Basic
College Mathematics

Basic College Mathematics

Fourth Edition

John Tobey

North Shore Community College
Danvers, Massachusetts

Jeffrey Slater

North Shore Community College
Danvers, Massachusetts

Prentice Hall

Prentice Hall
Upper Saddle River, NJ 07458

Library of Congress Cataloging-in-Publication Data

Tobey, John
    Basic college mathematics.--4th ed./John Tobey, Jeffrey Slater.
      p. cm.
    Slater's name appears first on the earlier eds.
    Includes index.
    ISBN 0-13-093229-9 -- ISBN 0-13-090954-8 (pbk.)
    1. Mathematics.   I. Slater, Jeffrey - II. Title.
  QA39.3 .T63 2002
  513.1--dc21
                                 2001021651

Executive Acquisition Editor: Karin E. Wagner
Project Manager: Mary Beckwith
Editor in Chief: Christine Hoag
Senior Managing Editor: Linda Mihatov Behrens
Executive Managing Editor: Kathleen Schiaparelli
Vice President/Director of Production and Manufacturing: David W. Riccardi
Production Management: Elm Street Publishing Services, Inc.
Manufacturing Buyer: Alan Fischer
Manufacturing Manager: Trudy Pisciotti
Executive Marketing Manager: Eilish Collins Main
Development Editor: Kathy Sessa Frederico
Editor in Chief, Development: Carol Trueheart
Media Project Manager, Developmental Math: Audra J. Walsh
Art Director: Maureen Eide
Assistant to the Art Director: John Christiana
Interior Designer: Studio Montage
Cover Design: Studio Montage
Managing Editor, Audio/Video Assets: Grace Hazeldine
Creative Director: Carole Anson
Director of Creative Services: Paul Belfanti
Photo Researcher: Christine Pullo
Photo Editor: Beth Boyd
Cover Photo: Peter Vanderwarker Photographs
Art Studio: Scientific Illustrators
Compositor: Preparé, Inc. / Emilcomp, S.r.l.

Photo credits appear on page P-1, which
constitutes a continuation of the copyright page.

© 2002, 1998, 1995, 1991 by Prentice-Hall, Inc.
Upper Saddle River, New Jersey 07458

Printed in the United States of America
10 9 8 7 6 5 4 3

ISBN 0-13-090954-8 (Student Edition paperback)
ISBN 0-13-093229-9 (Student Edition casebound)

Prentice-Hall International (UK) Limited, *London*
Prentice-Hall of Australia Pty. Limited, *Sydney*
Prentice-Hall Canada Inc., *Toronto*
Prentice-Hall Hispanoamericana, S.A., *Mexico City*
Prentice-Hall of India Private Limited, *New Delhi*
Prentice-Hall of Japan, Inc., *Tokyo*
Pearson Education Asia Pte. Ltd.
Editora Prentice-Hall do Brasil, Ltda., *Rio de Janeiro*

This book is dedicated to
Nancy Tobey
A loving wife for thirty-four years,
An outstanding mother of three children,
A dedicated but retired elementary teacher,
A helpful friend to so many,
A faithful administrative assistant who has
worked tirelessly to help produce this book

Contents

Chapter **1**

Whole Numbers   1

Chapter **2**

Fractions   107

Chapter 6

Measurement   367

Chapter 7

Geometry   415

Chapter 8

Statistics   513

Chapter 9

Signed Numbers   557

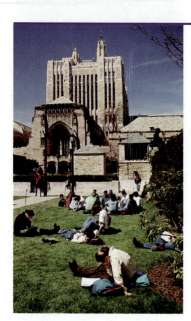

Chapter 10

Introduction to Algebra   607

Preface

To the Instructor

We share a partnership with you. For over thirty years we have taught mathematics courses at North Shore Community College. Each semester we join you in the daily task of sharing the knowledge of mathematics with students who often struggle with this subject. We enjoy teaching and helping students—and we are confident that you share these joys with us.

Mathematics instructors and students face many challenges today. *Basic College Mathematics* was written with these needs in mind. This textbook explains mathematics slowly, clearly, and in a way that is relevant to everyday life for the college student. As with previous editions, special attention has been given to problem solving in the fourth edition. This text is written to help students organize the information in any problem-solving situation, to reduce anxiety, and to provide a guide that enables students to become confident problem solvers.

One of the hallmark characteristics of *Basic College Mathematics* that makes the text easy to learn and teach from is the building-block organization. Each section is written to stand on its own, and each homework set is completely self-testing. Exercises are paired and graded and are of varying levels and types to ensure that all skills and concepts are covered. As a result, the text offers students an effective and proven learning program suitable for a variety of course formats—including lecture-based classes; discussion-oriented classes; distance learning centers; modular, self-paced courses; mathematics laboratories; and computer-supported centers.

Basic College Mathematics is the first text in a series that includes the following:

Tobey/Slater, *Basic College Mathematics*, Fourth Edition

Blair/Tobey/Slater, *Prealgebra*, Second Edition

Tobey/Slater, *Beginning Algebra*, Fifth Edition

Tobey/Slater, *Intermediate Algebra*, Fourth Edition

Tobey/Slater, *Beginning and Intermediate Algebra*

We have visited and listened to teachers across the country and have incorporated a number of suggestions into this edition to help you with the particular learning delivery system at your school. The following pages describe the key continuing features and changes in the fourth edition.

Key Features and Changes in the Fourth Edition

Developing Problem-Solving Abilities

We are committed as authors to producing a textbook that emphasizes mathematical reasoning and problem-solving techniques as recommended by AMATYC, NCTM, AMS, NADE, MAA, and other bodies. To this end, the problem sets are built on a wealth of real-life and real-data applications. Unique problems have been developed and incorporated into the exercise sets that help train students in data interpretation, mental mathematics, estimation, geometry and graphing, number sense, critical thinking, and decision making.

More Applied Problems

The exercises and applications have been extensively revised. Numerous real-world and real-data application problems show students the relevance of the math they are learning. The applications relate to everyday life, global issues beyond the borders of the United States, and other academic disciplines. Many include source citations. Nearly 50 percent of the applications are revised or new. The number of real-data applications has significantly increased. Roughly 30 percent of the applications have been contributed by actual students based on scenarios they have encountered in their home or work lives.

Math in the Media

New Math in the Media applications appear at the end of each chapter to offer students yet another opportunity to see why developing mastery of mathematical concepts enhances their understanding of the world around them. The applications are based on information from familiar media sources—either online or print. The exercises may ask students to interpret or verify information, perform calculations, make decisions or predictions, or provide a rationale for their response.

Putting Your Skills to Work

This highly successful feature has been revised and expanded in the fourth edition. There are 10 new Putting Your Skills to Work applications in the new edition. These nonroutine application problems challenge students to synthesize the knowledge they have gained and apply it to a totally new area. Each problem is specifically arranged for independent and cooperative learning or group investigation of mathematical problems that pique student interest. Students are given the opportunity to help one another discover mathematical solutions to extended problems. The investigations feature open-ended questions and extrapolation of data to areas beyond what is normally covered in such a course.

Internet Connections

As an integral part of each Putting Your Skills to Work problem, students are exposed to an interesting application of the Internet and encouraged to continue their investigation. This use of technology inspires students' confidence in their abilities to successfully use mathematics. In the fourth edition, the Internet Connections have been completely revised and updated.

The companion Web site now features annotated links to help students navigate the sites more efficiently and to provide a more user-friendly experience. `http://www.prenhall.com/tobey_basic`

Increased Integration and Emphasis on Geometry

Due to the emphasis on geometry on many statewide exams, geometry problems are integrated throughout the text. The new edition contains over 20 percent more geometry problems. Additionally, examples and exercises that incorporate a principle of geometry are now marked with a triangle icon for easy identification.

Mathematics Blueprint for Problem Solving

The successful Mathematics Blueprint for Problem Solving strengthens problem-solving skills by providing a consistent and interactive outline to help students organize their approach to problem solving. The blueprint now appears in at least one section of every chapter of the fourth edition. Once students fill in the blueprint, they can refer back to their plan as they do what is needed to solve the problem. Because of its flexibility, this feature can be used with single-step problems, multistep problems, applications, and nonroutine problems that require problem solving strategies. Students will not need to use the blueprint to solve every problem. It is available for those faced with a problem with which they are not familiar, to alleviate anxiety, to show them where to begin, and to assist them in the steps of reasoning.

Developing Your Study Skills

This highly successful feature has been retained in the new edition. The boxed notes are integrated throughout the text to provide students with techniques for improving their study skills and succeeding in math courses.

Graphs, Charts, and Tables

When students encounter mathematics in real-world publications, they often encounter data represented in a graph, chart, or table and are asked to make a reasonable conclusion based on the data presented. This emphasis on graphical interpretation is a continuing trend with today's expanding technology. The number of mathematical problems based on charts, graphs, and tables has been significantly increased in this edition. Students are asked to make simple interpretations, to solve medium-level problems, and to investigate challenging applied problems based on the data shown in a chart, graph, or table.

New Design

The fourth edition has a new design that enhances the accessible, student-friendly writing style. This new design includes new chapter opening applications and an improved and enhanced art program. See the walkthrough of features within this preface.

Mastering Mathematical Concepts

Chapter Pretests

Each chapter opens with a concise pretest to familiarize the student with the learning objectives for that particular chapter.

Learning Objectives

Concise learning objectives listed at the beginning of each section allow students to preview the goals of that section.

Examples and Exercises

The examples and exercises in this text have been carefully chosen to guide students through *Basic College Mathematics*. We have incorporated several different types of exercises and examples to assist your students in retaining the content of this course.

Practice Problems

Practice problems are found throughout the chapter, after the examples, and are designed to provide your students with immediate practice of the skills presented. The complete worked-out solution of each practice problem appears in the back of the book.

To Think About

These critical thinking questions may follow examples in the text and appear in the exercise sets. They extend the concept being taught, providing the opportunity for all students to stretch their minds, to look for patterns, and to make conclusions based on their previous experience. The number of these exercises has significantly increased in this edition.

Exercise Sets

Exercise sets are paired and graded. This design helps to ease the students into the problems; and the answers provide students with immediate feedback.

Cumulative Review Problems

Almost every exercise set concludes with a section of cumulative review problems. These problems review topics previously covered, and are designed to assist students in retaining the material. Many additional applied problems have been added to the cumulative review sections.

Calculator Problems

Calculator boxes are placed in the margin of the text to alert students to a scientific calculator application. In the exercise section a scientific calculator icon is used to indicate problems that are designed for solving with a calculator. There is also instruction on how to use a scientific calculator in the appendix.

Reviewing Mathematical Concepts

At the end of each chapter we have included problems and tests to provide your students with several different formats to help them review and reinforce the ideas that they have learned. This assists them not only with that specific chapter, but reviews previously covered topics as well.

Chapter Organizers

The concepts and mathematical procedures covered are reviewed at the end of each chapter in a unique chapter organizer. This device has been extremely popular with faculty and students alike. It not only lists concepts and methods, but provides a completely worked-out example for each type of problem. Students find that preparing a similar chapter organizer on their own in higher-level math courses becomes an invaluable way to master the content of a chapter of material.

Verbal and Writing Skills

The exercises provide students with the opportunity to extend a mathematical concept by allowing them to use their own words, to clarify their thinking, and to become familiar with mathematical terms. These exercises have been included at the beginning of exercise sets to set the stage for the practice that follows, or at the end of the practice as a summary. The number of these exercises has increased in the fourth edition.

Review Problems

These problems are grouped by section as a quick refresher at the end of the chapter. They can also be used by the student as a quiz of the chapter material.

Tests

Found at the end of the chapter, the chapter test is a representative review of the material from that particular chapter that simulates an actual testing format. This provides the students with a gauge to their preparedness for the actual examination.

Cumulative Tests

At the end of each chapter is a cumulative test. One-half of the content of each cumulative test is based on the math skills learned in previous chapters. By completing these tests for each chapter, the students build confidence that they have mastered not only the contents of the chapter but those of previous chapters as well.

Additional Content Changes in the Fourth Edition

- Throughout the text, explanations, definitions, and procedures have been carefully revised for clarity and precision.
- The coverage of fractions in Chapter 2 has been expanded as follows:

   Section 2.6 now begins with a discussion of least common multiples so that the subsequent coverage of least common denominators is more complete.

   A new lesson on order of operations in Section 2.8 offers students additional review of these rules and practice applying them to fractions.

   A new midchapter test on fractions appears after Section 2.5.

- Estimating is an important skill that allows students to check the reasonableness of their answers. The new edition contains increased coverage of estimating with fractions and decimals with new To Think About exercises in Sections 2.5, 2.8, and 3.3 and a new lesson in Section 3.7.
- Percent applications are now covered in two sections (Sections 5.4 and 5.5) to allow for a more relaxed presentation of these important topics.
- Because of the increased interest in geometry and the emphasis on this topic in many statewide exams, the coverage of geometry in Chapter 7 has been expanded and reorganized. The chapter now contains more coverage of angles and triangles as well as new coverage of rhombuses.
- The number of geometry problems integrated throughout the text has increased by over 20 percent. Examples and exercises that incorporate a principle of geometry are now marked by a triangle icon for easy identification.

Expanded and Enhanced Supplements Resource Package

The fourth edition is supported by a wealth of new supplements designed for added effectiveness and efficiency. New items include the MathPro 4.0 Explorer tutorial software together with a unique video clip feature, MathPro 5—the new online version of the popular tutorial program—providing online access anytime anywhere and enhanced course management for instructors; a new computerized testing system, TestGen-EQ with Quizmaster-EQ; all-new lecture videos; lecture videos digitized on CD-ROM; Prentice Hall Tutor Center; and options for online and distance learning courses. Please see the list of supplements and descriptions.

Options For Online and Distance Learning

For maximum convenience, Prentice Hall offers online interactivity and delivery options for a variety of distance learning needs. Instructors may access or adopt these in conjunction with this text, *Basic College Mathematics*, fourth edition.

Companion Web Site

Visit `http://www.prenhall.com/tobey_basic`
The companion Web site includes basic distance learning access to provide links to the text's Internet Connection activities. For the fourth edition the Internet Connection activities have been completely revised and updated. The Website now features annotated links to facilitate student navigation of the sites associated with the Internet Connection exercises. Links to additional sites of interest are also included.

This text-specific site offers students an online study guide via online self-quizzes. Questions are graded and students can e-mail their results. Syllabus Manager gives professors the option of creating their own online custom syllabus. Visit the Web site.

WebCT

Visit `http://www.prenhall.com/demo`
WebCT includes distance learning access to content found in the Tobey/Slater companion Web site plus more. WebCT provides tools to create, manage, and use online course materials. Save time and take advantage of items such as online help, communication tools, and access to instructor and student manuals. Your college may already have WebCT software installed on their server or you may choose to download it. Contact your local Prentice Hall sales representative for details.

BlackBoard

Visit `http://www.prenhall.com/demo`
For distance learning access to content and features from the Tobey/Slater companion Web site plus more. BlackBoard provides simple templates and tools to create, manage, and use online course materials. Take advantage of items such as online help, course management tools, communication tools, and access to instructor and student manuals. Contact your local Prentice Hall sales representative for details.

CourseCompass™ powered by BlackBoard

Visit `http://www.prenhall.com/demo`
For distance learning access to content and features from the Tobey/Slater companion Web site plus more. Prentice Hall content is preloaded in a customized version of BlackBoard 5. CourseCompass™ provides all of BlackBoard 5's powerful course management tools to create, manage, and use online course materials. Contact your local Prentice Hall sales representative for details.

Supplements for the Instructor

Printed Resources

Annotated Instructor's Edition (ISBN: 0-13-090955-6)

- Complete student text
- Answers appear in place on the same text page as exercises
- Teaching Tips placed in the margin at key points where students historically need extra help.
- Answers to all exercises in Pretests, Review Problems, Tests, Cumulative Tests, Diagnostic Pretest, and Practice Final

Instructor's Solutions Manual (ISBN: 0-13-092422-9)
- Detailed step-by-step solutions to the even-numbered exercises
- Solutions to every exercise (odd and even) in the Diagnostic Pretest, Pretests, Review Problems, Tests, Cumulative Tests, and Practice Final
- Solution methods reflect those emphasized in the text

Instructor's Resource Manual with Tests (ISBN: 0-13-092410-5)
- Nine test forms per chapter—6 free response, 3 multiple choice. Two of the free response tests are cumulative in nature.
- Four forms of Final Examination
- Answers to all items

Media Resources

New TestGen-EQ with QuizMaster-EQ (Windows/Macintosh) (ISBN: 0-13-092532-2)
- Algorithmically driven, text-specific testing program
- Networkable for administering tests and capturing grades online.
- The built-in Question Editor allows you to edit or add your own questions to create a nearly unlimited number of tests and worksheets
- Use the Function Plotter to create graphs
- Side-by-Side Testbank window and Test window show your test as you build it and as it will be printed.
- Extensive symbol palettes and expression templates assist professors in writing questions that include specialized tables and notation.
- Tests can be easily exported to HTML so they can be posted to the Web for student practice.
- QuizMaster-EQ Tests can be used for practice and graded tests. Instructors can set preferences to determine test availability, time limits, and number of tries.
- QuizMaster-EQ provides detailed reports of exam reports for individual students, classes, or the course.

New MathPro Explorer 4.0
Network Version for IBM/Macintosh (ISBN 0-13-092527-6)
- Enables instructors to create either customized or algorithmically generated practice tests from any section of a chapter, or a test of random items.
- Includes an e-mail function for network users, enabling instructors to send a message to a specific student or an entire group.
- Network based reports and summaries for a class or student and for cumulative or selected scores are available.

New MathPro 5
- The popular MathPro tutorial software available over the Internet.
- Online tutorial access–anytime, anywhere.
- Enhanced course management tools.

Companion Web Site

Visit `http://www.prenhall.com/tobey_basic`
- Internet Connection activities have been completely revised and updated. Annotated links facilitate student navigation of the sites associated with the Internet Connection exercises.

- Additional links provided to sites of interest or resources.
- Provides an online study guide via self-quizzes. Questions are graded and students can e-mail their results.
- Syllabus Manager gives professors the option of creating their own online custom syllabus. Visit the Web site to learn more.

Supplements for Students

Printed Resources

Student Solutions Manual (ISBN: 0-13-092419-9)

- Solutions to all odd-numbered exercises.
- Solutions to every (odd and even) exercise found in Pretests, Chapter Tests, Chapter Reviews, and Cumulative Reviews.
- Solution methods reflect those emphasized in the textbook.
- Ask your bookstore about ordering.

Media Resources

New MathPro Explorer 4.0 CD-ROM (Student version: 0-13-092533-0)

- Keyed to each section of the text for text-specific tutorial exercises and instruction.
- Warm-up exercises and graded practice problems.
- Video clips, providing a problem similar to one being attempted, show an explanation and worked-out solution on the board.
- Algorithmically generated exercises; includes bookmark, online help, glossary, and summary of scores for the exercises tried.
- Explorations enable students to explore concepts associated with each objective in more detail.

New MathPro 5 Anytime. Anywhere. With Assessment.

- The popular MathPro tutorial software available over the Internet.
- Online tutorial access–anytime, anywhere.

New Lecture Videos (ISBN: 0-13-092528-4)

- All new videotapes accompany the fourth edition.
- Keyed to each section of the text.
- Key concepts are explained step-by-step.

New Digitized Lecture Videos on CD-ROM (ISBN: 0-13-09252-0)

- The entire set of *Basic College Mathematics*, fourth edition lecture video tapes in digital form.
- Convenient access anytime to video tutorial support from a computer at home or on campus.
- Available shrink-wrapped with the text or stand-alone.

New Prentice Hall Tutor Center

- Staffed with developmental math instructors and open 5 days a week, 7 hours per day.
- Obtain help for examples and exercises in Tobey/Slater, *Basic College Mathematics*, fourth edition via toll-free telephone, fax, or e-mail.
- The Prentice Hall Tutor Center is accessed through a registration number that may be bundled with a new text or purchased separately with a used book.

- Contact your Prentice Hall sales representative for details, or visit
  `http://www.prenhall.com/tutorcenter`

Companion Web Site

Visit `http://www.prenhall.com/tobey_basic`

- Internet Connection activities have been completely revised and updated. Annotated links facilitate navigation of the sites associated with the Internet Connection exercises.
- Additional links provided to sites of interest or resources.
- Provides an online study guide via self-quizzes. Questions are graded and students can e-mail their results.

Additional Printed Material

Have your instructor contact the local Prentice Hall sales representative about the following resources:

- *How to Study Mathematics*
- *Math on the Internet: A Student's Guide*
- *Prentice Hall/New York Times, Theme of the Times Newspaper Supplement*

Acknowledgments

This book is the product of many years of work and many contributions from faculty and students across the country. We would like to thank the many reviewers and participants in focus groups and special meetings with the authors in preparation of previous editions.

Our deep appreciation to each of the following:

George J. Apostolopoulos, DeVry Institute of Technology

Katherine Barringer, Central Virginia Community College

Rita Beaver, Valencia Community College

Jamie Blair, Orange Coast College

Larry Blevins, Tyler Junior College

Brenda Callis, Rappahannock Community College

Joan P. Capps, Raritan Valley Community College

Robert Christie, Miami-Dade Community College

Mike Contino, California State University at Heyward

Judy Dechene, Fitchburg State University

Floyd L. Downs, Arizona State University

Barbara Edwards, Portland State University

Janice F. Gahan-Rech, University of Nebraska at Omaha

Colin Godfrey, University of Massachusetts, Boston

Carl Mancuso, William Paterson College

Janet McLaughlin, Montclair State College

Gloria Mills, Tarrant County Junior College

Norman Mittman, Northeastern Illinois University

Elizabeth A. Polen, County College of Morris

Ronald Ruemmler, Middlesex County College

Sally Search, Tallahassee Community College

Ara B. Sullenberger, Tarrant County Community College

Michael Trappuzanno, Arizona State University

Cora S. West, Florida Community College at Jacksonville

Jerry Wisnieski, Des Moines Community College

In addition, we want to thank the following individuals for providing splendid insight and suggestions for this new edition.

Sohrab Bakhtyari, St. Petersburg Junior College-Clearwater

Christine R. Bauman, Clark College at Larch

Vernon Bridges, Durham Technical Community College

Connie Buller, Metropolitan Community College

Oscar Caballero III, Laredo Community College

Nelson Collins, Joliet Junior College

Callie Jo Daniels, St. Charles County Community College

Ky Davis, Muskingum Area Technical College

Disa Enegren, Rose State College

Nancy Graham, Rose State College

Mary Beth Headlee, Manatee Community College

Sharon Louvier, Lee College

James A. Matovina, Community College of Southern Nevada

Doug Mace, Baker College

Beverly Meyers, Jefferson College

Wayne L. Miller, Lee College

Marcia Mollé, Metropolitan Community College

Jody E. Murphy, Lee College

Katrina Nichols, Delta College

Leticia M. Oropesa, University of Miami

Sandra Orr, West Virginia State Community College

Jim Osborn, Baker College

Linda Padilla, Joliet Junior College

Catherine Panik, Manatee Community College—South Campus

Gary Phillips, Clark College

Joel Rappaport, Miami Dade Community College

Dennis Runde, Manatee Community College

Richard Sturgeon, University of Southern Maine

Margie Thrall, Manatee Community College

We have been greatly helped by a supportive group of colleagues who not only teach at North Shore Community College but have also provided a number of ideas as well as extensive help on all of our mathematics books. Also, a special word of thanks to Hank Harmeling, Tom Rourke, Wally Hersey, Bob McDonald, Judy Carter, Bob Campbell, Rick Ponticelli, Russ Sullivan, Kathy LeBlanc, Lora Connelly, Sharyn Sharaf, Donna Stefano, and Nancy Tufo. Joan Peabody has done an excellent job of typing various materials for the manuscript and her help is gratefully acknowledged. Jenny Crawford provided new problems, new ideas, new answer keys, and great mental energy. Her excellent help was much appreciated.

We want to thank Louise Elton for providing several new applied problems and suggested applications. Error checking is a challenging task and few can do it well. So we especially want to thank the staff of Laurel Technical Services and Lauri Semarne for accuracy checking the content of the book at different stages of text preparation.

Additionally, Sherm Rosen researched the Internet Connections and provided splendid suggestions for improvements and helpful link annotations. Dave Nasby contributed the Math in the Media projects, and his hard work is much appreciated. Finally, we thank Richard Semmler for his careful work preparing the solutions manuals and accuracy checking.

Each textbook is a combination of ideas, writing, and revisions from the authors and wise editorial direction and assistance from the editors. We want to thank our Prentice Hall editor, Karin Wagner, for her helpful insight and perspective on each phase of the revision of the textbook. Her patience, her willingness to listen, and her flexibility to adapt to changing publishing decisions has been invaluable to the production of this book. Mary Beckwith, our project manager, provided daily support and encouragement as the book progressed. Her patient assistance with the art program and her attention to a variety of details was most appreciated. Kathy-Sessa Federico, our developmental editor, sifted through mountains of material and offered excellent suggestions for improvement and change. Gina Linko, our production director, kept things moving on schedule and cheerfully solved many crises.

Nancy Tobey retired from teaching and joined the team as our administrative assistant. Mailing, editing, photocopying, collating, and taping were cheerfully done each day. A special thanks goes to Nancy. We could not have finished the book without you.

Book writing is impossible for us without the loyal support of our families. Our deepest thanks and love to Nancy, Johnny, Melissa, Marcia, Shelley, Rusty, and Abby. Your understanding, your love and help, and your patience have been a source of great encouragement. Finally, we thank God for the strength and energy to write and the opportunity to help others through this textbook.

We have spent more than 30 years teaching mathematics. Each teaching day, we find that our greatest joy is helping students learn. We take a personal interest in ensuring that each student has a good learning experience in taking this course. If you have some personal comments, suggestions, or ideas for future editions of this textbook, please write to us at:

Prof. John Tobey and Prof. Jeffrey Slater
Prentice Hall Publishing
Office of the College Mathematics Editor
One Lake Street
Upper Saddle River, NJ 07458
or e-mail us at
jtobey@nscc.mass.edu.

We wish you success in this course and in your future life!

John Tobey
Jeffrey Slater

Enhanced, Student-Friendly Pedagogy

The Tobey/Slater series is a comprehensive learning system that features several pedagogical tools designed for ease-of-use and student success.

Chapter Organizer

*The Chapter Organizer appears at the end of each chapter and summarizes key concepts and mathematical procedures. It lists concepts and methods **and** provides a completely worked-out example for each type of problem.*

Chapter 2 Organizer

Topic	Procedure	Examples
Concept of a fractional part, p. 110.	The numerator is the number of parts selected. The denominator is the number of total parts.	What part of this sketch is shaded? $\frac{7}{10}$
Prime factorization, p. 117.	Prime factorization is the writing of a number as the product of prime numbers.	Write the prime factorization of 36. $36 = 4 \times 9$ $2 \times 2 \quad 3 \times 3$ $= 2 \times 2 \times 3 \times 3$

Page 188

Graphs, Charts, and Tables

Problems based on charts, graphs, and tables have been significantly increased. Students make simple interpretations, solve medium-level problems, and investigate challenging applied problems based on presented data.

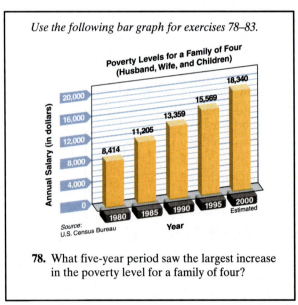

Use the following bar graph for exercises 78–83.

Poverty Levels for a Family of Four (Husband, Wife, and Children)

8,414 (1980)
11,205 (1985)
13,359 (1990)
15,569 (1995)
18,340 (2000 Estimated)

Annual Salary (in dollars) — Year

Source: U.S. Census Bureau

78. What five-year period saw the largest increase in the poverty level for a family of four?

Page 158

Developing Your Study Skills

Sprinkled throughout the text, these boxed notes provide students with techniques for improving their study skills and succeeding in math.

Developing Your Study Skills

Class Attendance

You will want to get started in the right direction by choosing to attend class every day, beginning with the first day of class. Statistics show that class attendance and good grades go together. Classroom activities are designed to enhance learning, and therefore you must be in class to benefit from them. Each day vital information and explanations are given that can help you understand concepts. Do not be deceived into thinking that you can just find out from a friend what went on in class. There is no good substitute for firsthand experience. Give yourself a push in the right direction by developing the habit of going to class every day.

Page 20

Math in the Media

Race for the Presidency Focuses on One-Third of the Country

Al Hunts Campaign Journal
THE WALL STREET JOURNAL

This is a national election but the real battle for the White House is focused on only one-third of the country.

With a week and a half to go, there are a dozen genuine tossup states. Three or four other states are clearly leaning to Vice President Al Gore or Gov. George W. Bush, but the deal may not be sealed yet. The three largest states—California, New York and Texas—aren't really in play.

The 12 states, with 132 electoral votes, that most experts and nonpartisan polls suggest could go either way are: Arkansas (with 6 electoral votes), Florida (25), Iowa (7), Michigan (18), Minnesota (10), Missouri (11), Nevada (4), New Mexico (5), Oregon (7), Pennsylvania (23), West Virginia (5) and Wisconsin (11).

Under this analysis, each candidate is clearly leading in enough states to have more that 200 electoral votes. It takes 270 to win. …

If the above analysis remains accurate, experts suggest that each can-

didate has approximately 200 electoral votes they can expect to win and the winner needs 70 additional electoral votes.

Source: October 26, 2000 WSJ.COM, The Wall Street Journal. © Dow Jones Company, Inc. Reprinted with permission.

Prior to the November 2000 presidential election, people were very fascinated in using mathematics to predict the results of the electoral vote. You may find this section quite interesting. Imagine yourself in the time during the week and a half before the election of November, 2000. Read the above article very carefully. Then try to solve these problems.

EXERCISES

1. What is the fewest number of states needed to get exactly 70 votes?

2. What is the maximum number of states possible to receive exactly 70 votes?

3. How does it appear on national TV that a particular state will carry the election for the new president?

🔴 Math in the Media

New Math in the Media exercises help students make connections between the real world and concepts learned in class.
A brief news clip or scenario based on the Web or print media appears at the end of each chapter. Related questions follow that may ask students to interpret the information, perform necessary calculations, or provide rationale for their decisions.

🔴 Chapter-Opening Application

A real-world application opens each chapter and links a specific situation to a "Putting Your Skills to Work" application that appears in that chapter to enhance students' awareness of the relevance of math.

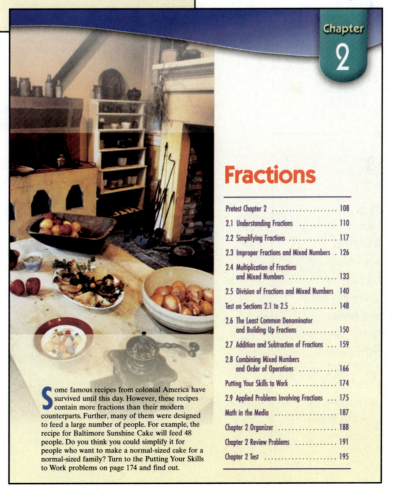

Chapter 2

Fractions

Some famous recipes from colonial America have survived until this day. However, these recipes contain more fractions than their modern counterparts. Further, many of them were designed to feed a large number of people. For example, the recipe for Baltimore Sunshine Cake will feed 48 people. Do you think you could simplify it for people who want to make a normal-sized cake for a normal-sized family? Turn to the Putting Your Skills to Work problems on page 174 and find out.

Integrated Problem Solving

🔴 Problem Solving

Problem Solving is thorough and easy to follow; key steps are highlighted with the pedagogical use of color. A clear problem-solving process is defined and reinforced throughout.

2.9 Applied Problems Involving Fractions

1 Solving Real-Life Problems with Fractions

All problem solving requires the same kind of thinking. In this section we will combine problem-solving skills with our new computational skills with fractions. Sometimes the difficulty is in figuring out what must be done. Sometimes it is in doing the computation. Remember that *estimating* is important in problem solving. We may use the following steps.

1. *Understand the problem.*
   (a) Read the problem carefully.
   (b) Draw a picture if this helps you.
   (c) Fill in the Mathematics Blueprint.

2. *Solve.*
   (a) Perform the calculations.
   (b) State the answer, including the unit of measure.

3. *Check.*
   Round fractions to the nearest whole number.
   swer with the estimate to see if your answer is

Student Learning Objectives

After studying this section, you will be able to:

1 Solve real-life problems with fractions.

SSM PH TUTOR CD & VIDEO MATH PRO WEB
CENTER

Page 175

Page 177

🔴 EXAMPLE 1 Change each mixed number to an improper fraction.

(a) $3\frac{2}{5}$ **(b)** $5\frac{4}{9}$ **(c)** $18\frac{3}{5}$

> Multiply the whole number by the denominator.   Add the numerator to the product.

(a) $3\frac{2}{5}$ $\dfrac{3 \times 5 + 2}{5} = \dfrac{15 + 2}{5} = \dfrac{17}{5}$ ← Write the sum over the denominator.

(b) $5\frac{4}{9} = \dfrac{9 \times 5 + 4}{9} = \dfrac{45 + 4}{9} = \dfrac{49}{9}$

(c) $18\frac{3}{5} = \dfrac{5 \times 18 + 3}{5} = \dfrac{90 + 3}{5} = \dfrac{93}{5}$

Practice Problem 1 Change the mixed numb

(a) $4\frac{3}{7}$ **(b)** $6\frac{2}{3}$ **(c)** $19\frac{4}{7}$

Page 127

▲ 🔴 **EXAMPLE 2** What is the inside diameter (distance across) of a cement storm drain pipe that has an outside diameter of $4\frac{1}{8}$ feet and is $\frac{3}{8}$ feet thick?

1. *Understand the problem.*
   Read the problem carefully. Draw a picture. The picture is in the margin on the left. Now fill in the Mathematics Blueprint.

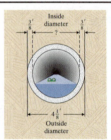

Mathematics Blueprint For Problem Solving

Gather The Facts	What Am I Asked To Do?	How Do I Proceed?	Key Points To Remember
Outside diameter is $4\frac{1}{8}$ feet. Thickness is $\frac{3}{8}$ foot on both ends of the diameter.	Find the *inside* diameter of the pipe.	Add the two measures of thickness. Then subtract this total from the outside diameter.	Since the LCD = 8, all fractions must have this denominator.

2. *Solve and state the answer.*
   Add the two thickness measurements together.
   Adding $\dfrac{3}{8} + \dfrac{3}{8} = \dfrac{6}{8}$ gives the total thickness of the pipe, $\dfrac{6}{8}$ foot. We will not reduce $\dfrac{6}{8}$ since the LCD is 8.
   We subtract the total of the two thickness measurements from the outside diameter.

$$
\begin{array}{rcr}
4\frac{1}{8} & = & 3\frac{9}{8} \\
-\ \frac{6}{8} & = & -\ \frac{6}{8} \\
\hline
& & 3\frac{3}{8}
\end{array}
$$

> We borrow 1 from 4 to get $3 + 1\frac{1}{8}$ or $3\frac{9}{8}$.

The inside diameter is $3\frac{3}{8}$ feet.

3. *Check.*
   We will work backward to check. We will use the exact values. If we have done our work correctly, $\frac{3}{8}$ foot $+\ 3\frac{3}{8}$ feet $+\ \frac{3}{8}$ foot should add up to the outside diameter, $4\frac{1}{8}$ feet.

$$\frac{3}{8} + 3\frac{3}{8} + \frac{3}{8} \stackrel{?}{=} 4\frac{1}{8}$$

🔴 Mathematics Blueprint for Problem Solving

Students begin the problem-solving process and plan the steps to be taken along the way using an outline to organize their approach to problem solving. Once students fill in the blueprint, they can refer back to their plan to solve the problem. The blueprint now appears in at least one section in every chapter.

Putting Your Skills to Work

The Thinning Ice Cap

According to top scientists, the great ice cap that stretches across the top of the earth has become significantly thinner than it was several decades ago.

From 1958 to 1976, the average thickness of the Arctic Sea ice was approximately 10 feet. During the period 1993 to 1997, it was about 6 feet. Scientists estimate that since that time, the ice has been thinning at the rate of about 4 inches a year.

This phenomenon is directly or indirectly related to global warming. In 2000, it was estimated that the temperature of the Earth had risen by 1.75 degrees Fahrenheit over the previous century. Most scientists (but not all) expect that the average temperature of the Earth will rise by another 3.5 degrees Fahrenheit by 2100 if emissions of heat-trapping gases such as carbon dioxide continue at present rates.

Problems for Individual Investigation

If the rate of thinning of the polar ice cap continues at the present rate, what will be the thickness of the ice in the Arctic Sea

1. in 2005?

2. in 2009?

Problems for Group Investigation and Cooperative Study

If the predictions for the rate of global warming over the next century are correct, what will be the amount of global warming

3. in 2015?

4. in 2068?

Internet Connections

 Netsite: http://www.prenhall.com/tobey_basic

This site contains facts and predictions related to global warming. Based on the information available on that site:

5. How many feet are the oceans expected to rise from 2000 to 2100?

6. The average temperature of New York City today is 54 degrees Fahrenheit. What will this be in 2100 if the present warming trend continues?

Page 237

Interesting and Diverse Exercises and Applications

Real-World Applications

Numerous real-world and real-data applications relate topics to everyday life, global issues, and other academic disciplines. An abundance of new real-world application problems show students the relevance of math in their daily lives.

Use the chart to answer questions 45–48.

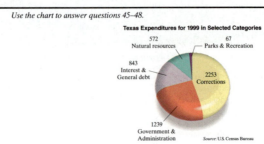

Texas Expenditures for 1999 in Selected Categories

572 Natural resources
67 Parks & Recreation
843 Interest & General debt
2253 Corrections
1239 Government & Administration

Source: U.S. Census Bureau

45. What percent of the total expenditures in these five categories was used for corrections?

46. What percent of the total expenditures in these five categories was used for government and administration?

Page 340

EXAMPLE 9
The level of sulfur dioxide emissions in the air has slowly been decreasing over the last 20 years, as can be seen in the accompanying bar graph. Find the average amount of sulfur dioxide emissions in the air over these five specific years.

First we take the sum of the five years.

$$9.37$$
$$9.30$$
$$8.68$$
$$7.37$$
$$+ \ 6.06$$
$$\overline{40.78}$$

Then we divide by five to obtain the average.

$$\begin{array}{r} 8.156 \\ 5{\overline{)40.780}} \\ \underline{40} \\ 7 \\ \underline{5} \\ 28 \\ \underline{25} \\ 30 \\ \underline{30} \\ \end{array}$$

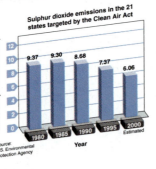

Sulphur dioxide emissions in the 21 states targeted by the Clean Air Act

Emissions (in millions of tons)

9.37 — 1980
9.30 — 1985
8.68 — 1990
7.37 — 1995
6.06 — 2000 Estimated

Year

Source: U.S. Environmental Protection Agency

Thus the yearly average is 8.156 million tons of sulfur dioxide emissions in these 21 states.

Practice Problem 9 Use the accompanying bar graph to find the average level of sulfur dioxide for the three years: 1985, 1990, and 1995. By how much does the three-year average differ from the five-year average?

Page 233

Page 144

Verbal and Writing Skills Exercises

These exercises ask students to use their own words. Writing helps students clarify their thinking and become familiar with mathematical terms.

Verbal and Writing Skills

1. In your own words explain how to remember that when you divide two fractions you invert the *second* fraction and multiply by the first. How can you be sure that you don't invert the *first* fraction by mistake?

2. Explain why $2 \div \frac{1}{3}$ is a larger number than $2 \div \frac{1}{2}$.

To Think About

Critical-thinking questions appear in the exercise sets to extend the concepts. They also give students the opportunity to look for patterns and draw conclusions.

To Think About

In countries outside the United States, very heavy items are measured in terms of **metric tons***. A metric ton is defined as 1000 kilograms (or 2205 pounds). Consider the following bar graph, which records the catch of fish (measured in millions of metric tons) by China for various years.*

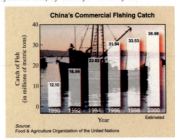

China's Commercial Fishing Catch

Catch of Fish (in millions of metric tons)

12.10 — 1990
18.56 — 1992
23.83 — 1994
31.94 — 1996
33.53 — 1998
36.98 — 2000 Estimated

Year

Source: Food & Agriculture Organization of the United Nations

60. By how many metric tons did the commercial catch of fish by China increase from 1990 to 1994?

61. By how many metric tons did the commercial catch of fish by China increase from 1996 to 2000?

Page 384

Pretest Chapter 2

Page 108

1. _____

2. _____

3. _____

4. _____

5. _____

6. _____

7. _____

If you are familiar with the topics in this chapter, take this test now. Check your answers with those in the back of the book. If an answer is wrong or you can't do a problem, study the appropriate section of the chapter.

If you are not familiar with the topics in this chapter, don't take this test now. Instead, study the examples, work the practice problems, and then take the test.

This test will help you identify which concepts you have mastered and which you need to study further. Make sure all fractions are simplified in the final answer.

Section 2.1

1. Use a fraction to represent the shaded part of the object.

2. Draw a sketch to show $\frac{3}{6}$ of an object.

3. An inspector checked 124 CD players. Of these, 5 were defective. Write a fraction that describes the part that was defective.

🔴 Chapter Pretests

A Chapter Pretest appears at the beginning of each chapter to familiarize students with the objectives in that chapter.

A Diagnostic Pretest with chapter references appears at the beginning of the text. A final exam with chapter references appears at the back of the text. All answers are included.

🔴 Emphasis on Geometry

A new geometry icon in the examples and exercises marks an increased emphasis on geometry.

▲ 🔴 **EXAMPLE 6** Find the area in square miles of a rectangle with width $1\frac{1}{3}$ miles and length $12\frac{1}{4}$ miles.

Length = $12\frac{1}{4}$ miles

Width = $1\frac{1}{3}$ miles

We find the area of a rectangle by multiplying the width times the length.

$$1\frac{1}{3} \times 12\frac{1}{4} = \frac{\cancel{4}}{3} \times \frac{49}{\cancel{4}} = \frac{49}{3} \quad \text{or} \quad 16\frac{1}{3}$$

The area is $16\frac{1}{3}$ square miles.

▲ **Practice Problem 6** Find the area in square meters of a rectangle with width $1\frac{1}{5}$ meters and length $4\frac{5}{6}$ meters. 🔴

Page 136

---

328 Chapter 5 Percent

Calculator

Percent of a Number
You can use a calculator to find 12% of 48. Enter

12 $\boxed{\%}$ $\boxed{\times}$ 48 $\boxed{=}$

The display should read

$\boxed{5.76}$

If your calculator does not have a percent key, use the keystrokes

0.12 $\boxed{\times}$ 48 $\boxed{=}$

What is 54% of 450?

Solving Percent Problems When We Do Not Know the Amount

🔴 **EXAMPLE 6** What is 45% of 590?

$\downarrow \quad \downarrow \quad \downarrow \quad \downarrow \quad \downarrow$

$n = 45\% \times 590$ Translate into an equation.

$n = (0.45)(590)$ Change the percent to decimal form.

$n = 265.5$ Multiply 0.45 × 590.

Practice Problem 6 What is 82% of 350? 🔴

🔴 **EXAMPLE 7** Find 160% of 500.

Find 160% of 500.   When you translate, remember that the word *find* is equiva-
$\downarrow \quad \downarrow \quad \downarrow \quad \downarrow$ lent to *what is.*

$n = 160\% \times 500$

$n = (1.60)(500)$ Change the percent to decimal form.

$n = 800$ Multiply 1.6 by 500.

Practice Problem 7 Find 230% of 400. 🔴

🔴 Calculator Notes

Calculator notes in the margin alert students to calculator applications. Calculator icons in the exercise sets mark problems that can be solved using a scientific calculator.

Page 328

To the Student

This book was written with your needs and interests in mind. The original manuscript and the first three editions of this book have been class-tested with students all across the country. Based on the suggestions of many students, the book has been refined and improved to maximize your learning while using this text.

We realize that students who enter college have sometimes never enjoyed math or never done well in a math course. You may find that you are anxious about taking this course. We want you to know that this book has been written to help you overcome those difficulties. Literally thousands of students across the country have found an amazing ability to learn math as they have used previous editions of this book. We have incorporated several learning tools and various types of exercises and examples to assist you in learning this material.

It helps to know that learning mathematics is going to help you in life. Perhaps you have entered this course feeling that little or no mathematics will be necessary for you in your future job. However, elementary school teachers, bus drivers, laboratory technicians, nurses, telephone operators, cable TV repair personnel, photographers, pharmacists, salespeople, doctors, architects, inspectors, counselors, and custodians who once believed that they needed little if any mathematics are finding that the mathematical skills presented in this course can help them. Mathematics, you will find, can help you too. In this book, a great number of examples and problems come from everyday life. You will be amazed at the number of ways mathematics is used in the world each day.

Our greatest wish is that you will find success and personal satisfaction in your mathematics course. We have written this book to help you accomplish that very goal.

Suggestions for Students

1. Be sure to take the time to read the boxes marked **Developing Your Study Skills**. These ideas come from faculty and students throughout the United States. They have found ways to succeed in mathematics and they want to share those ideas with you.

2. **Read** through each section of the book that is assigned by your instructor. You will be amazed at how much information you will learn as you read. It will "fill in the pieces" of mathematical knowledge. Reading a math book is a key part of learning.

3. Study carefully each **sample example**. Then work out the related practice problem. Direct involvement in doing problems and not just thinking about them is one of the greatest guarantees of success in this course.

4. Work out all assigned **homework problems**. Verify your answers in the back of the book. Ask questions when you don't understand or when you are not sure of an answer.

5. Be sure to take advantage of **end-of-chapter helps**. Study the chapter organizers. Do the chapter review problems and the practice test problems.

6. If you need help, remember that teachers and tutors can assist you. There is a **Student Solutions Manual** for this textbook that you will find most

valuable. This manual shows worked-out solutions for all the odd-numbered exercises as well as diagnostic pretest, chapter review, chapter test, and cumulative test problems in this book. (If your college bookstore does not carry the Student Solutions Manual, ask them to order a copy for you.)

7. Watch the **videotapes or digitized videos on CD-ROM.** Every section of this text is explained in detail on videotape. The work is solved using the same methods as explained in the text.

8. Use **MathPro Explorer 4.0** a tutorial software package that allows you to be tutored for any section in the book. It allows you to test yourself on your mastery of any section of any chapter of the text. Practice problem solving using the resources available to you.

9. Remember, learning mathematics takes time. However, the time spent is well worth the effort. **Take the time** to study, do homework, review, ask questions, and just reflect over what you have learned. The time you invest in learning mathematics will reap dividends in your future courses and your future life.

We encourage you to look over your textbook carefully. Many important features have been designed into the book to make learning mathematics a more enjoyable activity. There are some special features, unique to this textbook, that student throughout the country have told us that they found especially helpful.

Four Key Textbook Features to Help You

1. **Practice problems with worked-out solutions.** Immediately following every sample example is a similar problem called the Practice Problem. If you can work it out correctly by following the sample example as a general point of reference, then you will likely be able to do the homework exercises. If you encounter some difficulty, then you will find helpful the completely worked-out solutions to the practice problems that appear at the end of the text.

2. **Student-friendly application problems.** You will find realistic and interesting application problems in every chapter of this book. Many of the problems were actually written or suggested by students. As you develop your problem-solving and reasoning skills in this course, you will encounter a number of real-world situations that help you to see how very helpful mathematical skills are in today's complex world.

3. **Putting Your Skills to Work problems.** In each chapter are unique problem sets that ask you to analyze in depth some mathematical aspect of daily life. You will be asked to extend your knowledge, do some creative thinking, and work in small groups in a cooperative learning situation. You will even have a chance to explore some Web sites and see how the Internet can assist you in learning. These sections will awaken some new interests and help you to develop the critical thinking skills so necessary both in college and after you graduate.

4. **The Chapter Organizer.** Everything you need to learn in any one chapter of this book is readily available at your fingertips in the chapter organizer. This very popular chart summarizes all methods covered in the chapter, gives page references, and shows a sample example completely worked out for every major topic covered. Students have found this tool a most helpful way to master the content of any chapter of the book.

Diagnostic Pretest: Basic College Mathematics

Chapter 1

1. Add. $3846 + 527$

2. Divide. $58\overline{)1508}$

3. Subtract. $\begin{array}{r} 12{,}807 \\ -\,11{,}679 \\ \hline \end{array}$

4. The highway department used 115 truckloads of sand. Each truck held 8 tons of sand. How many tons of sand were used?

Chapter 2

5. Add. $\dfrac{3}{7} + \dfrac{2}{5}$

6. Multiply and simplify. $3\dfrac{3}{4} \times 2\dfrac{1}{5}$

7. Subtract. $2\dfrac{1}{6} - 1\dfrac{1}{3}$

8. Mike's car traveled 237 miles on $7\frac{9}{10}$ gallons of gas. How many miles per gallon did he achieve?

Chapter 3

9. Multiply. $\begin{array}{r} 51.06 \\ \times\,0.307 \\ \hline \end{array}$

10. Divide. $0.026\overline{)0.0884}$

11. The copper pipe was 24.375 centimeters long. Paula had to shorten it by cutting off 1.75 centimeters. How long will the copper pipe be when it is shortened?

12. Russ bicycled 20.5 miles on Monday, 5.8 miles on Tuesday, and 14.9 miles on Wednesday. How many miles did he bicycle on those three days?

1. _____

2. _____

3. _____

4. _____

5. _____

6. _____

7. _____

8. _____

9. _____

10. _____

11. _____

12. _____

13. _____

14. _____

15. _____

16. _____

17. _____

18. _____

19. _____

20. _____

21. _____

22. _____

23. _____

24. _____

25. _____

26. _____

27. _____

28. _____

Chapter 4

Solve each proportion problem. Round to the nearest tenth if necessary.

13. $\dfrac{3}{7} = \dfrac{n}{24}$

14. $\dfrac{0.5}{0.8} = \dfrac{220}{n}$

15. Wally's Landscape earned $600 for mowing lawns at 25 houses last week. At that rate, how much would he earn for doing 45 houses?

16. Two cities that are actually 300 miles apart appear to be 8 inches apart on the road map. How many miles apart are two cities that appear to be 6 inches apart on the map?

Chapter 5

Round to the nearest tenth if necessary.

17. Change to a percent: $\dfrac{3}{8}$

18. 138% of 5600 is what number?

19. At Mountainview College 53% of the students are women. There are 2067 women at the college. How many students are at the college?

20. At a manufacturing plant it was discovered that 9 out of every 3000 parts made were defective. What percent of the parts are defective?

Chapter 6

21. 15 qt = _____ gal

22. 3 cm = _____ meter

23. 1.56 tons = _____ lb

24. 4900 kg = _____ milligrams

Chapter 7

Round to the nearest hundredth when necessary. Use $\pi \approx 3.14$ when necessary.

25. Find the area of a triangle with a base of 34 meters and an altitude of 23 meters.

26. Find the cost to install carpet in a circular area with a radius of 5 yards at a cost of $35 per square yard.

27. In a right triangle the longest side is 15 meters and the shortest side is 9 meters. What is the length of the other side of the triangle?

28. How many pounds of fertilizer can be placed in a cylindrical tank that is 4 feet tall and has a radius of 5 feet if one cubic foot of fertilizer weighs 70 pounds?

Chapter 8

The following double bar graph indicates the sale of Dodge Neons for Westover County as reported by the district sales managers. Use this graph to answer questions 29–32.

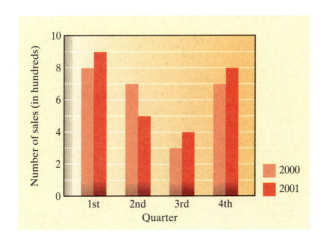

29. How many Dodge Neons were sold in the second quarter of 2001?

30. How many more Dodge Neons were sold in the fourth quarter of 2001 than were sold in the fourth quarter of 2000?

31. In which year were more Dodge Neons sold, in 2000 or 2001?

32. What is the *mean* number of Dodge Neons sold per quarter in 2000?

Chapter 9

Perform the following operations.

33. $-5 + (-2) + (-8)$ **34.** $-8 - (-20)$

35. $\left(-\dfrac{3}{4}\right) \div \left(\dfrac{5}{6}\right)$ **36.** $(-3)(2)(-1)(-3)$

Chapter 10

Simplify.

37. $9(x + y) - 3(2x - 5y)$

In exercises 38–39, solve for x.

38. $3x - 7 = 5x - 19$ **39.** $2(x - 3) + 4x = -2(3x + 1)$

40. A rectangle has a perimeter of 134 meters. The length of the rectangle is 4 meters longer than double the width of the rectangle. What is the length and the width of the rectangle?

29. _____

30. _____

31. _____

32. _____

33. _____

34. _____

35. _____

36. _____

37. _____

38. _____

39. _____

40. _____

Whole Numbers

Whenever we get a chance to take a vacation, we want to get the most for our money. However, sometimes making the decision as to what is the best buy is somewhat involved. Which vacation package would be the least expensive? How can the cost per person be determined? Do you think you could make a wise decision if you were faced with several possible choices for a ski vacation? Turn to the Putting Your Skills to Work problems on page 61 and find out.

Pretest Chapter 1

If you are familiar with the topics in this chapter, take this test now. Check your answers with those in the back of the book. If an answer is wrong or you can't answer a question, study the appropriate section of the chapter.

If you are not familiar with the topics in this chapter, don't take this test now. Instead, study the examples, work the practice problems, and then take the test.

This test will help you identify which concepts you have mastered and which you need to study further.

Section 1.1

1. Write in words. 78,310,436

2. Write in expanded notation. 38,247

3. Write in standard notation. five million, sixty-four thousand, one hundred twenty-two

Use the following table to answer questions 4 and 5.

Public High School Graduates (in thousands)

Source: U.S. Department of Education

1980	2,747
1995	2,273
2000*	2,583

*estimated

4. How many public school graduates were there in 1980?

5. How many public school graduates were there in 2000?

Section 1.2

Add.

6.
$$
\begin{array}{r}
13 \\
31 \\
88 \\
43 \\
+69 \\
\hline
\end{array}
$$

7.
$$
\begin{array}{r}
28,318 \\
5,039 \\
+17,213 \\
\hline
\end{array}
$$

8.
$$
\begin{array}{r}
7148 \\
500 \\
19 \\
+7062 \\
\hline
\end{array}
$$

Section 1.3

Subtract.

9.
$$
\begin{array}{r}
6439 \\
-4328 \\
\hline
\end{array}
$$

10.
$$
\begin{array}{r}
100,450 \\
-\ 24,139 \\
\hline
\end{array}
$$

11.
$$
\begin{array}{r}
45,861,413 \\
-43,879,761 \\
\hline
\end{array}
$$

Section 1.4

Multiply.

12. $9 \times 6 \times 1 \times 2$

13.
$$
\begin{array}{r}
2658 \\
\times\ \ \ 7 \\
\hline
\end{array}
$$

14.
$$
\begin{array}{r}
91 \\
\times 74 \\
\hline
\end{array}
$$

15.
$$
\begin{array}{r}
365 \\
\times 908 \\
\hline
\end{array}
$$

1. _____

2. _____

3. _____

4. _____

5. _____

6. _____

7. _____

8. _____

9. _____

10. _____

11. _____

12. _____

13. _____

14. _____

15. _____

Section 1.5

Divide. If there is a remainder, be sure to state it as part of the answer.

16. $8\overline{)84,840}$ **17.** $7\overline{)51,633}$ **18.** $42\overline{)5838}$

Section 1.6

19. Write in exponent form. $6 \times 6 \times 6 \times 6$ **20.** Evaluate. 3^4

Perform the operations in their proper order.

21. $2 \times 3^3 - (4 + 1)^2$ **22.** $2^4 + (6^2 + 36) \div 2$

23. $6 \times 7 - 2 \times 6 - 2^3 + 17 \div 17$

Section 1.7

24. Round to the nearest thousand.   270,612

25. Round to the nearest hundred.   26,539

26. Round to the nearest million.   59,540,000

Estimate by first rounding each number so that there is only one nonzero digit and then performing the calculation.

27. $187 + 458 + 375 + 803 + 666$ **28.** $56,784 \times 459,202$

Section 1.8

Solve each of the following applied problems.

29. Emily planned a trip of 1483 miles from her home to Chicago. She traveled 317 miles on her first day of the trip. How many more miles must she travel?

30. Betsey purchased 84 shares of stock at a cost of $1848.00. How much was the cost per share?

31. Dr. Alfonso bought three chairs at $46 each, three lamps at $29 each, and two tables at $37 each for the waiting room of his office. How much did his purchases cost?

Population for State of Alaska, U.S. Census Report

Year	1970	1980	1990	2000
Population	302,583	401,851	550,043	653,000

32. Using the preceding table, answer the following questions.
    (a) By how much did the population in Alaska increase from 1980 to 1990?
    (b) Some officials predict that between 2000 and 2010 the population of Alaska will increase by 134,890. If that figure is correct, what is the projected population of Alaska for the year 2010?

16. _____

17. _____

18. _____

19. _____

20. _____

21. _____

22. _____

23. _____

24. _____

25. _____

26. _____

27. _____

28. _____

29. _____

30. _____

31. _____

32. (a) _____

(b) _____

1.1 Understanding Whole Numbers

Student Learning Objectives

After studying this section, you will be able to:

1 Write numbers in expanded form.

2 Write a word name for a number and write a number for a word name.

3 Read numbers in tables.

SSM
PH TUTOR CENTER   CD & VIDEO   MATH PRO   WEB

1   *Writing Numbers in Expanded Form*

To count a number of objects or to answer the question "How many?" we use a set of numbers called **whole numbers.** These whole numbers are as follows.

$$0, 1, 2, 3, 4, 5, 6, 7, 8, 9, 10, 11, 12, 13, 14, 15, \ldots$$

There is no largest whole number. The three dots ... indicate that the set of whole numbers goes on indefinitely. Our number system is based on tens and ones and is called the **decimal system** (or the **base 10 system**). The numbers 0, 1, 2, 3, 4, 5, 6, 7, 8, 9 are called **digits.** The position, or placement, of the digits in the number tells the value of the digits. For example, in the number 521, the "5" means 5 hundreds (500). In the number 54, the "5" means 5 tens (50).

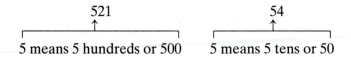

521   54

5 means 5 hundreds or 500   5 means 5 tens or 50

For this reason, our number system is called a **place-value system.**

Consider the number 5643. We will use a place-value chart to illustrate the value of each digit in the number 5643.

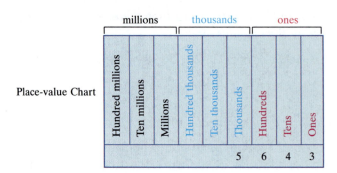

	millions			thousands			ones	
Place-value Chart	Hundred millions	Ten millions	Millions	Hundred thousands	Ten thousands	Thousands	Hundreds	Tens	Ones
						5	6	4	3

The value of the number is 5 thousands, 6 hundreds, 4 tens, 3 ones.

The place-value chart shows the value of each place, from ones on the right to hundred millions on the left. When we write very large numbers, we place a comma after every group of three digits called a **period,** moving from right to left. This makes the number easier to read. It is usually agreed that a four-digit number does not have a comma, but that numbers with five or more digits do. So 32,000 would be written with a comma but 7000 would not.

To show the value of each digit in a number, we sometimes write the number in expanded notation. For example, 56,327 is 5 ten thousands, 6 thousands, 3 hundreds, 2 tens, and 7 ones. In **expanded notation,** this is

$$50,000 + 6000 + 300 + 20 + 7.$$

EXAMPLE 1   Write each number in expanded notation.

(a) 2378   **(b)** 538,271   **(c)** 980,340,654

(a) Sometimes it helps to say the number to yourself.

	two thousand		three hundred		seventy		eight
2378 =	2000	+	300	+	70	+	8

4

(b)

Expanded notation

$$538,271 = 500,000 + 30,000 + 8000 + 200 + 70 + 1$$

(c) When 0 is used as a placeholder, you do not include it in the expanded form.

Expanded notation

$$980,340,654 = 900,000,000 + 80,000,000 + 300,000 + 40,000 + 600 + 50 + 4$$

Practice Problem 1   Write each number in expanded notation.

(a) 3182   **(b)** 520,890   **(c)** 709,680,059

   The number as you usually see it is called the **standard notation.** 980,340,654 is the standard notation for the number nine hundred eighty million, three hundred forty thousand, six hundred fifty-four.

EXAMPLE 2   Write each number in standard notation.

(a) 500 + 30 + 8   **(b)** 300,000 + 7000 + 40 + 7

(a) 538

(b) Be careful to keep track of the place value of each digit. You may need to use 0 as a placeholder.

3 hundred thousand

$$300,000 + 7000 + 40 + 7 = 307,047$$

We needed to use 0 in the ten thousands place and in the hundreds place.

7 thousand

Practice Problem 2   Write each number in standard notation.

(a) 400 + 90 + 2   **(b)** 80,000 + 400 + 20 + 7

EXAMPLE 3   Last year the population of Central City was 1,509,637. In the number 1,509,637

(a) How many ten thousands are there?   **(b)** How many tens are there?
(c) What is the value of the digit 5?   **(d)** In what place is the digit 6?

The place-value chart will help you identify the value of each place.

(a) Look at the digit in the ten thousands place. There are 0 ten thousands.
(b) Look at the digit in the tens place. There are 3 tens.
(c) The digit 5 is in the hundred thousands place. The value of the digit is 5 hundred thousand or 500,000.
(d) The digit 6 is in the hundreds place.

Practice Problem 3   The campus library has 904,759 books.

(a) What digit tells the number of hundreds?
(b) What digit tells the number of hundred thousands?
(c) What is the value of the digit 4?
(d) What is the value of the digit 9? Why does this question have two answers?

2 Writing Word Names for Numbers and Numbers for Word Names

A number has the same *value* no matter how we write it. For example, "a million dollars" means the same as "$1,000,000." In fact, any number in our number system can be written in several ways or forms:

- Standard notation   521
- Expanded notation   500 + 20 + 1
- Word name   five hundred twenty-one

You may want to write a number in any of these ways. To write a check, you need to use both standard notation and words.

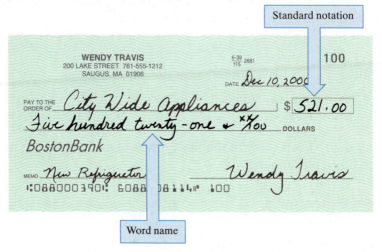

To write a word name, start from the left. Name the number in each period, followed by the name of the period, and a comma. The last period name, "ones" is not used of digits separated by a comma.

EXAMPLE 4   Write a word name for 364,128,957.

Place-value Chart

Billions			Millions			Thousands			Ones		
			3	6	4	1	2	8	9	5	7
Hundreds	Tens	Ones	Hundreds	Tens	Ones	Hundreds	Tens	Ones	Hundreds	Tens	Ones

We want to write a word name for 364, 128, 957.

three hundred sixty-four million,

one hundred twenty-eight thousand,

nine hundred fifty-seven

The answer is three hundred sixty-four million, one hundred twenty-eight thousand, nine hundred fifty-seven.

Practice Problem 4   Write a word name for 267,358,981.

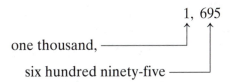 **EXAMPLE 5** Write the word name for each number.

(a) 1695   **(b)** 200,470   **(c)** 7,003,038

Look at the place-value chart if you need help identifying the place for each digit.

(a) To help us, we will put in the optional comma: 1,695.

$$1,\ 695$$

one thousand, ——————————↑   ↑

six hundred ninety-five ——————

The word name is one thousand, six hundred ninety-five.

(b) $$200,\ 470$$

two hundred thousand, ———↑   ↑

four hundred seventy ————

The word name is two hundred thousand, four hundred seventy.

(c) $$7,\ 003,\ 038$$

seven million, ——————↑   ↑   ↑

three thousand, ————————

thirty-eight ————————————

The word name is seven million, three thousand, thirty-eight.

Practice Problem 5   Write the word name for each number.

(a) 2736   **(b)** 980,306   **(c)** 12,000,021

🚫 **CAUTION: DO NOT USE THE WORD <u>AND</u> FOR WHOLE NUMBERS.** Many people use the word *and* when giving the word name for a whole number. For example, you might hear someone say the number 34,507 as "thirty-four thousand, five hundred *and* seven." However, this is not technically correct. In mathematics we do NOT use the word *and* when writing word names for whole numbers. In Chapter 3 we will use the word *and* to represent the decimal point. For example, 59.76 will have the word name "fifty-nine *and* seventy-six hundredths."

Very large numbers are used to measure quantities in some disciplines, such as distance in astronomy and the national debt in macroeconomics. We can extend the place-value chart to include these large numbers.

The national debt for the United States as of January 18, 2000, was $5,812,484,762,023. This number is indicated in the following place-value chart.

Place-value Chart														
Trillions			Billions			Millions			Thousands			Ones		
		5	8	1	2	4	8	4	7	6	2	0	2	3

EXAMPLE 6   Write the number for the national debt for the United States as of January 18, 2000, in the amount of $5,812,484,762,023 using a word name.

The national debt in January 2000 was five trillion, eight hundred twelve billion, four hundred eighty-four million, seven hundred sixty-two thousand, twenty-three dollars.

Practice Problem 6   As of July 6, 2000, the estimated population of the world was 6,153,584,805. Write this world population using a word name.

Occasionally you may want to write a word name as a number.

EXAMPLE 7   Write each number in standard notation.

(a) twenty-six thousand, eight hundred fifty-four

(b) two billion, three hundred eighty-six million, five hundred forty-seven thousand, one hundred ninety

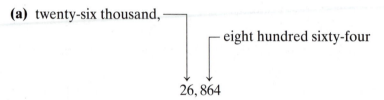
(a)   twenty-six thousand, ⎯

eight hundred sixty-four

26, 864

Thus we have 26,864.

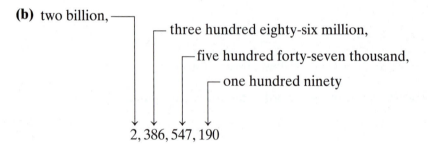
(b)   two billion, ⎯

three hundred eighty-six million,

five hundred forty-seven thousand,

one hundred ninety

2, 386, 547, 190

Thus we have 2,386,547,190.

Practice Problem 7   Write in standard notation.

(a) eight hundred three

(b) thirty thousand, two hundred twenty-nine

3   Reading Numbers in Tables

Sometimes numbers found in charts and tables are abbreviated. Look at the chart on the next page from the U.S. Bureau of the Census. Notice that the second line tells us the numbers represent thousands. To understand what these numbers mean, think "thousands." If the number 23 appears under 1740 for New Hampshire, the 23 represents 23 thousand. 23 thousand is 23,000. Note that census figures for some colonies are not available for certain years.

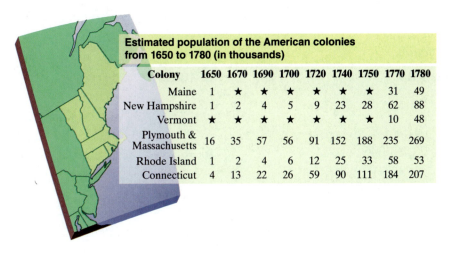

Estimated population of the American colonies from 1650 to 1780 (in thousands)

Colony	1650	1670	1690	1700	1720	1740	1750	1770	1780
Maine	1	★	★	★	★	★	★	31	49
New Hampshire	1	2	4	5	9	23	28	62	88
Vermont	★	★	★	★	★	★	★	10	48
Plymouth & Massachusetts	16	35	57	56	91	152	188	235	269
Rhode Island	1	2	4	6	12	25	33	58	53
Connecticut	4	13	22	26	59	90	111	184	207

EXAMPLE 8   Refer to the preceding chart to answer the following questions. Write each number in standard notation.

(a) What was the estimated population of Maine in 1780?

(b) What was the estimated population of Plymouth and Massachusetts in 1720?

(c) What was the estimated population of Rhode Island in 1700?

(a) To read the chart, first look for Maine on the left. Read across to the column under 1780. The number is 49. In this chart 49 means 49 thousands.

$$49 \text{ thousands} \Rightarrow 49,000$$

(b) Read the line of the chart for Plymouth and Massachusetts. The number for Plymouth and Massachusetts under 1720 is 91. This means 91 thousands. We will write this as 91,000.

(c) Read the line of the chart for Rhode Island. The number for Rhode Island under 1700 is 6. This means 6 thousands. We will write this as 6000.

To Think About   Why do you think Plymouth and Massachusetts had the largest population for the years shown in the table?

Practice Problem 8   Refer to the preceding chart to answer the following questions. Write each number in standard notation.

(a) What was the estimated population of Connecticut in 1670?

(b) What was the estimated population of New Hampshire in 1780?

(c) What was the estimated population of Vermont in 1770?

Write each number in expanded notation.

1. 6731

2. 9519

3. 108,276

4. 350,765

5. 23,761,345

6. 46,198,253

7. 103,260,768

8. 820,310,574

Write each number in standard notation.

9. 600 + 70 + 1

10. 500 + 90 + 6

11. 9000 + 800 + 60 + 3

12. 7000 + 600 + 50 + 2

13. 30,000 + 9000 + 700 + 30 + 3

14. 60,000 + 7000 + 200 + 4

15. 700,000 + 6000 + 200

16. 900,000 + 50,000 + 40 + 7

Verbal and Writing Skills

17. In the number 56,782
(a) What digit tells the number of hundreds?

(b) What is the value of the digit 5?

18. In the number 318,172
(a) What digit tells the number of thousands?

(b) What is the value of the digit 3?

19. In the number 945,302
(a) What digit tells the number of ten thousands?

(b) What is the value of the digit?

20. In the number 6,789,345
(a) What digit tells the number of thousands?

(b) What is the value of the digit?

Write a word name for each number.

21. 53

22. 46

23. 8936

24. 4629

25. 36,118   **26.** 55,742   **27.** 105,261   **28.** 370,258

29. 14,203,326   **30.** 71,189,530   **31.** 4,302,156,200   **32.** 7,436,210,400

Write each number in standard notation.

33. three hundred seventy-five

34. seven hundred thirty-six

35. twenty-seven thousand, three hundred eighty-two

36. ninety-two thousand, four hundred four

37. one hundred million, seventy-nine thousand, eight hundred twenty-six

38. four hundred fifty million, three hundred thousand, two hundred forty-nine

Applications

When writing a check, a person must write the word name for the dollar amount of the check.

ALEX J. WRITER
10 MAIN STREET   936-555-1212
WESTWOOD, TX 75862

$\frac{53-235}{113}$ 3254   6848

DATE April 1, 2001

PAY TO THE
ORDER OF _____ $ 1,965 00

_____ DOLLARS

Last
Chance Bank
Westwood, Texas

MEMO _____

⑈011302347⑈ 011081114⑈ 6848

39. Alex bought new equipment for his laboratory for $1965. What word name should he write on the check?

40. Alex later bought a new personal computer for $6383. What word name should he write on the check?

In exercises 41–44, use the following chart prepared with data from the U.S. Bureau of the Census. Notice that the second line tells us that the numbers represent millions. These values are only approximate values representing numbers written to the nearest million. They are not exact census figures.

**Estimated population of four states
from 1900 to 2000 (in millions)**

State	1900	1910	1920	1930	1940	1950	1960	1970	1980	1990	2000
California	1	2	3	6	7	11	16	20	24	30	33
Florida	1	1	1	1	2	3	5	7	10	13	15
Illinois	5	5	6	8	8	9	10	11	11	11	12
New York	7	9	10	13	13	15	17	18	18	18	18

Source: U.S. Bureau of the Census

41. What was the estimated population of New York in 1910?

42. What was the estimated population of Florida in 1970?

43. What was the estimated population of California in 2000?

44. What was the estimated population of Illinois in 1940?

In exercises 45–48, use the following chart prepared with data from the U.S. Department of Commerce. Notice that the title tells us the numbers represent billions of dollars. These values are only approximate values representing numbers written to the nearest billion. They are not exact financial figures.

**Personal Consumption Expenditures in Billions of Dollars
in the United States: Selected Years From 1990 to 1997**

Source: Bureau of Economic Analysis, U.S. Department of Commerce

Category	1990	1991	1995	1996	1997
Transportation	463	438	574	612	636
Medical care	615	656	875	912	957
Recreation	281	290	404	432	463
Religious and welfare activities	101	108	139	151	158

45. How much did people in the United States spend in 1990 on recreation?

46. How much did people in the United States spend in 1997 on medical care?

47. How much did people in the United States spend in 1995 on transportation?

48. How much did people in the United States spend in 1991 on religious and welfare activities?

49. The speed of light is approximately 29,979,250,000 centimeters per second.
(a) What digit tells the number of ten thousands?

(b) What digit tells the number of ten billions?

50. The radius of the earth is approximately 637,814,000 centimeters.
(a) What digit tells the number of thousands?

(b) What digit tells the number of millions?

51. In 1997 more than 798,400 immigrants came to the United States from other countries.
(a) Which digit tells the number of hundred thousands?
(b) Which digit tells the number of thousands?

52. The world's population is expected to reach 7,900,000,000 by the year 2020, according to the U.S. Bureau of the Census.
(a) Which digit tells the number of hundred millions?
(b) Which digit tells the number of billions?

53. Write in standard notation: six hundred thirteen trillion, one billion, thirty-three million, two hundred eight thousand, three.

To Think About

54. Write a word name for 3,682,968,009,931,960,747. (*Hint:* The digit 1 followed by 18 zeros represent the number *1 quintillion.* 1 followed by 15 zeros represents the number *1 quadrillion.*)

55. The number 50,000,000,000,000,000,000 is represented on some scientific calculators as 5 E 19. We will cover this in more detail in a later chapter. However, for the present we can see that this is a convenient notation that allows us to record very large whole numbers. Note that this number (50 quintillion) is a 5 followed by 19 zeros. Write in standard form the number that would be represented on a calculator as 6 E 22.

56. Think about the discussion in exercise 55. If the number 4 E 20 represented on a scientific calculator was divided by 2, what number would be the result? Write your answer in standard form.

57. Consider all the whole numbers between 200 and 800 that contain the digit 6. How many such numbers are there?

Student Learning Objectives

After studying this section, you will be able to:

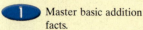 Master basic addition facts.

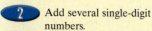 Add several single-digit numbers.

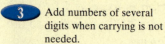 Add numbers of several digits when carrying is not needed.

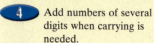 Add numbers of several digits when carrying is needed.

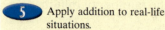 Apply addition to real-life situations.

SSM PH TUTOR CD & VIDEO MATH PRO WEB
CENTER

1 *Mastering Basic Addition Facts*

We see the addition process time and time again. Carpenters add to find the amount of lumber they need for a job. Auto mechanics add to make sure they have enough parts in the inventory. Bank tellers add to get cash totals.

What is addition? We do addition when we put sets of objects together.

$$5 \text{ objects} \quad + \quad 7 \text{ objects} \quad = \quad 12 \text{ objects}$$
$$5 + 7 = 12$$

Usually when we add numbers, we put one number under the other in a column. The numbers being added are called **addends.** The result is called the **sum.**

Suppose that we have four pencils in the car and we bring three more pencils from home. How many pencils do we have with us now? We add 4 and 3 to obtain a value of 7. In this case, the numbers 4 and 3 are the addends and the answer 7 is the sum.

$$
\begin{array}{r}
4 \quad \text{addend} \\
+ 3 \quad \text{addend} \\
\hline
7 \quad \text{sum}
\end{array}
$$

Think about what we do when we add 0 to another number. We are not making a change, so whenever we add zero to another number, that number will be the sum. Since this is always true, this is called a *property.* Since the sum is identical to the number added to zero, this is called the **identity property of zero.**

EXAMPLE 1 Add.

(a) $8 + 5$ **(b)** $3 + 7$ **(c)** $9 + 0$

$$
\textbf{(a)} \quad
\begin{array}{r}
8 \\
+ 5 \\
\hline
13
\end{array}
\qquad
\textbf{(b)} \quad
\begin{array}{r}
3 \\
+ 7 \\
\hline
10
\end{array}
\qquad
\textbf{(c)} \quad
\begin{array}{r}
9 \\
+ 0 \\
\hline
9
\end{array}
$$

Note: When we add zero to any other number, that number is the sum.

Practice Problem 1 Add.

$$
\textbf{(a)} \quad
\begin{array}{r}
7 \\
+ 5 \\
\hline
\end{array}
\qquad
\textbf{(b)} \quad
\begin{array}{r}
9 \\
+ 4 \\
\hline
\end{array}
\qquad
\textbf{(c)} \quad
\begin{array}{r}
3 \\
+ 0 \\
\hline
\end{array}
$$

The following table shows the basic addition facts. You should know these facts. If any of the answers don't come to you quickly, now is the time to learn them. To check your knowledge try Exercises 1.2, exercises 3 and 4.

Basic Addition Facts

+	0	1	2	3	4	5	6	7	8	9
0	0	1	2	3	4	5	6	7	8	9
1	1	2	3	4	5	6	7	8	9	10
2	2	3	4	5	6	7	8	9	10	11
3	3	4	5	6	7	8	9	10	11	12
4	4	5	6	7	8	9	10	11	12	13
5	5	6	7	8	9	10	11	12	13	14
6	6	7	8	9	10	11	12	13	14	15
7	7	8	9	10	11	12	13	14	15	16
8	8	9	10	11	12	13	14	15	16	17
9	9	10	11	12	13	14	15	16	17	18

To use the table to find the sum $4 + 7$, read across the top of the table to the 4 column, and then read down the left to the 7 row. The box where the 4 and 7 meet is 11, which means that $4 + 7 = 11$. Now read across the top to the 7 column and down the left to the 4 row. The box where these numbers meet is also 11. We can see that the order in which we add the numbers does not change the sum. $4 + 7 = 11$, and $7 + 4 = 11$. We call this the **commutative property of addition.**

This property does not hold true for everything in our lives. When you put on your socks and then your shoes, the result is not the same as if you put on your shoes first and then your socks! Can you think of any other examples where changing the order in which you add things would change the result?

2 Adding Several Single-Digit Numbers

If more than two numbers are to be added, we usually add from the first number to the next number and mentally note the sum. Then we add that sum to the next number, and so on.

EXAMPLE 2 Add. $3 + 4 + 8 + 2 + 5$

We rewrite the addition problem in a column format.

Mentally, we do these steps.

$$
\begin{array}{l}
\left.\begin{array}{l} 3 \\ 4 \end{array}\right\} 3 + 4 = 7 \\
8 \qquad\qquad\left.\right\} 7 + 8 = 15 \\
2 \qquad\qquad\qquad\left.\right\} 15 + 2 = 17 \\
\underline{+5} \qquad\qquad\qquad\qquad\left.\right\} 17 + 5 = 22 \\
22
\end{array}
$$

Practice Problem 2 Add. $7 + 6 + 5 + 8 + 2$

Because the order in which we add numbers doesn't matter, we can choose to add from the top down, from the bottom up, or in any other way. One shortcut is to add first any numbers that will give a sum of 10, or 20, or 30, and so on.

EXAMPLE 3 Add.

$$\begin{array}{r} 3 \\ 4 \\ 8 \\ 2 \\ +\,6 \\ \hline \end{array}$$

We mentally group the numbers into tens.

$$\begin{array}{l} 3 \\ 4 \leftarrow \\ 8 \leftarrow \\ 2 \leftarrow \quad 8 + 2 = 10 \quad \rightarrow 4 + 6 = 10 \\ 6 \leftarrow \end{array}$$

The sum is $10 + 10 + 3$ or 23.

Practice Problem 3 Add. $1 + 7 + 2 + 9 + 3$

3 Adding Numbers of Several Digits When Carrying is Not Needed

Of course, many numbers that we need to add have more than one digit. In such cases, we must be careful to first add the digits in the ones column, then the digits in the tens column, then those in the hundreds column, and so on. Notice that we move from *right to left*.

EXAMPLE 4 Add. $4304 + 5163$

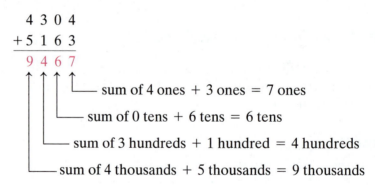

Practice Problem 4 Add.

$$\begin{array}{r} 8246 \\ +\,1702 \\ \hline \end{array}$$

4 Adding Numbers of Several Digits When Carrying Is Needed

When you add several whole numbers, often the sum in a column is greater than 9. However, we can only use *one* digit in any one place. What do we do with a two-digit sum? Look at the following example.

EXAMPLE 5 Add. $45 + 37$

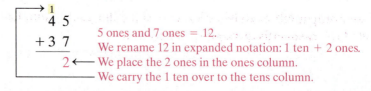

Note: Placing the 1 in the next column is often called "carrying the one."

$$\begin{array}{r} \overset{1}{4}\,5 \\ +\,3\,7 \\ \hline 8\,2 \end{array}$$ *Now we can add the digits in the tens column.*

Thus, $45 + 37 = 82$.

Practice Problem 5   Add. $\begin{array}{r} 56 \\ +\,36 \\ \hline \end{array}$

Often you must use carrying several times by bringing the left digit into the next column to the left.

EXAMPLE 6   Add. $257 + 688 + 94$

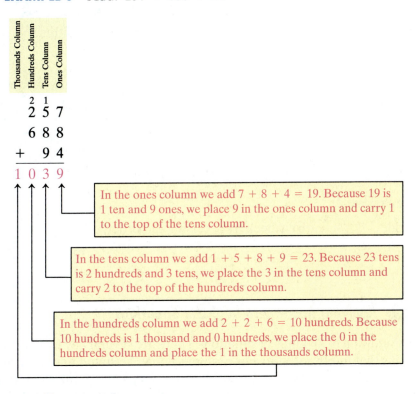

In the ones column we add $7 + 8 + 4 = 19$. Because 19 is 1 ten and 9 ones, we place 9 in the ones column and carry 1 to the top of the tens column.

In the tens column we add $1 + 5 + 8 + 9 = 23$. Because 23 tens is 2 hundreds and 3 tens, we place the 3 in the tens column and carry 2 to the top of the hundreds column.

In the hundreds column we add $2 + 2 + 6 = 10$ hundreds. Because 10 hundreds is 1 thousand and 0 hundreds, we place the 0 in the hundreds column and place the 1 in the thousands column.

Practice Problem 6   Add $789 + 63 + 297$

We can add numbers in more than one way. To add $5 + 3 + 7$ we can first add the 5 and 3. We do this by using parentheses to show the first operation to be done. This shows us that $5 + 3$ is to be grouped together.

$$5 + 3 + 7 = (5 + 3) + 7 = 15$$
$$= \quad 8 \quad + 7 = 15$$

We could add the 3 and 7 first. We use parentheses to show that we group $3 + 7$ together and that we will add these two numbers first.

$$5 + 3 + 7 = 5 + (3 + 7) = 15$$
$$= 5 + \quad 10 \quad = 15$$

The way we group numbers to be added does not change the sum. This property is called the **associative property of addition.**

Look again at the three properties of addition we have discussed in this section.

1. **Associative Property of Addition** When we add three numbers, we can group them in any way.	$(8 + 2) + 6 = 8 + (2 + 6)$ $10 + 6 = 8 + 8$ $16 = 16$
2. **Commutative Property of Addition** Two numbers can be added in either order with the same result.	$5 + 12 = 12 + 5$ $17 = 17$
3. **Identity Property of Zero** When zero is added to a number, the sum is that number.	$8 + 0 = 8$ $0 + 5 = 5$

Because of the commutative and associative properties of addition, we can check our addition by adding the numbers in the opposite order.

EXAMPLE 7 (a) Add the numbers. $39 + 7284 + 3132$

(b) Check by reversing the order of addition.

(a)
$$\begin{array}{r} \overset{1\,1}{39} \\ 7284 \\ + \ 3132 \\ \hline 10,455 \end{array}$$

Addition

(b)
$$\begin{array}{r} \overset{1\,1}{3132} \\ 7284 \\ + \quad 39 \\ \hline 10,455 \end{array}$$

Check by reversing the order.

The sum is the same in each case.

Practice Problem 7

(a) Add.
$$\begin{array}{r} 127 \\ 9876 \\ + \ 342 \\ \hline \end{array}$$

(b) Check by reversing the order.
$$\begin{array}{r} 342 \\ 9876 \\ + \ 127 \\ \hline \end{array}$$

5 Applying Addition to Real-Life Situations

We use addition in all kinds of situations. There are several key words in word problems that imply addition. For example, it may be stated that there are 12 math books, 9 chemistry books, and 8 biology books on a book shelf. To find the *total* number of books implies that we add the numbers $12 + 9 + 8$. Other key words are *how much, how many,* and *all.*

Sometimes a problem will have more information than you will need to answer the question. If you have too much information, to solve the problem you will need to separate out the facts that are not important. The following three steps are involved in the problem-solving process.

Step 1 Understand the problem.

Step 2 Calculate and state the answer.

Step 3 Check.

We may not write all of these steps down, but they are the steps we use to solve all problems.

EXAMPLE 8 The bookkeeper for Smithville Trucking was examining the following data for the company checking account.

Monday:	$23,416 was deposited and $17,389 was debited.
Tuesday:	$44,823 was deposited and $34,089 was debited.
Wednesday:	$16,213 was deposited and $20,057 was debited.

What was the total of all deposits during this period?

Step 1 *Understand the problem.*

Total implies that we will use addition. Since we don't need to know about the debits to answer this question, we use only the *deposit* amounts.

Step 2 *Calculate and state the answer.*

Monday:	$23,416 was deposited.	$\overset{11}{2}\overset{\,1}{3},416$
Tuesday:	$44,823 was deposited.	44,823
Wednesday:	$16,213 was deposited.	+ 16,213
		84,452

A total of $84,452 was deposited on those three days.

Step 3 *Check.*

You may add the numbers in reverse order to check. We leave the check up to you.

Practice Problem 8 North University has 23,413 men and 18,316 women. South University has 19,316 men and 24,789 women. East University has 20,078 men and 22,965 women. What is the total enrollment of *women* at the three universities?   ⬭

▲ ⬭ **EXAMPLE 9** Mr. Ortiz has a rectangular field whose length is 400 feet and whose width is 200 feet. What is the total number of feet of fence that would be required to fence in the field?

1. *Understand the problem.*

To help us to get a picture of what the field looks like, we will draw a diagram.

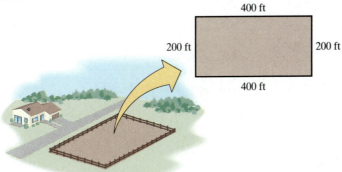

400 ft

200 ft   200 ft

400 ft

Note that ft is the abbreviation for feet. ft means feet.

2. *Calculate and state the answer.*

Since the fence will be along each side of the field, we add the lengths all around the field.

$$\begin{array}{r} 200 \\ 400 \\ 200 \\ + 400 \\ \hline 1200 \end{array}$$

The amount of fence that would be required is 1200 feet.

3. *Check.*

Regroup the addends and add.

$$\begin{array}{r} 200 \\ 200 \\ 400 \\ + 400 \\ \hline 1200 \end{array} \checkmark$$

▲ **Practice Problem 9** In Vermont, Gretchen fenced the rectangular field on which her sheep graze. The length of the field is 2000 feet and the width of the field is 1000 feet. What is the perimeter of the field? (*Hint:* The "distance around" an object [such as a field] is called the *perimeter.*)

Developing Your Study Skills

Class Attendance

You will want to get started in the right direction by choosing to attend class every day, beginning with the first day of class. Statistics show that class attendance and good grades go together. Classroom activities are designed to enhance learning, and therefore you must be in class to benefit from them. Each day vital information and explanations are given that can help you understand concepts. Do not be deceived into thinking that you can just find out from a friend what went on in class. There is no good substitute for firsthand experience. Give yourself a push in the right direction by developing the habit of going to class every day.

Verbal and Writing Skills

1. Explain in your own words.
   (a) the commutative property of addition   (b) the associative property of addition

2. When zero is added to any number, it does not change that number.
   Why do you think this is called the identity property of zero?

Complete the addition facts for each table. Strive for total accuracy, but work quickly. Allow a maximum of five minutes for each table.

3.

+	3	5	4	8	0	6	7	2	9	1
2										
7										
5										
3										
0										
4										
1										
8										
6										
9										

4.

+	1	6	5	3	0	9	4	7	2	8
3										
9										
4										
0										
2										
7										
8										
1										
6										
5										

Add.

5.
```
  4
  2
  8
+ 9
___
```

6.
```
  4
  6
  2
+ 7
___
```

7.
```
  3
  8
  9
+ 6
___
```

8.
```
  1
  4
  3
+ 8
___
```

9.
```
  10
  36
+  3
____
```

10.
```
  63
  11
+  6
____
```

11.
```
  63
  24
+ 12
____
```

12.
```
  54
  21
+ 23
____
```

13.
```
  2847
  1634
+   98
_____
```

14.
```
  5519
  1392
+  955
_____
```

15.
```
  7738
  1363
+ 3255
_____
```

16.
```
  5017
  2984
+ 1328
_____
```

17.
```
  8235
+ 5626
_____
```

18.
```
  5673
+ 3572
_____
```

19.
```
  26,108
+ 16,371
_____
```

20.
```
  78,315
+ 49,353
_____
```

Add from the top. Then check by adding in the reverse order.

21. 36
41
25
6
+ 13

22. 24
39
16
14
+ 9

23. 106
13
4
28
+ 981

24. 463
27
8
41
+ 507

Add.

25. 32
106
47
+ 7193

26. 189
7031
352
+ 1523

27. 1,362,214
7,002,316
+ 3,214,896

28. 4,002,983
2,134,702
+ 3,592,001

29. 837,241,000
+ 298,039,240

30. 982,306,000
+ 583,215,320

31. 516,208
24,317
+ 1,763,295

32. 32,500
763,420
+ 2,837,667

33. 12 + 8 + 156 + 72   **34.** 85 + 3 + 407 + 26   **35.** 15,216 + 485 + 5208 **36.** 26,002 + 599 + 3500

Applications

37. When Diana went shopping for holiday presents for her family, she spent $224 on Thursday, $387 on Friday, and $183 on Saturday. What is the total amount of money she spent on gifts?

38. Peter is redecorating the spare bedroom of the house to get ready for the arrival of a new baby. At the building supply store he made purchases of $421 on Friday, $218 on Saturday, and $79 on Sunday. How much did the new room cost to redecorate?

39. Paula's paycheck two months ago was $1184. Last month she made $1632, and this month she earned $1395. What is the total amount Paula made during this three-month period?

40. The taxes on Jack's house two years ago were $4658. Last year they were $5222. This year they are $6027. What is the total amount for three years?

▲ **41.** Nate wants to put a fence around his backyard. The sketch below indicates the length of each side of the yard. What is the total number of feet of fence he needs for his backyard?

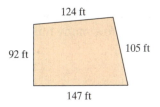

124 ft

92 ft   105 ft

147 ft

▲ **42.** Jessica has a field with the length of each side as labeled on the sketch. What is the total number of feet of fence that would be required to fence in the field? (Find the perimeter of the field.)

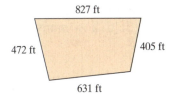

827 ft

472 ft   405 ft

631 ft

▲ **43.** The Pacific Ocean, the world's largest, has an area of 64,000,000 square miles. The Atlantic Ocean has an area of 31,800,000 square miles. The Indian Ocean has an area of 25,300,000 square miles. What is the total area for these oceans?

▲ **44.** The Arctic Ocean has an area of 5,400,000 square miles. The Mediterranean Sea has an area of 1,100,000 square miles. The Caribbean Sea has an area of 1,000,000 square miles. What is the total area for these bodies of water?

45. 2,144,856 people in California, 307,244 people in Oregon, and 470,239 people in Washington voted for Ross Perot for president in 1992. What was the total vote for Perot in those three states?

▲ **46.** Lake Huron, which is bordered by the United States and Canada, measures 23,010 square miles. Lake Tanganyika, which borders Tanzania and Zaire, measures 12,700 square miles. Lake Baikal in Russia measures 12,162 square miles. What is the total area for these lakes?

In exercises 47–48, be sure you understand the problem and then choose the numbers you need in order to answer each question. Then calculate the correct answer.

47. The admissions department of a competitive university is reviewing applications to see whether students are *eligible* or *ineligible* for student aid. On Monday, 415 were found eligible and 27 ineligible. On Tuesday, 364 were found eligible and 68 ineligible. On Wednesday, 159 were found eligible and 102 ineligible. On Thursday, 196 were found eligible and 61 ineligible.
 (a) How many students were eligible for student aid over the four days?

 (b) How many students were considered in all?

48. The quality control division of a motorcycle company classifies the final assembled bike as *passing* or *failing* final inspection. In January, 14,311 vehicles passed whereas 56 failed. In February, 11,077 passed and 158 failed. In March, 12,580 passed and 97 failed.
 (a) How many motorcycles passed the inspection during the three months?

 (b) How many motorcycles were assembled during the three months in all?

Use the following facts to answer exercises 49 and 50. It is 87 miles from Springfield to Weston. It is 17 miles from Weston to Boston. Driving directly, it is 98 miles from Springfield to Boston. It is 21 miles from Boston to Hamilton.

49. If Melissa drives from Springfield to Weston, then from Weston to Boston, and finally directly home to Springfield, how many miles does she drive?

50. If Marcia drives from Hamilton to Boston, then from Boston to Weston, and then from Weston to Springfield, how many miles does she drive?

51. Walter Swensen is examining the fences of a farm in Caribou, Maine. The shape of one field is in the shape of a four-sided figure with no sides equal. The field is enclosed with 2387 feet of wooden rail fence. The first side is 568 feet long, while the second side is 682 feet long. The third side is 703 feet long. How long is the fourth side?

52. Carlos Sontera is walking to examine the fences of a ranch in El Paso, Texas. The field he is examining is in the shape of a rectangle. The perimeter of the rectangle is 3456 feet. One side of the rectangle is 930 feet long. How long are the other sides? (*Hint:* The opposite sides of a rectangle are equal.)

53. Answer using the information in the following Western University expense chart for the current academic year.

Western University Yearly Expenses	In-State Student, U.S. Citizen	Out-of-State Student, U.S. Citizen	Foreign Student
Tuition	$3640	$5276	$8352
Room	1926	2437	2855
Board	1753	1840	1840

How much is the total cost for tuition, room, and board for
(a) an out-of-state U.S. citizen?
(b) an in-state U.S. citizen?
(c) a foreign student?

To Think About

In exercises 54–55, add.

54. 2,368,521,788 + 5,721,368,701 + 4,027,399,206

55. 89 + 166 + 23 + 45 + 72 + 190 + 203 + 77 + 18 + 93 + 46 + 73 + 66

56. What would happen if addition were not commutative?

57. What would happen if addition were not associative?

Cumulative Review Problems

Write the word name for each number.

58. 76,208,941

59. 121,000,374

Write each number in standard notation.

60. eight million, seven hundred twenty-four thousand, three hundred ninety-six

61. nine million, fifty-one thousand, seven hundred nineteen

Developing Your Study Skills

Class Participation

People learn mathematics through active participation, not through observation from the sidelines. If you want to do well in this course, get involved in classroom activities. Sit near the front where you can see and hear well, where your focus is on the instruction process. Ask questions, be ready to contribute toward solutions, and take part in all classroom activities. Your contributions are valuable to the class and to yourself. Class participation requires an investment of yourself in the learning process, which you will find pays huge dividends.

1.3 Subtraction of Whole Numbers

Student Learning Objectives

After studying this section, you will be able to:

1. Master basic subtraction facts.
2. Subtract whole numbers when borrowing is not necessary.
3. Subtract whole numbers when borrowing is necessary.
4. Check the answer to a subtraction problem.
5. Apply subtraction to real-life situations.

SSM PH TUTOR CENTER CD & VIDEO MATH PRO WEB

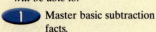 **Mastering Basic Subtraction Facts**

Subtraction is used day after day in the business world. The owner of a bakery placed an ad for his cakes in a local newspaper to see if this might increase his profits. To learn how many cakes had been sold, at closing time he subtracted the number of cakes remaining from the number of cakes the bakery had when it opened. To figure his profits, he subtracted his costs (including the cost of the ad) from his sales. Finally, to see if the ad paid off, he subtracted the profits he usually made in that period from the profits after advertising. He needed subtraction to see whether it paid to advertise.

What is subtraction? We do subtraction when we take objects away from a group. If you have 12 objects and take away 3 of them, 9 objects remain.

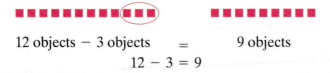

12 objects − 3 objects   =   9 objects

$$12 - 3 = 9$$

If you earn $400 per month, but have $100 taken out for taxes, how much do you have left?

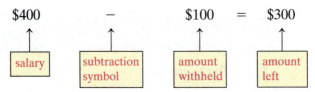

$400   − $100   = $300

salary | subtraction symbol | amount withheld | amount left

We can use addition to help with a subtraction problem.

To subtract: $200 - 196 =$ what number

We can think: $196 +$ what number $= 200$

Usually when we subtract numbers, we put one number under the other in a column. When we subtract one number from another, the answer is called the **difference.**

$$
\begin{array}{cccc}
9 & 8 & 12 & 17 \\
-2 & -3 & -6 & -9 \\
\hline
7 & 5 & 6 & 8
\end{array}
$$

Each of these is called the difference of the two numbers.

The other two parts of a subtraction problem have labels, although you will not often come across them. The number being subtracted is called the **subtrahend.** The number being subtracted from is called the **minuend.**

$$
\begin{array}{ll}
17 & \text{minuend} \\
-\;9 & \text{subtrahend} \\
\hline
8 & \text{difference}
\end{array}
$$

In this case; the number 17 is called the *minuend.* The number 9 is called the *subtrahend.* The number 8 is called the *difference.*

Quick Recall of Subtraction Facts

It is helpful if you can subtract quickly. See if you can do Example 1 correctly in 15 seconds or less. Repeat again with Practice Problem 1. Strive to obtain all answers correctly in 15 seconds or less.

 EXAMPLE 1   Subtract.

(a) $8 - 2$ **(b)** $13 - 5$ **(c)** $12 - 4$ **(d)** $15 - 8$ **(e)** $16 - 0$

(a) $\begin{array}{r} 8 \\ -2 \\ \hline 6 \end{array}$ **(b)** $\begin{array}{r} 13 \\ -5 \\ \hline 8 \end{array}$ **(c)** $\begin{array}{r} 12 \\ -4 \\ \hline 8 \end{array}$ **(d)** $\begin{array}{r} 15 \\ -8 \\ \hline 7 \end{array}$ **(e)** $\begin{array}{r} 16 \\ -0 \\ \hline 16 \end{array}$

Practice Problem 1   Subtract.

(a) $\begin{array}{r} 9 \\ -6 \\ \hline \end{array}$ **(b)** $\begin{array}{r} 12 \\ -5 \\ \hline \end{array}$ **(c)** $\begin{array}{r} 17 \\ -8 \\ \hline \end{array}$ **(d)** $\begin{array}{r} 14 \\ -0 \\ \hline \end{array}$ **(e)** $\begin{array}{r} 18 \\ -9 \\ \hline \end{array}$

2   *Subtracting Whole Numbers When Borrowing Is Not Necessary*

When we subtract numbers with more than two digits, in order to keep track of our work, we line up the ones column, the tens column, the hundreds column, and so on. Note that we begin with the ones column, and move from right to left.

 EXAMPLE 2   Subtract. $9867 - 3725$

$$\begin{array}{r} 9\ 8\ 6\ 7 \\ -\ 3\ 7\ 2\ 5 \\ \hline 6\ 1\ 4\ 2 \end{array}$$

7 ones − 5 ones = 2 ones

6 tens − 2 tens = 4 tens

8 hundreds − 7 hundreds = 1 hundred

9 thousands − 3 thousands = 6 thousands

Practice Problem 2   Subtract. $7695 - 3481$

3   *Subtracting Whole Numbers When Borrowing Is Necessary*

In the subtraction that we have looked at so far, each digit in the upper number (the minuend) has been greater than the digit in the lower number (the subtrahend) for each place value. Many times, however, a digit in the lower number is greater than the digit in the upper number for that place value.

$$\begin{array}{r} 42 \\ -28 \\ \hline \end{array}$$

The digit in the ones place in the lower number, the 8 of 28, is greater than the number in the ones place in the upper number, the 2 of 42. To subtract, we must *rename* 42, using place values. This is called **borrowing.**

EXAMPLE 3 Subtract. 42 − 28

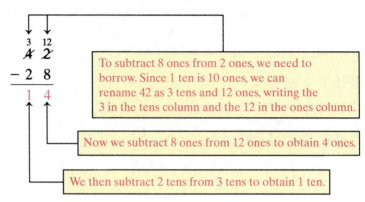

To subtract 8 ones from 2 ones, we need to borrow. Since 1 ten is 10 ones, we can rename 42 as 3 tens and 12 ones, writing the 3 in the tens column and the 12 in the ones column.

Now we subtract 8 ones from 12 ones to obtain 4 ones.

We then subtract 2 tens from 3 tens to obtain 1 ten.

Practice Problem 3 Subtract. 34 − 16

EXAMPLE 4 Subtract. 864 − 548

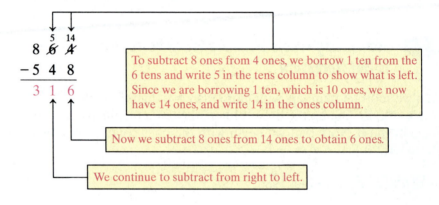

To subtract 8 ones from 4 ones, we borrow 1 ten from the 6 tens and write 5 in the tens column to show what is left. Since we are borrowing 1 ten, which is 10 ones, we now have 14 ones, and write 14 in the ones column.

Now we subtract 8 ones from 14 ones to obtain 6 ones.

We continue to subtract from right to left.

Practice Problem 4 Subtract.
$$693$$
$$- 426$$

EXAMPLE 5 Subtract. 8040 − 6375

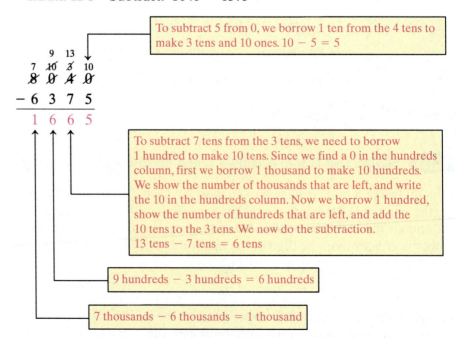

To subtract 5 from 0, we borrow 1 ten from the 4 tens to make 3 tens and 10 ones. 10 − 5 = 5

To subtract 7 tens from the 3 tens, we need to borrow 1 hundred to make 10 tens. Since we find a 0 in the hundreds column, first we borrow 1 thousand to make 10 hundreds. We show the number of thousands that are left, and write the 10 in the hundreds column. Now we borrow 1 hundred, show the number of hundreds that are left, and add the 10 tens to the 3 tens. We now do the subtraction. 13 tens − 7 tens = 6 tens

9 hundreds − 3 hundreds = 6 hundreds

7 thousands − 6 thousands = 1 thousand

Practice Problem 5   Subtract.
$$\begin{array}{r} 9070 \\ -\ 5886 \\ \hline \end{array}$$

EXAMPLE 6   Subtract.

(a) $9521 - 943$ **(b)** $40,000 - 29,056$

(a)
$$\begin{array}{r} {}^{8}\overset{14}{\cancel{4}}\ \overset{11}{\cancel{4}}\ \overset{11}{\cancel{1}} \\ \cancel{9}\ \cancel{5}\ \cancel{2}\ \cancel{1} \\ -\ \ \ 9\ 4\ 3 \\ \hline 8\ 5\ 7\ 8 \end{array}$$

(b)
$$\begin{array}{r} \overset{3}{\cancel{4}}\ \overset{9}{\cancel{0}},\ \overset{9}{\cancel{0}}\ \overset{9}{\cancel{0}}\ \overset{10}{\cancel{0}} \\ -\ 2\ 9,\ 0\ 5\ 6 \\ \hline 1\ 0,\ 9\ 4\ 4 \end{array}$$

Practice Problem 6   Subtract.

(a)
$$\begin{array}{r} 8964 \\ -\ 985 \\ \hline \end{array}$$
(b)
$$\begin{array}{r} 50,000 \\ -\ 32,508 \\ \hline \end{array}$$

4   Checking the Answer to a Subtraction Problem

We observe that when $9 - 7 = 2$ it follows that $7 + 2 = 9$. Each subtraction problem is equivalent to a corresponding addition problem. This gives us a convenient way to check our answers to subtraction.

EXAMPLE 7   Check this subtraction problem.

$$5829 - 3647 = 2182$$

$$\begin{array}{r} 5\ 8\ 2\ 9 \longleftarrow \\ -3\ 6\ 4\ 7 \\ \hline 2\ 1\ 8\ 2 \end{array} \quad \text{then} \quad \begin{array}{r} 3\ 6\ 4\ 7 \longleftarrow \\ +2\ 1\ 8\ 2 \\ \hline 5\ 8\ 2\ 9 \end{array}$$

The sum should equal 5829, which it does. We have checked our work, and it is correct.

Practice Problem 7   Check this subtraction problem by adding.

$$9763 - 5732 = 4031$$

EXAMPLE 8   Subtract and check your answers.

(a) $156,000 - 29,326$ **(b)** $1,264,308 - 1,057,612$

(a)
$$\begin{array}{r} 156,000 \longleftarrow \\ -\ \ 29,326 \\ \hline 126,674 \end{array} \qquad \begin{array}{r} 29,326 \\ +\ 126,674 \\ \hline 156,000 \longleftarrow \end{array}$$

It checks.

(b)
$$\begin{array}{r} 1,264,308 \longleftarrow \\ -1,057,612 \\ \hline 206,696 \end{array} \qquad \begin{array}{r} 1,057,612 \\ +\ \ 206,696 \\ \hline 1,264,308 \longleftarrow \end{array}$$

It checks.

Practice Problem 8   Subtract and check your answers.

(a)   284,000
    − 96,327

(b)   8,526,024
    − 6,397,518

Subtraction can be used to solve word problems. Some problems can be expressed (and solved) with an **equation.** An equation is a number sentence with an equal sign, such as

$$10 = 4 + x$$

Here we use the letter x to represent a number we do not know. When we write $10 = 4 + x$, we are stating that 10 is equal to 4 added to some other number. Since $10 − 4 = 6$, we would assume that the number is 6. If we substitute 6 for x in the equation, we have two values that are the same.

$$10 = 4 + x$$
$$10 = 4 + 6 \qquad \text{Substitute 6 for } x.$$
$$10 = 10 \qquad \text{Both sides of the equation are the same.}$$

We can write an equation when one of the addends is not known, then use subtraction to solve for the unknown.

EXAMPLE 9   The librarian knows that he has eight world atlases and that five of them are in full color. How many are not in full color?

We represent the number that we don't know as x and write an equation, or mathematical sentence.

$$8 = 5 + x$$

To solve an equation means to find those values that will make the equation true. We solve this equation by reasoning and by a knowledge of the relationship between addition and subtraction.

$$8 = 5 + x \text{ is equivalent to } 8 − 5 = x$$

We know that $8 − 5 = 3$. Then $x = 3$. We can check the answer by substituting 3 for x in the original equation.

$$8 = 5 + x$$
$$8 = 5 + 3 \qquad \text{True } \checkmark$$

We see that $x = 3$ checks, so our answer is correct. There are three atlases not in full color.

Practice Problem 9   Form an equation for each of the following problems and solve the equation in order to answer each question.

(a) The Salem Harbormaster's daily log noted that seventeen fishing vessels left the harbor yesterday during daylight hours. Walter was at the harbor all morning and saw twelve fishing vessels leave in the morning. How many vessels left in the afternoon? (Assume that sunset was at 6 P.M.)

(b) The Appalachian Mountain Club noted that twenty-two hikers left to climb Mount Washington during the morning. By 4 P.M., ten of them had returned. How many of the hikers were still on the mountain?

5   Applying Subtraction to Real-Life Situations

We use subtraction in all kinds of situations. There are several key words in word problems that imply subtraction. Words that involve comparison, such as *how much more, how much greater,* or how much a quantity *increased* or *decreased,* all imply subtraction. The *difference* between two numbers implies subtraction.

⬤ **EXAMPLE 10**   Look at the following population table.

**Estimated population of four states
from 1960 to 2000 (in millions)**

State	1960	1970	1980	1990	2000*
California	15,717,204	19,971,069	23,667,947	29,760,021	32,521,675
Texas	9,579,677	11,198,655	14,227,799	16,986,510	20,119,366
Arizona	1,302,161	1,775,399	2,716,598	3,665,228	4,798,037
New Mexico	951,023	1,017,055	1,303,302	1,515,069	1,860,429

Source: U.S. Bureau of the Census   *estimated

(a) In 1980, how much greater was the population of Texas than that of Arizona?

(b) How much did the population of California increase from 1960 to 2000?

(c) How much greater was the population of California in 1990 than that of the other three states combined?

(a)   14,227,799   1980 population of Texas
  − 2,716,598   1980 population of Arizona
  11,511,201   difference

The population of Texas was greater by 11,511,201.

(b)   32,521,675   2000 population of California
  − 15,717,204   1960 population of California
  16,804,471   difference

The population of California increased by 16,804,471 in those 40 years.

(c) First we need to find the total population in 1990 of Texas, Arizona, and New Mexico.

    16,986,510   1990 population of Texas
    3,665,228   1990 population of Arizona
  + 1,515,069   1990 population of New Mexico
    22,166,807

We use subtraction to compare this total with the population of California.

    29,760,021   1990 population of California
  − 22,166,807
    7,593,214

The population in California in 1990 was 7,593,214 more than the population in the other three states combined.

Practice Problem 10

(a) In 1980, how much greater was the population of California than the population of Texas?

(b) How much did the population of Texas increase from 1960 to 1970?

EXAMPLE 11 The number of real estate transfers in several towns during the years 1999 to 2001 is given in the following bar graph.

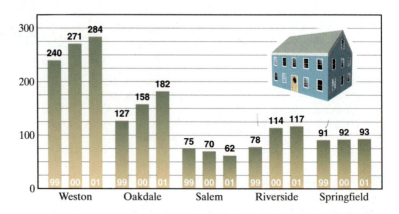

(a) What was the increase in homes sold in Weston from 2000 to 2001?

(b) What was the decrease in homes sold in Salem from 1999 to 2001?

(c) Between what two years did Oakdale have the greatest increase in sales?

(a) From the labels on the bar graph we see that 284 homes were sold in 2001 in Weston and 271 homes were sold in 2000. Thus the increase can be found by subtracting $284 - 271 = 13$. There was an increase of 13 homes sold in Weston from 2000 to 2001.

(b) In 1999, 75 homes were sold in Salem. In 2001, 62 homes were sold in Salem. The decrease in the number of homes sold is $75 - 62 = 13$. There was a decrease of 13 homes sold in Salem from 1999 to 2001.

(c) Here we will need to make two calculations in order to decide where the greatest increase occurs.

$$
\begin{array}{ll}
158 & \text{2000 sales} \\
-127 & \text{1999 sales} \\
\hline
31 & \text{Sales increase} \\
 & \text{from 1999 to 2000}
\end{array}
\qquad
\begin{array}{ll}
182 & \text{2001 sales} \\
-158 & \text{2000 sales} \\
\hline
24 & \text{Sales increase} \\
 & \text{from 2000 to 2001}
\end{array}
$$

The greatest increase in sales in Oakdale occurred from 1999 to 2000.

Practice Problem 11 Based on the preceding bar graph, answer the following questions.

(a) What was the increase in homes sold in Riverside from 1999 to 2000?

(b) How many more homes were sold in Springfield in 1999 than in Riverside in 1999?

(c) Between what two years did Weston have the greatest increase in sales?

1.3 Exercises

Verbal and Writing Skills

1. Explain how you can check a subtraction problem.

2. Explain how you use borrowing to calculate $107 - 88$.

3. Explain what number should be used to replace the question mark in the subtraction equation $32?5 - 1683 = 1592$.

4. Explain what steps need to be done to calculate 7 feet − 11 inches.

Try to do exercises 5–20 in one minute or less with no errors.

Subtract.

5.   6
  − 5
	6. 17
	  − 8
		7. 15
		  − 6
			8. 14
			  − 5
				9. 16
				  − 0
					10. 17
					  − 9

11.   18
  − 9
	12. 12
	  − 7
		13. 11
		  − 4
			14. 15
			  − 8
				15. 13
				  − 7
					16. 16
					  − 9

17.   11
  − 8
	18. 10
	  − 7
		19. 15
		  − 6
			20. 12
			  − 5		

Subtract. Check your answers by adding.

21.   47
  − 26
	22. 96
	  − 51
		23. 85
		  − 73
			24. 77
			  − 36
				25. 126
				  − 95

26.   193
  − 72
	27. 768
	  − 143
		28. 621
		  − 320
			29. 2893
			  − 572
				30. 5780
				  − 530

31.	24,396 − 13,205	32.	52,980 − 31,660	33.	986,302 − 433,201	34.	807,965 − 304,214

Check each subtraction. If the problem has not been done correctly, find the correct answer.

35.	129 − 19 110	36.	186 − 45 141	37.	7347 − 1213 6134	38.	9956 − 7254 2702

39.	6030 − 5020 1020	40.	7890 − 3200 7670	41.	99,583 − 41,181 58,402	42.	47,869 − 33,846 13,023

Subtract. Use borrowing if necessary.

43.	93 − 47	44.	86 − 33	45.	125 − 88	46.	136 − 95	47.	451 − 376	48.	706 − 435

49.	905 − 324	50.	861 − 345	51.	20,000 − 9,308	52.	50,000 − 36,670	53.	152,000 − 117,908

54.	361,000 − 121,520	55.	45,312 − 37,865	56.	64,381 − 29,997	57.	760,108 − 536,992	58.	580,092 − 349,905

Solve.

59. $x + 14 = 19$

60. $x + 35 = 50$

61. $34 = x + 13$

62. $25 = x + 18$

63. $100 + x = 127$

64. $86 + x = 120$

Applications

65. This year, for the college graduation ceremonies, the senior class voted for "Power of Example to Humanity." Martin Luther King received 765 votes, John F. Kennedy received 960 votes, and Mother Teresa received 778 votes. How many more votes did John F. Kennedy receive than Mother Teresa?

66. Three of the highest elevations on Earth are Mt. Everest in Asia, which is 8846 meters high; Mt. Aconcagua in Argentina, which is 6960 meters high; and Mt. McKinley in the United States (Alaska), which is 6194 meters high. How much higher is Mt. Everest than Mt. Aconcagua?

67. In 1999, the population of Ireland was approximately 3,632,944. In the same year, the population of Portugal was approximately 9,918,040. How much less than the population of Portugal was the population of Ireland in 1999?

68. The Nile River, the longest river in the world, is approximately 22,070,400 feet long. The Yangtze Kiang River, which is the longest river in China, is approximately 19,018,560 feet long. How much longer is the Nile River than the Yangtze Kiang River?

69. James earned $475 painting a few rooms for his neighbor. He gave $142 to his assistant and paid $85 to the hardware store for materials. How much money did he actually receive after paying expenses?

70. The Ski Club raised $1682 at a party for its favorite charity. The band received $350, and the club spent $265 on refreshments. How much money was actually raised for the charity after paying expenses?

In answering exercises 71–78, consider the following population table.

	1960	1970	1980	1990	2000
Illinois	10,081,158	11,110,285	11,427,409	11,430,602	12,051,683
Michigan	7,823,194	8,881,826	9,262,044	9,295,297	9,679,052
Indiana	4,662,498	5,195,392	5,490,212	5,544,159	6,045,521
Minnesota	3,413,864	3,806,103	4,075,970	4,375,099	4,830,784

Source: U.S. Census Bureau

71. How much did the population of Minnesota increase from 1960 to 2000?

72. How much did the population of Michigan increase from 1960 to 2000?

73. In 1970, how much greater was the population of Michigan than the population of Indiana?

74. In 1980, how much greater was the population of Indiana than the population of Minnesota?

75. How much did the population of Illinois increase from 1970 to 1990?

76. How much did the population of Michigan increase from 1970 to 1990?

77. Compare your answers to exercises 75 and 76. How much greater was the population increase of Michigan than the population increase of Illinois from 1970 to 1990?

78. In 1990, what was the difference in population between the state with the highest population and the state with the lowest population?

The number of real estate transfers in several towns during the years 1999 to 2001 is given in the following bar graph. Use the bar graph to answer exercises 79–86. The figures in the bar graph reflect sales of single-family detached homes only.

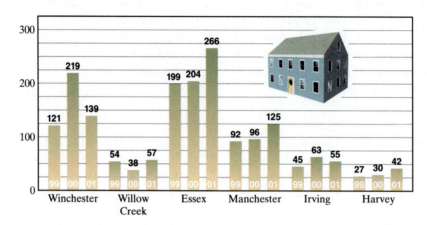

79. What was the increase in the number of homes sold in Manchester from 2000 to 2001?

80. What was the increase in the number of homes sold in Irving from 1999 to 2000?

81. What was the decrease in the number of homes sold in Winchester from 2000 to 2001?

82. What was the decrease in the number of homes sold in Willow Creek from 1999 to 2000?

83. Between what two years did the greatest change occur in the number of homes sold in Irving?

84. Between what two years did the greatest change occur in the number of homes sold in Winchester?

85. A real estate agent was trying to determine which two towns were closest to having the same number of sales of homes in 1999. What two towns should she select?

86. A real estate agent was trying to determine which two towns were closest to having the same number of sales of homes in 2000. What two towns should she select?

To Think About

87. In general, subtraction is not commutative. If a and b are whole numbers, $a - b \neq b - a$. For what types of numbers would it be true that $a - b = b - a$?

88. In general, subtraction is not associative. For example, $8 - (4 - 3) \neq (8 - 4) - 3$. In general, $a - (b - c) \neq (a - b) - c$. Can you find some numbers a, b, c for which $a - (b - c) = (a - b) - c$? (Remember, do operations inside the parentheses first.)

89. Walter Swensen wants to replace some of the fences on a farm in Caribou, Maine. The wooden rail fence costs about $60 for wood and $50 labor to install a fence that is 12 feet long. His son estimated he would need 276 feet of new fence. However, when he measured it he realized he would only need 216 new feet of fence. What is the difference in cost of his son's estimate versus his estimate with regard to how many feet of fence are needed?

90. Carlos Sontera is replacing some barbed-wire fence on a ranch in El Paso, Texas. The barbed wire and poles for 12 feet of fence cost about $80. The labor cost to install 12 feet of fence is about $40. A ranch hand reported that 300 new feet of fence were needed. However, when Carlos actually rode out there and measured it, he found that only 228 new feet of fence were needed. What is the difference in cost of the ranch hand's estimate versus Carlos's estimate of how many feet of fence are needed?

Cumulative Review Problems

91. Write in standard notation: eight million, four hundred sixty-six thousand, eighty-four

92. Write a word name for 296,308.

93. Add. $16 + 27 + 82 + 34 + 9$

94. Add. $156{,}325$
              $+ 963{,}209$

Developing Your Study Skills

How To Do Homework

Set aside time each day for your homework assignments. Make a weekly schedule and write down the times each day you will devote to doing math homework. Two hours spent studying outside class for each hour in class is usual for college courses. You may need more than that for mathematics.

Before beginning to solve your homework exercises, read your textbook very carefully. Expect to spend much more time reading a few pages of a mathematics textbook than several pages of another text. Read for complete understanding, not just for the general idea.

As you begin your homework assignments, read the directions carefully. You need to understand what is being asked. Concentrate on each exercise, taking time to solve it accurately. Rushing through your work usually causes errors. Check your answers with those given in the back of the textbook. If your answer is incorrect, check to see that you are doing the right problem. Redo the problem, watching for little errors. If it is still wrong, check with a friend. Perhaps the two of you can figure out where you are going wrong.

Work on your assignments every day and do as many exercises as it takes for you to know what you are doing. Begin by doing all the exercises that have been assigned. If there are more available in that section of your text, then do more. When you think you have done enough exercises to understand fully the topic at hand, do a few more to be sure. This may mean that you do many more exercises than the instructor assigns, but you can never practice mathematics too much. Practice improves your skills and increases your accuracy, speed, competence, and confidence.

Also, check the examples in the textbook or in your notes for a similar exercise. Can this one be solved in the same way? Give it some thought. You may want to leave it for a while by taking a break or doing a different exercise. But come back later and try again. If you are still unable to figure it out, ask your instructor for help during office hours or in class.

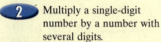
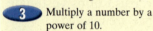
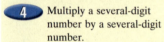
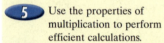
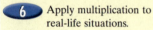

1 *Mastering Basic Multiplication Facts*

Like subtraction, multiplication is related to addition. Suppose that the pastry chef at the Gourmet Restaurant bakes croissants on a sheet that holds four croissants across, with room for three rows. How many croissants does the sheet hold?

We can add 4 + 4 + 4 to get the total, or we can use a shortcut: three rows of four is the same as 3 times 4, which equals 12. This is **multiplication,** a shortcut for repeated addition.

The numbers that we multiply are called **factors.** The answer is called the **product.** For now, we will use $\times$ to show multiplication. 3×4 is read "three times four."

$$
\underbrace{3}_{\text{factor}} \quad \times \quad \underbrace{4}_{\text{factor}} \quad = \quad \underbrace{12}_{\text{product}} \qquad \begin{array}{r} 3 \text{ factor} \\ \times\, 4 \text{ factor} \\ \hline 12 \text{ product} \end{array}
$$

Your skill in multiplication depends on how well you know the basic multiplication facts. Look at the table on page 39. You should learn these facts well enough to quickly and correctly give the products of any two factors in the table. To check your knowledge, try Exercises 1.4, exercises 3 and 4.

Study the table to see if you can discover any properties of multiplication. What do you see as results when you multiply zero by any number? When you multiply any number times zero, the result is zero. That is the **multiplication property of zero.**

$$2 \times 0 = 0 \qquad 5 \times 0 = 0 \qquad 0 \times 6 = 0 \qquad 0 \times 0 = 0$$

You may recall that zero plays a special role in addition. Zero is the *identity element* for addition. When we add any number to zero, that number does not change. Is there an identity element for multiplication? Look at the table. What is the identity element for multiplication? Do you see that it is 1? The **Identity element for multiplication** is 1.

$$5 \times 1 = 5 \qquad 1 \times 5 = 5$$

What other properties of addition hold for multiplication? Is multiplication commutative? Does the order in which you multiply two numbers change the results? Find the product of 3×4. Then find the product of 4×3.

$$3 \times 4 = 12$$
$$4 \times 3 = 12$$

The **commutative property of multiplication** tells us that when we multiply two numbers, changing the order of the numbers gives the same result.

Basic Multiplication Facts

×	0	1	2	3	4	5	6	7	8	9	10	11	12
0	0	0	0	0	0	0	0	0	0	0	0	0	0
1	0	1	2	3	4	5	6	7	8	9	10	11	12
2	0	2	4	6	8	10	12	14	16	18	20	22	24
3	0	3	6	9	12	15	18	21	24	27	30	33	36
4	0	4	8	12	16	20	24	28	32	36	40	44	48
5	0	5	10	15	20	25	30	35	40	45	50	55	60
6	0	6	12	18	24	30	36	42	48	54	60	66	72
7	0	7	14	21	28	35	42	49	56	63	70	77	74
8	0	8	16	24	32	40	48	56	64	72	80	88	96
9	0	9	18	27	36	45	54	63	72	81	90	99	108
10	0	10	20	30	40	50	60	70	80	90	100	110	120
11	0	11	22	33	44	55	66	77	88	99	110	121	132
12	0	12	24	36	48	60	72	84	96	108	120	132	144

Quick Recall of Multiplication Facts

It is helpful if you can multiply quickly. See if you can do Example 1 correctly in 15 seconds or less. Repeat again with Practice Problem 1. Strive to obtain all answers correctly in 15 seconds or less.

EXAMPLE 1 Multiply.

(a) 5×7 **(b)** 8×9 **(c)** 6×8 **(d)** 9×3 **(e)** 7×8

(a) $\begin{array}{r} 7 \\ \times 5 \\ \hline 35 \end{array}$ **(b)** $\begin{array}{r} 9 \\ \times 8 \\ \hline 72 \end{array}$ **(c)** $\begin{array}{r} 8 \\ \times 6 \\ \hline 48 \end{array}$ **(d)** $\begin{array}{r} 3 \\ \times 9 \\ \hline 27 \end{array}$ **(e)** $\begin{array}{r} 8 \\ \times 7 \\ \hline 56 \end{array}$

Practice Problem 1 Multiply.

(a) $\begin{array}{r} 8 \\ \times 8 \end{array}$ **(b)** $\begin{array}{r} 7 \\ \times 6 \end{array}$ **(c)** $\begin{array}{r} 5 \\ \times 8 \end{array}$ **(d)** $\begin{array}{r} 9 \\ \times 7 \end{array}$ **(e)** $\begin{array}{r} 9 \\ \times 9 \end{array}$

2 *Multiplying a Single-Digit Number by a Number with Several Digits*

EXAMPLE 2 Multiply. 4312×2

We first multiply the ones column, then the tens column, and so on, moving right to left.

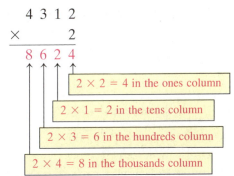

$$\begin{array}{r} 4\;3\;1\;2 \\ \times \quad\quad\; 2 \\ \hline 8\;6\;2\;4 \end{array}$$

$2 \times 2 = 4$ in the ones column

$2 \times 1 = 2$ in the tens column

$2 \times 3 = 6$ in the hundreds column

$2 \times 4 = 8$ in the thousands column

Practice Problem 2 Multiply. 3021×3

Usually, we will have to carry one digit of the result of some of the multiplication into the next left-hand column.

EXAMPLE 3   Multiply. 36×7

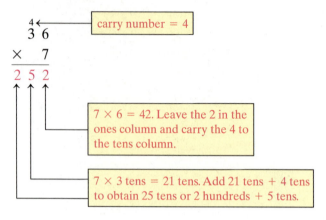

carry number = 4

$7 \times 6 = 42$. Leave the 2 in the ones column and carry the 4 to the tens column.

7×3 tens = 21 tens. Add 21 tens + 4 tens to obtain 25 tens or 2 hundreds + 5 tens.

Practice Problem 3   Multiply. 43×8

EXAMPLE 4   Multiply. 359×9

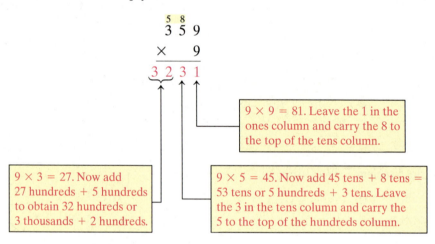

$9 \times 9 = 81$. Leave the 1 in the ones column and carry the 8 to the top of the tens column.

$9 \times 3 = 27$. Now add 27 hundreds + 5 hundreds to obtain 32 hundreds or 3 thousands + 2 hundreds.

$9 \times 5 = 45$. Now add 45 tens + 8 tens = 53 tens or 5 hundreds + 3 tens. Leave the 3 in the tens column and carry the 5 to the top of the hundreds column.

Practice Problem 4   Multiply.
$$\begin{array}{r} 579 \\ \times \quad 7 \\ \hline \end{array}$$

3   *Multiplying a Number by a Power of 10*

Observe what happens when a number is multiplied by 10, 100, 1000, 10,000, and so on.

one zero
$56 \times 10 \quad = \quad 560$

two zeros
$56 \times 100 \quad = \quad 5600$

three zeros
$56 \times 1000 \quad = \quad 56{,}000$

four zeros
$56 \times 10{,}000 \quad = \quad 560{,}000$

A **power of 10** is a whole number that begins with 1 and ends in one or more zeros. The numbers 10, 100, 1000, 10,000, and so on are powers of 10.

To multiply a whole number by a power of 10:

1. Count the number of zeros in the power of 10.

2. Attach that number of zeros to the right side of the other whole number to obtain the answer.

● **EXAMPLE 5** Multiply 358 by each number.

(a) 10   **(b)** 100   **(c)** 1000   **(d)** 100,000

(a) $358 \times 10 = 3580$ (one zero)
(b) $358 \times 100 = 35,800$ (two zeros)
(c) $358 \times 1000 = 358,000$ (three zeros)
(d) $358 \times 100,000 = 35,800,000$ (five zeros)

Practice Problem 5 Multiply 1267 by each number.

(a) 10   **(b)** 1000   **(c)** 10,000   **(d)** 1,000,000   ●

How can we handle zeros in multiplication involving a number that is not 10, 100, 1000, or any other power of 10? Consider 32×400. We can rewrite 400 as 4×100, which gives us $32 \times 4 \times 100$. We can simply multiply 32×4 and then attach two zeros for the factor 100. We find that $32 \times 4 = 128$. Attaching two zeros gives us 12,800, or $32 \times 400 = 12,800$.

● **EXAMPLE 6** Multiply.

(a) 12×3000 **(b)** 25×600 **(c)** 430×260

(a) $12 \times 3000 = 12 \times 3 \times 1000 = 36 \times 1000 = 36,000$
(b) $25 \times 600 = 25 \times 6 \times 100 = 150 \times 100 = 15,000$
(c) $430 \times 260 = 43 \times 26 \times 10 \times 10 = 1118 \times 100 = 111,800$

Practice Problem 6 Multiply.

(a) $9 \times 60,000$ **(b)** 15×400 **(c)** 270×800 ●

4 *Multiplying a Several-Digit Number by a Several-Digit Number*

● **EXAMPLE 7** Multiply. 234×21

We can consider 21 as 2 tens (20) and 1 one (1). First we multiply 234 by 1.

We also multiply 234×20. This gives us two **partial products.**

$$
\begin{array}{r}
234 \\
\times \quad 1 \\
\hline
234
\end{array}
\qquad
\begin{array}{r}
234 \\
\times \ 20 \\
\hline
4680
\end{array}
$$

Now we combine these two operations together by adding the two partial products to reach the final product, which is the solution.

$$
\begin{array}{r}
234 \\
\times \quad 21 \\
\hline
234 \\
4680 \\
\hline
4914
\end{array}
$$

234 ←——— Multiply 234×1.
4680 ←——— Multiply 234×20.
4914 ←——— Add the two partial products.

Practice Problem 7   Multiply.
$$
\begin{array}{r}
323 \\
\times \quad 32 \\
\hline
\end{array}
$$

EXAMPLE 8   Multiply. 671×35

$$
\begin{array}{r}
6\ 7\ 1 \\
\times \quad 3\ 5 \\
\hline
3\ 3\ 5\ 5 \\
2\ 0\ 1\ 3\ 0 \\
\hline
2\ 3\ 4\ 8\ 5
\end{array}
$$

3 3 5 5 ←——— First multiply 671×5.
2 0 1 3 0 ←——— Now multiply 671×30.
2 3 4 8 5 ←——— Now add the two partial products.

Note: We could omit zero on this line and leave the ones place blank.

Practice Problem 8   Multiply.
$$
\begin{array}{r}
385 \\
\times \quad 69 \\
\hline
\end{array}
$$

EXAMPLE 9   Multiply. 14×20

$$
\begin{array}{r}
14 \\
\times \quad 20 \\
\hline
0 \\
280 \\
\hline
\end{array}
$$

0 ←——— Multiply 14 by 0.
280 ←——— Multiply 14 by 2 tens.

Now add the ——→ 280
partial products.

Place 28 with the 8 in the tens column. To line up the digits for adding, we can insert a 0 in the ones column.

Notice that you will also get this result if you multiply $14 \times 2 = 28$ and then attach a zero to multiply it by 10: 280.

Practice Problem 9   Multiply. 34×20

EXAMPLE 10   Multiply. 120×40

$$
\begin{array}{r}
120 \\
\times \quad 40 \\
\hline
0 \\
4800 \\
\hline
\end{array}
$$

0 ←——— Multiply 120×0.
4800 ←——— Multiply 120 by 4 tens.

Now add the ——→ 4800
partial products.

The answer is 480 tens. We place the 0 of the 480 in the tens column. To line up the digits for adding, we can insert a 0 in the ones column.

Notice that this result is the same as $12 \times 4 = 48$ with two zeros attached: 4800.

Practice Problem 10   Multiply. 130×50

EXAMPLE 11 Multiply. 684×763

$$
\begin{array}{r}
6\ 8\ 4 \\
\times\ 7\ 6\ 3 \\
\hline
2\ 0\ 5\ 2 \\
4\ 1\ 0\ 4 \\
4\ 7\ 8\ 8 \\
\hline
5\ 2\ 1\ 8\ 9\ 2
\end{array}
$$

← Multiply 684×3.

← Multiply 684×60. Note that we omit the final zero.

← Multiply 684×700. Note that we omit the final two zeros.

Practice Problem 11 Multiply. 923×675

5 *Using the Properties of Multiplication to Perform Efficient Calculations*

When we add three numbers, we use the associative property. Recall that the associative property allows us to group the three numbers in any way. Thus to add $9 + 7 + 3$, we can group the numbers as $9 + (7 + 3)$ because it is easier to find the sum. $9 + (7 + 3) = 9 + 10 = 19$. We can demonstrate that multiplication is also associative.

Is this true? $2 \times (5 \times 3) = (2 \times 5) \times 3$

$$2 \times (15) = (10) \times 3$$

$$30 = 30$$

The final product is the same in both cases.

The way we group numbers to be multiplied does not change the product. This property is called the **associative property of multiplication.**

EXAMPLE 12 Multiply. $14 \times 2 \times 5$

Since we can group any two numbers together, let's take advantage of multiplying by 10.

$$14 \times 2 \times 5 = 14 \times (2 \times 5) = 14 \times 10 = 140$$

Practice Problem 12 Multiply. $25 \times 4 \times 17$

For convenience, we list the properties of multiplication that we have discussed in this section.

1. Associative Property of Multiplication. When we multiply three numbers, the multiplication can be grouped in any way.	$(7 \times 3) \times 2 = 7 \times (3 \times 2)$ $21 \times 2 = 7 \times 6$ $42 = 42$
2. Commutative Property of Multiplication. Two numbers can be multiplied in either order with the same result.	$9 \times 8 = 8 \times 9$ $72 = 72$
3. Identity Property of One. When one is multiplied by a number, the result is that number.	$7 \times 1 = 7$ $1 \times 15 = 15$
4. Multiplication Property of Zero. The product of any number and zero yields zero as a result.	$0 \times 14 = 0$ $2 \times 0 = 0$

Sometimes you can use several properties in one problem to make the calculation easier.

EXAMPLE 13 Multiply. $7 \times 20 \times 5 \times 6$

$$
\begin{aligned}
7 \times 20 \times 5 \times 6 &= 7 \times (20 \times 5) \times 6 & \text{Associative property} \\
&= 7 \times 6 \times (20 \times 5) & \text{Commutative property} \\
&= 42 \times 100 \\
&= 4200
\end{aligned}
$$

Practice Problem 13 Multiply. $8 \times 4 \times 3 \times 25$

Thus far we have discussed the properties of addition and the properties of multiplication. There is one more property that links both operations.

Before we discuss that property, we will illustrate several different ways of showing multiplication. The following are all the ways to show "3 times 4."

3×4	$(3)(4)$	$3(4)$ $(3)4$	$3 \cdot 4$	$3 * 4$
with an $\times$	with two sets of parentheses	with a single set of parentheses	with a dot	with a star

We will use parentheses to mean multiplication when we use the **distributive property.**

Sidelight

Why does our method of multiplying several-digit numbers work? Why can we say that 234×21 is the same as $234 \times 1 + 234 \times 20$?

The *distributive property of multiplication over addition* allows us to distribute the multiplication and then add the results. To illustrate, 234×21 can be written as $234(20 + 1)$. By the distributive property

$$
\begin{aligned}
234(20 + 1) &= (234 \times 20) &+ (234 \times 1) \\
&= \quad 4680 &+ \quad 234 \\
&= 4914
\end{aligned}
$$

This is what we actually do when we multiply.

$$
\begin{array}{r}
234 \\
\times\ 21 \\
\hline
234 \\
4680 \\
\hline
4914
\end{array}
$$

Distributive Property of Multiplication over Addition

Multiplication can be distributed over addition without changing the result.

$$5 \times (10 + 2) = (5 \times 10) + (5 \times 2)$$

6 **Applying Multiplication to Real-Life Situations**

To use multiplication in word problems, the number of items or the value of each item must be the same. Recall that multiplication is a quick way to do repeated addition where *each addend is the same*. In the beginning of the section we showed three rows of four croissants to illustrate 3 × 4. The number of croissants in each row was the same, 4. Look at another example. If we had six nickels, we could use multiplication to find the total value of the coins because the value of each nickel is the same, 5¢. Since 6 × 5 = 30, six nickels are worth 30¢.

In the following example the word *average* is used. The word *average* has several different meanings. In this example, we are told that the *average annual salary* of an employee at Software Associates is $21,342. This means that we can calculate the total payroll as if each employee made $21,342 even though we know that the president probably makes more than any other employee.

EXAMPLE 14   The average annual salary of an employee at Software Associates is $21,342. There are 38 employees. What is the annual payroll?

$$\begin{array}{r} \$21{,}342 \\ \times \qquad 38 \\ \hline 170\ 736 \\ 640\ 26 \\ \hline 810{,}996 \end{array}$$

The total annual payroll is $810,996.

Practice Problem 14   The average cost of a new car sold last year at Westover Chevrolet was $17,348. The dealership sold 378 cars. What were the total sales of cars at the dealership last year?

Another useful application of multiplication is area. The following example involves the area of a rectangle.

EXAMPLE 15   What is the area of a rectangular hallway that measures 4 feet wide and 8 feet long?

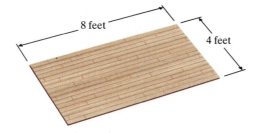

8 feet

4 feet

The **area** of a rectangle is the product of the length times the width. Thus for this hallway

$$\text{Area} = 8 \text{ feet} \times 4 \text{ feet} = 32 \text{ square feet}$$

The area of the hallway is 32 square feet.

Note: All measurements for area are given in square units such as square feet, square meters, square yards, and so on.

Practice Problem 15   What is the area of a rectangular rug that measures 5 yards by 7 yards?

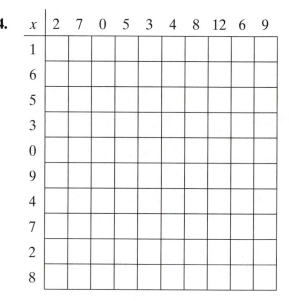

1.4 Exercises

Verbal and Writing Skills

1. Explain in your own words.
  (a) the commutative property of multiplication   **(b)** the associative property of multiplication

2. How does the distributive property of multiplication over addition help us to multiply 4×13?

Complete the multiplication facts for each table. Strive for total accuracy, but work quickly. (Allow a maximum of six minutes for each table.)

3.

x	6	2	3	8	0	5	7	9	12	4
5										
7										
1										
0										
6										
2										
3										
8										
4										
9										

4.

x	2	7	0	5	3	4	8	12	6	9
1										
6										
5										
3										
0										
9										
4										
7										
2										
8										

Multiply.

5. $\begin{array}{r} 32 \\ \times\ 3 \\ \hline \end{array}$

6. $\begin{array}{r} 21 \\ \times\ 4 \\ \hline \end{array}$

7. $\begin{array}{r} 203 \\ \times\ \ \ 3 \\ \hline \end{array}$

8. $\begin{array}{r} 313 \\ \times\ \ \ 3 \\ \hline \end{array}$

9. $\begin{array}{r} 6102 \\ \times\ \ \ \ \ 3 \\ \hline \end{array}$

10. $\begin{array}{r} 5203 \\ \times\ \ \ \ \ 2 \\ \hline \end{array}$

11. $\begin{array}{r} 20{,}103 \\ \times\ \ \ \ \ \ \ 4 \\ \hline \end{array}$

12. $\begin{array}{r} 422{,}102 \\ \times\ \ \ \ \ \ \ \ \ \ 3 \\ \hline \end{array}$

13. $\begin{array}{r} 14 \\ \times\ 5 \\ \hline \end{array}$

14. $\begin{array}{r} 12 \\ \times\ 4 \\ \hline \end{array}$

15. $\begin{array}{r} 87 \\ \times\ 6 \\ \hline \end{array}$

16. $\begin{array}{r} 95 \\ \times\ 7 \\ \hline \end{array}$

17. $\begin{array}{r} 326 \\ \times\ \ \ 5 \\ \hline \end{array}$

18. $\begin{array}{r} 732 \\ \times\ \ \ 6 \\ \hline \end{array}$

19. $\begin{array}{r} 2036 \\ \times\ \ \ \ \ 6 \\ \hline \end{array}$

20. 3215
$\times$ 6

21. 12,526
$\times$ 8

22. 48,761
$\times$ 7

23. 235,702
$\times$ 4

24. 127,054
$\times$ 6

Multiply by powers of 10.

25. 156
$\times$ 10

26. 278
$\times$ 10

27. 27,158
$\times$ 100

28. 89,361
$\times$ 100

29. 482
$\times$ 1000

30. 579
$\times$ 1000

31. 37,256
$\times$ 10,000

32. 614,260
$\times$ 10,000

33. 423
$\times$ 20

34. 134
$\times$ 20

35. 2120
$\times$ 30

36. 4230
$\times$ 20

37. 14,000
$\times$ 4000

38. 62,000
$\times$ 3000

Multiply.

39. 2103
$\times$ 32

40. 527
$\times$ 24

41. 146
$\times$ 54

42. 3021
$\times$ 12

43. 89
$\times$ 64

44. 68
$\times$ 49

45. 607
$\times$ 25

46. 817
$\times$ 84

47. 569
$\times$ 73

48. 915
$\times$ 47

49. 912
$\times$ 76

50. 498
$\times$ 39

51. 5123
$\times$ 29

52. 1268
$\times$ 38

53. 9053
$\times$ 91

54. 3078
$\times$ 72

55. 5536
$\times$ 224

56. 427
$\times$ 415

57. 678
$\times$ 132

58. 392
$\times$ 187

59.   2076
       × 105

60.   5092
       × 302

61.   3561
       × 403

62.   2074
       × 1003

63.   6035
       × 5006

64.   1298
       × 304

65.   260
       × 40

66.   403
       × 20

67.   403
       × 200

68.   150
       × 200

69. $7 \cdot 2 \cdot 5$

70. $8 \cdot 3 \cdot 2$

71. $11 \cdot 7 \cdot 4$

72. $15 \cdot 4 \cdot 4$

73. $10 \cdot 7 \cdot 10$

74. $20 \cdot 5 \cdot 3$

75. $11 \cdot 7 \cdot 5 \cdot 2$

76. $3 \cdot 6 \cdot 5 \cdot 8$

77. What is x if
$x = 8 \cdot 7 \cdot 6 \cdot 0$?

78. What is x if
$x = 3 \cdot 12 \cdot 0 \cdot 5$?

Applications

▲ **79.** Find the area of a rectangular university fitness center if the fitness center floor is 48 feet long and 67 feet wide.

▲ **80.** Find the area of a rectangular computer chip that is 17 millimeters long and 9 millimeters wide.

▲ **81.** Don Williams and his wife want to put down new carpet in the living room and the hallway of their house. The living room measures 12 feet by 14 feet. The hallway measures 9 feet by 3 feet. If the living room and the hallway are rectangular in shape, how many square feet of new carpet do Don and his wife need?

▲ **82.** Robert Tobey in Copper Center, Alaska, wants to put a field under helicopter surveillance because of a roving pack of wolves that are destroying other wildlife in the area. The field consists of two rectangular regions. The first one is 4 miles by 5 miles. The second one is 12 miles by 8 miles. How many square miles does he want to place under surveillance?

83. The student commons food supply needs to purchase espresso coffee. Find the cost of purchasing 240 pounds of espresso coffee beans at $5 per pound.

84. The music department of Wheaton College wishes to purchase 345 sets of headphones at the music supply store at a cost of $8 each. What will be the total amount of the purchase?

85. Helen pays $266 per month for her car payment on her new Honda Civic. What is her automobile payment cost for a one-year period?

86. A company rents a Ford Escort for a salesman at $276 per month for eight months. What is the cost for the car rental during this time?

87. Marcos has a Toyota Corolla that gets 34 miles per gallon during highway driving. Approximately how far can he travel if he has 18 gallons of gas in the tank?

88. Cheryl has a subcompact car that gets 48 miles per gallon on highway driving. Approximately how far can she travel if she has 12 gallons of gas in the tank?

89. During the holiday season, Borders Books has 25 employees working each day. One day, each employee sold 95 books. How many total books were sold that day?

90. Frank earns $245 each week at his part-time job as a waiter. How much does Frank make in one year? (There are 52 weeks in one year).

91. The country of Haiti has an average per capita (per person) income of $1070. If the approximate population of Haiti is 6,890,000, what is the approximate total yearly income of the entire country?

92. The kingdom of the Netherlands (Holland) has an approximate population of 15,800,000. The average per capita (per person) income is $22,000. What is the approximate total yearly income of the entire country?

To Think About

Use the following information to answer exercises 93–96. There are 98 puppies in a room, with an assortment of black and white ears and paws. 18 puppies have totally black ears and 2 white paws; 26 puppies have 1 black ear and 4 white paws; and 54 puppies have no black ears and 1 white paw.

93. How many black paws are in the room?

94. How many white paws are in the room?

95. How many black ears are in the room?

96. How many white ears are in the room?

In exercises 97–100, find the value of x in each equation.

97. $5(x) = 40$

98. $7(x) = 56$

99. $72 = 8(x)$

100. $63 = 9(x)$

101. Would the distributive property of multiplication be true for roman numerals such as (XII) × (IV)? Why or why not?

102. We saw that multiplication is distributive over addition. Is it distributive over subtraction? Why or why not? Give examples.

Cumulative Review Problems

103. Subtract.

$$\begin{array}{r} 34{,}084 \\ -\,27{,}328 \\ \hline \end{array}$$

104. Add.

$$\begin{array}{r} 263 \\ 27 \\ 891 \\ 5 \\ +\,63 \\ \hline \end{array}$$

105. Albert Gachet earns $156 per week. He has withheld weekly $12 for taxes, $2 for dues, and $3 for retirement. How much is left?

106. Mrs. May Washington retired at $743 per month. Her cost-of-living raise increased her payment to $802 per month. What was the increase?

Developing Your Study Skills

Why Is Homework Necessary?

Mathematics involves mastering a set of skills that you learn by practicing, not by watching someone else do it. Your instructor may make solving a mathematics problem look very easy, but for you to learn the necessary skills, you must practice them over and over again, just as your instructor once had to. There is no other way. Learning mathematics is like learning to play a musical instrument, to type, or to play a sport. No matter how much you watch someone else do it, no matter how many books you read on "how to" do it, no matter how easy it seems to be, the key to success is practice on a regular basis.

Homework provides this practice. The amount of practice needed varies for each individual, but usually students need to do most or all of the exercises provided at the end of each section in the text. The more exercises you do, the better you get. Some problems in a set are more difficult than others, and some stress different concepts. Only by working all the exercises will you cover the full range of difficulty.

1 *Mastering Basic Division Facts*

Suppose that we have eight quarters and want to divide them into two equal piles. We would discover that each pile contains four quarters.

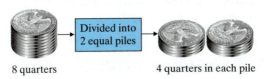

8 quarters   4 quarters in each pile

In mathematics we would express this thought by saying that

$$8 \div 2 = 4.$$

We know that this answer is right because two piles of four quarters is the same dollar amount as eight quarters. In other words, we know that $8 \div 2 = 4$ because $2 \times 4 = 8$. These two mathematical sentences are called **related sentences.** The division sentence $8 \div 2 = 4$ is related to the multiplication sentence $2 \times 4 = 8$.

In fact, in mathematics we usually define **division** in terms of multiplication. The answer to the division problem $12 \div 3$ is that number which when multiplied by 3 yields 12. Thus

$$12 \div 3 = 4 \text{ because } 3 \times 4 = 12.$$

Suppose that a surplus of $30 in the French Club budget at the end of the year is to be equally divided among the five club members. We would want to divide the $30 into five equal parts. We would write $30 \div 5 = 6$ because $5 \times 6 = 30$. Thus each of the five people would get $6 in this situation.

$30   5 piles of $6 each

As a mathematical sentence, $30 \div 5 = 6$.

The division problem $30 \div 5 = 6$ could also be written $\frac{30}{5} = 6$ or $5\overline{)30}^{\,6}$.

When referring to division, we sometimes use the words **divisor, dividend,** and **quotient** to identify the three parts.

$$\text{divisor} \overline{)\text{dividend}}^{\text{quotient}}$$

With $30 \div 5 = 6$, 30 is the dividend, 5 is the divisor, and 6 is the quotient.

$$\text{divisor} \to 5\overline{)30}^{\,6 \, \leftarrow \text{quotient}} \leftarrow \text{dividend}$$

So the quotient is the answer to a division problem. It is important that you be able to do short problems involving basic division facts quickly.

● EXAMPLE 1 Divide.

(a) $12 \div 4$ **(b)** $81 \div 9$ **(c)** $56 \div 8$ **(d)** $54 \div 6$

(a) $4\overline{)12}^{\,3}$ **(b)** $9\overline{)81}^{\,9}$ **(c)** $8\overline{)56}^{\,7}$ **(d)** $6\overline{)54}^{\,9}$

51

Practice Problem 1   Divide.

(a) $36 \div 4$ **(b)** $25 \div 5$ **(c)** $72 \div 9$ **(d)** $30 \div 6$ ⬤

Zero can be divided by any nonzero number, but division by zero is not possible. Why is this?

Suppose that we could divide by zero. Then $7 \div 0 =$ some number. Let us represent "some number" by the letter a.

$$\text{If } 7 \div 0 = a, \text{ then } 7 = 0 \times a,$$

because every division problem has a related multiplication problem. But zero times any number is zero, $0 \times a = 0$. Thus

$$7 = 0 \times a = 0.$$

That is, $7 = 0$, which we know is not true. Therefore, our assumption that $7 \div 0 = a$ is wrong. Thus we conclude that we cannot divide by zero. Mathematicians state this by saying, "Division by zero is **undefined.**"

It is helpful to remember the following basic concepts:

Division Problems Involving the Number 1 and the Number 0

1. Any nonzero number divided by itself is $1 (7 \div 7 = 1)$.

2. Any number divided by 1 remains unchanged $(29 \div 1 = 29)$.

3. Zero may be divided by any nonzero number; the result is always zero $(0 \div 4 = 0)$.

4. Zero can never be the divisor in a division problem $(3 \div 0$ is undefined; $0 \div 0$ is undefined).

⬤ **EXAMPLE 2**   Divide, if possible. If not possible, state why.

(a) $8 \div 8$ **(b)** $9 \div 1$ **(c)** $0 \div 6$ **(d)** $20 \div 0$ **(e)** $0 \div 0$

(a) $\dfrac{8}{8} = 1$ Any number divided by itself is 1.

(b) $\dfrac{9}{1} = 9$ Any number divided by 1 remains unchanged.

(c) $\dfrac{0}{6} = 0$ Zero divided by any nonzero number is zero.

(d) $\dfrac{20}{0}$ cannot be done   Division by zero is undefined.

(e) $\dfrac{0}{0}$ cannot be done   Division by zero is undefined.

Practice Problem 2   Divide, if possible.

(a) $7 \div 1$ **(b)** $\dfrac{9}{9}$ **(c)** $\dfrac{0}{5}$ **(d)** $12 \div 0$ **(e)** $\dfrac{7}{0}$ ⬤

② *Performing Division by a One-Digit Number*

Our accuracy with division is improved if we have a checking procedure. For each division fact, there is a related multiplication fact.

$$\text{If } 20 \div 4 = 5, \text{ then } 20 = 4 \times 5.$$
$$\text{If } 36 \div 9 = 4, \text{ then } 36 = 9 \times 4.$$

We will often use multiplication to check our answers.

When two numbers do not divide exactly, a number called the **remainder** is left over. For example, 13 cannot be divided exactly by 2. The number 1 is left over. We call this 1 the *remainder.*

$$\begin{array}{r} 6 \\ 2\overline{)13} \\ 12 \\ \hline 1 \end{array} \leftarrow \text{remainder}$$

Thus $13 \div 2 = 6$ with a remainder of 1. We can abbreviate this answer as

6 R 1.

To check this division, we multiply $2 \times 6 = 12$ and add the remainder: $12 + 1 = 13$. That is, $2 \times 6 + 1 = 13$. The result will be the dividend if the division was done correctly. The following box shows you how to check a division that has a remainder.

(divisor $\times$ quotient) + remainder = dividend

EXAMPLE 3 Divide. $33 \div 4$. Check your answer.

$$\begin{array}{r} 8 \\ 4\overline{)33} \\ 32 \\ \hline 1 \end{array}$$

$8 \rightarrow$ How many times can 4 be divided into 33?   8.

$32 \leftarrow$ What is 8×4?   32.

$1 \leftarrow$ What is 33 subtract 32?   1.

The answer is 8 with a remainder of 1. We abbreviate 8 R 1.

Check.

$$\begin{array}{r} 8 \\ \times\ 4 \\ \hline 32 \end{array}$$

Multiply. $8 \times 4 = 32$.

$$\begin{array}{r} 1 \\ \hline 33 \end{array}$$

Add the remainder. $32 + 1 = 33$.

Because the dividend is 33, the answer is correct.

Practice Problem 3 Divide. $45 \div 6$. Check your answer.

EXAMPLE 4 Divide. $158 \div 5$. Check your answer.

$$\begin{array}{r} 31 \\ 5\overline{)158} \\ 15 \\ \hline 08 \\ 5 \\ \hline 3 \end{array}$$

5 divided into 15?   3.

$15 \leftarrow$ What is 3×5?   15.

$08 \leftarrow$ 15 subtract 15?   0. Bring down 8.

$5 \leftarrow$ 5 divided into 8?   1. What is 1×5?   5.

$3 \leftarrow$ 8 subtract 5?   3.

The answer is 31 R 3.

Check.

$$\begin{array}{r} 31 \\ \times\ 5 \\ \hline 155 \end{array}$$

Multiply. $31 \times 5 = 155$.

$$\begin{array}{r} 3 \\ \hline 158 \end{array}$$

Add the remainder 3.

Because the dividend is 158, the answer is correct.

Practice Problem 4 Divide. $129 \div 6$. Check your answer.

EXAMPLE 5 Divide. 3672 ÷ 7

$$
\begin{array}{r}
524 \\
7\overline{)3672} \\
\underline{35} \\
17 \\
\underline{14} \\
32 \\
\underline{28} \\
4
\end{array}
$$

How many times can 7 be divided into 36?   5.

← What is 5 × 7?   35.

← 36 subtract 35?   1. Bring down 7.

← 7 divided into 17?   2. What is 2 × 7?   14.

← 17 subtract 14?   3. Bring down 2.

← 7 divided into 32?   4. What is 4 × 7?   28.

← 32 subtract 28?   4.

The answer is 524 R 4.

Practice Problem 5   Divide. 4237 ÷ 8

3 *Performing Division by a Two- or Three-Digit Number*

When the divisor has more than one digit, an estimation technique may help. Figure how many times the first digit of the divisor goes into the first two digits of the dividend. Try this answer as the first number in the quotient.

EXAMPLE 6   Divide. 283 ÷ 41

First guess:

$$
\begin{array}{r}
7 \\
41\overline{)283} \\
\underline{287} \text{ too large}
\end{array}
$$

How many times can the first digit of the divisor (4) be divided into the first two digits of the dividend (28)? 7. We try the answer 7 as the first number of the quotient. We multiply 7 × 41 = 287. We see that 287 is larger than 283.

Second guess:

$$
\begin{array}{r}
6 \\
41\overline{)283} \\
\underline{246} \\
37
\end{array}
$$

Because 7 is slightly too large, we try 6.

← 6 × 41?   246.

← 283 subtract 246?   37.

The answer is 6 R 37. (Note that the remainder must always be less than the divisor.)

Practice Problem 6   Divide. 229 ÷ 32

EXAMPLE 7   Divide. 33,897 ÷ 56

First guess:

$$
\begin{array}{r}
60 \\
56\overline{)33897} \\
\underline{336} \\
29
\end{array}
$$

How many times can 33 be divided by 5?   6.

← What is 6 × 56?   336.

338 subtract 336?   2.   Bring down 9.

56 cannot be divided into 29. Write 0 in quotient.

Second set of steps:

$$
\begin{array}{r}
605 \\
56\overline{)33897} \\
\underline{336} \\
297 \\
\underline{280} \\
17
\end{array}
$$

Bring down 7.

How many times can 5 be divided into 29?   5.

← What is 5 × 56?   280.   Subtract 297 − 280.

Remainder is 17.

The answer is 605 R 17.

Practice Problem 7   Divide. 42,183 ÷ 33

● **EXAMPLE 8**   Divide. 5629 ÷ 134

$$
\begin{array}{r}
42 \\
134\overline{)5629} \\
536 \\
\overline{269} \\
268 \\
\overline{1}
\end{array}
$$

How many times does 134 divide into 562?
We guess by saying that 1 divides into 5 five times,
but this is too large. ($5 \times 134 = 670$!)
So we try 4. What is 4×134?   536.
Subtract $562 - 536$. We obtain 26. Bring down 9.
How many times does 134 divide into 269?
We guess by saying that 1 divided into 2 goes 2 times.
What is 2×134?   268. Subtract $269 - 268$.
The remainder is 1.

The answer is 42 **R** 1.

Practice Problem 8   Divide. 3227 ÷ 128   ●

4   Applying Division to Real-Life Situations

When you solve a word problem that requires division, you will be given the total number and asked to calculate the number of items in each group or to calculate the number of groups. In the beginning of this section we showed eight quarters (the total number) and we divided them into two equal piles (the number of groups). Division was used to find how many quarters were in each pile (the number in each group). That is, $8 \div 2 = 4$. There were four quarters in each pile.

Let's look at another example. Suppose that $30 is to be divided equally among the members of a group. If each person receives $6, how many people are in the group? We use division, $30 \div 6 = 5$, to find that there are five people in the group.

● **EXAMPLE 9**   City Service Realty just purchased nine identical computers for the real estate agents in the office. The total cost for the nine computers was $25,848. What was the cost of one computer? Check your answer.

To find the cost of one computer, we need to divide the total cost by 9. Thus we will calculate $25,848 \div 9$.

$$
\begin{array}{r}
2872 \\
9\overline{)25848} \\
18 \\
\overline{78} \\
72 \\
\overline{64} \\
63 \\
\overline{18} \\
18 \\
\overline{0}
\end{array}
$$

Therefore, the cost of one computer is $2872. In order to check our work we will need to see if nine computers each costing $2872 will in fact result in a total of $25,848. We use multiplication to check division.

$$
\begin{array}{r}
2872 \\
\times \quad 9 \\
\hline
25848 \quad \checkmark
\end{array}
$$

We did obtain 25,848. Our answer is correct.

Practice Problem 9   The Dallas police department purchased seven identical police cars at a total cost of $117,964. Find the cost of one car. Check your answer.

In the following example you will see the word *average* used as it applies to division. The problem states that a car traveled 1144 miles in 22 hours. The problem asks you to find the average speed in miles per hour. This means that we will treat the problem as if the speed of the car were the same during each hour of the trip. We will use division to solve.

EXAMPLE 10   A car traveled 1144 miles in 22 hours. What was the average speed in miles per hour?

When doing distance problems, it is helpful to remember that distance ÷ time = rate. We need to divide 1144 miles by 22 hours to obtain the rate or speed in miles per hour.

$$
\begin{array}{r}
52 \\
22\overline{)1144} \\
110 \\
\hline
44 \\
44 \\
\hline
0
\end{array}
$$

The car traveled an average of 52 miles per hour.

Practice Problem 10   An airplane traveled 5138 miles in 14 hours. What was the average speed in miles per hour?

Developing Your Study Skills

Taking Notes in Class

An important part of mathematics studying is taking notes. In order to take meaningful notes, you must be an active listener. Keep your mind on what the instructor is saying, and be ready with questions whenever you do not understand something.

If you have previewed the lesson material, you will be prepared to take good notes. The important concepts will seem somewhat familiar. You will have a better idea of what needs to be written down. If you frantically try to write all that the instructor says or copy all the examples done in class, you may find your notes nearly worthless when you look at them at home. You may find that you are unable to make sense of what you have written.

Write down *important* ideas and examples as the instructor lectures, making sure that you are listening and following the logic. Include any helpful hints or suggestions that your instructor gives you or refers to in your text. You will be amazed at how easily you will forget these if you do not write them down. Try to review your notes the *same day* sometime after class. You will find the material in your notes easier to understand if you have attended class within the last few hours.

Successful note taking requires active listening and processing. Stay alert in class. You will realize the advantages of taking your own notes over copying those of someone else.

Verbal and Writing Skills

1. Explain in your own words what happens when you
   (a) divide a nonzero number by itself.   **(b)** divide a number by 1.

   (c) divide zero by a nonzero number.   **(d)** divide a number by 0.

Divide. See if you can work exercises 2–30 in three minutes or less.

2. $5\overline{)35}$ **3.** $6\overline{)42}$ **4.** $4\overline{)32}$ **5.** $8\overline{)24}$ **6.** $9\overline{)27}$ **7.** $8\overline{)40}$

8. $7\overline{)56}$ **9.** $9\overline{)36}$ **10.** $4\overline{)12}$ **11.** $7\overline{)21}$ **12.** $9\overline{)81}$ **13.** $8\overline{)56}$

14. $6\overline{)54}$ **15.** $7\overline{)63}$ **16.** $4\overline{)28}$ **17.** $8\overline{)72}$ **18.** $8\overline{)64}$ **19.** $9\overline{)63}$

20. $9\overline{)72}$ **21.** $6\overline{)24}$ **22.** $1\overline{)8}$ **23.** $1\overline{)7}$ **24.** $5\overline{)0}$ **25.** $6\overline{)0}$

26. $6\overline{)48}$ **27.** $54 \div 6$ **28.** $42 \div 7$ **29.** $6 \div 6$ **30.** $5 \div 5$

Divide. In exercises 31–42, check your answer.

31. $29 \div 6$ **32.** $42 \div 8$ **33.** $76 \div 8$ **34.** $39 \div 9$ **35.** $128 \div 5$

36. $6\overline{)103}$ **37.** $9\overline{)196}$ **38.** $8\overline{)424}$ **39.** $9\overline{)288}$ **40.** $7\overline{)126}$

41. $5\overline{)185}$ **42.** $8\overline{)224}$ **43.** $4\overline{)1289}$ **44.** $3\overline{)758}$ **45.** $6\overline{)763}$

46. $7\overline{)403}$ **47.** $8\overline{)6024}$ **48.** $9\overline{)2286}$ **49.** $3\overline{)3367}$ **50.** $6\overline{)8086}$

51. $7\overline{)12304}$ **52.** $6\overline{)18127}$ **53.** $5\overline{)12813}$ **54.** $8\overline{)32223}$

55. $185 \div 36$ **56.** $152 \div 48$ **57.** $267 \div 52$ **58.** $321 \div 53$ **59.** $427 \div 61$

60. $72\overline{)432}$ **61.** $12\overline{)1930}$ **62.** $13\overline{)6810}$ **63.** $30\overline{)1452}$ **64.** $40\overline{)1125}$

65. $12\overline{)3778}$ **66.** $13\overline{)9316}$ **67.** $36\overline{)7568}$ **68.** $32\overline{)3527}$ **69.** $18\overline{)3643}$

70. $19\overline{)2174}$ **71.** $174\overline{)700}$ **72.** $128\overline{)896}$ **73.** $322\overline{)2280}$ **74.** $41\overline{)205000}$

Solve.

75. $518 \div 14 = x$. What is the value of x?

76. $1572 \div 131 = x$. What is the value of x?

Applications

77. A *run* in skiing is going from the top of the ski lift to the bottom. If over seven days, 431,851 runs were made, what was the average number of ski runs per day?

78. Western Saddle Stable uses 21,900 pounds of feed per year to feed its 30 horses. How much does each horse eat per year?

79. A factory that produces silver earrings made 864 pairs of hoops in 36 hours. What was the average number of pairs of earrings produced per hour?

80. A telethon raised $3,677,880 over 15 hours. What was the average amount of money raised per hour?

81. A horse and carriage company in New York City bought seven new carriages at exactly the same price each. The total bill was $147,371. How much did each carriage cost?

82. A group of eight friends invested the same amount each in a beach property that sold for $369,432. How much did each friend pay?

83. Chung wishes to pay off a car loan of $4104 in 24 months. How large will his monthly payments be?

84. A new sorting machine sorts 26 letters per minute. The machine is fed 884 letters. How many minutes will it take to sort the letters?

85. The 2nd Avenue Delicatessen is making bagel sandwiches for a New York City Marathon party. The sandwich maker has 360 bagel halves, 340 slices of turkey, and 330 slices of swiss cheese. If he needs to make sandwiches each consisting of two bagel halves, two slices of turkey, and 2 slices of swiss cheese, what is the greatest number of sandwiches he can make?

86. In exercise 85, the sandwich maker finds an additional 100 slices of turkey and an additional 120 slices of swiss cheese. With these extra slices added to his original amount, now what is the greatest number of sandwiches he can make?

▲ **87.** Ace Landscaping is mowing a rectangular lawn that has an area of 2652 square feet. The company keeps a record of all lawn mowed in terms of length, width, square feet, and number of minutes it takes to mow the lawn. The width of the lawn is 34 feet. However, the page that lists the length of the lawn is soiled and the number cannot be read. Determine the length of the lawn.

▲ **88.** The space shuttle has recently gone through a number of repairs and improvements. NASA recently approved the use of a shuttle control panel that has an area of 3526 square centimeters. The control panel is rectangular. The width of the panel is 43 centimeters. What is the length of the panel?

To Think About

89. Division is not commutative. For example, $12 \div 4 \neq 4 \div 12$. If $a \div b = b \div a$, what must be true of the numbers a and b besides the fact that $b \neq 0$ and $a \neq 0$?

90. You can think of division as repeated subtraction. Show how $874 \div 138$ is related to repeated subtraction.

Cumulative Review Problems

Solve.

91. 128
     × 43
     ——

92. 7162
     × 145
     ——

93. 316,214 + 89,981

94. 1,360,000 − 1,293,156

Putting Your Skills to Work

Evaluating Actual Costs for Several Vacation Plans

You are arranging a vacation for you and three friends (four people total). You have agreed as a group, that you will try to make the most of your money and will take the best deal of three possible packages. The Jackson Hole Ski Resort in Jackson Hole, Wyoming, is offering three fabulous winter vacation packages, and you must figure out the best deal in terms of cost. Here are your choices:

1. An "Early Bird Special" ski package for four people costs $248 per person for four nights, including lodging and three days of ski passes.

2. A "New Year's" package for four people costs $510 per person for five nights, including lodging and four days of ski passes.

3. A "Presidents' Week" package for four people costs $933 per person for seven nights, including lodging and six days of ski passes.

Ski passes usually cost $58/day, $108 for two days, $156 for three days, $196 for four days, $240 for five days, $282 for six days, and $322 for seven days.

If you reserve a vacation package, the prices for ski passes are included in the package. The cost of meals is not included in the package.

Problems For Individual Investigation

(*Hint:* First Subtract the cost of the ski passes to obtain the answers.)

1. What is the cost of lodging per person per night for the "Early Bird Special"?

2. What is the cost of lodging per person per night for the "Presidents' Week" package?

Problems for Group Investigation and Cooperative Study

3. If the usual cost of lodging is $70 per night per person, how much does each person save on skiing if you take the "Early Bird Special"?

4. If the usual cost of lodging is $95 per night per person during Presidents' Week, how much will it cost per person to ski for six days if you take the "Presidents' Week" package?

5. How much will you save per person if you take the "Presidents' Week" package instead of paying the usual rates for lodging and ski passes for six days?

Internet Connections

  Netsite: http://www.prenhall.com/tobey_basic

This site contains links to two selected trip planners that can be used for travel virtually all over the world.

6. Plan the cost for you and a friend to travel by plane from Los Angeles to Hawaii and to stay three days in a hotel in Honolulu from June 1 to June 4. How much would you save if you flew with the same airline and stayed at the same hotel from January 1 to January 4?

7. Plan the cost for a family of four people to travel by plane from Chicago to Montreal and to stay for seven days in a hotel in Montreal from August 1 to August 8. How much would you save if you flew with the same airline and stayed at the same hotel from February 1 to February 8?

1.6 Exponents and Order of Operations

Student Learning Objectives

After studying this section, you will be able to:

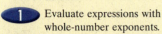

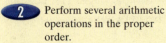

1 Evaluate expressions with whole-number exponents.

2 Perform several arithmetic operations in the proper order.

SSM
PH TUTOR   CD & VIDEO   MATH PRO   WEB
CENTER

1 *Evaluating Expressions with Whole-Number Exponents*

Sometimes a simple math idea comes "disguised" in technical language. For example, an **exponent** is just a "shorthand" number that saves writing multiplication of the same numbers.

10^3 The exponent 3 means $10 \times 10 \times 10$
(which takes longer to write).

The product 5×5 can be written as 5^2. The small number 2 is called the *exponent*. The exponent tells us how many factors are in the multiplication. The number 5 is called the **base.** The base is the number that is multiplied.

$$3 \times 3 \times 3 \times 3 = 3^4 \longleftarrow \text{exponent}$$
$$\underset{\text{base}}{\uparrow}$$

In 3^4 the base is 3 and the exponent is 4. (The 4 is sometimes called the *superscript.*) 3^4 is read as "three to the fourth power."

EXAMPLE 1   Write each product in exponent form.

(a) $15 \times 15 \times 15$ 　　　　　　　**(b)** $7 \times 7 \times 7 \times 7 \times 7$
(a) $15 \times 15 \times 15 = 15^3$ 　　　　**(b)** $7 \times 7 \times 7 \times 7 \times 7 = 7^5$

Practice Problem 1   Write each product in exponent form.

(a) $12 \times 12 \times 12 \times 12$ 　　　　**(b)** $2 \times 2 \times 2 \times 2 \times 2 \times 2$

EXAMPLE 2   Find the value of each expression.

(a) 3^3 　　**(b)** 7^2 　　**(c)** 2^5 　　**(d)** 1^8

(a) To find the value of 3^3, multiply the base 3 by itself 3 times.
$3^3 = 3 \times 3 \times 3 = 27$
(b) To find the value of 7^2, multiply the base 7 by itself 2 times.
$7^2 = 7 \times 7 = 49$
(c) $2^5 = 2 \times 2 \times 2 \times 2 \times 2 = 32$
(d) $1^8 = 1 \times 1 \times 1 \times 1 \times 1 \times 1 \times 1 \times 1 = 1$

Practice Problem 2   Find the value of each expression.

(a) 12^2 　　**(b)** 6^3 　　**(c)** 2^6 　　**(d)** 1^{10}

If a whole number does not have a visible exponent, the exponent is understood to be 1. Thus

$$3 = 3^1 \qquad \text{or} \qquad 10 = 10^1.$$

Large numbers are often expressed as a power of 10.

$10^1 = 10 = 1$ ten 　　　　　$10^4 = 10,000 = 1$ ten thousand
$10^2 = 100 = 1$ hundred 　　　$10^5 = 100,000 = 1$ hundred thousand
$10^3 = 1000 = 1$ thousand 　　$10^6 = 1,000,000 = 1$ million

What does it mean to have an exponent of zero? What is 10^0? Any whole number that is not zero can be raised to the zero power. The result is 1. Thus $10^0 = 1$, $3^0 = 1$, $5^0 = 1$, and so on. Why is this? Let's reexamine the powers of 10. As we go down one line at a time, notice the pattern that occurs.

$$10^5 = 100,000$$
$$10^4 = 10,000$$
$$10^3 = 1000$$
$$10^2 = 100$$
$$10^1 = 10$$
$$10^0 = 1$$

As we move down one line, we decrease the exponent by 1.

As we move down one line, we divide the previous number by 10.

Therefore, we present the following definition.

For any whole number a other than zero, $a^0 = 1$.

If numbers with exponents are added to other numbers, it is first necessary to **evaluate**, or find the value of, the number that is raised to a power. Then we may combine the results with another number.

 EXAMPLE 3   Find the value of each expression.

(a) $3^4 + 2^3$ **(b)** $5^3 + 7^0$ **(c)** $6^3 + 6$

(a) $3^4 + 2^3 = (3)(3)(3)(3) + (2)(2)(2) = 81 + 8 = 89$

(b) $5^3 + 7^0 = (5)(5)(5) + 1 = 125 + 1 = 126$

(c) $6^3 + 6 = (6)(6)(6) + 6 = 216 + 6 = 222$

Practice Problem 3   Find the value of each expression.

(a) $7^3 + 8^2$ **(b)** $9^2 + 6^0$ **(c)** $5^4 + 5$

② Performing Several Arithmetic Operations in the Proper Order

Sometimes the order in which we do things is not important. The order in which chefs hang up their pots and pans probably does not matter. The order in which they add and mix the elements in preparing food, however, makes all the difference in the world! If various cooks follow a recipe, though, they will get similar results. The recipe assures that the results will be consistent. It shows the **order of operations.**

In mathematics the order of operations is a list of priorities for working with the numbers in computational problems. This mathematical "recipe" tells how to handle certain indefinite computations. For example, how does a person solve $5 + 3 \times 2$?

A problem such as $5 + 3 \times 2$ sometimes causes students difficulty. Some people think $(5 + 3) \times 2 = 8 \times 2 = 16$. Some people think $5 + (3 \times 2) = 5 + 6 = 11$. Only one answer is right, 11. To obtain the right answer, follow the steps outlined in the following box.

Order of Operations

In the absence of grouping symbols:

Do first   **1.** Simplify any expressions with exponents.

⇕   **2.** Multiply or divide from left to right.

Do last   **3.** Add or subtract from left to right.

EXAMPLE 4   Evaluate. $3^2 + 5 - 4 \times 2$

$$3^2 + 5 - 4 \times 2 = 9 + 5 - 4 \times 2 \qquad \text{Evaluate the expression with exponents.}$$
$$= 9 + 5 - 8 \qquad \text{Multiply from left to right.}$$
$$= 14 - 8 \qquad \text{Add from left to right.}$$
$$= 6 \qquad \text{Subtract.}$$

Practice Problem 4   Evaluate. $7 + 4^3 \times 3$

EXAMPLE 5   Evaluate. $5 + 12 \div 2 - 4 + 3 \times 6$

There are no numbers to raise to a power, so we first do any multiplication or division in order from *left to right*.

$$5 + 12 \div 2 - 4 + 3 \times 6 \qquad \text{Multiply or divide from left to right.}$$
$$= 5 + 6 - 4 + 3 \times 6 \qquad \text{Divide.}$$
$$= 5 + 6 - 4 + 18 \qquad \text{Multiply.}$$
$$\qquad \text{Add or subtract from left to right.}$$
$$= 11 - 4 + 18 \qquad \text{Add.}$$
$$= 7 + 18 \qquad \text{Subtract.}$$
$$= 25 \qquad \text{Add.}$$

Practice Problem 5   Evaluate. $37 - 20 \div 5 + 2 - 3 \times 4$

EXAMPLE 6   Evaluate. $2^3 + 3^2 - 7 \times 2$

$$2^3 + 3^2 - 7 \times 2 = 8 + 9 - 7 \times 2 \qquad \text{Evaluate exponent expressions } 2^3 = 8 \text{ and } 3^2 = 9.$$
$$= 8 + 9 - 14 \qquad \text{Multiply.}$$
$$= 17 - 14 \qquad \text{Add.}$$
$$= 3 \qquad \text{Subtract.}$$

Practice Problem 6   Evaluate. $4^3 - 2 + 3^2$

You can change the order in which you compute by using grouping symbols. Place the numbers you want to calculate first within parentheses. This tells you to do those calculations first.

Order of Operations

With grouping symbols:

Do first   **1.** Perform operations inside the parentheses.

2. Simplify any expressions with exponents.

3. Multiply or divide from left to right.

Do last   **4.** Add or subtract from left to right.

EXAMPLE 7   Evaluate. $2 \times (7 + 5) \div 4 + 3 - 6$

First, we combine numbers inside the parentheses by adding the 7 to the 5. Next, because multiplication and division have equal priority, we work from left to right doing whichever of these operations comes first.

$$2 \times (7 + 5) \div 4 + 3 - 6$$
$$= 2 \times 12 \div 4 + 3 - 6$$
$$= 24 \div 4 + 3 - 6 \qquad \text{Multiply.}$$
$$= 6 + 3 - 6 \qquad \text{Divide.}$$
$$= 9 - 6 \qquad \text{Add.}$$
$$= 3 \qquad \text{Subtract.}$$

Practice Problem 7   Evaluate. $(17 + 7) \div 6 \times 2 + 7 \times 3 - 4$

EXAMPLE 8   Evaluate. $4^3 + 18 \div 3 - 2^4 - 3 \times (8 - 6)$

$$4^3 + 18 \div 3 - 2^4 - 3 \times (8 - 6)$$
$$= 4^3 + 18 \div 3 - 2^4 - 3 \times 2 \qquad \text{Work inside the parentheses.}$$
$$= 64 + 18 \div 3 - 16 - 3 \times 2 \qquad \text{Evaluate exponents.}$$
$$= 64 + 6 - 16 - 3 \times 2 \qquad \text{Divide.}$$
$$= 64 + 6 - 16 - 6 \qquad \text{Multiply.}$$
$$= 70 - 16 - 6 \qquad \text{Add.}$$
$$= 54 - 6 \qquad \text{Subtract.}$$
$$= 48 \qquad \text{Subtract.}$$

Practice Problem 8   Evaluate. $5^2 - 6 \div 2 + 3^4 + 7 \times (12 - 10)$

Verbal and Writing Skills

1. Explain what the expression 5^3 means. Evaluate 5^3.

2. In exponent notation, the _____ tells how many times to multiply the base.

3. In exponent notation, the _____ is the number that is multiplied.

4. 10^5 is read as _____ .

5. Explain the order in which we perform mathematical operations to ensure consistency.

6. Use the order of operations to solve $12 \times 5 + 3 \times 5 + 7 \times 5$. Is this the same as $5(12 + 3 + 7)$? Why or why not?

Write each number in exponent form.

7. $6 \times 6 \times 6 \times 6$ **8** $2 \times 2 \times 2 \times 2 \times 2$ **9.** $4 \times 4 \times 4 \times 4 \times 4$

10. $12 \times 12 \times 12 \times 12 \times 12 \times 12$ **11.** $8 \times 8 \times 8 \times 8$

12. $1 \times 1 \times 1 \times 1 \times 1 \times 1 \times 1$ **13.** 9   **14.** 27

Find the value of each expression.

15. 2^4 **16.** 3^3 **17.** 4^2 **18.** 5^2 **19.** 6^3 **20.** 10^3 **21.** 10^4 **22.** 1^{20}

23. 1^{17} **24.** 2^5 **25.** 2^6 **26.** 4^3 **27.** 3^5 **28.** 12^2 **29.** 15^2 **30.** 8^3

31. 7^3 **32.** 5^4 **33.** 4^4 **34.** 2^7 **35.** 9^0 **36.** 8^0 **37.** 25^2 **38.** 20^3

39. 10^6 **40.** 8^1 .   **41.** 4^5 **42.** 10^5 **43.** 9^1 **44.** 14^2 **45.** 7^4 **46.** 5^3

47. $2^4 + 1^8$ **48.** $7^0 + 4^3$ **49.** $6^3 + 3^2$

50. $7^3 + 4^2$ **51.** $8^3 + 8$ **52.** $9^2 + 9$

Work each exercise, using the correct order of operations.

53. $5 \times 6 - 3 + 10$ **54.** $8 \times 3 - 4 \times 2$ **55.** $3 \times 9 - 10 \div 2$ **56.** $20 \div 4 \times 6 - 12$

57. $48 \div 2^3 + 4$ **58.** $4^3 \div 4 - 11$ **59.** $7 \times 3^2 + 4 - 8$ **60.** $2^3 \times 4 + 6 - 9$

61. $4 \times 9^2 - 4 \times (7 - 2)$ **62.** $5 \times 2^5 - 4 \times (11 - 8)$ **63.** $(400 \div 20) \div 20$ **64.** $(600 \div 30) \div 20$

65. $950 \div (25 \div 5)$ **66.** $875 \div (35 \div 7)$ **67.** $(12)(5) - (12 + 5)$ **68.** $(3)(60) - (60 + 3)$

69. $3^2 + 4^2 \div 2^2$ **70.** $7^2 + 9^2 \div 3^2$ **71.** $(6)(7) - (12 - 8) \div 4$

72. $(8)(9) - (15 - 5) \div 5$ **73.** $100 - 3^2 \times 4$ **74.** $130 - 4^2 \times 5$

75. $5^2 + 2^2 + 3^3$ **76.** $2^3 + 3^2 + 4^3$ **77.** $72 \div 9 \times 3 \times 1 \div 2$

78. $120 \div 30 \times 2 \times 5 \div 8$ **79.** $12^2 - 6 \times 3 \times 4 \times 0$ **80.** $14^2 - 5 \times 2 \times 3 \times 0$

81. $4^2 \times 6 \div 3$

82. $7^2 \times 3 \div 3$

83. $5 + 6 \times 2 \div 4 - 1$

84. $7 + 5 \times 4 \div 10 - 2$

85. $3 + 3^2 \times 6 + 4$

86. $5 + 4^3 \times 2 + 7$

87. $12 \div 2 \times 3 - 2^4$

88. $30 \div 10 \times 5 - 3^2$

89. $3^2 \times 6 \div 9 + 4 \times 3$

90. $5^2 \times 3 \div 25 + 7 \times 6$

91. $4(10 - 6)^3 \div (2 + 6)$

92. $3(12 - 5)^2 \div (4 + 3)$

93. $1200 - 2^3(3) \div 6$

94. $2150 - 3^4(2) \div 9$

95. $250 \div 5 + 20 - 3^2$

96. $450 \div 25 - 8 + 2^5$

97. $20 \div 4 \times 3 - 6 + 2 \times (1 + 4)$

98. $50 \div 10 \times 2 - 8 + 3 \times (17 - 11)$

99. $2 \times 3 + (11 + 5) - 4 \times 5$

100. $(6 + 4) \times 10 - 5 \times 5 + 2 \times 4$

To Think About

101. The earth rotates once every 23 hours, 56 minutes, 4 seconds. How many seconds is that?

102. The planet Saturn rotates once every 10 hours, 12 minutes. How many minutes is that? How many seconds?

Cumulative Review Problems

103. Write in expanded notation.   156,312

104. Write in standard notation. two hundred million, seven hundred sixty-five thousand, nine hundred nine

105. Write in words.   261,763,002

▲ **106.** New Boston High School has an athletic field that needs to be enclosed by fencing. The rectangular field is 250 feet wide and 480 feet long. How many feet of fencing are needed to surround the field? Grass needs to be planted for a new playing field for next year. What is the area in square feet of the amount of grass that must be planted?

Rounding and Estimation

Student Learning Objectives

After studying this section, you will be able to:

 Round whole numbers.

 Estimate the answer to a problem involving calculation with whole numbers.

SSM

PH TUTOR   CD & VIDEO   MATH PRO   WEB
CENTER

1 *Rounding Whole Numbers*

Large numbers are often expressed to the nearest hundred or to the nearest thousand, because an approximate number is "good enough" for certain uses.

Distances from the earth to other galaxies are measured in light-years. Although light really travels at 5,865,696,000,000 miles a year, we usually **round** this number to the nearest trillion and say it travels at 6,000,000,000,000 miles a year. To round a number, we first determine the place we are rounding to—in this case, trillion. Then we find which value is closest to the number that we are rounding. In this case, the number we want to round is closer to 6 trillion than to 5 trillion. How do we know the number is closer to 6 trillion than to 5 trillion?

To see which is the closest value, we may picture a **number line,** where whole numbers are represented by points on a line. To show how to use the number line in rounding, we will round 368 to the nearest hundred. 368 is between 300 and 400. When we round off, we pick the hundred 368 is "closest to." We draw a number line to show 300 and 400. We also show the point midway between 300 and 400 to help us to determine which hundred 368 is closest to.

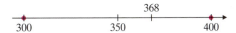

We find that the number 368 is closer to 400 than to 300, so we round 368 *up to* 400.

Let's look at another example. We will round 129 to the nearest hundred. 129 is between 100 and 200. We show this on the number line. We include the midpoint 150 as a guide.

We find that the number 129 is closer to 100 than to 200, so we round 129 *down to* 100.

This leads us to the following simple rule for rounding.

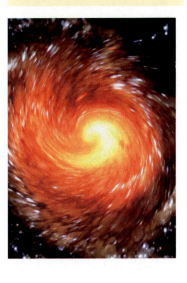

Rounding a Whole Number

1. If the first digit to the right of the round-off place is
   (a) *less than 5,* we make no change to the digit in the round-off place. (We know it is closer to the smaller number, so we round down.)
   (b) *5 or more,* we increase the digit in the round-off place by 1. (We know it is closer to the larger number, so we round up.)
2. Then we replace the digits to the right of the round-off place by zeros.

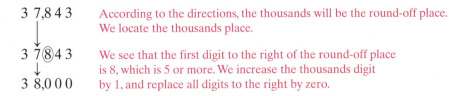

● **EXAMPLE 1**   Round 37,843 to the nearest thousand.

3 7,8 4 3   According to the directions, the thousands will be the round-off place. We locate the thousands place.

3 7⑧4 3   We see that the first digit to the right of the round-off place
              is 8, which is 5 or more. We increase the thousands digit
3 8,0 0 0   by 1, and replace all digits to the right by zero.

We have rounded 37,843 to the nearest thousand: 38,000. This means that 37,843 is closer to 38,000 than to 37,000.

Practice Problem 1 Round 65,528 to the nearest thousand.

EXAMPLE 2 Round 2,445,360 to the nearest hundred thousand.

2,4 4 5,3 6 0   Locate the hundred thousands round-off place.

2,4 ④ 5,3 6 0   The first digit to the right of this is less than 5, so round down. Do not change the hundred thousands digit.

2,4 0 0,0 0 0   Replace all digits to the right by zero.

Practice Problem 2 Round 172,963 to the nearest ten thousand.

EXAMPLE 3 Round as indicated.

(a) 561,328 to the nearest ten   **(b)** 3,798,152 to the nearest hundred
(c) 51,362,523 to the nearest million

(a) First locate the digit in the tens place.

↓
561,328   The digit to the right of the tens is greater than 5.
561,330   Round up.

561,328 rounded to the nearest ten is 561,330.

(b) 3,798,152   The digit to the right of the hundreds is 5.
3,798,200   Round up.
3,798,152 rounded to the nearest hundred is 3,798,200.

(c) 51,362,523   The digit to the right of the millions is less than 5.
51,000,000   Round down.
51,362,523 rounded to the nearest million is 51,000,000.

Practice Problem 3 Round as indicated.

(a) 53,282 to the nearest ten   **(b)** 164,485 to the nearest thousand
(c) 1,365,273 to the nearest hundred thousand

EXAMPLE 4 Round 763,571.

(a) To the nearest thousand
(b) To the nearest ten thousand
(c) To the nearest million

↓
(a) 763,571 = 764,000 to the nearest thousand. The digit to the right of the thousands is 5. We rounded up.

↓
(b) 763,571 = 760,000 to the nearest ten thousand. The digit to the right of the ten thousands is less than 5. We rounded down.

(c) 763,571 does not have any digits for millions. If it helps, you can think of this number as 0,763,571. Since the digit to the right of the millions place is 7, we round up to obtain one million or 1,000,000.

Practice Problem 4   Round 935,682 as indicated.

(a) To the nearest thousand

(b) To the nearest hundred thousand

(c) To the nearest million

EXAMPLE 5   Astronomers use the parsec as a measurement of distance. One parsec is approximately 30,900,000,000,000 kilometers. Round 1 parsec to the nearest trillion kilometers.

↓

30,900,000,000,000 km is 31,000,000,000,000 km or 31 trillion km to the nearest trillion kilometers.

Practice Problem 5   One light-year is approximately 9,460,000,000,000,000 meters. Round to the nearest hundred trillion meters.

2   *Estimating the Answer to a Problem Involving Calculation with Whole Numbers*

Often we need to quickly check the answer of a calculation to be reasonably assured that the answer is correct. If you expected your bill to be "around $40" for the groceries you had selected and the cashier's total came to $41.89, you would probably be confident that the bill is correct and pay it. If, however, the cashier rang up a bill of $367, you would not just assume that it is correct. You would know an error had been made. If the cashier's total came to $60, you might not be certain, but you would probably suspect an error and check the calculation.

In mathematics we often **estimate,** or determine the approximate value of a calculation, if we need to do a quick check. There are many ways to estimate, but in this book we will use one simple principle of estimation. We use the symbol ≈ to mean **approximately equal to.**

Principle of Estimation

1. Round the numbers so that there is one nonzero digit in each number.

2. Perform the calculation with the rounded numbers.

EXAMPLE 6   Estimate the sum. 163 + 237 + 846 + 922

We first determine where to round each number in our problem to leave only one nonzero digit in each. In this case, we round all numbers to the nearest hundred. Then we perform the calculation with the rounded numbers.

Actual Sum	*Estimated Sum*
163	200
237	200
846	800
+ 922	+ 900
	2100

We estimate the answer to be 2100. We say the sum ≈ 2100. If we calculate using the exact numbers, we obtain a sum of 2168, so our estimate is quite close to the actual sum.

Practice Problem 6   Estimate the sum. $3456 + 9876 + 5421 + 1278$

When we use the principle of estimation, we will not always round each number in a problem to the same place.

EXAMPLE 7   Phil and Melissa bought their first car last week. The selling price of this compact car was $8980. The dealer preparation charge was $289 and the sales tax was $449. Estimate the total cost of the car that Phil and Melissa had to pay.

We round each number to have only one nonzero digit, and add the rounded numbers.

8980	9000
289	300
+ 449	+ 400
	9700

The total cost $\approx$ $9700. (The exact answer is $9718, so we see that our answer is quite close.)

Practice Problem 7   Greg and Marcia purchased a new sofa for $697, plus $35 sales tax. The store also charged them $19 to deliver the sofa. Estimate their total cost.

Now we turn to a case where an estimate can help us discover an error.

EXAMPLE 8   Roberto added together four numbers and obtained the following result. Estimate the sum and determine if the answer seems reasonable.

$$12{,}456 + 17{,}976 + 18{,}452 + 32{,}128 \overset{?}{=} 61{,}012$$

We round each number so that there is one nonzero digit. In this case, we round them all to the nearest ten thousand.

12,456	10,000
17,976	20,000
18,452	20,000
+ 32,128	+ 30,000
	80,000 Our estimate is 80,000.

This is significantly different from 61,012, so we would suspect that an error has been made. In fact, Roberto did make an error. The exact sum is actually 81,012!

Practice Problem 8   Ming did the following calculation. Estimate to see if her sum appears to be correct or incorrect.

$$11,849 + 14,376 + 16,982 + 58,151 = 81,358$$

Next we look at a subtraction example where estimation is used.

EXAMPLE 9   The profit from Techno Industries for the first quarter of the year was \$642,987,000. The profit for the second quarter was \$238,890,000. Estimate how much less the profit was for the second quarter than for the first quarter.

We round each number so that there is one nonzero digit. Then we sub-tract, using the two rounded numbers.

$$
\begin{array}{r}
642,987,000 \\
-\ 238,890,000 \\
\hline
\end{array}
\qquad
\begin{array}{r}
600,000,000 \\
-\ 200,000,000 \\
\hline
400,000,000
\end{array}
$$

We estimate that the profit was \$400,000,000 less for the second quarter.

Practice Problem 9   The 1998 population of Florida was 14,916,235. The 1998 population of California was 32,667,584. Estimate how many more people lived in California in 1998 than in Florida.

We also use this principle to estimate results of multiplication and division.

EXAMPLE 10   Estimate the product. 56,789 × 529

We round each number so that there is one nonzero digit. Then we mul-tiply the rounded numbers to obtain our estimate.

$$
\begin{array}{r}
56,789 \\
\times\quad 529 \\
\hline
\end{array}
\qquad
\begin{array}{r}
60,000 \\
\times\quad 500 \\
\hline
30,000,000
\end{array}
$$

Therefore the product is ≈30,000,000. (This is reasonably close to the exact answer of 30,041,381.)

Practice Problem 10   Estimate the product. 8945 × 7317

EXAMPLE 11   Estimate the answer for the following division problem.

$$23\overline{)148,902}$$

We round each number to a number with one nonzero digit. Then we per-form the division, using the two rounded numbers.

$$
23\overline{)148,902}
\qquad
\begin{array}{r}
5000 \\
20\overline{)100,000}
\end{array}
$$

Our estimate is 5000. (The exact answer is 6474. We see that our estimate is "in the ballpark" but is not very close to the exact answer. Remember, an estimate is just a rough approximation of the exact answer.)

Practice Problem 11   Estimate the answer for the following division problem.

$$39\overline{)75,342}$$

Not all division estimates come out so easily. In some cases, you may need to carry out a long-division problem of several steps just to obtain the estimate.

EXAMPLE 12 Ralph drove his car a distance of 778 miles. He used 25 gallons of gas. Estimate how many miles he can travel on 1 gallon of gas.

In order to solve this problem, we need to divide 778 by 25 to obtain the number of miles Ralph gets with 1 gallon of gas. We round each number to a number with one nonzero digit and then perform the division, using the rounded numbers.

$$
25\overline{)778}
\qquad
\begin{array}{r}
26 \\
30\overline{)800} \\
\underline{60} \\
200 \\
\underline{180} \\
20 \quad \text{Remainder}
\end{array}
$$

We obtain an answer of 26 with a remainder of 20. For our estimate we will use the whole number 27. Thus we estimate that the number of miles Ralph's car obtained on 1 gallon of gas was 27 miles. (This is reasonably close to the exact answer, which is just slightly more than 31 miles per gallon of gas.)

Practice Problem 12 The highway department purchased 58 identical trucks at a total cost of $1,864,584. Estimate the cost for one truck.

1.7 Exercises

Verbal and Writing Skills

1. Explain the rule for rounding and provide examples.

2. What happens when you round 98 to the nearest ten?

Round to the nearest ten.

3. 83　　　　**4.** 45　　　　**5.** 65　　　　**6.** 57　　　　**7.** 92

8. 138　　　**9.** 526　　**10.** 2834　　**11.** 1672　　**12.** 7865

Round to the nearest hundred.

13. 247　　**14.** 661　　**15.** 2781　　**16.** 1258　　**17.** 7692　　**18.** 1643

Round to the nearest thousand.

19. 7621　　**20.** 3754　　**21.** 1672　　**22.** 515　　**23.** 27,863　　**24.** 94,671

Applications

25. The worst death rate from an earthquake was in Shaanxi, China, in 1556. That earthquake killed an estimated 832,400 people. Round this number to the nearest hundred thousand.

26. One light year (the distance light travels in 1 year) measures 5,878,612,843,000 miles. Round this figure to the nearest hundred million.

27. The Hubble Space Telescope's *Guide Star Catalogue* lists 15,169,873 stars. Round this figure to the nearest million.

28. The point of highest elevation in the world is Mt. Everest in the country of Nepal. Mt. Everest is 29,028 feet above sea level. Round this figure to the nearest ten thousand.

29. The total Native American population living on the Navajo and Trust Lands in the Arizona/New Mexico/Utah area numbered 163,298 in 2000. Round this figure to
   (a) the nearest thousand.
   (b) the nearest hundred.

30. A few years ago, there were 614,571 Catholic-school pupils in the United States. Round this figure to
   (a) the nearest thousand.
   (b) the nearest hundred.

▲ 31. The total area of mainland China is 3,705,392 square miles, or, 9,596,960 square kilometers. For *both* square miles and square kilometers, round this figure to
   (a) the nearest hundred thousand.

   (b) the nearest ten thousand.

32. Recently, in the United States 10,957,442,856 pounds of apples were produced.
   (a) Round this figure to the nearest hundred thousand.

   (b) Round this figure to the nearest ten thousand.

Use the principle of estimation to find an estimate for each calculation.

33. 613 + 252 + 137

34. 871 + 365 + 341

35. 34 + 78 + 59 + 31

36. 31 + 75 + 82 + 43

37. 146,270 + 47,566 + 96,112

38. 345,957 + 89,334 + 56,487

39. 567,984 − 129,562

40. 975,935 − 593,228

41. 78,945,000 − 61,076,500

42. 4,596,450 − 3,894,202

43. 33,261,378 − 18,199,276

44. 89,263,000 − 54,198,635

45. 47×62

46. 43×95

47. 1324×8

48. 5926×3

49. $631,540 \times 312$

50. $374,193 \times 193$

51. $24,318 \div 21$

52. $61,986 \div 32$

53. $156,721 \div 42$

54. $581,361 \div 28$

55. $3,885,720 \div 831$

56. $12,447,312 \div 497$

Estimate the result of each calculation. Some results are correct and some are incorrect. Which results appear to be correct? Which results appear to be incorrect?

57.
$$\begin{array}{r} 361 \\ 522 \\ 873 \\ + 164 \\ \hline 1320 \end{array}$$

58.
$$\begin{array}{r} 476 \\ 124 \\ 516 \\ + 389 \\ \hline 1505 \end{array}$$

59.
$$\begin{array}{r} 97,635 \\ 52,123 \\ + 41,986 \\ \hline 291,744 \end{array}$$

60.
$$\begin{array}{r} 26,181 \\ 47,998 \\ + 63,271 \\ \hline 137,450 \end{array}$$

61.
$$\begin{array}{r} 302,360 \\ - 89,518 \\ \hline 212,842 \end{array}$$

62.
$$\begin{array}{r} 735,128 \\ - 116,733 \\ \hline 518,395 \end{array}$$

63.
$$\begin{array}{r} 78,126,345 \\ - 48,972,103 \\ \hline 19,154,242 \end{array}$$

64.
$$\begin{array}{r} 42,765,317 \\ - 29,318,274 \\ \hline 23,447,043 \end{array}$$

65.
$$\begin{array}{r} 216 \\ \times\ \ 24 \\ \hline 6184 \end{array}$$

66.
$$\begin{array}{r} 578 \\ \times\ \ 32 \\ \hline 10,496 \end{array}$$

67.
$$\begin{array}{r} 5896 \\ \times\ \ 72 \\ \hline 424,512 \end{array}$$

68.
$$\begin{array}{r} 8076 \\ \times\ \ 89 \\ \hline 718,764 \end{array}$$

$$\begin{array}{r} 5286 \\ \textbf{69.}\ 78\overline{)412{,}308} \end{array} \qquad \begin{array}{r} 4793 \\ \textbf{70.}\ 58\overline{)277{,}994} \end{array} \qquad \begin{array}{r} 381 \\ \textbf{71.}\ 423\overline{)161{,}163} \end{array} \qquad \begin{array}{r} 612 \\ \textbf{72.}\ 781\overline{)477{,}972} \end{array}$$

Applications

▲ **73.** Larry and Nella are planning an outdoor wedding by the ocean. There is a beautiful meadow that is 35 feet wide and 62 feet long. Estimate the number of square feet in the meadow.

▲ **74.** A huge restaurant in New York City is 43 yards wide and 112 yards long. Estimate the number of square yards in the restaurant.

75. There are four salesmen at Oakridge Ford. The annual salaries of the four salesman last year were $17,894, $24,587, $37,892, and $59,643. Estimate the total amount that the four salesman earned in one year.

76. The highway departments in four towns in northwestern New York had the following budgets for snow removal for the year: $329,560, $672,940, $199,734, and $567,087. Estimate the total amount that the four towns spend for snow removal in one year.

77. The local pizzeria makes 267 pizzas on an average day. Estimate how many pizzas were made in the last 134 days.

78. To get ready for bathing suit weather, Wendy did 85 sit-ups daily for 63 days. Estimate how many sit-ups she did during that period.

79. U.S. air travel increased from 457,000,000 passengers in 1990 to 656,000,000 in 2000. Round each figure to the nearest ten million. Then estimate the increase.

80. The price of a Super Bowl ticket increased from $30 in 1980 to $280 in 2000. Estimate the increase. Determine the exact increase.

▲ **81.** The largest state of the United States is Alaska, with a land area of 586,412 square miles. The second largest state is Texas, with an area of 267,339 square miles. Round each figure to the nearest ten thousand. Then estimate how many square miles larger Alaska is than Texas.

82. Tina determined that she spent $9 on coffee each week last year. Estimate how much she spent on coffee over the entire year.

To Think About

83. A space probe travels at 23,560 miles per hour for a distance of 7,824,560,000 miles.
   (a) How many *hours* will it take the space probe to travel that distance? (Estimate.)

   (b) How many *days* will it take the space probe to travel that distance? (Estimate.)

84. A space probe travels at 28,367 miles per hour for a distance of 9,348,487,000 miles.
   (a) Estimate the number of *hours* it will take the space probe to travel that distance.

   (b) Estimate the number of *days* it will take the space probe to travel that distance.

Cumulative Review Problems

Evaluate.

85. $26 \times 3 + 20 \div 4$

86. $5^2 + 3^2 - (17 - 10)$

87. $3 \times (16 \div 4) + 8 \times 2$

88. $126 + 4 - (20 \div 5)^3$

1.8 Applied Problems

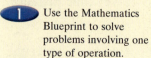

Student Learning Objectives

After studying this section, you will be able to:

1 Use the Mathematics Blueprint to solve problems involving one type of operation.

2 Use the Mathematics Blueprint to solve problems involving more than one type of operation.

SSM
PH TUTOR CENTER   CD & VIDEO   MATH PRO   WEB

1 *Solving Problems Involving One Type of Operation*

When a builder constructs a new home or office building, he or she often has a *blueprint*. This accurate drawing shows the basic structure of the building. It also shows the dimensions of the structure to be built. This blueprint serves as a useful reference throughout the construction process.

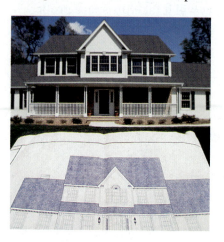

Similarly, when solving applied problems, it is helpful to have a "mathematics blueprint." This is a simple way to organize the information provided in the word problem. You record the facts you need to use and specify what you are solving for. You also record any other information that you feel will be helpful. We will use a Mathematics Blueprint for Problem Solving in the following situation.

Sometimes people feel totally lost when trying to solve a word problem. They sometimes say, "Where do I begin?" or "How in the world do you do this?" When you have this type of feeling, it sometimes helps to have a formal strategy or plan. Here is a plan you may find helpful:

1. *Understand the problem.*
   (a) Read the problem carefully.
   (b) Draw a picture if this helps you see the relationships more clearly.
   (c) Fill in the Mathematics Blueprint so that you have the facts and a method of proceeding in this situation.

2. *Solve and state the answer.*
   (a) Perform the calculations.
   (b) State the answer, including the unit of measure.

3. *Check.*
   (a) Estimate the answer.
   (b) Compare the exact answer with the estimate to see if your answer is reasonable.

Now exactly what does the Mathematics Blueprint for Problem Solving look like? It is a simple sheet of paper with four columns. Each column tells you something to do.

Gather the Facts—Find the numbers that you will need to use in your calculations.

What Am I Asked to Do?—Are you finding an area, a volume, a cost, the total number of people? What is it that you need to find?

How Do I Proceed?—Do you need to add items together? Do you need to multiply or divide? What types of calculations are required?

Key Points to Remember—Write down things you might forget. The length is in feet. The area is in square feet. We need the total number of something, not the intermediate totals. Whatever you need to help you, write it down in this column.

Mathematics Blueprint For Problem Solving

Gather the Facts	What Am I Asked to Do?	How Do I Proceed?	Key Points to Remember

● **EXAMPLE 1**   Gerald made deposits of $317, $512, $84, and $161 into his checking account. He also made out checks for $100 and $125. What was the total of his deposits?

1. *Understand the problem.*
First we read over the problem carefully and fill in the Mathematics Blueprint.

Mathematics Blueprint For Problem Solving

Gather the Facts	What Am I Asked to Do?	How Do I Proceed?	Key Points to Remember
We need only deposits—not checks. The **deposits** are $317, $512, $84, and $161.	Find the total of Gerald's four deposits.	I must add the four deposits to obtain the total.	Watch out! Don't use the **checks** of $100 and $125 in the calculation. We only want the total of the **deposits.**

2. *Solve and state the answer.*
We need to *add* to find the sum of the deposits.

$$
\begin{array}{r}
317 \\
512 \\
84 \\
+\ 161 \\
\hline
1074
\end{array}
$$

The total of the four deposits is $1074.

3. *Check.*

Reread the problem. Be sure you have answered the question that was asked. Did it ask for the total of the deposits? Yes. ✓

Is the calculation correct? You can use estimation to check. Here we round each of the deposits so that we have one nonzero digit.

$$
\begin{array}{rr}
317 & 300 \\
512 & 500 \\
84 & 80 \\
+161 & +200 \\
\hline
& 1080
\end{array}
$$

Our estimate is $1080. $1074 is close to our estimated answer of $1080. Our answer is reasonable. ✓

Thus we conclude that the total of the four deposits is $1074.

Practice Problem 1   Use the Mathematics Blueprint to solve the following problem. Diane's paycheck shows deductions of $135 for federal taxes, $28 for state taxes, $13 for FICA, and $34 for health insurance. Her gross pay (amount before deductions) is $1352. What is the total amount that is taken out of Diane's paycheck?

Mathematics Blueprint For Problem Solving

Gather the Facts	What Am I Asked to Do?	How Do I Proceed?	Key Points to Remember

Portland   028,353.0

Kansas   030,162.0

● **EXAMPLE 2**   Theofilos looked at his odometer before he began his trip from Portland, Oregon, to Kansas City, Kansas. He checked his odometer again when he arrived in Kansas City. The two readings are shown in the figure. How many miles did Theofilos travel?

1. *Understand the problem.*
Determine what information is given.

The mileage reading before the trip began and when the trip was over.
What do you need to find?
The number of miles traveled.

Mathematics Blueprint For Problem Solving

Gather the Facts	What Am I Asked to Do?	How Do I Proceed?	Key Points to Remember
At the start of the trip, the odometer read 28,353 miles. At the end of the trip, the odometer read 30,162 miles.	Find out how many miles Theofilos traveled.	I must subtract the two mileage readings.	Subtract the mileage at the start of the trip from the mileage at the end of the trip.

2. *Solve and state the answer.*

We need to subtract the two mileage readings to find the difference in the number of miles. This will give us the number of miles the car traveled on this trip alone.

$$30,162 - 28,353 = 1809 \quad \text{The trip totaled 1809 miles.}$$

3. *Check.*

We estimate and compare the estimate with the preceding answer.

Kansas City $30,162 \longrightarrow 30,000$ We subtract
Portland $28,353 \longrightarrow \underline{28,000}$ our rounded values.
$2,000$

Our estimate is 2000 miles. We compare this estimate with our answer. Our answer is reasonable. ✓

Practice Problem 2   The table on the right shows the results of the 1996 presidential race in the United States. By how many votes did the Democratic candidate beat the Republican candidate in that year?

1996 Presidential Race, Popular Votes

Source: Federal Election Commission

Candidate	Number of Votes
Clinton	47,402,357
Dole	39,198,755
Perot	8,085,402

Mathematics Blueprint For Problem Solving

Gather the Facts	What Am I Asked to Do?	How Do I Proceed?	Key Points to Remember

● **EXAMPLE 3**   One horsepower is the power needed to lift 550 pounds a distance of 1 foot in 1 second. How many pounds can be lifted 1 foot in 1 second by 7 horsepower?

1. *Understand the problem.*

    Simplify the problem. If 1 horsepower can lift 550 pounds, how many pounds can be lifted by 7 horsepower? We draw and label a diagram.

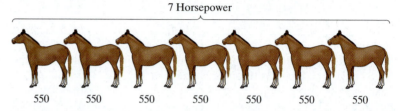

7 Horsepower

550   550   550   550   550   550   550

We use the mathematics blueprint to organize the information.

Mathematics Blueprint For Problem Solving

Gather the Facts	What Am I Asked to Do?	How Do I Proceed?	Key Points to Remember
One horsepower will lift 550 pounds.	Find how many pounds can be lifted by 7 horsepower.	I need to multiply 550 by 7.	I do not use the information about moving one foot in one second.

2. *Solve and state the answer.*

    To solve the problem we multiply the 7 horsepower by 550 pounds for each horsepower.

$$\begin{array}{r} 550 \\ \times\ \ \ 7 \\ \hline 3850 \end{array}$$

    We find that 7 horsepower moves 3850 pounds 1 foot in 1 second. We include 1 foot in 1 second in our answer because it is part of the unit of measure.

3. *Check.*

    We estimate our answer. We round 550 to 600 pounds.

$$600 \times 7 = 4200 \text{ pounds}$$

    Our estimate is 4200 pounds. Our calculations in step 2 gave us 3850. Is this reasonable? This answer is close to our estimate. Our answer is reasonable. ✓

Practice Problem 3   In a measure of liquid capacity, 1 gallon is 1024 fluid drams. How many fluid drams would be in 9 gallons?

Mathematics Blueprint For Problem Solving

Gather the Facts	What Am I Asked to Do?	How Do I Proceed?	Key Points to Remember

● **EXAMPLE 4**   Laura can type 35 words per minute. She has to type an English theme that has 5180 words. How many minutes will it take her to type the theme? How many hours and how many minutes will it take her to type the theme?

1. *Understand the problem.*

We draw a picture. Each "package" of 1 minute is 35 words. We want to know how many packages make up 5180 words.

We use the Mathematics Blueprint to organize the information.

Mathematics Blueprint For Problem Solving

Gather the Facts	What Am I Asked to Do?	How Do I Proceed?	Key Points to Remember
Laura can type 35 words per minute. She must type a paper with 5180 words.	Find out how many 35-word units are in 5180 words.	I need to divide 5180 by 35.	In converting minutes to hours, I will use the fact that 1 hour = 60 minutes.

2. *Solve and state the answer.*

$$
\begin{array}{r}
148 \\
35\overline{)5180} \\
35 \\
\hline
168 \\
140 \\
\hline
280 \\
280 \\
\hline
\end{array}
$$

It will take 148 minutes.

We will change this answer to hours and minutes. Since 60 minutes = 1 hour, we divide 148 by 60. The quotient will tell us how many hours. The remainder will tell us how many minutes.

$$
\begin{array}{r}
2\,\text{R}\,28 \\
60\overline{)148} \\
-\,120 \\
\hline
28
\end{array}
$$

Laura can type the theme in 148 minutes or 2 hours, 28 minutes.

3. *Check.*

The theme has 5180 words; she can type 35 words per minute. 5180 words is approximately 5000 words.

5180 words → 5000 words rounded to nearest thousand.

$$\begin{array}{r} 125 \\ 40\overline{)5000} \end{array}$$

35 words per minute → 40 words per minute rounded to nearest ten.

We divide our estimated values.

Our estimate is 125 minutes. This is close to our calculated answer. Our answer is reasonable. ✓

Practice Problem 4   Donna bought 45 shares of stock for $1620. How much did the stock cost her per share?

Mathematics Blueprint For Problem Solving

Gather the Facts	What Am I Asked to Do?	How Do I Proceed?	Key Points to Remember

2 Solving Problems Involving More Than One Type of Operation

Sometimes a chart, table, or bill of sale can be used to help us organize the data in an applied problem. In such cases, a blueprint may not be needed.

EXAMPLE 5   Cleanway Rent-A-Car bought four luxury sedans at $21,000 each, three compact sedans at $14,000 each, and seven subcompact sedans at $8000 each. What was the total cost of the purchase?

1. *Understand the problem.*
   We will make an imaginary bill of sale to help us to visualize the problem.
2. *Solve and state the answer.*
   We do the calculation and enter the results in the bill of sale.

Car Fleet Sales, Inc. Hamilton, Massachusetts

Customer: *Cleanway Rent-A-Car*			
Quantity	Type of Car	Cost per Car	Amount for This Type of Car
4	Luxury sedans	$21,000	$84,000 (4 × $21,000 = $84,000)
3	Compact sedans	$14,000	$42,000 (3 × $14,000 = $42,000)
7	Subcompact sedans	$ 8,000	$56,000 (7 × $8,000 = $56,000)
		TOTAL	$182,000 (sum of the three amounts)

The total cost of all 14 cars is $182,000.
3. *Check.*
   You may use estimation to check. The check is left to the student.

Practice Problem 5   Anderson Dining Commons purchased 50 tables at $200 each, 180 chairs at $40 each, and six moving carts at $65 each. What was the cost of the total purchase?

⬤ **EXAMPLE 6** Dawn had a balance of $410 in her checking account last month. She made deposits of $46, $18, $150, $379, and $22. She made out checks for $316, $400, and $89. What is her balance?

1. *Understand the problem.*

We want to *add* to get a total of all deposits and *add* to get a total of all checks.

| Old balance | + | total of deposits | − | total of checks | = | new balance |

Mathematics Blueprint For Problem Solving

Gather the Facts	What Am I Asked to Do?	How Do I Proceed?	Key Points to Remember
Old balance: $410. New deposits: $46, $18, $150, $379, and $22. New checks: $316, $400, and $89.	Find the amount of money in the checking account after deposits are made and checks are withdrawn.	(a) I need to calculate the total of the deposits and the total of the checks. (b) I add the total deposits to the old balance. (c) Then I subtract the total of the checks from that result.	Deposits are added to a checking account. Checks are subtracted from a checking account.

2. *Solve and state the answer.*

First we find the total sum of deposits:
```
   46
   18
  150
  379
+  22
$615
```
Then the total sum of checks:
```
  316
  400
+  89
$805
```

Add the deposits to the old balance and subtract the amount of the checks.

```
Old balance        410
+ total deposits  + 615
                  1025
− total checks    − 805
New balance        220
```

The new balance of the checking account is $220.

3. *Check.*

Work backward. You can add the total checks to the new balance and then subtract the total deposits. The result should be the old balance. Try it.

```
  410    Old balance  ✓
− 615
 1025
+ 805
  220    Work backward.
```

Practice Problem 6   Last month Bridget had $498 in a savings account. She made two deposits: one for $607 and one for $163. The bank credited her with $36 interest. Since last month, she has made four withdrawals: $19, $158, $582, and $74. What is her balance this month?

Mathematics Blueprint For Problem Solving

Gather the Facts	What Am I Asked to Do?	How Do I Proceed?	Key Points to Remember

EXAMPLE 7   When Lorenzo began his car trip, his gas tank was full and the odometer read 76,358 miles. He ended his trip at 76,668 miles and filled the gas tank with 10 gallons of gas. How many miles per gallon did he get with his car?

1. *Understand the problem.*

Mathematics Blueprint For Problem Solving

Gather the Facts	What Am I Asked to Do?	How Do I Proceed?	Key Points to Remember
Odometer reading at end of trip: 76,668 miles. Odometer reading at start of trip: 76,358 miles. Used on trip: 10 gallons of gas	Find the number of miles per gallon that the car obtained on the trip.	**(a)** I need to subtract the two odometer readings to obtain the number of miles traveled. **(b)** I divide the number of miles driven by the number of gallons of gas used to get the number of miles obtained per gallon of gas.	The gas tank was full at the beginning of the trip. 10 gallons fills the tank at the end of the trip.

2. *Solve and state the answer.*
    First we subtract the odometer readings to obtain the miles traveled.

$$\begin{array}{r} 76{,}668 \\ -\,76{,}358 \\ \hline 310 \end{array}$$

The trip was 310 miles.

Next we divide the miles driven by the number of gallons.

$$\begin{array}{r} 31 \\ 10\overline{)310} \\ \underline{30} \\ 10 \\ \underline{10} \\ 0 \end{array}$$

Thus Lorenzo obtained 31 miles per gallon on the trip.

3. *Check.*
We do not want to round to one nonzero digit here, because, if we do, the result will be zero when we subtract. Thus we will round to the nearest hundred for the values of mileage.

$$76,668 \longrightarrow 76,700$$

$$76,358 \longrightarrow 76,400$$

Now we subtract the estimated values.

$$\begin{array}{r} 76,700 \\ - 76,400 \\ \hline 300 \end{array}$$

Thus we estimate the trip to be 300 miles.
Then we divide.

$$\begin{array}{r} 30 \\ 10\overline{)300} \end{array}$$

We obtain 30 miles per gallon for our estimate. This is very close to our calculated value of 31 miles per gallon. ✓

Practice Problem 7   Deidre took a car trip with a full tank of gas. Her trip began with the odometer at 50,698 and ended at 51,118 miles. She then filled the tank with 12 gallons of gas. How many miles per gallon did her car get on the trip?

Mathematics Blueprint For Problem Solving

Gather the Facts	What Am I Asked to Do?	How Do I Proceed?	Key Points to Remember

Applications

You may want to use the Mathematics Blueprint for Problem Solving to help you to solve the word problems in exercises 1–34.

1. The Federal Nigeria game preserve has 24,111 animals, 327 full-time staff, and 793 volunteers. What is the total of these three groups? How many more volunteers are there than full-time staff?

2. The longest rivers in the world are the Nile River, the Amazon River, and the Mississippi River. Their lengths are 4132 miles, 3915 miles, and 3741 miles, respectively. How many total miles do these three rivers run? What is the difference in the lengths of the Nile and the Mississippi?

3. Wei Ling had a budget of $37,650 for home renovations for Dr. Smith's residence. When all the work was done, she had actually spent $42,318. How far over budget has she gone?

4. An airplane flying from Boston to Minneapolis is traveling at an altitude of 31,794 feet. At the same time, an airplane flying from Minneapolis to Boston is traveling at an altitude of 36,112 feet. How many feet apart are the planes at the moment the first plane is directly below the second?

5. There are 250 hors d'oeuvres in a box. The chef of Dexter's Hearthside ordered 15 boxes. How many hors d'oeuvres did she order?

6. There are 144 pencils in a gross. Mr. Jim Weston ordered 14 gross of pencils for the office. How many pencils did he order?

7. A 16-ounce can of beets costs 96¢. What is the unit cost of the beets? (How much do the beets cost per ounce?)

8. A 14-ounce can of chicken soup costs 98¢. What is the unit cost of the soup? (How much does the soup cost per ounce?)

9. Mike is a member of the Dartmouth College ski team. During the ski season, Mike had his skis tuned 14 times for a total of $322. How much did it cost to tune his skis each time?

10. There are approximately 50,000 bison living in the United States. If Northwest Trek, the animal preserve located in Mt. Rainier National Park, has 103 bison, how many bison are living elsewhere?

11. Sergio can bake 60 muffins in one hour. A large company ordered 300 muffins for a company breakfast. How many hours will Sergio need to fill the order? How many minutes is this?

▲ **12.** If a new amusement park covers 43 acres and there are 44,010 square feet in 1 acre, how many square feet of land does the amusement park cover?

13. Every 60 minutes, the world population increases by 100,000 people. How many people will be born during the next 480 minutes?

14. Roberto had $2158 in his savings account six months ago. In the last six months he made four deposits: $156, $238, $1119, and $866. The bank deposited $136 in interest over the six-month period. How much does he have in the savings account at present?

15. A games arcade has recently opened in a West Chicago neighborhood. The owners were nervous about whether it would be a success. Fortunately, the gross revenues over the last four weeks were $7356, $3257, $4777, and $4992. What was the gross revenue this month for the arcade?

16. The two largest cities in Saudi Arabia are Riyadh, the capital, with 1,250,000 people, and Jeddah, with 900,000 people. What is the difference in population between these two cities?

In exercises 17–34, more than one type of operation is required.

17. Harry just got promoted to assistant store manager of a Wal-Mart store. He bought two suits at $250 each, two shirts at $35 each, two pairs of shoes at $75 each, and three ties at $22 each. What was the total cost of his new work wardrobe?

18. Juanita bought six bath towels for $15 each, eight washcloths for $7 each, and three bathroom rugs for $20 each for her new bathroom. What was the total cost of these items?

19. Wei Mai Lee had a balance in her checking account of $61. During the last few months, she has made deposits of $385, $945, $732, and $144. She wrote checks against her account for $223, $29, $98, and $435. When all the deposits are recorded and all the checks clear, what balance will she have in her checking account?

20. Wagner had a balance in his checking account of $13. He made deposits of $786, $566, $415, and $50. He wrote out checks for $554, $351, $14, and $87. When all the deposits are recorded and all the checks clear, what balance will he have in his checking account?

21. Diana owns 85 acres of forest land in Oregon. She rents it to a timber grower for $250 per acre per year. Her property taxes are $57 per acre. How much profit does she make on the land each year?

22. Todd owns 13 acres of commercially zoned land in the city of Columbus, Ohio. He rents it to a construction company for $12,350 per acre per year. His property taxes to the city are $7362 per acre per year. How much profit does he make on the land each year?

23. Hanna wants to determine the miles-per-gallon rating of her Chevrolet Cavalier. She filled the tank when the odometer read 14,926 miles. She then drove her car on a trip. At the end of the trip, the odometer read 15,276 miles. It took 14 gallons to fill the tank. How many miles per gallon does her car deliver?

24. Gary wants to determine the miles-per-gallon rating of his Geo Metro. He filled the tank when the odometer read 28,862 miles. After ten days, the odometer read 29,438 miles and the tank required 18 gallons to be filled. How many miles per gallon did Gary's car achieve?

25. A beautiful piece of land in the Wilmot Nature Preserve has three times as many oak trees as birches, two times as many maples as oaks, and seven times as many pine trees as maples. If there are 18 birches on the land, how many of each of the other trees are there? How many trees are there in all?

26. The Cool Coffee Lounge in Albuquerque, New Mexico, has 27 tables, and each table has either two or four chairs. If there are a total of 94 chairs accompanying the 27 tables, how many tables have four chairs? How many tables have two chairs?

Use the following list to answer exercises 27–30.

The following is a partial list of the primary home languages of students attending Public School 139 in Queens, New York in the school year 2000-2001.

Source: Office of the Superintendent of Schools, Queens, New York.

Language	Number of Students
Russian	200
English	189
Spanish	174
Mandarin	53
Cantonese	44
Korean	40
Hindi	30
Chinese, other dialects	21
Filipino	10
Hebrew	8
Indonesian	8
Romanian	8
Urdu	8
Dari/Farsi/Persian	7
Albanian	6
Arabic	6
Bulgarian	6
Gujarati	5

28. How many students speak Korean, Hindi, or Filipino as the primary language in their homes?

29. How many more students speak Spanish or Russian rather than English as the primary language in their homes?

30. How many more students speak Indonesian or Romanian rather than Albanian as the primary language in their homes?

27. How many students speak Mandarin, Cantonese, or other Chinese dialects as the primary language in their homes?

Use the following bar graph to answer exercises 31–34.

31. How many more dollars were spent by state governments for highways in 1990 than in 1980?

32. How many more dollars were spent by state governments for highways in 2000 than in 1985?

33. If the exact same dollar increase occurs between 2000 and 2005 as occurred between 1995 and 2000, what will the expenditures by state governments for highways be in 2005?

34. If the amount of money expended by state governments for highways remains constant for the years 2000 to 2003, how much money will be spent for highways during that four-year period?

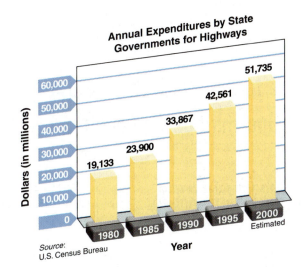

Annual Expenditures by State Governments for Highways

Source: U.S. Census Bureau

Cumulative Review Problems

35. Evaluate. 7^3

36. Perform in the proper order.
$3 \times 2^3 + 15 \div 3 - 4 \times 2$

37. Calculate. 126×38

38. Calculate. $12\overline{)3096}$

Math in the Media

Race for the Presidency Focuses on One-Third of the Country

Al Hunts Campaign Journal
THE WALL STREET JOURNAL

This is a national election but the real battle for the White House is focused on only one-third of the country.

With a week and a half to go, there are a dozen genuine tossup states. Three or four other states are clearly leaning to Vice President Al Gore or Gov. George W. Bush, but the deal may not be sealed yet. The three largest states—California, New York and Texas—aren't really in play.

The 12 states, with 132 electoral votes, that most experts and nonpartisan polls suggest could go either way are: Arkansas (with 6 electoral votes), Florida (25), Iowa (7), Michigan (18), Minnesota (10), Missouri (11), Nevada (4), New Mexico (5), Oregon (7), Pennsylvania (23), West Virginia (5) and Wisconsin (11).

Under this analysis, each candidate is clearly leading in enough states to have more that 200 electoral votes. It takes 270 to win. . . .

If the above analysis remains accurate, experts suggest that each candidate has approximately 200 electoral votes they can expect to win and the winner needs 70 additional electoral votes.

Source: October 26, 2000 WSJ.COM, The Wall Street Journal. © Dow Jones Company, Inc. Reprinted with permission.

Prior to the November 2000 presidential election, people were very fascinated in using mathematics to predict the results of the electoral vote. You may find this section quite interesting. Imagine yourself in the time during the week and a half before the election of November, 2000. Read the above article very carefully. Then try to solve these problems.

EXERCISES

1. What is the fewest number of states needed to get exactly 70 votes?

2. What is the maximum number of states possible to receive exactly 70 votes?

3. How does it appear on national TV that a particular state will carry the election for the new president?

Chapter 1 Organizer

Topic	Procedure	Examples
Place value of numbers, p. 4.	Each digit has a value depending on location.	In the number 2,896,341, what place value does 9 have? ten thousands
Writing expanded notation, p. 4.	Take the number of each digit and multiply it by one, ten, hundred, thousand, … according to its place.	Write in expanded notation. 46,235 $40,000 + 6000 + 200 + 30 + 5$
Writing whole numbers in words, p. 6.	Take the number in each period and indicate if they are (millions) (thousands) (ones) xxx, xxx, xxx	Write in words. 134,718,216. one hundred thirty-four million, seven hundred eighteen thousand, two hundred sixteen
Adding whole numbers, p. 14.	Starting with the right column, add each column separately. If a two-digit sum occurs, "carry" the first digit over to the next column to the left.	Add. $\begin{array}{r} {\scriptstyle 2\ 1} \\ 2\ 5\ 8 \\ 3\ 6\ 7 \\ 2\ 9\ 1 \\ +\ 4\ 5\ 3 \\ \hline 1\ 3\ 6\ 9 \end{array}$
Subtracting whole numbers, p. 26.	Starting with the right column, subtract each column separately. If necessary, borrow a unit from column to the left and bring it to right as a "10."	Subtract. $\begin{array}{r} {\scriptstyle 13} \\ {\scriptstyle 6\ 8\ 12} \\ 1\ 6,\cancel{7}\ \cancel{4}\ \cancel{2} \\ -\ 1\ 2,3\ 9\ 5 \\ \hline 4,3\ 4\ 7 \end{array}$
Multiplying several factors, p. 39.	Keep multiplying from left to right. Take each product and multiply by the next factor to the right. Continue until all factors are used once. (Since multiplication is commutative and associative, the factors can be multiplied in any order.)	Multiply. $\begin{aligned} 2 \times 9 \times 7 \times 6 \times 3 &= 18 \times 7 \times 6 \times 3 \\ &= 126 \times 6 \times 3 \\ &= 756 \times 3 \\ &= 2268 \end{aligned}$
Multiplying several-digit numbers, p. 41.	Multiply the top factor by the ones digit, then by the tens digit, then by the hundreds digit. Add the partial products together.	Multiply. $\begin{array}{r} 5\ 6\ 7 \\ \times\ 2\ 3\ 8 \\ \hline 4\ 5\ 3\ 6 \\ 1\ 7\ 0\ 1 \\ 1\ 1\ 3\ 4 \\ \hline 1\ 3\ 4,9\ 4\ 6 \end{array}$
Dividing by a two- or three-digit number, p. 54.	Figure how many times the first digit of the divisor goes into the first two digits of the dividend. To try this answer, multiply it back to see if it is too large or small. Continue each step of long division until finished.	Divide. $\begin{array}{r} 589 \\ 238\overline{)140182} \\ \underline{1190} \\ 2118 \\ \underline{1904} \\ 2142 \\ \underline{2142} \\ 0 \end{array}$

Topic	Procedure	Examples
Exponent form, p. 62.	To show in short form the repeated multiplication of the same number, write the number being multiplied. (This is the base.) Write in smaller print above the line the number of times it appears as a factor. (This is the exponent.) To evaluate the exponent form, write the factor the number of times shown in the exponent. Then multiply.	Write in exponent form. $10 \times 10 \times 10 \times 10 \times 10 \times 10 \times 10 \times 10$ 10^8 Evaluate. 6^3 $6 \times 6 \times 6 = 216$
Order of operations, p. 63.	1. Perform operations inside parentheses. 2. First raise to a power. 3. Then do multiplication and division in order from left to right. 4. Then do addition and subtraction in order from left to right.	Evaluate. $2^3 + 16 \div 4^2 \times 5 - 3$ Raise to a power first. $8 + 16 \div 16 \times 5 - 3$ Then do multiplication or division from left to right. $8 + 1 \times 5 - 3$ $8 + 5 - 3$ Then do addition and subtraction. $13 - 3 = 10$
Rounding, p. 69.	1. If the first digit to the right of the round-off place is less than 5, the digit in the round-off place is unchanged. 2. If the first digit to the right of the round-off place is 5 or more, the digit in the round-off place is increased by 1. 3. Digits to the right of the round-off place are replaced by zeros.	Round to the nearest hundred. 56,743 $\downarrow$ 5 6,7 ④ 3 The digit 4 is less than 5. 56,700 Round to the nearest thousand. 128,517 $\swarrow$ 1 2 8,5 1 7 The digit 5 is obviously 5 or greater. We increase the thousands digit by 1. 129,000
Estimating the answer to a calculation, p. 71.	1. Round each number so that there is one nonzero digit. 2. Perform the calculation with the rounded numbers.	Estimate the answer. $45,780 \times 9453$ First we round. $50,000 \times 9000$ Then we multiply. $\begin{array}{r} 50,000 \\ \times \quad 9,000 \\ \hline 450,000,000 \end{array}$ We estimate the answer to be 450,000,000.

Using the Mathematics Blueprint for Problem Solving, p. 80

In solving an applied problem, students may find it helpful to complete the following steps. You will not use all the steps all the time. Choose the steps that best fit the conditions of the problem.

1. Understand the problem.
   (a) Read the problem carefully.
   (b) Draw a picture if this helps you to visualize the situation. Think about what facts you are given and what you are asked to find.
   (c) Use the Mathematics Blueprint for Problem Solving to organize your work. Follow these four parts.
      1. Gather the facts (Write down specific values given in the problem.)
      2. What am I asked to do? (Identify what you must obtain for an answer.)
      3. How do I proceed? (What calculations need to be done.)
      4. Key points to remember (Record any facts, warnings, formulas, or concepts you think will be important as you solve the problem.)

2. Solve and state the answer.
   (a) Perform the necessary calculations.
   (b) State the answer, including the unit of measure.

3. Check.
   (a) Estimate the answer to the problem. Compare this estimate to the calculated value. Is your answer reasonable?
   (b) Repeat your calculations.
   (c) Work backward from your answer. Do you arrive at the original conditions of the problem?

Example   The Manchester highway department has just purchased two pickup trucks and three dump trucks. The cost of a pickup truck is $17,920. The cost of a dump truck is $48,670. What was the cost to purchase these five trucks?

Mathematics Blueprint For Problem Solving

Gather the Facts	What Am I Asked to Do?	How Do I Proceed?	Key Points to Remember
Buy 2 pickup trucks 3 dump trucks Cost Pickup: $17,920 Dump: $48,670	Find the total cost of the 5 trucks.	Find the cost of 2 pickup trucks. Find the cost of 3 dump trucks. Add to get final cost of all 5 trucks.	Multiply 2 times pickup truck cost. Multiply 3 times dump truck cost.

1. Understand the problem.

2. Solve.

Calculate cost of pickup trucks

$$\begin{array}{r} \$17{,}920 \\ \times\ 2 \\ \hline \$35{,}840 \end{array}$$

Calculate cost of dump trucks

$$\begin{array}{r} \$48{,}670 \\ \times\ 3 \\ \hline \$146{,}010 \end{array}$$

Find total cost. $35,840 + $146,010 = $181,850

The total cost of the five trucks is $181,850.

3. Check.   Estimate cost of pickup trucks $20,000 \times 2 = 40,000$

   Estimate cost of dump trucks $50,000 \times 3 = 150,000$

   Total estimate $40,000 + 150,000 = 190,000$

This is close to our calculated answer of $181,850. We determine that our answer is reasonable. ✓

Chapter 1 Review Problems

If you have trouble with a particular type of exercise, review the examples in the section indicated for that group of exercises. Answers to all exercises are located in the answer key.

1.1   *Write in words.*

1. 376

2. 5082

3. 109,276

4. 423,576,055

Write in expanded notation.

5. 4364

6. 27,986

7. 42,166,037

8. 1,305,128

Write in standard notation.

9. nine hundred twenty-four

10. six thousand ninety-five

11. one million, three hundred twenty-eight thousand, eight hundred twenty-eight

12. forty-five million, ninety-two thousand, six hundred fifty-one

1.2   *Add.*

13.   76
       + 39

14.   36
       + 94

15.   127
       + 563

16.   12
          28
          34
       + 76

17.   123
           61
            9
           84
       + 123

18.   937
          405
       + 256

19.   125
          364
       + 980

20.   28,364
       + 97,059

21.   1356
          2892
           561
            89
       + 9805

22.   26
           503
           935
          1257
       + 7861

1.3   *Subtract.*

23.   36
       − 19

24.   54
       − 48

25.   126
       − 99

26.   543
       − 372

27.   1296
       − 1137

28.   9821
       − 4993

29.   201,010
       − 137,864

30.   101,300
       − 98,274

31.   1,986,312
       − 1,761,555

32.   7,216,003
       − 5,985,312

1.4 *Multiply.*

33.   57
   $\times\ \ 2$

34.   12
   $\times\ \ 3$

35.   36
   $\times\ \ 0$

36.   24
   $\times\ \ 1$

37. $1 \times 3 \times 6$

38. $2 \times 4 \times 8$

39. $5 \times 7 \times 3$

40. $4 \times 6 \times 5$

41. $8 \times 1 \times 9 \times 2$

42. $7 \times 6 \times 0 \times 4$

43. $3 \cdot 4 \cdot 2 \cdot 2 \cdot 5$

44. $1 \cdot 2 \cdot 7 \cdot 3 \cdot 4$

45. $26{,}121 \times 100$

46. $84{,}312 \times 1000$

47. $832 \times 100{,}000$

48. $563 \times 1{,}000{,}000$

49.   58
   $\times\ \ 32$

50.   36
   $\times\ \ 24$

51.   150
   $\times\ \ 27$

52.   360
   $\times\ \ 38$

53.   709
   $\times\ \ 36$

54.   502
   $\times\ \ 48$

55.   123
   $\times\ 714$

56.   431
   $\times\ 623$

57.   1782
   $\times\ \ 305$

58.   2057
   $\times\ \ 124$

59.   300
   $\times\ 500$

60.   400
   $\times\ 600$

61.   1200
   $\times\ 6000$

62.   2500
   $\times\ 3000$

63.   100,000
   $\times\ \ 20{,}000$

64.   300,000
   $\times\ \ 40{,}000$

1.5 *Divide, if possible.*

65. $20 \div 10$

66. $40 \div 8$

67. $70 \div 5$

68. $36 \div 9$

69. $0 \div 8$

70. $12 \div 1$

71. $7 \div 1$

72. $0 \div 5$

73. $\dfrac{49}{7}$

74. $\dfrac{42}{6}$

75. $\dfrac{5}{0}$

76. $\dfrac{24}{6}$

77. $\dfrac{56}{8}$

78. $\dfrac{48}{8}$

79. $\dfrac{72}{9}$

80. $\dfrac{0}{0}$

Divide. Be sure to indicate the remainder, if one exists.

81. $6\overline{)750}$ **82.** $7\overline{)875}$ **83.** $5\overline{)1290}$ **84.** $4\overline{)1476}$

85. $3\overline{)77,622}$ **86.** $8\overline{)29,536}$ **87.** $6\overline{)221,748}$ **88.** $5\overline{)184,605}$

89. $8\overline{)127,890}$ **90.** $7\overline{)250,485}$ **91.** $67\overline{)490}$ **92.** $72\overline{)325}$

93. $21\overline{)666}$ **94.** $22\overline{)319}$ **95.** $68\overline{)2614}$ **96.** $76\overline{)4142}$

97. $45\overline{)4275}$ **98.** $35\overline{)9030}$ **99.** $132\overline{)7128}$ **100.** $204\overline{)3876}$

1.6 *Write in exponent form.*

101. 13×13 **102.** 24×24

103. $8 \times 8 \times 8 \times 8 \times 8$ **104.** $9 \times 9 \times 9 \times 9 \times 9$

Evaluate.

105. 2^6 **106.** 3^4 **107.** 2^7 **108.** 5^3

109. 7^2 **110.** 9^2 **111.** 6^3 **112.** 4^4

Perform each operation in proper order.

113. $7 + 2 \times 3 - 5$ **114.** $6 \times 2 - 4 + 3$ **115.** $2^5 + 4 - \left(5 + 3^2\right)$

116. $4^3 + 20 \div \left(2 + 2^3\right)$ **117.** $3^3 \times 4 - 6 \div 6$ **118.** $20 \div 20 + 5^2 \times 3$

119. $2^3 \times 5 \div 8 + 3 \times 4$ **120.** $3^2 \times 6 \div 3 + 5 \times 6$ **121.** $9 \times 2^2 + 3 \times 4 - 36 \div (4 + 5)$

122. $5 \times 3 + 5 \times 4^2 - 14 \div (6 + 1)$

1.7 *Round to the nearest ten.*

123. 1275   **124.** 5673   **125.** 15,305   **126.** 42,644

In exercises 127–130, round to the nearest thousand.

127. 12,350   **128.** 22,986   **129.** 675,800   **130.** 202,498

131. Round to the nearest hundred thousand.
5,668,243

132. Round to the nearest ten thousand. 9,995,312

Use the principle of estimation to find an estimate for each calculation.

133. 589 + 622 + 933 + 864

134. 25,981 + 36,782 + 73,125

135. 4,326,171 − 2,916,788

136. 29,378 − 17,924

137. 1763 × 5782

138. 2,965,372 × 893

139. 83,421 ÷ 24

140. 7,963,127 ÷ 378

Estimate the result of each calculation. Some results are correct and some are incorrect. Which results appear to be correct? Which results appear to be incorrect?

141. $87 + 36 + 94 + 55 \stackrel{?}{=} 272$

142. $938,526 - 398,445 \stackrel{?}{=} 540,081$

143.   176,394
        × 5216
    ──────────
    92,007,114

144.   27,911
    32)‾893,152‾

1.8   *Solve.*

145. Soft-drink cans come six to a package. There are 34 packages in the storage room. How many soft-drink cans are there?

146. Ward can type 25 words per minute on his computer. He typed for seven minutes at that speed. How many words did he type?

147. Alfonso drove 1362 kilometers last summer, 562 km during Christmas break, and 473 km during spring break. How many total kilometers were driven?

148. Applepickers, Inc. bought a truck for $26,300, a car for $14,520, and a minivan for $18,650. What was the total purchase price?

149. A plane was flying at 14,630 feet. It flew over a mountain 4329 feet high. How many feet was it from the plane to the top of the mountain?

150. Roberta was billed $11,658 for tuition. She received a $4630 grant. How much did she have to pay after the grant was deducted?

151. The expedition cost a total of $32,544 for 24 paying passengers, who shared the cost equally. What was the cost per passenger?

152. John bought 92 shares of stock for $5888. What was the cost per share?

153. Melissa's savings account balance last month was $810. The bank added $24 interest. Melissa deposited $105, $36, and $177. She made withdrawals of $18, $145, $250, and $461. What will be her balance this month?

154. Marcia's checking account balance last month was $436. She made deposits of $16, $98, $125, and $318. She made out checks of $29, $128, $100, and $402. What will be her balance this month?

155. Ali began a trip on a full tank of gas with the car odometer at 56,320 miles. He ended the trip at 56,720 miles and added 16 gallons of gas to refill the tank. How many miles per gallon did he get on the trip?

156. Amina began a trip on a full tank of gas with the car odometer at 24,396 miles. She ended the trip at 24,780 miles and added 16 gallons of gas to refill the tank. How many miles per gallon did she get on the trip?

157. The maintenance group bought three lawn mowers at $279, four power drills at $61, and two riding tractors at $1980. What was the total purchase price for these items?

158. The library bought 24 sets of shelves at $118 each, four desks at $120 each, and six chairs at $24 each. What was the total purchase price for these items?

Use the following bar graph to answer exercises 159–162.

159. How many more tons of solid waste were recovered and recycled in 1995 than in 1980?

160. What was the greatest increase in tons of solid waste recovered and recycled in a five-year period?

161. If the exact same increase in the number of tons recovered occurs from 2000 to 2005 as occurred from 1995 to 2000, how many tons of solid waste will be recovered and recycled in 2005?

162. If the exact same increase in the number of tons recovered occurs from 2000 to 2005 as occurred from 1995 to 2000, what will be the average increase in the number of tons per year during this five-year period?

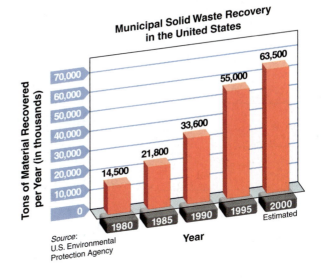

Municipal Solid Waste Recovery in the United States

Tons of Material Recovered per Year (in thousands)

63,500 (2000 Estimated)
55,000 (1995)
33,600 (1990)
21,800 (1985)
14,500 (1980)

Year

Source: U.S. Environmental Protection Agency

163. In 2001, the U.S. Marine Corps plans to buy 360 Osprey aircraft. The Air Force also intends to buy 50, while the Navy plans to buy 48. Each Osprey has an estimated cost of $42,900,000. What is the cost to buy all of these aircraft? (*Source:* U.S. Office of Management and Budget)

164. Marcia and Melissa had a big bowl of 1500 M&Ms in their dorm room. The bowl included yellow, red, green, brown, and blue candies. They knew they had 241 yellow, 407 green, and 117 blue pieces. The number of brown pieces was exactly double the number of blue pieces. How many red pieces were in the bowl?

Developing Your Study Skills

Exam Time: Getting Organized

Studying adequately for an exam requires careful preparation. Begin early so that you will be able to spread your review over several days. Even though you may still be learning new material at this time, you can be reviewing concepts previously learned in the chapter. Giving yourself plenty of time for review will take the pressure off. You need this time to process what you have learned and to tie concepts together.

Adequate preparation enables you to feel confident and to think clearly with less tension and anxiety.

Chapter 1 Test

Do these questions simulating test conditions.

1. _____

2. _____

3. _____

4. _____

5. _____

6. _____

7. _____

8. _____

9. _____

10. _____

11. _____

12. _____

13. _____

14. _____

15. _____

16. _____

17. _____

18. _____

1. Write in words. 44,007,635

2. Write in expanded notation. 26,859

3. Write in standard notation. three million, five hundred eighty-one thousand, seventy-six

Add.

4.
```
  156
   93
    8
  127
+ 593
```

5.
```
  470
  386
+ 189
```

6.
```
  135,484
    2,376
   81,004
+ 100,113
```

Subtract.

7.
```
  8961
-  894
```

8.
```
  300,523
- 262,182
```

9.
```
  18,400,100
- 13,174,332
```

Multiply.

10. $1 \times 6 \times 9 \times 7$

11.
```
    45
×   96
```

12.
```
  147
× 625
```

13.
```
  18,491
×      7
```

In problems 14–16, divide. If there is a remainder, be sure to state it as part of your answer.

14. $5\overline{)15,071}$

15. $8\overline{)18,264}$

16. $37\overline{)13,024}$

17. Write in exponent form. $11 \times 11 \times 11$

18. Evaluate. 2^6

In problems 19–21, perform each operation in proper order.

19. $5 + 6^2 - 2 \times (9 - 6)^2$

20. $2^3 + 4^3 + 18 \div 3$

21. $4 \times 6 + 3^3 \times 2 + 23 \div 23$

22. Round to the nearest ten.   88,073

23. Round to the nearest ten thousand.   6,462,431

24. Round to the nearest hundred thousand.   4,782,163

Estimate the answer.

25. $4,867,010 \times 27,058$ **26.** $1423 + 4287 + 4103 + 8549$

Solve.

27. A cruise for 15 people costs $32,220. If each person paid the same amount, how much will it cost each individual?

28. The river is 602 feet wide at Big Bend Corner. A boy is in the shallow water, 135 feet from the shore. How far is the boy from the other side of the river?

29. At the bookstore, Hector bought three notebooks at $2 each, one textbook for $45, two lamps at $21 each, and two sweatshirts at $17 each. What was his total bill?

30. Patricia is looking at her checkbook. She had a balance last month of $31. She deposited $902 and $399. She made out checks for $885, $103, $26, $17, and $9. What will be her new balance?

▲ **31.** The runway at Beverly Airport needs to be resurfaced. The rectangular runway is 6800 feet long and 110 feet wide. What is the area of the runway that needs to be resurfaced?

▲ **32.** Nancy Tobey planted a vegetable garden in the backyard. However, the deer and raccoons have been stealing all the vegetables. She asked John to fence in the garden. The rectangular garden measures 8 feet by 15 feet. How many feet of fence should John purchase if he wants to enclose the garden?

19. _____

20. _____

21. _____

22. _____

23. _____

24. _____

25. _____

26. _____

27. _____

28. _____

29. _____

30. _____

31. _____

32. _____

Fractions

Some famous recipes from colonial America have survived until this day. However, these recipes contain more fractions than their modern counterparts. Further, many of them were designed to feed a large number of people. For example, the recipe for Baltimore Sunshine Cake will feed 48 people. Do you think you could simplify it for people who want to make a normal-sized cake for a normal-sized family? Turn to the Putting Your Skills to Work problems on page 174 and find out.

Pretest Chapter 2

1. _____

2. _____

3. _____

4. _____

5. _____

6. _____

7. _____

8. _____

9. _____

10. _____

11. _____

12. _____

13. _____

14. _____

15. _____

16. _____

If you are familiar with the topics in this chapter, take this test now. Check your answers with those in the back of the book. If an answer is wrong or you can't do a problem, study the appropriate section of the chapter.

If you are not familiar with the topics in this chapter, don't take this test now. Instead, study the examples, work the practice problems, and then take the test.

This test will help you identify which concepts you have mastered and which you need to study further. Make sure all fractions are simplified in the final answer.

Section 2.1

1. Use a fraction to represent the shaded part of the object.

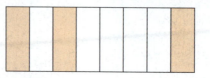

2. Draw a sketch to show $\frac{3}{6}$ of an object.

3. An inspector checked 124 CD players. Of these, 5 were defective. Write a fraction that describes the part that was defective.

Section 2.2

Reduce each fraction.

4. $\dfrac{3}{18}$ **5.** $\dfrac{13}{39}$ **6.** $\dfrac{16}{112}$ **7.** $\dfrac{175}{200}$ **8.** $\dfrac{44}{121}$

Section 2.3

Change to an improper fraction.

9. $3\dfrac{2}{3}$ **10.** $6\dfrac{1}{9}$

Change to a mixed number.

11. $\dfrac{97}{4}$ **12.** $\dfrac{29}{5}$ **13.** $\dfrac{36}{17}$

Section 2.4

Multiply.

14. $\dfrac{5}{11} \times \dfrac{1}{4}$ **15.** $\dfrac{3}{7} \times \dfrac{14}{9}$ **16.** $12\dfrac{1}{3} \times 5\dfrac{1}{2}$

Section 2.5

Divide.

17. $\dfrac{3}{7} \div \dfrac{3}{7}$

18. $\dfrac{7}{16} \div \dfrac{7}{8}$

19. $6\dfrac{4}{7} \div 1\dfrac{5}{21}$

20. $8 \div \dfrac{12}{7}$

Section 2.6

Find the least common denominator.

21. $\dfrac{1}{8}, \dfrac{3}{4}, \dfrac{1}{2}$

22. $\dfrac{2}{9}, \dfrac{4}{45}$

23. $\dfrac{4}{11}, \dfrac{2}{55}$

24. $\dfrac{5}{24}, \dfrac{7}{36}$

Section 2.7

Add or subtract.

25. $\dfrac{7}{18} - \dfrac{3}{24}$

26. $\dfrac{5}{24} + \dfrac{4}{9} + \dfrac{1}{36}$

Section 2.8

Perform the indicated operations.

27. $8 - 3\dfrac{2}{3}$

28. $1\dfrac{5}{6} + 2\dfrac{1}{7}$

29. $\dfrac{5}{9} \times \dfrac{9}{2} \div \dfrac{3}{2}$

Section 2.9

Answer each question.

30. Miguel and Lee set out to hike $13\frac{1}{2}$ miles from Arlington to Concord. During the first 5 hours, they covered $6\frac{1}{3}$ miles going from Arlington to Bedford. How many miles are left to be covered from Bedford to Concord?

31. Robert harvested $4\frac{3}{12}$ tons of wheat. His helper harvested $3\frac{1}{18}$ tons of wheat. How much did they harvest together?

32. The students of the fifth floor of Smith Dorm contributed money to make a stock purchase of one share of stock each. They paid $776 in all to buy the shares of stock. The cost of buying one share was $43\frac{1}{9}$. How many students shared in the purchase?

17. _____

18. _____

19. _____

20. _____

21. _____

22. _____

23. _____

24. _____

25. _____

26. _____

27. _____

28. _____

29. _____

30. _____

31. _____

32. _____

1 **Using a Fraction to Represent Part of a Whole**

In Chapter 1 we studied whole numbers. In this chapter we will study a fractional part of a whole number. One way to represent parts of a whole is with **fractions.** The word *fraction* (like the word *fracture*) suggests that something is being broken. In mathematics, fractions represent the part that is "broken off" from a whole. The whole can be a single object (like a whole pie) or a group (the employees of a company). Here are some examples.

Single object

$$\frac{1}{3}$$

The whole is the pie on the left. The fraction $\frac{1}{3}$ represents the shaded part of the pie, 1 of 3 pieces. $\frac{1}{3}$ is read "one-third."

A group: ACE company employs 150 men, 200 women.

$$\frac{150}{350}$$

The whole is the company of 350 people (150 men plus 200 women). The fraction $\frac{150}{350}$ represents that part of the company consisting of men.

Recipe: Applesauce
4 apples
1/2 cup sugar
1 teaspoon cinnamon

The whole is 1 whole cup of sugar. This recipe calls for $\frac{1}{2}$ cup of sugar. Notice that in many real-life situations $\frac{1}{2}$ is written as 1/2.

When we say "$\frac{3}{8}$ of a pizza has been eaten," we mean 3 of 8 equal parts of a pizza have been eaten. (See the figure.) When we write the fraction $\frac{3}{8}$, the number on the top, 3, is the **numerator,** and the number on the bottom, 8, is the **denominator.**

The numerator specifies how many parts → 3
The denominator specifies the total number of parts → 8

When we say, "$\frac{2}{3}$ of the marbles are red," we mean 2 marbles out of a total of 3 are red marbles.

Part we are interested in → 2   numerator
Total number in the group → 3   denominator

EXAMPLE 1 Use a fraction to represent the shaded or completed part of the whole shown.

(a)   **(b)**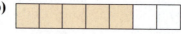

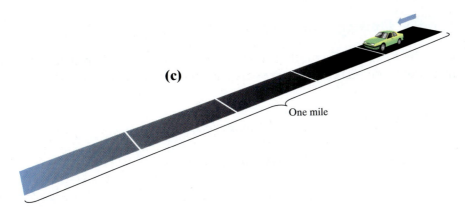

(c)

One mile

(a) Three out of four circles are red. The fraction is $\dfrac{3}{4}$.

(b) Five out of seven equal parts are shaded. The fraction is $\dfrac{5}{7}$.

(c) The mile is divided into five equal parts. The car has traveled 1 part out of 5 of the one mile distance. The fraction is $\dfrac{1}{5}$.

Practice Problem 1   Use a fraction to represent the shaded part of the whole.

(a)   **(b)**   **(c)**

We can also think of a fraction as a division problem.

$$\frac{1}{3} = 1 \div 3 \qquad \text{and} \qquad 1 \div 3 = \frac{1}{3}$$

The division way of looking at fractions asks the question:

> What is the result of dividing one whole into three equal parts?

Thus we can say the fraction $\frac{a}{b}$ means the same as $a \div b$. However, special care must be taken with the number 0.

Suppose that we had four equal parts and we wanted to take none of them. We would want $\frac{0}{4}$ of the parts. Since $\frac{0}{4} = 0 \div 4 = 0$, we see that $\frac{0}{4} = 0$. Any fraction with a 0 numerator equals zero.

$$\frac{0}{8} = 0 \qquad \frac{0}{5} = 0 \qquad \frac{0}{13} = 0$$

What happens when zero is in the denominator? $\frac{4}{0}$ means 4 out of 0 parts. Taking 4 out of 0 does not make sense. We say $\frac{4}{0}$ is **undefined.**

$$\frac{3}{0}, \frac{7}{0}, \frac{4}{0} \qquad \text{are \textbf{undefined}.}$$

We cannot have a fraction with 0 in the denominator. Since $\frac{4}{0} = 4 \div 0$, we say division by zero is *undefined*. We cannot divide by 0.

2   *Drawing a Sketch to Illustrate a Fraction*

Drawing a sketch of a mathematical situation is a powerful problem-solving technique. The picture often reveals information not always apparent in the words.

🔴 **EXAMPLE 2**   Draw a sketch to illustrate.

(a) $\dfrac{7}{11}$ of an object   **(b)** $\dfrac{2}{9}$ of a group

(a) The easiest figure to draw is a rectangular bar.

We divide the bar into 11 equal parts. We then shade in 7 parts to show $\dfrac{7}{11}$.

(b) We draw 9 circles of equal size to represent a group of 9.

We shade in 2 of the 9 circles to show $\dfrac{2}{9}$.

Practice Problem 2   Draw a sketch to illustrate.

(a) $\dfrac{4}{5}$ of an object   **(b)** $\dfrac{3}{7}$ of a group   🔴

Recall these facts about division problems involving the number 1 and the number 0.

Division Involving the Number 1 and the Number 0

1. Any nonzero number divided by itself is 1.

$$\frac{7}{7} = 1$$

2. Any number divided by 1 remains unchanged.

$$\frac{29}{1} = 29$$

3. Zero may be divided by any nonzero number; the result is always zero.

$$\frac{0}{4} = 0$$

4. Division by zero is undefined.

$$\frac{0}{0} \text{ is undefined} \qquad \frac{3}{0} \text{ is undefined}$$

3 *Using Fractions to Represent Real-Life Situations*

Several real-life situations can be described using fractions.

EXAMPLE 3 Use a fraction to describe each situation.

(a) A baseball player gets a hit 5 out of 12 times at bat.

(b) There are 156 men and 185 women taking psychology this semester. Describe the part of the class that consists of women.

(c) Robert Tobey found in the Alaska moose count that five-eighths of the moose observed were female.

(a) The baseball player got a hit $\frac{5}{12}$ of his times at bat.

(b) The total class is $156 + 185 = 341$. The fractional part that is women is 185 out of 341. Thus $\frac{185}{341}$ of the class is women.

156 men	185 women

Total class
341 students

(c) Five-eighths of the moose observed were female. The fraction is $\frac{5}{8}$.

Practice Problem 3 Use a fraction to describe each situation.

(a) 9 out of the 17 players on the basketball team are on the dean's list.

(b) The senior class has 382 men and 351 women. Describe the part of the class consisting of men.

(c) John needed seven-eighths of a yard of material.

EXAMPLE 4 Wanda made 13 calls, out of which she made five sales. Albert made 17 calls, out of which he made six sales. Write a fraction that describes for both people together the number of calls in which a sale was made compared with the total number of calls.

There are $5 + 6 = 11$ calls in which a sale was made.

There were $13 + 17 = 30$ total calls.

Thus $\dfrac{11}{30}$ of the calls resulted in a sale.

Practice Problem 4 An inspector found that one out of seven belts was defective. She also found that two out of nine shirts were defective. Write a fraction that describes what part of all the objects examined were defective.

Verbal and Writing Skills

1. A _____ can be used to represent part of a whole or part of a group.

2. In a fraction, the _____ tells the number of parts we are interested in.

3. In a fraction, the _____ tells the total number of parts in the whole or in the group.

4. Describe a real-life situation that involves fractions.

Name the numerator and the denominator in each fraction.

5. $\dfrac{3}{5}$ **6.** $\dfrac{9}{11}$ **7.** $\dfrac{2}{3}$ **8.** $\dfrac{3}{4}$ **9.** $\dfrac{1}{17}$ **10.** $\dfrac{1}{15}$

In exercises 11–30, use a fraction to represent the shaded part of the object or the shaded portion of the set of objects.

11. **12.** **13.** **14.**

15. **16.** **17.** **18.**

19. **20.** **21.** **22.**

23. **24.** **25.** ○○○○○○○ **26.**

27. **28.** **29.** **30.**

Draw a sketch to illustrate each fractional part.

31. $\frac{4}{5}$ of an object

32. $\frac{3}{7}$ of an object

33. $\frac{3}{8}$ of an object

34. $\frac{5}{12}$ of an object

35. $\frac{7}{10}$ of an object

36. $\frac{8}{15}$ of an object

Applications

37. Cecilia made 95 holiday wreaths to sell this year. Thirty-one of them were decorated with small silver bells, while the others had ribbon on them. What fractional part of the wreaths were made with silver bells?

38. The total purchase amount was 83¢, of which 5¢ was sales tax. What fractional part of the total purchase price was sales tax?

39. Lance bought a 100-CD jukebox for $750. Part of it was paid for with the $209 he earned parking cars for the valet service at a local wedding reception hall. What fractional part of the jukebox was paid for by his weekend earnings?

40. Charles drove for 47 minutes to go to the Laker game. 18 minutes of his trip were spent in bumper-to-bumper traffic. What fractional part of his time was spent in traffic?

41. The Democratic National Committee fund-raising event served 122 chicken dinners and 89 roast beef dinners to its contributors. What fractional part of the guests ate roast beef?

42. The East dormitory has 111 smokers and 180 nonsmokers. What fractional part of the dormitory has nonsmoking residents?

43. At Gold's Gym one day, 9 people were riding stationary bikes, 7 people were using rowing machines, and 13 people were running on treadmills. What fractional part of the people were using rowing machines?

44. At the local animal shelter there are 12 puppies, 25 adult dogs, 14 kittens, and 31 adult cats. What fractional part of the animals are either puppies or adult dogs?

45. The picnic table held two bowls of corn, three bowls of potato salad, four bowls of baked beans, and five bowls of ribs. What fractional part of the bowls on the table contains either ribs or beans?

46. A box of compact discs contains 5 classical CDs, 6 jazz CDs, 4 sound tracks, and 24 blues CDs. What fractional part of the total CDs is either jazz or blues?

47. The West Peabody Engine Company manufactured two items last week: 101 engines and 94 lawn mowers. It was discovered that 19 engines and 3 lawn mowers were defective. Of the engines that were not defective, 40 were properly constructed but 42 were not of the highest quality. Of the lawn mowers that were not defective, 50 were properly constructed but 41 were not of the highest quality.
(a) What fractional part of all items manufactured was of the highest quality?

(b) What fractional part of all items manufactured was defective?

48. A Chicago tour bus held 25 women and 33 men. 12 women wore jeans. 19 men wore jeans. In the group of 25 women, a subgroup of 8 women wore sandals. In the group of 19 men, a subgroup of 10 wore sandals.
(a) What fractional part of the people on the bus wore jeans?

(b) What fractional part of the women on the bus wore sandals?

To Think About

49. Illustrate a real-life example of the fraction $\frac{0}{6}$.

50. What happens when we try to illustrate a real-life example of the fraction $\frac{6}{0}$? Why?

Cumulative Review Problems

51. Add.   18
27
34
16
125
+ 21

52. Subtract.   38,114
− 27,008

53. Multiply.   4136
× 29

54. Divide. $12\overline{)2130}$

55. The annual marching band fund-raiser for Jefferson Valley High School has collected 2004 books. 282 are science fiction, 866 are novels, 42 are cookbooks, 317 are history books, 102 are biographies, 99 are foreign language books, and 115 are textbooks. The remaining books are reference books, such as encyclopedias and dictionaries. How many reference books are there?

Developing Your Study Skills

Previewing New Material

Part of your study time each day should consist of looking ahead to those sections in your text that are to be covered the following day. You do not necessarily have to study and learn the material on your own, but if you survey the concepts, terminology, diagrams, and examples, the new ideas will seem more familiar to you when the instructor presents them. You can take note of concepts that appear confusing or difficult and be ready to listen carefully for your instructor's explanations. You can be prepared to ask the questions that will increase your understanding. Previewing new material enables you to see what is coming and prepares you to be ready to absorb it.

1 *Writing a Number as a Product of Prime Factors*

A **prime number** is a whole number greater than 1 that cannot be evenly divided except by itself and 1. If you examine all the whole numbers from 1 to 50, you will find 15 prime numbers.

The First 15 Prime Numbers

$$2, 3, 5, 7, 11, 13, 17, 19, 23, 29, 31, 37, 41, 43, 47$$

A **composite number** is a whole number greater than 1 that can be divided by whole numbers other than 1 and itself. The number 12 is a composite number.

$$12 = 2 \times 6 \qquad \text{and} \qquad 12 = 3 \times 4$$

The number 1 is neither a prime nor a composite number. The number 0 is neither a prime nor a composite number.

Recall that factors are numbers that are multiplied together. Prime factors are prime numbers. To check to see if a number is prime or composite, simply divide the smaller primes (such as $2, 3, 5, 7, 11, \dots$) into the given number. If the number can be divided exactly without a remainder by one of the smaller primes, it is a composite and not a prime.

Some students find the following rules helpful when deciding if a number can be divided by 2, 3, or 5.

Divisibility Tests

1. A number is divisible by 2 if the last digit is 0, 2, 4, 6, or 8.

2. A number is divisible by 3 if the sum of the digits is divisible by 3.

3. A number is divisible by 5 if the last digit is 0 or 5.

To illustrate:

1. 478 is divisible by 2 since it ends in 8.

2. 531 is divisible by 3 since when we add the digits of 531 $(5 + 3 + 1)$, we get 9, which is divisible by 3.

3. 985 is divisible by 5 since it ends in 5.

EXAMPLE 1 Write each whole number as the product of prime factors.

(a) 12 **(b)** 60 **(c)** 168

(a) To start, write 12 as the product of any two factors. We will write 12 as 4×3.

$12 = \quad 4 \quad \times 3$ Now check whether the factors are prime. If not, factor these.

$\qquad\qquad 2 \times 2 \times 3$

$12 = 2 \times 2 \times 3$ Now all factors are prime, so 12 is completely factored.

117

Instead of writing $2 \times 2 \times 3$, we can write $2^2 \times 3$.

Note: To start, we could write 12 as 2×6. Begin this way and follow the preceding steps. Is the product of prime factors the same? Will this always be true?

(b) We follow the same steps as in (a).

$60 = $ 6 $\times$ 10

$3 \times 2 \times 2 \times 5$ Check that all factors are prime.

$60 = 2 \times 2 \times 3 \times 5$

Instead of writing $2 \times 2 \times 3 \times 5$, we can write $2^2 \times 3 \times 5$.

Note that in the final answer the prime factors are listed in order from least to greatest.

(c) Some students like to use a **factor tree** to help write a number as a product of prime factors as illustrated below.

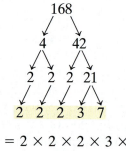

168

4   42

2 2 2 21

2 2 2 3 7

$168 = 2 \times 2 \times 2 \times 3 \times 7$

or $168 = 2^3 \times 3 \times 7$

Practice Problem 1   Write each whole number as a product of primes.

(a) 18   **(b)** 72   **(c)** 400

Suppose we started Example 1 (c) by writing $168 = 14 \times 12$. Would we get the same answer? Would our answer be correct? Let's compare.

Again we will use a factor tree.

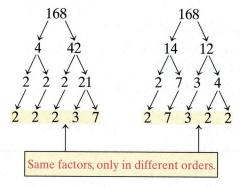

Same factors, only in different orders.

Thus $168 = 2 \times 2 \times 2 \times 3 \times 7$

or $= 2^3 \times 3 \times 7.$

The order of prime factors is not important because multiplication is commutative. No matter how we start, when we factor a composite number, we always get exactly the same prime factors.

The Fundamental Theorem of Arithmetic

Every composite number can be written in exactly one way as a product of prime numbers.

We have seen this in our Solution to Example 1(c).

You will be able to check this theorem again in Exercises 2.2, exercises 7–26. Writing a number as a product of prime factors is also called **prime factorization.**

2 ▸ Reducing a Fraction

You know that $5 + 2$ and $3 + 4$ are two ways to write the same number. We say they are *equivalent* because they are *equal* to the same *value*. They are both ways of writing the value 7.

Like whole numbers, fractions can be written in more than one way. For example, $\frac{2}{4}$ and $\frac{1}{2}$ are two ways to write the same number. The value of the fractions is the same. When we use fractions, we often need to write them in another form. If we make the numerator and denominator smaller, we *simplify* the fractions.

Compare the two fractions in the drawings on the right. In each picture the shaded part is the same size. The fractions $\frac{3}{4}$ and $\frac{6}{8}$ are called **equivalent fractions.** The fraction $\frac{3}{4}$ is in **simplest form.** To see how we can change $\frac{6}{8}$ to $\frac{3}{4}$, we look at a property of the number 1.

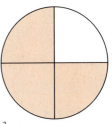

$\frac{3}{4}$ of the circle is shaded.

> Any nonzero number divided by itself is 1.
>
> $$\frac{5}{5} = \frac{17}{17} = \frac{c}{c} = 1.$$

Thus, if we multiply a fraction by $\frac{5}{5}$ or $\frac{17}{17}$ or $\frac{c}{c}$ (remember, c cannot be zero), the value of the fraction is unchanged because we are multiplying by a form of 1. We can use this rule to show that $\frac{3}{4}$ and $\frac{6}{8}$ are equivalent.

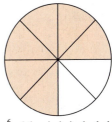

$\frac{6}{8}$ of the circle is shaded.

$$\frac{3}{4} \times \frac{2}{2} = \frac{6}{8}$$

In general, if b and c are not zero,

$$\frac{a}{b} = \frac{a \times c}{b \times c}.$$

To reduce a fraction, we find a **common factor** in the numerator and in the denominator and divide it out. In the fraction $\frac{6}{8}$, the common factor is 2.

$$\frac{6}{8} = \frac{3 \times 2}{4 \times 2} = \frac{3}{4}$$

$$\frac{6}{8} = \frac{3}{4}$$

For all fractions (where a, b, and c are not zero), if c is a common factor,

$$\frac{a}{b} = \frac{a \div c}{b \div c}.$$

A fraction is called **simplified, reduced,** or **in lowest terms** if the numerator and the denominator have only 1 *as a common factor.*

EXAMPLE 2   Simplify (write in lowest terms).

(a) $\dfrac{15}{25}$ **(b)** $\dfrac{42}{56}$

(a) $\dfrac{15}{25} = \dfrac{15 \div 5}{25 \div 5} = \dfrac{3}{5}$ The greatest common factor is 5. Divide the numerator and the denominator by 5.

(b) $\dfrac{42}{56} = \dfrac{42 \div 14}{56 \div 14} = \dfrac{3}{4}$ The greatest common factor is 14. Divide the numerator and the denominator by 14.

Perhaps 14 was not the first common factor you thought of. Perhaps you did see the common factor 2. Divide out 2. Then look for another common factor, 7. Now divide out 7.

$$\frac{42}{56} = \frac{42 \div 2}{56 \div 2} = \frac{21}{28} = \frac{21 \div 7}{21 \div 7} = \frac{3}{4}$$

If we do not see large factors at first, sometimes we can simplify a fraction by dividing both numerator and denominator by a smaller common factor several times, until no common factors are left.

Practice Problem 2   Simplify by dividing out common factors.

(a) $\dfrac{30}{42}$ **(b)** $\dfrac{60}{132}$

A second method to reduce or simplify fractions is called the *method of prime factors.* We factor the numerator and the denominator into prime numbers. We then divide the numerator and the denominator by any common prime factors.

EXAMPLE 3   Simplify the fractions by the method of prime factors.

(a) $\dfrac{35}{42}$ **(b)** $\dfrac{22}{110}$

(a) $\dfrac{35}{42} = \dfrac{5 \times 7}{2 \times 3 \times 7}$ We factor 35 and 42 into prime factors. The common prime factor is 7.

$$= \frac{5 \times \overset{1}{\cancel{7}}}{2 \times 3 \times \underset{1}{\cancel{7}}}$$ Now we divide out 7.

$$= \frac{5 \times 1}{2 \times 3 \times 1} = \frac{5}{6}$$ We multiply the factors in the numerator and denominator to write the reduced or simplified form.

Thus $\dfrac{35}{42} = \dfrac{5}{6}$, and $\dfrac{5}{6}$ is the simplified form.

(b) $\dfrac{22}{110} = \dfrac{2 \times 11}{2 \times 5 \times 11} = \dfrac{\overset{1}{\cancel{2}} \times \overset{1}{\cancel{11}}}{\underset{1}{\cancel{2}} \times 5 \times \underset{1}{\cancel{11}}} = \dfrac{1}{5}$

Practice Problem 3 Simplify the fractions by the method of prime factors.

(a) $\dfrac{120}{135}$ (b) $\dfrac{715}{880}$

3 ● Determining Whether Two Fractions Are Equal

After we simplify, how can we check that a reduced fraction is *equivalent* to the original fraction? If two fractions are equal, their diagonal products or **cross products** are equal. This is called the *equality test for fractions.* If $\frac{3}{4} = \frac{6}{8}$, then

$$\begin{array}{c}\dfrac{3}{4}\overset{?}{\underset{}{\times}}\dfrac{6}{8}\end{array} \longrightarrow \begin{array}{l} 4 \times 6 = 24 \\ 3 \times 8 = 24 \end{array} \longleftarrow \boxed{\text{Products are equal.}}$$

If two fractions are unequal (we use the symbol $\neq$), their *cross* products are unequal. If $\dfrac{5}{6} \neq \dfrac{6}{7}$, then

$$\dfrac{5}{6}\overset{?}{\times}\dfrac{6}{7} \longrightarrow \begin{array}{l} 6 \times 6 = 36 \\ 5 \times 7 = 35 \end{array} \longleftarrow \boxed{\text{Products are not equal.}}$$

Since $36 \neq 35$, we know that $\dfrac{5}{6} \neq \dfrac{6}{7}$. The test can be described in this way.

> **Equality Test for Fractions**
>
> For any two fractions where $a, b, c,$ and d are whole numbers and
> $b \neq 0, d \neq 0,$ if $\dfrac{a}{b} = \dfrac{c}{d}$, then $a \times d = b \times c$.

● **EXAMPLE 4** Are these fractions equal? Use the equality test.

(a) $\dfrac{2}{11} \overset{?}{=} \dfrac{18}{99}$ (b) $\dfrac{3}{16} \overset{?}{=} \dfrac{12}{62}$

(a) $\dfrac{2}{11}\overset{?}{=}\dfrac{18}{99} \longrightarrow \begin{array}{l} 11 \times 18 = 198 \\ 2 \times 99 = 198 \end{array} \longleftarrow \boxed{\text{Products are equal.}}$

Since $198 = 198$, we know that $\dfrac{2}{11} = \dfrac{18}{99}$.

(b) $\dfrac{3}{16}\overset{?}{=}\dfrac{12}{62} \longrightarrow \begin{array}{l} 16 \times 12 = 192 \\ 3 \times 62 = 186 \end{array} \longleftarrow \boxed{\text{Products are not equal.}}$

Since $192 \neq 186$, we know that $\dfrac{3}{16} \neq \dfrac{12}{62}$.

Practice Problem 4 Test whether the following fractions are equal.

(a) $\dfrac{84}{108} \overset{?}{=} \dfrac{7}{9}$ (b) $\dfrac{3}{7} \overset{?}{=} \dfrac{79}{182}$

Verbal and Writing Skills

1. Which of these whole numbers are prime?
   4, 12, 11, 15, 6, 19, 1, 41, 38, 24, 5, 46

2. A prime number is a whole number greater than 1 that cannot be evenly _____ except by itself and 1.

3. A _____ _____ is a whole number greater than 1 that can be divided by whole numbers other than itself and 1.

4. Every composite number can be written in exactly one way as a

   _____ of _____ numbers.

5. Give an example of a composite number written as a product of primes.

6. Give an example of equivalent (equal) fractions.

Write each number as a product of prime factors.

7. 15	**8.** 9	**9.** 6	**10.** 8	**11.** 49	**12.** 25	**13.** 64
14. 81	**15.** 55	**16.** 30	**17.** 63	**18.** 36	**19.** 75	**20.** 125
21. 54	**22.** 90	**23.** 84	**24.** 56	**25.** 98	**26.** 65	

Determine which of these whole numbers are prime. If a number is composite, write it as the product of prime factors.

27. 47	**28.** 31	**29.** 57	**30.** 51
31. 67	**32.** 71	**33.** 62	**34.** 91
35. 89	**36.** 97	**37.** 127	**38.** 119
39. 121	**40.** 95	**41.** 167	**42.** 169

Reduce each fraction by finding a common factor in the numerator and in the denominator and dividing by the common factor.

43. $\dfrac{18}{27}$
44. $\dfrac{16}{24}$
45. $\dfrac{32}{48}$
46. $\dfrac{30}{42}$

47. $\dfrac{30}{48}$
48. $\dfrac{48}{64}$
49. $\dfrac{35}{60}$
50. $\dfrac{42}{77}$

Reduce each fraction by the method of prime factors.

51. $\dfrac{3}{15}$
52. $\dfrac{7}{21}$
53. $\dfrac{66}{88}$
54. $\dfrac{42}{56}$

55. $\dfrac{18}{24}$
56. $\dfrac{65}{91}$
57. $\dfrac{27}{45}$
58. $\dfrac{28}{42}$

Mixed Practice

Reduce each fraction by any method.

59. $\dfrac{33}{36}$
60. $\dfrac{40}{96}$
61. $\dfrac{63}{108}$
62. $\dfrac{72}{132}$
63. $\dfrac{88}{121}$

64. $\dfrac{165}{180}$
65. $\dfrac{150}{200}$
66. $\dfrac{200}{300}$
67. $\dfrac{220}{260}$
68. $\dfrac{99}{189}$

Are these fractions equal? Why or why not?

69. $\dfrac{3}{17} \overset{?}{=} \dfrac{15}{85}$ **70.** $\dfrac{12}{13} \overset{?}{=} \dfrac{36}{39}$ **71.** $\dfrac{12}{40} \overset{?}{=} \dfrac{3}{13}$ **72.** $\dfrac{24}{72} \overset{?}{=} \dfrac{15}{45}$

73. $\dfrac{23}{27} \overset{?}{=} \dfrac{92}{107}$ **74.** $\dfrac{70}{120} \overset{?}{=} \dfrac{41}{73}$ **75.** $\dfrac{27}{57} \overset{?}{=} \dfrac{45}{95}$ **76.** $\dfrac{18}{24} \overset{?}{=} \dfrac{23}{28}$

77. $\dfrac{65}{70} \overset{?}{=} \dfrac{13}{14}$ **78.** $\dfrac{98}{182} \overset{?}{=} \dfrac{7}{13}$

Applications

Reduce the fractions in your answers.

79. Every December, Harold and Carolyn Bossel donate toys and games their children have outgrown to a local charity. This year, they found that 6 out of 10 board games were no longer age-appropriate. What fractional part of the games were donated?

80. Medical students frequently work long hours. James worked a 14-hour shift, spending 10 hours in the emergency room and 4 hours in surgery. What fractional part of his shift was he in the emergency room? What fractional part of his shift was he in surgery?

81. Professor Holbert found that 15 out of 95 students in the Introduction to Psychology class failed the midsemester exam. What fractional part of the class passed the exam?

82. William works for a wireless communications company that makes beepers and mobile phones. He inspected 315 beepers and found that 20 were defective. What fractional part of the beepers were not defective?

83. Rosie bought a used convertible for $16,000. She then paid $2000 to make it look and run like new. What fractional part of her total investment in the car went to make it look and run like new?

84. Monique's sister and her husband have been working two jobs each to put a down payment on a plot of land where they plan to build their house. The purchase price is $42,500. They have saved $5500. What fractional part of the cost of the land have they saved?

The following data was compiled on the students attending day classes at North Shore Community College.

Number of Students	Daily Distance Traveled from Home to College (miles)	Length of Commute
1100	0–6	Very short
1700	7–12	Short
900	13–18	Medium
500	19–24	Long
300	More than 24	Very long

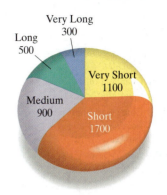

The number of students with each type of commute is displayed in the following circle graph.

Answer exercises 85–88 based on the preceding data. Reduce all fractions in your answers.

85. What fractional part of the student body has a short daily commute to the college?

86. What fractional part of the student body has a medium daily commute to the college?

87. What fractional part of the student body has a long or very long daily commute to the college?

88. What fractional part of the student body has a daily commute to the college that is considered less than long?

Cumulative Review Problems

89. Multiply. 386×425

90. Divide. $15{,}552 \div 12$

91. Multiply. 3600×1700

2.3 Improper Fractions and Mixed Numbers

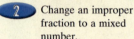

Student Learning Objectives

After studying this section, you will be able to:

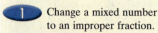 Change a mixed number to an improper fraction.

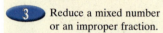 Change an improper fraction to a mixed number.

 Reduce a mixed number or an improper fraction.

SSM
PH TUTOR CENTER   CD & VIDEO   MATH PRO   WEB

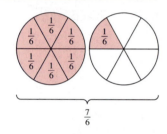

$$\frac{7}{6}$$

① *Changing a Mixed Number to an Improper Fraction*

We have names for different kinds of fractions. If the value of a fraction is less than 1, we say the fraction is proper.

$$\frac{3}{5}, \frac{5}{7}, \frac{1}{8} \quad \text{are called \textbf{proper fractions.}}$$

Notice that the numerator is less than the denominator. If the numerator is less than the denominator, the fraction is a proper fraction.

If the value of a fraction is greater than 1, the quantity can be written as an improper fraction or as a mixed number.

Suppose that we have 1 whole pizza and $\frac{1}{6}$ of a pizza. We could write this as $1\frac{1}{6}$. $1\frac{1}{6}$ is called a mixed number. A **mixed number** is the sum of a whole number greater than zero and a proper fraction. The notation $1\frac{1}{6}$ actually means $1 + \frac{1}{6}$. The plus sign is not usually shown.

Another way of writing $1\frac{1}{6}$ pizza is to write $\frac{7}{6}$ pizza. $\frac{7}{6}$ is called an improper fraction. Notice that the numerator is greater than the denominator. If the numerator is greater than or equal to the denominator, the fraction is an improper fraction.

$$\frac{7}{6}, \frac{6}{6}, \frac{5}{4}, \frac{8}{3}, \frac{2}{2} \quad \text{are \textbf{improper fractions.}}$$

The following chart will help you visualize these different fractions and their names.

Value Less Than 1	Value Equal To 1	Value Greater Than 1	
Proper Fraction	Improper Fraction	Improper Fraction or	Mixed Number
$\frac{3}{4}$	$\frac{4}{4}$	$\frac{5}{4}$ or $1\frac{1}{4}$	
$\frac{7}{8}$	$\frac{8}{8}$	$\frac{17}{8}$ or $2\frac{1}{8}$	
$\frac{3}{100}$	$\frac{100}{100}$	$\frac{109}{100}$ or $1\frac{9}{100}$	

126

Because improper fractions are easier to add, subtract, multiply, and divide than mixed numbers, we often change mixed numbers to improper fractions when we perform calculations with them.

Changing a Mixed Number to an Improper Fraction

1. Multiply the whole number by the denominator of the fraction.

2. Add the numerator of the fraction to the product found in step 1.

3. Write the sum found in step 2 over the denominator of the fraction.

EXAMPLE 1 Change each mixed number to an improper fraction.

(a) $3\dfrac{2}{5}$ **(b)** $5\dfrac{4}{9}$ **(c)** $18\dfrac{3}{5}$

> Multiply the whole number by the denominator.

> Add the numerator to the product.

(a) $3\dfrac{2}{5}$ $\dfrac{3 \times 5 + 2}{5} = \dfrac{15 + 2}{5} = \dfrac{17}{5}$

> Write the sum over the denominator.

(b) $5\dfrac{4}{9} = \dfrac{9 \times 5 + 4}{9} = \dfrac{45 + 4}{9} = \dfrac{49}{9}$

(c) $18\dfrac{3}{5} = \dfrac{5 \times 18 + 3}{5} = \dfrac{90 + 3}{5} = \dfrac{93}{5}$

Practice Problem 1 Change the mixed numbers to improper fractions.

(a) $4\dfrac{3}{7}$ **(b)** $6\dfrac{2}{3}$ **(c)** $19\dfrac{4}{7}$

2 *Changing an Improper Fraction to a Mixed Number*

We often need to change an improper fraction to a mixed number.

Changing an Improper Fraction to a Mixed Number

1. Divide the numerator by the denominator.

2. Write the quotient followed by the fraction with the remainder over the denominator.

$$\text{quotient } \dfrac{\text{remainder}}{\text{denominator}}$$

EXAMPLE 2 Write each improper fraction as a mixed number.

(a) $\dfrac{13}{5}$ **(b)** $\dfrac{29}{7}$ **(c)** $\dfrac{105}{31}$ **(d)** $\dfrac{85}{17}$

(a) We divide the denominator 5 into 13.

$$
\begin{array}{r}
2 \leftarrow \text{quotient} \\
5\overline{)13} \\
\underline{10} \\
3 \leftarrow \text{remainder}
\end{array}
$$

The answer is in the form quotient $\dfrac{\text{remainder}}{\text{denominator}}$

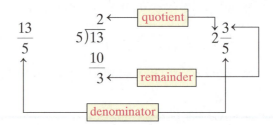

Thus $\dfrac{13}{5} = 2\dfrac{3}{5}$.

(b) $\begin{array}{r} 4 \\ 7\overline{)29} \\ \underline{28} \\ 1 \end{array}$ $\quad \dfrac{29}{7} = 4\dfrac{1}{7}$

(c) $\begin{array}{r} 3 \\ 31\overline{)105} \\ \underline{93} \\ 12 \end{array}$ $\quad \dfrac{105}{31} = 3\dfrac{12}{31}$

(d) $\begin{array}{r} 5 \\ 17\overline{)85} \\ \underline{85} \\ 0 \end{array}$ The remainder is 0, so $\dfrac{85}{17} = 5$, a whole number.

Practice Problem 2 Write as a mixed number or a whole number.

(a) $\dfrac{17}{4}$ **(b)** $\dfrac{36}{5}$ **(c)** $\dfrac{116}{27}$ **(d)** $\dfrac{91}{13}$

3 Reducing a Mixed Number or an Improper Fraction

Mixed numbers and improper fractions may need to be reduced if they are not in simplest form. Recall that we write the fraction in terms of prime factors. Then we look for common factors in the numerator and the denominator of the fraction. Then we divide the numerator and the denominator by the common factor.

EXAMPLE 3 Reduce the improper fraction. $\dfrac{22}{8}$

$$
\dfrac{22}{8} = \dfrac{\overset{1}{\cancel{2}} \times 11}{\underset{1}{\cancel{2}} \times 2 \times 2} = \dfrac{11}{4}
$$

Practice Problem 3 Reduce the improper fraction. $\dfrac{51}{15}$

● **EXAMPLE 4**   Reduce the mixed number. $4\frac{21}{28}$

We do not need to reduce the whole number 4, only the fraction $\frac{21}{28}$.

$$\frac{21}{28} = \frac{3 \times \overset{1}{\cancel{7}}}{4 \times \underset{1}{\cancel{7}}} = \frac{3}{4}$$

Therefore, $4\frac{21}{28} = 4\frac{3}{4}$.

Practice Problem 4   Reduce the mixed number. $3\frac{16}{80}$ ●

If an improper fraction contains a very large numerator and denominator, it is best to change the fraction to a mixed number before reducing.

● **EXAMPLE 5**   Reduce $\frac{945}{567}$ by first changing to a mixed number.

$$\begin{array}{r} 1 \\ 567\overline{)945} \\ \underline{567} \\ 378 \end{array} \qquad \text{so} \quad \frac{945}{567} = 1\frac{378}{567}$$

To reduce the fraction we write

$$\frac{378}{567} = \frac{3 \times 3 \times 3 \times 2 \times 7}{3 \times 3 \times 3 \times 3 \times 7} = \frac{\overset{1}{\cancel{3}} \times \overset{1}{\cancel{3}} \times \overset{1}{\cancel{3}} \times 2 \times \overset{1}{\cancel{7}}}{\underset{1}{\cancel{3}} \times \underset{1}{\cancel{3}} \times \underset{1}{\cancel{3}} \times 3 \times \underset{1}{\cancel{7}}} = \frac{2}{3}$$

So $\frac{945}{567} = 1\frac{378}{567} = 1\frac{2}{3}$.

Problems like Example 5 can be done in several different ways. It is not necessary to follow these exact steps when reducing this fraction.

Practice Problem 5   Reduce $\frac{1001}{572}$ by first changing to a mixed number.   ●

To Think About   A student concluded that just by looking at the denominator he could tell that the fraction $\frac{1655}{97}$ cannot be reduced unless $1655 \div 97$ is a whole number. How did he come to that conclusion?

Note that 97 is a prime number. The only factors of 97 are 97 and 1. Therefore, *any* fraction with 97 in the denominator can be reduced only if 97 is a factor of the numerator. Since $1655 \div 97$ is not a whole number (see the following division), it is therefore impossible to reduce $\frac{1655}{97}$.

$$\begin{array}{r} 17 \\ 97\overline{)1655} \\ \underline{97} \\ 685 \\ \underline{679} \\ 6 \end{array}$$

You may explore this idea in Exercises 2.3, exercises 81 and 82.

Verbal and Writing Skills

1. Describe in your own words how to change a mixed number to an improper fraction.

2. Describe in your own words how to change an improper fraction to a mixed number.

Change each mixed number to an improper fraction.

3. $4\dfrac{2}{3}$ **4.** $3\dfrac{5}{6}$ **5.** $2\dfrac{3}{7}$ **6.** $3\dfrac{3}{8}$ **7.** $6\dfrac{1}{7}$ **8.** $8\dfrac{3}{8}$

9. $10\dfrac{2}{3}$ **10.** $15\dfrac{3}{4}$ **11.** $21\dfrac{2}{3}$ **12.** $13\dfrac{1}{3}$ **13.** $9\dfrac{1}{6}$ **14.** $56\dfrac{1}{2}$

15. $28\dfrac{1}{6}$ **16.** $6\dfrac{6}{7}$ **17.** $10\dfrac{11}{12}$ **18.** $13\dfrac{5}{7}$ **19.** $7\dfrac{9}{10}$ **20.** $4\dfrac{1}{50}$

21. $8\dfrac{1}{25}$ **22.** $66\dfrac{2}{3}$ **23.** $105\dfrac{1}{2}$ **24.** $207\dfrac{2}{3}$ **25.** $164\dfrac{2}{3}$ **26.** $33\dfrac{1}{3}$

27. $8\dfrac{11}{15}$ **28.** $5\dfrac{19}{20}$ **29.** $5\dfrac{13}{25}$ **30.** $6\dfrac{18}{19}$

Change each improper fraction to a mixed number or a whole number.

31. $\dfrac{4}{3}$ **32.** $\dfrac{13}{4}$ **33.** $\dfrac{11}{4}$ **34.** $\dfrac{9}{5}$ **35.** $\dfrac{15}{6}$ **36.** $\dfrac{23}{6}$ **37.** $\dfrac{27}{8}$

38. $\dfrac{48}{16}$ **39.** $\dfrac{60}{12}$ **40.** $\dfrac{42}{13}$ **41.** $\dfrac{86}{9}$ **42.** $\dfrac{47}{2}$ **43.** $\dfrac{28}{13}$ **44.** $\dfrac{54}{17}$

45. $\dfrac{51}{16}$ **46.** $\dfrac{19}{3}$ **47.** $\dfrac{28}{3}$ **48.** $\dfrac{100}{3}$ **49.** $\dfrac{35}{2}$ **50.** $\dfrac{132}{11}$ **51.** $\dfrac{91}{7}$

52. $\dfrac{183}{7}$ **53.** $\dfrac{210}{15}$ **54.** $\dfrac{196}{9}$ **55.** $\dfrac{102}{17}$ **56.** $\dfrac{105}{8}$ **57.** $\dfrac{403}{11}$ **58.** $\dfrac{212}{9}$

Reduce each mixed number.

59. $2\dfrac{9}{12}$ **60.** $2\dfrac{10}{15}$ **61.** $4\dfrac{11}{66}$ **62.** $3\dfrac{15}{90}$ **63.** $12\dfrac{15}{40}$ **64.** $15\dfrac{12}{36}$

Reduce each improper fraction.

65. $\dfrac{24}{6}$ **66.** $\dfrac{36}{20}$ **67.** $\dfrac{36}{15}$ **68.** $\dfrac{63}{45}$ **69.** $\dfrac{78}{9}$ **70.** $\dfrac{143}{22}$

Change to a mixed number and reduce.

71. $\dfrac{340}{126}$ **72.** $\dfrac{386}{226}$ **73.** $\dfrac{986}{424}$ **74.** $\dfrac{764}{328}$ **75.** $\dfrac{508}{296}$ **76.** $\dfrac{2150}{1000}$

Applications

77. The Science Museum is hanging banners all over the building to commemorate the Apollo astronauts. The art department is using $360\frac{2}{3}$ yards of starry-sky parachute fabric. Change this number to an improper fraction.

78. For the Northwestern University alumni homecoming, the students studying sculpture have made a giant replica of the school using $244\frac{3}{4}$ pounds of clay. Change this number to an improper fraction.

79. A Cape Cod cranberry bog was contaminated by waste from an abandoned army base. Damage was done to $\frac{151}{3}$ acres of land. Write this as a mixed number.

80. Bill and Katie Naseth are adding a deck onto their new house. They need to purchase $\frac{119}{4}$ feet of lumber for part of the deck. Write this as a mixed number.

To Think About

81. Can $\dfrac{5687}{101}$ be reduced? Why or why not?

82. Can $\dfrac{9810}{157}$ be reduced? Why or why not?

Cumulative Review Problems

83. Subtract. $1,398,210 - 1,137,963$

84. Estimate the answer.

$78,964 \times 229,350$

85. Estimate the answer.

$872,365 \div 286$

86. Each semester college textbooks are often shipped to the bookstore in cardboard boxes that contain 24 textbooks. If 893 copies of *Basic College Mathematics* are shipped to the bookstore, how many full cartons are needed for the shipment? How many books are there in the carton that is not full?

2.4 Multiplication of Fractions and Mixed Numbers

1 *Multiplying Two Fractions That Are Proper or Improper*

Student Learning Objectives

After studying this section, you will be able to:

1 Multiply two fractions that are proper or improper.

2 Multiply a whole number by a fraction.

3 Multiply mixed numbers.

SSM · PH TUTOR CENTER · CD & VIDEO · MATH PRO · WEB

Fudge Squares

Ingredients:

2 cups sugar	1/4 teaspoon salt
4 oz chocolate	1 teaspoon vanilla
1/2 cup butter	1 cup all-purpose flour
4 eggs	1 cup nutmeats

Suppose you want to make an amount equal to half of what the recipe shown will produce. You would multiply the measure given for each ingredient by $\frac{1}{2}$.

$\frac{1}{2}$ of 2 cups sugar	$\frac{1}{2}$ of $\frac{1}{4}$ teaspoon salt
$\frac{1}{2}$ of 4 oz chocolate	$\frac{1}{2}$ of 1 teaspoon vanilla
$\frac{1}{2}$ of $\frac{1}{2}$ cup butter	$\frac{1}{2}$ of 1 cup all-purpose flour
$\frac{1}{2}$ of 4 eggs	$\frac{1}{2}$ of 1 cup nutmeats

We often use multiplication of fractions to describe taking a fractional part of something. To find $\frac{1}{2}$ of $\frac{3}{7}$ we multiply

$$\frac{1}{2} \times \frac{3}{7} = \frac{3}{14}$$

We begin with a bar that is $\frac{3}{7}$ shaded. To find $\frac{1}{2}$ of $\frac{3}{7}$ we divide the bar in half and take $\frac{1}{2}$ of the shaded section. $\frac{1}{2}$ of $\frac{3}{7}$ yields 3 out of 14 squares.

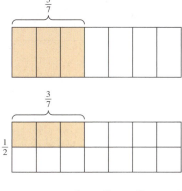

Thus $\dfrac{1}{2} \times \dfrac{3}{7} = \dfrac{3}{14}$

When you multiply two proper fractions together, you get a smaller fraction.

To multiply two fractions, we multiply the numerators and multiply the denominators.

$$\frac{2}{3} \times \frac{5}{7} = \frac{10}{21} \quad \begin{array}{l} \leftarrow 2 \times 5 = 10 \\ \leftarrow 3 \times 7 = 21 \end{array}$$

133

In general, for all positive whole numbers a, b, c, and d,

$$\frac{a}{b} \times \frac{c}{d} = \frac{a \times c}{b \times d}.$$

● EXAMPLE 1 Multiply.

(a) $\dfrac{3}{8} \times \dfrac{5}{7}$ **(b)** $\dfrac{1}{11} \times \dfrac{2}{13}$

(a) $\dfrac{3}{8} \times \dfrac{5}{7} = \dfrac{3 \times 5}{8 \times 7} = \dfrac{15}{56}$ **(b)** $\dfrac{1}{11} \times \dfrac{2}{13} = \dfrac{1 \times 2}{11 \times 13} = \dfrac{2}{143}$

Practice Problem 1 Multiply.

(a) $\dfrac{6}{7} \times \dfrac{3}{13}$ **(b)** $\dfrac{1}{5} \times \dfrac{11}{12}$

Some products may be reduced.

$$\frac{12}{35} \times \frac{25}{18} = \frac{300}{630} = \frac{10}{21}$$

By simplifying before multiplication, the reducing can be done more easily. For a multiplication problem, a factor in the numerator can be paired with a common factor in the denominator of the same or a different fraction. We can begin by finding the prime factors in the numerators and denominators. We then divide numerator and denominator by their common prime factors.

● EXAMPLE 2 Simplify first and then multiply. $\dfrac{12}{35} \times \dfrac{25}{18}$

$\dfrac{12}{35} \times \dfrac{25}{18} = \dfrac{2 \cdot 2 \cdot 3}{5 \cdot 7} \times \dfrac{5 \cdot 5}{3 \cdot 3 \cdot 2}$ First we find the prime factors.

$= \dfrac{2 \cdot 2 \cdot 3 \cdot 5 \cdot 5}{5 \cdot 7 \cdot 3 \cdot 3 \cdot 2}$ Write the product as one fraction.

$= \dfrac{\overset{1}{\cancel{2}} \cdot 2 \cdot \overset{1}{\cancel{3}} \cdot \overset{1}{\cancel{5}} \cdot 5}{\underset{1}{\cancel{2}} \cdot \underset{1}{\cancel{3}} \cdot 3 \cdot \underset{1}{\cancel{5}} \cdot 7}$ Arrange the factors in order and divide the numerator and denominator by the same numbers.

$= \dfrac{10}{21}$ Multiply the remaining factors.

Practice Problem 2 Simplify first and then multiply.

$$\frac{55}{72} \times \frac{16}{33}$$

Note: Although finding the prime factors of the numerators and denominators will help you avoid errors, you can also begin these problems by dividing the numerators and denominators by larger common factors. This method will be used for the remainder of the exercises in this section of the text.

2 *Multiplying a Whole Number by a Fraction*

When multiplying a fraction by a whole number, it is more convenient to express the whole number as a fraction with a denominator of 1. We know that $5 = \frac{5}{1}$, $7 = \frac{7}{1}$, and so on.

EXAMPLE 3 Multiply.

(a) $5 \times \dfrac{3}{8}$

(b) $\dfrac{22}{7} \times 14$

(a) $5 \times \dfrac{3}{8} = \dfrac{5}{1} \times \dfrac{3}{8} = \dfrac{15}{8}$ or $1\dfrac{7}{8}$

(b) $\dfrac{22}{7} \times 14 = \dfrac{22}{\overset{}{\underset{1}{7}}} \times \dfrac{\overset{2}{14}}{1} = \dfrac{44}{1} = 44$

Practice Problem 3 Multiply.

(a) $7 \times \dfrac{5}{13}$ **(b)** $\dfrac{13}{16} \times 8$

EXAMPLE 4 Mr. and Mrs. Jones found that $\frac{2}{7}$ of their income went to pay federal income taxes. Last year they earned $37,100. How much did they pay in taxes?

We need to find $\frac{2}{7}$ of $37,100. So we must multiply $\frac{2}{7} \times 37{,}100$.

$$\dfrac{2}{\overset{}{\underset{1}{7}}} \times \overset{5300}{\cancel{37{,}100}} = \dfrac{2}{1} \times 5300 = 10{,}600$$

They paid $10,600 in federal income taxes.

▲ **Practice Problem 4** Fred and Linda own 98,400 square feet of land. They found that $\frac{3}{8}$ of the land is in a wetland area and cannot be used for building. How many square feet of land are in the wetland area?

3 *Multiplying Mixed Numbers*

To multiply a fraction by a mixed number or to multiply two mixed numbers, first change each mixed number to an improper fraction.

EXAMPLE 5 Multiply.

(a) $\dfrac{5}{7} \times 3\dfrac{1}{4}$ **(b)** $20\dfrac{2}{5} \times 6\dfrac{2}{3}$ **(c)** $\dfrac{3}{4} \times 1\dfrac{1}{2} \times \dfrac{4}{7}$ **(d)** $4\dfrac{1}{3} \times 2\dfrac{1}{4}$

(a) $\dfrac{5}{7} \times 3\dfrac{1}{4} = \dfrac{5}{7} \times \dfrac{13}{4} = \dfrac{65}{28}$ or $2\dfrac{9}{28}$

(b) $20\dfrac{2}{5} \times 6\dfrac{2}{3} = \dfrac{\overset{34}{\cancel{102}}}{\underset{1}{\cancel{5}}} \times \dfrac{\overset{4}{\cancel{20}}}{\underset{1}{\cancel{3}}} = \dfrac{136}{1} = 136$

(c) $\dfrac{3}{4} \times 1\dfrac{1}{2} \times \dfrac{4}{7} = \dfrac{3}{\cancel{4}_{1}} \times \dfrac{3}{2} \times \dfrac{\cancel{4}^{1}}{7} = \dfrac{9}{14}$

(d) $4\dfrac{1}{3} \times 2\dfrac{1}{4} = \dfrac{13}{\cancel{3}_{1}} \times \dfrac{\cancel{9}^{3}}{4} = \dfrac{39}{4}$ or $9\dfrac{3}{4}$

Practice Problem 5 Multiply.

(a) $2\dfrac{1}{6} \times \dfrac{4}{7}$ **(b)** $10\dfrac{2}{3} \times 13\dfrac{1}{2}$ **(c)** $\dfrac{3}{5} \times 1\dfrac{1}{3} \times \dfrac{5}{8}$ **(d)** $3\dfrac{1}{5} \times 2\dfrac{1}{2}$ ●

▲ ● **EXAMPLE 6** Find the area in square miles of a rectangle with width $1\dfrac{1}{3}$ miles and length $12\dfrac{1}{4}$ miles.

Length = $12\dfrac{1}{4}$ miles

Width = $1\dfrac{1}{3}$ miles

We find the area of a rectangle by multiplying the width times the length.

$$1\dfrac{1}{3} \times 12\dfrac{1}{4} = \dfrac{\cancel{4}^{1}}{3} \times \dfrac{49}{\cancel{4}_{1}} = \dfrac{49}{3} \quad \text{or} \quad 16\dfrac{1}{3}$$

The area is $16\dfrac{1}{3}$ square miles.

▲ **Practice Problem 6** Find the area in square meters of a rectangle with width $1\dfrac{1}{5}$ meters and length $4\dfrac{5}{6}$ meters. ●

● **EXAMPLE 7** Find the value of x if

$$\dfrac{3}{7} \cdot x = \dfrac{15}{42}.$$

The variable x represents a fraction. We know that 3 times one number equals 15 and 7 times another equals 42.

Since $3 \cdot 5 = 15$ and $7 \cdot 6 = 42$ we know that $\dfrac{3}{7} \cdot \dfrac{5}{6} = \dfrac{15}{42}$

Therefore, $x = \dfrac{5}{6}$.

Practice Problem 7 Find the value of x if $\dfrac{8}{9} \cdot x = \dfrac{80}{81}$. ●

Multiply. Make sure all fractions are simplified in the final answer.

1. $\dfrac{3}{5} \times \dfrac{7}{11}$
 2. $\dfrac{1}{8} \times \dfrac{5}{11}$
 3. $\dfrac{3}{4} \times \dfrac{5}{13}$
 4. $\dfrac{4}{7} \times \dfrac{3}{5}$
 5. $\dfrac{6}{5} \times \dfrac{10}{12}$

6. $\dfrac{7}{8} \times \dfrac{16}{21}$
 7. $\dfrac{7}{36} \times \dfrac{30}{9}$
 8. $\dfrac{22}{45} \times \dfrac{5}{11}$
 9. $\dfrac{15}{28} \times \dfrac{7}{9}$
 10. $\dfrac{5}{24} \times \dfrac{18}{15}$

11. $\dfrac{9}{10} \times \dfrac{35}{12}$
 12. $\dfrac{12}{17} \times \dfrac{3}{24}$
 13. $8 \times \dfrac{3}{7}$
 14. $\dfrac{2}{11} \times 4$
 15. $\dfrac{5}{16} \times 8$

16. $5 \times \dfrac{7}{25}$
 17. $\dfrac{4}{9} \times \dfrac{3}{7} \times \dfrac{7}{8}$
 18. $\dfrac{8}{7} \times \dfrac{5}{12} \times \dfrac{3}{10}$
 19. $\dfrac{4}{5} \times \dfrac{1}{8} \times \dfrac{35}{7}$
 20. $\dfrac{10}{13} \times \dfrac{26}{15} \times \dfrac{2}{3}$

Multiply. Change any mixed number to an improper fraction before multiplying.

21. $2\dfrac{3}{4} \times \dfrac{8}{9}$
 22. $\dfrac{5}{6} \times 3\dfrac{3}{5}$
 23. $1\dfrac{1}{4} \times 3\dfrac{2}{3}$
 24. $2\dfrac{3}{5} \times 1\dfrac{4}{7}$

25. $2\dfrac{1}{2} \times 6$
 26. $4\dfrac{1}{3} \times 9$
 27. $2\dfrac{3}{10} \times \dfrac{3}{5}$
 28. $4\dfrac{3}{5} \times \dfrac{1}{10}$

29. $1\dfrac{3}{16} \times 0$
 30. $3\dfrac{7}{8} \times 1$
 31. $4\dfrac{1}{5} \times 12\dfrac{2}{9}$
 32. $5\dfrac{1}{4} \times 10\dfrac{3}{7}$

33. $6\dfrac{2}{5} \times \dfrac{1}{4}$
 34. $\dfrac{8}{9} \times 4\dfrac{1}{11}$
 35. $\dfrac{5}{5} \times 11\dfrac{5}{7}$
 36. $0 \times 6\dfrac{2}{3}$

Solve for x.

37. $\dfrac{2}{7} \cdot x = \dfrac{18}{35}$

38. $\dfrac{5}{8} \cdot x = \dfrac{35}{88}$

39. $\dfrac{7}{13} \cdot x = \dfrac{56}{117}$

40. $x \cdot \dfrac{11}{15} = \dfrac{77}{225}$

Applications

▲ **41.** A spy is running from his captors in a forest that is $8\frac{3}{4}$ miles long and $4\frac{1}{3}$ miles wide. Find the area of the forest where he is hiding. (*Hint:* The area of a rectangle is the product of the length times the width.)

▲ **42.** An area in the Midwest is a designated tornado danger zone. The land is $22\frac{5}{8}$ miles long and $16\frac{1}{2}$ miles wide. Find the area of the tornado danger zone. (*Hint:* The area of a rectangle is the product of the length times the width.)

43. A Lear jet airplane has 360 gallons of fuel. The plane averages $4\frac{1}{3}$ miles per gallon. How far can the plane go?

44. A jeep has $11\frac{1}{6}$ gallons of gas. The jeep averages 12 miles per gallon. How far will the jeep be able to go on what is in the tank?

45. A recipe from Nanette's French cookbook for a scalloped potato tart requires $90\frac{1}{2}$ grams of grated cheese. How many grams of cheese would she need if she made one tart for each of her 18 cousins?

46. We used $18\frac{1}{4}$ yards of wallpaper to decorate our TV room. If we were to wallpaper eight rooms with the exact same measurements, how many yards of wallpaper would we need?

47. Value Hardware Store is making a custom color paint for Susan's kitchen walls. The paint requires $12\frac{1}{4}$ ounces of pigment. How many ounces of pigment would be necessary to make $\frac{3}{4}$ of the formula?

48. The propeller on the Ipswich River Cruise Boat turns 320 revolutions per minute. How fast would it turn at $\frac{3}{4}$ of that speed?

49. Carlos has sent his résumé to 12,064 companies through an Internet job search service. If $\frac{1}{32}$ of the companies e-mail him with an invitation for an interview, how many companies will he have heard from?

50. Russ purchased a new Buick LeSabre for $26,500. After one year the car was worth $\frac{4}{5}$ of the purchase price. What was the car worth after one year?

51. Kathy walked $3\frac{1}{2}$ miles per hour for $1\frac{1}{4}$ hours. During $\frac{1}{2}$ of her walking time, she was walking in the rain. How many miles did she walk in the rain?

52. There were 1220 students at the Beverly campus of North Shore Community College during the spring 2001 semester. The registrar discovered that $\frac{2}{5}$ of these students live in the city of Beverly. He further discovered that $\frac{1}{4}$ of the students living in Beverly attend classes only on Monday, Wednesday, and Friday. How many students at the Beverly campus live in the city of Beverly and attend classes only on Monday, Wednesday, and Friday?

53. Of the students eligible to apply for grants, $\frac{3}{4}$ actually sent away for applications. Of the students who received applications, only $\frac{5}{6}$ sent their completed forms in.

(a) What fractional portion of the students actually sent in grant applications?

(b) If the school has 8600 eligible students, how many people actually sent in grant applications?

54. When Sue turned 13, her wealthy grandparents gave her an allowance and put $\frac{3}{5}$ of the money into a college fund. When she turned 18, her grandparents surprised her by telling her that $\frac{3}{8}$ of *that* money was put into a long-term fund.

(a) What portion of the original allowance went into the long-term fund?

(b) What amount was put into the long-term fund if the allowance was $150 per month?

To Think About

55. When we multiply two fractions, we look for opportunities to divide a numerator and a denominator by the same number. Why do we bother with that step? Why don't we just multiply the two numerators and the two denominators?

56. Suppose there is an unknown fraction that has *not* been simplified (it is not reduced). You multiply this unknown fraction by $\frac{2}{5}$ and you obtain a simplified answer of $\frac{6}{35}$. How many possible values could this unknown fraction be? Give at least three possible answers.

Cumulative Review Problems

57. A total of 16,399 cars used a toll bridge in January (31 days). What is the average number of cars using the bridge in one day?

58. The Office of Investors Services has 15,456 calls made per month by the sales personnel. There are 42 sales personnel in the office. What is the average number of calls made per month by one salesperson?

59. A computer printer can print 146 lines per minute. How many lines can it print in 12 minutes?

60. At cruising speed a new commercial jet uses 12,360 gallons of fuel per hour. How many gallons will be used in 14 hours of flying time?

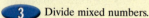

1 Dividing Two Proper or Improper Fractions

Why would you divide fractions? Consider this problem.

- A copper pipe that is $\frac{3}{4}$ of a foot long is to be cut into $\frac{1}{4}$-foot pieces. How many pieces will there be?

To find how many $\frac{1}{4}$'s are in $\frac{3}{4}$, we divide $\frac{3}{4} \div \frac{1}{4}$. We draw a sketch.

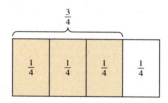

Notice that there are three $\frac{1}{4}$'s in $\frac{3}{4}$.

How do we divide two fractions? We **invert** the second fraction and multiply.

$$\frac{3}{4} \div \frac{1}{4} = \frac{3}{\cancel{4}_1} \times \frac{\cancel{4}^1}{1} = \frac{3}{1} = 3$$

When we invert a fraction, we interchange the numerator and the denominator. If we invert $\frac{5}{9}$, we obtain $\frac{9}{5}$. If we invert $\frac{6}{1}$, we obtain $\frac{1}{6}$. Numbers such as $\frac{5}{9}$ and $\frac{9}{5}$ are called **reciprocals** of each other.

> ### Rule for Division of Fractions
>
> To divide two fractions, we invert the second fraction and multiply.
>
> $$\frac{a}{b} \div \frac{c}{d} = \frac{a}{b} \times \frac{d}{c}$$
>
> (when b, c, and d are not zero).

EXAMPLE 1 Divide. (a) $\frac{3}{11} \div \frac{2}{5}$ (b) $\frac{5}{8} \div \frac{25}{16}$

(a) $\frac{3}{11} \div \frac{2}{5} = \frac{3}{11} \times \frac{5}{2} = \frac{15}{22}$ (b) $\frac{5}{8} \div \frac{25}{16} = \frac{\cancel{5}^1}{\cancel{8}_1} \times \frac{\cancel{16}^2}{\cancel{25}_5} = \frac{2}{5}$

Practice Problem 1 Divide. (a) $\frac{7}{13} \div \frac{3}{4}$ (b) $\frac{16}{35} \div \frac{24}{25}$

2 Dividing a Whole Number and a Fraction

When dividing with whole numbers, it is helpful to remember that for any whole number a, $a = \frac{a}{1}$.

EXAMPLE 2 Divide. (a) $\frac{3}{7} \div 2$ (b) $5 \div \frac{10}{13}$

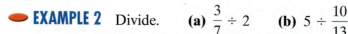

(a) $\dfrac{3}{7} \div 2 = \dfrac{3}{7} \div \dfrac{2}{1} = \dfrac{3}{7} \times \dfrac{1}{2} = \dfrac{3}{14}$

(b) $5 \div \dfrac{10}{13} = \dfrac{5}{1} \div \dfrac{10}{13} = \dfrac{\overset{1}{\cancel{5}}}{1} \times \dfrac{13}{\underset{2}{\cancel{10}}} = \dfrac{13}{2}$ or $6\dfrac{1}{2}$

Practice Problem 2 Divide. **(a)** $\dfrac{3}{17} \div 6$ **(b)** $14 \div \dfrac{7}{15}$

EXAMPLE 3 Divide, if possible.

(a) $\dfrac{23}{25} \div 1$ **(b)** $1 \div \dfrac{7}{5}$ **(c)** $0 \div \dfrac{4}{9}$ **(d)** $\dfrac{3}{17} \div 0$

(a) $\dfrac{23}{25} \div 1 = \dfrac{23}{25} \times \dfrac{1}{1} = \dfrac{23}{25}$

(b) $1 \div \dfrac{7}{5} = \dfrac{1}{1} \times \dfrac{5}{7} = \dfrac{5}{7}$

(c) $0 \div \dfrac{4}{9} = \dfrac{0}{1} \times \dfrac{9}{4} = \dfrac{0}{4} = 0$ Zero divided by any nonzero number is zero.

(d) $\dfrac{3}{17} \div 0$ Division by zero is undefined.

Practice Problem 3 Divide, if possible.

(a) $1 \div \dfrac{11}{13}$ **(b)** $\dfrac{14}{17} \div 1$ **(c)** $\dfrac{3}{11} \div 0$ **(d)** $0 \div \dfrac{9}{16}$

Sidelight

Why do we divide by inverting the second fraction and multiplying? What is really going on when we do this? We are actually multiplying by 1. To see why, consider the following.

$$\dfrac{3}{7} \div \dfrac{2}{3} = \dfrac{\dfrac{3}{7}}{\dfrac{2}{3}}$$ We write the division by using another fraction bar.

$$= \dfrac{\dfrac{3}{7}}{\dfrac{2}{3}} \times 1$$ Any fraction can be multiplied by 1 without changing the value of the fraction. This is the fundamental rule of fractions.

$$= \dfrac{\dfrac{3}{7}}{\dfrac{2}{3}} \times \dfrac{\dfrac{3}{2}}{\dfrac{3}{2}}$$ Any nonzero number divided by itself equals 1.

$$= \dfrac{\dfrac{3}{7} \times \dfrac{3}{2}}{\dfrac{2}{3} \times \dfrac{3}{2}}$$ Definition of multiplication of fractions.

$$= \dfrac{\dfrac{3}{7} \times \dfrac{3}{2}}{1} = \dfrac{3}{7} \times \dfrac{3}{2}$$ Any number can be written as a fraction with a denominator of 1 without changing its value.

Thus

$$\frac{3}{7} \div \frac{2}{3} = \frac{3}{7} \times \frac{3}{2} = \frac{9}{14}.$$

3 **Dividing Mixed Numbers**

If one or more mixed numbers are involved in the division, they should be converted to improper fractions first.

EXAMPLE 4 Divide.   **(a)** $3\frac{7}{15} \div 1\frac{1}{25}$ **(b)** $\frac{3}{5} \div 2\frac{1}{7}$

(a) $3\frac{7}{15} \div 1\frac{1}{25} = \frac{52}{15} \div \frac{26}{25} = \frac{\overset{2}{\cancel{52}}}{\underset{3}{\cancel{15}}} \times \frac{\overset{5}{\cancel{25}}}{\underset{1}{\cancel{26}}} = \frac{10}{3}$ or $3\frac{1}{3}$

(b) $\frac{3}{5} \div 2\frac{1}{7} = \frac{3}{5} \div \frac{15}{7} = \frac{\overset{1}{\cancel{3}}}{5} \times \frac{7}{\underset{5}{\cancel{15}}} = \frac{7}{25}$

Practice Problem 4 Divide.   **(a)** $1\frac{1}{5} \div \frac{7}{10}$ **(b)** $2\frac{1}{4} \div 1\frac{7}{8}$

The division of two fractions may be indicated by a wide fraction bar.

EXAMPLE 5 Divide.   **(a)** $\dfrac{10\frac{2}{9}}{2\frac{1}{3}}$ **(b)** $\dfrac{1\frac{1}{15}}{3\frac{1}{3}}$

(a) $\dfrac{10\frac{2}{9}}{2\frac{1}{3}} = 10\frac{2}{9} \div 2\frac{1}{3} = \frac{92}{9} \div \frac{7}{3} = \frac{92}{\underset{3}{\cancel{9}}} \times \frac{\overset{1}{\cancel{3}}}{7} = \frac{92}{21}$ or $4\frac{8}{21}$

(b) $\dfrac{1\frac{1}{15}}{3\frac{1}{3}} = 1\frac{1}{15} \div 3\frac{1}{3} = \frac{16}{15} \div \frac{10}{3} = \frac{\overset{8}{\cancel{16}}}{\underset{5}{\cancel{15}}} \times \frac{\overset{1}{\cancel{3}}}{\underset{5}{\cancel{10}}} = \frac{8}{25}$

Practice Problem 5 Divide.

(a) $\dfrac{5\frac{2}{3}}{7}$ **(b)** $\dfrac{1\frac{2}{5}}{2\frac{1}{3}}$

Some students may find Example 6 difficult at first. Read it slowly and carefully. It may be necessary to read it several times before it becomes clear.

EXAMPLE 6 Find the value of x if $x \div \frac{8}{7} = \frac{21}{40}$.

First we will change the division problem to an equivalent multiplication problem.

$$x \div \frac{8}{7} = \frac{21}{40}$$

$$x \cdot \frac{7}{8} = \frac{21}{40}$$

x represents a fraction.

In the numerator, we want to know what times 7 equals 21. In the denominator, we want to know what times 8 equals 40.

$$\frac{3}{5} \cdot \frac{7}{8} = \frac{21}{40}$$

Thus $x = \frac{3}{5}$.

Practice Problem 6   Find the value of x if $x \div \frac{3}{2} = \frac{22}{36}$.

EXAMPLE 7   There are 117 milligrams of cholesterol in $4\frac{1}{3}$ cups of milk. How much cholesterol is in 1 cup of milk?

We want to divide the 117 by $4\frac{1}{3}$ to find out how much is in 1 cup.

$$117 \div 4\frac{1}{3} = 117 \div \frac{13}{3} = \frac{\overset{9}{\cancel{117}}}{1} \times \frac{3}{\underset{1}{\cancel{13}}} = \frac{27}{1} = 27$$

Thus there are 27 milligrams of cholesterol in 1 cup of milk.

Practice Problem 7   A copper pipe that is $19\frac{1}{4}$ feet long will be cut into 14 equal pieces. How long will each piece be?

Developing Your Study Skills

Why Is Review Necessary?

You master a course in mathematics by learning the concepts one step at a time. There are basic concepts like addition, subtraction, multiplication, and division of whole numbers that are considered the foundation upon which all of mathematics is built. These must be mastered first. Then the study of mathematics is built step by step upon this foundation, each step supporting the next. The process is a carefully designed procedure, and so no steps can be skipped. A student of mathematics needs to realize the importance of this building process to succeed.

Because learning new concepts depends on those previously learned, students often need to take time to review. The reviewing process will strengthen the understanding and application of concepts that are weak due to lack of mastery or passage of time. Review at the right time on the right concepts can strengthen previously learned skills and make progress possible.

Timely, periodic review of previously learned mathematical concepts is absolutely necessary in order to master new concepts. You may have forgotten a concept or grown a bit rusty in applying it. Reviewing is the answer. Make use of any review sections in your textbook, whether they are assigned or not. Look back to previous chapters whenever you have forgotten how to do something. Study the examples and practice some exercises to refresh your understanding.

Be sure that you understand and can perform the computations of each new concept. This will enable you to be able to move successfully on to the next ones.

Make sure all fractions are simplified in the final answer.

Verbal and Writing Skills

1. In your own words explain how to remember that when you divide two fractions you invert the *second* fraction and multiply by the first. How can you be sure that you don't invert the *first* fraction by mistake?

2. Explain why $2 \div \frac{1}{3}$ is a larger number than $2 \div \frac{1}{2}$.

Divide, if possible.

3. $\frac{7}{8} \div \frac{2}{3}$

4. $\frac{3}{13} \div \frac{9}{26}$

5. $\frac{2}{3} \div \frac{4}{27}$

6. $\frac{5}{16} \div \frac{3}{8}$

7. $\frac{5}{9} \div \frac{10}{27}$

8. $\frac{8}{15} \div \frac{24}{35}$

9. $\frac{4}{5} \div 1$

10. $1 \div \frac{3}{7}$

11. $\frac{3}{11} \div 4$

12. $2 \div \frac{7}{8}$

13. $\frac{2}{9} \div \frac{1}{6}$

14. $\frac{3}{4} \div \frac{2}{3}$

15. $\frac{4}{15} \div \frac{4}{15}$

16. $\frac{2}{7} \div \frac{2}{7}$

17. $\frac{3}{7} \div \frac{7}{3}$

18. $\frac{11}{12} \div \frac{1}{5}$

19. $\frac{4}{3} \div \frac{7}{27}$

20. $\frac{8}{9} \div \frac{5}{81}$

21. $0 \div \frac{3}{17}$

22. $0 \div \frac{5}{16}$

23. $\frac{18}{19} \div 0$

24. $\frac{24}{29} \div 0$

25. $\frac{9}{16} \div \frac{3}{4}$

26. $\frac{3}{4} \div \frac{9}{16}$

27. $\frac{3}{7} \div \frac{15}{28}$

28. $\frac{5}{6} \div \frac{15}{18}$

29. $\frac{10}{25} \div \frac{20}{50}$

30. $\frac{3}{25} \div \frac{24}{125}$

31. $8 \div \frac{4}{5}$

32. $16 \div \frac{8}{11}$

33. $\dfrac{7}{8} \div 4$ **34.** $\dfrac{5}{6} \div 12$ **35.** $6000 \div \dfrac{6}{5}$ **36.** $8000 \div \dfrac{4}{7}$ **37.** $\dfrac{\frac{3}{5}}{6}$

38. $\dfrac{\frac{5}{9}}{10}$ **39.** $\dfrac{\frac{5}{8}}{\frac{25}{7}}$ **40.** $\dfrac{\frac{3}{16}}{\frac{5}{8}}$

Multiply or divide.

41. $3\dfrac{1}{5} \div \dfrac{3}{10}$ **42.** $2\dfrac{1}{8} \div \dfrac{1}{4}$ **43.** $2\dfrac{1}{3} \times \dfrac{1}{6}$ **44.** $6\dfrac{1}{2} \times \dfrac{1}{3}$

45. $5\dfrac{1}{4} \div 2\dfrac{5}{8}$ **46.** $1\dfrac{2}{9} \div 4\dfrac{1}{3}$ **47.** $5 \div 1\dfrac{1}{4}$ **48.** $7 \div 1\dfrac{2}{5}$

49. $12\dfrac{1}{2} \div 5\dfrac{5}{6}$ **50.** $14\dfrac{2}{3} \div 3\dfrac{1}{2}$ **51.** $8\dfrac{1}{4} \div 2\dfrac{3}{4}$ **52.** $2\dfrac{3}{8} \div 5\dfrac{3}{7}$

53. $3\dfrac{1}{2} \times \dfrac{9}{16}$ **54.** $1\dfrac{1}{8} \times \dfrac{3}{8}$ **55.** $3\dfrac{3}{4} \div 9$ **56.** $5\dfrac{5}{6} \div 7$

57. $\dfrac{3\frac{1}{2}}{\frac{1}{4}}$ **58.** $\dfrac{5\frac{3}{4}}{\frac{5}{8}}$ **59.** $\dfrac{1\frac{1}{4}}{1\frac{7}{8}}$ **60.** $\dfrac{2\frac{3}{5}}{1\frac{7}{10}}$

61. $\dfrac{\frac{7}{12}}{3\frac{2}{3}}$ **62.** $\dfrac{\frac{9}{10}}{3\frac{3}{5}}$ **63.** $\dfrac{5\frac{1}{3}}{2\frac{1}{2}}$ **64.** $\dfrac{6\frac{3}{4}}{3\frac{1}{2}}$

Review Example 6. Then find the value of x in each of the following.

65. $x \div \dfrac{4}{3} = \dfrac{21}{20}$

66. $x \div \dfrac{2}{5} = \dfrac{15}{16}$

67. $x \div \dfrac{9}{5} = \dfrac{20}{63}$

68. $x \div \dfrac{7}{3} = \dfrac{9}{28}$

Applications

Answer each question.

69. A leather factory in Morocco tans leather. In order to make the leather soft, it has to soak in a vat of uric acid and other ingredients. The main holding tank holds $20\frac{1}{4}$ gallons of the tanning mixture. If the mixture is distributed evenly into nine vats of equal size for the different colored leathers, how much will each vat hold?

70. A specially protected stretch of beach bordering the Great Barrier Reef in Australia is used for marine biology and ecological research. The beach, which is $7\frac{1}{2}$ miles long, has been broken up into 20 equal segments for comparison purposes. How long is each segment of the beach?

71. Bruce drove in a snowstorm to get to his favorite mountain to do some snowboarding. He traveled 125 miles in $3\frac{1}{3}$ hours. What was his average speed (in miles per hour)?

72. Roberto drove his truck to Cedarville, a distance of 200 miles, in $4\frac{1}{6}$ hours. What was his average speed (in miles per hour)?

73. The Patriotic Flag Company is making flags that need $3\frac{1}{8}$ yards of fabric each. The warehouse has $87\frac{1}{2}$ yards of flag fabric available. How many flags can the factory make?

74. The school cafeteria is making hamburgers for the annual Senior Day Festival. The cooks have decided that because hamburger shrinks on the grill, they will allow $\frac{2}{3}$ pound of meat for each student. If the kitchen has $38\frac{2}{3}$ pounds of meat, how many students will be fed?

75. Ellen bought 10 yards of lace to use on some dresses she is sewing. If each dress requires $1\frac{2}{3}$ yards of lace, how many dresses can she make?

76. A scientist needs to distribute 27 milliliters of a salt solution into small test tubes. If each test tube is to hold $\frac{3}{4}$ milliliter of solution, how many test tubes will he be able to fill?

77. In 1899, a time capsule was placed behind a steel wall measuring $4\frac{3}{4}$ inches thick. On December 22, 1999, a special drill was used to bore through the wall and extricate the time capsule. The drill could move only $\frac{5}{6}$ inch at a time. How many drill attempts did it take to reach the other side of the steel wall?

78. Imagination Ink supplies different colored inks for highlighter pens. Vat 1 has yellow ink, holds 150 gallons, and is $\frac{4}{5}$ full. Vat 2 has green ink, holds 50 gallons, and is $\frac{5}{8}$ full. One gallon of ink will fill 1200 pens. How many pens can be filled with the existing ink from Vats 1 and 2?

To Think About

When multiplying or dividing mixed numbers it is wise to estimate your answer by rounding each mixed number to the nearest whole number.

79. Estimate your answer to $14\frac{2}{3} \div 5\frac{1}{6}$ by rounding each mixed number to the nearest whole number. Then find the exact answer. How close was your estimate?

80. Estimate your answer to $18\frac{1}{4} \times 27\frac{1}{2}$ by rounding each mixed number to the nearest whole number. Then find the exact answer. How close was your estimate?

Cumulative Review Problems

81. Write in words. 39,576,304

82. Write in expanded form. 459,273

83. Add. $126 + 34 + 9 + 891 + 12 + 27$

84. Write in standard notation. eighty-seven million, five hundred ninety-five thousand, six hundred thirty-one

1. _____

2. _____

3. _____

4. _____

5. _____

6. _____

7. _____

8. _____

9. _____

10. _____

11. _____

12. _____

13. _____

14. _____

15. _____

16. _____

17. _____

18. _____

19. _____

20. _____

21. _____

22. _____

Solve. Make sure all fractions are simplified in the final answer.

1. Maria scored 23 out of 32 problems correct on the math final exam. Write a fraction that describes what part of the exam she completed correctly.

2. Carlos inspected the boxes that were shipped from the central warehouse. He found that 340 were the correct weight and 112 were not. Write a fraction that describes what part of the total number of the boxes were at the correct weight.

Reduce each fraction.

3. $\frac{19}{38}$

4. $\frac{12}{15}$

5. $\frac{24}{66}$

6. $\frac{125}{155}$

7. $\frac{39}{52}$

8. $\frac{51}{34}$

Change each mixed number to an improper fraction.

9. $3\frac{7}{12}$

10. $4\frac{1}{8}$

Change each improper fraction to a mixed number.

11. $\frac{45}{7}$

12. $\frac{33}{4}$

Multiply.

13. $\frac{4}{5} \times \frac{2}{7}$

14. $\frac{15}{7} \times \frac{3}{5}$

15. $18 \times \frac{5}{6}$

16. $\frac{3}{8} \times 44$

17. $2\frac{1}{3} \times 5\frac{3}{4}$

18. $1\frac{3}{7} \times 3\frac{1}{3}$

Divide.

19. $\frac{4}{7} \div \frac{3}{4}$

20. $\frac{8}{9} \div \frac{1}{6}$

21. $5\frac{1}{4} \div \frac{3}{4}$

22. $5\frac{3}{5} \div \frac{1}{2}$

Mixed Practice

Perform the indicated operations. Simplify your answers.

23. $2\dfrac{1}{4} \times 3\dfrac{1}{2}$

24. $6 \times 2\dfrac{1}{3}$

25. $5 \div 1\dfrac{7}{8}$

26. $5\dfrac{3}{4} \div 2$

27. $\dfrac{13}{20} \div \dfrac{4}{5}$

28. $\dfrac{4}{7} \div 8$

29. $\dfrac{9}{22} \times \dfrac{11}{16}$

30. $\dfrac{14}{25} \times \dfrac{65}{42}$

Solve. Simplify your answer.

▲ 31. A garden measures $5\dfrac{1}{4}$ feet by $8\dfrac{3}{4}$ feet. What is the area of the garden in square feet?

32. A recipe for two loaves of bread calls for $2\dfrac{2}{3}$ cups of flour. Sandra wants to make $1\dfrac{1}{2}$ times as much bread. How many cups of flour will she need?

33. Robert walked $4\dfrac{1}{2}$ miles yesterday. One-third of the trip was walking uphill. How many miles did he walk uphill?

34. The butcher prepared $12\dfrac{3}{8}$ pounds of lean ground round. He placed it in packages that held $\dfrac{3}{4}$ of a pound. How many full packages did he have? How much lean ground round was left over?

35. The college computer center has 136 computers. Samuel found that $\dfrac{3}{8}$ of them have Windows 2000 installed on them. How many computers have Windows 2000 installed on them?

36. The fire department has finished inspecting $\dfrac{3}{5}$ of the homes in the city to determine if the smoke detectors in these homes are functioning properly. 12,000 homes have been inspected. How many homes still need to be inspected?

37. Yung Kim was paid $132 last week at his part-time job. He was paid $8\dfrac{1}{4}$ per hour. How many hours did he work last week?

38. The Outdoor Shop is making some custom tents that are very light but totally waterproof. Each tent requires $8\dfrac{1}{4}$ yards of cloth. How many tents can be made from $56\dfrac{1}{2}$ yards of cloth? How much cloth will be left over?

39. Matthew has a bottle of barbecue sauce that holds 12 ounces. Matthew uses $\dfrac{9}{120}$ of an ounce every day. How many days will it take for Matthew to use up the bottle?

23.
24.
25.
26.
27.
28.
29.
30.
31.
32.
33.
34.
35.
36.
37.
38.
39.

2.6 The Least Common Denominator and Building Up Fractions

Student Learning Objectives

After studying this section, you will be able to:

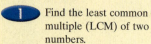

1 Find the least common multiple (LCM) of two numbers.

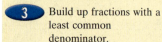

2 Find the least common denominator given two or three fractions.

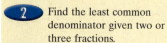

3 Build up fractions with a least common denominator.

SSM
PH TUTOR CENTER
CD & VIDEO
MATH PRO
WEB

1 Finding the Least Common Multiple (LCM) of Two Numbers

The idea of a multiple of a number is fairly straightforward. The **multiples** of a number are the products of that number and the numbers 1, 2, 3, 4, 5, 6, 7, ...

> For example, the multiples of 4 are   4, 8, 12, 16, 20, 24, 28, ...
> The multiples of 5 are   5, 10, 15, 20, 25, 30, 35, ...

The **least common multiple**, or **LCM**, of two natural numbers is the smallest number that is a multiple of both.

EXAMPLE 1 Find the least common multiple of 10 and 12.

> The multiples of 10 are   10,   20,   30,   40,   50,   60,   70, ...
> The multiples of 12 are   12,   24,   36,   48,   60,   72,   84, ...

The first multiple that appears on both lists is the least common multiple. Thus the number 60 is the least common multiple of 10 and 12.

Practice Problem 1   Find the least common multiple of 14 and 21.

EXAMPLE 2   Find the least common multiple of 6 and 8.

> The multiples of 6 are   6,   12,   18,   24,   30,   36,   42, ...
> The multiples of 8 are   8,   16,   24,   32,   40,   48,   56, ...

The first multiple that appears on both lists is the least common multiple. Thus the number 24 is the least common multiple of 6 and 8.

Practice Problem 2   Find the least common multiple of 15 and 25.

Now of course we can do the problem immediately if the larger number is a multiple of the smaller number. In such cases the larger number is the least common multiple.

EXAMPLE 3   Find the least common multiple of 7 and 35.

> Because $7 \times 5 = 35$, therefore 35 is a multiple of 7.

So we can state immediately that the least common multiple of 7 and 35 is 35.

Practice Problem 3   Find the least common multiple of 6 and 54.

2 Finding the Least Common Denominator Given Two or Three Fractions

We need some way to determine which of two fractions is larger. Suppose that Marcia and Melissa each have some left-over pizza.

Marcia's Pizza
$\frac{1}{3}$ of a pizza left

Melissa's Pizza
$\frac{1}{4}$ of a pizza left

Who has more pizza left? How much more? Comparing the amounts of pizza left would be easy if each pizza had been cut into equal-sized pieces. If the original pizzas had each been cut into 12 pieces, we would be able to see that Marcia had $\frac{1}{12}$ of a pizza more than Melissa had.

Marcia's Pizza
$$\left(\begin{array}{c} \text{We know that} \\ \dfrac{4}{12} = \dfrac{1}{3} \text{ by reducing.} \end{array} \right)$$

Melissa's Pizza
$$\left(\begin{array}{c} \text{We know that} \\ \dfrac{3}{12} = \dfrac{1}{4} \text{ by reducing.} \end{array} \right)$$

The denominator 12 is common to the fractions $\frac{4}{12}$ and $\frac{3}{12}$. We call the smallest denominator that allows us to compare fractions directly the *least common denominator*, abbreviated LCD. The number 12 is the least common denominator for the fractions $\frac{1}{3}$ and $\frac{1}{4}$.

Notice that 12 is the least common multiple of 3 and 4.

Definition

The **least common denominator (LCD)** of two or more fractions is the smallest number that can be divided evenly by each of the fractions' denominators.

How does this relate to least common multiples? The LCD of two fractions is the least common multiple of the two denominators.

In some problems you may be able to guess the LCD quite quickly. With practice, you can often find the LCD mentally. For example, you now know that if the denominators of two fractions are 3 and 4, the LCD is 12. For the fractions $\frac{1}{2}$ and $\frac{1}{4}$, the LCD is 4; for the fractions $\frac{1}{3}$ and $\frac{1}{6}$, the LCD is 6. We can see that if the denominator of one fraction divides without remainder into the denominator of another, the LCD of the two fractions is the larger of the denominators.

● **EXAMPLE 4**   Determine the LCD for each pair of fractions.

(a) $\dfrac{7}{15}$ and $\dfrac{4}{5}$ **(b)** $\dfrac{2}{3}$ and $\dfrac{5}{27}$

(a) Since 5 can be divided into 15, the LCD of $\dfrac{7}{15}$ and $\dfrac{4}{5}$ is 15. (Notice that the least common multiple of 5 and 15 is 15.)

(b) Since 3 can be divided into 27, the LCD of $\dfrac{2}{3}$ and $\dfrac{5}{27}$ is 27. (Notice that the least common multiple of 3 and 27 is 27.)

Practice Problem 4   Determine the LCD for each pair of fractions.

(a) $\dfrac{3}{4}$ and $\dfrac{11}{12}$ **(b)** $\dfrac{1}{7}$ and $\dfrac{8}{35}$ ●

In a few cases, the LCD is the product of the two denominators.

EXAMPLE 5   Find the LCD for $\frac{1}{4}$ and $\frac{3}{5}$.

We see that $4 \times 5 = 20$. Also, 20 is the *smallest* number that can be divided without remainder by 4 and by 5. We know this because the least common multiple of 4 and 5 is 20. So the LCD $= 20$.

Practice Problem 5   Find the LCD for $\frac{3}{7}$ and $\frac{5}{6}$.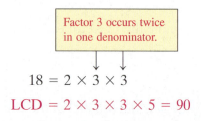

In cases where the LCD is not obvious, the following procedure will help us find the LCD.

Three-Step Procedure for Finding the Least Common Denominator

1. Write each denominator as the product of prime factors.
2. List all the prime factors that appear in either product.
3. Form a product of those prime factors, using each factor the greatest number of times it appears in any one denominator.

EXAMPLE 6   Find the LCD by the three-step procedure.

(a) $\frac{5}{6}$ and $\frac{4}{15}$ **(b)** $\frac{7}{18}$ and $\frac{7}{30}$ **(c)** $\frac{10}{27}$ and $\frac{5}{18}$

(a) Step 1 Write each denominator as a product of prime factors.

$$6 = 2 \times 3 \qquad 15 = 5 \times 3$$

Step 2 The LCD will contain the factors 2, 3, and 5.

$$6 = 2 \times 3 \qquad 15 = 5 \times 3$$

Step 3 LCD $= 2 \times 3 \times 5$ We form a product.

$$= 30$$

(b) Step 1 Write each denominator as a product of prime factors.

$$18 = 2 \times 9 = 2 \times 3 \times 3$$
$$30 = 3 \times 10 = 2 \times 3 \times 5$$

Step 2 The LCD will be a product containing 2, 3, and 5.

Step 3 The LCD will contain the factor 3 twice since it occurs twice in the denominator 18.

> Factor 3 occurs twice in one denominator.

$$18 = 2 \times 3 \times 3$$

$$\text{LCD} = 2 \times 3 \times 3 \times 5 = 90$$

(c) Write each denominator as a product of prime factors.

$$27 = 3 \times 3 \times 3 \qquad 18 = 3 \times 3 \times 2$$

Factor 3 occurs three times.

The LCD will contain the factor 2 once but the factor 3 three times.

$$LCD = 2 \times 3 \times 3 \times 3 = 54$$

Practice Problem 6 Find the LCD for each pair of fractions.

(a) $\dfrac{3}{14}$ and $\dfrac{1}{10}$ **(b)** $\dfrac{1}{15}$ and $\dfrac{7}{50}$ **(c)** $\dfrac{3}{16}$ and $\dfrac{5}{12}$

A similar procedure can be used for three fractions.

EXAMPLE 7 Find the LCD of $\dfrac{7}{12}, \dfrac{1}{15}$, and $\dfrac{11}{30}$.

$$
\begin{aligned}
12 &= 2 \times 2 \times 3 \\
15 &= \qquad\quad 3 \times 5 \\
30 &= \qquad 2 \times 3 \times 5 \\
LCD &= 2 \times 2 \times 3 \times 5 \\
&= 60
\end{aligned}
$$

Practice Problem 7 Find the LCD of $\dfrac{3}{49}, \dfrac{5}{21}$, and $\dfrac{6}{7}$.

3 ⬭ Building Up Fractions with a Least Common Denominator

We cannot add fractions with unlike denominators. To change denominators, we must (1) find the LCD and (2) build up the addends—the fractions being added—into equivalent fractions that have the LCD as the denominator. We know now how to find the LCD. Let's look at how we build up fractions. We know, for example, that

$$\frac{1}{2} = \frac{2}{4} = \frac{50}{100} \qquad \frac{1}{4} = \frac{25}{100} \qquad \text{and} \qquad \frac{3}{4} = \frac{75}{100}.$$

In these cases, we have mentally multiplied the given fraction by 1, in the form of a certain number, c, in the numerator and that same number, c, in the denominator.

$$\frac{1}{2} \times \boxed{\frac{c}{c}} = \frac{2}{4} \qquad \text{Here } c = 2, \frac{2}{2} = 1.$$

$$\frac{1}{2} \times \boxed{\frac{c}{c}} = \frac{50}{100} \qquad \text{Here } c = 50, \frac{50}{50} = 1.$$

This property is called the *building fraction property*.

Building Fraction Property

For whole numbers a, b, and c where $b \neq 0, c \neq 0$,

$$\frac{a}{b} = \frac{a}{b} \times 1 = \frac{a}{b} \times \boxed{\frac{c}{c}} = \frac{a \times c}{b \times c}.$$

EXAMPLE 8 Build each fraction to an equivalent fraction with the LCD.

(a) $\frac{3}{4}$, LCD = 28   **(b)** $\frac{4}{5}$, LCD = 45   **(c)** $\frac{1}{3}$ and $\frac{4}{5}$, LCD = 15

(a) $\frac{3}{4} \times \boxed{\frac{c}{c}} = \frac{?}{28}$ We know that $4 \times 7 = 28$, so the value c that we multiply numerator and denominator by is 7.

$\frac{3}{4} \times \frac{7}{7} = \frac{21}{28}$

(b) $\frac{4}{5} \times \boxed{\frac{c}{c}} = \frac{?}{45}$ We know that $5 \times 9 = 45$, so $c = 9$.

$\frac{4}{5} \times \frac{9}{9} = \frac{36}{45}$

(c) $\frac{1}{3} = \frac{?}{15}$ We know that $3 \times 5 = 15$, so we multiply numerator and denominator by 5.

$\frac{1}{3} \times \boxed{\frac{5}{5}} = \frac{5}{15}$

$\frac{4}{5} = \frac{?}{15}$ We know that $5 \times 3 = 15$, so we multiply numerator and denominator by 3.

$\frac{4}{5} \times \frac{3}{3} = \frac{12}{15}$

Thus $\frac{1}{3} = \frac{5}{15}$ and $\frac{4}{5} = \frac{12}{15}$.

Practice Problem 8 Build each fraction to an equivalent fraction with the LCD.

(a) $\frac{3}{5}$, LCD = 40   **(b)** $\frac{7}{11}$, LCD = 44   **(c)** $\frac{2}{7}$ and $\frac{3}{4}$, LCD = 28

⬬ EXAMPLE 9

(a) Find the LCD of $\dfrac{1}{32}$ and $\dfrac{7}{48}$.

(b) Build the fractions to equivalent fractions that have the LCD as their denominators.

(a) First we find the prime factors of 32 and 48.

$$32 = 2 \times 2 \times 2 \times 2 \times 2$$
$$48 = 2 \times 2 \times 2 \times 2 \times 3$$

Thus the LCD will require a factor of 2 five times and a factor of 3 one time.

$$\text{LCD} = 2 \times 2 \times 2 \times 2 \times 2 \times 3 = 96$$

(b) $\dfrac{1}{32} = \dfrac{?}{96}$ Since $32 \times 3 = 96$ we multiply by the fraction $\dfrac{3}{3}$.

$$\frac{1}{32} = \frac{1}{32} \times \boxed{\frac{3}{3}} = \frac{3}{96}$$

$\dfrac{7}{48} = \dfrac{?}{96}$ Since $48 \times 2 = 96$, we multiply by the fraction $\dfrac{2}{2}$.

$$\frac{7}{48} = \frac{7}{48} \times \boxed{\frac{2}{2}} = \frac{14}{96}$$

Practice Problem 9

(a) Find the LCD of $\dfrac{3}{20}$ and $\dfrac{11}{15}$.

(b) Build the fractions to equivalent fractions that have the LCD as their denominators.   ⬬

Find the least common multiple (LCM) for each pair of numbers.

1. 8 and 14 **2.** 6 and 20 **3.** 20 and 50 **4.** 22 and 55 **5.** 12 and 15

6. 18 and 30 **7.** 9 and 36 **8.** 8 and 72 **9.** 21 and 49 **10.** 25 and 35

Find the LCD for each pair of fractions.

11. $\dfrac{1}{5}$ and $\dfrac{3}{10}$ **12.** $\dfrac{3}{8}$ and $\dfrac{5}{16}$ **13.** $\dfrac{3}{7}$ and $\dfrac{1}{4}$ **14.** $\dfrac{5}{6}$ and $\dfrac{3}{5}$ **15.** $\dfrac{2}{5}$ and $\dfrac{3}{7}$

16. $\dfrac{1}{16}$ and $\dfrac{2}{3}$ **17.** $\dfrac{1}{6}$ and $\dfrac{5}{9}$ **18.** $\dfrac{1}{4}$ and $\dfrac{3}{14}$ **19.** $\dfrac{7}{12}$ and $\dfrac{14}{15}$ **20.** $\dfrac{7}{15}$ and $\dfrac{9}{25}$

21. $\dfrac{1}{16}$ and $\dfrac{3}{4}$ **22.** $\dfrac{2}{11}$ and $\dfrac{1}{44}$ **23.** $\dfrac{5}{10}$ and $\dfrac{11}{45}$ **24.** $\dfrac{13}{20}$ and $\dfrac{17}{30}$ **25.** $\dfrac{7}{12}$ and $\dfrac{7}{30}$

26. $\dfrac{5}{6}$ and $\dfrac{7}{15}$ **27.** $\dfrac{5}{21}$ and $\dfrac{8}{35}$ **28.** $\dfrac{1}{20}$ and $\dfrac{7}{8}$ **29.** $\dfrac{5}{18}$ and $\dfrac{11}{12}$ **30.** $\dfrac{1}{24}$ and $\dfrac{7}{40}$

Find the LCD for each set of three fractions.

31. $\dfrac{2}{3}, \dfrac{1}{2}, \dfrac{5}{6}$ **32.** $\dfrac{1}{5}, \dfrac{1}{3}, \dfrac{7}{10}$ **33.** $\dfrac{5}{24}, \dfrac{11}{15}, \dfrac{7}{30}$ **34.** $\dfrac{5}{18}, \dfrac{5}{12}, \dfrac{13}{20}$ **35.** $\dfrac{5}{11}, \dfrac{7}{12}, \dfrac{1}{6}$

36. $\dfrac{11}{16}, \dfrac{3}{20}, \dfrac{2}{5}$ **37.** $\dfrac{7}{12}, \dfrac{1}{21}, \dfrac{3}{14}$ **38.** $\dfrac{1}{30}, \dfrac{3}{40}, \dfrac{7}{8}$ **39.** $\dfrac{7}{15}, \dfrac{11}{12}, \dfrac{7}{8}$ **40.** $\dfrac{5}{36}, \dfrac{2}{48}, \dfrac{1}{24}$

Build each fraction to an equivalent fraction with the denominator specified. State the numerator.

41. $\dfrac{1}{3} = \dfrac{?}{9}$ **42.** $\dfrac{1}{5} = \dfrac{?}{35}$ **43.** $\dfrac{5}{6} = \dfrac{?}{54}$ **44.** $\dfrac{4}{7} = \dfrac{?}{28}$ **45.** $\dfrac{4}{11} = \dfrac{?}{55}$ **46.** $\dfrac{2}{13} = \dfrac{?}{39}$

47. $\dfrac{7}{24} = \dfrac{?}{48}$ **48.** $\dfrac{3}{50} = \dfrac{?}{100}$ **49.** $\dfrac{8}{9} = \dfrac{?}{108}$ **50.** $\dfrac{6}{7} = \dfrac{?}{147}$ **51.** $\dfrac{7}{20} = \dfrac{?}{180}$ **52.** $\dfrac{15}{32} = \dfrac{?}{192}$

The LCD of each pair of fractions is listed. Build each fraction to an equivalent fraction that has the LCD as the denominator.

53. LCD $= 36, \dfrac{7}{12}$ and $\dfrac{5}{9}$ **54.** LCD $= 20, \dfrac{9}{10}$ and $\dfrac{3}{4}$ **55.** LCD $= 200, \dfrac{3}{25}$ and $\dfrac{7}{40}$

56. LCD $= 72, \dfrac{5}{24}$ and $\dfrac{7}{36}$ **57.** LCD $= 160, \dfrac{9}{10}$ and $\dfrac{19}{20}$ **58.** LCD $= 240, \dfrac{13}{30}$ and $\dfrac{41}{80}$

Find the LCD. Build up the fractions to equivalent fractions having the LCD as the denominator.

59. $\dfrac{2}{5}$ and $\dfrac{9}{35}$ **60.** $\dfrac{7}{9}$ and $\dfrac{35}{54}$ **61.** $\dfrac{5}{12}$ and $\dfrac{1}{16}$ **62.** $\dfrac{7}{15}$ and $\dfrac{4}{25}$ **63.** $\dfrac{13}{20}$ and $\dfrac{11}{16}$

64. $\dfrac{9}{12}$ and $\dfrac{13}{18}$ **65.** $\dfrac{4}{15}$ and $\dfrac{5}{12}$ **66.** $\dfrac{9}{10}$ and $\dfrac{3}{25}$ **67.** $\dfrac{5}{24}, \dfrac{11}{36}, \dfrac{3}{72}$ **68.** $\dfrac{1}{30}, \dfrac{7}{15}, \dfrac{1}{45}$

69. $\dfrac{3}{56}, \dfrac{7}{8}, \dfrac{5}{7}$ **70.** $\dfrac{5}{9}, \dfrac{1}{6}, \dfrac{3}{54}$ **71.** $\dfrac{5}{63}, \dfrac{4}{21}, \dfrac{8}{9}$ **72.** $\dfrac{3}{8}, \dfrac{5}{14}, \dfrac{13}{16}$

Applications

73. Suppose that you wish to compare the lengths of the three portions of the given stainless steel pin that came out of a door.
(a) What is the LCD for the three fractions?

(b) Build up each fraction to an equivalent fraction that has the LCD as a denominator.

74. Suppose that you want to prepare a report on a plant that grew the given lengths during each week of a three-week experiment.
(a) What is the LCD for the three fractions?

(b) Build up each fraction to an equivalent fraction that has the LCD for a denominator.

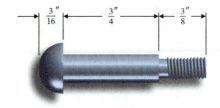

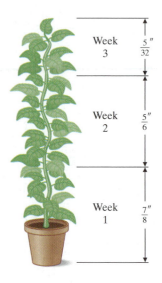

Cumulative Review Problems

75. Divide. $32\overline{)5699}$

76. Divide. $182\overline{)659,568}$

77. Multiply. 369×27

Use the following bar graph for exercises 78–83.

78. What five-year period saw the largest increase in the poverty level for a family of four?

79. What five-year period saw the smallest increase in the poverty level for a family of four?

80. If a husband earned $8100 a year in 1995, what would his wife have had to earn for the family income to be above the poverty level?

81. If a wife earned $9700 a year in 2000, what would her husband have had to earn for the family income to be above the poverty level?

82. If the same increase in the poverty level occurs between 2000 and 2005 as between 1995 and 2000, what would be the expected poverty level for a family of four in 2005?

83. If the same increase in the poverty level occurs between 2000 and 2010 as between 1990 and 2000, what would be the expected poverty level for a family of four in 2010?

2.7 Addition and Subtraction of Fractions

1 Adding and Subtracting Fractions with a Common Denominator

You must have common denominators (denominators that are alike) to add or subtract fractions.

If your problem has fractions without a common denominator or if it has mixed numbers, you must use what you already know about changing the form of each fraction (how the fraction looks). Only after all the fractions have a common denominator can you add or subtract.

An important distinction: You must have common denominators to add or subtract fractions, but you need not have common denominators to multiply or divide fractions.

To add two fractions that have the same denominator, add the numerators and write the sum over the common denominator.

To illustrate we use $\frac{1}{5} + \frac{2}{5} = \frac{3}{5}$. The figure shows that $\frac{1}{5} + \frac{2}{5} = \frac{3}{5}$.

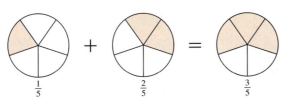

Student Learning Objectives

After studying this section, you will be able to:

 Add and subtract fractions with a common denominator.

 Add and subtract fractions without a common denominator.

SSM   PH TUTOR   CD & VIDEO   MATH PRO   WEB
CENTER

● EXAMPLE 1   Add. $\dfrac{5}{13} + \dfrac{7}{13}$

$$\frac{5}{13} + \frac{7}{13} = \frac{12}{13}$$

Practice Problem 1   Add. $\dfrac{3}{17} + \dfrac{12}{17}$

The answer may need to be reduced. Sometimes the answer may be written as a mixed number.

● EXAMPLE 2   Add.   **(a)** $\dfrac{4}{9} + \dfrac{2}{9}$ **(b)** $\dfrac{5}{7} + \dfrac{6}{7}$

(a) $\dfrac{4}{9} + \dfrac{2}{9} = \dfrac{6}{9} = \dfrac{2}{3}$ **(b)** $\dfrac{5}{7} + \dfrac{6}{7} = \dfrac{11}{7}$ or $1\dfrac{4}{7}$

Practice Problem 2   Add.   **(a)** $\dfrac{1}{12} + \dfrac{5}{12}$ **(b)** $\dfrac{13}{15} + \dfrac{7}{15}$

A similar rule is followed for subtraction, except that the numerators are subtracted and the result placed over a common denominator. Be sure to reduce all answers when possible.

● EXAMPLE 3   Subtract.   **(a)** $\dfrac{5}{13} - \dfrac{4}{13}$ **(b)** $\dfrac{17}{20} - \dfrac{3}{20}$

(a) $\dfrac{5}{13} - \dfrac{4}{13} = \dfrac{1}{13}$ **(b)** $\dfrac{17}{20} - \dfrac{3}{20} = \dfrac{14}{20} = \dfrac{7}{10}$

Practice Problem 3 Subtract. **(a)** $\dfrac{5}{19} - \dfrac{2}{19}$ **(b)** $\dfrac{21}{25} - \dfrac{6}{25}$

2 Adding and Subtracting Fractions without a Common Denominator

If the two fractions do not have a common denominator, we follow the procedure in Section 2.6: Find the LCD and then build up each fraction so that its denominator is the LCD.

EXAMPLE 4 Add. $\dfrac{7}{12} + \dfrac{1}{4}$

The LCD is 12. The fraction $\frac{7}{12}$ already has the least common denominator.

$$
\begin{array}{rcl}
\dfrac{7}{12} & = & \dfrac{7}{12} \\[2mm]
+\dfrac{1}{4} \times \dfrac{3}{3} & = +& \dfrac{3}{12} \\[2mm]
& & \dfrac{10}{12}
\end{array}
$$

We will need to reduce this fraction. Then we will have

$$\frac{7}{12} + \frac{1}{4} = \frac{7}{12} + \frac{3}{12} = \frac{10}{12} = \frac{5}{6}.$$

It is very important to remember to reduce our final answer.

Practice Problem 4 Add. $\dfrac{2}{15} + \dfrac{2}{5}$

EXAMPLE 5 Add. $\dfrac{7}{20} + \dfrac{4}{15}$
LCD $= 60$.

$$\frac{7}{20} \times \frac{3}{3} = \frac{21}{60} \qquad\qquad \frac{4}{15} \times \frac{4}{4} = \frac{16}{60}$$

Thus

$$\frac{7}{20} + \frac{4}{15} = \frac{21}{60} + \frac{16}{60} = \frac{37}{60}$$

Practice Problem 5 Add. $\dfrac{5}{12} + \dfrac{5}{16}$

A similar procedure holds for the addition of three or more fractions.

EXAMPLE 6 Add. $\dfrac{3}{8} + \dfrac{5}{6} + \dfrac{1}{4}$
LCD $= 24$.

$$\frac{3}{8} \times \frac{3}{3} = \frac{9}{24} \qquad \frac{5}{6} \times \frac{4}{4} = \frac{20}{24} \qquad \frac{1}{4} \times \frac{6}{6} = \frac{6}{24}$$

$$\frac{3}{8} + \frac{5}{6} + \frac{1}{4} = \frac{9}{24} + \frac{20}{24} + \frac{6}{24} = \frac{35}{24} \quad \text{or} \quad 1\frac{11}{24}$$

Practice Problem 6   Add. $\dfrac{3}{16} + \dfrac{1}{8} + \dfrac{1}{12}$

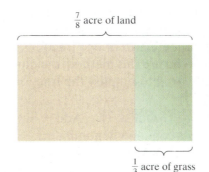

● **EXAMPLE 7**   Subtract. $\dfrac{17}{25} - \dfrac{3}{35}$

LCD = 175.

$$\dfrac{17}{25} \times \dfrac{7}{7} = \dfrac{119}{175} \qquad \dfrac{3}{35} \times \dfrac{5}{5} = \dfrac{15}{175}$$

Thus

$$\dfrac{17}{25} - \dfrac{3}{35} = \dfrac{119}{175} - \dfrac{15}{175} = \dfrac{104}{175}$$

Practice Problem 7   Subtract. $\dfrac{9}{48} - \dfrac{5}{32}$

● **EXAMPLE 8**   John and Nancy have a house on $\frac{7}{8}$ acre of land. They have $\frac{1}{3}$ acre of land planted with grass. How much of the land is not planted with grass?

1. *Understand the problem.*
   Draw a picture.

$\frac{7}{8}$ acre of land

$\frac{1}{3}$ acre of grass

   We need to subtract. $\dfrac{7}{8} - \dfrac{1}{3}$

2. *Solve and state the answer.*
   The LCD is 24.

$$\dfrac{7}{8} \times \dfrac{3}{3} = \dfrac{21}{24} \qquad \dfrac{1}{3} \times \dfrac{8}{8} = \dfrac{8}{24}$$

$$\dfrac{7}{8} - \dfrac{1}{3} = \dfrac{21}{24} - \dfrac{8}{24} = \dfrac{13}{24}$$

We conclude that $\dfrac{13}{24}$ acre of the land is not planted with grass.

3. *Check.*
   The check is left to the student.

Practice Problem 8   Leon had $\frac{9}{10}$ gallon of cleaning fluid in the garage. He used $\frac{1}{4}$ gallon to clean the garage floor. How much cleaning fluid is left?

Some students may find Example 9 difficult. Read it slowly and carefully.

🔴 **EXAMPLE 9**   Find the value of x in the equation $x + \frac{5}{6} = \frac{9}{10}$. Reduce your answer.

The LCD for the two fractions $\dfrac{5}{6}$ and $\dfrac{9}{10}$ is 30.

$$\frac{5}{6} \times \frac{5}{5} = \frac{25}{30} \qquad \frac{9}{10} \times \frac{3}{3} = \frac{27}{30}$$

Thus we can write the equation in the equivalent form.

$$x + \frac{25}{30} = \frac{27}{30}$$

The denominators are the same. Look at the numerators. We must add 2 to 25 to get 27.

$$\frac{2}{30} + \frac{25}{30} = \frac{27}{30}$$

So $x = \frac{2}{30}$ and we reduce the fraction to obtain $x = \frac{1}{15}$.

Practice Problem 9   Find the value of x in the equation $x + \frac{3}{10} = \frac{23}{25}$. 🔴

Alternate Method

In all the problems in this section so far, we have combined two fractions by first finding the least common denominator. However, there is an alternate approach. You are only required to find a common denominator, not necessarily the least common denominator. One way to quickly find a common denominator of two fractions is to multiply the two denominators. However, if you use this method, the numbers will usually be larger and you will usually need to simplify the fraction in your final answer.

🔴 **EXAMPLE 10**   Add $\frac{11}{12} + \frac{13}{30}$ by using the product of the two denominators as a common denominator.

Using this method we just multiply the numerator and denominator of each fraction by the denominator of the other fraction. Thus no steps are needed to determine what to multiply by.

$$\frac{11}{12} \times \frac{30}{30} = \frac{330}{360} \qquad \frac{13}{30} \times \frac{12}{12} = \frac{156}{360}$$

Thus $\quad \dfrac{11}{12} + \dfrac{13}{30} = \dfrac{330}{360} + \dfrac{156}{360} = \dfrac{486}{360}$

We must reduce the fraction: $\quad \dfrac{486}{360} = \dfrac{27}{20} \quad$ or $\quad 1\dfrac{7}{20}$

Practice Problem 10   Add $\frac{15}{16} + \frac{3}{40}$ by using the product of the two denominators as a common denominator. 🔴

Some students find this alternate method helpful because you do not have to find the LCD or the number each fraction must be multiplied by. Other students find this alternate method more difficult because of errors encountered when working with large numbers in reducing the final answer. You are encouraged to try a couple of the homework exercises by this method and make up your own mind.

Add or subtract. Simplify all answers.

1. $\dfrac{5}{9} + \dfrac{2}{9}$

2. $\dfrac{5}{8} + \dfrac{2}{8}$

3. $\dfrac{5}{18} + \dfrac{7}{18}$

4. $\dfrac{1}{25} + \dfrac{4}{25}$

5. $\dfrac{5}{24} - \dfrac{3}{24}$

6. $\dfrac{21}{23} - \dfrac{1}{23}$

7. $\dfrac{53}{88} - \dfrac{19}{88}$

8. $\dfrac{103}{110} - \dfrac{3}{110}$

Add or subtract. Simplify all answers.

9. $\dfrac{2}{7} + \dfrac{1}{2}$

10. $\dfrac{1}{4} + \dfrac{1}{3}$

11. $\dfrac{3}{10} + \dfrac{3}{20}$

12. $\dfrac{4}{9} + \dfrac{1}{6}$

13. $\dfrac{1}{8} + \dfrac{3}{4}$

14. $\dfrac{5}{16} + \dfrac{1}{2}$

15. $\dfrac{4}{5} + \dfrac{7}{20}$

16. $\dfrac{2}{3} + \dfrac{4}{7}$

17. $\dfrac{3}{10} + \dfrac{7}{100}$

18. $\dfrac{13}{100} + \dfrac{7}{10}$

19. $\dfrac{3}{25} + \dfrac{1}{35}$

20. $\dfrac{3}{15} + \dfrac{1}{25}$

21. $\dfrac{7}{8} + \dfrac{5}{12}$

22. $\dfrac{5}{6} + \dfrac{7}{8}$

23. $\dfrac{3}{8} + \dfrac{3}{10}$

24. $\dfrac{12}{35} + \dfrac{1}{10}$

25. $\dfrac{5}{12} - \dfrac{1}{6}$

26. $\dfrac{37}{20} - \dfrac{2}{5}$

27. $\dfrac{3}{7} - \dfrac{1}{5}$

28. $\dfrac{7}{8} - \dfrac{5}{6}$

29. $\dfrac{7}{9} - \dfrac{2}{27}$

30. $\dfrac{19}{40} - \dfrac{1}{10}$

31. $\dfrac{5}{12} - \dfrac{7}{30}$

32. $\dfrac{9}{24} - \dfrac{3}{8}$

33. $\dfrac{11}{12} - \dfrac{2}{3}$

34. $\dfrac{7}{10} - \dfrac{2}{5}$

35. $\dfrac{16}{25} - \dfrac{2}{5}$

36. $\dfrac{11}{18} - \dfrac{1}{9}$

37. $\dfrac{5}{12} - \dfrac{5}{16}$

38. $\dfrac{2}{5} - \dfrac{2}{15}$

39. $\dfrac{10}{16} - \dfrac{5}{8}$

40. $\dfrac{5}{6} - \dfrac{10}{12}$

41. $\dfrac{23}{36} - \dfrac{2}{9}$

42. $\dfrac{2}{3} - \dfrac{1}{16}$

43. $\dfrac{4}{5} + \dfrac{1}{20} + \dfrac{3}{4}$

44. $\dfrac{1}{3} + \dfrac{2}{24} + \dfrac{1}{6}$

45. $\dfrac{5}{30} + \dfrac{3}{40} + \dfrac{1}{8}$

46. $\dfrac{1}{12} + \dfrac{3}{14} + \dfrac{4}{21}$

47. $\dfrac{7}{30} + \dfrac{2}{5} + \dfrac{5}{6}$

48. $\dfrac{1}{12} + \dfrac{5}{36} + \dfrac{32}{36}$

Study Example 9 carefully. Then find the value of x in each equation.

49. $x + \dfrac{1}{7} = \dfrac{5}{14}$

50. $x + \dfrac{1}{8} = \dfrac{7}{16}$

51. $x + \dfrac{2}{3} = \dfrac{9}{11}$

52. $x + \dfrac{3}{4} = \dfrac{17}{18}$

53. $x - \dfrac{1}{5} = \dfrac{4}{12}$

54. $x - \dfrac{2}{3} = \dfrac{3}{11}$

Applications

55. Rita is baking a cake for a dinner party. The recipe calls for $\frac{2}{3}$ cup sugar for the frosting and $\frac{3}{4}$ cup sugar for the cake. How many total cups of sugar does she need?

56. Jackie started running on Wednesday. She ran $\frac{1}{2}$ mile. On Friday she ran $\frac{2}{3}$ mile. How many miles has she run so far this week?

57. Laurie purchased $\frac{2}{3}$ pound of bananas and $\frac{5}{6}$ pound of seedless grapes. How many pounds of fruit did she purchase?

58. Mandy purchased two new steel-belted all-weather radial tires for her car. The tread depth on the new tires measures $\frac{11}{32}$ of an inch. The dealer told her that when the tires have worn down and their tread depth measures $\frac{1}{8}$ of an inch, she should replace the worn tires with new ones. How much will the tread depth decrease over the useful life of the tire?

59. Travis typed $\frac{11}{12}$ of his book report on his computer. Then he printed out $\frac{3}{5}$ of his book report on his computer printer. Suddenly, there was a power outage, and he discovered that he hadn't saved his book report before the power went off. What fractional part of the book report was lost when the power failed?

60. An infant's father knows that straight apple juice is too strong for his daughter. Her bottle is $\frac{1}{2}$ full, and he adds $\frac{1}{3}$ of a bottle of water to dilute the apple juice.
(a) How much juice is in the bottle?

(b) If she drinks $\frac{2}{5}$ of the bottle, how much is left?

61. While he was at the grocery store, Raymond purchased a box of candy for himself. On the way back to the dorm he ate $\frac{1}{4}$ of the candy. As he was putting away the groceries he ate $\frac{1}{2}$ of what was left. There are now six chocolates left in the box. How many chocolates were in the box to begin with?

62. On Saturday morning, the gasoline storage tank at Dusty's station was $\frac{11}{12}$ full. During Saturday and Sunday the attendants pumped out $\frac{2}{3}$ of what was is the tank on Saturday morning. How much remained in the tank when they opened Monday morning?

63. The manager at Fit Factory Health Club was going through his files for 2000 and discovered that only $\frac{9}{14}$ of the members actually used the club. When he checked the numbers from the previous year of 1999, he found that $\frac{5}{6}$ of the members had used the club. What fractional part of the membership represents the decrease in club usage?

64. Consuela is a collage artist. She assembles different objects on canvases to present a message to the viewer. She is working on a collage that is attached to a dart board. She places the center of a computer chip $\frac{1}{3}$ inch from the center bull's-eye. Then, continuing in a straight line from the bull's-eye, she places the center of a tea bag $\frac{5}{9}$ inch from the center of the computer chip. The last item she uses is a chewing gum wrapper. The center of the wrapper is placed $\frac{3}{4}$ inch from the center of the tea bag. How far from the bull's-eye is the center of the chewing gum wrapper?

Cumulative Review Problems

65. Reduce to lowest terms. $\dfrac{15}{85}$

66. Reduce to lowest terms. $\dfrac{38}{57}$

67. Change to a mixed number. $\dfrac{123}{16}$

68. Change to an improper fraction. $14\frac{3}{7}$

69. Multiply. $2\frac{1}{2} \times 3\frac{3}{4}$

70. Divide. $4\frac{1}{3} \div 1\frac{1}{2}$

2.8 Combining Mixed Numbers and Order of Operations

Student Learning Objectives

After studying this section, you will be able to:

1 Add mixed numbers.

2 Subtract mixed numbers.

3 Evaluate expressions using the correct order of operations.

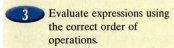

SSM
PH TUTOR CD & VIDEO MATH PRO WEB
CENTER

1 Adding Mixed Numbers

When adding mixed numbers, it is best to add the fractions together and then add the whole numbers together.

EXAMPLE 1   Add. $3\frac{1}{8} + 2\frac{5}{8}$

$$
\begin{array}{r}
3\ \dfrac{1}{8} \\[2mm]
+\,2\ \dfrac{5}{8} \\[2mm]
\hline
5\ \dfrac{6}{8}
\end{array}
$$

Add the whole numbers. $3 + 2 = 5$ → $5\dfrac{6}{8}$ ← Add the fractions. $\dfrac{1}{8} + \dfrac{5}{8} = \dfrac{6}{8}$

$= 5\ \dfrac{3}{4}$ ← Reduce $\dfrac{6}{8} = \dfrac{3}{4}$

Practice Problem 1   Add. $5\frac{1}{12} + 9\frac{5}{12}$

If the fraction portions of the mixed numbers do not have a common denominator, we must build up the fraction parts to obtain a common denominator before adding.

EXAMPLE 2   Add. $1\frac{2}{7} + 5\frac{1}{3}$

The LCD of $\dfrac{2}{7}$ and $\dfrac{1}{3}$ is 21.

$$\frac{2}{7} \times \frac{3}{3} = \frac{6}{21} \qquad \frac{1}{3} \times \frac{7}{7} = \frac{7}{21}$$

Thus $1\dfrac{2}{7} + 5\dfrac{1}{3} = 1\dfrac{6}{21} + 5\dfrac{7}{21}$.

$$
\begin{array}{r}
1\dfrac{2}{7} = \ \ 1\ \dfrac{6}{21} \\[2mm]
+\,5\dfrac{1}{3} = +5\ \dfrac{7}{21} \\[2mm]
\hline
6\ \dfrac{13}{21}
\end{array}
$$

Add the whole numbers. $1 + 5$ → $6\dfrac{13}{21}$ ← Add the fractions. $\dfrac{6}{21} + \dfrac{7}{21}$

Practice Problem 2   Add. $6\frac{1}{4} + 2\frac{2}{5}$

If the sum of the fractions is an improper fraction, we convert it to a mixed number and add the whole numbers together.

EXAMPLE 3 Add. $6\frac{5}{6} + 4\frac{3}{8}$

The LCD of $\frac{5}{6}$ and $\frac{3}{8}$ is 24.

$$
\begin{array}{rcl}
6 \;\boxed{\dfrac{5}{6} \times \dfrac{4}{4}} &=& 6 \;\boxed{\dfrac{20}{24}} \\[2ex]
+\,4 \;\boxed{\dfrac{3}{8} \times \dfrac{3}{3}} &=& +\,4 \;\boxed{\dfrac{9}{24}} \\[2ex]
\end{array}
$$

Add the whole numbers. $\rightarrow \quad 10 \;\boxed{\dfrac{29}{24}} \leftarrow$ Add the fractions.

$$= 10 + \boxed{1\frac{5}{24}} \quad \text{Since } \frac{29}{24} = 1\frac{5}{24}$$

$$= 11\frac{5}{24} \qquad \text{We add the whole numbers } 10 + 1 = 11.$$

Practice Problem 3 Add. $7\frac{1}{4} + 3\frac{5}{6}$

2 Subtracting Mixed Numbers

Subtracting mixed numbers is like adding.

EXAMPLE 4 Subtract. $8\frac{5}{7} - 5\frac{5}{14}$

The LCD of $\frac{5}{7}$ and $\frac{5}{14}$ is 14.

$$
\begin{array}{rcl}
8 \;\boxed{\dfrac{5}{7} \times \dfrac{2}{2}} &=& 8\dfrac{10}{14} \\[2ex]
-\,5\dfrac{5}{14} &=& -5\dfrac{5}{14} \\[2ex]
\end{array}
$$

Subtract the whole numbers. $\rightarrow 3\dfrac{5}{14} \leftarrow$ Subtract the fractions.

Practice Problem 4 Subtract. $12\frac{5}{6} - 7\frac{5}{12}$

Sometimes we must borrow before we can subtract.

EXAMPLE 5 Subtract.

(a) $9\frac{1}{4} - 6\frac{5}{14}$ **(b)** $15 - 9\frac{3}{16}$

The following example is fairly challenging. Read through each step carefully. Be sure to have paper and pencil handy and see if you can verify each step.

(a) The LCD of $\frac{1}{4}$ and $\frac{5}{14}$ is 28.

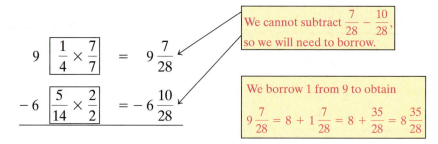

$$9 \;\boxed{\frac{1}{4} \times \frac{7}{7}} \;=\; 9\frac{7}{28}$$

We cannot subtract $\frac{7}{28} - \frac{10}{28}$, so we will need to borrow.

$$-6 \;\boxed{\frac{5}{14} \times \frac{2}{2}} \;=\; -6\frac{10}{28}$$

We borrow 1 from 9 to obtain
$$9\frac{7}{28} = 8 + 1\frac{7}{28} = 8 + \frac{35}{28} = 8\frac{35}{28}$$

We write this as

$$9\frac{7}{28} \;=\; 8\frac{35}{28}$$
$$-6\frac{10}{28} \;=\; -6\frac{10}{28}$$

$$\boxed{8 - 6 = 2} \longrightarrow 2\frac{25}{28} \longleftarrow \boxed{\frac{35}{28} - \frac{10}{28} = \frac{25}{28}}$$

(b) The LCD = 16.

$$15 \;=\; 14\frac{16}{16} \longleftarrow$$

We borrow 1 from 15 to obtain
$$15 = 14 + 1 = 14 + \frac{16}{16} = 14\frac{16}{16}$$

$$-\,9\frac{3}{16} \;=\; -\,9\frac{3}{16}$$

$$\boxed{14 - 9 = 5} \longrightarrow 5\frac{13}{16} \longleftarrow \boxed{\frac{16}{16} - \frac{3}{16} = \frac{13}{16}}$$

Practice Problem 5 Subtract. **(a)** $9\frac{1}{8} - 3\frac{2}{3}$ **(b)** $18 - 6\frac{7}{18}$

EXAMPLE 6 A plumber had a pipe $5\frac{3}{16}$ inches long for a fitting under the sink. He needed a pipe that was $3\frac{7}{8}$ inches long, so he cut the pipe down. How much of the pipe did he cut off?

We will need to subtract $5\frac{3}{16} - 3\frac{7}{8}$ to find the length that was cut off.

$$5\frac{3}{16} \;=\; 5\frac{3}{16}$$

$$-3\frac{7}{8} \times \frac{2}{2} \;=\; -3\frac{14}{16}$$

$$4\frac{19}{16} \longleftarrow$$

We borrow 1 from 5 to obtain
$$5\frac{3}{16} = 4 + 1\frac{3}{16} = 4 + \frac{19}{16}$$

$$-3\frac{14}{16}$$

$$\boxed{4 - 3 = 1} \longrightarrow 1\frac{5}{16} \longleftarrow \boxed{\frac{19}{16} - \frac{14}{16} = \frac{5}{16}}$$

The plumber had to cut off $1\frac{5}{16}$ inches of pipe.

Practice Problem 6 Hillary and Sam purchased $6\frac{1}{4}$ gallons of paint to paint the inside of their house. They used $4\frac{2}{3}$ gallons of paint. How much paint was left over?

Alternative Method

Can mixed numbers be added and subtracted as improper fractions? Yes. Recall Example 5(a).

$$9\frac{1}{4} - 6\frac{5}{14} = 2\frac{25}{28}$$

If we write $9\frac{1}{4} - 6\frac{5}{14}$ using improper fractions, we have $\frac{37}{4} - \frac{89}{14}$. Now we build up each of these improper fractions so that they both have the LCD for their denominators.

$$\frac{37}{4} \boxed{\times \frac{7}{7}} = \frac{259}{28}$$
$$-\frac{89}{14} \boxed{\times \frac{2}{2}} = -\frac{178}{28}$$
$$\frac{81}{28} = 2\frac{25}{28}$$

The same result is obtained as in Example 5(a). This method does not require borrowing. However, you do work with larger numbers. For more practice, see exercises 49–50.

Developing Your Study Skills

Problems with Accuracy

Strive for accuracy. Mistakes are often made because of human error rather than lack of understanding. Such mistakes are frustrating. A simple arithmetic or copying error can lead to an incorrect answer.

These five steps will help you cut down on errors.

1. Work carefully, and take your time. Do not rush through a problem just to get it done.

2. Concentrate on the problem. Sometimes problems become mechanical, and your mind begins to wander. You become careless and make a mistake.

3. Check your problem. Be sure that you copied it correctly from the book.

4. Check your computations from step to step. Check the solution to the problem. Does it work? Does it make sense?

5. Keep practicing new skills. Remember the old saying, "Practice makes perfect." An increase in practice results in an increase in accuracy. Many errors are due simply to lack of practice.

There is no magic formula for eliminating all errors, but these five steps will be a tremendous help in reducing them.

3 *Evaluating Expressions Using the Correct Order of Operations*

Recall that in Section 1.6 we discussed the order of operations when we were combining whole numbers. We will now encounter some similar problems involving fractions and mixed numbers. We will repeat here the four-step order of operations that we studied previously:

Order of Operations

With grouping symbols:

Do first   **1.** Perform operations inside the parentheses.

   2. Simplify any expressions with exponents.

   3. Multiply or divide from left to right.

Do last   **4.** Add or subtract from left to right.

EXAMPLE 7   Evaluate. $\dfrac{3}{4} - \dfrac{2}{3} \times \dfrac{1}{8}$

$$\dfrac{3}{4} - \dfrac{2}{3} \times \dfrac{1}{8} = \dfrac{3}{4} - \dfrac{1}{12}$$ First we must multiply $\dfrac{2}{3} \times \dfrac{1}{8}$.

$$= \dfrac{9}{12} - \dfrac{1}{12}$$ Now we subtract, but first we need to build $\dfrac{3}{4}$ to an equivalent fraction with a common denominator of 12.

$$= \dfrac{8}{12}$$ Now we can subtract $\dfrac{9}{12} - \dfrac{1}{12}$.

$$= \dfrac{2}{3}$$ Finally we reduce the fraction.

Practice Problem 7   Evaluate. $\dfrac{3}{5} - \dfrac{1}{15} \times \dfrac{10}{13}$

EXAMPLE 8   Evaluate. $\dfrac{2}{3} \times \dfrac{1}{4} + \dfrac{2}{5} \div \dfrac{14}{15}$

$$\dfrac{2}{3} \times \dfrac{1}{4} + \dfrac{2}{5} \div \dfrac{14}{15} = \dfrac{1}{6} + \dfrac{2}{5} \div \dfrac{14}{15}$$ First we multiply $\dfrac{2}{3} \times \dfrac{1}{4}$.

$$= \dfrac{1}{6} + \dfrac{2}{5} \times \dfrac{\overset{3}{\cancel{15}}}{\underset{7}{\cancel{14}}}$$ We express the division as a multiplication problem. We invert $\dfrac{14}{15}$ and multiply.

$$= \dfrac{1}{6} + \dfrac{3}{7}$$ Now we perform the multiplication.

$$= \dfrac{7}{42} + \dfrac{18}{42}$$ We obtain equivalent fractions with an LCD of 42.

$$= \dfrac{25}{42}$$ We add the two fractions.

Practice Problem 8   Evaluate. $\dfrac{1}{7} \times \dfrac{5}{6} + \dfrac{5}{3} \div \dfrac{7}{6}$

Add or subtract. Express the answer as a mixed number. Simplify all answers.

1. $7\frac{1}{8} + 2\frac{5}{8}$

2. $6\frac{3}{10} + 4\frac{1}{10}$

3. $15\frac{3}{14} - 11\frac{1}{14}$

4. $8\frac{3}{4} - 3\frac{1}{4}$

5. $12\frac{1}{3} + 5\frac{1}{6}$

6. $20\frac{1}{4} + 3\frac{1}{8}$

7. $5\frac{4}{5} + 10\frac{3}{10}$

8. $6\frac{3}{8} + 4\frac{1}{16}$

9. $1 - \frac{3}{7}$

10. $1 - \frac{9}{11}$

11. $1\frac{5}{6} + \frac{7}{8}$

12. $1\frac{2}{3} + \frac{5}{18}$

13. $7\frac{1}{2} + 8\frac{3}{4}$

14. $9\frac{1}{6} + 2\frac{2}{3}$

15. $9\frac{2}{3} - 7\frac{1}{8}$

16. $8\frac{11}{15} - 3\frac{3}{10}$

17. $12\frac{1}{3} - 7\frac{2}{5}$

18. $10\frac{10}{15} - 10\frac{2}{3}$

19. $30 - 15\frac{3}{7}$

20. $25 - 14\frac{2}{11}$

Add or subtract. Express the answer as a mixed number. Simplify all answers.

21. $\begin{aligned} 15\frac{4}{15} \\ + 26\frac{8}{15} \\ \hline \end{aligned}$

22. $\begin{aligned} 22\frac{1}{8} \\ + 14\frac{3}{8} \\ \hline \end{aligned}$

23. $\begin{aligned} 4\frac{1}{3} \\ + 2\frac{1}{4} \\ \hline \end{aligned}$

24. $\begin{aligned} 6\frac{1}{8} \\ + 7\frac{3}{4} \\ \hline \end{aligned}$

25. $\begin{aligned} 3\frac{3}{4} \\ + 4\frac{5}{12} \\ \hline \end{aligned}$

26. $\begin{aligned} 11\frac{5}{8} \\ + 13\frac{1}{2} \\ \hline \end{aligned}$

27. $\begin{aligned} 47\frac{3}{10} \\ + 26\frac{5}{8} \\ \hline \end{aligned}$

28. $\begin{aligned} 34\frac{1}{20} \\ + 45\frac{8}{15} \\ \hline \end{aligned}$

29. $\begin{aligned} 19\frac{5}{6} \\ - 14\frac{1}{3} \\ \hline \end{aligned}$

30. $\begin{aligned} 22\frac{7}{9} \\ - 16\frac{1}{4} \\ \hline \end{aligned}$

31. $\begin{aligned} 5\frac{5}{12} \\ - 5\frac{10}{24} \\ \hline \end{aligned}$

32. $\begin{aligned} 5\frac{1}{12} \\ - 3\frac{7}{18} \\ \hline \end{aligned}$

33. $12\frac{3}{20}$
$-\ 7\frac{7}{15}$

34. $8\frac{5}{12}$
$-\ 5\frac{9}{10}$

35. 12
$-\ 3\frac{7}{15}$

36. 19
$-\ 6\frac{3}{7}$

37. 120
$-\ 17\frac{3}{8}$

38. 98
$-\ 89\frac{15}{17}$

39. $3\frac{1}{8}$
$2\frac{1}{3}$
$+\ 7\frac{3}{4}$

40. $4\frac{2}{3}$
$3\frac{1}{5}$
$+\ 6\frac{3}{4}$

Applications

41. Lee Hong rode his mountain bike through part of the Sangre de Cristo Mountains in New Mexico. On Wednesday he rode $20\frac{3}{4}$ miles. On Thursday he rode $22\frac{3}{8}$ miles. What was his total biking distance during those two days?

42. Kathryn traveled on a flight to London that took $7\frac{5}{6}$ hours. She connected with another flight to Bombay that took $5\frac{4}{5}$ hours. What was her total time in the air?

43. Renaldo bought $6\frac{3}{8}$ pounds of cheese for a party. The guests ate $4\frac{1}{3}$ pounds of the cheese. How much cheese did Renaldo have left over?

44. Shanna purchased stock in 1985 at $\$21\frac{3}{8}$ per share. When her son was ready for college, she sold the stock in 1999 at $\$93\frac{5}{8}$ per share. How much did she make per share for her son's tuition?

45. Nina and Julie are the two tallest basketball players on their high school team. Nina is $69\frac{3}{4}$ inches tall and Julie is $72\frac{1}{2}$ inches tall. How many inches taller is Julie than Nina?

46. Christos bought $1\frac{3}{4}$ pounds of French vanilla coffee beans. He also purchased $2\frac{5}{6}$ pounds of hazelnut coffee beans. How many total pounds of coffee beans did Christos buy?

47. Jack's dog is too heavy. According to the veterinarian, the dog needs to lose 16 pounds over the next three months. During the first month he lost $5\frac{1}{8}$ pounds. The second month he lost $4\frac{2}{3}$ pounds.
   (a) How much weight did Jack's dog lose during the first two months?
   (b) How much weight does his dog need to lose in the third month to reach the goal advised by the veterinarian?

48. A young man has been under a doctor's care to lose weight. His doctor wanted him to lose 46 pounds in the first three months. He lost $17\frac{5}{8}$ pounds the first month and $13\frac{1}{2}$ pounds the second month.
   (a) How much did he lose during the first two months?
   (b) How much would he need to lose in the third month to reach the goal?

To Think About

Use improper fractions as discussed in the text to perform each calculation.

49. $\dfrac{379}{8} + \dfrac{89}{5}$

50. $\dfrac{151}{6} - \dfrac{130}{7}$

When adding or subtracting mixed numbers, it is wise to estimate your answer by rounding each mixed number to the nearest whole number.

51. Estimate your answer to $35\frac{1}{6} + 24\frac{5}{12}$ by rounding each mixed number to the nearest whole number. Then find the exact answer. How close was your estimate?

52. Estimate your answer to $102\frac{5}{7} - 86\frac{2}{3}$ by rounding each mixed number to the nearest whole number. Then find the exact answer. How close was your estimate?

Evaluate using the correct order of operation.

53. $\dfrac{6}{7} - \dfrac{4}{7} \times \dfrac{1}{3}$

54. $\dfrac{3}{5} - \dfrac{1}{3} \times \dfrac{6}{5}$

55. $\dfrac{1}{2} + \dfrac{3}{8} \div \dfrac{3}{4}$

56. $\dfrac{3}{4} + \dfrac{1}{4} \div \dfrac{5}{3}$

57. $\dfrac{5}{7} \times \dfrac{7}{2} \div \dfrac{3}{2}$

58. $\dfrac{2}{7} \times \dfrac{3}{4} \div \dfrac{1}{2}$

59. $\dfrac{3}{5} \times \dfrac{1}{2} + \dfrac{1}{5} \div \dfrac{2}{3}$

60. $\dfrac{5}{6} \times \dfrac{1}{2} + \dfrac{2}{3} \div \dfrac{4}{3}$

61. $\left(\dfrac{3}{5} - \dfrac{3}{20}\right) \times \dfrac{4}{5}$

62. $\left(\dfrac{1}{3} + \dfrac{1}{6}\right) \times \dfrac{5}{11}$

63. $\dfrac{8}{7} \div \left(\dfrac{2}{3} + \dfrac{1}{12}\right)$

64. $\dfrac{4}{3} \div \left(\dfrac{3}{5} - \dfrac{3}{10}\right)$

65. $\dfrac{1}{4} \times \left(\dfrac{2}{3}\right)^2$

66. $\dfrac{5}{8} \times \left(\dfrac{2}{5}\right)^2$

67. $\left(\dfrac{4}{3}\right)^2 \times \dfrac{9}{11}$

68. $\left(\dfrac{5}{3}\right)^2 \times \dfrac{7}{10}$

Cumulative Review Problems

Multiply.

69. $\begin{array}{r} 6737 \\ \times\ \ \ 76 \\ \hline \end{array}$

70. $\begin{array}{r} 4050 \\ \times\ 2106 \\ \hline \end{array}$

71. Suzanne is redecorating her family's house. She has a budget of $6300 for painting, new flooring, and kitchen appliances. The painter charges her $25/hour for 27 hours of work and $520 for materials. The flooring contractor charges her $30/hour for 8 hours of work and $2972 for materials. How much will she have left over for appliances?

Putting Your Skills to Work

Bringing a Colonial Recipe Down to Size

No one knows for sure the age of this old recipe for Baltimore Sunshine Cake. It may be as old as the colonial period. One thing is for sure—it makes a huge cake. It contains enough ingredients to make a cake that will serve 48 people.

Baltimore Sunshine Cake

20 eggs
5 cups milk
4 1/3 cups shortening
6 tablespoons ground orange peel
9 1/2 teaspoons ground lemon peel
16 1/3 cups sifted flour
15 3/4 teaspoons baking powder
4 1/2 teaspoons salt
12 tablespoons of lemon juice
7 1/4 cups sugar

Problems for Individual Investigation

1. Find the amounts needed for a cake for 24 people ($\frac{1}{2}$ of each ingredient).

2. Find the amounts needed for a cake for 16 people ($\frac{1}{3}$ of each ingredient).

Problems for Group Investigation and Cooperative Study

3. Find the amounts needed for a cake for 30 people ($\frac{5}{8}$ of each ingredient).

4. Find the amounts needed for a cake for 18 people ($\frac{3}{8}$ of each ingredient).

Internet Connections

   Netsite: http://www.prenhall.com/tobey_basic

This site contains a recipe converter. It will change the amount of each ingredient in a recipe to feed any number of people. Use the converter to solve the following:

5. Find the amounts needed for a Baltimore Sunshine Cake for seven people.

6. Find the amounts needed for a Baltimore Sunshine Cake for five people.

2.9 Applied Problems Involving Fractions

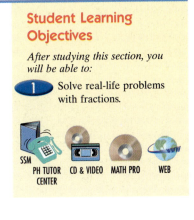

Student Learning Objectives

After studying this section, you will be able to:

1. Solve real-life problems with fractions.

SSM PH TUTOR CENTER CD & VIDEO MATH PRO WEB

1 Solving Real-Life Problems with Fractions

All problem solving requires the same kind of thinking. In this section we will combine problem-solving skills with our new computational skills with fractions. Sometimes the difficulty is in figuring out what must be done. Sometimes it is in doing the computation. Remember that *estimating* is important in problem solving. We may use the following steps.

1. *Understand the problem.*
   (a) Read the problem carefully.
   (b) Draw a picture if this helps you.
   (c) Fill in the Mathematics Blueprint.

2. *Solve.*
   (a) Perform the calculations.
   (b) State the answer, including the unit of measure.

3. *Check.*
   (a) Estimate the answer. Round fractions to the nearest whole number.
   (b) Compare the exact answer with the estimate to see if your answer is reasonable.

EXAMPLE 1 In designing a modern offshore speedboat, the designing engineer has determined that one of the oak frames near the engine housing needs to be $26\frac{1}{8}$ inches long. At the end of the oak frame there will be $2\frac{5}{8}$ inches of insulation. Finally there will be a steel mounting that is $3\frac{3}{4}$ inches long. When all three items are assembled, how long will the oak frame and insulation and steel mounting extend?

1. *Understand the problem.*
   We draw a picture to help us.
   Then we fill in the Mathematics Blueprint.

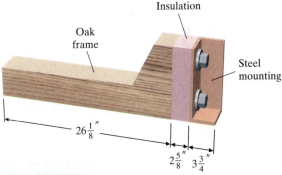

Mathematics Blueprint For Problem Solving

Gather The Facts	What Am I Asked To Do?	How Do I Proceed?	Key Points To Remember
Oak frame: $26\frac{1}{8}''$ Insulation: $2\frac{5}{8}''$ Steel mounting: $3\frac{3}{4}''$	Find the total length.	Add the lengths of the three items.	When adding mixed numbers, add the whole numbers first and then add the fractions.

2. *Solve and state the answer.*

Add the three amounts. $26\frac{1}{8} + 2\frac{5}{8} + 3\frac{3}{4}$

$$\text{LCD} = 8 \qquad 26\frac{1}{8} \qquad = \qquad 26\frac{1}{8}$$

$$2\frac{5}{8} \qquad = \qquad 2\frac{5}{8}$$

$$+ \; 3 \; \boxed{\frac{3}{4} \times \frac{2}{2}} \qquad = \qquad + \; 3\frac{6}{8}$$

$$31\frac{12}{8} = 32\frac{4}{8} = 32\frac{1}{2}$$

The entire assembly will be $32\frac{1}{2}$ inches.

3. *Check.*

Estimate the sum by rounding each fraction to one nonzero digit.

Thus $\qquad 26\frac{1}{8} + 2\frac{5}{8} + 3\frac{3}{4}$

becomes $\quad 26 + 3 + 4 = 33$

This is close to our answer, $32\frac{1}{2}$. Our answer seems reasonable.

One of the most important uses of estimation in mathematics is in the calculation of problems involving fractions. People find it easier to detect significant errors when working with whole numbers. However, the extra steps involved in the calculation with fractions and mixed numbers often distract our attention from an error that we should have detected.

Thus it is particularly critical to take the time to check your answer by estimating the results of the calculation with whole numbers. Be sure to ask yourself, is this answer reasonable? Does this answer seem realistic? Only by estimating our results with whole numbers will we be able to answer that question. It is this estimating skill that you will find more useful in your own life as a consumer and as a citizen.

Practice Problem 1 Nicole required the following amounts of gas for her farm tractor in the last three fill-ups: $18\frac{7}{10}$ gallons, $15\frac{2}{5}$ gallons, and $14\frac{1}{2}$ gallons. How many gallons did she need altogether?

The word *diameter* has two common meanings. First, it means a line segment that passes through the center of and intersects a circle twice. Second, it means the *length* of this segment.

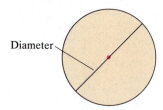
Diameter

▲ ⬭ **EXAMPLE 2**   What is the inside diameter (distance across) of a cement storm drain pipe that has an outside diameter of $4\frac{1}{8}$ feet and is $\frac{3}{8}$ feet thick?

1. *Understand the problem.*

Read the problem carefully. Draw a picture. The picture is in the margin on the left. Now fill in the Mathematics Blueprint.

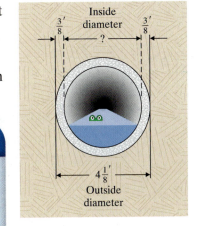

Mathematics Blueprint For Problem Solving

Gather The Facts	What Am I Asked To Do?	How Do I Proceed?	Key Points To Remember
Outside diameter is $4\frac{1}{8}$ feet. Thickness is $\frac{3}{8}$ foot on both ends of the diameter.	Find the *inside* diameter of the pipe.	Add the two measures of thickness. Then subtract this total from the outside diameter.	Since the LCD = 8, all fractions must have this denominator.

2. *Solve and state the answer.*

Add the two thickness measurements together.

Adding $\frac{3}{8} + \frac{3}{8} = \frac{6}{8}$ gives the total thickness of the pipe, $\frac{6}{8}$ foot. We will

not reduce $\frac{6}{8}$ since the LCD is 8.

We subtract the total of the two thickness measurements from the outside diameter.

$$
\begin{array}{rcl}
4\dfrac{1}{8} & = & 3\dfrac{9}{8} \\[2ex]
-\dfrac{6}{8} & = & -\dfrac{6}{8} \\[2ex]
\hline
& & 3\dfrac{3}{8}
\end{array}
$$

We borrow 1 from 4 to get $3 + 1\frac{1}{8}$ or $3\frac{9}{8}$.

The inside diameter is $3\frac{3}{8}$ feet.

3. *Check.*

We will work backward to check. We will use the exact values.
If we have done our work correctly, $\frac{3}{8}$ foot $+ 3\frac{3}{8}$ feet $+ \frac{3}{8}$ foot should add up to the outside diameter, $4\frac{1}{8}$ feet.

$$\frac{3}{8} + 3\frac{3}{8} + \frac{3}{8} \stackrel{?}{=} 4\frac{1}{8}$$

$$3\frac{9}{8} \stackrel{?}{=} 4\frac{1}{8}$$

$$4\frac{1}{8} = 4\frac{1}{8} \checkmark$$

Our answer of $3\frac{3}{8}$ feet is correct.

Practice Problem 2   A poster is $12\frac{1}{4}$ inches long. We want a $1\frac{3}{8}$-inch border on the top and a 2-inch border on the bottom. What is the length of the inside portion of the poster?

🔴 **EXAMPLE 3**   On Tuesday Michael earned $\$8\frac{1}{4}$ per hour working for eight hours. He also earned overtime pay, which is $1\frac{1}{2}$ times his regular rate of $\$8\frac{1}{4}$, for four hours on Tuesday. How much pay did he earn altogether on Tuesday?

1. *Understand the problem.*
    We draw a picture of the parts of Michael's pay on Tuesday.
    Michael's earnings on Tuesday are the sum of two parts:

Pay at regular pay rate		Pay at overtime pay rate		Total pay for the day
	$+$		$=$	

Now fill in the Mathematics Blueprint.

Mathematics Blueprint For Problem Solving

Gather The Facts	What Am I Asked To Do?	How Do I Proceed?	Key Points To Remember
He works eight hours at $\$8\frac{1}{4}$ per hour. He works four hours at the overtime rate, $1\frac{1}{2}$ times the regular rate.	Find his total pay for Tuesday.	Find out how much he is paid for regular time. Find out how much he is paid for overtime. Then add the two.	The overtime rate is $1\frac{1}{2}$ multiplied by the regular rate.

2. *Solve and state the answer.*
    Find his overtime pay rate.

$$1\frac{1}{2} \times 8\frac{1}{4} = \frac{3}{2} \times \frac{33}{4} = \frac{\$99}{8} \text{ per hour}$$

We leave our answer as an improper fraction because we will need to multiply it by another fraction.
How much was he paid for regular time? For overtime?

For eight regular hours, he earned $8 \times 8\frac{1}{4} = \overset{2}{\cancel{8}} \times \frac{33}{\underset{1}{\cancel{4}}} = \$66.$

For four overtime hours, he earned $\overset{1}{\cancel{4}} \times \frac{99}{\underset{2}{\cancel{8}}} = \frac{99}{2} = \$49\frac{1}{2}.$

Now we add to find the total pay.

$$\begin{array}{ll} \$66 & \text{Pay at regular pay rate} \\ \$49\frac{1}{2} & \text{Pay at overtime rate} \\ \hline \end{array}$$

Michael earned $\$115\frac{1}{2}$ working on Tuesday. (This is the same as $\$115.50$, which we will use in Chapter 3.)

3. *Check.*

We estimate his regular pay rate at $8 per hour.

We estimate his overtime pay rate at $1\frac{1}{2} \times 8 = \frac{3}{2} \times 8 = 12$ or $12 per hour.

$$8 \text{ hours} \times \$8 \text{ per hour} = \$64 \text{ regular pay}$$
$$4 \text{ hours} \times \$12 \text{ per hour} = \$48 \text{ overtime pay}$$

Estimated sum. $\$64 + \$48 \approx 60 + 50 = 110$

$110 is close to our calculated value, $115\frac{1}{2}$, so our answer is reasonable.   ✓

Practice Problem 3   A tent manufacturer uses $8\frac{1}{4}$ yards of waterproof duck cloth to make a regular tent. She uses $1\frac{1}{2}$ times that amount to make a large tent. How many yards of cloth will she need to make 6 regular tents and 16 large tents?

EXAMPLE 4   Alicia is buying some 8-foot boards for shelving. She wishes to make two bookcases, each with three shelves. Each shelf will be $3\frac{1}{4}$ feet long.

(a) How many boards does she need to buy?

(b) How many linear feet of shelving are actually needed to build the bookcases?

(c) How many linear feet of shelving will be left over?

1. *Understand the problem.*

Draw a sketch of a bookcase. Each bookcase will have three shelves. Alicia is making two such bookcases. (Alicia's boards are for the shelves, not the sides.)

Now fill in the Mathematics Blueprint.

Mathematics Blueprint For Problem Solving

Gather The Facts	What Am I Asked To Do?	How Do I Proceed?	Key Points To Remember
She needs three shelves for each bookcase. Each shelf is $3\frac{1}{4}$ feet long. She will make two bookcases. Shelves are cut from 8-foot boards.	Find out how many boards to buy. Find out how many feet of board are needed for shelves and how many feet will be left over.	First find out how many $3\frac{1}{4}$-foot shelves she can get from one board. Then see how many boards she needs to make all six shelves.	Each time she cuts up an 8-foot board, she will get some shelves and some leftover wood.

2. *Solve and state the answer.*
We want to know how many $3\frac{1}{4}$-foot boards are in an 8-foot board. By drawing a rough sketch, we would probably guess the answer is 2. To find exactly how many $3\frac{1}{4}$-foot long pieces are in 8 feet, we will use division.

$$8 \div 3\frac{1}{4} = \frac{8}{1} \div \frac{13}{4} = \frac{8}{1} \times \frac{4}{13} = \frac{32}{13} = 2\frac{6}{13} \text{ boards}$$

She will get two shelves from each board, and some wood will be left over.

(a) How many boards does Alicia need to build two bookcases? For two bookcases, she needs six shelves. She will get two shelves out of each board. $6 \div 2 = 3$. She will need three 8-foot boards.

(b) How many linear feet of shelving are actually needed to build the book-cases?
She needs 6 shelves at

$$3\frac{1}{4} \text{ feet} = 6 \times 3\frac{1}{4} = \overset{3}{\cancel{6}} \times \frac{13}{\underset{2}{\cancel{4}}} = \frac{39}{2} = 19\frac{1}{2}$$

linear feet. A total of $19\frac{1}{2}$ linear feet of shelving is needed.

(c) How many linear feet of shelving will be left over?
Each time she uses one board she will have

$$8 - 3\frac{1}{4} - 3\frac{1}{4} = 8 - 6\frac{1}{2} = 1\frac{1}{2}$$

feet left over. Each of the three boards will have $1\frac{1}{2}$ feet left over.

$$3 \times 1\frac{1}{2} = 3 \times \frac{3}{2} = \frac{9}{2} = 4\frac{1}{2}$$

linear feet. A total of $4\frac{1}{2}$ linear feet of shelving will be left over.

3. *Check.*
Work backward. See if you can check that with three 8-foot boards you
   (a) can make the six shelves for the two bookcases.
   (b) will use exactly $19\frac{1}{2}$ linear feet to make the shelves.
   (c) will have exactly $4\frac{1}{2}$ linear feet left over.
The check is left to you.

Practice Problem 4   Michael is purchasing 12-foot boards for shelving. He wishes to make two bookcases, each with four shelves. Each shelf will be $2\frac{3}{4}$ feet long.

(a) How many boards does he need to buy?

(b) How many linear feet of shelving are actually needed to build the bookcases?

(c) How many linear feet of shelving will be left over?

Another useful method for solving applied problems is called "Do a similar, simpler problem." When a problem seems difficult to understand because of the fractions, change the problem to an easier but similar problem. Then decide how to solve the simpler problem and use the same steps to solve the original problem. For example:

How many gallons of water can a tank hold if its volume is $58\frac{2}{3}$ cubic feet? (1 cubic foot holds about $7\frac{1}{2}$ gallons.)

A similar, easier problem would be: "If 1 cubic foot holds about 8 gallons and a tank holds 60 cubic feet, how many gallons of water does the tank hold?"

The easier problem can be read more quickly and seems to make more sense. Probably we will see how to solve the easier problem right away: "I can find the number of gallons by multiplying 8×60." Therefore we can solve the first problem by multiplying $7\frac{1}{2} \times 58\frac{2}{3}$ to obtain the number of gallons of water. See the next example.

● **EXAMPLE 5**   A fishing boat traveled $69\frac{3}{8}$ nautical miles in $3\frac{3}{4}$ hours. How many knots (nautical miles per hour) did the fishing boat average?

1. *Understand the problem.*
   Let us think of a simpler problem. If a boat traveled 70 nautical miles in 4 hours, how many knots did it average? We would divide distance by time.

$$70 \div 4 = \text{average speed}$$

Likewise in our original problem we need to divide distance by time.

$$69\frac{3}{8} \div 3\frac{3}{4} = \text{average speed}$$

Now fill in the Mathematics Blueprint.

Mathematics Blueprint For Problem Solving

Gather The Facts	What Am I Asked To Do?	How Do I Proceed?	Key Points To Remember
Distance is $69\frac{3}{8}$ nautical miles. Time is $3\frac{3}{4}$ hours.	Find the average speed of the boat.	Divide the distance in nautical miles by the time in hours.	You must change the mixed numbers to improper fractions before dividing.

2. *Solve and state the answer.*

Divide distance by time to get speed in knots.

$$69\frac{3}{8} \div 3\frac{3}{4} = \frac{555}{8} \div \frac{15}{4} = \frac{\overset{37}{\cancel{555}}}{\underset{2}{\cancel{8}}} \cdot \frac{\overset{1}{\cancel{4}}}{\underset{1}{\cancel{15}}}$$

$$= \frac{37}{2} \cdot \frac{1}{1} = \frac{37}{2} = 18\frac{1}{2} \text{ knots}$$

The speed of the boat was $18\frac{1}{2}$ knots.

3. *Check.*

We estimate $69\frac{3}{8} \div 3\frac{3}{4}$.

$$\text{Use } 70 \div 4 = 17\frac{1}{2} \text{ knots}$$

Our estimate is close to the calculated value.
Our answer is reasonable.   ✓

Practice Problem 5   Alfonso traveled $199\frac{3}{4}$ miles in his car and used $8\frac{1}{2}$ gallons of gas. How many miles per gallon did he get?

You may benefit from using the Mathematics Blueprint for Problem Solving when solving the following exercises.

Applications

▲ **1.** A triangle has three sides that measure $5\frac{1}{4}$ feet, $2\frac{5}{6}$ feet, and $8\frac{7}{12}$ feet. What is the perimeter (total distance around) of the triangle?

2. A deli used $8\frac{1}{4}$ pounds of roast beef on Monday, $7\frac{1}{2}$ pounds on Tuesday, and $6\frac{7}{10}$ pounds on Wednesday. How many pounds were used over the three days?

3. The coastal area of North Carolina experienced devastating erosion caused by fierce storms. The National Guard assisted by hauling sand to try to rebuild the dunes. During three days in September, the National Guard hauled $7\frac{5}{6}$ tons of sand on the first day, $8\frac{1}{8}$ tons on the second day, and $9\frac{1}{2}$ tons on the third. How many tons of sand were hauled over the three days?

4. Nancy wonders if a tree in her front yard will grow to be 8 feet tall. When the tree was planted, it was $2\frac{1}{6}$ feet above the ground in height. It grew $1\frac{1}{4}$ feet the first year after being planted. How many more feet does it need to grow to reach a height of 8 feet?

5. A bolt extends through $\frac{3}{4}$-inch-thick plywood, two washers that are each $\frac{1}{16}$ inch thick, and a nut that is $\frac{3}{16}$ inch thick. The bolt must be $\frac{1}{2}$ inch longer than the sum of the thicknesses of plywood, washers, and nut. What is the minimum length of the bolt?

6. A carpenter is using an 8-foot length of wood for a frame. The carpenter needs to cut a notch in the wood that is $4\frac{7}{8}$ feet from one end and $1\frac{2}{3}$ feet from the other end. How long does the notch need to be?

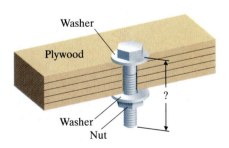

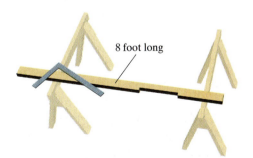

7. Hank is running the Boston Marathon, which is $26\frac{1}{5}$ miles long. At $6\frac{3}{4}$ miles from the start, he meets his wife, who is cheering him on. $9\frac{1}{2}$ miles further down the marathon course, he sees some friends from his running club volunteering at a water stop. Once he passes his friends, how many more miles does Hank have left to run?

8. Norman Olerud makes birdhouses as a hobby. He has a long piece of lumber that measures $14\frac{1}{4}$ feet. He needs to cut it into pieces that are $\frac{3}{4}$ foot long for the birdhouse floors. How many floors will he be able to cut from the long piece?

9. The following exaggerated story was repeated by certain basketball fans: Michael Jordan wasn't always 7 feet tall. When he was a little boy, at the height of $3\frac{1}{2}$ feet, no one knew that he would grow so quickly. The following year he grew $1\frac{5}{6}$ feet, and the year after that he grew $1\frac{1}{5}$ feet. How many more feet did he grow to reach 7 feet tall?

10. For a party of the British Literature Club using all "English foods," Nancy bought a $10\frac{2}{3}$-pound wheel of Stilton cheese, to go with the pears and the apples, at $\$8\frac{3}{4}$ per pound. How much did the wheel of Stilton cheese cost?

▲ **11.** How many gallons can a tank hold that has a volume of $36\frac{3}{4}$ cubic feet? (Assume that 1 cubic foot holds about $7\frac{1}{2}$ gallons.)

▲ **12.** A tank can hold a volume of $7\frac{1}{4}$ cubic feet. If it is filled with water, how much does the water weigh? (Assume that 1 cubic foot of water weighs $62\frac{1}{2}$ pounds.)

13. The night of the *Titanic* cruise ship disaster, the captain decided to run his ship at $22\frac{1}{2}$ knots (nautical miles per hour). The *Titanic* traveled at that speed for $4\frac{3}{4}$ hours before it met its tragic demise. How far did the *Titanic* travel at this excessive speed before the disaster?

14. Barbara earns $450 per week. She has $\frac{1}{5}$ of her income withheld for federal taxes, $\frac{1}{15}$ of her income withheld for state taxes, and $\frac{1}{25}$ of her income withheld for medical coverage. How much per week is left for Barbara after those three deductions?

15. Noriko earns $660 per week. She has $\frac{1}{5}$ of her income withheld for federal taxes, $\frac{1}{15}$ of her income withheld for state taxes, and $\frac{1}{20}$ of her income withheld for medical coverage. How much per week is left for Noriko after these three deductions?

16. Brenda and Luis are saving for a down payment on a house. Their total take-home pay is $870 per week. They have determined that $\frac{1}{5}$ of their weekly income is needed for car payments and insurance, $\frac{1}{10}$ is needed for groceries and monthly bills, and $\frac{1}{15}$ is needed for clothing and entertainment. How much per week is left to be saved for their down payment?

17. The crafts store bought $36\frac{3}{4}$ ounces of potpourri (a type of air freshener) for \$2. The shop divides the potpourri into $\frac{7}{8}$-ounce bags to put into drawers and closets, and sells each bag for $\$3\frac{1}{2}$.
  (a) How many $\frac{7}{8}$-ounces bags of potpourri can the shop make?
  (b) If all of the bags are sold-how much money will the store take in?
  (c) How much profit will it make?

▲ **18.** Jay and Pam are building a dog run for their collie. The fenced-in area measures $38\frac{1}{3}$ feet by $52\frac{3}{4}$ feet.
  (a) How many feet of coated fence wire will they need, if the wire is wrapped around the perimeter and $2\frac{1}{3}$ feet of wire is necessary to secure the end of the last wrap?
  (b) How many feet of fence wire would they need if they made the fence exactly one-half as long and one-half as wide?

19. Cecilia bought a loaf of sourdough bread that was made by a local gourmet bakery. The label said that the bread, plus its fancy box, weighed $18\frac{1}{2}$ ounces in total. Of this, $1\frac{1}{4}$ ounces turned out to be the weight of the ribbon. The box weighed $3\frac{1}{8}$ ounces.
  (a) How many ounces of bread did she actually buy?
  (b) The box stated its net weight as $14\frac{3}{4}$ ounces. (This means that she should have found $14\frac{3}{4}$ ounces of gourmet sourdough bread in the box.) How much in error was this measurement?

20. A special cookie recipe, requires $5\frac{1}{8}$ lbs of flour, $3\frac{2}{3}$ lbs of chocolate chips, $1\frac{1}{4}$ lbs of sugar, $2\frac{1}{2}$ lbs of molasses, and $2\frac{5}{6}$ lbs of oatmeal.
  (a) How much will one batch of this mixture weigh?
  (b) How much will 34 batches of this mixture weigh?

21. The largest Coast Guard boat stationed at San Diego can travel $160\frac{1}{8}$ nautical miles in $5\frac{1}{4}$ hours.
  (a) At how many knots is the boat traveling?
  (b) At this speed, how long would it take the Coast Guard boat to travel $213\frac{1}{2}$ nautical miles?

22. Russ and Norma's Mariah water ski boat can travel $72\frac{7}{8}$ nautical miles in $2\frac{3}{4}$ hours.
  (a) At how many knots is the boat traveling?
  (b) At this speed, how long would it take their water ski boat to travel $92\frac{3}{4}$ nautical miles?

▲ **23.** A Kansas wheat farmer has a storage bin with a capacity of $6856\frac{1}{4}$ cubic feet.
  (a) If a bushel of wheat is $1\frac{1}{4}$ cubic feet, how many bushels can the storage bin hold?
  (b) If a farmer wants to make a new storage bin $1\frac{3}{4}$ times larger, how many cubic feet will it hold?
  (c) How many bushels will the new bin hold?

24. A Greyhound bus made a long-distance run in upstate New York, stopping several times. The entire trip took $24\frac{1}{6}$ hours. The bus stopped a total of $2\frac{1}{3}$ hours for restroom and meal stops. The bus also stopped $1\frac{1}{2}$ hours for repairs. The bus traveled 854 miles. What was its average speed in miles per hour when it was moving? (Do not count stop times.)

To Think About

The Ortega family is having a birthday party for their 94-year-old grandmother. There are huge platters of cold cuts, salads, and cheese on the table. The table contains $9\frac{2}{3}$ lbs of roast beef, $16\frac{1}{2}$ lbs of honey-roasted ham, $4\frac{1}{4}$ lbs of turkey, $12\frac{3}{4}$ lbs of tuna salad, $10\frac{1}{3}$ lbs of taco salad, $6\frac{8}{9}$ lbs of corn salad, $10\frac{1}{8}$ lbs of swiss cheese, $8\frac{1}{3}$ lbs of cheddar cheese, and a wheel of brie cheese that weighs $6\frac{1}{5}$ lbs.

Grandpa Ortega remarked that this is exactly twice as much food in each category as they had at the party last year.

25. How many pounds of salad did they have at the party last year?

26. How many pounds of meat did they have at the party last year?

Cumulative Review Problems

27. Add.
$$\begin{array}{r} 16{,}846 \\ 19{,}321 \\ + \ 8{,}078 \\ \hline \end{array}$$

28. Subtract.
$$\begin{array}{r} 193{,}705 \\ - \ 165{,}891 \\ \hline \end{array}$$

29. Multiply.
$$\begin{array}{r} 1683 \\ \times \quad 27 \\ \hline \end{array}$$

30. Divide. $19\overline{)5282}$

Developing Your Study Skills

Why Study Mathematics?

Students often question the value of mathematics. They see little real use for it in their everyday lives. However, mathematics is often the key that opens the door to a better-paying job.

In our present-day technological world, many people use mathematics daily. Many vocational and professional areas—such as the fields of business, statistics, economics, psychology, finance, computer science, chemistry, physics, engineering, electronics, nuclear energy, banking, quality control, and teaching—require a certain level of expertise in mathematics. Those who want to work in these fields must be able to function at a given mathematical level. Those who cannot will not be able to enter these job areas.

So, whatever your field, be sure to realize the importance of mastering the basics of this course. It is very likely to help you advance to the career of your choice.

Math in the Media

Hour of Exercise Keeps Weight Off

Nancy Hellmich, USA TODAY

Researchers at the University of Pittsburgh School of Medicine and University of Colorado Health Sciences Center in Denver have compiled a National Weight Control Registry of people who've lost and kept off weight. Experts have analyzed detailed questionnaires from 2,800 people (80% women), who've lost an average of 65 pounds and kept at least 30 pounds off for $5\frac{1}{2}$ years.

Among the findings so far:

- People who've lost weight and kept if off burn an average of 2,800 calories a week with exercise—about 400 calories daily. (The average person burns about 100 calories every mile he or she walks.)

- On average, participants burn about 1,000 calories a week by walking.

- They burn 1,800 calories with a combination of activities, including aerobics, strength training and biking.

A separate study of 140 exercisers, mostly walkers, showed that those who maintained weight loss averaged 280 minutes of exercise a week, or 56 minutes five days a week.

EXERCISES

1. **(a)** Juan plans to exercise 280 minutes per week. If he plans to exercise 4 days of the week, how many minutes must he exercise each day?

   (b) If he plans to exercise 5 days of the week, what must be the time?

2. Using data from above, is it possible that Juan can burn 2,800 calories by walking? What speed must he maintain? Is it a reasonable speed for walking?

Chapter 2 Organizer

Topic	Procedure	Examples
Concept of a fractional part, p. 110.	The numerator is the number of parts selected. The denominator is the number of total parts.	What part of this sketch is shaded? $\frac{7}{10}$
Prime factorization, p. 117.	Prime factorization is the writing of a number as the product of prime numbers.	Write the prime factorization of 36. $36 = 4 \times 9$ $2 \times 2 \quad 3 \times 3$ $= 2 \times 2 \times 3 \times 3$
Reducing fractions, p. 119.	1. Factor numerator and denominator into prime factors. 2. Divide out factors common to numerator and denominator.	Reduce. $\frac{54}{90}$ $\frac{54}{90} = \frac{\overset{1}{\cancel{3}} \times \overset{1}{\cancel{3}} \times 3 \times \overset{1}{\cancel{2}}}{\underset{1}{\cancel{3}} \times \underset{1}{\cancel{3}} \times \underset{1}{\cancel{2}} \times 5} = \frac{3}{5}$
Changing a mixed number to an improper fraction, p. 126.	1. Multiply whole number by denominator. 2. Add product to numerator. 3. Place sum over denominator.	Write as an improper fraction. $7\frac{3}{4} = \frac{4 \times 7 + 3}{4} = \frac{28 + 3}{4} = \frac{31}{4}$
Changing an improper fraction to a mixed number, p. 127.	1. Divide denominator into numerator. 2. The quotient is the whole number. 3. The fraction is the remainder over the divisor.	Change to a mixed number. $\frac{32}{5}$ $\begin{array}{r} 6 \\ 5\overline{)32} \\ \underline{30} \\ 2 \end{array} = 6\frac{2}{5}$
Multiplying fractions, p. 133.	1. Divide out common factors from the numerators and denominators whenever possible. 2. Multiply numerators. 3. Multiply denominators.	Multiply. $\frac{3}{7} \times \frac{5}{13} = \frac{15}{91}$ Multiply. $\frac{\overset{1}{\cancel{5}}}{\underset{1}{\cancel{8}}} \times \frac{\overset{2}{\cancel{16}}}{\underset{3}{\cancel{15}}} = \frac{2}{3}$
Multiplying mixed and/or whole numbers, p. 135.	1. Change any whole numbers to fractions with a denominator of 1. 2. Change any mixed numbers to improper fractions. 3. Use multiplication rule for fractions.	Multiply. $7 \times 3\frac{1}{4}$ $\frac{7}{1} \times \frac{13}{4} = \frac{91}{4}$ $= 22\frac{3}{4}$
Dividing fractions, p. 140.	To divide two fractions, we invert the second fraction and multiply.	Divide. $\frac{3}{7} \div \frac{2}{9} = \frac{3}{7} \times \frac{9}{2} = \frac{27}{14}$ or $1\frac{13}{14}$
Dividing mixed numbers and/or whole numbers, p. 142.	1. Change any whole numbers to fractions with a denominator of 1. 2. Change any mixed numbers to improper fractions. 3. Use rule for division of fractions.	Divide. $8\frac{1}{3} \div 5\frac{5}{9} = \frac{25}{3} \div \frac{50}{9}$ $= \frac{\overset{1}{\cancel{25}}}{\underset{1}{\cancel{3}}} \times \frac{\overset{3}{\cancel{9}}}{\underset{2}{\cancel{50}}} = \frac{3}{2}$ or $1\frac{1}{2}$

Topic	Procedure	Examples
Finding the least common denominator, p. 152.	1. Write each denominator as the product of prime factors. 2. List all the prime factors that appear in both products. 3. Form a product of those factors, using each factor the greatest number of times it appears in any denominator.	Find LCD of $\frac{1}{10}, \frac{3}{8}$, and $\frac{7}{25}$. $10 = 2 \times 5$ $8 = 2 \times 2 \times 2$ $25 = 5 \times 5$ $LCD = 2 \times 2 \times 2 \times 5 \times 5 = 200$
Building fractions, p. 154.	1. Find how many times the original denominator can be divided into the new denominator. 2. Multiply that value by numerator and denominator of original fraction.	Build $\frac{5}{7}$ to an equivalent fraction with a denominator of 42. First we find $7\overline{)42}$ with quotient 6. Then we multiply the numerator and denominator by 6. $$\frac{5}{7} \times \frac{6}{6} = \frac{30}{42}$$
Adding or subtracting fractions with a common denominator, p. 159.	1. Add or subtract the numerators. 2. Keep the common denominator.	Add. $\frac{3}{13} + \frac{5}{13} = \frac{8}{13}$ Subtract. $\frac{15}{17} - \frac{12}{17} = \frac{3}{17}$
Adding or subtracting fractions without a common denominator, p. 160.	1. Find the LCD of the fractions. 2. Build up each fraction, if needed, to obtain the LCD in the denominator. 3. Follow the steps for adding and subtracting fractions with the same denominator.	Add. $\frac{1}{4} + \frac{3}{7} + \frac{5}{8}$ LCD = 56 $$\frac{1 \times 14}{4 \times 14} + \frac{3 \times 8}{7 \times 8} + \frac{5 \times 7}{8 \times 7}$$ $$= \frac{14}{56} + \frac{24}{56} + \frac{35}{56} = \frac{73}{56} = 1\frac{17}{56}$$
Adding mixed numbers, p. 166.	1. Change fractional parts to equivalent fractions with LCD as a denominator, if needed. 2. Add whole numbers and fractions separately. 3. If improper fractions occur, change to mixed numbers and simplify.	Add. $6\frac{3}{4} + 2\frac{5}{8}$ $6\boxed{\frac{3}{4} \times \frac{2}{2}} = 6\frac{6}{8}$ $+2\frac{5}{8} \qquad = 2\frac{5}{8}$ $\overline{\qquad\qquad\qquad 8\frac{11}{8} = 9\frac{3}{8}}$
Subtracting mixed numbers, p. 167.	1. Change fractional parts to equivalent fractions with LCD as a denominator, if needed. 2. If necessary, borrow from whole number to subtract fractions. 3. Subtract whole numbers and fractions separately.	Subtract. $8\frac{1}{5} - 4\frac{2}{3}$ $8\boxed{\frac{1}{5} \times \frac{3}{3}} = 8\frac{3}{15} = 7\frac{18}{15}$ $-4\boxed{\frac{2}{3} \times \frac{5}{5}} = -4\frac{10}{15} = -4\frac{10}{15}$ $\overline{\qquad\qquad\qquad\qquad\qquad 3\frac{8}{15}}$

Topic	Procedure	Examples

Order of Operations p. 170.	**Order of Operations** With grouping symbols: Do first **1.** Perform operations inside the parentheses. **2.** Simplify any expressions with exponents. **3.** Multiply or divide from left to right. Do last **4.** Add or subtract from left to right.	$\frac{5}{6} \div \left(\frac{4}{5} - \frac{7}{15}\right)$ First combine numbers inside the parentheses. $\frac{5}{6} \div \left(\frac{12}{15} - \frac{7}{15}\right)$ Transform $\frac{4}{5}$ to equivalent fraction $\frac{12}{15}$. $\frac{5}{6} \div \frac{1}{3}$ Subtract the two fractions inside the parentheses. $\frac{5}{6} \times \frac{3}{1}$ Invert the second fraction and multiply. $\frac{5}{2}$ or $2\frac{1}{2}$ Simplify.

Using the Mathematics Blueprint for Problem Solving, p. 175

In solving an applied problem with fractions, students may find it helpful to complete the following steps. You will not use the steps all of the time. Choose the steps that best fit the conditions of the problem.

1. *Understand the problem.*
   (a) Read the problem carefully.
   (b) Draw a picture if this helps you to visualize the situation. Think about what facts you are given and what you are asked to find.
   (c) It may help to write a similar, simpler problem to get started and to determine what operation to use.
   (d) Use the Mathematics Blueprint for Problem Solving to organize your work. Follow these four parts.
      1. Gather the facts (Write down specific values given in the problem.)
      2. What am I asked to do? (Identify what you must obtain for an answer.)
      3. How do I proceed? (What calculations need to be done.)
      4. Key points to remember (Record any facts, warnings, formulas, or concepts you think will be important as you solve the problem.)

2. *Solve and state the answer.*
   (a) Perform the necessary calculations.
   (b) State the answer, including the unit of measure.

3. *Check.*
   (a) Estimate the answer to the problem. Compare this estimate to the calculated value. Is your answer reasonable?
   (b) Repeat your calculations.
   (c) Work backward from your answer. Do you arrive at the original conditions of the problem?

Example

A wire is $95\frac{1}{3}$ feet long. It is cut up into smaller, equal-sized pieces, each $4\frac{1}{3}$ feet long. How many pieces will there be?

1. *Understand the problem.*
   Draw a picture of the situation.
   How will we find the number of pieces?
   Now we will use a simpler problem to clarify the idea. A wire 100 feet long is cut up into smaller pieces each 4 feet long. How many pieces will there be? We readily see that we would divide 100 by 4. Thus in our original problem we should divide $95\frac{1}{3}$ feet by $4\frac{1}{3}$ feet. This will tell us the number of pieces. Now we fill in the Mathematics Blueprint.

Mathematics Blueprint For Problem Solving

Gather The Facts	What Am I Asked To Do?	How Do I Proceed?	Key Points To Remember
Wire is $95\frac{1}{3}$ feet. It is cut into equal pieces $4\frac{1}{3}$ feet long.	Determine how many pieces of wire there will be.	Divide $95\frac{1}{3}$ by $4\frac{1}{3}$.	Change mixed numbers to improper fractions before carrying out the division.

2. *Solve and state the answer.*

We need to divide $95\dfrac{1}{3} \div 4\dfrac{1}{3}$

$$\frac{286}{3} \div \frac{13}{3} = \frac{\overset{22}{\cancel{286}}}{\underset{1}{\cancel{3}}} \times \frac{\overset{1}{\cancel{3}}}{\underset{1}{\cancel{13}}} = \frac{22}{1} = 22$$

There will be 22 pieces of wire.

3. *Check.*

Estimate. Rounded to the nearest ten, $95\dfrac{1}{3} \approx 100$.

Rounded to the nearest integer, $4\dfrac{1}{3} \approx 4$.

$$100 \div 4 = 25$$

This is close to our estimate. Our answer is reasonable. ✓

Chapter 2 Review Problems

Be sure to simplify all answers.

2.1 *Use a fraction to represent the shaded part of each object.*

1.

2.

In exercises 3 and 4, draw a sketch to illustrate each fraction.

3. $\dfrac{4}{7}$ of an object

4. $\dfrac{7}{9}$ of an object

5. The dean asked the 100 freshmen if they would be staying in the dorm over the holidays. A total of 87 said they would not. What fractional part of the freshmen said they would not?

6. An inspector looked at 31 parts and found 6 of them defective. What fractional part of these items was defective?

2.2 *Express each number as a product of prime factors.*

7. 54   **8.** 42   **9.** 168

Determine which of the following numbers are prime. If a number is composite, express it as the product of prime factors.

10. 59   **11.** 78   **12.** 167

Reduce each fraction.

13. $\dfrac{12}{42}$ **14.** $\dfrac{13}{52}$ **15.** $\dfrac{21}{36}$ **16.** $\dfrac{26}{34}$ **17.** $\dfrac{168}{192}$ **18.** $\dfrac{51}{105}$

2.3 *Change each mixed number to an improper fraction.*

19. $4\dfrac{3}{8}$ **20.** $2\dfrac{19}{23}$

In exercises 21 and 22, change each improper fraction to a mixed number.

21. $\dfrac{45}{8}$ **22.** $\dfrac{63}{13}$

23. Reduce and leave your answer as a mixed number.

$3\dfrac{15}{55}$

24. Reduce and leave your answer as an improper fraction.

$\dfrac{234}{16}$

25. Change to a mixed number and then reduce.

$\dfrac{385}{240}$

2.4 *In exercises 26–33, multiply.*

26. $\dfrac{4}{7} \times \dfrac{5}{11}$ **27.** $\dfrac{7}{9} \times \dfrac{21}{35}$ **28.** $12 \times \dfrac{3}{7} \times 0$ **29.** $\dfrac{3}{5} \times \dfrac{2}{7} \times \dfrac{10}{27}$

30. $12 \times 8\dfrac{1}{5}$ **31.** $5\dfrac{1}{4} \times 4\dfrac{6}{7}$ **32.** $5\dfrac{3}{8} \times 3\dfrac{4}{5}$ **33.** $35 \times \dfrac{7}{10}$

34. In 1999, one share of stock cost $\$37\frac{5}{8}$. How much money did 18 shares cost?

▲ **35.** Find the area of a rectangle that is $18\frac{1}{5}$ inches wide and $26\frac{3}{4}$ inches long.

2.5 *In exercises 36–43, divide, if possible.*

36. $\dfrac{3}{7} \div \dfrac{2}{5}$ **37.** $\dfrac{9}{17} \div \dfrac{18}{5}$ **38.** $1200 \div \dfrac{5}{8}$ **39.** $\dfrac{2\frac{1}{4}}{3\frac{1}{3}}$

40. $\dfrac{\frac{20}{4}}{4\frac{4}{5}}$ **41.** $2\dfrac{1}{8} \div 20\dfrac{1}{2}$ **42.** $0 \div 3\dfrac{7}{5}$ **43.** $4\dfrac{2}{11} \div 3$

44. A dress requires $3\frac{1}{8}$ yards of fabric. Amy has $21\frac{7}{8}$ yards available. How many dresses can she make?

45. There are 420 calories in $2\frac{1}{4}$ cans of grape soda. How many calories are in 1 can of soda?

2.6 *Find the LCD for each pair of fractions.*

46. $\dfrac{7}{14}$ and $\dfrac{3}{49}$

47. $\dfrac{7}{40}$ and $\dfrac{11}{30}$

48. $\dfrac{5}{18}, \dfrac{1}{6}, \dfrac{7}{45}$

Build each fraction to an equivalent fraction with the specified denominator.

49. $\dfrac{3}{7} = \dfrac{?}{56}$

50. $\dfrac{11}{24} = \dfrac{?}{72}$

51. $\dfrac{9}{43} = \dfrac{?}{172}$

52. $\dfrac{17}{18} = \dfrac{?}{198}$

2.7 *Add or subtract.*

53. $\dfrac{3}{7} - \dfrac{5}{14}$

54. $\dfrac{1}{2} + \dfrac{1}{3} + \dfrac{1}{4}$

55. $\dfrac{4}{7} + \dfrac{7}{9}$

56. $\dfrac{7}{8} - \dfrac{3}{5}$

57. $\dfrac{7}{30} + \dfrac{2}{21}$

58. $\dfrac{5}{18} + \dfrac{5}{12}$

59. $\dfrac{15}{16} - \dfrac{13}{24}$

60. $\dfrac{14}{15} - \dfrac{3}{25}$

2.8 *In exercises 61–68, evaluate using the correct order of operations.*

61. $1 - \dfrac{17}{23}$

62. $6 - \dfrac{5}{9}$

63. $3 + 5\dfrac{2}{3}$

64. $8 + 12\dfrac{5}{7}$

65. $3\dfrac{1}{4} + 1\dfrac{5}{8}$

66. $7\dfrac{3}{16} - 2\dfrac{5}{6}$

67. $\dfrac{3}{5} \times \dfrac{1}{2} + \dfrac{2}{5} \div \dfrac{2}{3}$

68. $\left(\dfrac{3}{7} - \dfrac{1}{14}\right) \times \dfrac{8}{9}$

69. Bob jogged $1\frac{7}{8}$ miles on Monday, $2\frac{3}{4}$ miles on Tuesday, and $4\frac{1}{10}$ miles on Wednesday. How many miles did he jog on these three days?

70. When it was new, Ginny Sue's car got $28\frac{1}{6}$ miles per gallon. It now gets $1\frac{5}{6}$ miles per gallon less. How far can she drive now if the car has $10\frac{3}{4}$ gallons in the tank?

71. A recipe calls for $3\frac{1}{3}$ cups of sugar and $4\frac{1}{4}$ cups of flour. How much sugar and how much flour would be needed for $\frac{1}{2}$ of that recipe?

72. Rafael travels in a car that gets $24\frac{1}{4}$ miles per gallon. He has $8\frac{1}{2}$ gallons of gas in the gas tank. Approximately how far can he drive?

73. How many lengths of pipe $3\frac{1}{5}$ inches long can be cut from a pipe 48 inches long?

74. A car radiator holds $15\frac{3}{4}$ liters. If it contains $6\frac{1}{8}$ liters of antifreeze and the rest is water, how much is water?

75. Delbert types 366 words in $12\frac{1}{5}$ minutes. How many words per minute can he type?

76. Alicia earns $\$4\frac{1}{2}$ per hour for regular pay and $1\frac{1}{2}$ times that rate of pay for overtime. On Monday she worked eight hours at regular pay and three hours at overtime. How much did she earn on Monday?

77. George purchased stock in 1998 at $\$88\frac{3}{8}$ a share. He sold it in 1999 at $\$79\frac{5}{8}$ a share. How much did the value of the stock decrease during that time period?

78. A 3-inch bolt passes through $1\frac{1}{2}$ inches of pine board, a $\frac{1}{16}$-inch washer, and a $\frac{1}{8}$-inch nut. How many inches does the bolt extend beyond the board, washer, and nut if the head of the bolt is $\frac{1}{4}$ inch long?

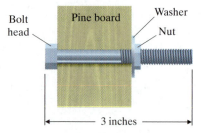

79. Francine has a take-home pay of $880 per month. She gives $\frac{1}{10}$ of it to her church, spends $\frac{1}{2}$ of it for rent and food, and spends $\frac{1}{8}$ of it on electricity, heat, and telephone. How many dollars per month does she have left for other things?

80. Manuel's new car used $18\frac{2}{5}$ gallons of gas on a 460-mile trip.
   (a) How many miles can his car travel on 1 gallon of gas?
   (b) How much did his trip cost him in gasoline expense if the average cost of gasoline was $\$1\frac{1}{5}$ per gallon?

1. Use a fraction to represent the shaded part of the object.

2. A basketball star shot at the hoop 388 times. The ball went in 311 times. Write a fraction that describes the part of the time that his shots went in.

In problems 3–5, reduce each fraction.

3. $\dfrac{18}{42}$

4. $\dfrac{15}{70}$

5. $\dfrac{225}{50}$

6. Change to an improper fraction. $6\dfrac{4}{5}$

7. Change to a mixed number. $\dfrac{114}{14}$

Multiply.

8. $42 \times \dfrac{2}{7}$

9. $\dfrac{7}{9} \times \dfrac{2}{5}$

10. $4\dfrac{1}{3} \times 7\dfrac{1}{5}$

Divide.

11. $\dfrac{7}{8} \div \dfrac{5}{11}$

12. $\dfrac{12}{31} \div \dfrac{8}{13}$

13. $7\dfrac{1}{5} \div 1\dfrac{1}{25}$

14. $5\dfrac{1}{7} \div 3$

In problems 15–17, find the least common denominator of each set of fractions.

15. $\dfrac{5}{24}$ and $\dfrac{7}{18}$

16. $\dfrac{3}{16}$ and $\dfrac{1}{24}$

17. $\dfrac{1}{4}, \dfrac{3}{8}, \dfrac{5}{6}$

18. Build the fraction to an equivalent fraction with the specified denominator. $\dfrac{5}{12} = \dfrac{?}{72}$

Evaluate using the correct order of operations.

19. $\dfrac{7}{9} - \dfrac{5}{12}$

20. $\dfrac{2}{15} + \dfrac{5}{12}$

21. $\dfrac{1}{4} + \dfrac{3}{7} + \dfrac{3}{14}$

22. $8\dfrac{3}{5} + 5\dfrac{4}{7}$

1. _____
2. _____
3. _____
4. _____
5. _____
6. _____
7. _____
8. _____
9. _____
10. _____
11. _____
12. _____
13. _____
14. _____
15. _____
16. _____
17. _____
18. _____
19. _____
20. _____
21. _____
22. _____

23. _____

24. _____

25. _____

26. _____

27. _____

28. _____

29. _____

30. (a) _____

(b) _____

31. (a) _____

(b) _____

(c) _____

23. $18\frac{6}{7} - 13\frac{13}{14}$

24. $\frac{6}{7} - \frac{5}{7} \times \frac{1}{4}$

25. $\left(\frac{1}{2} + \frac{1}{3}\right) \times \frac{7}{5}$

Answer each question.

▲**26.** A hallway measures $8\frac{1}{6}$ yards by $5\frac{1}{7}$ yards. How many square yards is the area of the hallway?

27. A butcher has $18\frac{2}{3}$ pounds of steak that he wishes to place into packages that average $2\frac{1}{3}$ pounds each. How many packages can he make?

28. From central parking it is $\frac{9}{10}$ of a mile to the science building. Bob started at central parking and walked $\frac{1}{5}$ of a mile toward the science building. He stopped for coffee. When he finished, how much farther did he have to walk to reach the science building?

29. Robin jogged $4\frac{1}{8}$ miles on Monday, $3\frac{1}{6}$ miles on Tuesday, and $6\frac{3}{4}$ miles on Wednesday. How far did she jog on those three days?

30. Mr. and Mrs. Samuel visited Florida and purchased 120 oranges. They gave $\frac{1}{4}$ of them to relatives, ate $\frac{1}{12}$ of them in the hotel, and gave $\frac{1}{3}$ of them to friends. They shipped the rest home to Illinois.
   (a) How many oranges did they ship?
   (b) If it costs 24¢ for each orange to be shipped to Illinois, what was the total shipping bill?

31. A candle company purchased $48\frac{1}{8}$ pounds of wax to make specialty candles. It takes $\frac{5}{8}$ pound of wax to make one candle. The owners of the business plan to sell the candles for $12 each. The specialty wax cost them $2 per pound.
   (a) How many candles can they make?
   (b) How much does it cost to make one candle?
   (c) How much profit will they make if they sell all of the candles?

One-half of this test is based on Chapter 1 material. The remainder is based on material covered in Chapter 2.

1. Write in words.   84,361,208

2. Add.   128
452
178
34
+ 77

3. Add.   156,200
364,700
+ 198,320

4. Subtract.   5718
− 3643

5. Subtract.   1,000,361
− 983,145

6. Multiply.   126
× 38

7. Multiply.   16,908
×   12

8. Divide. $7\overline{)32,606}$

9. Divide. $18\overline{)6642}$

10. Evaluate. 7^2

11. Round to the nearest thousand.
6,037,452

12. Perform the operations in their proper order. $4 \times 3^2 + 12 \div 6$

13. Roone bought three shirts at $26 and two pairs of pants at $48. What was his total bill?

14. Leslie had a balance of $64 in her checking account. She deposited $1160. She made checks out for $516, $199, and $203. What will be her new balance?

15. Eighty-four students enrolled in psychology. Fifty-five were women. Write a fraction that describes the part of the class that was made up of women.

16. Reduce. $\dfrac{28}{52}$

17. Write as an improper fraction. $18\dfrac{3}{4}$

1. _____

2. _____

3. _____

4. _____

5. _____

6. _____

7. _____

8. _____

9. _____

10. _____

11. _____

12. _____

13. _____

14. _____

15. _____

16. _____

17. _____

18. _____

19. _____

20. _____

21. _____

22. _____

23. _____

24. _____

25. _____

26. _____

27. _____

28. _____

29. _____

18. Write as a mixed number. $\dfrac{100}{7}$

19. Multiply. $3\dfrac{7}{8} \times 2\dfrac{5}{6}$

20. Divide. $\dfrac{44}{49} \div 2\dfrac{13}{21}$

21. Find the least common denominator of $\dfrac{6}{13}$ and $\dfrac{5}{39}$.

Evaluate using the correct order of operations.

22. $\dfrac{7}{18} + \dfrac{5}{27}$

23. $2\dfrac{1}{8} + 6\dfrac{3}{4}$

24. $12\dfrac{1}{5} - 4\dfrac{2}{3}$

25. $\dfrac{1}{3} + \dfrac{5}{8} \div \dfrac{5}{4}$

Answer each question.

26. A truck hauled $9\dfrac{1}{2}$ tons of gravel on Monday. On Tuesday, it hauled $6\dfrac{3}{8}$ tons and on Wednesday, $7\dfrac{1}{4}$ tons. How many tons were hauled on the three days?

27. Melinda traveled $221\dfrac{2}{5}$ miles on 9 gallons of gas. How many miles per gallon did her car get?

28. A recipe requires $3\dfrac{1}{4}$ cups of sugar and $2\dfrac{1}{3}$ cups of flour. Marcia wants to make $2\dfrac{1}{2}$ times that recipe. How many cups of each will she need?

29. A space probe travels at 28,356 miles per hour for 2142 hours. Estimate how many miles it travels.

Decimals

Recent expeditions to the Arctic and submarines traveling under the North Pole have confirmed that the layer of ice that covers the Arctic region is getting thinner. How fast is this change taking place? Can you predict how thin the ice layer will become in the near future? Turn to page 237 and try the Putting Your Skills to Work problems to find out.

Pretest Chapter 3

If you are familiar with the topics in this chapter, take this test now. Check your answers with those in the back of the book. If an answer is wrong or you can't answer a question, study the appropriate section of the chapter.

If you are not familiar with the topics in this chapter, don't take this test now. Instead, study the examples, work the practice problems, and then take the test.

This test will help you identify which concepts you have mastered and which you need to study further.

Section 3.1

1. Write a word name for the decimal. 47.813

2. Express as a decimal. $\dfrac{567}{10,000}$

Write as a fraction or a mixed number. Reduce whenever possible.

3. 2.11

4. 0.525

Section 3.2

5. Place the set of numbers in the proper order from smallest to largest. 1.6, 1.59, 1.61, 1.601

6. Round to the nearest tenth. 123.49268

7. Round to the nearest thousandth. 1.053458

Section 3.3

Add.

8.
$$\begin{array}{r} 5.12 \\ 4.7 \\ 8.03 \\ +\ 1.6 \\ \hline \end{array}$$

9. 24.613 + 0.273 + 2.305

Subtract.

10.
$$\begin{array}{r} 42.16 \\ -\ 31.57 \\ \hline \end{array}$$

11. 26 − 18.329

Section 3.4

Multiply.

12.
$$11.67$$
$$\times\ \ 0.03$$

13. 4.7805×1000

14. 0.0003796×10^5

Section 3.5

Divide.

15. $0.09\overline{)0.03186}$

16. $0.3328 \div 2.6$

Section 3.6

17. Write as a decimal. $\dfrac{11}{16}$

18. Write as a repeating decimal. $\dfrac{5}{22}$

19. Perform the operations in the correct order. $(0.2)^2 + 8.7 \times 0.3 - 1.68$

Section 3.7

20. When Marcia began her trip, the odometer on her car read 57,124.8. When she ended the trip, it read 57,312.8. She used 10.5 gallons of gas. How many miles per gallon did her car get? Round your answer to the nearest tenth.

▲ **21.** A rectangular room is 10.5 yards long and 3.6 yards wide. How much will it cost to install carpeting for the entire room at $12.95 per square yard?

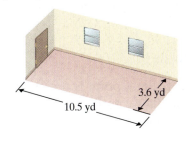

3.6 yd

10.5 yd

22. Tony worked 35 hours last week. He was paid $298.55. How much was he paid per hour?

12.	
13.	
14.	
15.	
16.	
17.	
18.	
19.	
20.	
21.	
22.	

3.1 Decimal Notation

Student Learning Objectives

After studying this section, you will be able to:

 1 Write a word name for a decimal fraction.

 2 Change from fractional notation to decimal notation.

 3 Change from decimal notation to fractional notation.

SSM

PH TUTOR   CD & VIDEO   MATH PRO   WEB
CENTER

1 Writing a Word Name for a Decimal Fraction

In Chapter 2 we discussed *fractions*—the set of numbers such as $\frac{1}{2}, \frac{2}{3}, \frac{1}{10}, \frac{6}{7}, \frac{18}{100}$, and so on. Now we will take a closer look at **decimal fractions**—that is, fractions with 10, 100, 1000, and so on, in the denominator, such as $\frac{1}{10}, \frac{18}{100}$, and $\frac{43}{1000}$.

Why, of all fractions, do we take special notice of these? Our hands have 10 digits. Our U.S. money system is based on the dollar, which has 100 equal parts, or cents. And the international system of measurement called the *metric system* is based on 10 and powers of 10.

As with other numbers, these decimal fractions can be written in different ways (forms). For example, the shaded part of the whole in the following drawing can be written:

in words (one-tenth)

in fractional form $\left(\frac{1}{10}\right)$

in decimal form (0.1)

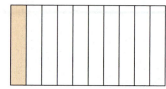

All mean the same quantity, namely 1 out of 10 equal parts of the whole. We'll see that when we use decimal notation, computations can be easily done based on the old rules for whole numbers and a few new rules about where to place the decimal point. In a world where calculators and computers are commonplace, many of the fractions we encounter are decimal fractions. A decimal fraction is a fraction whose denominator is a power of 10.

$$\frac{7}{10} \text{ is a decimal fraction.} \qquad \frac{89}{10^2} = \frac{89}{100} \text{ is a decimal fraction.}$$

Decimal fractions can be written with numerals in two ways: fractional form or decimal form. Some decimal fractions are shown in decimal form below.

Fractional Form		Decimal Form
$\frac{3}{10}$	=	0.3
$\frac{59}{100}$	=	0.59
$\frac{171}{1000}$	=	0.171

The zero in front of the decimal point is not actually required. We place it there simply to make sure that we don't miss seeing the decimal point. A number written in decimal notation has three parts.

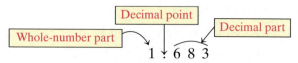

When a number is written in decimal form, the first digit to the right of the decimal point represents tenths, the next digit hundredths, the next digit thousandths, and so on. 0.9 means nine tenths and is equivalent to $\frac{9}{10}$. 0.51 means fifty-one hundredths and is equivalent to $\frac{51}{100}$. Some decimals are larger than 1. For example, 1.683 means one and six hundred eighty-three thousandths.

It is equivalent to $1\frac{683}{1000}$. Note that the word *and* is used to indicate the decimal point. A place-value chart is helpful.

Decimal Place Values

Hundreds	Tens	Ones	Decimal point	Tenths	Hundredths	Thousandths	Ten-thousandths
1	5	6	.	2	8	7	4
100	10	1	"and"	$\frac{1}{10}$	$\frac{1}{100}$	$\frac{1}{1,000}$	$\frac{1}{10,000}$

So, we can write 156.2874 in words as one hundred fifty-six and two thousand eight hundred seventy-four ten-thousandths. We say ten-thousandths because it is the name of the last decimal place on the right.

EXAMPLE 1 Write a word name for each decimal.

(a) 0.79   **(b)** 0.5308   **(c)** 1.6   **(d)** 23.765

(a) 0.79 = seventy-nine hundredths
(b) 0.5308 = five thousand three hundred eight ten-thousandths
(c) 1.6 = one and six tenths
(d) 23.765 = twenty-three and seven hundred sixty-five thousandths

Practice Problem 1 Write a word name for each decimal.

(a) 0.073   **(b)** 4.68   **(c)** 0.0017   **(d)** 561.78

Sometimes, decimals are used where we would not expect them. For example, we commonly say that there are 365 days in a year, with 366 days in every fourth year (or leap year). However, this is not quite correct. In fact, from time to time further adjustments need to be made to the calendar to adjust for these inconsistencies. Astronomers know that a more accurate measure of a year is called a **tropical year** (measured from one equinox to the next). Rounded to the nearest hundred-thousandth, 1 tropical year = 365.24122 days. This is read "three hundred sixty-five and twenty-four thousand, one hundred twenty-two hundred-thousandths." This approximate value is a more accurate measurement of the amount of time it takes the earth to complete one orbit around the sun.

Note the relationship between fractions and their equivalent numbers' decimal forms.

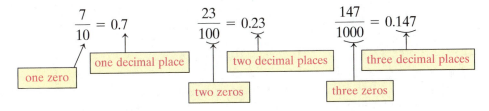

Decimal notation is commonly used with money. When writing a check, we often write the amount that is less than 1 dollar, such as 23¢, as $\frac{23}{100}$ dollar.

ALICE J. JENNINGTON
208 BARTON SPRINGS 512-555-1212
AUSTIN, TX 78704

37-86
110

3680

DATE _March 18, 2000_

PAY TO THE
ORDER OF _Rosetta Ramirez_ | $ | 59.23 |

Fifty-nine and $\frac{23}{100}$ ———————————— DOLLARS

ACB Austin Central Bank
Austin, Texas

MEMO ___ _textbooks_

⑈063000420⑈ 800⑈3649 28⑈ 3680

● EXAMPLE 2 Write a word name for the amount on a check made out for $672.89.

six hundred seventy-two and $\frac{89}{100}$ dollars

Practice Problem 2 Write a word name for the amount of a check made out for $7863.04. ●

2 *Changing from Fractional Notation to Decimal Notation*

It is helpful to be able to write decimals in both decimal notation and fractional notation. First we illustrate changing a fraction with a denominator of 10, 100, or 1000 into decimal form.

EXAMPLE 3 Write as a decimal.

(a) $\frac{8}{10}$ **(b)** $\frac{74}{100}$ **(c)** $1\frac{3}{10}$ **(d)** $2\frac{56}{1000}$

(a) $\frac{8}{10} = 0.8$ **(b)** $\frac{74}{100} = 0.74$ **(c)** $1\frac{3}{10} = 1.3$ **(d)** $2\frac{56}{1000} = 2.056$

Note: We need to add a zero before the digits 56. Since there are three zeroes in the denominator, we need three decimal places in the decimal number.

Practice Problem 3 Write as a decimal.

(a) $\frac{9}{10}$ **(b)** $\frac{136}{1000}$ **(c)** $2\frac{56}{100}$ **(d)** $34\frac{86}{1000}$ ●

3 *Changing from Decimal Notation to Fractional Notation*

● EXAMPLE 4 Write in fractional notation.

(a) 0.51 **(b)** 18.1 **(c)** 0.7611 **(d)** 1.363

(a) $0.51 = \frac{51}{100}$ **(b)** $18.1 = 18\frac{1}{10}$

(c) $0.7611 = \frac{7611}{10,000}$ **(d)** $1.363 = 1\frac{363}{1000}$

Practice Problem 4   Write in fractional notation.

(a) 0.37   **(b)** 182.3   **(c)** 0.7131   **(d)** 42.019

When we convert from decimal form to fractional form, we reduce whenever possible.

EXAMPLE 5   Write in fractional notation. Reduce whenever possible.

(a) 2.6   **(b)** 0.38   **(c)** 0.525   **(d)** 361.007

(a) $2.6 = 2\dfrac{6}{10} = 2\dfrac{3}{5}$ **(b)** $0.38 = \dfrac{38}{100} = \dfrac{19}{50}$

(c) $0.525 = \dfrac{525}{1000} = \dfrac{105}{200} = \dfrac{21}{40}$

(d) $361.007 = 361\dfrac{7}{1000}$ (cannot be reduced)

Practice Problem 5   Write in fractional notation. Reduce whenever possible.

(a) 8.5   **(b)** 0.58   **(c)** 36.25   **(d)** 106.013

EXAMPLE 6   A chemist found that the concentration of lead in a water sample was 5 parts per million. What fraction would represent the concentration of lead?

Five parts per million means 5 parts out of 1,000,000. As a fraction, this is $\frac{5}{1,000,000}$. We can reduce this by dividing numerator and denominator by 5. Thus

$$\frac{5}{1,000,000} = \frac{1}{200,000}.$$

The concentration of lead in the water sample is $\frac{1}{200,000}$.

Practice Problem 6   A chemist found that the concentration of PCBs in a water sample was 2 parts per billion. What fraction would represent the concentration of PCBs?

Developing Your Study Skills

Steps Toward Success In Mathematics

Mathematics is a building process, mastered one step at a time. The foundation of this process is formed by a few basic requirements. Those who are successful in mathematics realize the absolute necessity for building a study of mathematics on the firm foundation of these six minimum requirements.

1. Attend class every day.

2. Read the textbook.

3. Take notes in class.

4. Do assigned homework every day.

5. Get help immediately when needed.

6. Review regularly.

Verbal and Writing Skills

1. Describe a decimal fraction and provide examples.

2. What word is used to describe the decimal point when writing the word name for a decimal that is greater than one?

3. What is the name of the last decimal place on the right for the decimal 132.45678?

4. When writing $82.75 on a check, we write 75¢ as

_____.

Write a word name for each decimal.

5. 0.57

6. 0.78

7. 3.8

8. 5.2

9. 5.283

10. 3.117

11. 28.0037

12. 54.0013

Write a word name as you would on a check.

13. $124.20

14. $239.55

15. $1236.08

16. $7652.02

17. $10,000.76

18. $20,000.67

Write in decimal notation.

19. seven tenths

20. twelve hundredths

21. forty-five hundredths

22. seventy-five thousandths

23. twenty-two thousandths

24. twenty-nine ten-thousandths

25. two hundred eighty-six millionths

26. seven hundred sixteen millionths

Write each fraction as a decimal.

27. $\dfrac{7}{10}$

28. $\dfrac{3}{10}$

29. $\dfrac{76}{100}$

30. $\dfrac{84}{100}$

31. $\dfrac{1}{100}$

32. $\dfrac{6}{100}$

33. $\dfrac{53}{1000}$

34. $\dfrac{42}{1000}$

35. $\dfrac{2403}{10,000}$

36. $\dfrac{7794}{10,000}$

37. $8\dfrac{3}{10}$

38. $3\dfrac{1}{10}$

39. $84\dfrac{13}{100}$ **40.** $52\dfrac{77}{100}$ **41.** $1\dfrac{19}{1000}$ **42.** $2\dfrac{23}{1000}$ **43.** $126\dfrac{571}{10,000}$ **44.** $198\dfrac{333}{10,000}$

Write in fractional notation. Reduce whenever possible.

45. 0.02   **46.** 0.05   **47.** 3.6   **48.** 7.4

49. 8.24   **50.** 15.75   **51.** 12.625   **52.** 29.875

53. 7.0015   **54.** 4.0016   **55.** 235.1254   **56.** 581.2406

57. 0.0187   **58.** 0.0209   **59.** 8.0108   **60.** 7.0605

Applications

61. American bald eagles have been fighting extinction due to environmental hazards such as DDT, PCBs, and dioxin. The problem is with the food chain. Fish or rodents consume contaminated food and/or water. Then the eagles ingest the poison, which in turn affects the durability of the eagles' eggs. It takes only 4 parts per million of certain chemicals to ruin an eagle egg; write this number as a fraction in lowest terms. (In 1994 the bald eagle was removed from the endangered species list.)

62. Every year turtles lay eggs on the islands of South Carolina. Unfortunately, due to illegal polluting, a lot of the eggs are contaminated. If the turtle eggs contain more than 2 parts per one hundred million of chemical pollutants, they will not hatch and the population will continue to head toward extinction. Write the preceding amount of chemical pollutants as a fraction in the lowest terms.

Cumulative Review Problems

63. Add.
$$\begin{array}{r} 156 \\ 84 \\ 39 \\ 463 \\ +\ \ 76 \\ \hline \end{array}$$

64. Subtract.
$$\begin{array}{r} 12,843 \\ -\ 11,905 \\ \hline \end{array}$$

65. Round to the nearest *hundred*. 56,758

66. Round to the nearest *thousand*. 8,069,482

3.2 Comparing, Ordering, and Rounding Decimals

Student Learning Objectives

After studying this section, you will be able to:

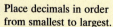 **1** Compare decimals.

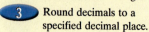 **2** Place decimals in order from smallest to largest.

3 Round decimals to a specified decimal place.

SSM

PH TUTOR CENTER   CD & VIDEO   MATH PRO   WEB

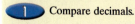

0 1 2 3 4 5 6 7 8 9 10

Comparing Decimals

All of the numbers we have studied have a specific order. To illustrate this order, we can place the numbers on a **number line.** Look at the number line on the left. Each number has a specific place on it. The arrow points in the direction of increasing value. Thus, if one number is to the right of a second number, it is larger, or greater, than that number. Since 5 is to the right of 2 on the number line, we say that 5 is greater than 2. We write $5 > 2$.

Since 4 is to the left of 6 on the number line, we say that 4 is less than 6. We write $4 < 6$.

The symbols " $>$ " and " $<$ " are called **inequality symbols.**

$a < b$ is read "*a* is less than *b*."

$a > b$ is read "*a* is greater than *b*."

We can assign exactly one point on the number line to each decimal number. When two decimal numbers are placed on a number line, the one farther to the right is the larger. Thus we can say that $3.4 > 2.7$ and $4.3 > 4.0$. We can also say that $0.5 < 1.0$ and $1.8 < 2.2$. Why?

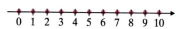

To compare or order decimals without using a number line, we compare each digit.

Comparing Two Numbers in Decimal Notation

1. Start at the left and compare corresponding digits. If the digits are the same, move one place to the right.

2. When two digits are different, the larger number is the one with the larger digit.

EXAMPLE 1   Write an inequality statement with 0.167 and 0.166. The numbers in the tenths place are the same. They are both 1.

0.**1** 6 7   0.**1** 6 6

The numbers in the hundredths place are the same. They are both 6.

0.1 **6** 7   0.1 **6** 6

The numbers in the thousandths place differ.

0. 1 6 **7**   0.1. 6 **6**

Since $7 > 6$, we know that $0.167 > 0.166$.

Practice Problem 1   Write an inequality statement with 5.74 and 5.75.

208

Whenever necessary, extra zeros can be written to the right of the last digit—that is, to the right of the decimal point—without changing the value of the decimal. Thus

$$0.56 = 0.56000 \quad \text{and} \quad 0.7768 = 0.77680.$$

The zero to the left of the decimal point is optional. Thus $0.56 = .56$. Both notations are used. You are encouraged to place a zero to the left of the decimal point so that you don't miss the decimal point when you work with decimals.

● **EXAMPLE 2**   Fill in the blank with one of the symbols $<, =,$ or $>$.

$$0.77 _____ 0.777$$

We begin by adding a zero to the first decimal.

$$0.77\underline{0} \qquad 0.77\underline{7}$$

We see that the tenths and hundredths digits are equal. But the thousandths digits differ. Since $0 < 7$, we have $0.770 < 0.777$.

Practice Problem 2   Fill in the blank with one of the symbols $<, =,$ or $>$.

$$0.894 _____ 0.89$$

● **2** ● *Placing Decimals in Order from Smallest to Largest*

Which is the heaviest—a puppy that weighs 6.2 ounces, a puppy that weighs 6.28 ounces, or a puppy that weighs 6.028 ounces? Did you choose the puppy that weighs 6.28 ounces? You are correct.

You can place three or more decimals in order. If you are asked to order the decimals from smallest to largest, look for the smallest decimal and place it first.

● **EXAMPLE 3**   Place the following five decimal numbers in order from smallest to largest.

$$1.834, \quad 1.83, \quad 1.381, \quad 1.38, \quad 1.8$$

First we add zeros to make the comparison easier.

$$1.834, \quad 1.830, \quad 1.381, \quad 1.380, \quad 1.800$$

Now we rearrange with smallest first.

$$1.380, \quad 1.381, \quad 1.800, \quad 1.830, \quad 1.834$$

Practice Problem 3   Place the following five decimal numbers in order from smallest to largest.

$$2.45, \quad 2.543, \quad 2.46, \quad 2.54, \quad 2.5$$

● **3** ● *Rounding Decimals to a Specified Decimal Place*

Sometimes in calculations involving money, we see numbers like $386.432 and $29.5986. To make these useful, we usually round them to the nearest cent. $386.432 is rounded to $386.43. $29.5986 is rounded to $29.60. A general rule for rounding decimals follows.

Rounding Decimals

1. Find the decimal place (units, tenths, hundredths, and so on) for which rounding off is required.
2. If the first digit to the right of the given place value is less than 5, drop it and all digits to the right of it.
3. If the first digit to the right of the given place value is 5 or greater, increase the number in the given place value by one. Drop all digits to the right of this place.

EXAMPLE 4   Round 156.37 to the nearest tenth.

156.3 7
└─────── We find the tenths place.

Note that 7, the next place to the right, is greater than 5. We round up to 156. **4** and drop the digits to the right. The answer is 156.4.

Practice Problem 4   Round 723.88 to the nearest tenth.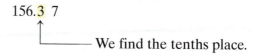

EXAMPLE 5   Round to the nearest thousandth.

(a) 0.06358   **(b)** 128.37448

(a) 0.06 **3** 58
    └─────── We locate the thousandths place.

Note that the digit to the right of the thousandths place is 5. We round up to 0.064 and drop all the digits to the right.

(b) 128.37 **4** 48
     └─────── We locate the thousandths place.

Note that the digit to the right of the thousandths place is less than 5. We round to 128.374 and drop all the digits to the right.

Practice Problem 5   Round to the nearest thousandth.

(a) 12.92647   **(b)** 0.007892

Remember that rounding up to the next digit in a position may result in several digits being changed.

EXAMPLE 6   Round to the nearest hundredth. Fred and Linda used 203.9964 kilowatt hours of electricity in their house in May.

203.9 **9** 64
    └─────── We locate the hundredths place.

Since the digit to the right of hundredths is greater than 5, we round up. This affects the next two positions. Do you see why? The result is 204.00 kilowatt hours.

Practice Problem 6   Round to the nearest tenth. Last month the college gymnasium used 15,699.953 kilowatt hours of electricity.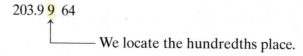

Sometimes we round a decimal to the nearest whole number. For example, when writing figures on income tax forms, a taxpayer may round all figures to the nearest dollar.

● EXAMPLE 7 To complete her income tax return, Marge needs to round these figures to the nearest whole dollar.

Medical bills $779.86   Taxes $563.49

Retirement contributions $674.38   Contributions to charity $534.77

Round the amounts.

	Original Figure	Rounded To Nearest Dollar
Medical bills	779.86	780
Taxes	563.49	563
Retirement	674.38	674
Charity	534.77	535

Practice Problem 7 Round the following figures to the nearest whole dollar.

Medical bills $375.50   Taxes $981.39

Retirement contributions $980.49   Contributions to charity $817.65 ●

⊘ **WARNING** Why is it so important to consider only *one* digit to the right of the desired round-off position? What is wrong with rounding in steps? Suppose that Mark rounds 1.349 to the nearest tenth in steps. First he rounds 1.349 to 1.35 (nearest hundredth). Then he rounds 1.35 to 1.4 (nearest tenth). What is wrong with this reasoning?

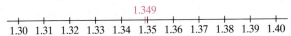

To round 1.349 to the nearest tenth, We ask if 1.349 is closer to 1.3 or to 1.4. It is closer to 1.3. Mark got 1.4, so he is not correct. He "rounded in steps" by first moving to 1.35, thus increasing the error and moving in the wrong direction. To control rounding errors, we consider *only* the first digit to the right of the decimal place to which we are rounding.

Fill in the blank with one of the symbols < , = , or >.

1. 1.3 ___ 1.29

2. 2.6 ___ 2.58

3. 0.68 ___ 0.681

4. 72.54 ___ 72.56

5. 18.92 ___ 18.93

6. 0.460 ___ 0.46

7. 0.0006 ___ 0.0005

8. 0.0037 ___ 0.003

9. 1.002 ___ 1.0021

10. 2.0056 ___ 2.006

11. 126.34 ___ 125.35

12. 406.78 ___ 407.75

13. $\dfrac{72}{1000}$ ___ 0.072

14. $\dfrac{54}{1000}$ ___ 0.054

15. $\dfrac{8}{10}$ ___ 0.08

16. $\dfrac{5}{100}$ ___ 0.005

Arrange each set of decimals from smallest to largest.

17. 12.6, 12.8, 12.65

18. 18.32, 18.038, 18.04

19. 0.0071, 0.05, 0.007

20. 0.0025, 0.0052, 0.002

21. 5.2, 5.23, 5.3, 5.12

22. 4.67, 4.1, 4.73, 4.7

23. 26.034, 26.003, 26.04, 26.033

24. 33.082, 33.02, 33.088, 33.079

25. 18.006, 18.060, 18.066, 18.606, 18.065

26. 15.020, 15.002, 15.001, 15.018, 15.0019

Round to the nearest tenth.

27. 5.67

28. 8.35

29. 29.43

30. 47.94

31. 578.064

32. 311.091

33. 2176.83

34. 4082.74

Round to the nearest hundredth.

35. 26.032

36. 47.071

37. 5.76582

38. 2.98613

39. 156.1749

40. 283.8441

41. 2786.706

42. 4609.285

Round to the nearest given place.

43. 1.06132   thousandths

44. 8.10263   thousandths

45. 0.047357   ten-thousandths

46. 0.063148   ten-thousandths

47. 5.00761238   hundred-thousandths

48. 4.01062378   hundred-thousandths

49. 135.564   nearest whole number

50. 208.7302   nearest whole number

Round to the nearest dollar.

51. $2536.85   **52.** $5319.62   **53.** $10,098.47   **54.** $20,159.48

Round to the nearest cent.

55. $56.9832   **56.** $28.7619   **57.** $5783.716   **58.** $3928.649

Applications

59. The average man consumes 0.095 kilogram of protein per day, while the average woman consumes 0.066 kilogram of protein per day. Round these values to the nearest hundredth.

60. In the 1960s, each person in the U.S. ate an average of 10.62 pounds of chocolate per year. In 2000, the average person ate 26.38 pounds of chocolate per year. Round these values to the nearest tenth.

61. The number of days in a year is 365.24122. Round this value to the nearest hundredth.

62. The numbers π and e (used in higher mathematics) are approximately equal to 3.14159 and 2.71828, respectively. Round these values to the nearest hundredth.

To Think About

63. Arrange in order from smallest to largest.

$$0.61, 0.062, \frac{6}{10}, 0.006, 0.0059,$$

$$\frac{6}{100}, 0.0601, 0.0519, 0.0612$$

64. Arrange in order from smallest to largest.

$$1.05, 1.512, \frac{15}{10}, 1.0513, 0.049,$$

$$\frac{151}{100}, 0.0515, 0.052, 1.051$$

65. A person wants to round 86.23498 to the nearest hundredth. He first rounds 86.23498 to 86.2350. He then rounds to 86.235. Finally, he rounds to 86.24. What is wrong with his reasoning?

66. Fred is checking the calculations on his monthly bank statement. An interest charge of $16.3724 was rounded to $16.38. An interest charge of $43.7214 was rounded to $43.73. What rule does the bank use for rounding off to the nearest cent?

Cumulative Review Problems

67. Add. $3\frac{1}{4} + 2\frac{1}{2} + 6\frac{3}{8}$

68. Subtract. $27\frac{1}{5} - 16\frac{3}{4}$

69. Mary drove her Dodge Caravan on a trip. At the start of the trip, the odometer (which measures distance) read 46,381. At the end of the trip, it read 47,073. How many miles long was the trip?

70. The cost of four pieces of equipment for the college computer center was $1736, $2714, $892, and $4316. What was their total cost?

3.3 Addition and Subtraction of Decimals

Student Learning Objectives

After studying this section, you will be able to:

 Add decimals.

 Subtract decimals.

SSM PH TUTOR CD & VIDEO MATH PRO WEB
        CENTER

1 Adding Decimals

We often add decimals when we check the addition of our bill at a restaurant or at a store. We can relate addition of decimals to addition of fractions. For example,

$$\frac{3}{10} + \frac{6}{10} = \frac{9}{10} \quad \text{and} \quad 1\frac{1}{10} + 2\frac{8}{10} = 3\frac{9}{10}.$$

These same problems can be written more efficiently as decimals.

$$\begin{array}{r} 0.3 \\ + 0.6 \\ \hline 0.9 \end{array} \qquad \begin{array}{r} 1.1 \\ + 2.8 \\ \hline 3.9 \end{array}$$

The steps to follow when adding decimals are listed in the following box.

Adding Decimals

1. Write the numbers to be added vertically and line up the decimal points. Extra zeros may be placed to the right of the decimal points if needed.

2. Add all the digits with the same place value, starting with the right column and moving to the left.

3. Place the decimal point of the sum in line with the decimal points of the numbers added.

● EXAMPLE 1   Add.

(a) 2.8 + 5.6 + 3.2   **(b)** 158.26 + 200.07 + 315.98

(c) 5.3 + 26.182 + 0.0007 + 624

$$\textbf{(a)} \quad \begin{array}{r} \overset{1}{2.8} \\ 5.6 \\ + \ 3.2 \\ \hline 11.6 \end{array} \qquad \textbf{(b)} \quad \begin{array}{r} \overset{11\ 2}{158.26} \\ 200.07 \\ + 315.98 \\ \hline 674.31 \end{array} \qquad \textbf{(c)} \quad \begin{array}{r} 5.3000 \\ 26.1820 \\ 0.0007 \\ + 624.0000 \\ \hline 655.4827 \end{array}$$

Extra zeros have been added to make the problem easier. *Note:* The decimal point is understood to be to the right of the digit 4.

Practice Problem 1   Add.

(a) $\begin{array}{r} 9.8 \\ 3.6 \\ + 5.4 \\ \hline \end{array}$ **(b)** $\begin{array}{r} 300.72 \\ 163.75 \\ + 291.08 \\ \hline \end{array}$ **(c)** 8.9 + 37.056 + 0.0023 + 945

Sidelight

When we add decimals like 3.1 + 2.16 + 4.007, we may write in zeros, as shown:

$$\begin{array}{r} 3.100 \\ 2.160 \\ + 4.007 \\ \hline 9.267 \end{array}$$

What are we really doing here? What is the advantage of adding these extra zeros?

"Decimals" means "decimal fractions." If we look at the number as fractions, we see that we are actually using the property of multiplying a fraction by 1 in order to obtain common denominators. Look at the problem this way:

$$3.1 \quad = 3\frac{1}{10}$$

$$2.16 \quad = 2\frac{16}{100}$$

$$4.007 = 4\frac{7}{1000}$$

The least common denominator is 1000. To obtain the common denominator for the first two fractions, we multiply.

$$3 \; \frac{1}{10} \times \frac{100}{100} \quad = 3\frac{100}{1000}$$

$$2 \; \frac{16}{100} \times \frac{10}{10} \quad = 2\frac{160}{1000}$$

$$+4 \; \frac{7}{1000} \qquad\quad = 4\frac{7}{1000}$$

Once we obtain a common denominator, we can add the three fractions.

$$9\frac{267}{1000} = 9.267$$

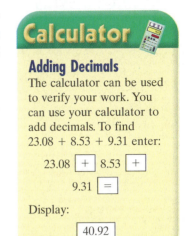

Calculator

Adding Decimals
The calculator can be used to verify your work. You can use your calculator to add decimals. To find $23.08 + 8.53 + 9.31$ enter:

23.08 [+] 8.53 [+]
9.31 [=]

Display:

40.92

This is the answer we arrived at earlier using the decimal form for each number. Thus writing in zeros in a decimal fraction is really an easy way to transform fractions to equivalent fractions with a common denominator. Working with decimal fractions is easier than working with other fractions.

The final digit of most odometers measures tenths of a mile. The odometer reading shown in the odometer on the right is 38,516.2 miles.

EXAMPLE 2 Barbara checked her odometer before the summer began. It read 49,645.8 miles. She traveled 3852.6 miles that summer in her car. What was the odometer reading at the end of the summer?

$$\begin{array}{r} \overset{1\,1\quad\;1}{49,645.8} \\ +\;\;3,852.6 \\ \hline 53,498.4 \end{array}$$

The odometer read 53,498.4 miles.

Practice Problem 2 A car odometer read 93,521.8 miles before a trip of 1634.8 miles. What was the final odometer reading?

EXAMPLE 3 During his first semester at Tarrant County Community College, Kelvey deposited checks into his checking account in the amounts of $98.64, $157.32, $204.81, $36.07, and $229.89. What was the sum of his five checks?

$$\begin{array}{r} \overset{2\,3\,2\;2}{\$\;\;98.64} \\ 157.32 \\ 204.81 \\ 36.07 \\ +\;\;229.89 \\ \hline \$726.73 \end{array}$$

Practice Problem 3 During the spring semester, Will deposited the following checks into his account: $80.95, $133.91, $256.47, $53.08, and $381.32. What was the sum of his five checks?

2 Subtracting Decimals

It is important to see the relationship between the decimal form of a mixed number and the fractional form of a mixed number. This relationship helps us understand why calculations with decimals are done the way they are. Recall that when we subtract mixed numbers with common denominators, sometimes we must borrow from the whole number.

$$5\frac{4}{10} = 4\frac{14}{10}$$
$$-2\frac{7}{10} = -2\frac{7}{10}$$
$$2\frac{7}{10}$$

We could write the same problem in decimal form:

$$
\begin{array}{r}
\overset{4}{\cancel{5}}.\overset{14}{\cancel{4}} \\
-\ 2.7 \\
\hline
2.7
\end{array}
$$

Subtraction of decimals is thus similar to subtraction of fractions (we get the same result), but it's usually easier to subtract with decimals than to subtract with fractions.

Subtracting Decimals

1. Write the decimals to be subtracted vertically and line up the decimal points. Additional zeros may be placed to the right of the decimal point if not all numbers have the same number of decimal places.

2. Subtract all digits with the same place value, starting with the right column and moving to the left. Borrow when necessary.

3. Place the decimal point of the difference in line with the decimal point of the two numbers being subtracted.

EXAMPLE 4 Subtract. **(a)** 84.8   **(b)** 1076.320
                                  − 27.3   − 983.518

(a)
$$
\begin{array}{r}
\overset{7}{\cancel{8}}\ \overset{14}{\cancel{4}}.8 \\
-\ 2\ 7.3 \\
\hline
5\ 7.5
\end{array}
$$

(b)
$$
\begin{array}{r}
\cancel{1}\ \overset{9}{\cancel{0}}\ \overset{\overset{10}{\cancel{10}}}{}\ \overset{17}{\cancel{7}}\ \overset{5}{\cancel{6}}.\ \overset{13}{\cancel{3}}\ \overset{1}{\cancel{2}}\ \overset{10}{\cancel{0}} \\
-\ \ \ 9\ 8\ 3.5\ 1\ 8 \\
\hline
9\ 2.8\ 0\ 2
\end{array}
$$

Practice Problem 4   Subtract.

(a)   38.8   **(b)**   2034.908
        − 26.9   − 1986.325

When the two numbers being subtracted do not have the same number of decimal places, write in zeros as needed.

EXAMPLE 5 Subtract. **(a)** $12 - 8.362$ **(b)** $156.381 - 99.82$

(a)
$$
\begin{array}{r}
\overset{\underset{\displaystyle 11}{\cancel{1}} \ \overset{9}{\cancel{\overset{10}{\cancel{0}}}} \ \overset{9}{\cancel{\overset{10}{\cancel{0}}}} \ \overset{}{\cancel{\overset{10}{\cancel{0}}}}}{\cancel{1}\,\cancel{2}\,.\,\cancel{0}\,\cancel{0}\,\cancel{0}} \\
-\quad 8\,.\,3\ 6\ 2 \\
\hline
3\,.\,6\ 3\ 8
\end{array}
$$

(b)
$$
\begin{array}{r}
\overset{\underset{}{\ \ \ \overset{14}{\cancel{\overset{4}{\cancel{1}}}} \ \overset{15}{\cancel{\overset{8}{\cancel{5}}}} \ \overset{13}{\cancel{3}}}}{\cancel{1}\,\cancel{5}\,\cancel{6}\,.\,\cancel{3}\,8\ 1} \\
-\quad 9\ 9\,.\,8\ 2\ 0 \\
\hline
5\ 6\,.\,5\ 6\ 1
\end{array}
$$

Practice Problem 5 Subtract. **(a)** $19 - 12.579$ **(b)** $283.076 - 96.38$

EXAMPLE 6 On Tuesday, Don Ling filled the gas tank in his car. The odometer read 56,098.5. He drove for four days. The next time he filled the tank, the odometer read 56,420.2. How many miles had he driven?

$$
\begin{array}{r}
\overset{\underset{}{\ \ \ \ \ \overset{11}{\ } \ \ \overset{9}{\ }}}{\ \ \ \ \overset{3}{\ } \ \overset{1}{\cancel{4}} \ \overset{10}{\cancel{0}} \ \overset{12}{\cancel{2}}} \\
5\ 6,\cancel{4}\ \cancel{2}\ \cancel{0}.\cancel{2} \\
-\ 5\ 6,0\ \ 9\ \ 8.5 \\
\hline
3\ \ 2\ \ 1.7
\end{array}
$$

He had driven 321.7 miles.

Practice Problem 6 Abdul had his car oil changed when his car odometer read 82,370.9 miles. When he changed the oil again, the odometer read 87,160.1 miles. How many miles did he drive between oil changes?

EXAMPLE 7 Find the value of x if $x + 3.9 = 14.6$.

Recall that the letter x is a variable. It represents a number that is added to 3.9 to obtain 14.6. We can find the number x if we calculate $14.6 - 3.9$.

$$
\begin{array}{r}
\overset{\underset{}{\ \ \overset{3}{\ }\ \overset{16}{\ }}}{1\,\cancel{4}.\cancel{6}} \\
-\quad 3.9 \\
\hline
1\,0.7
\end{array}
$$

Thus $x = 10.7$.

Check. Is this true? If we replace x by 10.7, do we get a true statement?

$$
\begin{aligned}
x + 3.9 &= 14.6 \\
10.7 + 3.9 &\overset{?}{=} 14.6 \\
14.6 &= 14.6 \ \checkmark
\end{aligned}
$$

Practice Problem 7 Find the value of x if $x + 10.8 = 15.3$.

Developing Your Study Skills

Making a Friend in the Class

Attempt to make a friend in your class. You may find that you enjoy sitting together and drawing support and encouragement from each other. Exchange phone numbers so you can call each other whenever you get stuck in your study. Set up convenient times to study together on a regular basis, to do homework, and to review for exams.

You must not depend on a friend or fellow student to tutor you, do your work for you, or in any way be responsible for your learning. However, you will learn from each other as you seek to master the course. Studying with a friend and comparing notes, methods, and solutions can be very helpful. And it can make learning mathematics a lot more fun!

3.3 Exercises

Add.

1. 57.1 + 19.7

2. 78.3 + 29.4

3. 718.98 + 496.57

4. 813.47 + 629.86

5.
```
   13.4
    7.6
+ 275.2
```

6.
```
  176.5
    8.4
+  22.5
```

7.
```
   5.6
  9.23
+ 8.17
```

8.
```
  2.65
   3.2
+ 7.76
```

9.
```
  4.9637
  28.12
+  3.645
```

10.
```
  7.0276
  3.451
+ 16.98
```

11.
```
12.
    3.62
+  51.8
```

12.
```
13.
    4.52
+  63.7
```

13. 753.61 + 28.75 + 162.3 + 100.5 + 67

14. 432.51 + 16.08 + 892.1 + 301.2 + 84

Applications

In exercises 15 and 16, calculate the perimeter of each triangle.

15.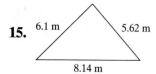

6.1 m   5.62 m   8.14 m

16.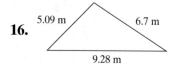

5.09 m   6.7 m   9.28 m

17. Lamar is losing weight by walking each evening after dinner. During the first week in February he lost 1.75 pounds. During the second, third, and fourth weeks, he lost 2.5 pounds, 1.55 pounds, and 2.8 pounds, respectively. How many total pounds did Lamar lose in February?

18. Teresa knows she needs to drink more water while at work. One day during her morning break she drank 7.15 ounces. At lunch she drank 12.45 ounces and throughout the afternoon she drank 10.75 ounces. How many total ounces of water did she drink?

19. Mick and Keith have arrived in Miami and are going to the beach. They buy sunblock for $4.99, beverages for $12.50, sandwiches for $11.85, towels for $28.50, bottled water for $3.29, and two novels for $16.99. After they got what they needed, what was Mick and Keith's bill for their day at the beach?

20. Brent purchased items at a grocery store just off-campus. They cost $7.18, $5.29, $0.61, $1.15, and $1.49. What was the total of Brent's grocery bill?

In exercises 21 and 22, a portion of a bank checking account deposit slip is shown. Add the numbers to determine the total deposit. The line drawn between the dollars and the cents column serves as the decimal point.

21.

22.

Subtract.

23. $6.8 - 2.9$

24. $3.6 - 2.8$

25. $123 - 96.34$

26. $161 - 89.29$

27.
$$\begin{array}{r} 132.2 \\ -\ 16.67 \\ \hline \end{array}$$

28.
$$\begin{array}{r} 473.6 \\ -\ 79.7 \\ \hline \end{array}$$

29.
$$\begin{array}{r} 586.513 \\ -\ 78.2 \\ \hline \end{array}$$

30.
$$\begin{array}{r} 243.967 \\ -\ 84.2 \\ \hline \end{array}$$

31.
$$\begin{array}{r} 3.00269 \\ -\ 0.80368 \\ \hline \end{array}$$

32.
$$\begin{array}{r} 5.00735 \\ -\ 1.05921 \\ \hline \end{array}$$

33.
$$\begin{array}{r} 24.0079 \\ -\ 19.3614 \\ \hline \end{array}$$

34.
$$\begin{array}{r} 52.0708 \\ -\ 41.9312 \\ \hline \end{array}$$

35.
$$\begin{array}{r} 8 \\ -\ 1.263 \\ \hline \end{array}$$

36.
$$\begin{array}{r} 12 \\ -\ 7.981 \\ \hline \end{array}$$

37.
$$\begin{array}{r} 7362.14 \\ -\ 6173.07 \\ \hline \end{array}$$

38.
$$\begin{array}{r} 4986.71 \\ -\ 3615.93 \\ \hline \end{array}$$

39.
$$\begin{array}{r} 1.5 \\ -\ 0.0365 \\ \hline \end{array}$$

40.
$$\begin{array}{r} 2.8 \\ -\ 0.07763 \\ \hline \end{array}$$

Applications

41. The average length of an Emerald Tree boa constrictor is 1.8 meters. A native found one that measured 3.264 meters long. How much longer was this snake than the average Emerald Tree boa constrictor?

42. During the three months of August, September, and October, baby Kathryn gained 7.675 kilograms. During the three months of November, December, and January, she gained another 9.986 kilograms. How much weight has the baby gained in the last six months?

43. A child's beginner telescope is priced at $79.49. The price of a certain professional telescope is $37,026.65. How much more does the professional telescope cost?

44. Tamika drove on a summer trip. When she began, the odometer read 26,052.3 miles. At the end of the trip, the odometer read 28,715.1 miles. How long was the trip?

45. Malcolm took a taxi from John F. Kennedy Airport in New York to his hotel in the city. His fare was $47.70 and he tipped the driver $7.00. How much change did Malcolm get back if he gave the driver a $100 bill?

46. Nathan spent $28.54 at Pet Express and $39.38 at Super Stop and Shop. If he had $100 before he went shopping, how much does he have left?

47. An insulated wire measures 12.62 centimeters. The last 0.98 centimeter of the wire is exposed. How long is the part of the wire that is not exposed?

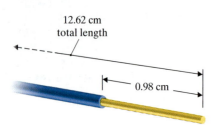

12.62 cm
total length

0.98 cm

▲ **48.** The outside radius of a pipe is 9.39 centimeters. The inside radius is 7.93 centimeters. What is the thickness of the pipe?

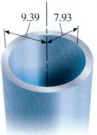

9.39   7.93

49. A cancer researcher is involved in an important experiment. She is trying to determine how much of an anticancer drug is necessary for a Stage I (nonhuman or animal) test. She pours 2.45 liters of the experimental anticancer formula in one container and 1.35 liters of a reactive liquid in another. She then pours the contents of one container into the other. If 0.85 liters is expected to evaporate during the process, how much liquid will be left?

50. Everyone is becoming aware of the rapid loss of the Earth's rainforests. Between 1981 and 1999, tropical South America lost a substantial amount of natural resources due to deforestation and development. In 1981, there were 797,100,000 hectares of rainforest. In 1999 there were 683,400,000 hectares. How much rainforest, in hectares, was destroyed? *(Source:* United Nations Statistics Division*)*

The federal water safety standard requires that drinking water contain no more than 0.015 milligrams of lead per liter of water. (Source: *Environmental Protection Agency*)

51. Carlos and Maria had the well that supplies their home analyzed for safety. A sample of well water contained 0.0089 milligrams of lead per liter of water. What is the difference between their sample and the federal safety standard? Is it safe for them to drink the water?

52. Fred and Donna use water provided by the city for the drinking water in their home. A sample of their tap water contained 0.023 milligrams of lead per liter of water. What is the difference between their sample and the federal safety standard? Is it safe for them to drink the water?

The following table shows the income of the United States by industry. Use this table for exercises 53–56. Write each answer as a decimal and as a whole number. The table values are recorded in billions of dollars.

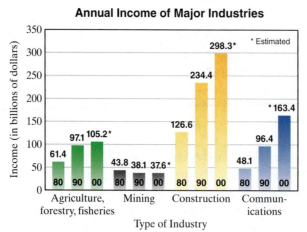

Annual Income of Major Industries

Income (in billions of dollars) / Type of Industry

* Estimated

298.3*

234.4

*163.4

126.6

105.2*
97.1
96.4

61.4

43.8 38.1 37.6*

48.1

Agriculture, forestry, fisheries   Mining   Construction   Communications

80 90 00

Source: Bureau of Labor Statistics

53. How many more dollars were earned in mining in 1980 than in 2000?

54. How many more dollars were earned in construction in 2000 than in 1980?

55. In 1980, how many more dollars were earned in agriculture, forestry, and fisheries than in communications?

56. In 1990, how many more dollars were earned in communication than in mining?

To Think About

Mr. Jensen made up the following shopping list of items he needs and the cost of each item. Use the list to answer exercises 57 and 58.

can cranberry sauce	$0.99
hot dog relish	$0.79
ranch salad dressing	$1.47
large can solid white tuna	$2.29
can tomato soup	$0.68
can sliced peaches	$1.26
large jar tomato sauce	$1.65
large box Cheerios	$3.79
medium box Raisin Bran	$2.63
medium jar peanut butter	$2.19

57. Mr. Jensen goes to the store to buy the following items from his list: Raisin Bran, ranch salad dressing, sliced peaches, hot dog relish, and peanut butter. He has a ten-dollar bill. Round each item to the nearest ten cents and estimate the cost of buying these items by first rounding the cost of each item to the nearest ten cents. Does he have enough money to buy all of them? Find the exact cost of these items. How close was your estimate?

58. The next day the Jensens' daughter, Brenda, goes to the store to buy the following items from the list: Cheerios, tomato sauce, peanut butter, white tuna, tomato soup, and cranberry sauce. She has fifteen dollars. Estimate the cost of buying these items by first rounding the cost of each item to the nearest ten cents. Does she have enough money to buy all of them? Find the exact cost of these items. How close was your estimate?

Find the value of x.

59. $x + 7.1 = 15.5$

60. $x + 4.8 = 23.1$

61. $156.9 + x = 200.6$

62. $210.3 + x = 301.2$

63. $4.162 = x + 2.053$

64. $7.076 = x + 5.602$

Cumulative Review Problems

Multiply.

65. $\begin{array}{r} 2536 \\ \times\ \ \ 8 \\ \hline \end{array}$

66. $\begin{array}{r} 467 \\ \times\ \ 39 \\ \hline \end{array}$

67. $\dfrac{22}{7} \times \dfrac{49}{50}$

68. $2\dfrac{1}{3} \times 3\dfrac{3}{4}$

3.4 Multiplication of Decimals

1 Multiplying a Decimal by a Decimal or a Whole Number

We learned previously that the product of two fractions is the product of the numerators over the product of the denominators. For example,

$$\frac{3}{10} \times \frac{7}{100} = \frac{21}{1000}$$

In decimal form this product would be written

$$0.3 \times 0.07 = 0.021$$

one decimal place   two decimal places   three decimal places

---

Multiplication of Decimals

1. Multiply the numbers just as you would multiply whole numbers.

2. Find the sum of the decimal places in the two factors.

3. Place the decimal point in the product so that the product has the same number of decimal places as the sum in step 2. You may need to write zeros to the left of the number found in step 1.

Now use these steps to do the preceding multiplication problem.

EXAMPLE 1   Multiply. 0.07×0.3

$$
\begin{array}{rl}
0.07 & \text{2 decimal places} \\
\times\ 0.3 & \text{1 decimal place} \\
\hline
0.021 & \text{3 decimal places in product } (2 + 1 = 3)
\end{array}
$$

Practice Problem 1   Multiply. 0.09×0.6

When performing the calculation, it is usually easier to place the factor with the smallest number of nonzero digits underneath the other factor.

EXAMPLE 2   Multiply.

(a) 0.38×0.26

(b) 12.64×0.572

(a)
$$
\begin{array}{rl}
0.38 & \text{2 decimal places} \\
\times 0.26 & \text{2 decimal places} \\
\hline
228 & \\
76 & \\
\hline
0.0988 & \text{4 decimal places} \\
& (2 + 2 = 4)
\end{array}
$$

Note that we need to insert a zero before the 988.

(b)
$$
\begin{array}{rl}
12.64 & \text{2 decimal places} \\
\times 0.572 & \text{3 decimal places} \\
\hline
2528 & \\
8848 & \\
6\,320 & \\
\hline
7.23008 & \text{5 decimal places} \\
& (2 + 3 = 5)
\end{array}
$$

222

Practice Problem 2   Multiply.

(a) 0.47×0.28

(b) 0.436×18.39

When multiplying decimal fractions by a whole number, you need to remember that a whole number has no decimal places.

EXAMPLE 3   Multiply. 5.261×45

$$
\begin{array}{rl}
5.261 & \text{3 decimal places} \\
\times\quad 45 & \text{0 decimal places} \\
\hline
26\,305 & \\
210\,44 & \\
\hline
236.745 & \text{3 decimal places } (3 + 0 = 3)
\end{array}
$$

Calculator

Multiplying Decimals
You can use your calculator to multiply a decimal by a decimal. To find 0.08×1.53 enter:

$$0.08 \;\boxed{\times}\; 1.53 \;\boxed{=}$$

Display:

$$\boxed{0.1224}$$

Practice Problem 3   Multiply. 0.4264×38

EXAMPLE 4   Uncle Roger's rectangular front lawn measures 50.6 yards wide and 71.4 yards long. What is the area of the lawn in square yards?

Since the lawn is rectangular, we will use the fact that to find the area of a rectangle we multiply the length by the width.

$$
\begin{array}{rl}
71.4 & \text{1 decimal place} \\
\times\, 50.6 & \text{1 decimal place} \\
\hline
42\,84 & \\
3570\,0 & \\
\hline
3612.84 & \text{2 decimal places}
\end{array}
$$

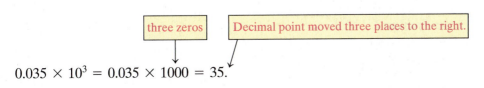

71.4 yards   50.6 yards

The area of the lawn is 3612.84 square yards.

Practice Problem 4   A rectangular computer chip measures 1.26 millimeters wide and 2.3 millimeters long. What is the area of the chip in square millimeters?

2   Multiplying a Decimal by a Power of 10

Observe the following pattern.

one zero   Decimal point moved one place to the right.

$0.035 \times 10^1 = 0.035 \times 10 = 0.35$

two zeros   Decimal point moved two places to the right.

$0.035 \times 10^2 = 0.035 \times 100 = 3.5$

three zeros   Decimal point moved three places to the right.

$0.035 \times 10^3 = 0.035 \times 1000 = 35.$

Multiplication of a Decimal by a Power of 10

To multiply a decimal by a power of 10, move the decimal point to the right the same number of places as the number of zeros in the power of 10.

EXAMPLE 5   Multiply.   **(a)** 2.671×10 **(b)** 37.85×100

(a) 2.671×10 $= 26.71$

one zero   Decimal point moved one place to the right.

(b) 37.85×100 $= 3785.$

two zeros   Decimal point moved two places to the right.

Practice Problem 5   Multiply.   **(a)** 0.0561×10 **(b)** 1462.37×100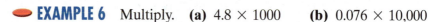

Sometimes it is necessary to add extra zeros before placing the decimal point in the answer.

EXAMPLE 6   Multiply.   **(a)** 4.8×1000 **(b)** $0.076 \times 10{,}000$

(a) 4.8×1000 $= 4800.$

(three zeros)   Decimal point moved three places to the right. Two extra zeros were needed.

(b) $0.076 \times 10{,}000$ $= 760.$

(four zeros)   Decimal point moved four places to the right. One extra zero was needed.

Practice Problem 6   Multiply.   **(a)** 0.26×1000 **(b)** $5862.89 \times 10{,}000$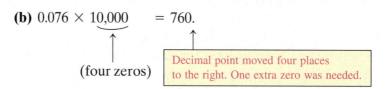

If the number that is a power of 10 is in exponent form, move the decimal point to the right the same number of places as the number that is the exponent.

EXAMPLE 7   Multiply. 3.68×10^3

Exponent of 3   Decimal point moved three places to the right.

$3.68 \times 10^3 = 3680.$

Practice Problem 7   Multiply. 7.684×10^4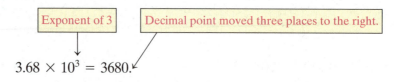

Sidelight

Can you devise a quick rule to use when multiplying a decimal fraction by $\frac{1}{10}, \frac{1}{100}, \frac{1}{1000}$, and so on? How is it like the rules developed in this section? Consider a few examples.

Original Problem	Change Fraction to Decimal	Decimal Multiplication	Observation
$86 \times \dfrac{1}{10}$	86×0.1	$\begin{array}{r} 86 \\ \times 0.1 \\ \hline 8.6 \end{array}$	Decimal point moved one place to the left.
$86 \times \dfrac{1}{100}$	86×0.01	$\begin{array}{r} 86 \\ \times 0.01 \\ \hline 0.86 \end{array}$	Decimal point moved two places to the left.
$86 \times \dfrac{1}{1000}$	86×0.001	$\begin{array}{r} 86 \\ \times 0.001 \\ \hline 0.086 \end{array}$	Decimal point moved three places to the left.

Can you think of a way to describe a rule that you could use in solving this type of problem without going through all the foregoing steps?

You use multiplying by a power of 10 when you convert a larger unit of measure to a smaller unit of measure in the metric system.

EXAMPLE 8   Change 2.96 kilometers to meters.

Since we are going from a larger unit of measure to a smaller one, we multiply. There are 1000 meters in 1 kilometer. Multiply 2.96 by 1000.

$$2.96 \times 1000 = 2960$$

2.96 kilometers is equal to 2960 meters.

Practice Problem 8   Change 156.2 kilometers to meters.

Names Used to Describe Large Numbers   Often when reading the newspaper or watching television news shows, we hear words like 3.46 trillion or 67.8 billion. These are abbreviated notations that are used to describe large numbers. When you encounter these numbers, you can change them to standard notation by multiplication of the appropriate value.

For example, if someone says that the population of China is 1.31 billion people, we can write 1.31 billion = 1.31 × 1 billion = 1.31 × 1,000,000,000 = 1,310,000,000. If someone says the population of Chicago is 2.92 million people, we can write

2.92 million = 2.92 × 1 million = 2.92 × 1,000,000 = 2,920,000.

3.4 Exercises

Multiply.

1. 0.6
     × 0.2

2. 0.9
     × 0.3

3. 0.12
     × 0.5

4. 0.17
     × 0.4

5. 0.0036
     ×   0.8

6. 0.067
     × 0.07

7. 0.079
     × 0.09

8. 0.0034
     ×   0.5

9. 0.043
     × 0.012

10. 0.037
      × 0.011

11. 10.97
      × 0.06

12. 18.07
      × 0.05

13. 5167
      × 0.19

14. 7986
      × 0.32

15. 2.163
      × 0.008

16. 1.892
      × 0.007

17. 0.7613
      ×   1009

18. 0.6178
      ×   5004

19. 2350
      × 3.6

20. 3720
      × 8.1

21. 4.57 × 11.8

22. 73.2 × 2.45

23. 0.001 × 6523.7

24. 0.01 × 826.75

Applications

25. Kenny is making payments on his Ford Escort of $155.40 per month for the next 60 months. How much will he have spent in car payments after he sends in his final payment?

26. Melissa earns $5.85 per hour for a 40-hour week. How much does she make in one week?

27. Mei Lee works for a company that manufactures electric and electronic equipment and earns $9.55 per hour for a 40-hour week. How much does she earn in one week? (The average wage in 1999 for U.S. electric/electronic equipment manufacturers was $10.18 per hour for an average of 41.3 hours, for a total of $420.43 per week.) (*Source*: Bureau of Labor Statistics)

28. Elva works for a company that manufactures textile products. She earns $7.20 per hour for a 40-hour week. How much does she earn in one week? (The average wage in 1999 for U.S. textile mill industries was $7.38 per hour for an average of 40.6 hours, for a total of $299.63 per week. (*Source*: Bureau of Labor Statistics)

▲ **29.** Ralph and Darlene are getting new carpet in their bedroom and need to find how many square feet they need to purchase. The dimensions of their rectangular bedroom are 15.5 feet and 19.2 feet. What is the area of the room in square feet?

▲ **30.** Lisa has a framed rectangular poster in her living room. Its dimensions are 18.3 inches and 27.8 inches. What is the area of the poster in square inches?

31. Dwight is paying off a student loan at Westmont College with payments of $36.90 per month for the next 18 months. How much will he pay off during the next 18 months?

32. Marcia is making car payments to Highfield Center Chevrolet of $230.50 per month for 16 more months. How much will she pay for car payments in the next 16 months?

33. Steve's car gets approximately 26.4 miles per gallon. His gas tank holds 19.5 gallons. Approximately how many miles can he travel on a full tank of gas?

34. Jim's 4 × 4 truck gets approximately 18.6 miles per gallon. His gas tank holds 19.5 gallons. Approximately how many miles can he travel on a full tank of gas? Compare this to your answer in exercise 33.

Multiply.

35. 2.86×10

36. 1.98×10

37. 43.0×10

38. 98.0×10

39. 128.65×1000

40. 204.37×1000

41. $5.60982 \times 10,000$

42. $1.27986 \times 10,000$

43. $280,560.2 \times 10^2$

44. 7163.241×10^2

45. 816.32×10^3

46. 763.49×10^4

Applications

47. To convert from meters to centimeters, multiply by 100. How many centimeters are in 5.932 meters?

48. One meter is about 39.36 inches. About how many inches are in 100 meters?

49. To convert from kilometers to meters, multiply by 1000. How many meters are in 2.98 kilometers?

50. Macy's purchased 100 men's shirts at $20.53 each. How much did Macy's pay for the order?

51. In June, Gabrielle receives $820.00 on her tax return. She decides to spend her money on family holiday gifts, six months early, so that she doesn't have to worry about it later. She spends $124.00 on a gift for her parents, $110.00 on a gift for her sister, $83.60 on a gift for her brother, $76.00 on a gift for her grandmother, and $44.60 for a gift for each of her four cousins. How much money does she have left over.

52. Tomba is a beautiful orange tabby cat. When he was found by the side of the road, he was three weeks old and weighed 0.95 lb. At the age of three months, he weighed 2.85 lb. At the age of nine months, he weighed 6.30 lb; at one year, he weighed 11.7 lb. Today, Tomba the cat is $1\frac{1}{2}$ years old, and weighs 15.75 lb.
(a) How much weight did he gain?
(b) If the veterinarian wants him to lose 0.25 lb per week until he weighs 13.5 lb, how long will it take?

▲ **53.** The college is purchasing new carpeting for the learning center. What is the price of a carpet that is 19.6 yards wide and 254.2 yards long if the cost is $12.50 per square yard?

54. A jewelry store purchased long lengths of gold chain, which will be cut and made into necklaces and bracelets. The store purchased 3220 grams of gold chain at $3.50 per gram.
(a) How much did the jewelry store spend?

(b) If they sell a 28-gram gold necklace for $17.75 per gram, how much profit will they make on the necklace?

To Think About

55. State in your own words a rule for mental multiplication by 0.1, 0.01, 0.001, 0.0001, and so on.

56. State in your own words a rule for mental multiplication by 0.2, 0.02, 0.002, 0.0002, and so on.

Cumulative Review Problems

Divide. Be sure to include any remainder as part of your answer.

57. $12\overline{)1176}$

58. $14\overline{)1204}$

59. $37\overline{)4629}$

60. $29\overline{)3745}$

The relief costs for the five most devastating hurricanes to hit the United States are represented in the following bar graph. Use the bar graph to answer exercises 61–64.

61. How much more were the relief costs for Hurricane Floyd than Hurricane Fran?

62. How much more were the relief costs for Hurricane Hugo than Hurricane Floyd?

63. FEMA has projected a contingency budget for 2001 that is $0.876 billion more that the relief budget for Hurricane George. What is that total amount of money?

64. FEMA has projected a contingency budget for 2002 that is $1.78 billion more than the relief budget for Hurricane Hugo. What is that total amount of money?

3.5 Division of Decimals

1 Dividing a Decimal by a Whole Number

When you divide a decimal by a whole number, place the decimal point for the quotient directly above the decimal point in the dividend. Then divide as if the numbers were whole numbers.

To divide 26.8 by 4, we place the decimal point of our answer (the quotient) directly *above* the decimal point in the dividend.

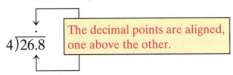

$$4\overline{)26.8}$$ The decimal points are aligned, one above the other.

Then we divide as if we were dividing whole numbers.

$$
\begin{array}{r}
6.7 \\
4\overline{)26.8} \\
\underline{24} \\
28 \\
\underline{28} \\
0
\end{array}
$$

The quotient is 6.7.

The quotient to a problem may have all digits to the right of the decimal point. In some cases you will have to put a zero in the quotient as a "place holder." Let's divide 0.268 by 4,

$$
\begin{array}{r}
0.067 \\
4\overline{)0.268} \\
\underline{24} \\
28 \\
\underline{28} \\
0
\end{array}
$$

Note that we must have a zero after the decimal point in 0.067.

EXAMPLE 1 Divide.

(a) $9\overline{)0.3204}$ **(b)** $14\overline{)36.12}$

(a)
$$
\begin{array}{r}
0.0356 \\
9\overline{)0.3204} \\
\underline{27} \\
50 \\
\underline{45} \\
54 \\
\underline{54} \\
0
\end{array}
$$

Note the zero *after* the decimal point.

(b)
$$
\begin{array}{r}
2.58 \\
14\overline{)36.12} \\
\underline{28} \\
81 \\
\underline{70} \\
112 \\
\underline{112} \\
0
\end{array}
$$

Practice Problem 1 Divide.

(a) $7\overline{)1.806}$ **(b)** $16\overline{)0.0928}$

Student Learning Objectives

After studying this section, you will be able to:

1 Divide a decimal by a whole number.

2 Divide a decimal by a decimal.

SSM   PH TUTOR   CD & VIDEO   MATH PRO   WEB
CENTER

229

Some division problems do not yield a remainder of zero. In such cases, we may be asked to round off the answer to a specified place. ==To round off, we carry out the division until our answer contains a digit that is one place to the right of that to which we intend to round. Then we round our answer to the specified place.== For example, to round to the nearest thousandth, we carry out the division to the ten-thousandths place. In some division problems, you will need to write in zeros at the end of the dividend so that this division can be carried out.

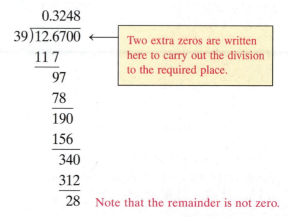 **EXAMPLE 2**   Divide and round the quotient to the nearest thousandth.

$$12.67 \div 39$$

We will carry out our division to the ten-thousandths place. Then we will round our answer to the nearest thousandth.

$$
\begin{array}{r}
0.3248 \\
39\overline{)12.6700} \\
\underline{11\ 7} \\
97 \\
\underline{78} \\
190 \\
\underline{156} \\
340 \\
\underline{312} \\
28
\end{array}
$$

Two extra zeros are written here to carry out the division to the required place.

Note that the remainder is not zero.

Now we round 0.3248 to 0.325. The answer is rounded to the nearest thousandth.

Practice Problem 2   Divide and round the quotient to the nearest hundredth. 23.82 ÷ 46

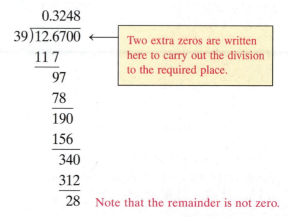 **EXAMPLE 3**   Maria paid $5.92 for 16 pounds of tomatoes. How much did she pay per pound?

The cost of one pound of tomatoes equals the total cost, $5.92, divided by 16 pounds. Thus we will divide.

$$
\begin{array}{r}
0.37 \\
16\overline{)5.92} \\
\underline{4\ 8} \\
112 \\
\underline{112}
\end{array}
$$

Maria paid $0.37 per pound for the tomatoes.

Practice Problem 3   Won Lin will pay off his auto loan for $3538.75 over 19 months. If the monthly payments are equal, how much will he pay each month?

2   Dividing a Decimal by a Decimal

When the divisor is not a whole number, we can convert the division problem to an equivalent problem that has a whole number as a divisor. Think about the reasons why this procedure will work. We will ask you about it after you study Examples 4 and 5.

Dividing by a Decimal

1. Make the divisor a whole number by moving the decimal point to the right. Mark that position with a caret ($_\wedge$). Count the number of places the decimal point moved.

2. Move the decimal point in the dividend to the right the same number of places. Mark that position with a caret.

3. Place the decimal point of your answer directly above the caret marking the decimal point of the dividend.

4. Divide as with whole numbers.

🔴 **EXAMPLE 4**   **(a)** Divide. $0.08\overline{)1.632}$ **(b)** Divide. $1.352 \div 0.026$

(a) $0.08.\overline{)1.63.2}$ Move each decimal point two places to the right.

Place the decimal point of the answer directly above the caret.

$0.08_\wedge\overline{)1.63_\wedge2}$ Mark the new position by a caret ($_\wedge$).

$$
\begin{array}{r}
20\,.\,4 \\
0.08_\wedge\overline{)1.63_\wedge2} \\
\underline{16} \\
3\ \ 2 \\
\underline{3\ \ 2} \\
0
\end{array}
$$

The answer is 20.4.

Perform the division.

(b)
$$
\begin{array}{r}
52\,. \\
0.026_\wedge\overline{)1.352_\wedge} \\
\underline{1\ 30} \\
52 \\
\underline{52} \\
0
\end{array}
$$

Move each decimal point three places to the right and mark the new position by a caret .

The answer is 52.

Practice Problem 4   Divide.   **(a)** $0.09\overline{)0.1008}$ **(b)** $1.702 \div 0.037$ 🔴

To Think About   Why do we move the decimal point to the right in the divisor and the dividend? What rule allows us to do this? How do we know the answer will be valid? We are actually using the property that multiplication of a fraction by 1 leaves the fraction unchanged. This is called the *multiplication identity*. Let us examine Example 4(b) again. We will write $1.352 \div 0.026$ as a fraction.

$\dfrac{1.352}{0.026} \times 1$ Multiplication of a fraction by 1 does not change the value of the fraction.

$= \dfrac{1.352}{0.026} \times \dfrac{1000}{1000}$ We know that $\dfrac{1000}{1000} = 1$.

$= \dfrac{1352}{26}$ Multiplication by 1000 can be done by moving the decimal point three places to the right.

$= 52$ Divide the whole numbers.

Thus in Example 4(b) when we moved the decimal point three places to the right in the divisor and the dividend, we were actually creating an equivalent fraction where the numerator and the denominator of the original fraction were multiplied by 1000.

 EXAMPLE 5 Divide. **(a)** $1.7 \overline{)0.0323}$ **(b)** $0.0032 \overline{)7.68}$

(a)
$$\begin{array}{r} 0.019 \\ 1.7_\wedge \overline{)0.0_\wedge 323} \\ \underline{17} \\ 153 \\ \underline{153} \\ 0 \end{array}$$

Move the decimal point in the divisor and dividend one place to the right and mark that position with a caret.

(b)
$$\begin{array}{r} 2400. \\ 0.0032_\wedge \overline{)7.6800_\wedge} \\ \underline{64} \\ 128 \\ \underline{128} \\ 000 \end{array}$$

Note that two extra zeros are needed in the dividend as we move the decimal point four places to the right.

Practice Problem 5 Divide. **(a)** $1.8 \overline{)0.0414}$ **(b)** $0.0036 \overline{)8.316}$

 EXAMPLE 6 **(a)** Find $2.9 \overline{)431.2}$ rounded to the nearest tenth.
(b) Find $2.17 \overline{)0.08}$ rounded to the nearest thousandth.

(a)
$$\begin{array}{r} 148.68 \\ 2.9_\wedge \overline{)431.2_\wedge 00} \\ \underline{29} \\ 141 \\ \underline{116} \\ 252 \\ \underline{232} \\ 200 \\ \underline{174} \\ 260 \\ \underline{232} \\ 28 \end{array}$$

Calculate to the hundredths place and round the answer to the nearest tenth.

The answer rounded to the nearest tenth is 148.7.

(b)
$$\begin{array}{r} 0.0368 \\ 2.17_\wedge \overline{)0.08_\wedge 0000} \\ \underline{651} \\ 1490 \\ \underline{1302} \\ 1880 \\ \underline{1736} \\ 144 \end{array}$$

Calculate to the ten-thousandths place and then round the answer. Rounding 0.0368 to the nearest thousandth, we obtain 0.037.

Practice Problem 6 **(a)** Find $3.8 \overline{)521.6}$ rounded to the nearest tenth.
(b) Find $8.05 \overline{)0.17}$ rounded to the nearest thousandth.

 EXAMPLE 7 John drove his 1997 Cavalier 420.5 miles to Chicago. He used 14.5 gallons of gas on the trip. How many miles per gallon did his car get on the trip?

To find miles per gallon we need to divide the number of miles, 420.5, by the number of gallons, 14.5.

$$\begin{array}{r} 29. \\ 14.5_\wedge \overline{)420.5_\wedge} \\ \underline{290} \\ 1305 \\ \underline{1305} \end{array}$$

Calculator

Dividing Decimals
You can use your calculator to divide a decimal by a decimal. To find $21.83 \overline{)54.53}$ rounded to the nearest hundredth, enter:

54.53 ÷ 21.38 =

Display:

2.5505145

This is an approximation. Some calculators will round to eight digits. The answer rounded to the nearest hundredth is 2.55.

John's car achieved 29 miles per gallon on the trip to Chicago.

Practice Problem 7   Sarah rented a large truck to move to Boston. She drove 454.4 miles yesterday. She used 28.5 gallons of gas on the trip. How many miles per gallon did the rental truck get?

EXAMPLE 8   Find the value of n if $0.8 \times n = 2.68$.

Here 0.8 is multiplied by some number n to obtain 2.68. What is this number n? If we divide 2.68 by 0.8, we will find the value of n.

$$
\begin{array}{r}
3.35 \\
0.8_\wedge \overline{)2.6_\wedge 80} \\
\underline{2\,4} \\
2\;8 \\
\underline{2\;4} \\
40 \\
\underline{40}
\end{array}
$$

Thus the value of n is 3.35.

Check.   Is this true? Are we sure the value of $n = 3.35$?
We substitute the value of $n = 3.35$ into the equation to see if it makes the statement true.

$$
\begin{array}{rcl}
0.8 \times n & = & 2.68 \\
0.8 \times 3.35 & \overset{?}{=} & 2.68 \\
2.68 & = & 2.68 \quad \checkmark \quad \text{Yes, it is true.}
\end{array}
$$

Practice Problem 8   Find the value of n if $0.12 \times n = 0.696$.

EXAMPLE 9   The level of sulfur dioxide emissions in the air has slowly been decreasing over the last 20 years, as can be seen in the accompanying bar graph. Find the average amount of sulfur dioxide emissions in the air over these five specific years.

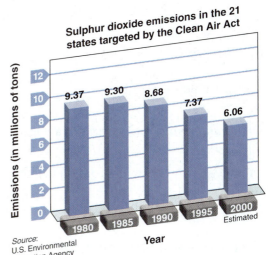

First we take the sum of the five years.

$$
\begin{array}{r}
9.37 \\
9.30 \\
8.68 \\
7.37 \\
+\;6.06 \\
\hline
40.78
\end{array}
$$

Then we divide by five to obtain the average.

$$
\begin{array}{r}
8.156 \\
5\overline{)40.780} \\
\underline{40} \\
7 \\
\underline{5} \\
28 \\
\underline{25} \\
30 \\
\underline{30}
\end{array}
$$

Thus the yearly average is 8.156 million tons of sulfur dioxide emissions in these 21 states.

Practice Problem 9   Use the accompanying bar graph to find the average level of sulfur dioxide for the three years: 1985, 1990, and 1995. By how much does the three-year average differ from the five-year average?

3.5 Exercises

Divide until there is a remainder of zero.

1. $6\overline{)12.6}$
2. $8\overline{)17.28}$
3. $4\overline{)71.32}$
4. $6\overline{)83.16}$

5. $7\overline{)128.17}$
6. $8\overline{)972}$
7. $6\overline{)819}$
8. $0.5\overline{)32.15}$

9. $0.4\overline{)47.28}$
10. $0.09\overline{)0.7209}$
11. $0.8113 \div 0.07$
12. $75.6 \div 3.6$

13. $68.4 \div 3.8$
14. $728 \div 5.6$
15. $40.30 \div 0.31$

Divide and round your answer to the nearest tenth.

16. $8\overline{)44}$
17. $9\overline{)47.31}$
18. $1.8\overline{)4.16}$

19. $1.9\overline{)2.36}$
20. $0.95\overline{)32.067}$
21. $0.85\overline{)41.901}$

Divide and round your answer to the nearest hundredth.

22. $4\overline{)263.82}$
23. $5\overline{)471.03}$
24. $1.7\overline{)20.8}$

Divide and round your answer to the nearest thousandth.

25. $9\overline{)241}$
26. $6\overline{)409.387}$
27. $0.69\overline{)8.45}$
28. $0.87\overline{)79.40}$

Divide and round your answer to the nearest whole number.

29. $12\overline{)1396}$
30. $0.065\overline{)4.398}$
31. $0.55\overline{)7.00}$
32. $0.39\overline{)5.00}$

Applications

33. The college cafeteria staff is portioning out butterscotch pudding for the next meal. The cook has 235.6 ounces of pudding and wants to end up with 38 equal portions. How many ounces will each portion have if the cook uses up the entire batch of butterscotch pudding?

34. The Miller family wants to use the latest technology to access the Internet from their home television system. The equipment needed to upgrade their existing equipment will cost $992.76. If the Millers make 12 equal monthly payments, how much will they pay per month?

35. Four students sit down to their weekly lasagna dinner. At one end of the table, there is a bottle containing 67.6 ounces of a popular soft drink. At the other end of the table is a bottle that contains 33.6 ounces of water.
  (a) If the students share the soft drink and water equally, how many ounces of liquid will each student drink?
  (b) At the last minute, another student is asked to join the group. How many ounces of liquid will each of the five students share?

36. Wally owns a Plymouth Breeze that travels 360 miles on 13.2 gallons of gas. How many miles per gallon does it achieve? (Round your answer to the nearest tenth.)

37. The local concert promoter paid a total of $3388.50 for nine bands to play in the outdoor festival. Each band received an equal amount. How much did he pay per band?

38. Andrea makes Mother's Day bouquets each year for extra income. This year her goal is to make $300. If she sells each bouquet for $12.50, how many bouquets must she sell to reach her goal?

39. Demitri had a contractor build an outdoor deck for his back porch. He now has $1131.75 to pay off, and he agreed to pay $125.75 per month. How many more payments on the outdoor deck must he make?

40. For their wedding reception, Sharon and Richard spent $1865.50 on food and drinks. If the caterer charged them $10.25 per person, how many guests did they have?

41.

Year	U.S. Annual Per Capita Turkey Consumption (Boneless Weight)
1980	8.1 lb
1985	9.2 lb
1990	13.9 lb
1995	14.3 lb
2000*	14.6 lb

*estimated
Source: U.S. Department of Agriculture

  (a) Using the preceding chart, find the average number of pounds of turkey consumed per person for the years 1980 and 1985.
  (b) What is the average increase in turkey consumption per person per year over this 20-year period?

42. Yoshi is working as an inspector for a company that makes snowboards. A Mach 1 snowboard weighs 3.8 kilograms. How many of these snowboards are contained in a box in which the contents weigh 87.40 kilograms? If the box is labeled CONTENTS: 24 SNOWBOARDS, how great an error was made in packing the box?

Find the value of n.

43. $0.7 \times n = 0.0861$

44. $0.6 \times n = 3.948$

45. $1.6 \times n = 110.4$

46. $1.4 \times n = 821.8$

47. $n \times 0.063 = 2.835$

48. $n \times 0.098 = 4.312$

To Think About

Multiply the numerator and denominator of each fraction by 10,000. Then divide the numerator by the denominator. Is the result the same if we divided the original numerator by the original denominator? Why?

49. $\dfrac{3.8702}{0.0523}$

50. $\dfrac{2.9356}{0.0716}$

Cumulative Review Problems

51. Add. $\dfrac{3}{8} + 1\dfrac{2}{5}$

52. Subtract. $2\dfrac{13}{16} - 1\dfrac{7}{8}$

53. Multiply. $3\dfrac{1}{2} \times 2\dfrac{1}{6}$

54. Divide. $4\dfrac{1}{3} \div 2\dfrac{3}{5}$

Putting Your Skills to Work

The Thinning Ice Cap

According to top scientists, the great ice cap that stretches across the top of the earth has become significantly thinner than it was several decades ago.

From 1958 to 1976, the average thickness of the Arctic Sea ice was approximately 10 feet. During the period 1993 to 1997, it was about 6 feet. Scientists estimate that since that time, the ice has been thinning at the rate of about 4 inches a year.

This phenomenon is directly or indirectly related to global warming. In 2000, it was estimated that the temperature of the Earth had risen by 1.75 degrees Fahrenheit over the previous century. Most scientists (but not all) expect that the average temperature of the Earth will rise by another 3.5 degrees Fahrenheit by 2100 if emissions of heat-trapping gases such as carbon dioxide continue at present rates.

Problems for Individual Investigation

If the rate of thinning of the polar ice cap continues at the present rate, what will be the thickness of the ice in the Arctic Sea

1. in 2005?

2. in 2009?

Problems for Group Investigation and Cooperative Study

If the predictions for the rate of global warming over the next century are correct, what will be the amount of global warming

3. in 2015?

4. in 2068?

Internet Connections

  Netsite: http://www.prenhall.com/tobey_basic

This site contains facts and predictions related to global warming. Based on the information available on that site:

5. How many feet are the oceans expected to rise from 2000 to 2100?

6. The average temperature of New York City today is 54 degrees Fahrenheit. What will this be in 2100 if the present warming trend continues?

SSM · PH TUTOR CENTER · CD & VIDEO · MATH PRO · WEB

Springfield $2\frac{1}{2}$ miles

Springfield 2.5 miles

1 Converting a Fraction to a Decimal

A number can be expressed in two equivalent forms: as a fraction and as a decimal.

Fraction $2\frac{1}{2}$

two and one-half

2.5 Decimal

two and five-tenths

Same quantity, different appearance

Every decimal can be expressed as an equivalent fraction. For example,

Decimal form ⇒ fraction form

$$0.75 = \frac{75}{100} \quad \text{or} \quad \frac{3}{4}$$

$$0.5 = \frac{5}{10} \quad \text{or} \quad \frac{1}{2}$$

$$2.5 = 2\frac{5}{10} = 2\frac{1}{2} \quad \text{or} \quad \frac{5}{2}$$

And every fraction can be expressed as an equivalent decimal, as we will learn in this section. For example,

Fraction form ⇒ decimal form

$$\frac{1}{5} = 0.20 \quad \text{or} \quad 0.2$$

$$\frac{3}{8} = 0.375$$

$$\frac{5}{11} = 0.4545 \dots \quad \text{(The "45" keeps repeating.)}$$

Some of these decimal equivalents are so common that people find it helpful to memorize them. You would be wise to memorize the following equivalents:

$$\frac{1}{2} = 0.5 \qquad \frac{1}{4} = 0.25 \qquad \frac{1}{5} = 0.2 \qquad \frac{1}{10} = 0.1$$

We previously studied how to convert some fractions with a denominator of 10, 100, 1000, and so on to decimal form. For example, $\frac{3}{10} = 0.3$ and $\frac{7}{100} = 0.07$. We need to develop a procedure to write other fractions, such as $\frac{3}{8}$ and $\frac{5}{16}$, in decimal form.

Converting a Fraction to an Equivalent Decimal

Divide the denominator into the numerator until
   (a) the remainder becomes zero, or
   (b) the remainder repeats itself, or
   (c) the desired number of decimal places is achieved.

EXAMPLE 1 Write as an equivalent decimal.

(a) $\frac{3}{8}$ **(b)** $\frac{31}{40}$ of a second

Divide the denominator into the numerator until the remainder becomes zero.

(a)
$$
\begin{array}{r}
0.375 \\
8\overline{)3.000} \\
\underline{2\,4} \\
60 \\
\underline{56} \\
40 \\
\underline{40} \\
0
\end{array}
$$

Therefore, $\dfrac{3}{8} = 0.375$.

(b)
$$
\begin{array}{r}
0.775 \\
40\overline{)31.000} \\
\underline{28\,0} \\
3\,00 \\
\underline{2\,80} \\
200 \\
\underline{200} \\
0
\end{array}
$$

Therefore, $\dfrac{31}{40} = 0.775$ of a second.

Practice Problem 1   Write as an equivalent decimal. **(a)** $\dfrac{5}{16}$ **(b)** $\dfrac{11}{80}$

Athletics' times in Olympic events, such as the 100-meter dash, are measured to the nearest hundredth of a second. Future Olympic athletics' times will be measured to the nearest thousandth of a second.

Decimals such as 0.375 and 0.775 are called **terminating decimals.** When converting $\frac{3}{8}$ to 0.375 or $\frac{31}{40}$ to 0.775, the division operation eventually yields a remainder of zero. Other fractions yield a repeating pattern. For example, $\frac{1}{3} = 0.3333\ldots$ and $\frac{2}{3} = 0.6666\ldots$ have a pattern of repeating digits. Decimals that have a digit or a group of digits that repeats are called **repeating decimals.** We often indicate the repeating pattern with a bar over the repeating group of digits:

$$0.3333\ldots = 0.\overline{3} \qquad 0.74\,74\,74\ldots = 0.\overline{74}$$
$$0.\,218\,\,218\,\,218\ldots = 0.\overline{218} \qquad 0.8942\,8942\ldots = 0.\overline{8942}$$

If when converting fractions to decimal form the remainder repeats itself, we know that we have a repeating decimal.

EXAMPLE 2   Write as an equivalent decimal.

(a) $\dfrac{5}{11}$ **(b)** $\dfrac{13}{22}$ **(c)** $\dfrac{5}{37}$

(a)
$$
\begin{array}{r}
0.4545 \\
11\overline{)5.0000} \\
\underline{4\,4} \\
60 \\
\underline{55} \\
50 \\
\underline{44} \\
60 \\
\underline{55} \\
5
\end{array}
$$

repeating remainders

Thus $\dfrac{5}{11} = 0.4545\ldots = 0.\overline{45}$

(b)
$$
\begin{array}{r}
0.59090 \\
22\overline{)13.00000} \\
\underline{11\,0} \\
2\,00 \\
\underline{1\,98} \\
2\,00 \\
\underline{1\,98} \\
20
\end{array}
$$

repeating remainders

Thus $\dfrac{13}{22} = 0.5909090\ldots = 0.5\overline{90}$

Notice that the bar is over the digits 9 and 0 but *not* over the digit 5.

(c)
$$
\begin{array}{r}
0.1351 \\
37\overline{)5.0000} \\
\underline{37} \\
130 \\
\underline{111} \\
190 \\
\underline{185} \\
50 \\
\underline{37} \\
13
\end{array}
$$

repeating remainders

Thus $\dfrac{5}{37} = 0.135135\ldots = 0.\overline{135}$

Calculator

Fraction to Decimal

You can use a calculator to change $\frac{5}{8}$ to a decimal.

Enter:

5 ÷ 8 =

The display should read

| 0.625 |

Try the following.

(a) $\frac{17}{25}$ (b) $\frac{2}{9}$

(c) $\frac{13}{10}$ (d) $\frac{15}{19}$

Note: 0.78947368 is an approximation for $\frac{15}{19}$.

Some calculators round to only eight places.

Practice Problem 2   Write as an equivalent decimal.

(a) $\frac{7}{11}$ (b) $\frac{8}{15}$ (c) $\frac{13}{44}$

 EXAMPLE 3   Write as an equivalent decimal.

(a) $3\frac{7}{15}$ (b) $\frac{20}{11}$

(a) $3\frac{7}{15}$ means $3 + \frac{7}{15}$

$$
\begin{array}{r}
0.466 \\
15\overline{)7.000} \\
\underline{60} \\
100 \\
\underline{90} \\
100 \\
\underline{90} \\
10
\end{array}
$$

(b)
$$
\begin{array}{r}
1.818 \\
11\overline{)20.000} \\
\underline{11} \\
90 \\
\underline{88} \\
20 \\
\underline{11} \\
90 \\
\underline{88} \\
2
\end{array}
$$

Thus $\frac{7}{15} = 0.4\overline{6}$ and $3\frac{7}{15} = 3.4\overline{6}$ Thus $\frac{20}{11} = 1.818181\ldots = 1.\overline{81}$

Practice Problem 3   Write as an equivalent decimal.

(a) $2\frac{11}{18}$ (b) $\frac{28}{27}$

In some cases, the pattern of repeating is quite long. For example,

$$\frac{1}{7} = 0.142857142857\ldots = 0.\overline{142857}$$

Such problems are often rounded to a certain value.

 EXAMPLE 4   Express $\frac{5}{7}$ as a decimal rounded to the nearest thousandth.

$$
\begin{array}{r}
0.7142 \\
7\overline{)5.0000} \\
\underline{49} \\
10 \\
\underline{7} \\
30 \\
\underline{28} \\
20 \\
\underline{14} \\
6
\end{array}
$$

Rounding to the nearest thousandth, we round 0.7142 to 0.714. (In repeating form, $\frac{5}{7} = 0.714285714285\ldots = 0.\overline{714285}$.)

Practice Problem 4   Express $\dfrac{19}{24}$ as a decimal rounded to the nearest thousandth.

Recall that we studied placing two decimals in order in Section 3.2. If we are required to place a fraction and a decimal in order, it is usually easiest to change the fraction to decimal form and then compare the two decimals.

EXAMPLE 5   Fill in the blank with one of the symbols $<$, $=$, or $>$.

$$\frac{7}{16} \underline{\hspace{1cm}} 0.43$$

Now we divide to find the decimal equivalent of $\frac{7}{16}$.

$$
\begin{array}{r}
0.4375 \\
16\overline{)7.0000} \\
\underline{6\,4} \\
60 \\
\underline{48} \\
120 \\
\underline{112} \\
80 \\
\underline{80} \\
0
\end{array}
$$

Now in the thousandths place $7 > 0$, so we know

$$0.43\,7\,5 > 0.43\,0\,0.$$

Therefore, $\frac{7}{16} > 0.43$.

Practice Problem 5   Fill in the blank with one of the symbols $<$, $=$, or $>$.

$$\frac{5}{8} \underline{\hspace{1cm}} 0.63$$

2 Using the Correct Order of Operations with Decimals

The rules for order of operations that we discussed in Section 1.6 apply to operations with decimals.

Order of Operations

Do first   **1.** Perform operations inside parentheses.

2. Simplify any expressions with exponents.

3. Multiply or divide from left to right.

Do last   **4.** Add or subtract from left to right.

Sometimes exponents are used with decimals. In such cases, we merely evaluate using repeated multiplication.

$$(0.2)^2 = 0.2 \times 0.2 = 0.04$$
$$(0.2)^3 = 0.2 \times 0.2 \times 0.2 = 0.008$$
$$(0.2)^4 = 0.2 \times 0.2 \times 0.2 \times 0.2 = 0.0016$$

EXAMPLE 6   Evaluate. $(0.3)^3 + 0.6 \times 0.2 + 0.013$

First we need to evaluate $(0.3)^3 = 0.3 \times 0.3 \times 0.3 = 0.027$. Thus

$(0.3)^3 + 0.6 \times 0.2 + 0.013$

$= 0.027 + 0.6 \times 0.2 + 0.013$

$= 0.027 + 0.12 + 0.013$ ⟵   When addends have a different number of decimal places, writing the problem in column form makes adding easier.

$$\begin{array}{r} 0.027 \\ 0.120 \\ +\ 0.013 \\ \hline 0.160 \end{array}$$

$= 0.16$

Practice Problem 6   Evaluate. $0.3 \times 0.5 + (0.4)^3 - 0.036$

In the next example all four steps of the rules for order of operations will be used.

EXAMPLE 7   Evaluate. $(8 - 0.12) \div 2^3 + 5.68 \times 0.1$

$(8 - 0.12) \div 2^3 + 5.68 \times 0.1$

$= 7.88 \div 2^3 + 5.68 \times 0.1$ First do subtraction inside the parentheses.

$= 7.88 \div 8 + 5.68 \times 0.1$ Simplify the expression with exponents.

$= 0.985 + 0.568$ From left to right do division and multiplication.

$= 1.553$ Add the final two numbers.

Practice Problem 7   Evaluate. $6.56 \div (2 - 0.36) + (8.5 - 8.3)^2$

Developing Your Study Skills

Keep Trying

We live in a highly technical world, and you cannot afford to give up on the study of mathematics. Dropping mathematics may prevent you from entering certain career fields that you may find interesting. You may not have to take math courses as high-level as calculus, but such courses as intermediate algebra, finite math, college algebra, and trigonometry may be necessary. Learning mathematics can open new doors for you.

Learning mathematics is a process that takes time and effort. You will find that regular study and daily practice are necessary to strengthen your skills and to help you grow academically. This process will lead you toward success in mathematics. Then, as you become more successful, your confidence in your ability to do mathematics will grow.

3.6 Exercises

Verbal and Writing Skills

1. 0.75 and $\frac{3}{4}$ are different ways to express the _____.

2. To convert a fraction to an equivalent decimal, divide the _____ into the numerator.

3. Why is $0.\overline{8942}$ called a repeating decimal?

4. The order of operations for decimals is the same as the order of operations for whole numbers. Write the steps for the order of operations.

Write as an equivalent decimal. If a repeating decimal is obtained, use notation such as $0.\overline{7}$, $0.\overline{16}$, or $0.\overline{245}$.

5. $\frac{1}{4}$

6. $\frac{3}{4}$

7. $\frac{4}{5}$

8. $\frac{2}{5}$

9. $\frac{7}{16}$

10. $\frac{3}{16}$

11. $\frac{7}{20}$

12. $\frac{3}{40}$

13. $\frac{31}{50}$

14. $\frac{23}{25}$

15. $\frac{9}{4}$

16. $\frac{14}{5}$

17. $2\frac{1}{8}$

18. $3\frac{13}{16}$

19. $1\frac{7}{16}$

20. $1\frac{1}{40}$

21. $\frac{14}{15}$

22. $\frac{1}{6}$

23. $\frac{5}{11}$

24. $\frac{7}{11}$

25. $3\frac{7}{12}$

26. $7\frac{7}{11}$

27. $2\frac{5}{18}$

28. $6\frac{1}{6}$

Write as an equivalent decimal or a decimal approximation. Round your answer to the nearest thousandth if needed.

29. $\frac{4}{7}$

30. $\frac{13}{27}$

31. $\frac{19}{21}$

32. $\frac{20}{21}$

33. $\frac{7}{48}$

34. $\frac{5}{48}$

35. $\frac{35}{27}$

36. $\frac{37}{23}$

243

37. $\dfrac{21}{52}$ **38.** $\dfrac{1}{38}$ **39.** $\dfrac{17}{18}$ **40.** $\dfrac{5}{13}$

41. $\dfrac{22}{7}$ **42.** $\dfrac{17}{14}$ **43.** $\dfrac{9}{19}$ **44.** $\dfrac{11}{17}$

Applications

45. Your fingernails grow approximately $\frac{1}{8}$ inch each month. Write the thickness as a decimal.

46. The hair on your head grows approximately $\frac{9}{20}$ inch a month. Write the thickness as a decimal.

47. An ice climber had the local mountaineering shop install boot heaters to the back of his hiking boots. The installer drilled a hole $\frac{3}{8}$ inch in diameter. The hole for the boot heater should have been 0.5 inch in diameter. Is the hole too large or too small? By how much?

48. A master carpenter is re-creating a room for the set of a movie being filmed. He is using a burled maple veneer $\frac{7}{16}$ inch thick. The designer specified maple veneer 0.45 inch thick. Is the veneer he is using too thick or too thin? By how much?

49. A machinist in a factory is using sheet metal $\frac{17}{32}$ inch thick. The plans call for sheet metal 0.53 inch thick. Is the metal he is using too thick or too thin? By how much?

50. To manufacture a circuit board, Rick must program a computer to place a piece of thin plastic atop a circuit board. For the current to flow through the circuit, the top plastic piece must form a border of exactly $\frac{1}{16}$ inch with the circuit board. A few circuit boards were made with a border of 0.055 inch by accident. Is this border too small or too large? By how much?

Evaluate.

51. $2.4 + (0.5)^2 - 0.35$

52. $9.6 + 3.6 - (0.4)^2$

53. $12.2 \times 9.4 - 2.68 + 1.6 \div 0.8$

54. $2.3 \times 12.6 - 1.98 + 12.6 \div 1.4$

55. $12 \div 0.03 - 50 \times (0.5 + 1.5)^3$

56. $61.95 \div 1.05 - 2 \times (1.7 + 1.3)^3$

57. $(1.1)^3 + 2.6 \div 0.13 + 0.083$

58. $(1.1)^3 + 8.6 \div 2.15 - 0.086$

59. $(14.73 - 14.61)^2 \div (1.18 + 0.82)$

60. $(32.16 - 32.02)^2 \div (2.24 + 1.76)$

61. $(0.5)^3 + (3 - 2.6) \times 0.5$

62. $(0.6)^3 + (7 - 6.3) \times 0.07$

63. $(0.76 + 4.24) \div 0.25 + 8.6$

64. $(2.4)^2 + 3.6 \div (1.2 - 0.7)$

Evaluate.

65. $(1.6)^3 + (2.4)^2 + 18.666 \div 3.05 + 4.86$

66. $5.9 \times 3.6 \times 2.4 - 0.1 \times 0.2 \times 0.3 \times 0.4$

Write as a decimal. Round your answer to six decimal places.

 67. $\dfrac{5236}{8921}$

 68. $\dfrac{17,359}{19,826}$

To Think About

69. Subtract. $0.\overline{16} - 0.00\overline{16}$
   (a) What do you obtain?

   (b) Now subtract $0.\overline{16} - 0.01\overline{6}$. What do you obtain?

   (c) What is different about these results?

70. Subtract. $1.\overline{89} - 0.01\overline{89}$
   (a) What do you obtain?

   (b) Now subtract $1.\overline{89} - 0.18\overline{9}$. What do you obtain?

   (c) What is different about these results?

Cumulative Review Problems

▲ **71.** What is the area of a rectangle that measures 12 feet by 26 feet?

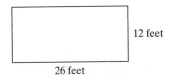

12 feet

26 feet

72. In the 2000 presidential election, 1,213,714 voters in Virginia voted for Al Gore and 1,242,987 voters in Wisconsin voted for Al Gore. What was the difference in the number of votes for Al Gore between these two states? (*Source:* Federal Election Commission)

73. If a person makes deposits of $56, $81, $42, and $198, what is the total amount deposited in the account?

74. If Central Texas Catering equally divides $3320 in tips from this weekend among its 40 employees, how much will each person receive?

3.7 Estimating and Solving Applied Problems Involving Decimals

Student Learning Objectives

After studying this section, you will be able to:

1. Estimate sums, differences, products, and quotients of decimals.

2. Solve applied problems using operations with decimals.

SSM

PH TUTOR CD & VIDEO MATH PRO WEB
CENTER

---

1 Estimating Sums, Differences, Products, and Quotients of Decimals

When we encounter real-life applied problems, it is important to know if an answer is reasonable. A car may get 21.8 miles per gallon. However, a car will not get 218 miles per gallon. Neither will a car get 2.18 miles per gallon. To avoid making an error in solving applied problems, it is wise to make an estimate. The most useful time to make an estimate is at the end of solving the problem, in order to see if the answer is reasonable.

There are several different rules for estimating. Not all mathematicians agree what is the best method for estimating in each case. Most students find that a very quick and simple method to estimate is to round each number so that there is one nonzero digit. Then perform the calculation. We will use that approach in this section of the book. However, you should be aware that there are other valid approaches. Your instructor may wish you to use another method.

EXAMPLE 1 Estimate.

(a) $184{,}987.09 + 676{,}393.95$

(b) $56.98876 - 48.87447$

(c) 145.87×78.323

(d) $138.85 \div 5.887$

In each case we will round to one nonzero digit to estimate.

(a) $184{,}987.09 + 676{,}393.95 \approx 200{,}000 + 700{,}000 = 900{,}000$

(b) $56.98876 - 48.87447 \approx 60 - 50 = 10$

(c) $145.87 \times 78.323 \approx$

$$
\begin{array}{r}
100 \\
\times \quad 80 \\
\hline
8000
\end{array}
$$

Thus $145.87 \times 78.323 \approx 8000$

(d) $138.85 \div 5.887 \approx 6\overline{)100} = 16\frac{4}{6} \approx 17$

$$
\begin{array}{r}
16 \\
6\overline{)100} \\
\underline{6} \\
40 \\
\underline{36} \\
4
\end{array}
$$

Thus $138.85 \div 5.887 \approx 17$

Here we round the answer to the nearest whole number.

Practice Problem 1 Round to one nonzero digit. Then estimate the result of the indicated calculation.

(a) $385.98 + 875.34$

(b) $623{,}999 - 212{,}876$

(c) 5876.34×0.087

(d) $46{,}873 \div 8.456$

Take a few minutes to review Example 1. Be sure you can perform these estimation steps. We will use this type of estimation to check our work in the applied problems in this section.

2   *Solving Applied Problems Using Operations with Decimals*

We use the basic plan of solving applied problems that we discussed in Section 1.8 and Section 2.9. Let us review how we analyze applied-problem situations.

1. *Understand the problem.*

2. *Solve and state the answer.*

3. *Check.*

In the United States for almost all jobs where you are paid an hourly wage, if you work more than 40 hours in one week, you should be paid overtime, at a rate of 1.5 times the normal hourly rate, for the extra hours worked in that week. The next problem deals with overtime wages.

EXAMPLE 2   A laborer is paid $7.38 per hour for a 40-hour week and 1.5 times that wage for any hours worked beyond the standard 40. If he works 47 hours in a week, what will he earn?

1. *Understand the problem.*

Mathematics Blueprint For Problem Solving

Gather the Facts	What Am I Asked to Do?	How Do I Proceed?	Key Points to Remember
He works 47 hours. He gets paid $7.38 per hour for 40 hours. He gets paid 1.5 × $7.38 per hour for 7 hours.	Find the earnings of the laborer if he works 47 hours in one week.	Add the earnings of 40 hours at $7.38 per hour to the earnings of 7 hours at overtime pay.	Multiply 1.5 × $7.38 to find the pay he earns for overtime.

2. *Solve and state the answer.*

We want to compute his regular pay and his overtime pay and add the results.

$$\text{Regular pay} + \text{Overtime pay} = \text{Total pay}$$

Regular pay: Calculate his pay for 40 hours of work.

$$\begin{array}{r} 7.38 \\ \times\ 40 \\ \hline 295.20 \end{array}$$ He earns $295.20 at $7.38 per hour.

Overtime pay: Calculate his overtime pay rate. This is 7.38 × 1.5.

$$\begin{array}{r} 7.38 \\ \times\ 1.5 \\ \hline 3\,690 \\ 7\,38 \\ \hline 11.070 \end{array}$$ He earns $11.07 per hour in overtime.

Calculate how much he earned doing 7 hours of overtime work.

$$\begin{array}{r} 11.07 \\ \times\ \ \ 7 \\ \hline 77.49 \end{array}$$ For 7 hours overtime he earns $77.49.

Total pay: Add the two amounts.

$$\begin{array}{r} \$295.20 \\ +\quad 77.49 \\ \hline \$372.69 \end{array}$$
    Regular 40-hour week earnings
    Overtime earnings
    Total earnings

The total earnings of the laborer for a 47-hour workweek will be $372.69.

3. *Check.*

Estimate his regular pay.

$$40 \times \$7 = \$280$$

Estimate his overtime rate of pay, and then his overtime pay.

$$1.5 \times \$7 = \$10.50$$
$$7 \times \$11 = \$77$$

Then add.

$$\begin{array}{r} \$280 \\ +\quad 77 \\ \hline \$357 \end{array}$$

Our estimate of $357 is close to our answer of $372.69. Our answer is reasonable. ✓

Practice Problem 2   Melinda works for the phone company as a line repair technician. She earns $9.36 per hour. She worked 51 hours last week. If she gets time and a half for all hours worked above 40 hours per week, how much did she earn last week?

EXAMPLE 3   A chemist is testing 36.85 liters of cleaning fluid. She wishes to pour it into several smaller containers that each hold 0.67 liter of fluid. **(a)** How many containers will she need?   **(b)** If each liter of this fluid costs $3.50, how much does the cleaning fluid in one container cost? (Round your answer to the nearest cent.)

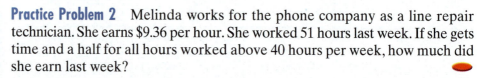

Mathematics Blueprint For Problem Solving

Gather the Facts	What Am I Asked to Do?	How Do I Proceed?	Key Points to Remember
The total amount of cleaning fluid is 36.85 liters. Each small container holds 0.67 liter. Each liter of fluid costs $3.50.	**(a)** Find out how many containers the chemist needs. **(b)** Find the cost of cleaning fluid in each small container.	**(a)** Divide the total, 36.85 liters, by the amount in each small container, 0.67 liter, to find the number of containers. **(b)** Multiply the cost of one liter, $3.50, by the amount of liters in one container, 0.67.	If you are not clear as to what to do at any stage of the problem, then do a similar, simpler problem.

(a) How many containers will the chemist need?

She has 36.85 liters of cleaning fluid and she wants to put it into several equal-sized containers each holding 0.67 liter. Suppose we are not sure what to do. Let's do a similar, simpler problem. If we had 40 liters of cleaning fluid and we wanted to put it into little containers each holding 2 liters, what would we do? Since the little containers would only hold 2 liters, we would need 20 containers. We know that $40 \div 2 = 20$. So we see that, in general, we divide the total number of liters by the amount in the small container. Thus $36.85 \div 0.67$ will give us the number of containers in this case.

$$
\begin{array}{r}
55\,. \\
0.67_\wedge \overline{)36.85_\wedge} \\
\underline{33\ 5} \\
3\ 35 \\
\underline{3\ 35} \\
\end{array}
$$

The chemist will need 55 containers to hold this amount of cleaning fluid.

(b) How much does the cleaning fluid in each container cost? Each container will hold only 0.67 liter. If one liter costs $3.50, then to find the cost of one container we multiply $0.67 \times \$3.50$.

$$
\begin{array}{r}
3.50 \\
\times\ 0.67 \\
\hline
2450 \\
2100 \\
\hline
2.3450 \\
\end{array}
$$

We round our answer to the nearest cent. Thus each container would cost $2.35.

Check.

(a) Is it really true that 55 containers each holding 0.67 liter will hold a total of 36.85 liters? To check we multiply.

$$
\begin{array}{r}
55 \\
\times\ 0.67 \\
\hline
385 \\
330 \\
\hline
36.85 \quad \checkmark \\
\end{array}
$$

(b) One liter of cleaning fluid costs $3.50. We would expect the cost of 0.67 liter to be less than $3.50. $2.35 is less than $3.50. ✓

We use estimation to check more closely.

$$
\begin{array}{rcl}
\$3.50 & \longrightarrow & \$4.00 \\
\times\ \ 0.67 & \longrightarrow & \times\ \ \ 0.7 \\
\hline
 & & \$2.800 \\
\end{array}
$$

$2.80 is fairly close to $2.35. Our answer is reasonable. ✓

Practice Problem 3   A butcher divides 17.4 pounds of prime steak into small equal-sized packages. Each package contains 1.45 pounds of prime steak. **(a)** How many packages of steak will he have?   **(b)** Prime steak sells for $4.60 per pound. How much will each package of prime steak cost?

Developing Your Study Skills

Applications or Word Problems

Applications or word problems are the very life of mathematics! They are the reason for doing mathematics, because they teach you how to put into use the mathematical skills you have developed. Learning mathematics without ever doing word problems is similar to learning all the skills of a sport without ever playing a game or learning all the notes on an instrument without ever playing a song.

The key to success is practice. Make yourself do as many problems as you can. You may not be able to do them all correctly at first, but keep trying. Do not give up whenever you reach a difficult one. If you cannot solve it, just try another one. Then come back and try it again later. Ask for help from your teacher or the tutoring lab. Ask other classmates how they solved the problem.

A misconception among students when they begin studying word problems is that each problem is different. At first the problems may seem this way, but as you practice more and more, you will begin to see the similarities, the different "types." You will see patterns in solving problems, which will enable you to solve problems of a given type more easily.

3.7 Exercises

In exercises 1–10, round to one nonzero digit. Then estimate the result of the calculation.

1. 238,598,980 + 487,903,870

2. 5,927,000 + 9,983,000

3. 56,789.345 − 33,875.125

4. 6949.45 − 1432.88

5. 445,000 × 0.7634

6. 68,976 × 0.875

7. 879,654 ÷ 5682

8. 34.5684 ÷ 0.55

9. The average family of four people spends $512.34 on cereal in one year. Estimate how much the family spends per week.

10. Last year the sales of boats in Massachusetts totaled $865,987,273.45. If this represented a purchase of 55,872 boats, estimate the average price per boat.

Applications

11. Carlos is taking a trip to Denmark. Before he leaves, he checks the newspaper and finds that every U.S. dollar is equal to 7.5 kroners (Danish currency). If Carlos takes $650 on his trip, how many kroners will he receive when he does the exchange?

▲ **12.** A parking lot in front of the Orlando Outback Steak House measures 103.4 meters long by 76.3 meters wide. What is the area in square meters of this parking lot?

▲ **13.** Juan and Gloria are having their roof reshingled and need to determine its area in square feet. The dimensions of the roof are 48.3 feet by 56.9 feet. What is the area of the roof in square feet?

14. Cynthia is making holiday cookies, which she will give as gifts to her friends and co-workers. She has made 34.5 pounds of cookies, which need to be divided up into equal packages containing 0.75 pound of cookies. How many packages can she make?

15. Hans is making gourmet chocolate in Switzerland. He has 11.52 liters of liquid white chocolate that will be poured into molds that hold 0.12 liter each. How many individual molds can Hans make with his 11.52 liters of liquid white chocolate?

16. David bought MacIntosh apples and Anjou pears at the grocery store for a fruit salad. At the checkout counter, the apples weighed 2.7 pounds and the pears weighed 1.8 pounds. If the apples cost $1.29 per pound and the pears cost $1.49 per pound, how much did David spend on fruit? (Round your answer to the nearest cent.)

17. One year in Mount Waialeale, Hawaii, considered the "rainiest place in the world," the yearly rainfall totaled 11.68 meters. The next year, the yearly rainfall on this mountain totaled 10.42 meters. The third year it was 12.67 meters. On average, how much rain fell on Mount Waialeale, Hawaii, per year?

18. Emma and Jennie took a trip in their Ford Taurus from Saskatoon, Saskatchewan, to Calgary, Alberta, in Canada to check out the glacier lakes. When they left, their odometer read 54,089. When they returned home, the odometer read 55,401. They used 65.6 gallons of gas. How many miles per gallon did they get on the trip?

19. The York Steak House uses 3.5 boxes of customized paper towels per day for its customers. If the master box holds 42 regular boxes, how many days will the master box last?

20. Sylvia's telephone company offers a special rate of $0.23 per minute on calls made to the Philippines during certain parts of the day. If Sylvia makes a 28.5-minute call to the Philippines at this special rate, how much will it cost?

21. The local Police Athletic League raised enough money to renovate the local youth hall and turn it into a coffeehouse/activity center so that there is a safe place to hang out. The room that holds the Ping-Pong table needs 43.9 square yards of new carpeting. The entryway needs 11.3 square yards, and the stage/seating area needs 63.4 yards. The carpeting will cost $10.65 per square yard. What will be the total bill for carpeting these three areas of the coffeehouse?

22. Kevin has a job as a house painter. One family needs its kitchen, family room, and hallway painted. The respective amounts needed are 2.7 gallons, 3.3 gallons, and 1.8 gallons. If paint costs $7.40 per gallon, how much will Kevin need to spend on paint to do the job?

23. Shirley works at the local pizza franchise and is paid $6.20 per hour for a 40-hour week. She is paid time and a half for overtime (1.5 times the hourly wage) for every hour more than 40 hours worked in the same week. If she works 52 hours in one week, what will she earn for that week?

24. An electrician is paid $14.30 per hour for a 40-hour week. She is paid time and a half for overtime (1.5 times the hourly wage) for every hour more than 40 hours worked in the same week. If she works 48 hours in one week, what will she earn for that week?

25. Barbara had $420.13 in her savings account last month. Since then she has made deposits of $116.32 and $318.57. The bank credited her with interest of $1.86 for the month. She wrote three checks, for $16.50, $36.89, and $376.94. What is her new balance? (Assume that no bank fees have been charged.)

26. Larry had $79.50 in cash on Sunday. Throughout the week he spent $12.42 on soda and snacks and $26.75 on lunches. On Saturday Larry's friend paid him $30 for helping him move, and another friend paid Larry the $17.25 she owed him. How much money did Larry have at the end of the week?

27. Charlie borrowed $11,500 to purchase a new car. His loan requires him to pay $288.65 each month over the next 60 months (five years). How much will he pay over the five years? How much more will he pay back than the amount of the loan?

28. Hector and Junita borrowed $80,000 to buy their new home. They make monthly payments to the bank of $450.25 to repay the loan. They will be making these monthly payments for the next 30 years. How much money will they pay to the bank in the next 30 years? How much more will they pay back than they borrowed?

29. The EPA standard for safe drinking water is a maximum of 1.3 milligrams of copper per liter of water. A study was conducted on a sample of 7 liters of water drawn from Jeff Slater's house. The analysis revealed 8.06 milligrams of copper in the sample. Is the water safe or not? By how much?

30. The EPA standard for safe drinking water is a maximum of 0.015 milligram of lead per liter of water. A study was conducted on 6 liters of water from West Towers Dormitory. The analysis revealed 0.0795 milligram of lead in the sample. Is the water safe or not? By how much?

31. A jet fuel tank containing 17,316.8 gallons is being emptied at the rate of 126.4 gallons per minute. How many minutes will it take to empty the tank?

32. In a New Jersey mall, the average price of a Parker Brothers Monopoly game is $11.50. The Alfred Dunhill Company made a special commemorative set for $25,000,000.00. Instead of plastic houses and hotels, you can buy and trade gold houses and silver hotels! How many regular Monopoly games could you purchase for the price of one special commemorative set?

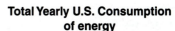

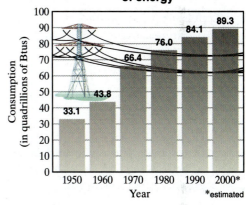

Total Yearly U.S. Consumption of energy

Source: U.S. Department of Energy

Use the preceding bar graph to answer exercises 33–36.

33. How many more Btu were consumed in the United States during 2000 than in 1970?

34. What was the greatest increase in consumption of energy in a 10-year period? When did it occur?

35. What was the average consumption of energy per year in the United States for the years 1950, 1960, and 1970? Write your answer in quadrillion Btu and then write your answer in Btu. (Remember that a quadrillion is 1000 trillion)

36. What was the average consumption of energy per year in the United States for the years 1970, 1980, and 1990? Write your answer in quadrillion Btu and then write your answer in Btu. (Remember that a quadrillion is 1000 trillion.)

Cumulative Review Problems

Calculate.

37. $\dfrac{1}{5} + \dfrac{3}{7} \times \dfrac{1}{2}$

38. $\dfrac{10}{17} + \dfrac{1}{34} - \dfrac{5}{17}$

39. $\dfrac{5}{12} \times \dfrac{36}{27}$

40. $2\dfrac{2}{3} \times 8\dfrac{1}{4}$

Math in the Media

Rise in Health Care Costs Rests Largely on Drug Prices

By Robert Pear
THE NEW YORK TIMES ON THE WEB, November 14, 2000

"Washington, Nov. 13—Prescription drugs accounted for 44 percent of the increase in health costs last year, researchers said today.

In a report published in the journal *Health Affairs*, the researcher said overall health costs for services covered by private insurance rose by 6.6 percent last year, while drug spending increased by 18.4 percent. The study did not separately examine costs for people without insurance."

. . .

"The report in *Health Affairs* said that about one-third of the increase in drug spending last year was attributable to higher prices. The remainder, it said, was attributable to a higher volume of sales, reflecting the advent of new medicines and the increased use of existing drugs.

While prescription drugs accounted for 44 percent of the increase in health costs last year, the report said, doctors' services accounted for 32 percent, and outpatient hospital care accounted for 21 percent, while inpatient hospital care was responsible for only 3 percent."

Read over the article very carefully. Assume that overall health costs for services covered by private insurance includes all doctor services. Also assume that the costs are increased at the same rate for people who had insurance and for people who did not have insurance.

EXERCISES

Using the information quoted answer questions 1–4. Round your expense values to the nearest cent. Round all percents to the nearest tenth.

1. Suppose that your prescription drugs last year cost $200, that you spent $300 for doctor services and were fortunate to have no hospital expenses. That is, your total medical expenses last year were $500. Based on the excerpts quoted above, what should you expect next year's expenses to be when figuring your budget?

2. What is your anticipated percent increase in medical costs?

3. Suppose an opportunity exists to buy health insurance at $50 per month. The insurance policy does not cover prescription drugs, but covers 80 percent of all other medical expenses. What should be the estimate for next year's medical expenses?

4. Why might spending more money to have insurance be a good idea?

Chapter 3 Organizer

Topic	Procedure	Examples
Word names for decimals, p. 203.		The word name for 341.6783 is three hundred forty-one and six thousand seven hundred eighty-three ten-thousandths.

	Hundreds	Tens	Ones	Decimal point	Tenths	Hundredths	Thousandths	Ten-thousandths
	3	4	1	.	6	7	8	3

Topic	Procedure	Examples
Writing a decimal as a fraction, p. 204.	1. Read the decimal in words. 2. Write it in fraction form. 3. Reduce if possible.	Write 0.36 as a fraction. 1. 0.36 is read "thirty-six hundredths." 2. Write the fractional form. $\dfrac{36}{100}$ 3. Reduce. $\dfrac{36}{100} = \dfrac{9}{25}$
Determining which of two decimals is larger, p. 208.	1. Start at the left and compare corresponding digits. Write in extra zeros if needed. 2. When two digits are different, the larger number is the one with the larger digit.	Which is larger? 0.138 or 0.13 0.13**8** ? 0.13**0** $8 > 0$ So, $0.138 > 0.130$.
Rounding decimals, p. 210.	1. Locate the place (units, tenths, hundredths, etc.) for which the round-off is required. 2. If the first digit to the right of the given place value is less than 5, drop it and all the digits to the right of it. 3. If the first digit to the right of the given place value is 5 or greater, increase the number in the given place value by one. Drop all digits to the right.	Round to the nearest hundredth. 0.8652 0.87 Round to the nearest thousandth. 0.21648 0.216
Adding and subtracting decimals, pp. 214.	1. Write the numbers vertically and line up the decimal points. Extra zeros may be written to the right of the decimal points if needed. 2. Add or subtract all the digits with the same place value, starting with the right column, moving to the left. Use carrying or borrowing as needed. 3. Place the decimal point of the result in line with the decimal points of all the numbers added or subtracted.	Add. Subtract. $36.3 + 8.007 + 5.26$ $82.5 - 36.843$ $\begin{array}{r} \overset{1}{36.300} \\ 8.007 \\ +\ 5.260 \\ \hline 49.567 \end{array}$ $\begin{array}{r} {\scriptstyle 7\ \ 11\ 14\ \ 9\ \ 10} \\ 8\ 2.5\ 0\ 0 \\ -\ 3\ 6.8\ 4\ 3 \\ \hline 4\ 5.6\ 5\ 7 \end{array}$
Multiplying decimals, p. 222.	1. Multiply the numbers just as you would multiply whole numbers. 2. Find the sum of the decimal places in the two factors. 3. Place the decimal point in the product so that the product has the same number of decimal places as the sum in step 2. You may need to insert zeros to the left of the number found in step 1.	Multiply. $\begin{array}{r} 0.2 \\ \times\ 0.6 \\ \hline 0.12 \end{array}$ $\begin{array}{r} 0.3174 \\ \times\ \ \ \ 0.8 \\ \hline 0.25392 \end{array}$ $\begin{array}{r} 0.0064 \\ \times\ \ \ 0.21 \\ \hline 64 \\ 128 \\ \hline 0.001344 \end{array}$ $\begin{array}{r} 1364 \\ \times\ \ \ 0.7 \\ \hline 954.8 \end{array}$

Topic	Procedure	Examples
Multiplying a decimal by a power of 10, p. 224.	Move the decimal point to the right the same number of places as there are zeros in the power of 10. (Sometimes it is necessary to write extra zeros before placing the decimal point in the answer.)	Multiply. $5.623 \times 10 = 56.23$ $0.597 \times 10^4 = 5970$ $0.0082 \times 1000 = 8.2$ $0.075 \times 10^6 = 75,000$ $28.93 \times 10^2 = 2893$
Dividing by a decimal, p. 231.	1. Make the divisor a whole number by moving the decimal point to the right. Mark that position with a caret ($\wedge$). 2. Move the decimal point in the dividend to the right the same number of places. Mark that position with a caret. 3. Place the decimal point of your answer directly above the caret in the dividend. 4. Divide as with whole numbers.	Divide. **(a)** $0.06\overline{)0.162}$ **(b)** $0.003\overline{)85.8}$ **(a)** $0.06_\wedge\overline{)0.16_\wedge2}$ gives 2.7 with 12, 42, 42, 0 **(b)** $0.003_\wedge\overline{)85.800_\wedge}$ gives $28\,600.$ with 6, 25, 24, 18, 18, 0
Converting a fraction to a decimal, p. 238.	Divide the denominator into the numerator until 1. the remainder is zero, or 2. the decimal repeats itself, or 3. the desired number of decimal places is achieved.	Find the decimal equivalent. **(a)** $\dfrac{13}{22}$ **(b)** $\dfrac{5}{7}$, rounded to the nearest ten-thousandth **(a)** $22\overline{)13.0000}$ gives 0.5909 with 110, 200, 198, 200, 198, 2 $\dfrac{13}{22} = 0.59\overline{0}$ or $0.5909090\ldots$ **(b)** $7\overline{)5.00000}$ gives 0.71428 with 49, 10, 7, 30, 28, 20, 14, 60, 56, 4 0.71428 rounded to the nearest ten thousandth is 0.7143.
Order of operations with decimal numbers, p. 241.	Same as order of operations of whole numbers. 1. Perform operations inside parentheses. 2. Simplify any expressions with exponents. 3. Multiply or divide from left to right. 4. Add or subtract from left to right.	Evaluate. $(0.4)^3 + 1.26 \div 0.12 - 0.12 \times (1.3 - 1.1)$ $= (0.4)^3 + 1.26 \div 0.12 - 0.12 \times 0.2$ $= 0.064 + 1.26 \div 0.12 - 0.12 \times 0.2$ $= 0.064 + 10.5 - 0.024$ $= 10.564 - 0.024$ $= 10.54$

Using the Mathematics Blueprint for Problem Solving, p. 247

In solving a real-life problem with decimals, students may find it helpful to complete the following steps. You will not use all the steps all of the time. Choose the steps that best fit the conditions of the problem.

1. *Understand the problem.*
   (a) Read the problem carefully.
   (b) Draw a picture if it helps you visualize the situation. Think about what facts you are given and what you are asked to find.
   (c) It may help to write a similar, simpler problem to get started and to determine what operation to use.
   (d) Use the Mathematics Blueprint for Problem Solving to organize your work. Follow these four parts.
      1. Gather the Facts (Write down specific values given in the problem.)
      2. What Am I Asked to Do? (Identify what you must obtain for an answer.)
      3. How Do I Proceed? (Determine what calculations need to be done).
      4. Key Points to Remember (Record any facts, warnings, formulas, or concepts you think will be important as you solve the problem.)

2. *Solve and state the answer.*
   (a) Perform the necessary calculations.
   (b) State the answer, including the unit of measure.

3. *Check.*
   (a) Estimate the answer to the problem. Compare this estimate to the calculated value. Is your answer reasonable?
   (b) Repeat your calculations.
   (c) Work backward from your answer. Do you arrive at the original conditions of the problem?

Example

▲ Fred has a rectangular living room that is 3.5 yards wide and 6.8 yards long. He has a hallway that is 1.8 yards wide and 3.5 yards long. He wants to carpet each area using carpeting that costs $12.50 per square yard. What will the carpeting cost him?

1. *Understand the problem.*
   It is helpful to draw a sketch.

6.8 yd

Living Room | 3.5 yd

3.5 yd

Hallway | 1.8 yd

Mathematics Blueprint For Problem Solving

Gather the Facts	What Am I Asked to Do?	How Do I Proceed?	Key Points to Remember
Living room: 6.8 yards by 3.5 yards Hallway: 3.5 yards by 1.8 yards Cost of carpet: $12.50 per square yard	Find out what the carpeting will cost Fred.	Find the area of each room. Add the two areas. Multiply the total area by $12.50.	Multiply the length by the width to get the area of the room. Remember, area is measured in square yards.

continues on next page

To find the area of each room, we multiply the dimensions for each room.

Living room 6.8 × 3.5 = 23.80 square yards   Hallway 3.5 × 1.8 = 6.30 square yards

Add the two areas.

$$\begin{array}{r} 23.80 \\ +\ 6.30 \\ \hline 30.10 \end{array} \text{ square yards}$$

Multiply the total area by the cost per square yard.

30.1 × 12.50 = $376.25

Estimate to check. You may be able to do some of this mentally.

7 × 4 = 28 square yards   4 × 2 = 8 square yards

$$\begin{array}{r} 28 \\ +\ 8 \\ \hline 36 \end{array} \text{ square yards}$$

36 × 10 = $360 $360 is close to $376.25. ✓

Chapter 3 Review Problems

3.1 *Write a word name for each decimal.*

1. 13.672

2. 0.00084

Write as a decimal.

3. $\dfrac{7}{10}$

4. $\dfrac{81}{100}$

5. $1\dfrac{523}{1000}$

6. $\dfrac{79}{10,000}$

Write as a fraction or a mixed number.

7. 0.17

8. 0.365

9. 34.24

10. 1.00025

3.2 *Fill in the blank with* < *,* = *, or* > *.*

11. $\dfrac{13}{100}$ _____ 0.103

12. 0.716 _____ 0.706

In exercises 13 and 14, arrange each set of decimal numbers from smallest to largest.

13. 0.981, 0.918, 0.98, 0.901

14. 5.62, 5.2, 5.6, 5.26, 5.59

15. Round to the nearest tenth. 0.613

16. Round to the nearest hundredth. 19.2076

17. Round to the nearest thousandth. 1.09952

18. Round to the nearest dollar. $156.48

3.3

19. Add.
$$\begin{array}{r} 9.6 \\ 11.5 \\ 21.8 \\ + 34.7 \\ \hline \end{array}$$

20. Add.
$$\begin{array}{r} 1.8 \\ 2.603 \\ 0.52 \\ + 1.716 \\ \hline \end{array}$$

21. Subtract.
$$\begin{array}{r} 5.19 \\ - 1.296 \\ \hline \end{array}$$

22. Subtract.
$$\begin{array}{r} 352.806 \\ - 195.992 \\ \hline \end{array}$$

3.4 *In exercises 23–28, multiply.*

23.
$$\begin{array}{r} 0.098 \\ \times\ 0.032 \\ \hline \end{array}$$

24.
$$\begin{array}{r} 126.83 \\ \times\ \ \ 7 \\ \hline \end{array}$$

25.
$$\begin{array}{r} 78 \\ \times\ 5.2 \\ \hline \end{array}$$

26.
$$\begin{array}{r} 7053 \\ \times\ 0.34 \\ \hline \end{array}$$

27. 0.000613×10^3

28. 1.2354×10^5

29. Hillary purchased 3.6 pounds of bananas at $0.35 per pound. How much did the purchase cost?

3.5 *In exercises 30–32, divide until there is a remainder of zero.*

30. $0.07\overline{)0.0001806}$

31. $5.2\overline{)191.36}$

32. $8\overline{)1863.2}$

33. Divide and round your answer to the nearest tenth.

$$1.3\overline{)746.75}$$

34. Divide and round your answer to the nearest thousandth.

$$0.06\overline{)0.003539}$$

3.6 *Write as an equivalent decimal.*

35. $\dfrac{5}{18}$

36. $\dfrac{7}{40}$

37. $1\dfrac{5}{6}$

38. $\dfrac{19}{16}$

Write as a decimal rounded to the nearest thousandth.

39. $\dfrac{11}{14}$

40. $2\dfrac{5}{17}$

Evaluate by doing the operations in proper order.

41. $1.6 \times 2.3 + 0.4 - 0.6 \times 0.8$

42. $0.03 + (1.2)^2 - 5.3 \times 0.06$

43. $(1.02)^3 + 5.76 \div 1.2 \times 0.05$

44. $6.63 + 8.24 \div (5.76 - 5.68) - 22.5$

Mixed Practice

Calculate.

45. $2398.26 - 1959.07$

46. 67.036×0.006

47. $0.061 + 0.0023 + 0.777$

48. $110.72 \div 1.6$

49. $8 \div 0.4 + 0.1 \times (0.2)^2$

50. $(3.8 - 2.8)^3 \div (0.5 + 0.3)$

Applications

3.7 *Solve each problem.*

51. At a large football stadium there are 2,600 people in line for tickets. In the first two minutes the computer is running slowly and tickets 228 people. Then the computer stops. For the next 2.5 minutes, the computer runs at medium speed and tickets 388 people per minute. For the next three minutes the computer runs at full speed and tickets 430 people per minute. Then the computer stops. How many people still have not received their tickets?

52. Dan drove to the mountains. His odometer read 26,005.8 miles at the start, and 26,325.8 miles at the end of the trip. He used 12.9 gallons of gas on the trip. How many miles per gallon did his car get? (Round your answer to the nearest tenth.)

53. Robert is considering buying a car and making installment payments of $189.60 for 48 months. The cash price of the car is $6930.50. How much extra does he pay if he uses the installment plan instead of buying the car with one payment?

54. Mr. Zeno has a choice of working as an assistant manager at ABC Company at $315.00 per week or receiving an hourly salary of $8.26 per hour at the XYZ company. He learned from several previous assistant managers at both companies that they usually worked 38 hours per week. At which company will he probably earn more money?

55. The EPA standard for safe drinking water is a maximum of 0.002 milligram of mercury in one liter of water. The town wells at Winchester were tested. The test was done on 12 liters of water. The entire 12-liter sample contained 0.03 milligram of mercury. Is the water safe or not? By how much does it differ from the standard?

56. A chemist wishes to take a sample of 0.322 liter of acid and place it in small test tubes each containing 0.046 liter. How many test tubes will be needed?

▲ **57.** Dick Wright's new rectangular garden measures 18.3 feet by 9.6 feet. He needs to install wire fence on all four sides. How many feet of fence does he need?

▲ **58.** Bill Tupper's rectangular driveway needs to be resurfaced. It is 75.5 feet long and 18.5 feet wide. How large is the area of the driveway?

59. The following strip map shows the distances in miles between several local towns in Pennsylvania. Find the distance from Coudersport to Wellsboro.

▲ **60.** A farmer in Vermont has a field with an irregular shape. The distances are marked on the diagram. There is no fence but there is a path on the edge of the field. How long is the walking path around the field?

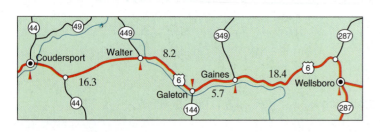

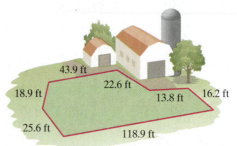

61. Marcia and Greg purchased a new car. For the next five years they will be making monthly payments of $212.50. Their bank has offered to give them a loan at a smaller interest rate so that they would make monthly payments of only $199.50. The bank would charge them $285.00 to reissue their car loan. How much would it cost them to keep their original loan? How much would it cost them if they took the new loan from the bank? Should they make the change or keep the original loan?

Use the following bar graph to answer exercises 62–67. Round all answers to the nearest cent.

62. How much did the average monthly social security benefit increase from 1980 to 1990?

63. How much did the average monthly social security benefit increase from 1990 to 2000?

64. What was the average daily social security benefit in 1985? (Assume 30 days in a month.)

65. What was the average daily social security benefit in 1995? (Assume 30 days in a month.)

66. If the average daily social security benefit increases by the same amount from 2000 to 2005 as it did from 1995 to 2000, what will be the average daily social security benefit in 2005?

67. If the average daily social security benefit increases by the same amount from 2000 to 2010 as it did from 1990 to 2000, what will be the average daily social security benefit in 2010?

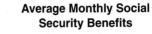

Average Monthly Social Security Benefits

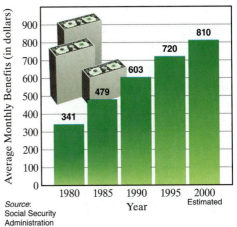

Source: Social Security Administration

Developing Your Study Skills

Exam Time: How To Review

Reviewing adequately for an exam enables you to bring together the concepts you have learned over several sections. For your review, you will need to do the following:

1. Reread your textbook. Make a list of any terms, rules, or formulas you need to know for the exam. Be sure you understand them all.

2. Reread your notes. Go over returned homework and quizzes. Redo the problems you missed.

3. Practice some of each type of problem covered in the chapter(s) you are to be tested on. In fact, it is a good idea to construct a practice test of your own and then discuss it with a friend from class.

4. Use the end-of-chapter materials provided in your textbook. Read carefully through the Chapter Organizer. Do the Chapter Review Problems. Take the Chapter Test. When you are finished, check your answers. Redo any problems you missed.

5. Get help if any concepts give you difficulty.

1. Write a word name for the decimal. 0.157

2. Write as a decimal. $\dfrac{3,977}{10,000}$

In questions 3 and 4, write in fractional notation. Reduce whenever possible.

3. 7.15   **4.** 0.261

5. Arrange from smallest to largest.   2.19, 2.91, 2.9, 2.907

6. Round to the nearest hundredth.   78.6562

7. Round to the nearest ten-thousandth.   0.0341752

Add.

8. 96.2
     1.348
  + 2.15

9. 17 + 2.1 + 16.8 + 0.04 + 1.59

Subtract.

10. 1.0075
    − 0.9096

11. 72.3 − 1.145

Multiply.

12. 8.31
    × 0.07

13. 2.189 × 100

Divide.

14. $0.004\overline{)0.1028}$

15. $0.69\overline{)32.43}$

Write as a decimal. **16.** $\dfrac{11}{9}$ **17.** $\dfrac{9}{16}$

In questions 18 and 19, perform the operations in the proper order.

18. $(0.3)^3 + 1.02 \div 0.5 - 0.58$

19. $19.36 \div (0.24 + 0.26) \times (0.4)^2$

20. A beef roast weighing 7.8 pounds costs $4.25 per pound. How much will the roast cost?

21. Frank traveled from the city to the shore. His odometer read 42,620.5 miles at the start and 42,780.5 at the end of the trip. He used 8.5 gallons of gas. How many miles per gallon did his car achieve? Round to the nearest tenth.

22. The rainfall for March in Central City was 8.01 centimeters; for April, 5.03 centimeters; and for May, 8.53 centimeters. The normal rainfall for these three months is 25 centimeters. How much less rain fell during these three months than usual; that is, how does this year's figure compare with the figure for normal rainfall?

23. Wendy is earning $7.30 per hour in her new job as a teller trainee at the Springfield National Bank. She earns 1.5 times that amount for every hour over 40 hours she works in one week. She was asked to work 49 hours last week. How much did she earn last week?

1. _____

2. _____

3. _____

4. _____

5. _____

6. _____

7. _____

8. _____

9. _____

10. _____

11. _____

12. _____

13. _____

14. _____

15. _____

16. _____

17. _____

18. _____

19. _____

20. _____

21. _____

22. _____

23. _____

Approximately one-half of this test is based on Chapter 3 material. The remainder is based on material covered in Chapters 1 and 2.

1. Write in words. 38,056,954

2. Add.   156,028
             301,579
          + 21,980

3. Subtract.   1,091,000
                  − 1,036,520

4. Multiply.   589
                 × 67

5. Divide. $17\overline{)4386}$

6. Evaluate. $20 \div 4 + 2^5 - 7 \times 3$

7. Reduce. $\dfrac{33}{88}$

8. Add. $4\dfrac{1}{3} + 3\dfrac{1}{6}$

9. Subtract. $\dfrac{23}{35} - \dfrac{2}{5}$

10. Evaluate. $\dfrac{7}{10} \times \dfrac{5}{3} - \dfrac{5}{12} \times \dfrac{1}{2}$

11. Divide. $52 \div 3\dfrac{1}{4}$

12. Divide. $1\dfrac{3}{8} \div \dfrac{5}{12}$

13. Estimate. $58{,}216 \times 438{,}207$

14. Write as a decimal. $\dfrac{571}{1000}$

15. Arrange from smallest to largest. 2.1, 20.1, 2.01, 2.12, 2.11

16. Round to the nearest thousandth. 26.07984

17. Add.   1.9
             2.36
            15.2
          + 0.08

18. Subtract.   28.007
                  − 19.368

19. Multiply.   56.8
                  × 0.02

20. Multiply. 365.123×100

21. Divide. $0.06\overline{)0.06348}$

22. Write as a decimal. $\dfrac{13}{16}$

23. Perform the operations in the correct order.

$1.44 \div 0.12 + (0.3)^3 + 1.57$

24. A car gets 28.5 miles per gallon. Its gas tank holds 16.0 gallons of gas. Approximately how many miles can the car travel on one tank of gas?

25. Sue's savings account balance is $199.36. This month she earned interest of $1.03. She deposited $166.35 and $93.50. She withdrew money three times, in the amounts of $90.00, $37.49, and $137.18. What will her balance be at the start of next month?

26. Russ and Norma Camp borrowed some money from the bank to purchase a new Dodge Caravan for their family. They are paying off the car loan at the rate of $320.50 per month. At the end of the loan period they will have paid $19,230.00 to the bank. How many months will it take to pay off this car loan?

1. _____
2. _____
3. _____
4. _____
5. _____
6. _____
7. _____
8. _____
9. _____
10. _____
11. _____
12. _____
13. _____
14. _____
15. _____
16. _____
17. _____
18. _____
19. _____
20. _____
21. _____
22. _____
23. _____
24. _____
25. _____
26. _____

Ratio and Proportion

T he state of Alaska maintains a count of the number of moose and other wildlife in the state. This close tracking of animals in the wild has been valuable in preserving a proper balance of animal species in the environment and in keeping a variety of animals from becoming endangered species. The scientists and biologists responsible for maintaining these records use proportions to determine future populations. Do you think you could perform these kinds of calculations? Turn to page 297 and try the Putting Your Skills to Work problems to find out.

Pretest Chapter 4

1. _____

2. _____

3. _____

4. _____

5. (a) _____

   (b) _____

6. _____

7. _____

8. _____

9. _____

10. _____

11. _____

12. _____

13. _____

If you are familiar with the topics in this chapter, take this test now. Check your answers with those in the back of the book. If an answer is wrong or you can't answer a question, study the appropriate section of the chapter.

If you are not familiar with the topics in this chapter, don't take this test now. Instead, study the examples, work the practice problems, and then take the test.

This test will help you identify which concepts you have mastered and which you need to study further.

Section 4.1

In questions 1–4, write each ratio in simplest form.

1. 13 to 18

2. 44 to 220

3. $72 to $16

4. 121 kilograms to 132 kilograms

5. Sam's take-home pay is $240 per week. $70 per week is withheld for federal taxes and $22 per week is withheld for state taxes.
(a) Find the ratio of federal withholding to take-home pay.
(b) Find the ratio of state withholding to take-home pay.

Write each rate in simplest form.

6. 9 flight attendants for 300 passengers

7. 620 gallons of water for each 840 square feet of lawn

Write as a unit rate. Round to the nearest tenth if necessary.

8. 122 miles are traveled in 4 hours. What is the rate in miles per hour?

9. 15 CD players are purchased for $435. What is the cost per CD player?

Section 4.2

Write a proportion.

10. 13 is to 40 as 39 is to 120

11. 116 is to 158 as 29 is to 37

Determine whether each equation is a proportion.

12. $\dfrac{14}{31} = \dfrac{42}{93}$

13. $\dfrac{17}{33} = \dfrac{19}{45}$

Section 4.3

Solve for n in each equation.

14. $9 \times n = 153$

15. $234 = 13 \times n$

Solve for n in each proportion.

16. $\dfrac{36}{16} = \dfrac{9}{n}$

17. $\dfrac{3}{144} = \dfrac{n}{336}$

18. $\dfrac{n}{900} = \dfrac{15}{22.5}$

19. $\dfrac{\frac{1}{2}}{n} = \dfrac{\frac{3}{4}}{3}$

Section 4.4

Solve using a proportion.

20. A recipe for six portions calls for 1.5 cups of flour. How many cups of flour are needed to make 14 portions?

21. Maria's car can travel 81 miles on 2 gallons of gas. How far can she travel on 9 gallons of gas?

22. Two cities are 5 inches apart on a map, but the actual distance between them is 365 miles. What is the actual distance between two other cities that are 2 inches apart on the map?

23. A shipment of 121 light bulbs had 6 defective bulbs. How many defective bulbs would we expect, at the same rate, in a shipment of 1089 light bulbs?

24. Last year one of the Red Sox pitchers gave up 67 runs in 245 innings. If he normally gives up runs at this rate, how many runs would you expect him to give up if he pitched a nine-inning game? (Round your answer to the nearest tenth.)

25. A recent survey showed that 3 out of every 10 people in Massachusetts read the *Boston Globe*. In a Massachusetts town of 45,600 people, how many people would you expect read the *Boston Globe*?

14. _____

15. _____

16. _____

17. _____

18. _____

19. _____

20. _____

21. _____

22. _____

23. _____

24. _____

25. _____

4.1 Ratios and Rates

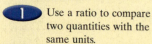

1 Using a Ratio to Compare Two Quantities with the Same Units

Assume that you earn 13 dollars an hour and your friend earns 10 dollars per hour. The *ratio* 13 : 10 compares what you and your friend make. This ratio means that for every 13 dollars you earn, your friend earns 10. The *rate* you are paid is 13 dollars per hour, which compares 13 dollars to 1 hour. In this section we see how to use both rates and ratios to solve many everyday problems.

Suppose that we want to compare an object weighing 20 pounds to an object weighing 23 pounds. The ratio of their weights would be 20 to 23. We may also write this as $\frac{20}{23}$. A **ratio** is the comparison of two quantities that have the *same units*.

A commonly used video display for a computer has a horizontal dimension of 14 inches and a vertical dimension of 10 inches. The ratio of the horizontal dimension to the vertical dimension is 14 to 10. In reduced form we would write that as 7 to 5. We can express the ratio three ways.

We can write "the ratio of 7 to 5."

We can write 7 : 5 using a colon.

We can write $\frac{7}{5}$ using a fraction.

All three notations are valid ways to compare 7 to 5. Each is read as "7 to 5."

> We always want to write a ratio in simplest form. A ratio is in **simplest form** when the two numbers do not have a common factor and both numbers are whole numbers.

EXAMPLE 1 Write in simplest form. Express your answer as a fraction.

(a) the ratio of 15 hours to 20 hours   **(b)** the ratio of 36 hours to 30 hours
(c) 125 : 150

(a) $\dfrac{15}{20} = \dfrac{3}{4}$ **(b)** $\dfrac{36}{30} = \dfrac{6}{5}$ **(c)** $\dfrac{125}{150} = \dfrac{5}{6}$

Notice that in each case the two numbers *do* have a common factor. When we form the fraction—that is, the ratio—we take the extra step of *reducing* the fraction.

Practice Problem 1 Write in simplest form. Express your answer as a fraction.

(a) the ratio of 36 feet to 40 feet   **(b)** the ratio of 18 feet to 15 feet   **(c)** 220 : 270

EXAMPLE 2 Martin earns $350 weekly. However, he takes home only $250 per week in his paycheck.

$350.00 gross pay   (what Martin earns)

45.00 withheld for federal tax
20.00 withheld for state tax
35.00 withheld for retirement   } (what is taken out of Martin's earnings)

$250.00 take-home pay   (what Martin has left)

(a) What is the ratio of the amount withheld for federal tax to gross pay?

(b) What is the ratio of the amount withheld for state tax to the amount withheld for federal tax?

(a) The ratio of the amount withheld for federal tax to gross pay is

$$\frac{45}{350} = \frac{9}{70}.$$

(b) The ratio of the amount withheld for state tax to the amount withheld for federal tax is

$$\frac{20}{45} = \frac{4}{9}.$$

Practice Problem 2   Recently President Reeves conducted a survey of students at North Shore Community College who use the Internet. He wanted to determine how many of the students use the college Internet provider versus how many use AOL, Compuserve, or other commercial Internet providers. The results of his survey are shown in the circle graph.

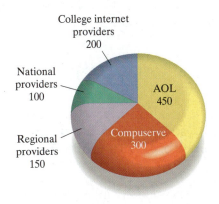

(a) Write the ratio of the number of students who use the college Internet provider to the number of students who use AOL.

(b) Write the ratio of the number of students who use Compuserve to the total number of students who use the Internet.

To Think About   Perhaps you have heard statements like "a certain jet plane travels at Mach 2.2." What does that mean? A Mach number is a ratio that compares the velocity (speed) of an object to the velocity of sound. Sound travels at about 330 meters per second. The Mach number is written in decimal form.

What is the Mach number of a jet traveling at 690 meters per second?

$$\text{Mach number of jet} = \frac{690 \text{ meters per second}}{330 \text{ meters per second}}$$

$$= 2.09090909\ldots \text{ or } 2.\overline{09}$$

Rounded to the nearest tenth, the Mach number of the jet is 2.1.

Exercises 69 and 70 in Exercises 4.1 deal with Mach numbers.

2 *Using a Rate to Compare Two Quantities with Different Units*

A **rate** is a comparison of two quantities with *different units*. Usually, to avoid misunderstanding, we express a rate as a fraction with the units included.

EXAMPLE 3 Recently an automobile manufacturer spent $946,000 for a 48-second television commercial shown on a national network. What is the rate of dollars spent to seconds of commercial time?

$$\text{The rate is} \quad \frac{946{,}000 \text{ dollars}}{48 \text{ seconds}} = \frac{59{,}125 \text{ dollars}}{3 \text{ seconds}}.$$

Practice Problem 3 A farmer is charged a $44 storage fee for every 900 tons of grain he stores. What is the rate of the storage fee in dollars to tons of grain?

Often we want to know the rate for a single unit, which is the unit rate. A **unit rate** is a rate in which the denominator is the number 1.

EXAMPLE 4 A car traveled 301 miles in seven hours. Find the unit rate.
$\frac{301}{7}$ can be simplified. We find $301 \div 7 = 43$.
Thus

$$\frac{301 \text{ miles}}{7 \text{ hours}} = \frac{43 \text{ miles}}{1 \text{ hour}}$$

The denominator is 1. We write our answer as 43 miles/hour. The fraction line is read as the word "per," so our answer here is read "43 miles per hour." *Per* means "for every," so a rate of 43 miles per hour means 43 miles traveled for every hour traveled.

Practice Problem 4 A car traveled 212 miles in four hours. Find the unit rate.

EXAMPLE 5 A grocer purchased 200 pounds of apples for $68. He sold the 200 pounds of apples for $86. How much profit did he make per pound of apples?

$$
\begin{array}{rl}
\$86 & \text{selling price} \\
-\ 68 & \text{cost} \\
\hline
\$18 & \text{profit}
\end{array}
$$

The rate that compares profit to pounds of apples sold is $\dfrac{18 \text{ dollars}}{200 \text{ pounds}}$. We will find $18 \div 200$.

$$
\begin{array}{r}
0.09 \\
200\overline{)18.00} \\
\underline{18\ 00} \\
0
\end{array}
$$

The unit rate of profit is $0.09 per pound.

Practice Problem 5 A retailer purchased 120 nickel-cadmium batteries for flashlights for $129.60. She sold them for $170.40. What was her profit per battery?

EXAMPLE 6 Hamburger at a local butcher is packaged in large and extra-large packages. A large package costs $7.86 for 6 pounds and an extra-large package is $10.08 for 8 pounds.

(a) What is the unit rate in dollars per pound for each size package?

(b) How much per pound does a consumer save by buying the extra-large package?

(a) $\dfrac{7.86 \text{ dollars}}{6 \text{ pounds}} = \$1.31/\text{pound for the large package}$

$\dfrac{10.08 \text{ dollars}}{8 \text{ pounds}} = \$1.26/\text{pound for the extra-large package}$

(b) $\begin{array}{r} \$1.31 \\ -\ 1.26 \\ \hline \$0.05 \end{array}$ A person saves $0.05/pound by buying the extra-large package.

Practice Problem 6 A 12-ounce package of Fred's favorite cereal costs $2.04. A 20-ounce package of the same cereal costs $2.80.

(a) What is the cost per ounce of each size of cereal?

(b) How much per ounce would Fred save by buying the larger size?

Verbal and Writing Skills

1. A _____ is a comparison of two quantities that have the same units.

2. A rate is a comparison of two quantities that have _____ units.

3. The ratio $5:8$ is read _____ .

4. Marion compares the number of loaves of bread she bakes to the number of pounds of flour she needs to make the bread. Is this a ratio or a rate? Why?

Write in simplest form. Express your answer as a fraction.

5. $6:18$

6. $8:20$

7. $21:18$

8. $42:51$

9. $36:132$

10. $75:135$

11. $150:225$

12. $360:480$

13. 60 to 64

14. 33 to 57

15. 28 to 42

16. 21 to 98

17. 32 to 20

18. 90 to 54

19. 8 ounces to 12 ounces

20. 50 years to 85 years

21. 39 kilograms to 26 kilograms

22. $86 to $120

23. $54 to $63

24. 255 meters to 15 meters

25. 312 yards to 24 yards

26. 91 tons to 133 tons

27. $2\frac{1}{2}$ pounds to $4\frac{1}{4}$ pounds

28. $4\frac{1}{3}$ feet to $5\frac{2}{3}$ feet

Use the following table to answer exercises 29–32.

ROBIN'S WEEKLY PAYCHECK

Total (Gross) Pay	Federal Withholding	State Withholding	Retirement	Insurance	Take-Home Pay
$285	$35	$20	$28	$16	$165

29. What is the ratio of take-home pay to total (gross) pay?

30. What is the ratio of retirement to insurance?

31. What is the ratio of federal withholding to take-home pay?

32. What is the ratio of retirement to total (gross) pay?

An automobile insurance company prepared the following analysis for its clients. Use this table for exercises 33–36.

33. What is the ratio of sedans that lasted two years or less to the total number of sedans?

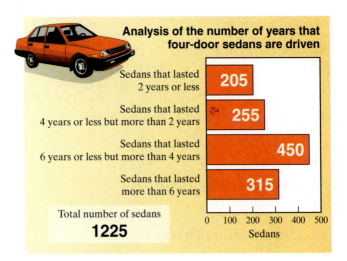

Analysis of the number of years that four-door sedans are driven

Sedans that lasted 2 years or less **205**

Sedans that lasted 4 years or less but more than 2 years **255**

Sedans that lasted 6 years or less but more than 4 years **450**

Sedans that lasted more than 6 years **315**

Total number of sedans **1225**

Sedans 0 100 200 300 400 500

34. What is the ratio of sedans that lasted more than six years to the total number of sedans?

35. What is the ratio of the number of sedans that lasted six years or less but more than four years to the number of sedans that lasted two years or less?

36. What is the ratio of the number of sedans that lasted more than six years to the number of sedans that lasted four years or less but more than two years?

37. A basketball team scored a total of 672 points during one season. Of these, 42 points were scored by making one-point free throws. What is the ratio of free-throw points to total points?

38. When Shawn bought his stereo, he paid $40 in tax. The total cost was $760. What is the ratio of tax to total cost?

Write as a rate in simplest form.

39. $40 for 16 magazines

40. $189 for 18 picture frames

41. $170 for 12 bushes

42. 98 pounds for 22 people

43. 310 gallons of water for every 625 square feet of lawn

44. 460 gallons of water for every 825 square feet of lawn

45. 6150 revolutions for every 15 miles

46. 9540 revolutions for every 18 miles

47. $330,000 for 12 employees

48. 28 pies for 168 people

Write as a unit rate.

49. Earn $520 in 40 hours

50. Earn $304 in 38 hours

51. Travel 192 miles on 12 gallons of gas

52. Travel 322 miles on 14 gallons of gas

53. Pump 2480 gallons in 16 hours

54. Pump 2880 gallons in 24 hours

55. 2250 pencils in 18 boxes

56. 1122 books in 11 crates

57. Travel 276 miles in 4 hours

58. Travel 325 miles in 5 hours

Applications

59. $3870 was spent for 129 shares of Mattel stock. Find the cost per share.

60. $6150 was spent for 150 shares of Polaroid stock. Find the cost per share.

61. A bookstore owner purchased 80 calendars for $760. He sold them for $1200. How much profit did he make per calendar?

62. A retailer purchased 24 compact disc players for $3380. She sold them for $4030. How much profit did she make per compact disc player? Round your answer to the nearest dollar.

63. A 16-ounce box of dry pasta costs $1.28. A 24-ounce box of the same pasta costs $1.68.
(a) What is the cost per ounce of each box of pasta?

(b) How much does the educated consumer save by buying the larger box?

(c) How much does the consumer save by buying 2 large boxes instead of 3 small boxes?

64. A 16-ounce can of beef stew costs $2.88. A 26-ounce can of the same beef stew costs $4.16.
(a) What is the cost per ounce of each can of stew?

(b) How much does the consumer save per ounce by buying the larger can?

65. Dr. Robert Tobey completed a count of two herds of moose in the central regions of Alaska. He recorded 3978 moose on the North Slope and 5520 moose on the South Slope. There are 306 acres on the North Slope and 460 acres on the South Slope.
  (a) How many moose per acre were found on the North Slope?
  (b) How many moose per acre were found on the South Slope?
  (c) In which region are the moose more closely crowded together?

66. In Melbourne, Australia, 27,900 people live in the suburb of St. Kilda and 38,700 live in the suburb of Caulfield. The area of St. Kilda is 6500 acres. The area of Caulfield is 9200 acres. Round your answers to the nearest tenth.
  (a) How many people per acre live in St. Kilda?
  (b) How many people per acre live in Caulfield?
  (c) Which suburb is more crowded?

67. Mr. Jenson bought 525 shares of Zenith stock for $12,876.50. How much did he pay per share? Round your answer to the nearest cent.

68. In 1999 Mark McGwire of the St. Louis Cardinals hit 65 home runs with 521 "at-bats." In the same year, Sammy Sosa of the Chicago Cubs hit 63 home runs with 625 at-bats.
  (a) What are the rates of at-bats per home run for Mark and for Sammy? Round to the nearest tenth.
  (b) Which person hits home runs more often?

To Think About

For exercises 69 and 70, recall that the speed of sound is about 330 meters per second. (See the To Think About discussion on page 269.) Round your answers to the nearest tenth.

69. A jet plane was originally designed to fly at 750 meters per second. It was modified to fly at 810 meters per second. By how much was its Mach number increased?

70. A rocket was first flown at 1960 meters per second. It proved unstable and unreliable at that speed. It is now flown at a maximum of 1920 meters per second. By how much was its Mach number decreased?

Cumulative Review Problems

Calculate.

71. $2\frac{1}{4} + \frac{3}{8}$

72. $\frac{5}{7} \div \frac{3}{21}$

73. $\frac{3}{5} \times \frac{5}{8} - \frac{2}{3} \times \frac{1}{4}$

74. $3\frac{1}{16} - 2\frac{1}{24}$

▲ **75.** A room 12 yards × 5.2 yards had a carpet installed. The bill was $764.40. What was the cost of the installed carpet per square yard?

76. An electronics superstore bought 1050 computer games for $23 each. How much did the store pay in all for these games? The store sold the games for $39 each. How much profit did the store make?

1 Writing a Proportion

A **proportion** states that two ratios or two rates are equal. For example, $\frac{5}{8} = \frac{15}{24}$ is a proportion and $\frac{7 \text{ feet}}{8 \text{ dollars}} = \frac{35 \text{ feet}}{40 \text{ dollars}}$ is also a proportion. A proportion can be read two ways. The proportion $\frac{5}{8} = \frac{15}{24}$ can be read "five eighths equals fifteen twenty-fourths," or it can be read "five *is to* eight *as* fifteen *is to* twenty-four."

EXAMPLE 1   Write the proportion 5 is to 7 as 15 is to 21.

$$\frac{5}{7} = \frac{15}{21}$$

Practice Problem 1   Write the proportion 6 is to 8 as 9 is to 12.

EXAMPLE 2   Write a proportion to express the following: If four rolls of wallpaper measure 300 feet, then eight rolls of wallpaper will measure 600 feet.

When you write a proportion, order is important. Be sure that the similar units for the rates are in the same position in the fractions

$$\frac{4 \text{ rolls}}{300 \text{ feet}} = \frac{8 \text{ rolls}}{600 \text{ feet}}$$

Practice Problem 2   Write a proportion to express the following: If it takes two hours to drive 72 miles, then it will take three hours to drive 108 miles.

2 Determining Whether a Statement is a Proportion

By definition, a proportion states that two ratios are equal. $\frac{2}{7} = \frac{4}{14}$ is a proportion because $\frac{2}{7}$ and $\frac{4}{14}$ are equivalent fractions. You might say that $\frac{2}{7} = \frac{4}{14}$ is a *true* statement. It is easy enough to see that $\frac{2}{7} = \frac{4}{14}$ is true. However, is $\frac{4}{14} = \frac{6}{21}$ true? Is $\frac{4}{14} = \frac{6}{21}$ a proportion? To determine whether a statement is a proportion, we use the equality test for fractions.

Equality Test for Fractions

For any two fractions where $b \neq 0$ and $d \neq 0$,

$$\text{if and only if } \frac{a}{b} = \frac{c}{d}, \text{ then } a \times d = b \times c.$$

Thus, to see if $\dfrac{4}{14} = \dfrac{6}{21}$, we can multiply.

$$\dfrac{4}{14} \overset{\times}{=} \dfrac{6}{21} \qquad \begin{array}{l} 14 \times 6 = 84 \leftarrow \\ 4 \times 21 = 84 \leftarrow \end{array} \qquad \boxed{\text{The cross products are equals.}}$$

$\dfrac{4}{14} = \dfrac{6}{21}$ is true. $\dfrac{4}{14} = \dfrac{6}{21}$ is a proportion.

This method is called finding **cross products**.

EXAMPLE 3　Determine which equations are proportions.

(a) $\dfrac{14}{18} \overset{?}{=} \dfrac{35}{45}$ 　　　　**(b)** $\dfrac{16}{21} \overset{?}{=} \dfrac{174}{231}$

(a) $\dfrac{14}{18} \overset{?}{=} \dfrac{35}{45}$

$$18 \times 35 = \boxed{630}$$

$\dfrac{14}{18} \overset{\times}{} \dfrac{35}{45}$ 　The cross products are equal.　Thus $\dfrac{14}{18} = \dfrac{35}{45}$. This is a proportion.

$$14 \times 45 = \boxed{630}$$

(b) $\dfrac{16}{21} \overset{?}{=} \dfrac{174}{231}$

$$21 \times 174 = \boxed{3654}$$

$\dfrac{16}{21} \overset{\times}{} \dfrac{174}{231}$ 　The cross products are not equal.　Thus $\dfrac{16}{21} \neq \dfrac{174}{231}$. This is not a proportion.

$$16 \times 231 = \boxed{3696}$$

Practice Problem 3　Determine which of the following equations are proportions.

(a) $\dfrac{10}{18} \overset{?}{=} \dfrac{25}{45}$ 　　　　**(b)** $\dfrac{42}{100} \overset{?}{=} \dfrac{22}{55}$

Proportions may involve fractions or decimals.

EXAMPLE 4　Determine which equations are proportions.

(a) $\dfrac{5.5}{7} \overset{?}{=} \dfrac{33}{42}$ 　　　　**(b)** $\dfrac{5}{8\frac{3}{4}} \overset{?}{=} \dfrac{40}{72}$

(a) $\dfrac{5.5}{7} \overset{?}{=} \dfrac{33}{42}$

$$7 \times 33 = \boxed{231}$$

$\dfrac{5.5}{7} \overset{\times}{} \dfrac{33}{42}$ 　The cross products are equal.　Thus $\dfrac{5.5}{7} = \dfrac{33}{42}$. This is a proportion.

$$5.5 \times 42 = \boxed{231}$$

(b) $\dfrac{5}{8\frac{3}{4}} \overset{?}{=} \dfrac{40}{72}$

First we multiply $8\frac{3}{4} \times 40 = \dfrac{35}{\cancel{4}} \times \overset{10}{\cancel{40}} = 35 \times 10 = 350$

$$8\frac{3}{4} \times 40 = \boxed{350}$$

$\dfrac{5}{8\frac{3}{4}} \diagtimes \dfrac{40}{72}$ The cross products are not equal. Thus $\dfrac{5}{8\frac{3}{4}} \neq \dfrac{40}{72}$. This is not a proportion.

$$5 \times 72 = \boxed{360}$$

Practice Problem 4   Determine which equations are proportions.

(a) $\dfrac{2.4}{3} \overset{?}{=} \dfrac{12}{15}$ **(b)** $\dfrac{2\frac{1}{3}}{6} \overset{?}{=} \dfrac{14}{38}$

● **EXAMPLE 5**   **(a)** Is the rate $\dfrac{\$86}{13 \text{ tons}}$ equal to the rate $\dfrac{\$79}{12 \text{ tons}}$?

(b) Is the rate $\dfrac{3 \text{ American dollars}}{2 \text{ British pounds}}$ equal to the rate $\dfrac{27 \text{ American dollars}}{18 \text{ British pounds}}$?

(a) We want to know whether $\dfrac{86}{13} = \dfrac{79}{12}$.

$$13 \times 79 = \boxed{1027}$$

$\dfrac{86}{13} \diagtimes \dfrac{79}{12}$ The cross products are not equal.   Thus the two rates are not equal. This is not a proportion.

$$86 \times 12 = \boxed{1032}$$

(b) We want to know whether $\dfrac{3}{2} = \dfrac{27}{18}$.

$$2 \times 27 = \boxed{54}$$

$\dfrac{3}{2} \diagtimes \dfrac{27}{18}$ The cross products are equal.   Thus the two rates are equal. This is a proportion.

$$3 \times 18 = \boxed{54}$$

Practice Problem 5

(a) Is the rate $\dfrac{1260 \text{ words}}{7 \text{ pages}}$ equal to the rate $\dfrac{3530 \text{ words}}{20 \text{ pages}}$?

(b) Is the rate $\dfrac{2 \text{ American dollars}}{11 \text{ French francs}}$ equal to the rate $\dfrac{16 \text{ American dollars}}{88 \text{ French francs}}$? ●

Verbal and Writing Skills

1. A proportion states that two ratios or rates are _____ .

2. Explain in your own words how we use the equality test for fractions to determine if a statement is a proportion. Give an example.

Write a proportion.

3. 18 is to 9 as 2 is to 1.   **4.** 16 is to 12 as 4 is to 3.   **5.** 20 is to 36 as 5 is to 9.

6. 42 is to 48 as 14 is to 16.   **7.** 84 is to 105 as 12 is to 15.   **8.** $1\frac{1}{2}$ is to 6 as $7\frac{3}{4}$ is to 31.

9. $5\frac{1}{2}$ is to 16 as $7\frac{2}{3}$ is to 23.   **10.** 5.5 is to 10 as 11 is to 20.   **11.** 6.5 is to 14 as 13 is to 28.

Applications

Write a proportion.

12. When Jenny makes rice in her steamer, she mixes 2 cups of rice with 3 cups of water. To make 8 cups of rice, she needs 12 cups of water.

13. A cartographer (a person who makes maps) uses a scale of 3 inches to represent 40 miles. 27 inches would then represent 360 miles.

14. If Adam can read 18 pages of his biology book in two hours, he can read 45 pages in five hours.

15. If Ty hit 10 home runs in 45 baseball games, then he should hit 36 home runs in 162 games.

16. If 20 pounds of pistachio nuts cost $75, then 30 pounds will cost $112.50.

17. If three credit hours at El Paso Community College costs $525, then seven credit hours should cost $1225.

18. Last year the Little League Baseball finals held 2200 people on 100 benches. This year we will have 2750 people on 125 benches.

19. There are 3 teaching assistants for every 40 children in the elementary school. If we have 280 children, then we will have 21 teaching assistants.

▲ **20.** If 16 pounds of fertilizer cover 1520 square feet of lawn, then 19 pounds of fertilizer should cover 1805 square feet of lawn.

21. When New City had 4800 people, it had three restaurants. Now New City has 11,200 people, so it should have seven restaurants.

Determine which equations are proportions.

22. $\dfrac{8}{6} \overset{?}{=} \dfrac{20}{15}$

23. $\dfrac{10}{25} \overset{?}{=} \dfrac{6}{15}$

24. $\dfrac{12}{7} \overset{?}{=} \dfrac{15}{9}$

25. $\dfrac{11}{7} \overset{?}{=} \dfrac{20}{13}$

26. $\dfrac{99}{100} \overset{?}{=} \dfrac{49}{50}$

27. $\dfrac{17}{75} \overset{?}{=} \dfrac{22}{100}$

28. $\dfrac{315}{2100} \overset{?}{=} \dfrac{15}{100}$

29. $\dfrac{102}{120} \overset{?}{=} \dfrac{85}{100}$

30. $\dfrac{6}{14} \overset{?}{=} \dfrac{4.5}{10.5}$

31. $\dfrac{7}{9} \overset{?}{=} \dfrac{10.5}{13.5}$

32. $\dfrac{11}{12} \overset{?}{=} \dfrac{9.5}{10}$

33. $\dfrac{3}{17} \overset{?}{=} \dfrac{4.5}{24.5}$

34. $\dfrac{2}{4\frac{1}{3}} \overset{?}{=} \dfrac{6}{13}$

35. $\dfrac{2\frac{1}{3}}{3} \overset{?}{=} \dfrac{7}{15}$

36. $\dfrac{8}{19} \overset{?}{=} \dfrac{2}{4\frac{3}{4}}$

37. $\dfrac{9}{22} \overset{?}{=} \dfrac{3}{7\frac{1}{3}}$

38. $\dfrac{75 \text{ miles}}{5 \text{ hours}} \overset{?}{=} \dfrac{105 \text{ miles}}{7 \text{ hours}}$

39. $\dfrac{135 \text{ miles}}{3 \text{ hours}} \overset{?}{=} \dfrac{225 \text{ miles}}{5 \text{ hours}}$

40. $\dfrac{286 \text{ gallons}}{12 \text{ acres}} \overset{?}{=} \dfrac{429 \text{ gallons}}{18 \text{ acres}}$

41. $\dfrac{166 \text{ gallons}}{14 \text{ acres}} \overset{?}{=} \dfrac{249 \text{ gallons}}{21 \text{ acres}}$

42. $\dfrac{48 \text{ points}}{56 \text{ games}} \overset{?}{=} \dfrac{40 \text{ points}}{45 \text{ games}}$

43. $\dfrac{27 \text{ points}}{45 \text{ games}} \overset{?}{=} \dfrac{22 \text{ points}}{40 \text{ games}}$

44. At the Michael W. Smith concert on Friday there were 9600 female fans and 8200 male fans. The concert on Saturday had 12,480 female fans and 10,660 male fans. Is the ratio of female fans to male fans the same for both nights of the concert?

45. In the beginner scuba diving course, there are 96 male students and 54 female students. In the intermediate scuba diving course, there are 144 male students and 81 female students. Is the ratio of male students to female students the same in both scuba classes?

46. A machine folds 730 boxes in six hours. Another machine folds 1090 boxes in nine hours. Do they fold boxes at the same rate?

47. A car traveled 550 miles in 15 hours. A bus traveled 230 miles in 6 hours. Did they travel at the same rate?

▲ **48.** A common size for a color television screen is 22 inches wide by 16 inches tall. Does a smaller color television screen that is 11 inches wide by 8.5 inches tall have the same ratio of width to length?

▲ **49.** A common size for a driveway in suburban Wheaton, Illinois, is 75 feet long by 20 feet wide. Does a larger driveway that is 105 feet long by 28 feet wide have the same ratio of width to length?

75 feet

20 feet

To Think About

50. Determine whether $\dfrac{63}{161} = \dfrac{171}{437}$

    (a) by reducing each side to lowest terms.

    (b) by using the equality test for fractions.

    (c) Which method was faster? Why?

51. Determine whether $\dfrac{169}{221} = \dfrac{247}{323}$

    (a) by reducing each side to lowest terms.

    (b) by using the equality test for fractions.

    (c) Which method was faster? Why?

Cumulative Review Problems

Calculate.

52. $9.6 + 7.8 + 2.56 + 3.004 + 0.1765$

53. 2.83×5.002

54. $\begin{array}{r} 29{,}366.215 \\ -\,28{,}963.807 \\ \hline \end{array}$

55. $7.03\overline{)181.374}$

56. Xavier's car holds $16\frac{1}{8}$ gallons of gas. He used $3\frac{3}{4}$ gallons today, $5\frac{1}{3}$ gallons yesterday, and $2\frac{5}{6}$ gallons the day before. Assuming he started with a full tank, determine how many gallons are left in the tank.

4.3 Solving Proportions

1 *Solving for the variable n in an Equation of the Form a × n = b*

Consider this expression: "3 times a number yields 15. What is the number?" We could write this as

$$3 \times \boxed{?} = 15$$

and guess that the number $\boxed{?} = 5$. There is a better way of solving this problem, a way that eliminates the guesswork. We will begin by using a **variable**. That is, we will use a letter to represent a number we do not yet know. We briefly used variables in Chapters 1–3. Now we use them more extensively.

Let the letter *n* represent the unknown number. We write

$$3 \times n = 15.$$

This is called an **equation**. An equation has an equal sign. This indicates that the values on each side of it are equivalent. We want to find the number *n* in this equation without guessing. We will not change the value of *n* in the equation if we divide both sides of the equation by 3. Thus if

$$3 \times n = 15,$$

we can say

$$\frac{3 \times n}{3} = \frac{15}{3},$$

which is

$$\frac{3}{3} \times n = 5$$

or

$$1 \times n = 5.$$

Since $1 \times$ any number is the same number, we know that $n = 5$. Any equation of the form $a \times n = b$ can be solved in this way. We divide both sides of an equation of the form $a \times n = b$ by the number that is multiplied by *n*. (This method will not work for $3 + n = 15$, since here the 3 is added to *n* and not multiplied by *n*.)

● EXAMPLE 1 Solve for *n*.   **(a)** $16 \times n = 80$ **(b)** $24 \times n = 240$

(a) $16 \times n = 80$

$$\frac{16 \times n}{16} = \frac{80}{16} \qquad \text{Divide each side by 16.}$$

$$n = 5 \qquad \text{because } 16 \div 16 = 1 \text{ and } 80 \div 16 = 5.$$

(b) $24 \times n = 240$

$$\frac{24 \times n}{24} = \frac{240}{24} \qquad \text{Divide each side by 24.}$$

$$n = 10 \qquad \text{because } 24 \div 24 = 1 \text{ and } 240 \div 24 = 10.$$

Practice Problem 1 Solve for *n*.

(a) $5 \times n = 45$ **(b)** $7 \times n = 84$

The same procedure is followed if the variable n is on the right side of the equation.

● EXAMPLE 2 Solve for n.

(a) $66 = 11 \times n$

(b) $143 = 13 \times n$

(a) $66 = 11 \times n$

$$\frac{66}{11} = \frac{11 \times n}{11} \quad \text{Divide each side by 11.}$$

$$6 = n$$

(b) $143 = 13 \times n$

$$\frac{143}{13} = \frac{13 \times n}{13} \quad \text{Divide each side by 13.}$$

$$11 = n$$

Practice Problem 2 Solve for n.

(a) $108 = 9 \times n$

(b) $210 = 14 \times n$

The numbers in the equations are not always whole numbers, and the answer to an equation is not always a whole number.

● EXAMPLE 3 Solve for n.

(a) $16 \times n = 56$

(b) $18.2 = 2.6 \times n$

(a) $16 \times n = 56$

$$\frac{16 \times n}{16} = \frac{56}{16} \quad \text{Divide each side by 16.}$$

$$n = 3.5$$

$$\begin{array}{r} 3.5 \\ 16\overline{)56.0} \\ \underline{48} \\ 8\,0 \\ \underline{8\,0} \\ 0 \end{array}$$

(b) $18.2 = 2.6 \times n$

$$\frac{18.2}{2.6} = \frac{2.6 \times n}{2.6} \quad \text{Divide each side by 2.6}$$

$$7 = n$$

$$\begin{array}{r} 7. \\ 2.6_\wedge\overline{)18.2_\wedge} \\ \underline{18\,2} \\ 0 \end{array}$$

Practice Problem 3 Solve for n.

(a) $15 \times n = 63$

(b) $39.2 = 5.6 \times n$

② *Finding the Missing Number in a Proportion*

Sometimes one of the pieces of a proportion is unknown. We can use an equation such as $a \times n = b$ and solve for n to find the unknown quantity. Suppose we want to know the value of n in the proportion

$$\frac{5}{12} = \frac{n}{144}.$$

Since this is a proportion, we know that $5 \times 144 = 12 \times n$. Simplifying, we have

$$720 = 12 \times n.$$

Next we divide both sides by 12.

$$\frac{720}{12} = \frac{12 \times n}{12}$$

$$60 = n$$

We check to see if this is correct. Do we have a true proportion?

$$\frac{5}{12} \overset{?}{=} \frac{60}{144}$$

$$\frac{5}{12} \diagup\!\!\!\!\diagdown \frac{60}{144} \qquad 12 \times 60 = 720 \leftarrow$$
$$5 \times 144 = 720 \leftarrow \qquad \text{The cross products are equal.}$$

Thus $\dfrac{5}{12} = \dfrac{60}{144}$ is true. We have checked our answer.

To Solve for a Missing Number in a Proportion

1 Find the cross products.
2 Divide each side of the equation by the number multiplied by n.
3 Simplify the result.
4 Check your answer.

● **EXAMPLE 4** Find the value of n in $\dfrac{25}{4} = \dfrac{n}{12}$.

$$25 \times 12 = 4 \times n \qquad \text{Find the cross products.}$$
$$300 = 4 \times n$$
$$\frac{300}{4} = \frac{4 \times n}{4} \qquad \text{Divide each side by 4.}$$
$$75 = n$$

Check: Is this a proportion?

$$\frac{25}{4} \overset{?}{=} \frac{75}{12}$$
$$25 \times 12 \overset{?}{=} 4 \times 75$$
$$300 = 300 \qquad \checkmark$$

It is a proportion. The answer $n = 75$ is correct.

Practice Problem 4 Find the value of n in $\dfrac{24}{n} = \dfrac{3}{7}$. ●

Some answers will be exact values. In other cases we will obtain answers that are rounded to a certain decimal place. Recall that we sometimes use the $\approx$ symbol, which means "is approximately equal to."

The answer to the next problem is not a whole number.

● **EXAMPLE 5** Find the value of n in $\dfrac{125}{2} = \dfrac{150}{n}$.

$$125 \times n = 2 \times 150 \quad \text{Find the cross products.} \qquad \textit{Check.} \qquad \frac{125}{2} \overset{?}{=} \frac{150}{2.4}$$

$$125 \times n = 300 \qquad\qquad\qquad\qquad\qquad\qquad 125 \times 2.4 \overset{?}{=} 2 \times 150$$

$$\frac{125 \times n}{125} = \frac{300}{125} \qquad \text{Divide each side by 125.} \qquad\qquad 300 = 300 \quad \checkmark$$

$$n = 2.4$$

Practice Problem 5 Find the value of n in $\dfrac{176}{4} = \dfrac{286}{n}$.

In real-life situations it is helpful to write the units of measure in the proportion. Remember, order is important. The same units should be in the same position in the fractions.

EXAMPLE 6 If 5 grams of a non-icing additive are placed in 8 liters of diesel fuel, how many grams n should be added to 12 liters of diesel fuel?

We need to find the value of n in $\dfrac{n \text{ grams}}{12 \text{ liters}} = \dfrac{5 \text{ grams}}{8 \text{ liters}}$.

$$8 \times n = 12 \times 5$$
$$8 \times n = 60$$
$$\frac{8 \times n}{8} = \frac{60}{8}$$
$$n = 7.5$$

The answer is 7.5 grams. 7.5 grams of the additive should be added to 12 liters of the diesel fuel.

Check. $\dfrac{7.5 \text{ grams}}{12 \text{ liters}} \overset{?}{=} \dfrac{5 \text{ grams}}{8 \text{ liters}}$

$$7.5 \times 8 \overset{?}{=} 12 \times 5$$
$$60 = 60 \checkmark$$

Practice Problem 6 Find the value of n in $\dfrac{80 \text{ dollars}}{5 \text{ tons}} = \dfrac{n \text{ dollars}}{6 \text{ tons}}$.

Some proportions contain decimals.

EXAMPLE 7 Find the value of n in $\dfrac{141.75 \text{ miles}}{4.5 \text{ hours}} = \dfrac{63 \text{ miles}}{n \text{ hours}}$.

$$141.75 \times n = 63 \times 4.5$$
$$141.75 \times n = 283.5$$
$$\frac{141.75 \times n}{141.75} = \frac{283.5}{141.75}$$
$$n = 2$$

The answer is $n = 2$. The check is up to you.

Practice Problem 7 Find the value of n in $\dfrac{264.25 \text{ meters}}{3.5 \text{ seconds}} = \dfrac{n \text{ meters}}{2 \text{ seconds}}$.

To Think About Suppose that the proportion contains fractions or mixed numbers. Could you still follow all the steps? For example, find n when

$$\frac{n}{3\frac{1}{4}} = \frac{5\frac{1}{6}}{2\frac{1}{3}}$$

We have $2\frac{1}{3} \times n = 5\frac{1}{6} \times 3\frac{1}{4}$.

This can be written as

$$\frac{7}{3} \times n = \frac{31}{6} \times \frac{13}{4}$$

$$\boxed{\frac{7}{3} \times n = \frac{403}{24}} \quad \color{red}{\text{equation (1)}}$$

Now we divide each side of equation (1) by $\frac{7}{3}$. Why?

$$\frac{\frac{7}{3} \times n}{\frac{7}{3}} = \frac{\frac{403}{24}}{\frac{7}{3}}$$

Be careful here. The right-hand side means $\frac{403}{24} \div \frac{7}{3}$, which we evaluate by *inverting* the second fraction and multiplying.

$$\frac{403}{\underset{8}{\cancel{24}}} \times \frac{\overset{1}{\cancel{3}}}{7} = \frac{403}{56}$$

Thus $n = \frac{403}{56}$ or $7\frac{11}{56}$. Think about all the steps to solving this problem. Can you follow them? There is another way to do the problem. We could multiply each side of equation (1) by $\frac{3}{7}$. Try it. Why does this work? Now try exercises 56–59 in Exercises 4.3.

Verbal and Writing Skills

1. Suppose you have an equation of the form $a \times n = b$, where the letters a and b represent whole numbers. Explain in your own words how you would solve the equation.

2. Suppose you have an equation of the form $\frac{n}{a} = \frac{b}{c}$, where a, b, and c represent whole numbers. Explain in your own words how you would solve the equation.

Solve for n.

3. $12 \times n = 132$

4. $14 \times n = 126$

5. $3 \times n = 16.8$

6. $2 \times n = 19.6$

7. $n \times 3.8 = 95$

8. $n \times 11.4 = 57$

9. $40.6 = 5.8 \times n$

10. $50.4 = 6.3 \times n$

11. $\frac{3}{4} \times n = 26$

12. $\frac{5}{6} \times n = 32$

Find the value of n. Check your answer.

13. $\frac{n}{20} = \frac{3}{4}$

14. $\frac{n}{28} = \frac{3}{7}$

15. $\frac{6}{n} = \frac{3}{8}$

16. $\frac{4}{n} = \frac{2}{7}$

17. $\frac{12}{40} = \frac{n}{25}$

18. $\frac{13}{30} = \frac{n}{15}$

19. $\frac{25}{100} = \frac{8}{n}$

20. $\frac{18}{150} = \frac{6}{n}$

21. $\frac{n}{6} = \frac{150}{12}$

22. $\frac{n}{22} = \frac{25}{11}$

23. $\frac{15}{4} = \frac{n}{6}$

24. $\frac{16}{10} = \frac{n}{9}$

25. $\frac{18}{n} = \frac{3}{11}$

26. $\frac{16}{n} = \frac{2}{7}$

Find the value of n. Round your answer to the nearest tenth when necessary.

27. $\dfrac{21}{n} = \dfrac{2}{3}$ **28.** $\dfrac{58}{n} = \dfrac{11}{6}$ **29.** $\dfrac{9}{26} = \dfrac{n}{52}$ **30.** $\dfrac{12}{8} = \dfrac{21}{n}$ **31.** $\dfrac{15}{12} = \dfrac{10}{n}$

32. $\dfrac{n}{18} = \dfrac{3.5}{1}$ **33.** $\dfrac{n}{36} = \dfrac{4.5}{1}$ **34.** $\dfrac{2.5}{n} = \dfrac{0.5}{10}$ **35.** $\dfrac{1.5}{n} = \dfrac{0.3}{8}$ **36.** $\dfrac{3}{4} = \dfrac{n}{3.8}$

37. $\dfrac{7}{8} = \dfrac{n}{4.2}$ **38.** $\dfrac{12.5}{16} = \dfrac{n}{12}$ **39.** $\dfrac{13.8}{15} = \dfrac{n}{6}$ **40.** $\dfrac{27\frac{3}{11}}{100} = \dfrac{3}{n}$ **41.** $\dfrac{22\frac{2}{9}}{100} = \dfrac{2}{n}$

Applications

Find the value of n. Round to the nearest hundredth when necessary.

42. $\dfrac{n \text{ grams}}{10 \text{ liters}} = \dfrac{7 \text{ grams}}{25 \text{ liters}}$ **43.** $\dfrac{n \text{ grams}}{14 \text{ liters}} = \dfrac{6 \text{ grams}}{22 \text{ liters}}$ **44.** $\dfrac{105 \text{ miles}}{2 \text{ hours}} = \dfrac{n \text{ miles}}{5 \text{ hours}}$

45. $\dfrac{128 \text{ miles}}{4 \text{ hours}} = \dfrac{80 \text{ miles}}{n \text{ hours}}$ **46.** $\dfrac{50 \text{ gallons}}{12 \text{ acres}} = \dfrac{36 \text{ gallons}}{n \text{ acres}}$ **47.** $\dfrac{80 \text{ gallons}}{26 \text{ acres}} = \dfrac{42 \text{ gallons}}{n \text{ acres}}$

48. $\dfrac{10 \text{ miles}}{16.1 \text{ kilometers}} = \dfrac{n \text{ miles}}{7 \text{ kilometers}}$ **49.** $\dfrac{3 \text{ kilometers}}{1.86 \text{ miles}} = \dfrac{n \text{ kilometers}}{4 \text{ miles}}$

50. $\dfrac{12 \text{ quarters}}{3 \text{ dollars}} = \dfrac{87 \text{ quarters}}{n \text{ dollars}}$ **51.** $\dfrac{35 \text{ dimes}}{3.5 \text{ dollars}} = \dfrac{n \text{ dimes}}{8 \text{ dollars}}$

52. $\dfrac{2\frac{1}{2} \text{ acres}}{3 \text{ people}} = \dfrac{n \text{ acres}}{5 \text{ people}}$ **53.** $\dfrac{3\frac{1}{4} \text{ feet}}{8 \text{ pounds}} = \dfrac{n \text{ feet}}{12 \text{ pounds}}$

▲ **54.** A photographic negative is 3.5 centimeters wide and 2.5 centimeters tall. If you want to make a color print that is 6 centimeters tall, how wide will the print be?

▲ **55.** A color photograph is 5 inches wide and 3 inches tall. If you want to make an enlargement of this photograph that is 5 inches tall, how wide will the enlargement be?

To Think About

Study the "To Think About" example in the text. Then solve for n in exercises 56–59. Express n as a mixed number.

56. $\dfrac{n}{2\frac{1}{3}} = \dfrac{4\frac{5}{6}}{3\frac{1}{9}}$

57. $\dfrac{n}{7\frac{1}{4}} = \dfrac{2\frac{1}{5}}{4\frac{1}{8}}$

58. $\dfrac{8\frac{1}{6}}{n} = \dfrac{5\frac{1}{2}}{7\frac{1}{3}}$

59. $\dfrac{9\frac{3}{4}}{n} = \dfrac{8\frac{1}{2}}{4\frac{1}{3}}$

Cumulative Review Problems

Evaluate by doing each operation in the proper order.

60. $4^3 + 20 \div 5 + 6 \times 3 - 5 \times 2$

61. $(1.6)^2 - 0.12 \times 3.5 + 36.8 \div 2.5$

62. Write a word name for the decimal 0.563.

63. Write thirty-four ten-thousandths in decimal notation.

64. The North Bend Women's Soccer League has eight teams. If each team plays all the others twice, how many games will have been played?

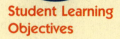 **Applied Problems Involving Proportions**

Student Learning Objectives

After studying this section, you will be able to:

 Solve applied problems using proportions.

SSM

PH TUTOR CENTER

CD & VIDEO

MATH PRO

WEB

① Solving Applied Problems Using Proportions

Let us examine a variety of applied problems that can be solved by proportions.

● **EXAMPLE 1** A company that makes eyeglasses conducted a recent survey using a quality control test. It was discovered that 37 pairs of eyeglasses in a sample of 120 pairs of eyeglasses were defective. If this rate remains the same each year, how many of the 36,000 pairs of eyeglasses made by this company each year are defective?

Mathematics Blueprint For Problem Solving

Gather the Facts	What Am I Asked to Do?	How Do I Proceed?	Key Points to Remember
Sample: 37 defective pairs in a total of 120 pairs 36,000 pairs were made by the company.	Find how many of the 36,000 pairs of eyeglasses are defective.	Set up a proportion comparing defective eyeglasses to total eyeglasses.	Make sure one fraction represents the sample and one fraction represents the total number of eyeglasses made by the company.

We will use the letter *n* to represent the number of defective eyeglasses in the total.

We compare the sample   to   the total number

$$\frac{37 \text{ defective pairs}}{120 \text{ total pairs of eyeglasses}} = \frac{n \text{ defective pairs}}{36,000 \text{ total pairs of eyeglasses}}$$

$37 \times 36,000 = 120 \times n$ Form the cross products.

$1,332,000 = 120 \times n$ Simplify.

$\dfrac{1,332,000}{120} = \dfrac{120 \times n}{120}$ Divide each side by 120.

$11,100 = n$

Thus, if the rate of defective eyeglasses holds steady, there are about 11,100 defective pairs of eyeglasses made by the company each year.

Practice Problem 1 Yesterday an automobile assembly line produced 243 engines, of which 27 were defective. If the same rate is true each day, how many of the 4131 engines produced this month are defective?   ●

Looking back at Example 1, perhaps it occurred to you that the fractions in the proportion could be set up in an alternative way. You can set up this problem in several different ways as long as the units are in correctly corresponding positions. It would be correct to set up the problem in the form

$$\frac{\text{defective pairs in sample}}{\text{total defective pairs}} = \frac{\text{total glasses in sample}}{\text{total glasses made by company}}$$

or

$$\frac{\text{total glasses in sample}}{\text{defective pairs in sample}} = \frac{\text{total glasses made by company}}{\text{total defective pairs}}.$$

But we **cannot** set up the problem this way.

$$\frac{\text{defective pairs in sample}}{\text{total glasses made by company}} = \frac{\text{total defective pairs}}{\text{total glasses in sample}}$$

This is *not* correct. Do you see why?

EXAMPLE 2   Ted's car can go 245 miles on 7 gallons of gas. Ted wants to take a trip of 455 miles. Approximately how many gallons of gas will this take?

Let n = the unknown number of gallons.

$$\frac{245 \text{ miles}}{7 \text{ gallons}} = \frac{455 \text{ miles}}{n \text{ gallons}}$$

$245 \times n = 7 \times 455$	Form the cross products.
$245 \times n = 3185$	Simplify.
$\dfrac{245 \times n}{245} = \dfrac{3185}{245}$	Divide both sides by 245.
$n = 13$	

Ted will need approximately 13 gallons of gas for the trip.

Practice Problem 2   Cindy's car travels 234 miles on 9 gallons of gas. How many gallons of gas will Cindy need to take a 312-mile trip?

EXAMPLE 3   In a certain gear, Alice's 10-speed bicycle has a gear ratio of three revolutions of the pedal for every two revolutions of the bicycle wheel. If her bicycle wheel is turning at 65 revolutions per minute, how many times must she pedal per minute?

Let n = the number of revolutions of the pedal.

$$\frac{3 \text{ revolutions of the pedal}}{2 \text{ revolutions of the wheel}} = \frac{n \text{ revolutions of the pedal}}{65 \text{ revolutions of the wheel}}$$

$3 \times 65 = 2 \times n$	Cross-multiply.
$195 = 2 \times n$	Simplify.
$\dfrac{195}{2} = \dfrac{2 \times n}{2}$	Divide both sides by 2.
$97.5 = n$	

Alice will pedal at the rate of 97.5 revolutions in 1 minute.

Practice Problem 3   Alicia must pedal at 80 revolutions per minute to ride her bicycle at 16 miles per hour. If she pedals at 90 revolutions per minute, how fast will she be riding?

EXAMPLE 4 Tim operates a bicycle rental center during the summer months on the island of Martha's Vineyard. He discovered that when the ferryboats brought 8500 passengers a day to the island, his center rented 340 bicycles a day. Next summer the ferryboats plan to bring 10,300 passengers a day to the island. How many bicycles a day should Tim plan to rent?

Two important cautions are necessary before we solve the proportion. We need to be sure that the bicycle rentals are directly related to the number of people on the ferryboat. (Presumably, people who fly to the island or who take small pleasure boats to the island also rent bicycles.)

Next we need to be sure that the people who represent the increase in passengers per day would be as likely to rent bicycles as the present number of passengers do. For example, if the new visitors to the island are all senior citizens, they are not as likely to rent bicycles as younger people. If we assume those two conditions are satisfied, then we can solve the problem as follows.

$$\frac{8500 \text{ passengers per day now}}{340 \text{ bike rentals per day now}} = \frac{10{,}300 \text{ passengers per day later}}{n \text{ bike rentals per day later}}$$

$$8500 \times n = 340 \times 10{,}300$$

$$8500 \times n = 3{,}502{,}000$$

$$\frac{8500 \times n}{8500} = \frac{3{,}502{,}000}{8500}$$

$$n = 412$$

If the two conditions are satisfied, we would predict 412 bicycle rentals.

Practice Problem 4   For every 4050 people who walk into Tom's Souvenir Shop, 729 make a purchase. Assuming the same conditions, if 5500 people walk into Tom's Souvenir Shop, how many people may be expected to make a purchase?

Biologists and others who observe or protect wildlife sometimes use the capture-mark-recapture method to determine how many animals are in a certain region. In this approach some animals are caught and tagged in a way that does not harm them. They are then released into the wild, where they mix with their kind.

It is assumed (usually correctly) that the tagged animals will mix throughout the entire population in that region, so that when they are recaptured in a future sample, the biologists can use them to make reasonable estimates about the total population. We will employ the capture-mark-recapture method in the next example.

EXAMPLE 5   A biologist catches 42 fish in a lake and tags them. She then quickly returns them to the lake. In a few days she catches a new sample of 50 fish. Of those 50 fish, 7 have her tag. Approximately how many fish are in the lake?

$$\frac{\text{42 fish tagged in 1st sample}}{\text{n fish in lake}} = \frac{\text{7 fish tagged in 2nd sample}}{\text{50 fish caught in 2nd sample}}$$

$$42 \times 50 = 7 \times n$$

$$2100 = 7 \times n$$

$$\frac{2100}{7} = \frac{7 \times n}{7}$$

$$300 = n$$

Assuming that no tagged fish died and that the tagged fish mixed throughout the population of fish in the lake, we estimate that there are 300 fish in the lake.

Practice Problem 5   A park ranger in Alaska captures and tags 50 bears. He then releases them to range through the forest. Sometime later he captures 50 bears. Of the 50, 4 have tags from the previous capture. Estimate the number of bears in the forest.

Developing Your Study Skills

Getting Help

Getting the right kind of help at the right time can be a key ingredient in being successful in mathematics. When you have gone to class on a regular basis, taken careful notes, methodically read your textbook, and diligently done your homework—in other words, when you have made every effort possible to learn the mathematics—you may still find that you are having difficulty. If this is the case, then you need to seek help. Make an appointment with your instructor to find out what help is available to you. The instructor, tutoring services, a mathematics lab, videotapes, and computer software may be among the resources you can draw on.

4.4 Exercises

Applications

1. The policy at the Colonnade Hotel is to have 19 desserts for every 16 people if a buffet is being served. If the Saturday buffet has 320 people, how many desserts must be available?

2. An automobile dealership has found that for every 140 cars sold, 23 will be brought back to the dealer for major repairs. If the dealership sells 980 cars this year, approximately how many cars will be brought back for major repairs?

3. The directions on a bottle of bleach say to use $\frac{3}{4}$ cup bleach for every 1 gallon of soapy water. Ron needs a 4-gallon mixture to mop his floors. How many cups of bleach will he need?

4. There are approximately $2\frac{1}{2}$ centimeters in one inch. Approximately how many centimeters are in one foot?

5. One day in January 2000, the exchange rate between U.S. and Swiss currency was 10 U.S. dollars for every 15 Swiss francs. If on a trip to Switzerland on that day, you exchanged $430 how many francs did you receive?

6. When Douglas flew to Sweden, the exchange rate was 65 Swedish kronor for every 8 U.S. dollars. If Douglas brought 320 U.S. dollars for spending money, how many kronor did he receive?

In exercises 7 and 8 two nearby objects cast shadows at the same time of day. The ratio of the height of one of the objects to the length of its shadow is equal to the ratio of the height of the other object to the length of its shadow.

▲ 7. A pro football offensive tackle who stands 6.5 feet tall casts a 5-foot shadow. At the same time, the football stadium, which he is standing next to, casts a shadow of 152 feet. How tall is the stadium? Round your answer to the nearest tenth.

▲ 8. A souvenir life-sized cardboard cutout of Queen Elizabeth (for the tourists) sits outside a London office building. It is 5 feet tall and casts a shadow that is 3 feet long. The office building casts a shadow that is 165 feet long. How tall is the building?

9. On Dr. Jennings's map of Antarctica, the scale states that 4 inches represent 250 miles of actual distance. Two Antarctic mountains are 5.7 inches apart on the map. What is the approximate distance in miles between the two mountains? Round your answer to the nearest mile.

10. On a tour guide map of Madagascar, the scale states that 3 inches represent 125 miles. Two beaches are 5.2 inches apart on the map. What is the approximate distance in miles between the two beaches? Round your answer to the nearest mile.

11. Jenny's blueberry pancake recipe uses 5 cups of blueberries for every 12 people. How many cups of blueberries does she need for the same recipe prepared for 28 people?

12. In his lasagna recipe, Giovanni uses 2 cups of marinara sauce for every 7 people. How much sauce will he need to make lasagna for 62 people? Round to the nearest tenth.

13. During a basketball game against the Miami Heat, the Denver Nuggets made 17 out of 25 free throws attempted. If they attempt 150 free throws in the remaining games of the season, how many will they make if their success rate remains the same?

14. A baseball pitcher gave up 52 earned runs in 260 innings of pitching. At that rate, how many runs would he give up in a 9-inning game? (This decimal is called the pitcher's *earned run average*.)

15. The Chicago-O'Hare Taxi Drivers Association has discovered that for every 7 plane flights arriving at O'Hare Airport, 79 people need a taxi ride from the airport. If tomorrow 434 plane flights are scheduled to arrive at O'Hare, how many people would you estimate will need a taxi ride from the airport?

16. The employees at a local coffee shop know that during their morning rush, 6 out of every 15 people who buy coffee will also buy a muffin. They use this fact to make sure they bake enough muffins. If they are expecting 300 customers in the next few days, how many muffins will they need to have prepared?

17. In Kenya, a worker at the game preserve captures 26 giraffes, tags them, and then releases them back into the preserve. The next month, he captures 18 giraffes and finds that 6 of them have already been tagged. Estimate the number of giraffes on the preserve.

18. An ornithologist is studying hawks in the Adirondack Mountains. She catches 24 hawks over a period of one month, tags them, and releases them back into the wild. The next month, she catches 20 hawks and finds that 12 are already tagged. Estimate the number of hawks in this part of the mountains.

19. Bill and Shirley Grant are raising tomatoes to sell at the local co-op. The farm has a yield of 425 pounds of tomatoes for every 3 acres. The farm has 14 acres of good tomatoes. The crop this year should bring $1.80 per pound for each pound of tomatoes. How much will the Grants get from the sale of the tomato crop?

▲ **20.** A paint manufacturer suggests 2 gallons of flat latex paint for every 750 square feet of wall. A painter is going to paint 7875 square feet of wall in a Dallas office building with paint that costs $8.50 per gallon. How much will the painter spend for the paint?

21. A company that manufactures computer chips expects 5 out of every 100 made to be defective. In a shipment of 5400 chips, how many are expected to be defective?

22. The editor of a small-town newspaper conducted a survey to find out how many customers are satisfied with their delivery service. Of the 100 people surveyed, 88 customers said they were satisfied. If 1700 people receive the newspaper, how many are satisfied?

To Think About

The following chart is used for several brands of instant mashed potatoes. Use this chart in answering exercises 23–26.

TO MAKE	WATER	MARGARINE OR BUTTER	SALT (optional)	MILK	FLAKES
2 servings	2/3 cup	1 tablespoon	1/8 teaspoon	1/4 cup	2/3 cup
4 servings	1-1/3 cups	2 tablespoons	1/4 teaspoon	1/2 cup	1-1/3 cups
6 servings	2 cups	3 tablespoons	1/2 teaspoon	3/4 cup	2 cups
Entire box	5 cups	1/2 cup	1 teaspoon	2-1/2 cups	Entire box

23. How many cups of water and how many cups of milk are needed to make enough mashed potatoes for three people?

24. How many cups of water and how many cups of milk are needed to make enough mashed potatoes for five people?

25. The instructions on the box say that in high altitude (above 5000 feet) the amount of water should be reduced by $\frac{1}{4}$. If you had to make enough mashed potatoes for eight people in a high-altitude city, how many cups of water and how many cups of milk would be needed?

26. If you used the entire box, how many servings would you obtain?

During the 1999 playing season, Ken Griffey, Jr. of the Seattle Mariners hit 48 home runs and was paid an annual salary of $8,510,532. During the same season, Jay Bell of the Arizona Diamondbacks hit 38 home runs and was paid an annual salary of $6,000,000. (Source: U.S. Bureau of Labor Statistics)

27. Express the salary of each player as a unit rate in terms of dollars paid to home runs hit.

28. Which player hit more home runs per dollar?

During the 1999 playing season, Roger Clemens of the New York Yankees pitched in 30 games and was paid an annual salary of $8,250,000. During the same season, Pedro Martinez of the Boston Red Sox pitched in 31 games and was paid an annual salary of $11,000,000. (Source: U.S. Bureau of Labor Statistics)

29. Express the salary of each player as a unit rate in terms of dollars paid to number of games played.

30. Which player pitched more games per dollar?

Cumulative Review Problems

31. Round to the nearest hundred. 56,179

32. Round to the nearest ten thousand. 196,379,910

33. Round to the nearest tenth. 56.148

34. Round to the nearest hundredth. 1.96341

▲ **35.** An eyewear company makes very expensive carbon fiber sunglasses. The material is made into long sheets, and the basic eyeglass frame is punched out by a machine. It takes a section measuring $1\frac{3}{16}$ feet by $\frac{4}{5}$ foot to make one pair of glasses.
  (a) How many square feet of this material are needed for one frame?
  (b) How many square feet of this material are needed for 1500 frames?

Putting Your Skills to Work

Studying the Moose Population

In November 2000, it was estimated that Alaska had a population of 163,400 moose. Each year hunters usually kill between 6000 and 8000 moose. Since 1990, the moose population has grown by 25,000. (*Source*: Alaska Department of Wildlife Management)

Problems for Individual Investigation

1. The ratio of females to males in the moose population is usually 23 to 27. What was the estimated number of males in the moose population as of November 2000?

2. Assume that the next year 20,000 females gave birth to twin calves, and 15,000 females had only one calf. Also assume that the next year hunters killed 6000 moose while 19,000 died from natural causes or natural predators. What was the moose population in 2001?

Problems for Group Investigation and Cooperative Study

3. Assume that over the ten years from 2000 to 2010, the total moose population grows by 3000 per year. How many females will be in the moose population in November 2010?

4. If the total moose population grows by 3500 per year for the five years from 2000 to 2005 and then, due to a very severe winter and lack of natural food, the moose population declines by 5000 per year for two years, how many males will be in the moose population in November 2007?

Internet Connections

  Netsite: http://www.prenhall.com/tobey_basic

This site contains information about the actual count of various animals in Alaska over the last several years and a projected count for the future.

5. Using the harvest figures as an indication of population and considering only areas where a moose harvest occurred, which GMU (Game Management Unit) appears to have the highest density of moose population (i.e. individuals per square mile)? The lowest?

6. If you were interested in photographing bears, which area would you choose as probably having the greatest density of black bears and brown bears combined?

Math in the Media

Digital Camera Sales Click

Digital Camera Sales (projected in millions)	
1996	$386
1997	$518
1998	$573
1999	$819

Digital Camera Sales Click

"An increasing number of consumers —eager to see their photographs instantly—are buying digital cameras."

From the table above it is easy to compute the amount of change in sales (in millions) from year to year. Thus from 1996 to 1997 the amount of change would be $518 − $386 = $132 (in millions). This is also known as the average (yearly) rate of change from 1996 to 1997 and it's represented as $132/1 which represents the rate as $132 (in millions) per year.

EXERCISES

The average rate of change of sales over other periods of time = change in sales/change in time. For example: ($573 − $386)/2 = $187/2 = $93.5 (in millions) per year from 1996 to 1998. Refer to the table for questions 1 and 2.

1. Find the yearly rate of change from 1997 to 1998 and 1998 to 1999.

2. Find the average rates of change from 1997 to 1999 and 1996 to 1999.

3. What is the relationship between the rates of change computed in exercise 1 and the average rate of change for two years (from 1997 to 1999) computed in exercise 2?

Chapter 4 Organizer

Topic	Procedure	Examples
Forming a ratio, p. 268.	A *ratio* is the comparison of two quantities that have the same units. A ratio is usually expressed as a fraction. The fractions should be in reduced form.	**1.** Find the ratio of 20 books to 35 books. $$\frac{20}{35} = \frac{4}{7}$$ **2.** Find the ratio in simplest form of $88:99$. $$\frac{88}{99} = \frac{8}{9}$$ **3.** Bob earns \$250 each week, but \$15 is withheld for medical insurance. Find the ratio of medical insurance to total pay. $$\frac{\$15}{\$250} = \frac{3}{50}$$
Forming a rate, p. 270.	A *rate* is a comparison of two quantities that have different units. A rate is usually expressed as a fraction in reduced form.	A college has 2520 students with 154 faculty. What is the rate of students to faculty? $$\frac{2520 \text{ students}}{154 \text{ faculty}} = \frac{180 \text{ students}}{11 \text{ faculty}}$$
Forming a unit rate, p. 270.	A *unit rate* is a rate with a denominator of 1. Divide the denominator into the numerator to obtain the unit rate.	A car traveled 416 miles in 8 hours. Find the unit rate. $$\frac{416 \text{ miles}}{8 \text{ hours}} = 52 \text{ miles/hour}$$ Bob spread 50 pounds of fertilizer over 1870 square feet of land. Find the unit rate of square feet per pound. $$\frac{1870 \text{ square feet}}{50 \text{ pounds}} = 37.4 \text{ square feet/pound}$$
Writing proportions, p. 276.	A *proportion* is a statement that two rates or two ratios are equal. A proportion statement a is to b as c is to d can be written $$\frac{a}{b} = \frac{c}{d}.$$	Write a proportion for 17 is to 34 as 13 is to 26. $$\frac{17}{34} = \frac{13}{26}$$
Determining whether a relationship is a proportion, p. 277.	For any two fractions where $b \neq 0$ and $d \neq 0$, $\frac{a}{b} = \frac{c}{d}$ if and only if $a \times d = b \times c$. A proportion is a statement that two rates or two ratios are equal.	**1.** Is this a proportion? $\frac{7}{56} = \frac{3}{24}$ $$7 \times 24 \stackrel{?}{=} 56 \times 3$$ $$168 = 168 \checkmark$$ It is a proportion. **2.** Is this a proportion? $$\frac{64 \text{ gallons}}{5 \text{ acres}} = \frac{89 \text{ gallons}}{7 \text{ acres}}$$ $$64 \times 7 \stackrel{?}{=} 5 \times 89$$ $$448 \neq 445$$ It is not a proportion.

Topic	Procedure	Examples
Solving a proportion, p. 284.	To solve a proportion where the value n is not known: **1.** Cross-multiply. **2.** Divide both sides of the equation by the number multiplied by n.	**1.** Solve for n. $$\frac{17}{n} = \frac{51}{9}$$ $17 \times 9 = 51 \times n$ Cross-multiply. $153 = 51 \times n$ Simplify. $$\frac{153}{51} = \frac{51 \times n}{51}$$ Divide by 51. $3 = n$
Solving applied problems, p. 290.	**1.** Write a proportion with n representing the unknown value. **2.** Solve the proportion.	Bob purchased eight notebooks for \$19. How much would 14 notebooks cost? $$\frac{8 \text{ notebooks}}{\$19} = \frac{14 \text{ notebooks}}{n}$$ $8 \times n = 19 \times 14$ $8 \times n = 266$ $$\frac{8 \times n}{8} = \frac{266}{8}$$ $n = 33.25$ The 14 notebooks would cost \$33.25.

Chapter 4 Review Problems

4.1   *Write in simplest form. Express your answer as a fraction.*

1. $88:40$ **2.** $65:39$ **3.** $28:35$ **4.** $250:475$ **5.** $2\frac{1}{3}$ to $4\frac{1}{4}$ **6.** 27 to 81

7. 280 to 651   **8.** 156 to 441   **9.** 26 tons to 65 tons

Bob earns \$215 per week and has \$35 per week withheld for federal taxes and \$20 per week withheld for state taxes.

10. Write the ratio of federal taxes withheld to earned income.

11. Write the ratio of state taxes withheld to earned income.

Write as a rate in simplest form.

12. 10 gallons of water for every 18 people

13. 44 revolutions every 121 minutes

14. 188 vibrations every 16 seconds

15. 12 cups of flour for every 38 people

In exercises 16–21, write as a unit rate. Round to the nearest tenth when necessary.

16. $2125 was paid for 125 shares of stock. Find the cost per share.

17. $2244 was paid for 132 chairs. Find the cost per chair.

▲**18.** $742.50 was spent for 55 square yards of carpet. Find the cost per square yard.

19. The Baseball Boosters Club spent $768.80 for 62 tickets to the ball game. Find the cost per ticket.

20. A 12.5-ounce can of white tuna costs $2.75. A 7.0-ounce can of the same brand of white tuna costs $1.75.
   (a) What is the cost per ounce of the large can?

   (b) What is the cost per ounce of the small can?

   (c) How much per ounce do you save by buying the larger can?

21. A 4-ounce jar of instant coffee costs $2.96. A 9-ounce jar of the same brand of instant coffee costs $5.22.
   (a) What is the cost per ounce of the 4-ounce jar?

   (b) What is the cost per ounce of the 9-ounce jar?

   (c) How much per ounce do you save by buying the larger jar?

4.2 *Write as a proportion.*

22. 12 is to 48 as 7 is to 28

23. $1\frac{1}{2}$ is to 5 as 4 is to $13\frac{1}{3}$

24. 7.5 is to 45 as 22.5 is to 135

25. If 15 pounds cost $4.50, then 27 pounds will cost $8.10.

26. If three buses can transport 138 passengers, then five buses can transport 230 passengers.

Determine whether each equation is a proportion.

27. $\dfrac{16}{48} = \dfrac{2}{12}$

28. $\dfrac{20}{25} = \dfrac{8}{10}$

29. $\dfrac{24}{20} = \dfrac{18}{15}$

30. $\dfrac{84}{48} = \dfrac{14}{8}$

31. $\dfrac{37}{33} = \dfrac{22}{19}$

32. $\dfrac{15}{18} = \dfrac{18}{22}$

33. $\dfrac{84 \text{ miles}}{7 \text{ gallons}} = \dfrac{108 \text{ miles}}{9 \text{ gallons}}$

34. $\dfrac{156 \text{ revolutions}}{6 \text{ minutes}} = \dfrac{181 \text{ revolutions}}{7 \text{ minutes}}$

4.3   *Solve for n.*

35. $7 \times n = 161$ **36.** $8 \times n = 42$ **37.** $558 = 18 \times n$ **38.** $663 = 39 \times n$

Solve. Round to the nearest tenth when necessary.

39. $\dfrac{3}{11} = \dfrac{9}{n}$ **40.** $\dfrac{2}{7} = \dfrac{12}{n}$ **41.** $\dfrac{n}{28} = \dfrac{6}{24}$ **42.** $\dfrac{n}{32} = \dfrac{15}{20}$ **43.** $\dfrac{2\frac{1}{4}}{9} = \dfrac{4\frac{3}{4}}{n}$ **44.** $\dfrac{3\frac{1}{3}}{2\frac{2}{3}} = \dfrac{7}{n}$

45. $\dfrac{54}{72} = \dfrac{n}{4}$ **46.** $\dfrac{45}{135} = \dfrac{n}{3}$ **47.** $\dfrac{6}{n} = \dfrac{2}{29}$ **48.** $\dfrac{8}{n} = \dfrac{2}{81}$ **49.** $\dfrac{25}{7} = \dfrac{60}{n}$ **50.** $\dfrac{60}{9} = \dfrac{31}{n}$

51. $\dfrac{35 \text{ miles}}{28 \text{ gallons}} = \dfrac{15 \text{ miles}}{n \text{ gallons}}$

52. $\dfrac{8 \text{ defective parts}}{100 \text{ perfect parts}} = \dfrac{44 \text{ defective parts}}{n \text{ perfect parts}}$

4.4   *Solve using a proportion. Round your answer to the nearest hundredth when necessary.*

53. The school volunteers used 3 gallons of paint to paint two rooms. How many gallons would they need to paint 10 rooms of the same size?

54. Several recent surveys show that 49 out of every 100 adults in America drink coffee. If a computer company employs 3450 people, how many of those employees would you expect would drink coffee? Round to the nearest whole number.

55. When Marguerite traveled, the rate of French francs to American dollars was 24 francs to 5 dollars. How many francs did Marguerite receive for 420 dollars?

56. A catering service recommends providing 7.5 pounds of cold cuts for every 18 people at the reception. The men's glee club is having a reception for 120 people. How many pounds of cold cuts should they order?

57. Two cities located 225 miles apart appear 3 inches apart on a map. If two other cities appear 8 inches apart on the map, how many miles apart are the cities?

58. If Melissa pedals her bicycle at 84 revolutions per minute, she travels at 14 miles per hour. How fast does she go if she pedals at 96 revolutions per minute?

▲ **59.** In the setting sun, a 6-foot man casts a shadow 16 feet long. At the same time a building casts a shadow of 320 feet. How tall is the building?

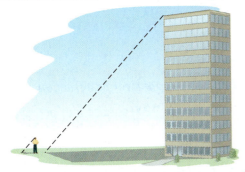

60. During the first 680 miles of a trip, Cindy and Melinda used 26 gallons of gas. They need to travel 200 more miles. Assume that the car will have the same rate of gas consumption.
(a) How many more gallons of gas will they need?
(b) If gas costs $1.15 per gallon, what will fuel cost them for the last 200 miles?

▲ **61.** A film negative is 3.5 centimeters wide and 2.5 centimeters tall. If you want to make a color print that is 8 centimeters wide, how tall will the print be?

62. The dosage of a certain medication is 3 grams for every 50 pounds of body weight. If a person weighs 125 pounds, how many grams of this medication should she take?

63. Maria is making a brick walkway for the front of her new house. She needs 33 bricks to make a walkway section that is 2 feet long. How many bricks will she need to buy to make a walkway section that is 11 feet long?

64. In a survey of a town prior to the presidential election, 8 out of every 13 people surveyed said they would vote on election day. If the town has 15,600 registered voters, how many people will probably vote on election day in the town?

▲ **65.** When Carlos was painting his apartment he found that he used 3 gallons of paint to cover 500 square feet of wall space. He is planning to paint his sister's apartment. She said there are 1400 square feet of wall space that need to be painted. How many gallons of paint will Carlos need?

66. Melinda recently purchased some shares of stock in a new high-tech company. The next week, she found that the company declared a stock split of five shares for every three owned. If Melinda purchased 51 shares of stock before the split, how many shares did she have after the split?

▲ **67.** A scale model of a new church sanctuary has a length of 14 centimeters. When the church is built, the actual length will be 145 feet. In the scale model, the width measures 11 centimeters. What will be the actual width of the church sanctuary?

68. Jeff checked the time on his new watch. In 40 days his watch gained 3 minutes. How much time will the watch gain in a year? (Assume it is not a leap year.)

69. Jean, the top soccer player for the Springfield Comets, scored a total of 68 goals during the season. During the season the team played 32 games, but Jean played in only 27 of them due to a leg injury. The league has been expanded and next season the team will play 34 games. If Jean scores goals at the same rate and is able to play in every game, how many goals might she be expected to score? Round your answer to the nearest whole number.

70. Hank found out there were 345 calories in the 10-ounce chocolate milkshake that he purchased yesterday. Today he decided to order the 16-ounce milkshake. How many calories would you expect to be in the 16-ounce milkshake?

1. _____

2. _____

3. _____

4. _____

5. _____

6. _____

7. _____

8. _____

9. _____

10. _____

11. _____

12. _____

13. _____

14. _____

15. _____

16. _____

17. _____

18. _____

19. _____

20. _____

Write as a ratio in simplest form.

1. $18:52$

2. 70 to 185

Write as a rate in simplest form. Express your answer as a fraction.

3. 784 miles per 24 gallons

4. 2100 square feet per 45 pounds

Write as a unit rate. Round to the nearest hundredth when necessary.

5. 19 tons in five days

6. $57.96 for seven hours

7. 5400 feet per 22 telephone poles

8. $9373 for 110 shares of stock

Write as a proportion.

9. 17 is to 29 as 51 is to 87

10. $3\frac{1}{2}$ is to 5 as 18 is to $25\frac{5}{7}$

11. 490 miles is to 21 gallons as 280 miles is to 12 gallons

12. 5 tablespoons of flour is to 18 people as 15 tablespoons of flour is to 54 people

Determine whether each equation is a proportion.

13. $\dfrac{50}{24} = \dfrac{34}{16}$

14. $\dfrac{18.4}{20} = \dfrac{46}{50}$

15. $\dfrac{32 \text{ smokers}}{46 \text{ nonsmokers}} = \dfrac{160 \text{ smokers}}{230 \text{ nonsmokers}}$

16. $\dfrac{\$0.74}{16 \text{ ounces}} = \dfrac{\$1.84}{40 \text{ ounces}}$

Solve. Round to the nearest tenth when necessary.

17. $\dfrac{n}{20} = \dfrac{4}{5}$

18. $\dfrac{9}{2} = \dfrac{63}{n}$

19. $\dfrac{2\frac{2}{3}}{8} = \dfrac{6\frac{1}{3}}{n}$

20. $\dfrac{4.2}{11} = \dfrac{n}{77}$

21. $\dfrac{45 \text{ women}}{15 \text{ men}} = \dfrac{n \text{ women}}{40 \text{ men}}$

22. $\dfrac{3.5 \text{ ounces}}{4.2 \text{ grams}} = \dfrac{7 \text{ ounces}}{n \text{ grams}}$

23. $\dfrac{n \text{ inches of snow}}{14 \text{ inches of rain}} = \dfrac{12 \text{ inches of snow}}{1.4 \text{ inches of rain}}$

24. $\dfrac{28 \text{ pounds of bananas}}{\$n} = \dfrac{3 \text{ pounds of bananas}}{\$0.55}$

Solve using a proportion. Round your answer to the nearest hundredth when necessary.

25. Bob's recipe for pancakes calls for three eggs and will serve 11 people. If he wants to feed 22 people, how many eggs will he need?

26. A steel cable 42 feet long weighs 170 pounds. How much will 20 feet of this cable weigh?

27. If 9 inches on a map represents 57 miles, what distance does 3 inches represent?

▲ **28.** John and Nancy found it would cost $240 per year to fertilize their front lawn of 4000 square feet. How much would it cost to fertilize 6000 square feet?

29. If Jenny's car uses 1.5 quarts of oil every 3000 miles, how many quarts will it use in 8000 miles?

30. Stephen traveled 570 kilometers in 9 hours. At this rate, how far could he go in 11 hours?

31. Peter found out that his lawn mower uses 0.6 gallon of gas every 2 hours. He figured that mowing the lawn during the summer, he ran the lawn mower for 46 hours. How many gallons of gas did he use running the lawn mower during the summer?

32. On the tri-city softball league, Wilma got seven hits in 34 times at bat last week. During the entire playing season, she was at bat 155 times. If she gets hits at the same rate all season as during last week's game, how many hits would she have for the entire season? Round to the nearest whole number.

21. _____

22. _____

23. _____

24. _____

25. _____

26. _____

27. _____

28. _____

29. _____

30. _____

31. _____

32. _____

Cumulative Test for Chapters 1–4

Approximately one-half of this test is based on Chapter 4 material. The remainder is based on material covered in Chapters 1–3.

1. Write in words. 26,597,089

2. Divide. $23\overline{)1564}$

3. Combine. $\dfrac{1}{4} + \dfrac{1}{8} \times \dfrac{3}{4}$

4. Subtract. $2\dfrac{1}{5} - 1\dfrac{3}{7}$

5. Multiply. $4\dfrac{1}{2} \times 3\dfrac{1}{4}$

6. Subtract.
$$\begin{array}{r} 12.1 \\ -\ \ 3.8416 \end{array}$$

7. Multiply.
$$\begin{array}{r} 0.8163 \\ \times\ \ \ 0.22 \end{array}$$

8. Write as a unit rate in simplest form. $1.68 for 12 bananas

9. Write as a unit rate in simplest form. 12 yen for 3 pesos

Determine whether each equation is a proportion.

10. $\dfrac{12}{17} = \dfrac{30}{42.5}$

11. $\dfrac{4\frac{1}{3}}{13} = \dfrac{2\frac{2}{3}}{8}$

Solve. Round to the nearest tenth when necessary.

12. $\dfrac{9}{2.1} = \dfrac{n}{0.7}$

13. $\dfrac{50}{20} = \dfrac{5}{n}$

14. $\dfrac{n}{56} = \dfrac{16}{7}$

15. $\dfrac{7}{n} = \dfrac{28}{36}$

16. $\dfrac{n}{11} = \dfrac{5}{16}$

17. $\dfrac{3\frac{1}{3}}{7} = \dfrac{10}{n}$

Solve. Round to the nearest hundredth when necessary.

18. Two cities that are located 300 miles apart appear 4 inches apart on a map. If two other cities are 625 miles apart, how far apart will they appear on the same map?

19. Last week Bob earned $84. He had $7 withheld for federal income tax. He earned $9000 last year. Assuming the same ratio, how much did he have withheld for federal income tax last year?

20. Emily Robinson's lasagna recipe feeds 14 people and calls for 3.5 pounds of sausage. If she wants to feed 20 people, how much sausage does she need?

21. Loring Kerr in Nova Scotia produces his own maple syrup. He has found that 39 gallons of maple sap produce 2 gallons of maple syrup. How much sap is needed to produce 11 gallons of syrup?

1. _____

2. _____

3. _____

4. _____

5. _____

6. _____

7. _____

8. _____

9. _____

10. _____

11. _____

12. _____

13. _____

14. _____

15. _____

16. _____

17. _____

18. _____

19. _____

20. _____

21. _____

Chapter 5

Percent

Experts in the area of nutrition carefully measure the contents of the foods we eat to determine how many calories and grams of fat each item contains. A summary of these nutrition facts is contained on the label that appears on most foods sold in the United States. However, sometimes we need information that is not provided on the label. Could you use your knowledge of percents to find this kind of information? Turn to the Putting Your Skills to Work problems in page 354 to find out.

Pretest Chapter 5

1. _____

2. _____

3. _____

4. _____

5. _____

6. _____

7. _____

8. _____

9. _____

10. _____

11. _____

12. _____

13. _____

14. _____

15. _____

16. _____

17. _____

18. _____

19. _____

20. _____

21. _____

22. _____

23. _____

24. _____

If you are familiar with the topics in this chapter, take this test now. Check your answers with those in the back of the book. If an answer is wrong or you can't answer a question, study the appropriate section of the chapter.

If you are not familiar with the topics in this chapter, don't take this test now. Instead, study the examples, work the practice problems, and then take the test.

This test will help you identify which concepts you have mastered and which you need to study further.

Section 5.1

Write as a percent.

1. 0.17

2. 0.387

3. 1.34

4. 8.94

5. 0.006

6. 0.004

7. $\dfrac{17}{100}$

8. $\dfrac{27}{100}$

9. $\dfrac{13.4}{100}$

10. $\dfrac{19.8}{100}$

11. $\dfrac{6\frac{1}{2}}{100}$

12. $\dfrac{1\frac{3}{8}}{100}$

Section 5.2

Change to a percent. Round to the nearest hundredth of a percent when necessary.

13. $\dfrac{8}{10}$

14. $\dfrac{1}{40}$

15. $\dfrac{52}{20}$

16. $\dfrac{17}{16}$

17. $\dfrac{5}{7}$

18. $\dfrac{2}{7}$

19. $\dfrac{22}{23}$

20. $\dfrac{13}{19}$

21. $4\dfrac{2}{5}$

22. $2\dfrac{3}{4}$

23. $\dfrac{1}{300}$

24. $\dfrac{1}{400}$

Write as a fraction in simplified form.

25. 22%   **26.** 53%   **27.** 150%   **28.** 160%

29. $6\frac{1}{3}\%$ **30.** $4\frac{2}{3}\%$ **31.** $51\frac{1}{4}\%$ **32.** $43\frac{3}{4}\%$

Section 5.3

Solve. Round to the nearest hundredth when necessary.

33. Find 24% of 230.

34. What is 78% of 62?

35. 68 is what percent of 72?

36. What percent of 76 is 34?

37. 8% of what number is 240?

38. 354 is 40% of what number?

Section 5.4

Solve. Round to the nearest hundredth when necessary.

39. The home team won 24 of 38 basketball games. What percent of the basketball games did the home team win?

40. Robert paid sales tax of $0.72 on his dinner. The sales tax rate is 5%. What was the cost of his dinner without the tax?

41. A television set that sold originally for $480 was sold for $336. What was the percent of decrease in the selling price?

Section 5.5

42. A salesperson earns a commission rate of 24%. How much commission would be paid if the salesperson sold $22,500 worth of goods?

43. Find the simple interest on a loan of $1700 that is borrowed at 12% annual interest for a period of two years.

25. _____

26. _____

27. _____

28. _____

29. _____

30. _____

31. _____

32. _____

33. _____

34. _____

35. _____

36. _____

37. _____

38. _____

39. _____

40. _____

41. _____

42. _____

43. _____

5.1 Understanding Percent

Student Learning Objectives

After studying this section, you will be able to:

1. Express a fraction whose denominator is 100 as a percent.

2. Write a percent as a decimal.

3. Write a decimal as a percent.

SSM   PH TUTOR   CD & VIDEO   MATH PRO   WEB
CENTER

1 Expressing a Fraction Whose Denominator Is 100 as a Percent

"My raise came through. I got a 6% increase!"

"The leading economic indicators show inflation rising at a rate of 1.3%."

"Mark McGwire and Babe Ruth each hit quite a few home runs. But I wonder who has the higher percentage of home runs per at-bat?"

We use percents often in our everyday lives. In business, in sports, in shopping, and in many areas of life, percentages play an important role. In this section we introduce the idea of percent, which means "*per centum*" or "per hundred." We then show how to use percentages.

In previous chapters, when we described parts of a whole, we used fractions or decimals. Using a percent is another way to describe a part of a whole. Percents can be described as ratios whose denominators are 100. The word **percent** means per 100. This sketch has 100 rectangles.

Of 100 rectangles, 11 are shaded. We can say that 11 percent of the whole is shaded. We use the symbol % for percent. It means "parts per 100." When we write 11 percent as 11%, we understand that it means 11 parts per one hundred, or, as a fraction, $\frac{11}{100}$.

EXAMPLE 1   Recently 100 college students were surveyed about their intentions for voting in the next presidential election. 39 students intended to vote for the Republican candidate, 28 students intended to vote for the Democratic candidate, and 22 students were undecided about which candidate to vote for. The remaining 11 students admitted that they were not planning to vote.

(a) What percent of the students intended to vote for the Democratic candidate?

(b) What percent of the students intended to vote for the Republican candidate?

(c) What percent of the students were undecided as to which candidate they would vote for?

(d) What percent of the students were not planning to vote?

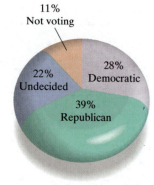

(a) $\frac{28}{100} = 28\%$ **(b)** $\frac{39}{100} = 39\%$ **(c)** $\frac{22}{100} = 22\%$ **(d)** $\frac{11}{100} = 11\%$

Percent notation is often used in circle graphs or pie charts.

Practice Problem 1   Write as a percent.

(a) 51 out of 100 students in the class were women.

(b) 68 out of 100 cars in the parking lot have front-wheel drive.

(c) 7 out of 100 students in the dorm quit smoking.

(d) 26 out of 100 students did not vote in class elections.

Some percents are larger than 100. Consider the following situations.

EXAMPLE 2

(a) Write $\dfrac{386}{100}$ as a percent.

(b) Twenty years ago, four car tires for a full-sized car cost $100. Now the average price for four car tires for a full-sized car is $270. Write the present cost as a percent of the cost 20 years ago.

(a) $\dfrac{386}{100} = 386\%$

(b) The ratio is $\dfrac{\$270 \text{ for four tires now}}{\$100 \text{ for four tires then}}$. $\dfrac{270}{100} = 270\%$

The present cost of four car tires for a full-sized car is 270% of the cost 20 years ago.

Practice Problem 2

(a) Write $\dfrac{238}{100}$ as a percent.

(b) Last year 100 students tried out for varsity baseball. This year 121 students tried out. Write this year's number as a percent of last year's number. .

Some percents are smaller than 1%.

$\dfrac{0.7}{100}$ can be written as 0.7%. $\dfrac{0.3}{100}$ can be written as 0.3%.

$\dfrac{0.04}{100}$ can be written as 0.04%.

Decimal to Percent

EXAMPLE 3   Write as a percent.

(a) $\dfrac{0.9}{100}$ **(b)** $\dfrac{0.002}{100}$ **(c)** $\dfrac{0.07}{100}$

(a) $\dfrac{0.9}{100} = 0.9\%$ **(b)** $\dfrac{0.002}{100} = 0.002\%$ **(c)** $\dfrac{0.07}{100} = 0.07\%$

Practice Problem 3   Write as a percent.

(a) $\dfrac{0.5}{100}$ **(b)** $\dfrac{0.06}{100}$ **(c)** $\dfrac{0.003}{100}$

2 *Writing a Percent as a Decimal*

Suppose we have a percent such as 59%. What would be the equivalent in decimal form? Using our definition of percent, $59\% = \frac{59}{100}$. This fraction could be written in decimal form as 0.59. In a similar way, we could write 21% as $\frac{21}{100} = 0.21$. This pattern allows us to quickly change the form of a number from a percent to a fraction whose denominator is 100 to a decimal.

EXAMPLE 4 Write as a decimal.

(a) 38% **(b)** 6%

(a) $38\% = \dfrac{38}{100} = 0.38$ **(b)** $6\% = \dfrac{6}{100} = 0.06$

Practice Problem 4 Write as a decimal.

(a) 47% **(b)** 2%

The results of Example 4 suggest that when you remove a percent sign (%) you are dividing by 100. When we divide by 100 this moves the decimal point of a number two places to the left. Now that you understand this process we can abbreviate it with the following rule.

> **Changing a Percent to a Decimal**
>
> **1.** Drop the % symbol.
> **2.** Move the decimal point two places to the left.

EXAMPLE 5 Write as a decimal.

(a) 26.9% **(b)** 7.2% **(c)** 0.13% **(d)** 158%

In each case, we drop the percent symbol and move the decimal point two places to the left.

(a) $26.9\% = 0.269 = 0.269$

(b) $7.2\% = 0.072 = 0.072$ Note that we need to add an extra zero to the left of the seven.

(c) $0.13\% = 0.0013 = 0.0013$ Here we added zeros to the left of the 1.

(d) $158\% = 1.58 = 1.58$

Practice Problem 5 Write as a decimal.

(a) 80.6% **(b)** 2.5% **(c)** 0.29% **(d)** 231%

3 *Writing a Decimal as a Percent*

In Example 4(a) we changed 38% to $\frac{38}{100}$ to 0.38. We can start with 0.38 and reverse the process. We obtain $0.38 = \frac{38}{100} = 38\%$. Study all the parts of Examples 4 and 5. You will see that the steps are reversible. Thus $0.38 = 38\%$, $0.06 = 6\%$, $0.70 = 70\%$, $0.269 = 26.9\%$, $0.072 = 7.2\%$, $0.0013 = 0.13\%$, and $1.58 = 158\%$.

In each part we are multiplying by 100. So to change a decimal number to a percent we are multiplying the number by 100. In each part the decimal point is moved two places to the right. Then the percent symbol is written after the number.

> **Changing a Decimal to a Percent**
>
> **1.** Move the decimal point two places to the right.
> **2.** Then write the % symbol at the end of the number.

EXAMPLE 6   Write as a percent.

(a) 0.47   **(b)** 0.08   **(c)** 6.31   **(d)** 0.055   **(e)** 0.001

In each part we move the decimal point two places to the right and write the percent symbol at the end of the number.

(a) $0.47 = 47\%$ **(b)** $0.08 = 8\%$ **(c)** $6.31 = 631\%$

(d) $0.055 = 5.5\%$ **(e)** $0.001 = 0.1\%$

Practice Problem 6   Write as a percent.

(a) 0.78   **(b)** 0.02   **(c)** 5.07   **(d)** 0.029   **(e)** 0.006

To Think About   What is really happening when we change a decimal to a percent? Suppose that we wanted to change 0.59 to a percent.

$$0.59 = \frac{59}{100} \qquad \text{\textcolor{red}{Definition of a decimal.}}$$

$$= 59 \times \frac{1}{100} \qquad \text{\textcolor{red}{Definition of multiplying fractions.}}$$

$$= 59 \text{ percent} \qquad \text{\textcolor{red}{Because "per 100" means percent.}}$$

$$= 59\% \qquad \text{\textcolor{red}{Writing the symbol for percent.}}$$

Can you see why each step is valid? Since we know the reason behind each step, we know we can always move the decimal point two places to the right and write the percent symbol. See Exercises 5.1, exercises 75 and 76.

Calculator

Percent to Decimal
You can use a calculator to change 52% to a decimal.
Enter

52 $\boxed{\%}$

The display should read

$\boxed{0.52}$

Try the following.

(a) 46%   **(b)** 137%
(c) 9.3%   **(d)** 6%

Note: The calculator divides by 100 when the percent key is pressed. If you do not have a $\boxed{\%}$ key then you can use the key-strokes $\boxed{\div}$ 100 $\boxed{=}$.

Verbal and Writing Skills

1. In this section we introduced percent, which means "per centum" or "per _____."

2. The number 1 written as a percent is _____.

3. To change a percent to a decimal, move the decimal point _____ places to the _____. _____ the % symbol.

4. To change a decimal to a percent, move the decimal point _____ places to the _____. _____ the % symbol at the end of the number.

Write as a percent.

5. $\dfrac{48}{100}$

6. $\dfrac{45}{100}$

7. $\dfrac{9}{100}$

8. $\dfrac{7}{100}$

9. $\dfrac{80}{100}$

10. $\dfrac{60}{100}$

11. $\dfrac{245}{100}$

12. $\dfrac{110}{100}$

13. $\dfrac{5.3}{100}$

14. $\dfrac{8.3}{100}$

15. $\dfrac{0.07}{100}$

16. $\dfrac{0.019}{100}$

Applications

Write a percent to express each of the following.

17. 13 out of 100 loaves of bread had gone stale.

18. 54 out of 100 dog owners have attended an obedience class.

19. 28 out of 100 cars sold were Toyota Camrys.

20. 64 out of 100 college sophomores are confident of their selection of a major.

Write as a decimal.

21. 51%   22. 42%   23. 7%   24. 6%   25. 20%   26. 40%

27. 43.6%   **28.** 81.5%   **29.** 0.03%   **30.** 0.09%   **31.** 0.72%   **32.** 0.61%

33. 126%   **34.** 175%   **35.** 366%   **36.** 398%

Write as a percent.

37. 0.74   **38.** 0.66   **39.** 0.50   **40.** 0.40   **41.** 0.08   **42.** 0.03

43. 0.563   **44.** 0.408   **45.** 0.002   **46.** 0.009   **47.** 0.0057   **48.** 0.0026

49. 1.35   **50.** 1.86   **51.** 2.72   **52.** 3.04

53. Bob paid 0.27 of his income for federal income tax.

54. Sally spends 0.31 of her income for housing.

55. Lou donates 0.09 of his income to charity.

56. Sally invests 0.05 of her income in the stock market.

Mixed Practice

Write as a percent.

57. 0.94   **58.** 0.25   **59.** 2.31   **60.** 1.48

61. $\dfrac{10}{100}$ **62.** $\dfrac{40}{100}$ **63.** 0.009   **64.** 0.005

Write as a decimal.

65. 62%   **66.** 49%   **67.** $\dfrac{138}{100}$ **68.** $\dfrac{210}{100}$

69. $\dfrac{0.3}{100}$ **70.** $\dfrac{0.8}{100}$ **71.** $\dfrac{80}{100}$ **72.** $\dfrac{60}{100}$

Applications

Write as a percent.

73. For every 100 people in Texas who voted in the 2000 presidential election, 59 voted for George W. Bush.

74. For every 100 people in Tennessee who voted in the 2000 presidential election, 48 voted for Al Gore.

To Think About

75. Suppose that we want to change 36% to 0.36 by moving the decimal point two places to the left and dropping the % symbol. Explain the steps to show what is really involved in changing 36% to 0.36. Why does the rule work?

76. Suppose that we want to change 10.65 to 1065%. Give a complete explanation of the steps.

Write the given value (a) as a decimal, (b) as a fraction with a denominator of 100, and (c) as a reduced fraction.

77. 55,562%

78. 60,724%

Cumulative Review Problems

Write as a fraction in simplest form.

79. 0.56

80. 0.78

Write as a decimal.

81. $\dfrac{11}{16}$

82. $\dfrac{7}{8}$

83. A very successful commercial ceramics studio makes beautiful vases for gift stores. In one corner of the warehouse, the storage area has 24 shelves. Three shelves have 246 vases each, seven shelves have 380 vases each, five shelves have 168 vases each, and nine shelves have 122 vases each. How many vases are there in this corner of the studio?

5.2 Changing Between Percents, Decimals, and Fractions

1 *Changing a Percent to a Fraction*

By using the definition of percent, we can write any percent as a fraction whose denominator is 100. Thus when we change a percent to a fraction, we remove the percent symbol and write the number over 100. To write a number over 100 means that we are dividing by 100. If possible, we then simplify the fraction.

EXAMPLE 1 Write as a fraction in simplest form.

(a) 37% **(b)** 75% **(c)** 2%

(a) $37\% = \dfrac{37}{100}$ **(b)** $75\% = \dfrac{75}{100} = \dfrac{3}{4}$ **(c)** $2\% = \dfrac{2}{100} = \dfrac{1}{50}$

Practice Problem 1 Write as a fraction in simplest form.

(a) 71% **(b)** 25% **(c)** 8%

In some cases, it may be helpful to write the percent as a decimal before you write it as a fraction in simplest form.

EXAMPLE 2 Write as a fraction in simplest form.

(a) 43.5% **(b)** 36.75%

(a) $43.5\% = 0.435$ Change the percent to a decimal.

$= \dfrac{435}{1000}$ Change the decimal to a fraction.

$= \dfrac{87}{200}$ Reduce the fraction.

(b) $36.75\% = 0.3675 = \dfrac{3675}{10,000} = \dfrac{147}{400}$

Practice Problem 2 Write as a fraction in simplest form.

(a) 8.4% **(b)** 28.5%

If the percent is greater than 100%, the simplified fraction is usually changed to a mixed number.

EXAMPLE 3 Write as a mixed number.

(a) 225% **(b)** 138%

(a) $225\% = 2.25 = 2\dfrac{25}{100} = 2\dfrac{1}{4}$ **(b)** $138\% = 1.38 = 1\dfrac{38}{100} = 1\dfrac{19}{50}$

Practice Problem 3 Write as a mixed number.

(a) 170% **(b)** 288%

Sometimes a percent is not a whole number, such as 9% or 10%. Instead, it contains a fraction, such as $9\frac{1}{12}\%$ or $9\frac{3}{8}\%$. Extra steps will be needed to write such a percent as a simplified fraction.

48-month certificates
of deposit now earn
9 ³/₈% interest
Buy one now!
Invest for the future.

🔴 **EXAMPLE 4**   Convert $9\frac{3}{8}\%$ to a fraction in simplest form.

$$9\frac{3}{8}\% = \frac{9\frac{3}{8}}{100} \qquad \text{Change the percent to a fraction.}$$

$$= 9\frac{3}{8} \div \frac{100}{1} \qquad \text{Write the division horizontally. } \frac{9\frac{3}{8}}{100} \text{ means } 9\frac{3}{8} \text{ divided by 100.}$$

$$= \frac{75}{8} \div \frac{100}{1} \qquad \text{Write } 9\frac{3}{8} \text{ as an improper fraction.}$$

$$= \frac{75}{8} \times \frac{1}{100} \qquad \text{Use the definition of division of fractions.}$$

$$= \frac{\overset{3}{\cancel{75}}}{8} \times \frac{1}{\underset{4}{\cancel{100}}} \qquad \text{Simplify.}$$

$$= \frac{3}{32} \qquad \text{Reduce the fraction.}$$

Practice Problem 4   Convert $7\frac{5}{8}\%$ to a fraction in simplest form.   🔴

🔴 **EXAMPLE 5**   In the fiscal 2000 budget of the United States, approximately $15\frac{3}{8}\%$ of the budget was designated for defense. (*Source*: U.S. Office of Management and Budget). Write this percent as a fraction.

$$15\frac{3}{8}\% = \frac{15\frac{3}{8}}{100} = 15\frac{3}{8} \div 100 = \frac{123}{8} \times \frac{1}{100} = \frac{123}{800}$$

Thus we could say $\frac{123}{800}$ of the fiscal 2000 budget was designated for defense. That is, for every \$800 in the budget, \$123 was spent for defense.

Practice Problem 5   In the fiscal 2000 budget of the United States, approximately $23\frac{1}{6}\%$ was designated for social security. (*Source:* U.S. Office of Management and Budget). Write this percent as a fraction.   🔴

Certain percents occur very often, especially in money matters. Here are some common equivalents that you may already know. If not, be sure to memorize them.

$$25\% = \frac{1}{4} \qquad 33\frac{1}{3}\% = \frac{1}{3} \qquad 10\% = \frac{1}{10}$$

$$50\% = \frac{1}{2} \qquad 66\frac{2}{3}\% = \frac{2}{3}$$

$$75\% = \frac{3}{4}$$

2   *Changing a Fraction to a Percent*

A convenient way to change a fraction to a percent is to write the fraction in decimal form first and then convert the decimal to a percent.

● **EXAMPLE 6**   Write $\dfrac{3}{8}$ as a percent.

We see that $\dfrac{3}{8} = 0.375$ by calculating $3 \div 8$.

$$
\begin{array}{r}
0.375 \\
8\overline{)3.000} \\
\underline{24} \\
60 \\
\underline{56} \\
40 \\
\underline{40} \\
0
\end{array}
$$

Thus $\dfrac{3}{8} = 0.375 = 37.5\%$.

Practice Problem 6   Write $\dfrac{5}{8}$ as a percent.

●

● **EXAMPLE 7**   Write as a percent.   **(a)** $\dfrac{7}{40}$ **(b)** $\dfrac{39}{50}$

(a) $\dfrac{7}{40} = 0.175 = 17.5\%$ **(b)** $\dfrac{39}{50} = 0.78 = 78\%$

Practice Problem 7   Write as a percent.

(a) $\dfrac{21}{25}$ **(b)** $\dfrac{7}{16}$

●

 Changing some fractions to decimal form results in infinitely repeating decimals. In such cases, we usually round to the nearest hundredth of a percent.

● **EXAMPLE 8**   Write as a percent. Round to the nearest hundredth of a percent.

(a) $\dfrac{1}{6}$ **(b)** $\dfrac{15}{33}$

(a) We find that $\dfrac{1}{6} = 0.16666\ldots$ by calculating $1 \div 6$.

$$
\begin{array}{r}
0.166 \\
6\overline{)1.000} \\
\underline{6} \\
40 \\
\underline{36} \\
40
\end{array}
$$

The repeating quotient can be written as $0.1\overline{6}$. If we round the decimal to the nearest ten-thousandth, we have $\frac{1}{6} \approx 0.1667$. If we change this to a percent, we have

$$
\frac{1}{6} \approx 16.67\%.
$$

This is correct to the nearest hundredth of a percent.

(b) By calculating $15 \div 33$, we see that $\dfrac{15}{33} = 0.45454545\ldots$. This can be written as $0.\overline{45}$. If we round to the nearest ten-thousandth, we have

$$\frac{15}{33} \approx 0.4545 = 45.45\%.$$

This rounded value is correct to the nearest hundredth of a percent.

Practice Problem 8   Write as a percent. Round to the nearest hundredth of a percent.

(a) $\dfrac{7}{9}$ **(b)** $\dfrac{19}{30}$

Recall that sometimes percents are written with fractions.

EXAMPLE 9   Express $\dfrac{11}{12}$ as a percent containing a fraction.

We will stop the division after two steps and write the remainder in fraction form.

$$
\begin{array}{r}
0.91 \\
12\overline{)11.00} \\
\underline{108} \\
20 \\
\underline{12} \\
8
\end{array}
$$

This division tells us that we can write

$$\frac{11}{12} \quad \text{as} \quad 0.91\frac{8}{12} \quad \text{or} \quad 0.91\frac{2}{3}$$

We now have a decimal with a fraction. When we express this decimal as a percent, we move the decimal point two places to the right. We do not write the decimal point in front of the fraction.

$$0.91\frac{2}{3} = 91\frac{2}{3}\%$$

Note that our answer in Example 9 is an *exact answer*. We have not rounded off or approximated in any way.

Practice Problem 9   Express $\dfrac{7}{12}$ as a percent containing a fraction.

3 Changing a Percent, a Decimal, or a Fraction to Equivalent Forms

We have seen so far that a fraction, a decimal, and a percent are three different forms (notations) for the same number. We can illustrate this in a chart.

EXAMPLE 10   Complete the following table of equivalent notations. Round decimals to the nearest ten-thousandth. Round percents to the nearest hundredth of a percent.

Fraction	Decimal	Percent
$\frac{11}{16}$		
	0.265	
		$17\frac{1}{5}\%$

Begin with the first row. The number is written as a fraction. We will change the fraction to a decimal and then to a percent.

The <mark>fraction</mark> is changed to a <mark>decimal</mark> is changed to a <mark>percent.</mark>

$$\frac{11}{16} \longrightarrow 16\overline{)11.0000}^{\,0.6875} \longrightarrow 68.75\%$$

In the second row the number is written as a decimal. This can easily be written as a percent.

$$0.265 \longrightarrow 26.5\%$$

Now write 0.265 as a fraction and simplify.

$$0.265 \\ \downarrow \\ \frac{53}{200} \longleftarrow \frac{265}{1000}$$

In the third row the number is written as a percent. Proceed from right to left—that is, write the number as a decimal and then as a fraction.

$$\frac{17\frac{1}{5}}{100} \longleftarrow 17\frac{1}{5}\% \\ \downarrow \\ \boxed{\frac{86}{5} \times \frac{1}{100}} \\ \downarrow \\ 0.172 \longleftarrow \frac{86}{500} \quad \text{Divide.} \quad 500\overline{)86.000}^{\,0.172}$$

and

$$\frac{43}{250} \longleftarrow \frac{86}{500}$$

Thus the completed table is as follows.

Fraction	Decimal	Percent
$\frac{11}{16}$	0.6875	68.75%
$\frac{53}{200}$	0.265	26.5%
$\frac{43}{250}$	0.172	$17\frac{1}{5}\%$

Calculator

Fraction to Decimal

You can use a calculator to change $\frac{3}{5}$ to a decimal. Enter

$$3 \;\boxed{\div}\; 5 \;\boxed{=}$$

The display should read

$$\boxed{\qquad 0.6}$$

Try the following.

(a) $\dfrac{17}{25}$ **(b)** $\dfrac{2}{9}$

(c) $\dfrac{13}{10}$ **(d)** $\dfrac{15}{19}$

Note: 0.78947368 is an approximation for $\frac{15}{19}$. Some calculators round to only eight places.

Practice Problem 10 Complete the following table of equivalent notations. Round decimals to the nearest ten-thousandth. Round percents to the nearest hundredth of a percent.

Fraction	Decimal	Percent
$\dfrac{23}{99}$		
	0.516	
		$38\dfrac{4}{5}\%$

Alternative Method

Another way to convert a fraction to a percent is to use a proportion. To change $\frac{7}{8}$ to a percent, write the proportion

$$\frac{7}{8} = \frac{n}{100}$$

$$7 \times 100 = 8 \times n \qquad \text{Cross-multiply.}$$

$$700 = 8 \times n \qquad \text{Simplify.}$$

$$\frac{700}{8} = \frac{8 \times n}{8} \qquad \text{Divide each side by 8.}$$

$$87.5 = n \qquad \text{Simplify.}$$

Thus $\frac{7}{8} = 87.5\%$. You will use this approach in Exercises 5.2, exercises 85 and 86.

Developing Your Study Skills

Reading the Textbook

Homework time each day should begin with the careful reading of the section(s) assigned in your textbook. Much time and effort have gone into the selection of a particular text, and your instructor has chosen a book that will help you become successful in this mathematics class. Expensive textbooks can be a wise investment if you take advantage of them by reading them.

Reading a mathematics textbook is unlike reading many other types of books that you may use in your literature, history, psychology, or sociology courses. Mathematics texts are technical books that provide you with exercises to practice on. Reading a mathematics text requires slow and careful reading of each word, which takes time and effort.

Begin reading your textbook with a paper and pencil in hand. As you come across a new definition, or concept, underline it in the text and/or write it down in your notebook. Whenever you encounter an unfamiliar term, look it up and make a note of it. When you come to an example, work through it step by step. Be sure to read each word and to follow directions carefully.

Notice the helpful hints the author provides to guide you to correct solutions and prevent you from making errors. Take advantage of these pieces of expert advice.

Be sure that you understand what you are reading. Make a note of any of those things that you do not understand and ask your instructor about them. Do not hurry through the material. Learning mathematics takes time.

Verbal and Writing Skills

1. Explain in your own words how to change a percent to a fraction.

2. Explain in your own words how to change a fraction to a percent.

Write as a fraction or as a mixed number.

3. 10% **4.** 8% **5.** 33% **6.** 47% **7.** 55% **8.** 35%

9. 75% **10.** 25% **11.** 20% **12.** 40% **13.** 9.5% **14.** 6.5%

15. 17.5% **16.** 77.5% **17.** 64.8% **18.** 12.2% **19.** 71.25% **20.** 38.75%

21. 176% **22.** 228% **23.** 340% **24.** 420% **25.** 1200% **26.** 3600%

27. $2\frac{1}{6}\%$ **28.** $7\frac{5}{6}\%$ **29.** $12\frac{1}{2}\%$ **30.** $37\frac{1}{2}\%$ **31.** $8\frac{4}{5}\%$ **32.** $9\frac{3}{5}\%$

Applications

33. During the 1998–1999 television season, the number-three-ranked regularly scheduled network program was *Frasier*, with $15\frac{5}{9}\%$ of all households watching the show. Write this percent as a fraction.

34. During an inspection, $17\frac{3}{4}\%$ of the labels printed had the wrong colors. What *fraction* of the labels was in the wrong color?

35. In the 2000 budget of the United States, approximately $2\frac{4}{11}\%$ was designated for veterans' benefits and Services. (*Source*: U.S. Office of Management and Budget). Write this percent as a fraction.

36. In the 2000 budget of the United States, approximately $12\frac{3}{11}\%$ was designated for Medicare. (*Source*: U.S. Office of Management and Budget). Write this percent as a fraction.

Write as a percent. Round to the nearest hundredth of a percent when necessary.

37. $\dfrac{3}{4}$ **38.** $\dfrac{1}{4}$ **39.** $\dfrac{1}{3}$ **40.** $\dfrac{2}{3}$ **41.** $\dfrac{7}{20}$ **42.** $\dfrac{11}{20}$ **43.** $\dfrac{7}{25}$ **44.** $\dfrac{9}{25}$

45. $\dfrac{11}{40}$ **46.** $\dfrac{13}{40}$ **47.** $\dfrac{5}{12}$ **48.** $\dfrac{8}{12}$ **49.** $\dfrac{18}{5}$ **50.** $\dfrac{7}{4}$ **51.** $2\dfrac{5}{6}$ **52.** $3\dfrac{1}{6}$

53. $4\dfrac{1}{8}$ **54.** $2\dfrac{5}{8}$ **55.** $\dfrac{3}{7}$ **56.** $\dfrac{2}{7}$ **57.** $\dfrac{31}{16}$ **58.** $\dfrac{43}{16}$ **59.** $\dfrac{26}{50}$ **60.** $\dfrac{43}{50}$

Applications

Round to the nearest hundredth of a percent.

61. The brain represents approximately $\frac{1}{40}$ of an average person's weight. Express this fraction as a percent.

62. To calculate your maximum monthly house payment, a real estate agent multiplies your monthly income by $\frac{7}{25}$. Express this fraction as a percent.

63. The U.S. mail service estimates that people do not send as many personal letters through the mail as they did 20 years ago. In fact, in 2000 only $\frac{35}{3000}$ of all mail delivered consisted of personal letters. Express this fraction as a percent.

64. During waking hours, a person blinks $\frac{9}{2000}$ of the time. Express this fraction as a percent.

Express as a percent containing a fraction. (See Example 9.)

65. $\dfrac{5}{6}$ **66.** $\dfrac{1}{6}$ **67.** $\dfrac{5}{14}$ **68.** $\dfrac{9}{14}$ **69.** $\dfrac{3}{8}$ **70.** $\dfrac{5}{8}$ **71.** $\dfrac{3}{40}$ **72.** $\dfrac{11}{90}$

Mixed Practice

In exercises 73–82, complete the table of equivalents. Round decimals to the nearest ten-thousandth. Round percents to the nearest hundredth of a percent.

	Fraction	Decimal	Percent
73.	$\dfrac{11}{12}$		
75.		0.56	
77.	$\dfrac{3}{200}$		
79.	$\dfrac{5}{9}$		
81.			$3\frac{1}{8}\%$

	Fraction	Decimal	Percent
74.	$\dfrac{1}{12}$		
76.		0.85	
78.	$\dfrac{9}{200}$		
80.	$\dfrac{7}{9}$		
82.			$2\frac{5}{8}\%$

83. Write $28\frac{15}{16}\%$ as a fraction.

84. Write $18\frac{7}{12}\%$ as a fraction.

Change each fraction to a percent by using a proportion.

85. $\dfrac{123}{800}$

86. $\dfrac{417}{600}$

To Think About

87. What can you say about a decimal number if it is equivalent to a percent that is less than 0.1%?

88. What can you say about a decimal number if it is equivalent to a percent that is less than 0.01%?

Cumulative Review Problems

Solve for n.

89. $\dfrac{15}{n} = \dfrac{8}{3}$

90. $\dfrac{32}{24} = \dfrac{n}{3}$

91. The law firm of Dewey, Cheatham & Howe was required to review 54 years of documents of one of its clients. The first file contained 10,041 documents. The second file contained 986 documents. The third file contained 4,283 documents. The last file contained 533,855 documents. How many total documents were there?

5.3A Solving Percent Problems Using an Equation

1 Translating a Percent Problem into an Equation

In word problems like the ones in this section, we can translate from words to mathematical symbols and back again. After we have the mathematical symbols arranged in an *equation*, we solve the equation. When we find the values that make the equation true, we have also found the answer to our word problem.

To solve a percent problem, we express it as an equation with an unknown quantity. We use the letter n to represent the number we do not know. The following table is helpful when translating from a percent problem to an equation.

Word	Mathematical Symbol
of	Any multiplication symbol: $\times$ or ( ) or $\cdot$
is	$=$
what	Any letter; for example, n
find	$n =$

In Examples 1–5 we show how to translate words into an equation. Please do **not** solve the problem. Translate into an equation only.

EXAMPLE 1 Translate into an equation.

What is 5% of 19.00?

$$n = 5\% \times 19.00$$

Practice Problem 1 Translate into an equation. What is 26% of 35?

EXAMPLE 2 Translate into an equation.

Find 0.6% of 400.

Notice here that the words *what is* are missing. The word *find* is equivalent to *what is.*

Find 0.6% of 400.

$$n = 0.6\% \times 400$$

Practice Problem 2 Translate into an equation. Find 0.08% of 350.

The unknown quantity, n, does not always stand alone in an equation.

EXAMPLE 3 Translate into an equation.

(a) 35% of what is 60?   **(b)** 7.2 is 120% of what?

(a) 35% of what is 60?

$$35\% \times n = 60$$

(b) 7.2 is 120% of what?

$$7.2 = 120\% \times n$$

Practice Problem 3 Translate into an equation.

(a) 58% of what is 400?   **(b)** 9.1 is 135% of what?

EXAMPLE 4 Translate into an equation.

$$\underbrace{\text{What percent}}_{n} \text{ of } \underset{\downarrow}{50} \underset{\downarrow}{\text{ is }} \underset{\downarrow}{10?}$$
$$n \qquad \times\ 50\ =\ 10$$

We see here that the words *what percent* are represented by the letter *n*.

Practice Problem 4 Translate into an equation. What percent of 250 is 36?

EXAMPLE 5 Translate into an equation.

(a) 30 is what percent of 16?   **(b)** What percent of 3000 is 2.6?

(a) 30 is what percent of 16?   **(b)** What percent of 3000 is 2.6?

$$30\ =\quad n \quad\times\ 16 \qquad\qquad n \quad\times\ 3000\ =\ 2.6$$

Practice Problem 5 Translate into an equation.

(a) 50 is what percent of 20?   **(b)** What percent of 2000 is 4.5?

2 Solving a Percent Problem by Solving an Equation

The percent problems we have translated are of three types. Consider the equation $60 = 20\% \times 300$. This problem has the form

amount = percent × base.

Any one of these quantities—amount, percent, or base—may be unknown.

1. When *we do not know the amount*, we have an equation like

$$n = 20\% \times 300$$

2. When *we do not know the base*, we have an equation like

$$60 = 20\% \times n$$

3. When *we do not know the percent*, we have an equation like

$$60 = n \times 300$$

We will study each type separately. It is not necessary to memorize the three types, but it is helpful to look carefully at the examples we give of each. In each example, do the computation in a way that is easiest for you. This may be using a pencil and paper, using a calculator, or, in some cases, doing the problem mentally.

Percent of a Number
You can use a calculator to find 12% of 48.
Enter

12 % × 48 =

The display should read

5.76

If your calculator does not have a percent key, use the keystrokes

0.12 × 48 =

What is 54% of 450?

Solving Percent Problems When We Do Not Know the Amount

EXAMPLE 6 What is 45% of 590?

$$
\begin{aligned}
n &= 45\% \times 590 & &\text{Translate into an equation.}\\
n &= (0.45)(590) & &\text{Change the percent to decimal form.}\\
n &= 265.5 & &\text{Multiply } 0.45 \times 590.
\end{aligned}
$$

Practice Problem 6 What is 82% of 350?

EXAMPLE 7 Find 160% of 500.

Find 160% of 500.

$$
\begin{aligned}
n &= 160\% \times 500 \\
n &= (1.60)(500) & &\text{Change the percent to decimal form.}\\
n &= 800 & &\text{Multiply 1.6 by 500.}
\end{aligned}
$$

When you translate, remember that the word *find* is equivalent to *what is*.

Practice Problem 7 Find 230% of 400.

EXAMPLE 8 When Rick bought a new Dodge Neon, he had to pay a sales tax of 5% on the cost of the car, which was $12,000. What was the sales tax?
This problem is asking

What is 5% of $12,000?

$$
\begin{aligned}
n &= 5\% \times \$12{,}000 \\
n &= 0.05 \times 12{,}000 \\
n &= \$600
\end{aligned}
$$

The sales tax was $600.

Practice Problem 8 When Oprah bought an airplane ticket, she had to pay a tax of 8% on the cost of the ticket, which was $350. What was the tax?

Solving Percent Problems When We Do Not Know the Base

If a number is multiplied by the letter n, this can be indicated by a multiplication sign, parentheses, a dot, or placing the number in front of the letter. Thus $3 \times n = 3(n) = 3 \cdot n = 3n$. In this section we use equations like $3n = 9$ and $0.5n = 20$. To solve these equations we use the procedures developed in Chapter 4. We divide each side by the number multiplied by n.

EXAMPLE 9 12 is 0.6% of what?

$$
\begin{aligned}
12 &= 0.6\% \times n & &\text{Translate into an equation.}\\
12 &= 0.006n & &\text{Change 0.6\% to a decimal.}\\
\frac{12}{0.006} &= \frac{0.006n}{0.006} & &\text{Divide each side of the equation by 0.006.}\\
2000 &= n & &\text{Divide } 12 \div 0.006.
\end{aligned}
$$

Practice Problem 9 32 is 0.4% of what?

 **EXAMPLE 10**   Dave and Elsie went out to dinner. They gave the waiter a tip that was 15% of the total bill. The waiter received $6. What was the total bill (not including the tip)?

This problem is asking

$$
\begin{array}{cccc}
15\% \text{ of} & \text{what} & \text{is} & \$6? \\
\downarrow \quad \downarrow & \downarrow & \downarrow & \downarrow \\
15\% \times & n & = & 6
\end{array}
$$

$$0.15n = 6$$

$$\frac{0.15n}{0.15} = \frac{6}{0.15}$$

$$n = 40$$

The total bill for the meal (not including the tip) was $40.

Practice Problem 10   The coach of the university baseball team said that 30% of the players on his team are left-handed. Six people on the team are left-handed. How many people are on the team?

Solving Percent Problems When We Do Not Know the Percent

In solving these problems, we notice that there is no % symbol in the problem. The percent is what we are trying to find. Therefore, our answer for this type of problem will always have a percent symbol.

 **EXAMPLE 11**
$$
\begin{array}{ccccc}
\underbrace{\text{What percent}} & \text{of} & 5000 & \text{is} & 3.8? \\
\downarrow & \downarrow & \downarrow & \downarrow & \downarrow \\
n & \times & 5000 & = & 3.8
\end{array}
$$
Translate into an equation.

$$5000n = 3.8$$
Multiplication is commutative. $n \times 5000 = 5000 \times n$.

$$\frac{5000n}{5000} = \frac{3.8}{5000}$$
Divide each side by 5000.

$$n = 0.00076$$ Divide 3.8 by 5000.

$$n = 0.076\%$$ Express the decimal as a percent.

Practice Problem 11   What percent of 9000 is 4.5?

 **EXAMPLE 12**
$$
\begin{array}{ccccc}
90 & \text{is} & \underbrace{\text{what percent}} & \text{of} & 20? \\
\downarrow & \downarrow & \downarrow & \downarrow & \downarrow \\
90 & = & n & \times & 20
\end{array}
$$
Translate into an equation.

$$90 = 20n$$
Multiplication is commutative. $n \times 20 = 20 \times n$.

$$\frac{90}{20} = \frac{20n}{20}$$
Divide each side by 20.

$$4.5 = n$$
Divide 90 by 20.

$$450\% = n$$
Express the decimal as a percent.

Practice Problem 12   198 is what percent of 33?

Calculator

Finding the Percent
You can use a calculator to find a missing percent. What percent of 95 is 19?

1. Enter as a fraction. Enter the number after "is," and then the division key. Then enter the number after the word "of."

$$19 \;\boxed{\div}\; 95$$

2. Change to a percent.

$$19 \;\boxed{\div}\; 95 \;\boxed{\times}\; 100 \;\boxed{=}$$

The display should read

$$\boxed{20}$$

This means 20%. What percent of 625 is 250?

EXAMPLE 13   In a recent basketball game for the New York Knicks, Patrick Ewing made 10 of his 24 shots. What percent of his shots did he make? (Round to the nearest tenth of a percent.)

This is equivalent to

10 is what percent of 24?

$$10 = \qquad n \qquad \times 24$$

$$10 = 24n$$

$$\frac{10}{24} = \frac{24n}{24}$$

$$0.41666\ldots = n$$

To the nearest tenth of a percent we have

$$n = 41.7\%$$

Patrick Ewing made 41.7% of his shots in this game.

Practice Problem 13   In a basketball game for the Los Angeles Lakers, A.C. Green made 5 of his 16 shots. What percent of his shots did he make? (Round to the nearest tenth of a percent.)

*Translate into a mathematical equation in exercises 1–6. Use the letter n for the unknown quantity. Do **not** solve, but rather just obtain the equation.*

1. What is 5% of 90?

2. What is 9% of 65?

3. 70% of what is 2?

4. 55% of what is 34?

5. 17 is what percent of 85?

6. 24 is what percent of 144?

Solve.

7. What is 20% of 140?

8. What is 30% of 210?

9. Find 152% of 600.

10. Find 136% of 500.

Applications

11. Clarice rented a new Ford Escort for one week. Her bill before the 6% tax on rental cars was $106.50. What was the amount of her tax?

12. The Center City check-cashing service charges a fee that is 7% of the amount of the check they cash. How much is the fee for cashing a check for $240?

Solve.

13. 25% of what is 14?

14. 40% of what is 30?

15. 52 is 4% of what?

16. 36 is 6% of what?

Applications

17. In Australia, all general sales (except for food) have a hidden tax of 22% built into the final price. Walter is planning to purchase a camera while in Australia. He wants to know the before-tax price that the dealer is charging before he adds on the hidden tax of $33. Can you determine the amount of the before-tax price?

18. Jocelyn is teaching her golden retriever puppy how to fetch. Jack, her puppy, fetched the ball 72% of the time today. Jack fetched the ball 54 times today. How many times did Jocelyn actually throw the ball for Jack?

Solve.

19. What percent of 64 is 32?

20. What percent of 64 is 16?

21. 56 is what percent of 200?

22. 45 is what percent of 300?

Applications

23. The total number of points scored in a basketball game was 120. The winning team scored 78 of those points. What percent of the points were scored by the winning team?

24. Randy's bill for car repairs was $140. Of this amount, $28 was charged for labor and $112 was charged for parts. What percent of the bill was for labor?

Mixed Practice

Solve.

25. 20% of 155 is what?

26. 60% of 215 is what?

27. 170% of what is 144.5?

28. 160% of what is 152?

29. 84 is what percent of 700?

30. 72 is what percent of 900?

31. Find 0.4% of 820.

32. Find 0.3% of 540.

33. What percent of 35 is 22.4?

34. What percent of 45 is 16.2?

35. 89 is 20% of what?

36. 42 is 40% of what?

37. 42 is what percent of 120?

38. 26 is what percent of 160?

39. What is 16.5% of 240?

40. What is 18.5% of 360?

41. Scoring 44 problems out of 55 problems correctly on a test is what percent?

42. Scoring 27 problems out of 45 problems correctly on a test is what percent?

Applications

43. The special effects office at Magic Movie Studio fabricated 80 minimodels of office buildings for the disaster sequence of a new film. The director of photography discovered that on camera, only 68 of them really looked authentic. What percent of the models were acceptable?

44. An Olympic equestrian rider practiced jumping over a water hazard. In 400 attempts, she and her horse touched the water 15 times. What percent of her jump attempts were not perfect?

45. At Brookstone State College 44% of the freshman class is employed part-time. There are 1260 freshmen this year. How many of them are employed part-time? Round to the nearest whole number.

46. A recent study indicates that 15% of all middle school students do not eat a proper breakfast. If Pineridge Middle School has 420 students, how many do not eat a proper breakfast?

47. The swim team at Stonybrook College has gone on to the state championships 24 times over the years. If that translates to 60% of the time in which the team has qualified for the finals, how many years has the swim team qualified for the finals?

48. North Shore Community College found that 60% of its graduates go on for further education. Last year 570 of the graduates went on for further education. How many students graduated from the college last year?

49. Find 12% of 30% of $1600.

50. Find 90% of 15% of 2700.

To Think About

▲ **51.** A farmer has a rectangular field. The width of the field is 30% of the length of the field. The perimeter of the field is 1040 feet. What is the width of the field? What is the length of the field?

▲ **52.** The back portion of the schoolyard is used as a playground for elementary children. The playground has a length that is 200% of the width. The area of the playground is 3200 square feet. What is the length and width of the playground?

Cumulative Review Problems

Multiply.

53. 1.36
     $\times$ 1.8

54. 2.04
     $\times$ 7.3

Divide.

55. $0.06\overline{)170.04}$

56. $0.08\overline{)233.36}$

1 *Identifying the Parts of the Percent Proportion*

In Section 5.3A we showed you how to use an equation to solve a percent problem. Some students find it easier to use proportions to solve percent problems. We will show you how to use proportions in this section. The two methods work equally well. Using percent proportions allows you to see another of the many uses of the proportions that we studied in Chapter 4.

Suppose your math class of 25 students has 19 right-handed students and 6 left-handed students. You could say that $\frac{19}{25}$ of the class or 76% is right-handed. Consider the following relationship.

$$\frac{19}{25} = 76\%$$

This can be written as

$$\frac{19}{25} = \frac{76}{100}$$

As a rule, we can write this relationship using the **percent proportion**

$$\frac{\text{amount}}{\text{base}} = \frac{\text{percent number}}{100}.$$

To use this equation effectively, we need to find the amount, base, and percent number in a word problem. The easiest of these three parts to find is the percent number. We use the letter p (a variable) to represent the **percent number.**

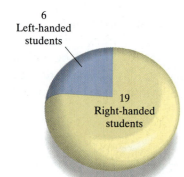

6
Left-handed students

19
Right-handed students

 **EXAMPLE 1** Identify the percent number p.

(a) Find 16% of 370

(b) 28% of what is 25?

(c) What percent of 18 is 4.5?

(a) Find 16% of 370.
The value of p is 16.

(b) 28% of what is 25?
The value of p is 28.

(c) What percent of 18 is 4.5?
↓
p

We let p represent the unknown percent number.

Practice Problem 1 Identify the percent number p.

(a) Find 83% of 460.   **(b)** 18% of what number is 90?

(c) What percent of 64 is 8?

We use the letter b to represent the base number. The **base** is the entire quantity or the total involved. The number that is the base usually appears after the word *of.* The **amount**, which we represent by the letter a, is the part being compared to the whole.

⬤ **EXAMPLE 2** Identify the base b and the amount a.

(a) 20% of 320 is 64.　　　　**(b)** 12 is 60% of what?

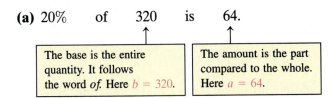

(a) 20%　　of　　320　　is　　64.

The base is the entire quantity. It follows the word *of*. Here $b = 320$.	The amount is the part compared to the whole. Here $a = 64$.

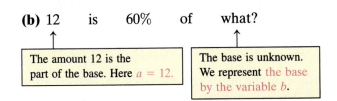

(b) 12　　is　　60%　　of　　what?

The amount 12 is the part of the base. Here $a = 12$.	The base is unknown. We represent the base by the variable b.

Practice Problem 2 Identify the base b and the amount a.

(a) 30% of 52 is 15.6.　　　　**(b)** 170 is 85% of what?　　⬤

When identifying p, b, and a in a problem, it is easiest to identify p and b first. The remaining quantity or variable is a.

⬤ **EXAMPLE 3** Find p, b, and a.

(a) What is 52% of 300?　　　　**(b)** What percent of 30 is 18?

The value of p is 52.

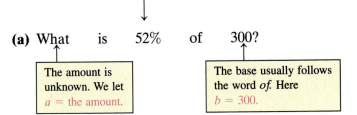

(a) What　　is　　52%　　of　　300?

The amount is unknown. We let a = the amount.	The base usually follows the word *of*. Here $b = 300$.

(b)

The value of p is not known. We let p represent the unknown percent.

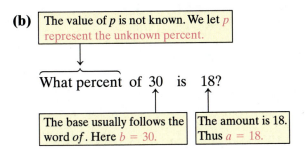

What percent　of　30　is　18?

The base usually follows the word *of*. Here $b = 30$.	The amount is 18. Thus $a = 18$.

Practice Problem 3 Find p, b, and a.

(a) What is 18% of 240?　　　　**(b)** What percent of 64 is 4?　　⬤

2 Using the Percent Proportion to Solve Percent Problems

When we solve the percent proportion, we will have enough information to state the numerical value for two of the three variables a, b, p in the equation

$$\frac{a}{b} = \frac{p}{100}.$$

We first identify those two values, and then substitute those values into the equation. Then we will use the skills that we acquired for solving proportions in Chapter 4 to find the value we do not know.

EXAMPLE 4 Find 260% of 40.

The percent $p = 260$. The number that is the base usually appears after the word *of.* The base $b = 40$. The amount is unknown. We use the variable a. Thus

$$\frac{a}{b} = \frac{p}{100} \qquad \text{becomes} \qquad \frac{a}{40} = \frac{260}{100}.$$

If we reduce the fraction on the right-hand side, we have

$$\frac{a}{40} = \frac{13}{5}$$

$$5a = (40)(13) \qquad \text{Cross multiply.}$$

$$5a = 520 \qquad \text{Simplify.}$$

$$\frac{5a}{5} = \frac{520}{5} \qquad \text{Divide each side of the equation by 5.}$$

$$a = 104$$

Thus 260% of 40 is 104.

Practice Problem 4 Find 340% of 70.

EXAMPLE 5 85% of what is 221?

The percent $p = 85$. The base is unknown. We use the variable b. The amount a is 221. Thus

$$\frac{a}{b} = \frac{p}{100} \qquad \text{becomes} \qquad \frac{221}{b} = \frac{85}{100}.$$

If we reduce the fraction on the right-hand side, we have

$$\frac{221}{b} = \frac{17}{20}$$

$$(221)(20) = 17b \qquad \text{Cross multiply.}$$

$$4420 = 17b \qquad \text{Simplify.}$$

$$\frac{4420}{17} = \frac{17b}{17} \qquad \text{Divide each side by 17.}$$

$$260 = b. \qquad \text{Divide 4420 by 17.}$$

Thus 85% of 260 is 221.

Practice Problem 5 68% of what is 476?

EXAMPLE 6 George and Barbara purchased some no-load mutual funds. The account manager charged a service fee of 0.2% of the value of the mutual funds. George and Barbara paid this fee, which amounted to $53. What was the value of the mutual funds that they purchased?

The basic situation here is that 0.2% of some number is $53. This is equivalent to saying $53 is 0.2% of what? If we want to answer the question "53 is 0.2% of what?", we need to identify a, b, and p.

The percent $p = 0.2$. The base is unknown. We use the variable b. The amount $a = 53$. Thus

$$\frac{a}{b} = \frac{p}{100} \qquad \text{becomes} \qquad \frac{53}{b} = \frac{0.2}{100}.$$

When we cross multiply, we obtain

$$(53)(100) = 0.2b$$
$$5300 = 0.2b$$
$$\frac{5300}{0.2} = \frac{0.2b}{0.2}$$
$$26{,}500 = b$$

Thus $53 is 0.2% of $26,500. Therefore the value of the mutual funds was $26,500.

Practice Problem 6 Everett Hatfield recently exchanged U.S. dollars to Canadian dollars for his company, Nova Scotia Central Trucking, Ltd. The bank charged a fee of 0.3% of the total U.S. dollars exchanged. The fee amounted to $216 in U.S. money. How many U.S. dollars were exchanged?

EXAMPLE 7 What percent of 4000 is 160?

The percent is unknown. We use the variable p. The base $b = 4000$. The amount $a = 160$. Thus

$$\frac{a}{b} = \frac{p}{100} \qquad \text{becomes} \qquad \frac{160}{4000} = \frac{p}{100}.$$

If we reduce the fraction on the left-hand side, we have

$$\frac{1}{25} = \frac{p}{100}$$
$$100 = 25p \qquad \text{\color{red}Cross multiply.}$$
$$\frac{100}{25} = \frac{25p}{25} \qquad \text{\color{red}Divide each side by 25.}$$
$$4 = p \qquad \text{\color{red}Divide 100 by 25.}$$

Thus 4% of 4000 is 160.

Practice Problem 7 What percent of 3500 is 105?

Identify p, b, and a. Do not solve for the unknown.

	p	*b*	*a*
1. 75% of 660 is 495.	_____	_____	_____
2. 65% of 820 is 532.	_____	_____	_____
3. What is 22% of 60?	_____	_____	_____
4. What is 35% of 95?	_____	_____	_____
5. 49% of what is 2450?	_____	_____	_____
6. 38% of what is 2280?	_____	_____	_____
7. 30 is what percent of 50?	_____	_____	_____
8. 50 is what percent of 250?	_____	_____	_____

Solve using the percent proportion

$$\frac{a}{b} = \frac{p}{100}.$$

In exercises 9–14, the amount a is not known.

9. 30% of 120 is what?

10. 80% of 320 is what?

11. Find 280% of 70.

12. Find 340% of 90.

13. 0.7% of 8000 is what?

14. 0.8% of 9000 is what?

In exercises 15–20, the base b is not known.

15. 20 is 25% of what?

16. 45 is 60% of what?

17. 250% of what is 200?

18. 175% of what is 350?

19. 3000 is 0.5% of what?

20. 6000 is 0.4% of what?

In exercises 21–24, the percent p is not known.

21. 56 is what percent of 280?

22. 70 is what percent of 1400?

23. What percent of 260 is 10.4?

24. What percent of 350 is 10.5?

Mixed Practice

25. 18% of 150 is what?

26. 26% of 250 is what?

27. 150% of what is 120?

28. 200% of what is 120?

29. 82 is what percent of 500?

30. 75 is what percent of 600?

31. Find 0.7% of 520.

32. Find 0.4% of 650.

33. What percent of 66 is 16.5?

34. What percent of 49 is 34.3?

35. 68 is 40% of what?

36. 52 is 40% of what?

Applications

37. Victor purchased a used car for $9500. He made a down payment of 24% of the purchase price. How much was his down payment?

38. Last year Priscilla had 21% of her salary withheld for taxes. If the total amount withheld was $4200 for the year, what was her annual salary?

39. Ed and Suzie went out to eat at Pizzeria Uno. The dinner check was $26.00. They left a tip of $3.90. What percent of the check was the tip?

40. Nathaniel got 27 problems correct on a 30-problem test. What percent of the test problems did he do correctly?

41. Trudy took a biology test with 40 problems. She got 8 of the problems wrong and 32 of the problems right. What percent of the test problems did she do incorrectly?

42. The Super Shop and Save store had 120 gallons of milk placed on the shelf one night. During the next morning's inspection, the manager found that 15% of the milk had passed the expiration date. How many gallons of milk had passed the expiration date?

43. Michael and Susan Baley spend 18% of their combined net income on a monthly mortgage payment. Their monthly mortgage payment is $720. What is their combined monthly net income?

44. During June the Wenham police stopped 250 drivers for speed violations. It was found that 8% of the people who were stopped had outstanding warrants for their arrest. How many people had outstanding warrants for their arrest?

To Think About

Recently a review was made of the expenditures in Texas during 1999 for the following five categories: corrections, government and administration, interest on general debt, natural resources, and parks and recreation. The amount of money (in millions of dollars) expended in each of these five categories is listed in the following pie chart.

Round all answers to the nearest tenth.

Use the chart to answer questions 45–48.

Texas Expenditures for 1999 in Selected Categories

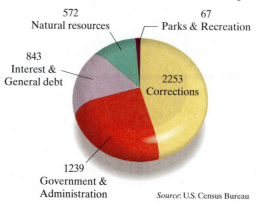

Source: U.S. Census Bureau

45. What percent of the total expenditures in these five categories was used for corrections?

46. What percent of the total expenditures in these five categories was used for government and administration?

47. Suppose that compared to 1999, expenditures in 2004 for corrections were 25% larger, expenditures for government and administration were 30% larger, and the other three categories remained at the same dollar figure. What percent of the total expenditures in 2004 in these five categories would be used for natural resources?

48. Suppose that compared to 1999, expenditures in 2007 for corrections were 40% larger, expenditures for government and administration were 35% larger, expenditures for interest on general debt were 20% larger, and the other two categories remained at the same dollar figure. What percent of the total expenditures in 2007 would be used for interest on general debt?

Cumulative Review Problems

Simplify.

49. $\dfrac{4}{5} + \dfrac{8}{9}$

50. $\dfrac{7}{13} - \dfrac{1}{2}$

51. $\left(2\dfrac{4}{5}\right)\left(1\dfrac{1}{2}\right)$

52. $1\dfrac{2}{5} \div \dfrac{3}{4}$

5.4 Solving Applied Percent Problems

1 Solving General Applied Percent Problems

Student Learning Objectives

After studying this section, you will be able to:

1 Solve general applied percent problems.

2 Solve problems when percents are added.

3 Solve discount problems.

 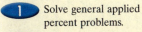

SSM PH TUTOR CD & VIDEO MATH PRO WEB
CENTER

In Sections 5.3A and 5.3B, we learned the three types of applied percent problems. Some problems ask you to find a percent of a number. Some problems give you an amount and a percent and ask you to find the base (or whole). Other problems give an amount and a base and ask you to find the percent. We will now see how the three types of percent problems occur in real life.

● **EXAMPLE 1** Of all the computers manufactured last month, an inspector found 18 that were defective. This is 2.5% of all the computers manufactured last month. How many computers were manufactured last month?

Method A Translate to an equation.
The problem is equivalent to: 2.5% of the number of computers is 18.
Let n = the number of computers.

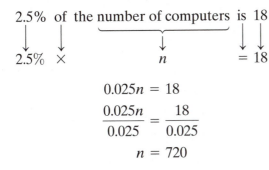

$$0.025n = 18$$
$$\frac{0.025n}{0.025} = \frac{18}{0.025}$$
$$n = 720$$

720 computers were manufactured last month.

Method B Use the percent proportion $\frac{a}{b} = \frac{p}{100}$.

The percent p = 2.5. The base is unknown. We will use the variable b. The amount a = 18. Thus

$$\frac{a}{b} = \frac{p}{100} \quad \text{becomes} \quad \frac{18}{b} = \frac{2.5}{100}.$$

Using cross multiplication, we have

$$(18)(100) = 2.5b$$
$$1800 = 2.5b$$
$$\frac{1800}{2.5} = \frac{2.5b}{2.5}$$
$$720 = b$$

720 computers were manufactured last month.
By either Method A or Method B, we obtain the same number of computers, 720.
Substitute 720 into the original problem to check.

2.5% of 720 computers are defective.
$$(0.025)(720) = 18 \checkmark$$

Practice Problem 1 4800 people, or 12% of all passengers holding tickets for American Airlines flights in one month, did not show up for their flights. How many people held tickets that month? ●

341

EXAMPLE 2 How much sales tax will you pay on a color television priced at $299 if the sales tax is 5%?

Method A Translate to an equation.

$$\text{What is } 5\% \text{ of } \$299?$$
$$\downarrow \ \downarrow \ \downarrow \quad \downarrow \quad \downarrow$$
$$n = 5\% \times 299$$
$$n = (0.05)(299)$$
$$n = 14.95 \qquad \text{The tax is } \$14.95.$$

Method B Use the percent proportion $\dfrac{a}{b} = \dfrac{p}{100}$.

The percent $p = 5$. The base $b = 299$. The amount is unknown. We use the variable a. Thus

$$\frac{a}{b} = \frac{p}{100} \qquad \text{becomes} \qquad \frac{a}{299} = \frac{5}{100}.$$

If we reduce the fraction on the right-hand side, we have

$$\frac{a}{299} = \frac{1}{20}.$$

We then cross multiply to obtain

$$20a = 299$$
$$\frac{20a}{20} = \frac{299}{20}$$
$$a = 14.95 \quad \text{The tax is } \$14.95.$$

Thus, by either method, the amount of the sales tax is $14.95.

Let's see if our answer is reasonable. Is 5% of $299 really $14.95? If we round $299 to one nonzero digit, we have $300. Thus we have 5% of 300 = 15. Since 15 is quite close to our value of $14.95, our answer seems reasonable.

Practice Problem 2 A salesperson rented a hotel room for $62.30 per night. The tax in her state is 8%. What tax does she pay for one night at the hotel?

EXAMPLE 3 A failing student attended class 39 times out of the 45 times the class met last semester. What percent of the classes did he attend? Round to the nearest tenth of a percent.

Method A Translate to an equation.

This problem is equivalent to:

$$39 \ \text{ is } \text{what percent} \ \text{ of } \ 45?$$
$$\downarrow \ \downarrow \qquad\qquad \downarrow \qquad\quad \downarrow$$
$$39 = \qquad\qquad n \qquad\quad \times 45$$
$$39 = 45n$$
$$\frac{39}{45} = \frac{45n}{45}$$
$$0.8666\ldots = n$$

To the nearest tenth of a percent we have $n = 86.7\%$.

Method B Use the percent proportion $\dfrac{a}{b} = \dfrac{p}{100}$.

The percent is unknown. We use the variable p. The base b is 45. The amount a is 39. Thus

$$\frac{a}{b} = \frac{p}{100} \quad \text{becomes} \quad \frac{39}{45} = \frac{p}{100}.$$

When we cross multiply, we get

$$(39)(100) = 45p$$
$$3900 = 45p$$
$$\frac{3900}{45} = \frac{45p}{45}$$
$$86.666\ldots = p$$

To the nearest tenth, the answer is 86.7%.

By using either method, we discover that the failing student attended approximately 86.7% of the classes.

Verify by estimating that the answer is reasonable.

Practice Problem 3 Of the 130 flights at Orange County Airport yesterday, only 105 of them were on time. What percent of the flights were on time? (Round to the nearest tenth of a percent.)

Now you have some experience solving the three types of percent problems in real-life applications. You can use either Method A or Method B to solve applied percent problems. In the following pages we will present more percent applications. We will not list all the steps of Method A or Method B. Most students will find after a careful study of Examples 1–3 that they do not need to write out all the steps of Method A or Method B when solving applied percent problems.

2 Solving Problems When Percents Are Added

Percents can be added if the base (whole) is the same. For example, 50% of your salary added to 20% of your salary = 70% of your salary. 100% of your cost added to 15% of your cost = 115% of your cost. Problems like this are often called **markup problems.** If we add 15% of the cost of an item to the original cost, the markup is 15%. We will add percents in some applied situations.

The following example is interesting, but it is a little challenging. So please read it very carefully. A lot of students find it difficult at first.

EXAMPLE 4 Walter and Mary Ann are going out to a restaurant. They have a limit of $63.25 to spend for the evening. They want to tip the waitress 15% of the cost of the meal. How much money can they afford to spend on the meal itself? (Assume there is no tax.)

In some of the problems in this section, it may help you to use the Mathematics Blueprint. We will use it here for Example 4.

Mathematics Blueprint For Problem Solving

Gather the Facts	What Am I Asked to Do?	How Do I Proceed?	Key Points to Remember
They have a spending limit of $63.25. They want to tip the waitress 15% of the cost of the meal.	Find the amount of money that the meal will cost.	Separate the $63.25 into two parts: the cost of meal and the tip. Add these two parts to get $63.25.	We are not taking 15% of $63.25, but rather 15% of the cost of meal.

Let n = the cost of the meal. 15% of the cost = the amount of the tip. We want to add the percents of the meal.

$$\boxed{\begin{array}{c}\text{Cost of}\\\text{meal } n\end{array}} + \boxed{\begin{array}{c}\text{tip of 15\%}\\\text{of the cost}\end{array}} = \boxed{\$63.25}$$

$$100\% \text{ of } n \ + \ 15\% \text{ of } n \ = \ \$63.25$$

Note that 100% of n added to 15% of n is 115% of n.

$$115\% \text{ of } n = \$63.25$$
$$1.15 \times n = 63.25$$
$$\frac{1.15 \times n}{1.15} = \frac{63.25}{1.15} \qquad \text{Divide both sides by 1.15.}$$
$$n = 55$$

They can spend up to $55.00 on the meal itself.

Does this answer seem reasonable?

Practice Problem 4 Sue and Sam have $46.00 to spend at a restaurant, including a 15% tip. How much can they spend on the meal itself? (Assume there is no tax.)

3 Solving Discount Problems

Frequently, we see signs urging us to buy during a sale when the list price is discounted by a certain percent.

The amount of a **discount** is the product of the discount rate and the list price.

> Discount = discount rate × list price

EXAMPLE 5 Jeff purchased a compact disc player on sale at a 35% discount. The list price was $430.00.

(a) What was the amount of the discount?

(b) How much did Jeff pay for the CD player?

(a) Discount = discount rate × list price

$$= 35\% \times 430$$
$$= 0.35 \times 430$$
$$= 150.5$$

The discount was $150.50.

(b) We subtract the discount from the list price to get the selling price.

$$\begin{array}{ll}\$430.00 & \text{list price}\\ -\$150.50 & \text{discount}\\ \hline \$279.50 & \text{selling price}\end{array}$$

Jeff paid $279.50 for the CD player.

Practice Problem 5 Betty bought a car that lists for $13,600 at a 7% discount.

(a) What was the discount?   **(b)** What did she pay for the car?

Exercises

Applications

Exercises 1–18 present the three types of percent problems. They are similar to Examples 1–3. Take the time to master exercises 1–18 before going on to the next ones. Round to the nearest hundredth when necessary.

1. No graphite was found in 4500 pencils shipped to Sureway School Supplies. This was 2.5% of the total number of pencils received by Sureway. How many pencils in total were in the order?

2. A high-jumper on the track and field team hit the bar 58 times last week. This means that he did not succeed in 29% of his jump attempts. How many total attempts did he make last week?

3. Scott's phone bill averages $80.50 per month. This is 115% of his average monthly bill last year. What was his average monthly phone bill last year?

4. Renata now earns $9.50 per hour. This is 125% of what she earned last year. What did she earn per hour last year?

5. The Incredible Chocolate Chip Company has discovered that 36 out of 400 chocolate chip cookies do not contain enough chocolate chips. What percent of the chocolate chip cookies do not have enough chips?

6. Every day this year, Sam ordered either cappuccino or espresso from the coffee bar downstairs. He had 85 espressos and 280 cappuccinos. What percent of the coffees were espressos?

7. In a survey taken by the Department of Agriculture of 450 people, 74% admitted that they disliked brussels sprouts. How many people in this survey do not care for this vegetable?

8. At the backstage entrance to the Hollywood Bowl Concert Arena, 85% of the people who came to the backstage door after a recent concert requested autographs. If 510 people asked for autographs, how many people came to the backstage door?

9. Dean bought a new mountain bike. The sales tax in his state is 4%, and he paid $9.60 in tax. What was the price of the mountain bike before the tax?

10. Maha received a raise of 8% of her monthly salary. The amount of her raise was $48.16 per month. What was her monthly salary before the raise?

11. Peter and Judy together earn $5060 per month. Their mortgage payment is $1265 per month. What percent of their household income goes toward paying the mortgage?

12. Jackie puts aside $65.50 per week for her monthly car payment. She earns $327.50 per week. What percent of her income is set aside for car payments.

13. The Children's Wish Charity raised 75% of its funds from sporting promotions. Last year the charity received $7,200,000 from its sporting promotions. What was the charity's total income last year?

14. Shannon paid $8400 in federal and state income taxes as a lab technician, which amounted to 28% of her annual income. What was her income last year?

15. Of the 206 bones in the human body, 26 are in the foot. What percent of the bones in your body are in your foot?

16. The nutrition label on a bag of cookies reads "Calories 180; Calories from Fat 40." What percent of the calories in a serving of cookies is from fat?

17. In Flagstaff, Arizona, 0.9% of all babies walk before they reach the age of 11 months. If 24,000 babies were born in Flagstaff in the last 20 years, how many of them will have walked before they reached the age of 11 months?

18. At a Boston Celtics basketball game scheduled for 8:30 P.M., 0.8% of the spectators were children under age 12. If 28,000 people showed up for the game, how many children under age 12 were in the stands?

Exercises 19–34 include percents that are added and discounts. Solve. Round to the nearest hundredth when necessary.

19. Manuel is taking Cynthia out to dinner. He has $66 to spend. If he wants to tip the server 20%, how much can be afford to spend on their meal?

20. Belinda asked Martin out to dinner. She has $47.50 to spend. She wants to tip the waitress 15% of the cost of their meal. How much money can she afford to spend on the meal itself?

21. John and Chris Maney are building a new house. When finished, the house will cost $163,500. The price of the house is 9% higher than the price when the original plans were made. What was the price of the house when the original plans were made?

22. Dan and Connie Lacorazza purchased a new Ford Windstar. The purchase price was $23,320. The price was 6% higher than the price of a similar Windstar three years ago. What was the price of the Windstar three years ago?

23. When a new computer case is made, there is some waste of the plastic material used for the front of the computer case. Approximately 3% of the plastic is of poor quality and is thrown away, while 8% of the plastic is waste material that is thrown away as the front pieces are created by a giant stamping machine. If 20,000 pounds of plastic are used to make the front of computer cases each month, how many pounds are thrown away?

24. In a recent survey of planes landing at Logan Airport in Boston, it was observed that 12% of the flights were delayed less than an hour and 7% of the flights were delayed an hour or more but less than two hours. If 9000 flights arrive at Logan Airport in a day, how many flights are delayed less than two hours?

25. The Democratic National Committee has a budget of $33,000,000 to spend on the inauguration of the new president. 15% of the costs will be paid to personnel, 12% of the costs will go toward food, and 10% will go to decorations. How much money will go for personnel, food, and decorations? How much will be left over to cover security, facility rental, and all other expenses?

26. A major research facility has developed an experimental drug to treat Alzheimer's disease. Twenty percent of the research costs was paid to the staff. Sixteen percent of the research costs was paid to rent the building where the research was conducted. The company has spent $6,000,000 in research on this new drug. How much was paid to cover the cost of staff and rental of the building?

7:13

27. Melinda purchased a new blouse, jeans, and a sweater. All of the clothes were discounted 35%. Before the sale, the total purchase price would have been $190 for these three items. How much did she pay for them with the discount?

28. Juan went to purchase two new radial tires for his Honda Accord. The set of two tires normally costs $130. However, he bought them on sale at a discount of 30%. How much did he pay for the tires with the discount?

29. Carol and David went to purchase a discount coupon book for several local restaurants in their area. The coupon book costs $30 and is good for one year. It has coupons that are worth 15% off all meals served before 6 P.M. at 20 local restaurants. They usually spend $400 a year dining at these restaurants. How many dollars would they save each year if they purchased this book?

30. Anne wanted to join TJ's Discount Shopping Club. The annual membership fee is $45. With her membership card she will get a discount of 8% on any item in the store. She estimates that she would probably purchase around $900 worth of store items each year. How much will she save over a period of one year if she purchases these items at the store but also pays the annual membership fee?

31. Susie bought her first Harley-Davidson motorcycle. The list price was $16,000, but the dealer gave her a discount of 8%.
(a) What was the discount?

(b) How much did she pay for the motorcycle?

32. Jane bought a leather sofa that had been used as the floor model. The list price of the sofa was $1200, but the dealer gave her a discount of 35%.
(a) What was the discount?

(b) How much did she pay for the sofa?

33. Old Tyme Antiques needs to clear space for a new shipment. They are offering 30% off all items in one area of the store. Angela collects antique clocks and finds a mantel clock priced at $1110.
(a) What is the discount?

(b) How much will she pay for the mantel clock?

34. The Smiths are having a swimming pool and sauna installed in their backyard. They saw an advertisement that Swim & Spa was offering 15% off the total cost if you buy both items. When they went to the store, the sign read "10% off." Mrs. Smith and the store's owner agreed to average the discounts, and came up with a savings of 12.5%. If the total cost is $13,000, how much do the Smiths save with the 12.5% discount?

Cumulative Review Problems

35. Round to the nearest thousand. 1,698,481

36. Round to the nearest hundred. 2,452,399

37. Round to the nearest hundredth. 1.63474

38. Round to the nearest thousandth. 0.793468

39. Round to the nearest ten-thousandth. 0.055613

40. Round to the nearest ten-thousandth. 0.079152

5.5 Solving Commission, Percent of Increase or Decrease, and Interest Problems

1 Solving Commission Problems

If you work as a salesperson, your earnings may be in part or in total a certain percentage of the sales you make. The amount of money you get that is a percentage of the value of your sales is called your **commission.** It is calculated by multiplying the percentage (called the **commission rate**) by the value of the sales.

Commission = commission rate × value of sales

EXAMPLE 1 A salesperson has a commission rate of 17%. She sells $32,500 worth of goods in a department store in two months. What is her commission?

$$\text{Commission} = \text{commission rate} \times \text{value of sales}$$
$$\text{Commission} = 17\% \times \$32,500$$
$$= 0.17 \times 32,500$$
$$= 5525$$

Her commission is $5525.00.

Does this answer seem reasonable? Check by estimating.

Practice Problem 1 A real estate salesperson earns a commission rate of 6% when he sells a $156,000 home. What is his commission?

In some problems, the unknown quantity will be the commission rate or the value of sales. However, the same equation is used:

Commission = commission rate × value of sales

2 Solving Percent-of-Increase or Percent-of-Decrease Problems

We sometimes need to find the percent by which a number increases or decreases. If a car costs $7000 and the price decreases $1750, we say that the percent of decrease is $\frac{1750}{7000} = 0.25 = 25\%$.

$$\textbf{Percent of decrease} = \frac{\text{amount of decrease}}{\text{original amount}}$$

Similarly, if a population of 12,000 people increases by 1920 people, we say that the percent of increase is $\frac{1920}{12,000} = 0.16 = 16\%$.

$$\textbf{Percent of increase} = \frac{\text{amount of increase}}{\text{original amount}}$$

Note that for these types of problems the base is always the *original amount*.

EXAMPLE 2 The population of Center City increased from 50,000 to 59,500. What was the percent of increase?

For this problem as well as others in this section, you may find it helpful to use the Mathematics Blueprint.

Mathematics Blueprint For Problem Solving

Gather the Facts	What Am I Asked to Do?	How Do I Proceed?	Key Points to Remember
The population increased from 50,000 to 59,500.	We must find the percent of increase.	First subtract to find the amount of increase. Then divide the amount of increase by the original amount.	Always divide by the original amount.

Amount of increase

$$
\begin{array}{r}
59,500 \\
- 50,000 \\
\hline
9,500
\end{array}
$$

$$
\text{Percent of increase} = \frac{\text{amount of increase}}{\text{original amount}} = \frac{9500}{50,000}
$$

$$
= 0.19 = 19\%
$$

The percent of increase is 19%.

Practice Problem 2   A new car is sold for $15,000. A year later its price had decreased to $10,500. What is the percent of decrease?

3   Solving Simple Interest Problems

Interest is money paid for the use of money. If you deposit money in a bank, the bank uses that money and pays you interest. If you borrow money, you pay the bank interest for the use of that money. The **principal** is the amount deposited or borrowed. Interest is usually expressed as a percent rate of the principal. The **interest rate** is assumed to be per year, unless otherwise stated. The formula used in business to compute simple interest is

$$\text{Interest} = \text{principal} \times \text{rate} \times \text{time}$$
$$I = P \times R \times T$$

If the interest rate is *per year*, the time T *must* be in *years*.

EXAMPLE 3   Find the simple interest on a loan of $7500 borrowed at 13% for one year.

$$I = P \times R \times T$$

$P = \text{principal} = \$7500$ $R = \text{rate} = 13\%$ $T = \text{time} = 1 \text{ year}$

$$I = 7500 \times 13\% \times 1 = 7500 \times 0.13 = 975$$

The interest is $975.

Practice Problem 3   Find the simple interest on a loan of $5600 borrowed at 12% for one year.

Our formula is based on a yearly interest rate. Time periods of more than one year or a fractional part of a year are sometimes needed.

EXAMPLE 4   Find the simple interest on a loan of $2500 that is borrowed at 9% for

(a) three years.   **(b)** three months.

(a)
$$I = P \times R \times T$$
$$P = \$2500 \qquad R = 9\% \qquad T = 3 \text{ years}$$
$$I = 2500 \times 0.09 \times 3 = 225 \times 3 = 675$$

The interest for three years is $675.

(b) Three months $= \dfrac{1}{4}$ year. The period must be in years to use the formula.

Since $T = \dfrac{1}{4}$ year, we have

$$I = 2500 \times 0.09 \times \frac{1}{4}$$
$$= 225 \times \frac{1}{4}$$
$$= \frac{225}{4} = 56.25$$

The interest for $\dfrac{1}{4}$ year is $56.25.

Practice Problem 4   Find the simple interest on a loan of $1800 that is borrowed at 11% for

(a) four years.   **(b)** six months.

Applications

Exercises 1–18, are problems involving commissions, percent of increase or decrease, and simple interest.

1. Walter works as an appliance salesman in a department store. Last month he sold $170,000 worth of appliances. His commission rate is 2%. How much money did he earn in commission last month?

2. Susan works at the Acura dealership in Winchester. Last month she had car sales totaling $230,000. Her commission rate is 3%. How much money did she earn in commission last month?

3. Allison works in the local Verizon office selling mobile phones. She is paid $300 per month plus 4% of her total sales in mobile phones. Last month she sold $96,000 worth of mobile phones. What was her total income for the month?

4. Matthew is a stockbroker. He is paid $500 per month plus 0.5% of the total sales of stocks that he sells. Last month he sold $340,000 worth of stock. What was his total income for the month?

5. The daycare enrollment in the town of Harvey increased from 5000 to 7600 children per day. What was the percent of increase?

6. The cost of a certain Chevrolet Camaro increased from $16,000 last year to $17,280 this year. What was the percent of increase?

7. A stereo originally priced at $985 was sold for $394. What was the percent of decrease?

8. Tay weighed 285 pounds two years ago. After careful supervision at a weight loss center, he reduced his weight to 171 pounds. What was the percent of decrease in his weight?

9. Phil placed $2000 in a one-year CD at the bank. The bank is paying simple interest of 7% for one year on the CD. How much interest will Phil earn in one year?

10. Charlotte has a checking account that pays her simple interest of 2% on the average balance in her checking account. Last year her average balance was $450. How much interest did she earn in her checking account

11. Melinda has a MasterCard account with Centerville Bank. She has to pay a monthly interest rate of 1.5% on the average daily balance of the amount she owes on her credit card. Last month her average daily balance was $500. How much interest was she charged last month? (*Hint:* The formula $I = P \times R \times T$ can be used if the interest rate is *per month* and the time is in *months.*)

12. Walter borrowed $3000 for a student loan to finish college this year. Next year he will need to pay 7% simple interest on the amount he borrowed. How much interest will he need to pay next year?

13. Abe had to borrow $12,000 for a house construction loan for three months. The interest rate was 16% per year. How much interest did he have to pay for borrowing the money for three months?

14. Joan needed to borrow $8000 for two months to finance some renovations to her beauty salon. The interest rate she was charged was 18% per year. How much interest did she have to pay for borrowing the money for two months?

15. Robert sells life insurance for a major insurance company for a commission. Last year he sold $12,000,000 worth of insurance. He earned $72,000 in commissions. What was his commission rate?

16. Hillary sells medical supplies to doctors' offices for a major medical supply company. Last year she sold $9,000,000 worth of medical supplies. She works on a commission basis and last year she earned $63,000 in commissions. What was her commission rate?

17. Jennifer sells furniture for a major department store. Last year she was paid $48,000 for commissions. If her commission rate is 3%, what was the sales total of the furniture that she sold last year?

18. Michael sells used cars for Beltway Motors. Last year he was paid $42,000 in commissions. Beltway Motors pays the salespeople a commission rate of 6%. What was the sales total of the cars that Michael sold last year?

Mixed Applications

Exercises 19–30 are a variety of percent problems. They involve commissions, percent of increase or decreases, and simple interest. There are also some of each kind of percent problem encountered in the chapter.

19. Ted is trying to decrease his spending on entertainment. He earns $265 per week and is allowing himself to spend only 15% per week on movies, dining out, and so on. How much can Ted spend per week on entertainment?

20. The maximum capacity of your lungs is 4.58 liters of air. In a typical breath, you breathe in 12% of the maximum capacity. How many liters of air do you breathe in a typical breath?

21. Of all the boxes of Girl Scout cookies sold, 25% are Thin Mints. If a Girl Scout troop sells 156 boxes of cookies, how many are Thin Mints?

22. 11% of the Scottish population have red hair. If the population of Scotland is 5,600,000, how many people have red hair?

23. An inspector found that 32 computer parts in a shipment were defective. If this is 4% of the total number of parts inspected, how many parts were inspected?

24. Babies are born with 350 soft bones. When bone fusion is complete (at 20–25 years of age), the body has 206 bones. What is the percentage of decrease in the number of bones?

25. Adam deposited $3700 in his savings account for one year. His savings account earns 3.2% interest annually. He did not add any more money within the year, and at the end of that time, he withdrew all funds.
   (a) How much interest did he earn?
   (b) How much money did he withdraw from the bank?

26. Nikki had $1258 outstanding on her Master-Card, which charges 2% monthly interest. At the end of this month Nikki paid off the loan.
   (a) How much interest did Nikki pay for one month?
   (b) How much did it cost to pay off the loan totally?

27. Bryce went shopping and bought a pair of sandals for $52, swimming trunks for $38, and sunglasses for $26. The tax in Bryce's city is 6%.
   (a) What is the total sales tax?
   (b) What is the total of the purchases?

28. Finn bought a used Honda Accord for $10,500. The tax in his state is 8%.
   (a) What is the sales tax?
   (b) What is the final price of the Honda Accord?

Solve. Round to the nearest cent.

 29. How much sales tax would you pay to purchase a new Honda Accord that costs $18,456.82 if the sales tax rate is 4.6%

 30. The Hartling family purchased a new living room set. The list price was $1249.95. However, they got a discount of 29%. How much did they pay for the new living room set?

Cumulative Review Problems

Perform the following calculations using the correct order of operations.

31. $3(12 - 6) - 4(12 \div 3)$

32. $7 + 4^3 \times 2 - 15$

33. $\left(\dfrac{5}{3}\right)\left(\dfrac{3}{8}\right) - \left(\dfrac{1}{2} - \dfrac{1}{3}\right)$

34. $(6.8 - 6.6)^2 + 2(1.8)$

Putting Your Skills to Work

Nutrition Facts and Percents

Almost all packaged foods in the United States are required to have posted a Nutrition Facts label, such as the one shown from a box of raisin bran. The information on the label is helpful, but sometimes people desire additional information and the use of percents is required. Study the label and answer the following questions.

Problems for Individual Investigation

1. Melinda discovered that when she pours a bowl of raisin bran for breakfast, she uses only $\frac{3}{4}$ of a cup of cereal with her $\frac{1}{2}$ cup of milk. What percent of the daily recommended values of potassium and total carbohydrates will she obtain from her breakfast if she is on a 2000-calorie diet?

2. Doug discovered that when he pours a bowl of raisin bran for breakfast, he uses $1\frac{1}{2}$ cups of cereal with the $\frac{1}{2}$ cup of milk. What percent of the daily recommended values of potassium and total carbohydrates will he obtain from his breakfast if he is on a 2500-calorie diet?

Problems for Group Investigation and Cooperative Study

3. Fred is the starting center for the college basketball team. Fred is 6 feet 3 inches tall and weighs 280 pounds. His doctor has examined him and said he is totally healthy and is the right weight for his body type and height. His doctor has told him that he should consume 3000 calories each day. What percent of his suggested daily amount of calories and total fat is provided by 1 cup of raisin bran and $\frac{1}{2}$ cup of fat-free milk? (Assume that the daily recommended amount of fat in a 3000-calorie diet is proportional to the daily recommended amount of fat in a 2000-calorie diet.)

4. Lucy is a member of the college gymnastics team. She is 4 feet 10 inches tall and weighs 95 pounds. Her doctor has examined her and said she is totally healthy and is the right weight for her body type and height. Her doctor has told her that she should consume 1800 calories each day. What percent of her suggested daily amount of calories and total fat is provided by 1 cup of raisin bran and $\frac{1}{2}$ cup of fat-free milk? (Assume that the daily recommended amount of fat in an 1800-calorie diet is proportional to the daily recommended amount of fat in a 2000-calorie diet.)

Nutrition Facts

Serving Size 1 Cup (59g/2.1 oz.)
Servings per Package About 9

Amount Per Serving	Cereal	Cereal with 1/2 cup Vitamins A&D Fat Free Milk
Calories	190	230
Calories from Fat	15	15

	% Daily Value**	
Total Fat 1.5g*	2%	2%
Saturated Fat 0g	0%	0%
Cholesterol 0 mg	0%	0%
Sodium 350 mg	15%	17%
Potassium 360 mg	10%	16%
Total Carbohydrate 45g	15%	17%
Dietary Fiber 8g	32%	32%
Sugars 18g		
Other Carbohydrate 19g		
Protein 5g		

Vitamin A	15%	20%
Vitamin C	0%	2%
Calcium	2%	15%
Iron	25%	25%
Vitamin D	10%	25%
Thiamin	25%	30%
Riboflavin	25%	35%
Niacin	25%	25%
Vitamin B_6	25%	25%
Folic Acid	25%	25%
Vitamin B_{12}	25%	35%
Phosphorus	25%	35%
Magnesium	20%	25%
Zinc	25%	30%
Copper	10%	10%

* Amount in cereal. One half cup of fat free milk contributes an additional 40 calories, 65mg sodium, 6g total carbohydrate (6g sugars), 200mg potassium, and 4g protein.

** Percent Daily Values are based on a 2,000 calorie diet. Your daily values may be higher or lower depending on your calorie needs:

	Calories	2,000	2,500
Total Fat	Less than	65g	80g
Sat. Fat	Less than	20g	25g
Cholesterol	Less than	300mg	300mg
Sodium	Less than	2,400mg	2,400mg
Potassium		3,500mg	3,500mg
Total Carbohydrate		300g	375g
Dietary Fiber		25g	30g

Calories per gram: Fat 9 · Carbohydrate 4 · Protein 4

Internet Connections

  Netsite: http://www.prenhall.com/tobey_basic

This site contains a nutrition counter that covers fat and calories for a variety of foods.

Choose a meal and make your choices. Then click Submit at the bottom. After reading the calorie and fat count, click Order another meal, or return here and click this link again to do another meal.

5. Assume that for lunch you had a $1\frac{1}{2}$ cup chef salad with four tablespoons of non-diet French dressing, a dill pickle, and an 8-ounce glass of lowfat 2% milk. After the salad you ate an apple. Find the number of calories and grams of fat for your lunch.

6. Assume that for dinner you had six ounces of broiled ground beef, one baked potato, a serving of cooked carrots, an 8-ounce glass of skim milk, and a slice of apple pie. Find the number of calories and grams of fat for your lunch.

Math in the Media

The Law of Zero Return

Paul Cox, THE GLOSSARY OF MATHEMATICAL MISTAKES, April, 2000, www.mathmistakes.com

Reprinted with Permission

Paul Cox defines the Law of Zero Return by the formula: Return on Investment = Inflation + Taxes. In other words, an investment has zero return if all profits are consumed by inflation and taxes leaving no return which could be used for discretionary items, for example entertainment, a new car, etc.

EXERCISES

To illustrate, consider an investment of $10,000 which returns $500 in one year in questions 1–3 below.

1. If the inflation rate was 3.5% and an income tax rate was 13%, was the investment one of zero return? Hint: First, determine the base of each percent. Are they the same?

2. What change in the inflation rate would make it a zero return investment?

3. What change in the tax rate would make it a zero return investment?

Good financial planning requires an understanding of policies and issues that may influence future earnings. Consider the following statement:

"Inflation affects your income in three ways. First, your dollars are worth less over time, even though you may have more of them. Second, the money you gain just to keep up with inflation is subjected to additional taxes. Third, because income tax is progressive—the more money you earn the higher your tax burden—inflation is pushing people into higher tax brackets than they cannot afford. Tax rates are not adjusted by inflation. That 30.6% top tax rate that the middle class did not care about in 1993 because it was only 'rich,' will eventually become the tax rate of the middle class as inflation raises the cost of living."

4. What evidence can you find that indicates the federal government is monitoring inflation and the tax burden? Have any policies been proposed or actions taken to control or corrct this?

Chapter 5 Organizer

Topic	Procedure	Examples
Converting a decimal to a percent, p. 313.	1. Move the decimal point two places to the right. 2. Add the percent sign.	$0.19 = 19\%$ $0.516 = 51.6\%$ $0.04 = 4\%$ $1.53 = 153\%$ $0.006 = 0.6\%$
Converting a fraction with a denominator of 100 to a percent, p. 311.	1. Use the numerator only. 2. Add the percent sign.	$\dfrac{29}{100} = 29\% \qquad \dfrac{5.6}{100} = 5.6\%$ $\dfrac{3}{100} = 3\% \qquad \dfrac{7\frac{1}{3}}{100} = 7\frac{1}{3}\%$ $\dfrac{231}{100} = 231\%$
Changing a fraction (whose denominator is not 100) to a percent, p. 319.	1. Divide the numerator by the denominator and obtain a decimal. 2. Change the decimal to a percent.	$\dfrac{13}{50} = 0.26 = 26\%$ $\dfrac{1}{20} = 0.05 = 5\%$ $\dfrac{3}{800} = 0.00375 = 0.375\%$ $\dfrac{312}{200} = 1.56 = 156\%$
Changing a mixed number to a percent, p. 319.	1. Change the mixed number to a decimal. 2. Transform the decimal to an equivalent percent.	$3\frac{1}{4} = 3.25 = 325\% \qquad 6\frac{4}{5} = 6.8 = 680\%$ $7\frac{3}{8} = 7.375 = 737.5\%$
Changing a percent to a decimal, p. 312.	1. Drop the percent sign. 2. Move the decimal point two places to the left.	$49\% = 0.49$ $2\% = 0.02$ $0.5\% = 0.005$ $196\% = 1.96$ $1.36\% = 0.0136$
Changing a percent to a fraction, p. 317.	1. If the percent does not contain a decimal point, remove the % and write the number over a denominator of 100. Reduce the fraction if possible. 2. If the percent contains a decimal point, change the percent to a decimal by removing the % and moving the decimal point two places to the left. Then write the decimal as a fraction and reduce if possible. 3. If the percent contains a fraction, remove the % and write the number over a denominator of 100. If the numerator is a mixed number, change the numerator to an improper fraction. Next simplify by the "invert and multiply" rule. Then reduce the fraction if possible.	$25\% = \dfrac{25}{100} = \dfrac{1}{4}$ $38\% = \dfrac{38}{100} = \dfrac{19}{50}$ $130\% = \dfrac{130}{100} = \dfrac{13}{10}$ $5.8\% = 0.058$ $\quad = \dfrac{58}{1000} = \dfrac{29}{500}$ $2.72\% = 0.0272$ $\quad = \dfrac{272}{10{,}000} = \dfrac{17}{625}$ $7\frac{1}{8}\% = \dfrac{7\frac{1}{8}}{100}$ $\quad = 7\frac{1}{8} \div \dfrac{100}{1}$ $\quad = \dfrac{57}{8} \times \dfrac{1}{100} = \dfrac{57}{800}$

Topic	Procedure	Examples
Solving percent problems by translating to equations, p. 326.	**1.** Translate by replacing "of" with $\times$ "is" with $=$ "what" with n "find" with $n =$ **2.** Solve the resulting equation.	**(a)** What is 3% of 56? $\downarrow\downarrow\ \downarrow\ \downarrow\ \downarrow$ $n = 3\% \times 56$ $n = (0.03)(56)$ $n = 1.68$ **(b)** 16% of what is 208? $\downarrow\ \ \downarrow\ \ \ \downarrow\ \downarrow\ \downarrow$ $16\% \times\ \ n\ = 208$ $0.16n = 208$ $\dfrac{0.16n}{0.16} = \dfrac{208}{0.16}$ $n = 1300$ **(c)** What percent of 70 is 30? $\underbrace{}$ $\ \downarrow\ \downarrow\ \downarrow\ \downarrow$ $\quad\ n\qquad \times\ 70 = 30$ $70n = 30$ $\dfrac{70n}{70} = \dfrac{30}{70}$ $n = 0.428571\ldots$ $n \approx 42.86\%.$
Solving percent problems by using proportions, p. 336.	**1.** Identify the parts of the percent proportion. $a =$ the amount $b =$ the base (the whole; it usually appears after the word "of") $p =$ the percent number **2.** Write the percent proportion $\dfrac{a}{b} = \dfrac{p}{100}$ using the values obtained in step 1 and solve.	**(a)** What is 28% of 420? The percent $p = 28$. The base $b = 420$. The amount a is unknown. We use the variable a. $\dfrac{a}{b} = \dfrac{p}{100}$ becomes $\dfrac{a}{420} = \dfrac{28}{100}$ If we reduce the fraction on the right-hand side, we have $\dfrac{a}{420} = \dfrac{7}{25}$ $25a = (7)(420)$ $25a = 2940$ $\dfrac{25a}{25} = \dfrac{2940}{25}$ $a = 117.6$ Thus, 28% of 420 is 117.6. **(b)** 64% of what is 320? The percent $p = 64$. The base is unknown. We use the variable b. The amount $a = 320$. $\dfrac{a}{b} = \dfrac{p}{100}$ $\dfrac{320}{b} = \dfrac{64}{100}$ If we reduce the fraction on the right-hand side, we have $\dfrac{320}{b} = \dfrac{16}{25}$ $(320)(25) = 16b$ $8000 = 16b$ $\dfrac{8000}{16} = \dfrac{16b}{16}$ $500 = b$ Thus, 64% of 500 is 320.

(continues on next page)

Topic	Procedure	Examples
(continued) **Solving percent problems by using proportions, p. 337.**	**1.** Identify the parts of the percent proportion. a = the amount b = the base (the whole; it usually appears after the word "of") p = the percent number **2.** Write the percent proportion $\dfrac{a}{b} = \dfrac{p}{100}$ using the values obtained in step 1 and solve.	**(c)** What percent of 140 is 105? The percent is unknown. The base $b = 140$, the amount $a = 105$. $\dfrac{a}{b} = \dfrac{p}{100}$ becomes $\dfrac{105}{140} = \dfrac{p}{100}$ If we reduce the fraction on the left side, we have $$\dfrac{3}{4} = \dfrac{p}{100}$$ $$(100)(3) = 4p$$ $$300 = 4p$$ $$\dfrac{300}{4} = \dfrac{4p}{4}$$ $$75 = p$$ Thus 105 is 75% of 140.
Solving discount problems, p. 344.	Discount = discount rate × list price	Carla purchased a color TV set that lists for $350 at an 18% discount. **(a)** How much was the discount? **(b)** How much did she pay for the TV set? **(a)** Discount = $(0.18)(350)$ = \$63 **(b)** $350 - 63 = 287$ She paid $287 for the color TV set.
Solving commission problems, p. 348.	Commission = commission rate × value of sales	A housewares salesperson gets a 16% commission on sales he makes. How much commission does he earn if he sells $12,000 in housewares? $$\text{Commission} = (0.16)(12{,}000)$$ $$= \$1920$$
Solving simple interest problems, p. 349.	Interest = principal × rate × time $$I = P \times R \times T$$	Hector borrowed $3000 for 4 years at a simple interest rate of 12%. How much interest did he owe after 4 years? $$I = P \times R \times T$$ $$I = (3000)(0.12)(4)$$ $$= (360)(4)$$ $$= 1440$$ Hector owed $1440 in interest.
Percent-of-increase or percent-of-decrease problems, p. 348.	Percent of increase or decrease = amount of increase or decrease ÷ base	A car that costs $16,500 now cost only $15,000 last year. What is the percent of increase? $\begin{array}{r} 16{,}500 \\ -\,15{,}000 \\ \hline 1{,}500 \text{ increase} \end{array}$ $\dfrac{1500}{15{,}000} = 0.10$ Percent of increase = 10%

Chapter 5 Review Problems

5.1 *Write as a percent. Round to the nearest hundredth of a percent when necessary.*

1. 0.62　　**2.** 0.43　　**3.** 0.372　　**4.** 0.529　　**5.** 0.0828　　**6.** 0.0719

7. 2.52　　**8.** 4.37　　**9.** 1.036　　**10.** 1.052　　**11.** 0.006　　**12.** 0.002

13. $\dfrac{0.52}{100}$　　**14.** $\dfrac{0.67}{100}$　　**15.** $\dfrac{4\frac{1}{12}}{100}$　　**16.** $\dfrac{3\frac{5}{12}}{100}$　　**17.** $\dfrac{317}{100}$　　**18.** $\dfrac{225}{100}$

5.2 *Change to a percent. Round to the nearest hundredth of a percent when necessary.*

19. $\dfrac{19}{25}$　　**20.** $\dfrac{13}{25}$　　**21.** $\dfrac{11}{20}$　　**22.** $\dfrac{9}{40}$　　**23.** $\dfrac{5}{11}$　　**24.** $\dfrac{4}{9}$　　**25.** $2\frac{1}{4}$

26. $3\frac{3}{4}$　　**27.** $4\frac{3}{7}$　　**28.** $5\frac{5}{9}$　　**29.** $\dfrac{152}{80}$　　**30.** $\dfrac{165}{90}$　　**31.** $\dfrac{3}{800}$　　**32.** $\dfrac{5}{800}$

Change to decimal form.

33. 0.8%　　**34.** 0.6%　　**35.** 82.7%　　**36.** 59.6%

37. 236%　　**38.** 177%　　**39.** $32\frac{1}{8}$%　　**40.** $26\frac{3}{8}$%

Change to fractional form.

41. 72%　　**42.** 92%　　**43.** 185%　　**44.** 225%　　**45.** 16.4%

46. 30.5%　　**47.** $31\frac{1}{4}$%　　**48.** $43\frac{3}{4}$%　　**49.** 0.05%　　**50.** 0.06%

Complete the following chart.

	Fraction	Decimal	Percent
51.	$\dfrac{3}{5}$		
52.	$\dfrac{7}{8}$		
53.			37.5%
54.			56.25%
55.		0.008	
56.		0.45	

5.3 *Solve. Round to the nearest hundredth when necessary.*

57. What is 83% of 400?

58. What is 45% of 900?

59. 18 is 20% of what number?

60. 70 is 40% of what number?

61. 50 is what percent of 125?

62. 70 is what percent of 175?

63. Find 162% of 60.

64. Find 124% of 80.

65. 92% of what number is 147.2?

66. 68% of what number is 95.2?

67. What percent of 70 is 14?

68. What percent of 200 is 116?

5.4 and 5.5 *Solve. Round your answer to the nearest hundredth when necessary.*

69. Professor Wonson found that 34% of his class is left-handed. He has 150 students in his class. How many are left-handed?

70. A Vermont truck dealer found that 64% of all the trucks he sold had four-wheel drive. If he sold 150 trucks, how many had four-wheel drive?

71. Today Yvonne's car has 61% of the value that it had two years ago. Today it is worth $6832. What was it worth two years ago?

72. A charity organization spent 12% of its budget for administrative expenses. It spent $9624 on administrative expenses. What was the total budget?

73. In Seattle it rained 20 days in February, 18 days in March, and 16 days in April. What percent of those three months did it rain? (Assume it was a leap year.)

74. Moorehouse Industries received 600 applications and hired 45 of the applicants. What percent of the applicants obtained a job?

75. The average temperature in Winchester during the last 10 years has been 45.8°F. In the previous 10 years the average temperature was 44°F. What percent of increase is this?

76. Gary purchased a car for $18,600. The sales tax in his state is 3%. What did he pay in sales tax?

77. Joan and Michael budget 38% of their income for housing. They spend $684 per month for housing. What is their monthly income?

78. Beachfront property is very expensive. A real estate agent in South Carolina earned $26,000 in commissions. The property she sold was worth $650,000. What was her commission rate?

79. Adam sold encyclopedias last summer to raise tuition money. He sold $83,500 worth of encyclopedias and was paid $5010 in commissions. What commission rate did he earn?

80. Roberta earns a commission at the rate of 7.5%. Last month she sold $16,000 worth of goods. How much commission did she make last month?

81. Irene purchased a dining room set at a 20% discount. The list price was $1595.
  (a) What was the discount?
  (b) What did she pay for the set?

82. A Compaq laptop computer listed for $2125. This week, Lisa heard that the manufacturer is offering a rebate of 12%.
  (a) What is the rebate?
  (b) How much will Lisa pay for the computer?

83. Mark and Julie wanted to buy a prefabricated log cabin to put on their property in the Colorado Rockies. The price of the kit is listed at $24,000. At the after-holiday cabin sale, a discount of 14% was offered.
  (a) What was the discount?
  (b) How much did they pay for the cabin?

84. Sally invested $6000 in mutual funds earning 11% simple interest. How much interest will she earn in
  (a) six months?
  (b) two years?

85. Reed took out a college loan of $3000. He will be charged 8% simple interest on the loan.
  (a) How much interest will be due on the loan in three months?
  (b) How much interest will be due on the loan in three years?

Chapter 5 Test

1. _____

2. _____

3. _____

4. _____

5. _____

6. _____

7. _____

8. _____

9. _____

10. _____

11. _____

12. _____

13. _____

14. _____

15. _____

16. _____

Write as a percent. Round to the nearest hundredth of a percent when necessary.

1. 0.57

2. 0.01

3. 0.008

4. 0.139

5. 3.56

6. $\dfrac{71}{100}$

7. $\dfrac{1.8}{100}$

8. $\dfrac{3\frac{1}{7}}{100}$

Change to a percent. Round to the nearest hundredth of a percent when necessary.

9. $\dfrac{19}{40}$

10. $\dfrac{180}{450}$

11. $\dfrac{225}{75}$

12. $1\dfrac{3}{4}$

Write as a percent.

13. 0.1713

14. 3.024

Write as a fraction in simplified form.

15. 152%

16. $7\dfrac{3}{4}\%$

Solve. Round to the nearest hundredth if necessary.

17. What is 17% of 157?

18. 33.8 is 26% of what number?

19. What percent of 72 is 40?

20. Find 0.8% of 25,000.

21. 16% of what number is 800?

22. 92 is what percent of 200?

23. 132% of 530 is what number?

24. What percent is 8 of 350?

Solve. Round to the nearest hundredth if necessary.

25. A real estate agent sells a house for $152,300. She gets a commission of 4% on the sale. What is her commission?

26. Julia and Charles bought a new dishwasher at a 33% discount. The list price was $457.
   (a) What was the discount?
   (b) How much did they pay for the dishwasher?

27. An inspector found that 75 out of 84 parts were not defective. What percent of the parts were not defective?

28. At Cedars University a three-credit college course meets for 51 hours in a semester. Twenty years ago a three-credit course met for 56 hours. What is the percent of decrease?

29. A total of 5160 people voted in the city election. This was 43% of the registered voters. How many registered voters are in the city?

30. Wanda borrowed $3000 at a simple interest rate of 16%.
   (a) How much interest did she pay in six months?
   (b) How much interest did she pay in two years?

17. _____

18. _____

19. _____

20. _____

21. _____

22. _____

23. _____

24. _____

25. _____

26. (a) _____

(b) _____

27. _____

28. _____

29. _____

30. (a) _____

(b) _____

1. _____

2. _____

3. _____

4. _____

5. _____

6. _____

7. _____

8. _____

9. _____

10. _____

11. _____

12. _____

13. _____

14. _____

15. _____

16. _____

Approximately one-half of this test is based on Chapter 5 material. The remainder is based on material covered in Chapters 1–4.

Solve. Simplify your answer.

1. Add.   38
   196
   $+ 2007$

2. Subtract.   23,007
   $- 14,563$

3. Multiply.   126
   $\times\ \ 42$

4. Divide. $36\overline{)3204}$

5. Add. $2\dfrac{1}{4} + 3\dfrac{1}{3}$

6. Subtract. $\dfrac{11}{12} - \dfrac{5}{6}$

7. Multiply. $3\dfrac{17}{36} \times \dfrac{21}{25}$

8. Divide. $\dfrac{5}{12} \div 1\dfrac{3}{4}$

9. Round to the nearest tenth. 5731.652

10. Add.   5.6
   3.21
   18.3
   $+\ \ 7.008$

11. Multiply.   5.62
   $\times\ \ 0.3$

12. Divide. $1.4\overline{)0.5152}$

13. Write as a rate in simplest form. 78 pounds to 130 square feet

14. Is this equation a proportion? $\dfrac{20}{25} = \dfrac{300}{375}$

15. Solve the proportion. $\dfrac{8}{2.5} = \dfrac{n}{7.5}$

16. A college has a ratio of 3 faculty members for every 19 students. The student body presently has 4263 students. How many faculty members are there? Round to the nearest whole number.

In questions 17–30, round to the nearest hundredth when necessary.

Write as a percent.

17. 0.023

18. $\frac{46.8}{100}$

19. 1.98

20. $\frac{3}{80}$

In questions 21 and 22, write as a decimal.

21. 243%

22. $6\frac{3}{4}\%$

23. What percent of 214 is 38?

24. Find 1.7% of 6740.

25. 219 is 73% of what number?

26. 114% of 630 is what number?

27. Alice bought a new car. She got a 7% discount. The car listed for $9000. How much did she pay for the car?

28. A total of 896 freshmen were admitted to King Frederich College. Freshmen make up 28% of the student body. How big is the student body?

29. The air pollution level in Centerville is 8.86 parts per million. Ten years ago it was 7.96 parts per million. What is the percent of increase of the air pollution level?

30. Fred borrowed $1600 for two years. He was charged simple interest at a rate of 11%. How much interest did he pay?

17.	
18.	
19.	
20.	
21.	
22.	
23.	
24.	
25.	
26.	
27.	
28.	
29.	
30.	

Measurement

O n land, we use miles as our basic unit to measure long distances. However, at sea, distances are measured in nautical miles and speeds are measured in knots, or nautical miles per hour. Do you think you can adjust to these nautical terms and solve problems where measurements are made in nautical miles? Turn to the Putting Your Skills to Work problems on page 399 to find out.

Pretest Chapter 6

1. _____   17. _____

2. _____   18. _____

3. _____   19. _____

4. _____   20. _____

5. _____   21. _____

6. _____   22. _____

7. _____   23. _____

8. _____   24. _____

9. _____   25. _____

10. _____   26. _____

11. _____   27. _____

12. _____   28. (a) _____

13. _____   (b) _____

14. _____   29. _____

15. _____   30. _____

16. _____

If you are familiar with the topics in this chapter, take this test now. Check your answers with those in the back of the book. If an answer is wrong or you can't answer a question, study the appropriate section of the chapter.

If you are not familiar with the topics in this chapter, don't take this test now. Instead, study the examples, work the practice problems, and then take the test.

This test will help you identify which concepts you have mastered and which you need to study further.

Section 6.1

Convert. When necessary, express your answer as a decimal rounded to the nearest hundredth.

1. 19 ft = ___ in.   **2.** 5 gal = ___ pt   **3.** 3 mi = ___ yd

4. 3.2 tons = ___ lb   **5.** 22 min = ___ sec   **6.** 6 gal = ___ qt

Section 6.2

Perform each conversion.

7. 6.75 km = ___ m   **8.** 73.9 m = ___ cm   **9.** 986 mm = ___ cm

10. 27 mm = ___ m   **11.** 5296 mm = ___ cm   **12.** 482 m = ___ km

Convert to meters and add.

13. 1.2 km + 192 m + 984 m   **14.** 3862 cm + 9342 mm + 46.3 m

Section 6.3

Perform each conversion.

15. 5.66 L = ___ mL   **16.** 7835 g = ___ kg   **17.** 56.3 kg = ___ t

18. 4.8 kL = ___ L   **19.** 568 mg = ___ g   **20.** 8.9 L = ___ cm^3

Section 6.4

Perform each conversion. Round to the nearest hundredth when necessary.

21. 14 cm = ___ in.   **22.** 4.2 ft = ___ m   **23.** 96 km = ___ mi

24. 482 gal = ___ L   **25.** 1.4 oz = ___ g   **26.** 54 kg = ___ lb

Section 6.5

Solve. Round to the nearest hundredth when necessary.

▲ **27.** Find the perimeter of the triangle at right. Express your answer in **feet.**

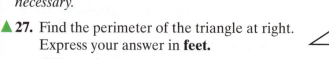

$7\frac{4}{5}$ yd $4\frac{1}{5}$ yd $6\frac{3}{5}$ yd

28. The radio reported the temperature today as 35°C. The record high temperature for this day is 98°F.
   (a) What was the Fahrenheit temperature today?
   (b) Did the temperature today set a new record?

29. Juanita traveled in Mexico for two hours at 95 kilometers per hour. She had to travel a distance of 130 miles. How far does she still need to travel? (Express your answer in **miles**.)

30. A pump is running at 1.5 quarts per minute. What is this rate in gallons per hour?

6.1 American Units

1 Identifying the Basic Unit Equivalencies in the American System

Student Learning Objectives

After studying this section, you will be able to:

1 Identify the basic unit equivalencies in the American system.

2 Convert from one unit of measure to another.

SSM

PH TUTOR CD & VIDEO MATH PRO WEB
CENTER

We often ask questions about measurements. How far is it to work? How much does this bottle hold? What is the weight of this box? How long will it be until exams? To answer these questions we need to agree on a unit of measure for each type of measurement.

At present there are two main systems of measurement, each with its own set of units: **the metric system** and the **American system**. Nearly all countries in the world use the metric system. In the United States, however, except in science, most measurements are made in American units.

The United States is using the metric system more and more frequently, however, and may eventually convert to metric units as the standard. But for now we need to be familiar with both systems. We cover American units in this section.

One of the most familiar measuring devices is a ruler that measures lengths as great as 1 foot. It is divided into 12 inches; that is,

1 foot = 12 inches.

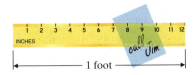

There are several other important relationships you need to know. Your instructor may require you to memorize the following facts.

Length
12 inches = 1 foot
3 feet = 1 yard
5280 feet = 1 mile
1760 yards = 1 mile

Time
60 seconds = 1 minute
60 minutes = 1 hour
24 hours = 1 day
7 days = 1 week

Note that time is measured in the same units in both the metric and the American systems.

Weight
16 ounces = 1 pound
2000 pounds = 1 ton

Volume
8 fluid ounces = 1 cup
2 cups = 1 pint
2 pints = 1 quart
4 quarts = 1 gallon

We can choose to measure an object—say, a bridge—using a small unit (an inch), a larger unit (a foot), or a still larger unit (a mile). We may say that the bridge spans 7920 inches, 660 feet, or an eighth of a mile. Although we probably would not choose to express our measurement in inches because this is not a convenient measurement to work with for an object as long as a bridge, the bridge length is the same whatever unit of measurement we use.

Notice that the smaller the measuring unit, the larger the number of those units in the final measurement. The inch is the smallest unit in our example, and the inch measurement has the greatest number of units (7920). The mile is the largest unit, and it has the smallest number of units (an eighth equals 0.125). Whatever measuring system you use, and whatever you measure (length, volume, and so on), the smaller the unit of measurement you use, the greater the number of those units.

After studying the values in the length, time, weight, and volume tables, see if you can quickly do Example 1.

EXAMPLE 1 Answer rapidly the following questions.

(a) How many inches in a foot?   **(b)** How many yards in a mile?
(c) How many seconds in a minute?   **(d)** How many hours in a day?
(e) How many pounds in a ton?   **(f)** How many cups in a pint?

(a) 12   **(b)** 1760   **(c)** 60   **(d)** 24   **(e)** 2000   **(f)** 2

Practice Problem 1 Answer rapidly the following questions.

(a) How many feet in a yard?   **(b)** How many feet in a mile?
(c) How many minutes in an hour?   **(d)** How many days in a week?
(e) How many ounces in a pound?   **(f)** How many pints in a quart?
(g) How many quarts in a gallon?

2 **Converting from One Unit of Measure to Another**

To convert or change one measurement to another, we simply multiply by 1 since multiplying by 1 does not change the value of a quantity. For example, to convert 180 inches to feet, we look for a name for 1 that has inches and feet.

$$1 = \frac{1 \text{ foot}}{12 \text{ inches}}$$

A ratio of measurements for which the measurement in the numerator is equivalent to the measurement in the denominator is called a **unit fraction**. We now use the unit fraction $\dfrac{1 \text{ foot}}{12 \text{ inches}}$ to convert 180 inches to feet.

$$180 \text{ inches} \times \frac{1 \text{ foot}}{12 \text{ inches}} = \frac{180 \text{ feet}}{12} = 15 \text{ feet}$$

Notice that when we multiplied, the inches divided out. We are left with the unit feet.

What name for 1 (that is, unit fraction) should we choose if we want to change from feet to inches? Convert 4 feet to inches.

$$1 = \frac{12 \text{ inches}}{1 \text{ foot}}$$

$$4 \text{ feet} \times \frac{12 \text{ inches}}{1 \text{ foot}} = \frac{48 \text{ inches}}{1} = 48 \text{ inches}$$

> When multiplying by a unit fraction, the unit we want to change to should be in the *numerator*. The unit we start with should be in the *denominator*. This unit will divide out.

EXAMPLE 2 Convert.   8800 yards to miles

$$8800 \text{ yards} \times \frac{1 \text{ mile}}{1760 \text{ yards}} = \frac{8800}{1760} \text{ miles} = 5 \text{ miles}$$

Practice Problem 2 Convert.   15,840 feet to miles

Some conversions involve fractions or decimals. Whether you want to measure the area of a living room or the dimensions of a piece of property, it is helpful to be able to make conversions like these.

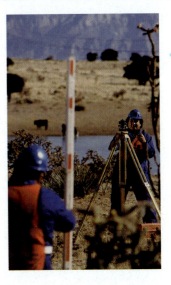

● **EXAMPLE 3**   Convert.   **(a)** 26.48 miles to yards   **(b)** $3\frac{2}{3}$ feet to yards

(a) $26.48 \text{ miles} \times \dfrac{1760 \text{ yards}}{1 \text{ mile}} = 46{,}604.8 \text{ yards}$

(b) $3\frac{2}{3} \text{ feet} \times \dfrac{1 \text{ yard}}{3 \text{ feet}} = \dfrac{11}{3} \times \dfrac{1}{3} \text{ yard} = \dfrac{11}{9} \text{ yards} = 1\frac{2}{9} \text{ yards}$

Practice Problem 3   Convert.   **(a)** 18.93 miles to feet

(b) $16\frac{1}{2}$ inches to yards

● **EXAMPLE 4**   Lynda's new car weighs 2.43 tons. How many pounds is that?

$$2.43 \text{ tons} \times \dfrac{2000 \text{ pounds}}{1 \text{ ton}} = 4860 \text{ pounds}$$

Practice Problem 4   A package weighs 760.5 pounds. How many ounces does it weigh?

● **EXAMPLE 5**   The chemistry lab has 34 quarts of weak hydrochloric acid. How many gallons of this acid are in the lab? (Express your answer as a decimal.)

$$34 \text{ quarts} \times \dfrac{1 \text{ gallon}}{4 \text{ quarts}} = \dfrac{34}{4} \text{ gallons} = 8.5 \text{ gallons}$$

Practice Problem 5   19 pints of milk is the same as how many quarts? (Express your answer as a decimal.)

● **EXAMPLE 6**   A window is 4 feet 5 inches wide. How many inches is that?
4 feet 5 inches means 4 feet and 5 inches. Change the 4 feet to inches and add the 5 inches.

$$4 \text{ feet} \times \dfrac{12 \text{ inches}}{1 \text{ foot}} = 48 \text{ inches}$$

48 inches + 5 inches = 53 inches   The window is 53 inches wide.

Practice Problem 6   A bag of potatoes weighs 7 pounds 12 ounces. How many ounces is that?

● **EXAMPLE 7**   The Charlotte all-night garage charges $1.50 per hour for parking both day and night. A businessman left his car there for $2\frac{1}{4}$ days. How much was he charged?

1. *Understand the problem.*
   Here it might help to look at a simpler problem. If the businessman had left his car for two hours, we would multiply.

The fraction bar means "per." $\rightarrow$ $\dfrac{1.50 \text{ dollars}}{1 \text{ hour}} \times 2 \text{ hours} = 3.00 \text{ dollars or } \$3.00$

Thus, if the businessman had left his car for two hours, he would have been charged $3. We see that we need to multiply $1.50 by the number of hours the car was in the garage to solve the problem.

Since the original problem gave the time in days, not hours, we will need to change the days to hours.

2. Solve and state the answer.
Now that we know that the way to solve the problem is to multiply by hours, we will begin. To make our calculations easier we will write $2\frac{1}{4}$ as 2.25. Change days to hours. Then multiply by $1.50 per hour.

$$2.25 \; \cancel{\text{days}} \times \frac{24 \; \cancel{\text{hours}}}{1 \; \cancel{\text{day}}} \times \frac{1.50 \text{ dollars}}{1 \; \cancel{\text{hour}}} = 81 \text{ dollars or } \$81$$

The businessman was charged $81.

3. Check.
Is our answer in the desired unit? Yes. The answer is in dollars and we would expect it to be in dollars. ✓
The check is up to you.

Practice Problem 7 A businesswoman parked her car at a garage for $1\frac{3}{4}$ days. The garage charges $1.50 per hour. How much did she pay to park the car?

Alternative Method Using Proportions

How did people first come up with the idea of multiplying by a unit fraction? What mathematical principles are involved here? Actually, this is the same as solving a proportion. Consider Example 5, where we changed 34 quarts to 8.5 gallons by multiplying.

$$34 \; \cancel{\text{quarts}} \times \frac{1 \text{ gallon}}{4 \; \cancel{\text{quarts}}} = \frac{34}{4} \text{ gallons} = 8.5 \text{ gallons}$$

What we were actually doing is setting up the proportion:

1 gallon is to 4 quarts as n gallons is to 34 quarts

$$\frac{1 \text{ gallon}}{4 \text{ quarts}} = \frac{n \text{ gallons}}{34 \text{ quarts}}$$

$1 \text{ gallon} \times 34 \text{ quarts} = 4 \text{ quarts} \times n \text{ gallons}$	Cross multiplying
$\dfrac{1 \text{ gallon} \times 34 \; \cancel{\text{quarts}}}{4 \; \cancel{\text{quarts}}} = \dfrac{\cancel{4} \; \cancel{\text{quarts}} \times n \text{ gallons}}{\cancel{4} \; \cancel{\text{quarts}}}$	Dividing both sides of the equation by 4 quarts
$1 \text{ gallon} \times \dfrac{34}{4} = n \text{ gallons}$	Simplifying
$8.5 \text{ gallons} = n \text{ gallons}$	

Thus the number of gallons is 8.5. Using proportions takes a little longer, so multiplying by a fractional name for 1 is the more popular method.

Verbal and Writing Skills

1. Explain in your own words how you would use a unit fraction to change 23 miles to inches.

2. Explain in your own words how you would use a unit fraction to change 27 days to minutes.

From memory, write the equivalent value

3. _____ yards = 1 mile

4. _____ feet = 1 mile

5. 1 ton = _____ pounds

6. 1 pound = _____ ounces

7. _____ quarts = 1 gallon

8. _____ cups = 1 pint

9. 1 quart = _____ pints

10. _____ minutes = 1 hour

Convert. When necessary, express your answer as a decimal.

11. 21 feet = _____ yards

12. 63 feet = _____ yards

13. 108 inches = _____ feet

14. 180 inches = _____ feet

15. 10,560 feet = _____ miles

16. 5280 yards = _____ miles

17. 7 miles = _____ yards

18. 21,120 feet = _____ miles

19. 12 feet = _____ inches

20. 16 feet = _____ inches

21. 16 cups = _____ fluid ounces

22. 40 fluid ounces = _____ cups

23. 75 inches = _____ feet

24. 87 inches = _____ feet

25. 192 ounces = _____ pounds

26. 144 ounces = _____ pounds

27. 12,000 pounds = _____ tons

28. 17 tons = _____ pounds

29. 2.25 pounds = _____ ounces

30. 4.25 pounds = _____ ounces

31. 7 gallons = _____ quarts

32. 5 gallons = _____ quarts

33. 48 quarts = _____ gallons

34. 24 pints = _____ quarts

35. 31 pints = _____ cups

36. 38 cups = _____ pints

37. 8 gallons = _____ pints

38. 6 gallons = _____ pints

39. 12 weeks = _____ days

40. 7 weeks = _____ days

41. 18 hours = _____ seconds

42. 15 hours = _____ seconds

Applications

43. Judy is making a wild mushroom sauce for pasta tonight with a large group of friends. She bought 26 ounces of wild mushrooms at $6.00 per pound. How much were the mushrooms?

44. Kurt is trying to eat a low-fat diet. He finds a store that sells 1% fat ground white-meat turkey breast at $6.00 per pound. He buys one packet weighing 18 ounces and another weighing 22 ounces. How much does he pay?

45. Candace Cable is a champion wheelchair racer. In Grandma's Marathon, she covered the 26.2-mile course in 1:46 (1 hour and 46 minutes). How many feet did she travel in the race?

46. Mount Whitney in California is approximately 2.745 miles high. How many feet is that? Round to the nearest 10 feet.

47. Jane ran a 10-kilometer race in 44:25 (44 minutes and 25 seconds). How many seconds did the race take her?

48. Curtis threw the shot put $39\frac{1}{2}$ feet at the high school track meet. How many inches is that?

▲ **49.** A window in Brianna and Nelson's house needs to be replaced. The rectangular window is 2 feet 3 inches wide and 3 feet 9 inches tall. Change each of the measurements to inches.
(a) Find the perimeter of the window in inches.

(b) If the perimeter needs to be sealed with insulation that is $0.85 per inch, how much will it cost to insulate the perimeter?

▲ **50.** The cellar in Jeff and Shelley's house needs to be sealed along the edge of the concrete floor. The rectangular floor measures 7 yards 2 feet wide and 12 yards 1 foot long. Change each of the measurements to feet.
(a) Find the perimeter of the cellar in feet.

(b) If the perimeter (edge) of the cellar floor needs to be sealed with waterproof sealer that costs $1.75 per foot, how much will it cost to seal the perimeter?

51. Every day, your heart pumps 7200 quarts of blood through your body. How many cups is that?

52. A seedling peach tree grew for seven years until it produced its first fruit. How many hours was that if you assume that there are 365 days in a year?

To Think About

There are approximately 6080 feet in a nautical mile and 5280 feet in a land mile. On sea and in the air, distance and speed are often measured in nautical miles. The ratio of land miles to nautical miles is approximately 38 to 33.

53. A windjammer sailboat was used as a floating school. During one school year, the ship traveled 12,800 nautical miles. What would be the equivalent in land miles? Round to the nearest whole mile.

54. A merchant ship took some tourists on board for extra income. The ship carried them the equivalent of 850 land miles. How many nautical miles did they travel? Round to the nearest whole mile.

Cumulative Review Problems

55. Melinda and Robert are refinancing their house. They were making payments of $560 per month for the next 20 years. They obtained a new mortgage for which their payments will be $515 per month for the next 20 years. Starting with the date they began payments on the new mortgage, how much will they save with the new monthly payments during this entire period?

56. Michael has a computer disk that stores 650 MB of data. He stored three programs that each require 123 MB of storage space. He then stored two programs that each require 69 MB of storage space. What percent of his computer disk is still free for storage of future programs?

57. Johnny and Phil took a bicycle trip covering 115 miles in five days last summer. This summer they hope to take a similar bicycle trip lasting seven days. How many miles are they likely to cover?

58. The dean observed that 150 students enrolled in mathematics courses at the University of Hawaii during the last summer semester and that 12 of them dropped the course during the first two weeks. If this ratio has remained consistent over the past seven summers and if 1300 students enrolled in the summer mathematics courses during this time, how many of them dropped the course during the first two weeks of the semester?

6.2 Metric Measurements: Length

Student Learning Objectives

After studying this section, you will be able to:

 1 Understand prefixes in metric units.

 2 Convert from one metric unit of length to another.

    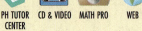

SSM | PH TUTOR CENTER | CD & VIDEO | MATH PRO | WEB

---

1 **Understanding Prefixes in Metric Units**

The **metric system** of measurement is used in most industrialized nations of the world. As we move toward a global economy, it is important to be familiar with the metric system. The metric system is designed for ease in calculating and in converting from one unit to another.

In the metric system, the basic unit of length measurement is the **meter**. A meter is just slightly longer than a yard. To be more precise, the meter is approximately 39.37 inches long.

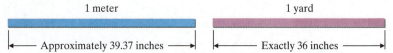

1 meter	1 yard
Approximately 39.37 inches	Exactly 36 inches

Units that are larger or smaller than the meter are based on the meter and powers of 10. For example, the unit *deka*meter is *ten* meters. The unit *deci*meter is *one-tenth* of a meter. The prefix *deka* means 10. What does the prefix *deci* mean? All the prefixes in the metric system are names for multiples of 10. A list of metric prefixes and their meanings follows.

Prefix	Meaning
kilo-	thousand
hecto-	hundred
deka-	ten
deci-	tenth
centi-	hundredth
milli-	thousandth

The most commonly used prefixes are *kilo-*, *centi-*, and *milli-*. *Kilo-* means thousand, so a *kilo*meter is a thousand meters. Similarly, *centi-* means one hundredth, so a *centi*meter is one hundredth of a meter. And *milli-* means one thousandth, so a *milli*meter is one thousandth of a meter.

The kilometer is used to measure distances much larger than the meter. How far did you travel in a car? The centimeter is used to measure shorter lengths. What are the dimensions of this textbook? The millimeter is used to measure very small lengths. What is the width of the lead in a pencil?

Amiens
Le Havre
120 km
180 km
Reims
130 km
Paris
200 km
110 km
Le Mans   Orléans   250 km
France   Dijon

 EXAMPLE 1 Write the prefixes that mean **(a)** thousand and **(b)** tenth.

(a) The prefix *kilo-* is used for thousand.

(b) The prefix *deci-* is used for tenth.

Practice Problem 1 Write the prefixes that mean **(a)** ten and **(b)** thousandth.

---

2 **Converting from One Metric Unit of Length to Another**

How do we convert from one metric unit to another? For example, how do we change 5 kilometers into an equivalent number of meters?

Recall from Chapter 3 that when we multiply by 10 we move the decimal point one place to the right. When we divide by 10 we move the decimal point one place to the left. Let's see how we use this idea to change from one metric unit to another.

Changing from Larger Metric Units to Smaller Ones

When you change from one metric prefix to another by moving to the **right** on this prefix chart, move the decimal point to the **right** the same number of places.

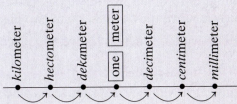

Thus 1 meter = 100 centimeters because we move two places to the right on the chart of prefixes and we also move the decimal point (1.00) two places to the right.

Now let us examine the four most commonly used metric measurements of length and their abbreviations.

Commonly Used Metric Lengths

1 kilometer (km) = 1000 meters
1 meter (m) (the basic unit of length in the metric system)
1 centimeter (cm) = 0.01 meter
1 millimeter (mm) = 0.001 meter

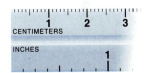

Now let us see how we can change a measurement stated in a larger unit to an equivalent measurement stated in smaller units.

⬤ EXAMPLE 2

(a) Change 5 kilometers to meters. **(b)** Change 20 meters to centimeters.

(a) To go from *kilometer* to meter, we move three places to the right on the prefix chart. So we move the decimal point three places to the right.

5 kilometers = 5.000 meters (move three places) = 5000 meters

(b) To go from meter to *centimeter*, we move two places to the right on the prefix chart. Thus we move the decimal point two places to the right.

20 meters = 20.00 centimeters (move two places) = 2000 centimeters

Practice Problem 2

(a) Change 4 meters to centimeters.
(b) Change 30 centimeters to millimeters. ⬤

Changing from Smaller Metric Units to Larger Ones

When you change from one metric prefix to another by moving to the **left** on this prefix chart, move the decimal point to the **left** the same number of places.

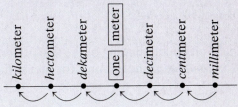

● EXAMPLE 3

(a) Change 163 centimeters to meters.

(b) Change 56 millimeters to kilometers.

(a) To go from *centi*meter to meter, we move *two* places to the left on the prefix chart. Thus we move the decimal point two places to the left.

$$163 \text{ centimeters} = 1.63. \quad \text{meters (move two places to the left)}$$

$$= 1.63 \text{ meters}$$

(b) To go from *milli*meter to *kilo*meter, we move six places to the left on the prefix chart. Thus we move the decimal point six places to the left.

$$56 \text{ millimeters} = 0.000056. \text{ kilometer} \quad \text{(move six places to the left)}$$

$$= 0.000056 \text{ kilometer}$$

Practice Problem 3

(a) Change 3 *milli*meters to meters.

(b) Change 47 *centi*meters to *kilo*meters.

Thinking Metric

Long distances are customarily measured in kilometers. A kilometer is about 0.62 mile. It takes about 1.6 kilometers to make a mile. The following drawing shows the relationship between the kilometer and the mile.

> 1 kilometer
>
> 1 mile (about 1.6 kilometers)

Many small distances are measured in centimeters. A centimeter is about 0.394 inch. It takes 2.54 centimeters to make an inch. You can get a good idea of their size by looking at a ruler marked in both inches and centimeters.

> 1 centimeter
>
> 1 inch   (2.54 centimeters)

Try to visualize how many centimeters long or wide this book is. It is 21 centimeters wide and 27.5 centimeters long.

A millimeter is a very small unit of measurement, often used in manufacturing.

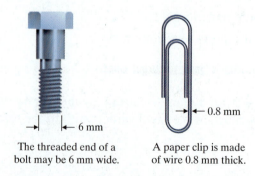

←‖←– 6 mm	→‖←– 0.8 mm
The threaded end of a bolt may be 6 mm wide.	A paper clip is made of wire 0.8 mm thick.

Now that you have an understanding of the size of these metric units, let's try to select the most convenient metric unit for measuring the length of an object.

EXAMPLE 4   Bob measured the width of a doorway in his house. He wrote down "73." What unit of measurement did he use?

(a) 73 kilometers   **(b)** 73 meters   **(c)** 73 centimeters

The most reasonable choice is **(c)**, 73 centimeters. The other two units would be much too long. A meter is close to a yard, and a doorway would not be 73 yards wide! A kilometer is even larger than a meter.

Practice Problem 4   Joan measured the length of her car. She wrote down 3.8. Which unit of measurement did she use?

(a) 3.8 kilometers   **(b)** 3.8 meters   **(c)** 3.8 centimeters

The most frequent metric conversions in length are done between kilometers, meters, centimeters, and millimeters. Their abbreviations are km, m, cm, and mm, and you should be able to use them correctly.

EXAMPLE 5   Convert.   **(a)** 982 cm to m   **(b)** 5.2 m to mm

In the first case, we move the decimal point to the left because we are going from a smaller unit to a larger unit.

(a) 982 cm = 9.82 m   (two places to left)
            = 9.82 m

In the last case, we need to move the decimal point to the right because we are going from a larger unit to a smaller unit.

(b) 5.2 m = 5200. mm   (three places to left)
            = 5200 mm

Practice Problem 5   Convert.   **(a)** 375 cm to m   **(b)** 46 m to mm

The other metric units of length are the hectometer, the dekameter, and the decimeter. These are not used very frequently, but it is good to understand how their lengths relate to the basic unit, the meter. A complete list of the metric lengths we have discussed appears in the following table.

Metric Lengths with Abbreviations

1 kilometer (km) = 1000 meters
1 hectometer (hm) = 100 meters
1 dekameter (dam) = 10 meters
1 meter (m)
1 decimeter (dm) = 0.1 meter
1 centimeter (cm) = 0.01 meter
1 millimeter (mm) = 0.001 meter

EXAMPLE 6 Convert.

(a) 426 decimeters to kilometers   **(b)** 9.47 hectometers to meters

(a) We are converting from a smaller unit, dm, to a larger one, km. Therefore, there will be fewer kilometers than decimeters. (The number we get will be smaller than 426.) We move the decimal point four places to the left.

$$426 \text{ dm} = 0.0426. \text{ km} \quad \text{(four places to left)}$$
$$= 0.0426 \text{ km}$$

(b) We are converting from a larger unit, hm, to a smaller one, m. Therefore, there will be more meters than hectometers. (The number will be larger than 9.47.) We move the decimal point two places to the right.

$$9.47 \text{ hm} = 9.47 \text{ m} \quad \text{(two places to right)}$$
$$= 947 \text{ m}$$

Practice Problem 6 Convert.

(a) 389 millimeters to dekameters   **(b)** 0.48 hectometer to centimeters

When several metric measurements are to be added, we change them to a convenient common unit.

EXAMPLE 7 Add. 125 m + 1.8 km + 793 m

First we change the kilometer measurement to a measurement in meters.

$$1.8 \text{ km} = 1800 \text{ m}$$

Then we add.

$$
\begin{array}{r}
125 \text{ m} \\
1800 \text{ m} \\
+ \quad 793 \text{ m} \\
\hline
2718 \text{ m}
\end{array}
$$

Practice Problem 7 Add. 782 cm + 2 m + 537 m

Sidelight: Extremely Large Metric Distances

Is the biggest length in the metric system a kilometer? Is the smallest length a millimeter? No. The system extends to very large units and very small ones. Usually, only scientists use these units.

1 gigameter = 1,000,000,000 meters
1 megameter = 1,000,000 meters
1 kilometer = 1000 meters
1 meter
1 millimeter = 0.001 meter
1 micrometer = 0.000001 meter
1 nanometer = 0.000000001 meter

For example, 26 megameters equals a length of 26,000,000 meters. A length of 31 micrometers equals a length of 0.000031 meter.

Sidelight: Metric Measurements for Computers

A **byte** is the amount of computer memory needed to store one alphanumeric character. When referring to computers you may hear the following words: **kilobytes, megabytes,** and **gigabytes**. The following chart may help you.

> 1 gigabyte (GB) = one billion bytes = 1,000,000,000 bytes
> 1 megabyte (MB) = one million bytes* = 1,000,000 bytes
> 1 kilobyte (KB) = one thousand bytes† = 1,000 bytes

A Challenge for You   Before we move on to the exercises, see if you can use the large metric distance and use computer memory size to convert the following measurements.

1. 18 megameters = _____ kilometers

2. 26 millimeters = _____ micrometers

3. 17 nanometers = _____ millimeter

4. 38 meters = _____ megameter

5. 1.2 gigabytes = _____ bytes

6. 528 megabytes = _____ bytes

7. 78.9 kilobytes = _____ bytes

8. 24.9 gigabytes = _____ bytes

---

*Sometimes in computer science 1 megabyte is considered to be 1,048,576 bytes.
†Sometimes in computer science 1 kilobyte is considered to be 1024 bytes.

Verbal and Writing Skills

Write the prefix for each of the following.

1. Hundred

2. Hundredth

3. Tenth

4. Thousandth

5. Thousand

6. Ten

The following conversions involve metric units that are very commonly used. You should be able to perform each conversion without any notes and without consulting your text.

7. 46 centimeters = _____ millimeters

8. 79 centimeters = _____ millimeters

9. 5.2 kilometers = _____ meters

10. 8.3 kilometers = _____ meters

11. 1670 millimeters = _____ meters

12. 2540 millimeters = _____ meters

13. 7.32 cm = _____ meters

14. 9.14 cm = _____ meters

15. 2 kilometers = _____ centimeters

16. 7 kilometers = _____ centimeters

17. 78,000 millimeters = _____ kilometer

18. 96,000 millimeters = _____ kilometer

Abbreviations are used in the following. Fill in the blanks with the correct values.

19. 35 mm = _____ cm = _____ m

20. 83 mm = _____ cm = _____ m

21. 3582 mm = _____ m = _____ km

22. 7812 mm = _____ m = _____ km

Applications

23. Peter measured the length of his dining room table. Choose the most reasonable measurement.
(a) 2 m   **(b)** 2 km   **(c)** 2 cm

24. Wally measured the length of his college identification card. Choose the most reasonable measurement.
(a) 10 km   **(b)** 10 m   **(c)** 10 cm

25. Eddie measured the width of a compact disc case. Choose the most appropriate measurement.
(a) 80.5 km   **(b)** 80.5 m   **(c)** 80.5 mm

26. The Botanical Gardens are located in the center of the city. If it takes approximately 10 minutes to walk directly from the east entrance to the west entrance, which would be the most likely measurement of the width of the Gardens?
(a) 1.4 km   **(b)** 1.4 m   **(c)** 1.4 mm

27. Joan measured her math book so she could buy the right size book cover. Choose the most appropriate measurement for the length.
(a) 32 mm   **(b)** 32 cm   **(c)** 32 km

28. Miles ran in a road race to help raise money for a local charity. Which would be the most likely measurement of the length of the race?
(a) 8 cm   **(b)** 8 m   **(c)** 8 km

The following conversions involve metric units that are not used extensively. You should be able to perform each conversion, but it is not necessary to do it from memory.

29. 390 decimeters = _____ meters

30. 270 decimeters = _____ meters

31. 198 millimeters = _____ decimeters

32. 2.93 decimeters = _____ hectometer

33. 48.2 meters = _____ hectometer

34. 435 hectometers = _____ kilometers

Change to a convenient unit of measure and add.

35. 243 m + 2.7 km + 312 m

36. 845 m + 5.79 km + 701 m

37. 5.2 cm + 361 cm + 968 mm

38. 4.8 cm + 607 cm + 834 mm

39. 15 mm + 2 dm + 42 cm

40. 8 dm + 21 mm + 38 cm

41. The outside casing of a stereo cabinet is built of plywood 0.95 centimeter thick attached to plastic 1.35 centimeters thick and a piece of mahogany veneer 2.464 millimeters thick. How thick is the stereo casing?

42. A plywood board is 2.2 centimeters thick. A layer of tar paper is 3.42 millimeters thick. A layer of false brick siding is 2.7 centimeters thick. A house wall consists of these three layers. How thick is the wall?

Mixed Practice

43. 4.8 cm + 607 cm + 834 mm

44. 82 m + 471 cm + 0.32 km

45. 46 m + 986 cm + 0.884 km

46. 56.3 centimeters = _____ meter

47. 96.4 centimeters = _____ meter

Write true *or* false *for each statement.*

48. 1 meter = 0.01 centimeter

49. 1 kilometer = 0.001 meter

50. 1000 meters = 1 kilometer

51. 10 millimeters = 1 centimeter

52. An airport runway might be 2 kilometers long.

53. A man might be 2 meters tall.

54. A centimeter is larger than an inch.

55. The glass in a drinking glass is usually about 2 millimeters thick.

Applications

56. The world's longest train run is on the Trans-Siberian line in Russia, from Moscow to Nakhodka on the Sea of Japan. The length of the run, which has 97 stops, measures 94,380,000 centimeters. The run takes 8 days, 4 hours, and 25 minutes to travel.
 (a) How many meters is the run?

 (b) How many kilometers is the run?

57. The highest railroad line in the world is a track on the Morococha branch of the Peruvian State Railways at La Cima. The track is 4818 meters high.
 (a) How many centimeters high is the track?

 (b) How many kilometers high is the track?

58. The world's highest dam, the Rogun dam in Tajikistan, is 335 meters high.
(a) How many kilometers high is the dam?

(b) How many centimeters high is the dam?

59. A typical virus of the human body measures just 0.000000254 centimeter in diameter. How many meters in diameter is a typical virus?

To Think About

In countries outside the United States, very heavy items are measured in terms of **metric tons**. *A metric ton is defined as 1000 kilograms (or 2205 pounds). Consider the following bar graph, which records the catch of fish (measured in millions of metric tons) by China for various years.*

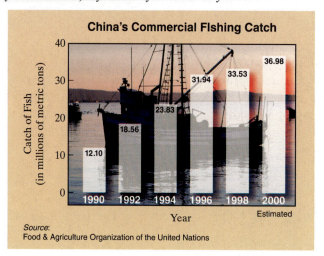

China's Commercial Fishing Catch

Catch of Fish (in millions of metric tons)

12.10 — 1990
18.56 — 1992
23.83 — 1994
31.94 — 1996
33.53 — 1998
36.98 — 2000 (Estimated)

Year

Source:
Food & Agriculture Organization of the United Nations

60. By how many metric tons did the commercial catch of fish by China increase from 1990 to 1994?

61. By how many metric tons did the commercial catch of fish by China increase from 1996 to 2000?

62. How many kilograms of fish were caught by China in 1998?

63. How many kilograms of fish were caught by China in 1992?

64. If the same *percent of increase* that occurred from 1994 to 2000 also occurs from 2000 to 2006, how many metric tons of fish will China catch in 2006?

65. If the same *percent of increase* that occurred from 1990 to 2000 also occurs from 2000 to 2010, how many metric tons of fish will China catch in 2010?

Cumulative Review Problems

66. 57% of what number is 2850?

67. Find 0.03% of 5900.

6.3 Metric Measurements: Volume and Weight

1 Converting between Metric Units of Volume

As products are distributed worldwide, more and more of them are being sold in metric units. Soft drinks come in 1-, 2-, or 3-liter bottles. Labels often contain these amounts in both American units and metric units. Get used to looking at these labels to gain a sense of the size of metric units of volume.

The basic metric unit for volume is the liter. A **liter** is defined as the volume of a box 10 cm × 10 cm × 10 cm, or 1000 cm³. A cubic centimeter may be written as cc, so we sometimes see 1000 cc = 1 liter. A liter is slightly larger than a quart; 1 liter of liquid is 1.057 quarts of that liquid.

The most common metric units of volume are the milliliter, the liter, and the kiloliter. Often a capital letter L is used as an abbreviation for *liter*.

> **Common Metric Volume Measurements**
>
> 1 kiloliter (kL) = 1000 liters
>
> 1 liter (L)
>
> 1 milliliter (mL) = 0.001 liter

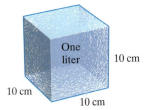

One liter — 10 cm × 10 cm × 10 cm

We know that 1000 cc = 1 liter. Dividing each side of that equation by 1000, we get 1 cc = 1 mL.

> Use this prefix chart as a guide when you change one metric prefix to another. Move the decimal point in the same direction and the same number of places.
>
>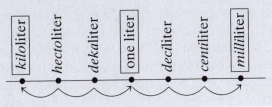
>
> kiloliter   hectoliter   dekaliter   one liter   deciliter   centiliter   milliliter

1 liter   2 liters

The prefixes for liter follow the pattern we have seen for meter. Note that *kilo-* is three places to the left of the liter and *milli-* is three places to the right. Because the kiloliter, the liter, and the milliliter are the most commonly used units of volume, we will focus exclusively on them in this text.

● EXAMPLE 1   Convert.

(a) 3 L = _____ mL   **(b)** 24 kL = _____ L   **(c)** 0.084 L = _____ mL

(a) The prefix *milli-* is three places to the right. We move the decimal point three places to the right.

$$3 \text{ L} = 3.000 = 3000 \text{ mL}$$

(b) $24 \text{ kL} = 24.000 = 24{,}000 \text{ L}$

(c) $0.084 \text{ L} = 0.084 = 84 \text{ mL}$

Practice Problem 1   Convert.

(a) 5 L = _____ mL   **(b)** 84 kL = _____ L   **(c)** 0.732 L = _____ mL

EXAMPLE 2 Convert.

(a) 26.4 mL = _____ L   **(b)** 5982 mL = _____ L   **(c)** 6.7 L = _____ kL

(a) The unit L is three places to the left. We move the decimal three places to the left. 26.4 mL = 0.0264 L

(b) 5982 mL = 5.982 L   **(c)** 6.7 L = 0.0067 kL

Practice Problem 2 Convert.

(a) 15.8 mL = _____ L   **(b)** 12,340 mL = _____ L   **(c)** 86.3 L = _____ kL

The cubic centimeter is often used in medicine. Recall 1 mL = 1 cm^3 or 1 cc.

EXAMPLE 3 Convert.

(a) 26 mL = _____ cm^3 **(b)** 0.82 L =_____ cm^3

(a) A milliliter and a cubic centimeter are equivalent. 26 mL = 26 cm^3

(b) We use the same rule to convert liters to cubic centimeters as we do to convert liters to milliliters. 0.82 L = 820 cm^3

Practice Problem 3 Convert.

(a) 396 mL = _____ cm^3 **(b)** 0.096 L = _____ cm^3

2 Converting Between Metric Units of Weight

One Gram of Water

1 cm
1 cm
1 cm

In a science class we make a distinction between weight and **mass**. Mass is the amount of material in an object. Weight is a measure of the pull of gravity on an object. The farther you are from the center of the earth, the less you weigh. If you were in an astronaut suit floating in outer space you would be weightless. The mass of your body, however, would not change. Throughout this book we will refer to *weight* only. The technical difference between weight and mass is not something we need to pay attention to in everyday life.

In the metric system the basic unit of weight is the gram. A **gram** is the weight of the water in a box that is 1 centimeter on each side. To get an idea of how small a gram is, we note that two small paper clips weigh about 1 gram. A gram is only about 0.035 ounce.

One kilogram is 1000 times larger than a gram. A kilogram weighs about 2.2 pounds. Some of the measures of weight in the metric system are shown in the following chart.

Common Metric Weight Measurements

1 metric ton (t) = 1,000,000 grams

1 kilogram (kg) = 1000 grams

1 gram (g)

1 milligram (mg) = 0.001 gram

Use this prefix chart as a guide when you change one metric prefix to another. Move the decimal point in the same direction and the same number of places.

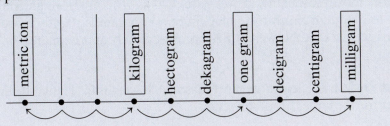

metric ton | kilogram | hectogram | dekagram | one gram | decigram | centigram | milligram

We will focus exclusively on the metric ton, kilogram, gram, and milligram. We convert weight measurements the same way we convert volume and length measurements.

EXAMPLE 4 Convert.

(a) 2 t = _____ kg

(b) 0.42 kg = _____ g

(a) 2 t = 2000 kg

(b) 0.42 kg = 420 g

Practice Problem 4 Convert.

(a) 3.2 t = _____ kg

(b) 7.08 kg = _____ g

A kilogram is slightly more than 2 pounds. A 3-week-old miniature pinscher weighs about 1 kilogram.

EXAMPLE 5 Convert.

(a) 283 kg = _____ t

(b) 7.98 mg = _____ g

(a) 283 kg = 0.283 t

(b) 7.98 mg = 0.00798 g

Practice Problem 5 Convert.

(a) 59 kg = _____ t

(b) 28.3 mg = _____ g

EXAMPLE 6 If a chemical costs $0.03 per gram, what will it cost per kilogram?

Since there are 1000 grams in a kilogram, a chemical that costs $0.03 per gram would cost 1000 times as much per kilogram.

$$1000 \times \$0.03 = \$30.00$$

The chemical would cost $30.00 per kilogram.

Practice Problem 6 If coffee costs $10.00 per kilogram, what will it cost per gram?

EXAMPLE 7 Select the most reasonable weight for Tammy's Toyota.

(a) 820 t   **(b)** 820 g   **(c)** 820 kg   **(d)** 820 mg

The most reasonable answer is **(c)** 820 kg. The other weight values are much too large or much too small. Since a kilogram is slightly more than 2 pounds, we see that this weight, 820 kg, most closely approximates the weight of a car.

Practice Problem 7 Select the most reasonable weight for Hank, starting linebacker for the college football team.

(a) 120 kg   **(b)** 120 g   **(c)** 120 mg

Sidelight

When dealing with very small particles or atomic elements, scientists sometimes use units smaller than a gram.

> **Small Weight Measurements**
>
> 1 milligram = 0.001 gram
>
> 1 microgram = 0.000001 gram
>
> 1 nanogram = 0.000000001 gram
>
> 1 picogram = 0.000000000001 gram

We could make the following conversions.

2.6 picograms = 0.0026 nanogram

29.7 micrograms = 0.0297 milligram

58 nanograms = 58,000 picograms

58 nanograms = 0.058 microgram

See Exercises 6.3, exercises 63 and 64 for such conversion problems.

Verbal and Writing Skills

Write the metric unit that represents each measurement.

1. one thousand liters

2. one thousandth of a liter

3. one thousandth of a gram

4. one thousand grams

5. one thousandth of a kilogram

6. one thousand milliliters

Perform each conversion.

7. 58 kL = _____ L

8. 45 kL = _____ L

9. 2.6 L = _____ mL

10. 9.4 L = _____ mL

11. 18.9 mL = _____ L

12. 31.5 mL = _____ L

13. 752 L = _____ kL

14. 493 L = _____ kL

15. 2.43 kL = _____ mL

16. 1.76 kL = _____ mL

17. 82 mL = _____ cm^3

18. 152 mL = _____ cm^3

19. 5261 mL = _____ kL

20. 28,156 mL = _____ kL

21. 74 L = _____ cm^3

22. 122 L = _____ cm^3

23. 1620 g = _____ kg

24. 2940 g = _____ kg

25. 35 mg = _____ g

26. 13 mg = _____ g

27. 6328 mg = _____ g

28. 986 mg = _____ g

29. 2.92 kg = _____ g

30. 14.6 kg = _____ g

31. 17 t = _____ kg

32. 36 t = _____ kg

Fill in the blanks with the correct values.

33. 7 mL = _____ L = _____ kL

34. 18 mL = _____ L = _____ kL

35. 128 cm^3 = _____ L = _____ kL

36. 199.8 cm^3 = _____ L = _____ kL

37. 0.033 kg = _____ g = _____ mg

38. 0.098 kg = _____ g = _____ mg

39. 2.58 metric tons = _____ kg = _____ g

40. 7183 g = _____ kg = _____ t

41. Alice bought a jar of apple juice at the store. Choose the most reasonable measurement for its contents.
(a) 0.32 kL (b) 0.32 L (c) 0.32 mL

42. A nurse gave an injection of insulin to a diabetic patient. Choose the most reasonable measurement for the dose.
(a) 4 kL (b) 4 L (c) 4 mL

43. A ship owner purchased a new oil tanker. Choose the most reasonable measurement for its weight.
(a) 4000 t (b) 4000 kg (c) 4000 g

44. Robert bought a new psychology textbook. Choose the most reasonable measurement for its weight.
(a) 0.49 t (b) 0.49 kg (c) 0.49 g

Find the convenient unit of measure and add.

45. 83 L + 822 mL + 30.1 L

46. 152 L + 473 mL + 77.3 L

47. 20 g + 52 mg + 1.5 kg

48. 2 kg + 42 mg + 120 g

Mixed Practice

Write true *or* false *for each statement.*

49. 1 milliliter = 0.001 liter

50. 1 kiloliter = 1000 liters

51. Milk can be purchased in kiloliter jugs at the food store.

52. Small amounts of medicine are often measured in liters.

53. 1 metric ton = 1000 grams

54. 1 kL = 100 mL

55. A nickel weighs about 5 grams.

56. A convenient size for a family purchase is 2 kilograms of ground beef.

Applications

Convert the units.

57. A wholesaler sells peanut butter in canisters of 5000 g for $7.25. An elementary school needs to buy 75 kg to restock its pantry. How much will the school spend on peanut butter?

58. Randy's Premier Pizza needs to order tomato sauce from the distributor. The sauce comes in 4000-gram jars for $6.80. If Randy orders 56 kg, how much will he pay for the tomato sauce?

59. A very rare essence of an almost extinct flower found in the Amazon jungle of South America is extracted by a biogenetic company trying to copy and synthesize it. The company estimates that if the procedure is successful, the product will cost the company $850 per milliliter to produce. How much will it cost the company to produce 0.4 liter of the engineered essence?

60. In a recent year the price of gold was $9.40 per gram. At that price how much would a kilogram of gold cost?

61. A government rocket is carrying cargo into space for a private company. The charge for the freight shipment is $22,450 per kilogram of the special cargo, and the cargo weighs 0.45 metric ton. How much will the private company have to pay?

62. The individual coal-cutting record is 45.4 metric tons per person in one shift (six hours), by five Soviet miners. If the coal is sold for $2.00 per kilogram, what is the value of the coal?

Convert.

63. 5632 picograms = _____ micrograms

64. 0.076182 milligram = _____ nanograms

To Think About

The following chart shows the production in weight and value of platinum in the United States for various years. Use this chart to answer the following questions. Round to the nearest dollar when necessary.

Production of Platinum Metal in the U.S.

Source: U.S. Geological Survey Annual Reports

Year	1990	1995	1997	1998	2000
Production in kg	1810	1590	2610	3500	3860*
Value in millions	$27	$21	$24	$33	$36*

*estimated

65. How many more kilograms were produced in 2000 than in 1990?

66. How many fewer kilograms were produced in 1995 than in 1990?

67. What was the value per kilogram of the platinum mined in 1997? What was the value per gram?

68. What was the value per kilogram of the platinum mined in 2000? What was the value per gram?

69. In what year was the value per kilogram of the platinum mined the lowest?

70. In what year was the value per kilogram of the platinum mined the highest?

Cumulative Review Questions

71. 14 out of 70 is what percent?

72. What is 23% of 250?

73. What is 1.7% of $18,900?

74. A salesperson earns a commission of 8%. She sold furniture worth $8960. How much commission did she earn?

6.4 Conversion of Units (Optional)

Student Learning Objectives

After studying this section, you will be able to:

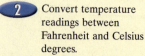

1 Convert units of length, volume, or weight between the metric and American systems.

2 Convert temperature readings between Fahrenheit and Celsius degrees.

SSM
PH TUTOR   CD & VIDEO   MATH PRO   WEB
CENTER

1 Converting Units of Length, Volume, or Weight Between the Metric and American Systems

So far we've seen how to convert units when working *within* either the American or the metric system. Many people, however, work in *both* the metric and the American systems. If you study such fields as chemistry, electromechanical technology, business, X-ray technology, nursing, or computers, you will probably need to convert measurements between the two systems.

To convert between American units and metric units, it is helpful to have equivalent values. The most commonly used equivalents are listed in the following table. Most of these are approximate.

Equivalent Measures

	American to Metric	Metric to American
Units of length	1 mile ≈ 1.61 kilometers 1 yard ≈ 0.914 meter 1 foot ≈ 0.305 meter 1 inch = 2.54 centimeters*	1 kilometer ≈ 0.62 mile 1 meter ≈ 3.28 feet 1 meter ≈ 1.09 yards 1 centimeter ≈ 0.394 inch
Units of volume	1 gallon ≈ 3.79 liter 1 quart ≈ 0.946 liter	1 liter ≈ 0.264 gallon 1 liter ≈ 1.06 quarts
Units of weight	1 pound ≈ 0.454 kilogram 1 ounce ≈ 28.35 gram	1 kilogram ≈ 2.2 pounds 1 gram ≈ 0.0353 ounce

*exact value

Remember that to convert from one unit to another you multiply by a fraction that is equivalent to 1. Create a fraction from the equivalent measures table so that the unit in the denominator cancels the unit you are changing.

To change 5 miles to kilometers, we look in the table and find that 1 mile ≈ 1.61 kilometers. We will use the unit fraction

$$\frac{1.61 \text{ kilometers}}{1 \text{ mile}}$$

because we want to have miles in the denominator.

$$5 \text{ miles} \times \frac{1.61 \text{ kilometers}}{1 \text{ mile}} = 5 \times 1.61 \text{ kilometers} = 8.05 \text{ kilometers}$$

Thus 5 miles ≈ 8.05 kilometers.

Is this the only way to do the problem? No. To make the previous conversion, we also could have used the relationship that 1 kilometer ≈ 0.62 mile. Again, we want to have miles in the denominator, so we use

$$\frac{1 \text{ kilometer}}{0.62 \text{ mile}}$$

$$5 \text{ miles} \times \frac{1 \text{ kilometers}}{0.62 \text{ mile}} = \frac{5}{0.62} \approx 8.06 \text{ kilometers}$$

Using this approach, we find that 5 miles ≈ 8.06 kilometers. This is not the same result we obtained before. The discrepancy between the results of the two conversions has occurred because the numbers given in the equivalent measures tables are approximations.

Often we have to make conversions in order to make comparisons. For example, is the 6 inches of attic insulation commonly used in the northeastern United States more or less than the 16 centimeters of attic insulation commonly used in Sweden? In order to find out we would have to convert 6 inches to centimeters. As we will see in Example 1(b) 6 inches is somewhat less than 16 centimeters.

● EXAMPLE 1   Convert 3 feet to meters.

$$3 \text{ feet} \times \frac{0.305 \text{ meter}}{1 \text{ foot}} = 0.915 \text{ meter}$$

Practice Problem 1   Convert 7 feet to meters.   ●

Unit abbreviations are quite common, so we will use them for the remainder of this section. We list them here for your reference.

American Measure (Alphabetical Order)	Standard Abbreviation
feet	ft
gallon	gal
inch	in.
mile	mi
ounce	oz
pound	lb
quart	qt
yard	yd

Metric Measure	Standard Abbreviation
centimeter	cm
gram	g
kilogram	kg
kilometer	km
liter	L
meter	m
millimeter	mm

● EXAMPLE 2

(a) Convert 26 m to yd. **(b)** Convert 1.9 km to mi.
(c) Convert 14 gal to L. **(d)** Convert 2.5 L to qt.

(a) $26 \text{ m} \times \dfrac{1.09 \text{ yd}}{1 \text{ m}} = 28.34 \text{ yd}$ **(b)** $1.9 \text{ km} \times \dfrac{0.62 \text{ mi}}{1 \text{ km}} = 1.178 \text{ mi}$

(c) $14 \text{ gal} \times \dfrac{3.79 \text{ L}}{1 \text{ gal}} = 53.06 \text{ L}$ **(d)** $2.5 \text{ L} \times \dfrac{1.06 \text{ qt}}{1 \text{ L}} = 2.65 \text{ qt}$

Practice Problem 2

(a) Convert 17 m to yd. **(b)** Convert 29.6 km to mi.
(c) Convert 26 gal to L. **(d)** Convert 6.2 L to qt.   ●

Some conversions require more than one step.

● EXAMPLE 3   Convert 235 cm to ft. Round to the nearest hundredth of a foot.

Our first fraction converts centimeters to inches. Our second fraction converts inches to feet.

$$235 \text{ cm} \times \frac{0.394 \text{ in.}}{1 \text{ cm}} \times \frac{1 \text{ ft}}{12 \text{ in.}} = \frac{92.59}{12} \text{ ft}$$

$$\approx 7.72 \text{ ft (rounded to the nearest hundredth)}$$

Practice Problem 3 Convert 180 cm to ft.

The same rules can be followed for a rate such as 50 miles per hour.

EXAMPLE 4 Convert 100 km/hr to mi/hr.

$$\frac{100 \text{ km}}{\text{hr}} \times \frac{0.62 \text{ mi}}{1 \text{ km}} = 62 \text{ mi/hr}$$

Thus 100 km/hr is approximately equal to 62 mi/hr.

Practice Problem 4 Convert 88 km/hr to mi/hr.

Sometimes we need more than one unit fraction to make the conversion of two rates. We will see how this is accomplished in Example 5.

EXAMPLE 5 A rocket carrying a communication satellite is launched from a rocket launch pad. It is traveling at 700 miles per hour. How many feet per second is the rocket traveling? Round to the nearest whole number.

$$\frac{700 \text{ miles}}{\text{hr}} \times \frac{5280 \text{ ft}}{1 \text{ mile}} \times \frac{1 \text{ hr}}{60 \text{ min}} \times \frac{1 \text{ min}}{60 \text{ sec}}$$

$$= \frac{700 \times 5280 \text{ ft}}{60 \times 60 \text{ sec}} = \frac{3,696,000 \text{ ft}}{3600 \text{ sec}} \approx 1027 \text{ ft/sec}$$

The missile is traveling at 1027 feet per second.

Practice Problem 5 A Concorde jet flew at 900 miles per hour. What was the speed of the Concorde jet in feet per second?

▲ **Sidelight**

Suppose we consider a rectangle that measures 2 yards wide by 4 yards long. The area would be 2 yards × 4 yards = 8 square yards. How could you change 8 square yards to square meters? Suppose that we look at 1 square yard. Each side is 1 yard long, which is equivalent to 0.9144 meter.

$$\text{Area} = 1 \text{ yard} \times 1 \text{ yard} \approx 0.9144 \text{ meter} \times 0.9144 \text{ meter}$$

$$\text{Area} = 1 \text{ square yard} \approx 0.8361 \text{ square meter}$$

Thus 1 yd^2 ≈ 0.8361 m^2. Therefore

$$8 \text{ yd}^2 \times \frac{0.8361 \text{ m}^2}{1 \text{ yd}^2} = 6.6888 \text{ m}^2.$$

8 square yards ≈ 6.6888 square meters.

2 *Converting Temperature Readings Between Fahrenheit and Celsius Degrees*

In the metric system, temperature is measured on the **Celsius scale**. Water boils at 100° (100°C) and freezes at 0°(0°C) on the Celsius scale. In the **Fahrenheit system**, water boils at 212°(212°F) and freezes at 32°(32°F).

To convert Celsius to Fahrenheit, we can use the formula

$$F = 1.8 \times C + 32,$$

where C is the number of Celsius degrees and F is the number of Fahrenheit degrees.

 EXAMPLE 6 When the temperature is 35°C, what is the Fahrenheit reading?

$$F = 1.8 \times C + 32$$
$$= 1.8 \times 35 + 32$$
$$= 63 + 32$$
$$= 95$$

The temperature is 95°F.

Practice Problem 6 Convert 20°C to Fahrenheit temperature.

To convert Fahrenheit temperature to Celsius, we can use the formula

$$C = \frac{5 \times F - 160}{9}$$

where F is the number of Fahrenheit degrees and C is the number of Celsius degrees.

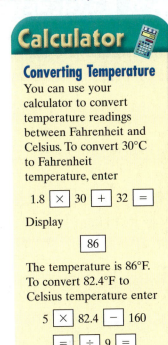

Calculator

Converting Temperature
You can use your calculator to convert temperature readings between Fahrenheit and Celsius. To convert 30°C to Fahrenheit temperature, enter

1.8 $\times$ 30 $+$ 32 $=$

Display

86

The temperature is 86°F. To convert 82.4°F to Celsius temperature enter

5 $\times$ 82.4 $-$ 160

$=$ $\div$ 9 $=$

Display

28

The temperature is 28°C.

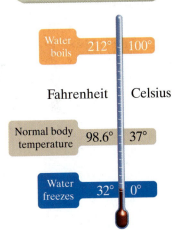

Water boils 212° 100°

Fahrenheit   Celsius

Normal body temperature 98.6° 37°

Water freezes 32° 0°

Perform each conversion. Round to the nearest hundredth when necessary.

1. 7 ft to m

2. 11 ft to m

3. 9 in. to cm

4. 13 in. to cm

5. 14 m to yd

6. 18 m to yd

7. 30.8 yd to m

8. 42.5 yd to m

9. 82 mi to km

10. 68 mi to km

11. 16 ft to m

12. 12 ft to m

13. 17.5 cm to in.

14. 19.6 cm to in.

15. 200 m to yd

16. 350 m to yd

17. 65 in. to cm

18. 80 in. to cm

19. 280 gal to L

20. 140 gal to L

21. 23 qt to L

22. 28 qt to L

23. 19 L to gal

24. 15 L to gal

25. 4.5 L to qt

26. 6.5 L to qt

27. 130 lb to kg

28. 155 lb to kg

29. 126 g to oz

30. 186 g to oz

31. 1260 cm to ft

32. 1872 cm to ft

33. 800 mi/hr to ft/sec (Round to the nearest whole number.)

34. 500 mi/hr to ft/sec (Round to the nearest whole number.)

35. 400 ft/sec to mi/hr (Round to the nearest whole number.)

36. 300 ft/sec to mi/hr (Round to the nearest whole number.)

37. A wire that is 13 mm wide is how many inches wide?

38. A bolt that is 7 mm wide is how many inches wide?

39. 85°C to Fahrenheit   **40.** 105°C to Fahrenheit   **41.** 12°C to Fahrenheit   **42.** 21°C to Fahrenheit

43. 140°F to Celsius   **44.** 131°F to Celsius   **45.** 40°F to Celsius   **46.** 52°F to Celsius

Applications

Solve. Round to the nearest hundredth when necessary.

47. Mr. and Mrs. Weston have traveled 67 miles on a boat cruise from Seattle, Washington, to Victoria Island, Vancouver, B.C., Canada. They have 36 kilometers until their rendezvous point with another boat. How many kilometers in total will they have traveled?

48. John and Sandy Westphal traveled to Spain for a bike tour. The woman leading the tour told the participants they would be biking 75 km the first day, 83 km the second day, and 78 km the third day. How many miles did John and Sandy bike in those three days?

49. Pierre had a Jeep imported into France. During a trip from Paris to Lyon, he used 38 liters of gas. The tank, which he filled before starting the trip, holds 15 gallons of gas. How many liters of gas were left in the tank when he arrived?

50. A surgeon is irrigating an abdominal cavity after a cancerous growth is removed. There is a supply of 3 gallons of distilled water in the operating room. The surgeon uses a total of 7 liters of the water during the procedure. How many liters of water are left over after the operation?

51. One of the heaviest males documented in medical records weighed 635 kg in 1978. What would have been his weight in pounds?

52. The average weight for a 7-year-old girl is 22.2 kilograms. What is the average weight in pounds?

53. Tourists who visit Ayers Rock in central Australia in the summer begin climbing at four o'clock in the morning, when the temperature is 19° Celsius. They do this because after seven o'clock in the morning, the temperature can reach 45°C and can cause climbers to die of dehydration. What are equivalent Fahrenheit temperatures?

54. A holiday turkey in Buenos Aires, Argentina, was roasted at 200° Celsius for 4 hours. What would have been the Fahrenheit roasting temperature in Joplin, Missouri?

 55. There are 96,550 km of blood vessels in the human body. How many miles of blood vessels is this?

56. The distance around the Earth at the equator (the circumference) is approximately 40,325 km. How many miles is this?

Round to four decimal places.

▲ **57.** 28 square inches = ? square centimeters

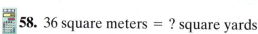 **58.** 36 square meters = ? square yards

To Think About

▲ **59.** Frederick Himlein is planning to carpet his rectangular living room, which measures 8 yards by 4 yards. He has found some carpet in New York City that he likes that costs $28 per square yard. While visiting family friends in Germany, his wife Gertrude found some carpet that costs $30 per square meter. The company has a New York City office and can sell it in America for the same price. Frederick says that the German carpet is much too expensive. How much would it cost to carpet the living room with the American carpet? How much would it cost to carpet the living room with the German carpet? How much difference in cost is there between these two choices? (Round to the nearest dollar.)

▲ **60.** Phillipe Bertoude is planning to carpet his rectangular family room, which measures 7 yards by 5 yards. His wife found some carpet in Boston that she likes, and it costs $24 per square yard. While Phillipe was visiting his father in Paris, he found some carpet that costs $26 per square meter. The company has a Boston office and can sell it in America for the same price. Phillipe told his wife that the carpet in Paris was a better buy. She does not agree. How much would it cost to carpet the family room with the French carpet? How much would it cost to carpet the family room with the American carpet? How much difference in cost is there between these two choices? (Round to the nearest dollar.)

Cumulative Review Problems

Do the operations in the correct order.

61. $2^3 \times 6 - 4 + 3$

62. $5 + 2 - 3 + 5 \times 3^2$

63. $\dfrac{1}{2} \cdot \dfrac{3}{4} - \dfrac{1}{5}\left(\dfrac{1}{2}\right)^2$

64. $\dfrac{5}{6} - \dfrac{1}{2}\left(\dfrac{5}{6} - \dfrac{1}{3}\right)$

Putting Your Skills to Work

Ships and Nautical Miles

Although land speeds of trains and cars are almost always measured in miles per hour, speeds of boats and airplanes are usually measured in knots, or nautical miles per hour. A nautical mile contains approximately 6080 feet, compared to 5280 feet in a land mile. To understand the properties of boats and ships, we will need to be more familiar with nautical miles and speeds measured in knots.

The *Queen Elizabeth II* is considered one of the finest passenger ships in the world. It measures 963 feet long, weighs 70,327 tons, and carries 1778 passengers. See if you can answer the following questions.

Problems for Individual Investigation

1. The maximum speed of the *Queen Elizabeth II* is 32.5 knots. Approximately how many miles per hour is the maximum speed?

2. The normal cruising speed of the *Queen Elizabeth II* is 28 knots. Approximately how many miles per hour is the normal cruising speed?

Problems for Group Investigation and Cooperative Study

3. At cruising speed, the *Queen Elizabeth II* moves only 6 inches for each gallon of diesel fuel that it burns. How many gallons of fuel will it require to travel the distance from Southampton, England, to New York City at cruising speed? (It is 2977 nautical miles from Southampton to New York City).

4. At maximum speed, the *Queen Elizabeth II* moves only 5 inches for each gallon of diesel fuel that it burns. How many gallons of fuel will it require to travel the distance from New York City to Cape Town, South Africa, at maximum speed? (It is 6772 nautical miles from New York City to Cape Town.)

Internet Connections

  Netsite: http://www.prenhall.com/tobey_basic

This site gives the distance between any two cities in the world.

5. If the *Queen Elizabeth II* travels from San Francisco, California, to Sydney, Australia, at cruising speed, how many gallons of fuel will be needed?

6. If the *Queen Elizabeth II* travels from Sydney, Australia, to Singapore at cruising speed, how many gallons of fuel will be needed?

Student Learning Objectives

After studying this section, you will be able to:

1. Solve applied problems involving metric and American units.

SSM PH TUTOR CENTER CD & VIDEO MATH PRO WEB

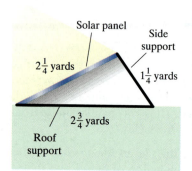

Solar panel

$2\frac{1}{4}$ yards

Side support

$1\frac{1}{4}$ yards

$2\frac{3}{4}$ yards

Roof support

1 **Solving Applied Problems Involving Metric and American Units**

Once again, we will be using the Mathematics Blueprint for solving applied problems that we used previously.

▲ 🔴 **EXAMPLE 1** A triangular support piece holds a solar panel. The sketch shows the dimensions of the triangle. Find the perimeter of this triangle. Express the answer in *feet*.

Mathematics Blueprint For Problem Solving

Gather the Facts	What Am I Asked to Do?	How Do I Proceed?	Key Points to Remember
The triangle has three sides: $2\frac{1}{4}$ yd, $1\frac{1}{4}$ yd, and $2\frac{3}{4}$ yd.	Find the perimeter.	Add the lengths of the three sides. Then change the answer from yards to feet.	Be sure to change the number of yards to an improper fraction before multiplying by 3 to obtain feet.

1. **Understand the problem.**

   The perimeter is the sum of the lengths of the sides. We are asked to express the answer in feet. Remember to convert the yards to feet.

2. **Solve and state the answer.**

   Add the three sides.

$$2\frac{1}{4} \text{ yards}$$

$$1\frac{1}{4} \text{ yards}$$

$$+ 2\frac{3}{4} \text{ yards}$$

$$\overline{5\frac{5}{4} \text{ yards} = 6\frac{1}{4} \text{ yards}}$$

   Convert $6\frac{1}{4}$ yards to feet using the fact that 1 yard = 3 feet. To make the calculation easier, we will change $6\frac{1}{4}$ to $\frac{25}{4}$.

$$6\frac{1}{4} \text{ yards} \times \frac{3 \text{ feet}}{1 \text{ yard}} = \frac{25}{4} \text{ yards} \times \frac{3 \text{ feet}}{1 \text{ yard}} = \frac{75}{4} \text{ feet} = 18\frac{3}{4} \text{ feet}$$

   The perimeter of the triangle is $18\frac{3}{4}$ feet.

3. **Check.**

   We will check by estimating the answer.

$$2\frac{1}{4} \text{ yd} \approx 2 \text{ yd} \qquad 1\frac{1}{4} \text{ yd} \approx 1 \text{ yd} \qquad 2\frac{3}{4} \text{ yd} \approx 3 \text{ yd}$$

   Now we add the three sides, using our estimated values.

$$2 + 1 + 3 = 6 \text{ yards} \qquad 6 \text{ yards} = 18 \text{ feet}$$

   Our estimated answer, 18 feet, is close to our calculated answer, $18\frac{3}{4}$ feet. Thus our answer seems reasonable. ✓

▲ **Practice Problem 1**   Find the perimeter of the rectangle on the right. Express the answer in *feet*.

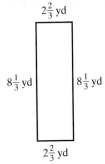

$2\frac{2}{3}$ yd

$8\frac{1}{3}$ yd $8\frac{1}{3}$ yd

$2\frac{2}{3}$ yd

EXAMPLE 2   How many 210-*liter* gasoline barrels can be filled from a tank of 5.04 *kiloliters* of gasoline?

1. Understand the problem.

Mathematics Blueprint For Problem Solving

Gather the Facts	What Am I Asked to Do?	How Do I Proceed?	Key Points to Remember
We have 5.04 kiloliters of gasoline. We are going to divide the gasoline into smaller barrels that hold 210 liters each.	Find out how many of these smaller 210-liter barrels can be filled.	We need to get all measurements in the same units. We choose to convert 5.04 kiloliters to liters. Then we divide that result by 210 to find out how many barrels can be filled.	To convert 5.04 kiloliters to liters, we move the decimal point three places to the right.

2. Solve and state the answer.
First we convert 5.04 kiloliters to liters.

$$5.04 \text{ kiloliters} = 5040 \text{ liters}$$

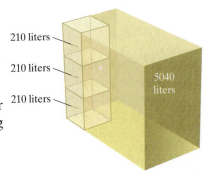

210 liters

210 liters

210 liters

5040 liters

Now we find out how many barrels can be filled. How many 210-liter barrels will 5040 liters fill? Visualize fitting 210-liter barrels into a big barrel that holds 5040 liters. (Use rectangular barrels.)
    We need to divide

$$\frac{5040 \text{ liters}}{210 \text{ liters}} = 24.$$

Thus we can fill 24 of the 210-liter barrels.

3. Check.
Estimate each value. First 5.04 kiloliters is approximately 5 kiloliters or 5000 liters. 210-liter barrels hold approximately 200 liters. How many times does 200 fit into 5000?

$$\frac{5000}{200} = 25$$

We estimate that 25 barrels can be filled. This is very close to our calculated value of 24 barrels. Thus our answer is reasonable. ✓

Practice Problem 2   A lab assistant must use 18.06 liters of solution to fill 42 jars. How many milliliters of the solution will go into each jar?

Applications

Solve.

▲ **1.** Find the perimeter of the triangle. Express your answer in feet.

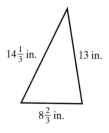

$14\frac{1}{3}$ in.   13 in.

$8\frac{2}{3}$ in.

▲ **2.** Find the perimeter of the triangle. Express your answer in feet

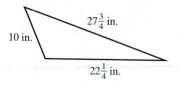

$27\frac{3}{4}$ in.

10 in.

$22\frac{1}{4}$ in.

▲ **3.** The triangular giraffe enclosure at the zoo is being re-fenced. One side is 86 yards long and a second side is 77 yards long. If the zoo has 522 feet of fencing, how much will be left over for the third side? Express your answer in yards.

▲ **4.** A farmer has 600 feet of fencing to fence in a triangular region. One side is to be 75 yards and a second 65 yards. How much fencing will the farmer have left for the third side? Express your answer in yards.

▲ **5.** A rectangular picture window measures 87 centimeters × 152 centimeters. Window insulation is applied along all four sides. The insulation costs $7.00 per meter. What will it cost to insulate the window?

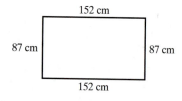

152 cm

87 cm   87 cm

152 cm

▲ **6.** A rectangular doorway measures 90 centimeters × 200 centimeters. Weatherstripping is applied on the top and the two sides. The weatherstripping costs $6.00 per meter. What did it cost to weatherstrip the door?

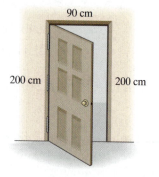

90 cm

200 cm   200 cm

7. A tack supply company, which makes saddles, fittings, bridles, and other equipment for horses and riders, has a length of braided leather 12.4 meters long. The leather must be cut into 4 equal pieces. How many centimeters long will each piece be?

8. A stretch of road 1.863 kilometers long has 230 parking spaces of equal length painted in white on the pavement. How many meters long is each parking space?

9. The innermost lane on an outdoor running track is 400 meters long. The relay exchange zones make up $\frac{1}{10}$ of the lane. How many feet of the lane are designated as exchange zones for relay races?

10. An antenna 186 meters tall is on the roof of a building. The top $\frac{1}{8}$ of the antenna has a special rubberized insulation. How many centimeters long is the portion of the antenna with this insulation?

11. One serving of raisin bran contains 340 mg of potassium. If you eat a serving every day for breakfast, how many grams of potassium will you have consumed in one week?

12. A snack size bag of Doritos chips contains 350 mg of sodium. If you eat 1 bag each day for a week, how many grams of sodium will you have consumed?

13. A bottle of superglue states on its label that it should not be used at temperatures above 85°F. The sign on the display in the store reads NOT TO BE USED AT TEMPERATURES ABOVE 27°C. What is the discrepancy in degrees Fahrenheit between the two suggested temperatures?

14. The temperature in Dakar, Senegal, today is 33°C. The temperature on this date last year was 94°F. What is the difference in degrees Fahrenheit between the temperature in Dakar today and the temperature one year ago?

15. A Swedish flight attendant is heating passenger meals for the new American airline he is working for. He is used to heating meals at 180°C. His co-worker tells him that the food must be heated at 350°F. What is the difference in temperature in degrees Fahrenheit between the two temperatures? Which temperature is hotter?

16. A French cook wishes to bake potatoes in an oven at 195°C. The American assistant set the oven at 400°F. By how many degrees Fahrenheit was the oven temperature off? Was the oven too hot or too cold?

17. Sharon and James Hanson traveled from Arizona to Acapulco, Mexico. The last day of their trip, they traveled 520 miles and it took them eight hours. The maximum speed limit is 110 kilometers/hour.
    (a) How many kilometers per hour did they average on the last day of the trip?
    (b) Did they break the speed limit?

18. A small corporate jet travels at 600 kilometers/hour for 1.5 hours. The pilot said the plane will be on time if it travels at 350 miles per hour.
    (a) What was the jet's speed in miles per hour?
    (b) Will it arrive on time?

19. The concession stand at the high school sells large 1-pint sodas. On an average night, two sodas are sold each minute. How many gallons of soda are sold per hour?

20. A leaky faucet drips 1 pint of water per hour. How many gallons of water is this per day?

21. 3.4 tons of bananas are off-loaded from a ship that has just arrived from Costa Rica. The port taxes imported fruit at $0.015 per pound. What is the tax on the entire shipment?

22. Trucks in Sam's home state are taxed each year at $0.03 per pound. Sam's empty truck weighs 1.8 tons. What is his annual tax?

23. A can of peaches contains 16 ounces. Fred discovered that 5 ounces were syrup and the rest was fruit. How many grams of fruit were in the can?

24. A box of raisin bran contains 14 ounces of cereal. Of this, 11 ounces are bran flakes and the rest is raisins. How many grams of raisins are in the box?

25. Carlos is mixing a fruit tree spray to retard the damage caused by beetles. He has 16 quarts of spray concentrate available. The old recipe he used in Mexico called for 11 liters of spray concentrate.
 (a) How many extra quarts of spray concentrate does he have?
 (b) If the spray costs $2.89 per quart, how much will it cost him to prepare the old recipe?

26. Maria bought 18 quarts of motor oil for $1.39 per quart. Her uncle from Mexico is visiting this summer. He said he will use 12 liters of oil while driving his car this summer.
 (a) How many extra quarts of oil did Maria buy?
 (b) How much did this extra oil cost her?

27. Jackson's small motorcycle gets 56 kilometers per liter. He drives 392 kilometers to Quebec City, Canada, and gas is $0.78 per liter.
 (a) How much does the gas used for the trip cost?
 (b) How many miles per gallon does Jackson's motorcycle get?

28. Ryan's Volvo station wagon runs on diesel fuel. He gets 20 miles per gallon on the highway.
 (a) If he drives 275 kilometers to Mexico City, and diesel fuel costs $1.30 per gallon, how much does the fuel used for the trip cost him? Round to the nearest cent.
 (b) How many kilometers per liter does he get?

29. The flow rate of a safety discharge pipe at a dam in Lowell, Massachusetts, is rated for a maximum of 240,000 gallons per hour. The inspector asked if this flow rate could have handled the floods of 1927. During the floods of 1927 the flow rate at the dam was measured as 440 pints per second. Could the safety discharge pipe safely handle a flow rate of 440 pints per second? Why?

30. The old main lines that run water from the Quabbin Reservoir to Boston have some leaks. It is estimated that the lines leak approximately 36,000 gallons per hour. A local newspaper said that the lines leak 80 pints per second. Did the newspaper have the correct information? Why?

Cumulative Review Problems

Solve for n.

31. $\dfrac{n}{16} = \dfrac{2}{50}$

32. The highest known mountain in the solar system is the Martian volcano Olympus Mons, which rises approximately three times higher than Mt. Everest. (Mt. Everest is 8846 meters high.)
   (a) How high is the Martian volcano in meters?

   (b) How high is the Martian volcano in feet? (Round to the nearest foot.)

33. Justin is going to India. While reading his atlas he sees that Bombay is 6 inches from where he will begin his travel. The scale shows that 3 inches represents 7.75 miles on the ground. If he travels by train on a straight track, how far must the train go to take him to his destination?

34. Thompson has a scale model of a famous fishing schooner. Every 2 centimeters on the model represents an actual length of 7.5 yards. The model is 11 centimeters long. How long is the famous fishing schooner?

Math in the Media

Converting Between the Different Temperature Scales

Chad Palmer
USA TODAY, 11/01/00

Measurement of temperature is found in every weather report on radio, tv, or in the newspaper. In the United States we use the Fahrenheit scale. In Canada, our close neighbor, the Celsius scale is used. In situations where only positive temperatures are desired, the Fahrenheit and Celsius scales are modified to become the Kelvin and Rankin scales, respectively.

The following formulae relate the different temperature scales. The subscripts F, C, K and R indicate temperatures in Fahrenheit, Celsius, Kelvin, and Rankin scales, respectively.

$$T_F = \frac{9}{5}(T_C) + 32 \qquad T_C = \frac{5}{9}(T_F - 32)$$

$$T_K = T_C + 273 \qquad T_R = T_F + 460$$

Another method for conversion back and forth between Fahrenheit and Celsius scales is the following: Regardless of which direction you want to convert, Fahrenheit to Celsius or Celsius to Fahrenheit, always first add 40 to the number. Next multiply by $\frac{5}{9}$ or $\frac{9}{5}$, then always substract out the 40 you just added. For example:

Convert Celsius zero, freezing, to Fahrenheit freezing

$$(0 + 40)\left(\frac{9}{5}\right) - 40 = \frac{360}{5} - \frac{200}{5}$$

$$= \frac{160}{5} = 32$$

Convert Fahrenheit boiling, 212, to Celsius boiling (water at sea level)

$$(212 + 40)\left(\frac{5}{9}\right) - 40 = 252\left(\frac{5}{9}\right) - \frac{360}{9}$$

$$= \left(\frac{1260 - 360}{9}\right) = \frac{900}{9} = 100$$

EXERCISES

Use the formulas for Fahrenheit, Celsius, Kelvin and Rankin for questions 1–4.

1. Find the temperature for freezing water at sea level in every scale.

2. Find the temperature for boiling water at sea level in every scale.

3. In which scales will a one-degree change in temperature feel the same?

4. In which scales will a one-degree measure change represent the smallest temperature change?

Use the alternative conversion method for Fahrenheit-Celsius for questions 5 and 6.

5. Normal human temperature in Fahrenheit is 98.6. Convert to Celsius.

6. How does one remember which to use in multiplication?

Chapter 6 Organizer

Topic	Procedure	Examples
Changing from one American unit to another, p. 370.	1. Find the equality statement that relates what you want to find and what you know. 2. Form a unit fraction. The denominator will contain the units of the original measurement. 3. Multiply by the unit fraction and simplify.	Convert 210 inches to feet. 1. Use 12 inches = 1 foot. 2. Unit fraction = $\dfrac{1 \text{ foot}}{12 \text{ inches}}$ 3. $210 \text{ in.} \times \dfrac{1 \text{ ft}}{12 \text{ in.}} = \dfrac{210}{12} \text{ ft}$ $\qquad = 17.5 \text{ ft}$ Convert 86 yards to feet. 1. Use 1 yard = 3 feet. 2. Unit fraction = $\dfrac{3 \text{ feet}}{1 \text{ yard}}$ 3. $86 \text{ yd} \times \dfrac{3 \text{ ft}}{1 \text{ yd}} = 86 \times 3 \text{ ft} = 258 \text{ ft}$
Changing from one metric unit to another, pp. 376, 385, 386.	When you change from one prefix to another by moving to the *left* in the prefix guide, move the decimal point to the *left* the same number of places. kilo- = 1000 hecto- = 100 deka- = 10 one unit = 1 deci- = 0.1 centi- = 0.01 milli- = 0.001 When you change from one prefix to another by moving to the *right* in the prefix guide, move the decimal point to the *right* the same number of places.	Change 7.2 meters to kilometers. 1. Move three decimal places to the left. $$0.007.2$$ 2. 7.2 m = 0.0072 km Change 196 centimeters to meters. 1. Move two places to the left. $$196.$$ 2. 196 cm = 1.96 m Change 17.3 liters to milliliters. 1. Move three decimal places to the right. $$17.300$$ 2. 17.3 L = 17,300 mL
Changing from American units to metric units, p. 392.	1. From the list of approximate equivalent measures, pick an equality statement that begins with the unit in the original measurement. 1 mi ≈ 1.61 km 1 yd ≈ 0.914 m 1 ft ≈ 0.305 m 1 in. = 2.54 cm (exact) 1 gal ≈ 3.79 L 1 qt ≈ 0.946 L 1 lb ≈ 0.454 kg 1 oz ≈ 28.35 g 2. Multiply by a unit fraction.	Convert 7 gallons to liters. 1. 1 gal ≈ 3.79 L 2. $7 \text{ gal} \times \dfrac{3.79 \text{ L}}{1 \text{ gal}} = 26.53 \text{ L}$ Convert 18 pounds to kilograms. 1. 1 lb ≈ 0.454 kg 2. $18 \text{ lb} \times \dfrac{0.454 \text{ kg}}{1 \text{ lb}} = 8.172 \text{ kg}$

Topic	Procedure	Examples
Changing from metric units to American units, p. 393.	1. From the list of approximate equivalent measures, pick an equality statement that begins with the unit in the original measurement and ends with the unit you want. $\qquad$ 1 km ≈ 0.62 mi $\qquad$ 1 m ≈ 3.28 ft $\qquad$ 1 m ≈ 1.09 yd $\qquad$ 1 cm ≈ 0.394 in. $\qquad$ 1 L ≈ 0.264 gal $\qquad$ 1 L ≈ 1.06 qt $\qquad$ 1 kg ≈ 2.2 lb $\qquad$ 1 g ≈ 0.0353 oz 2. Multiply by a unit fraction.	Convert 605 grams to ounces. 1. 1 g ≈ 0.0353 oz 2. $605 \ g \times \dfrac{0.0353 \text{ oz}}{1 \ g} = 21.3565 \text{ oz}$ Convert 80 km/hr to mi/hr. 1. 1 km ≈ 0.62 mi 2. $80 \dfrac{km}{hr} \times \dfrac{0.62 \text{ mi}}{1 \ km} = 49.6 \text{ mi/hr}$
Changing from Celsius to Fahrenheit temperature, p. 395.	1. To convert Celsius to Fahrenheit, we use the formula $$F = 1.8 \times C + 32.$$ 2. We replace C by the Celsius temperature. 3. We calculate to find the Fahrenheit temperature.	Convert 65°C to Fahrenheit. 1. $F = 1.8 \times C + 32$ 2. $F = 1.8 \times 65 + 32$ 3. $F = 117 + 32 = 149$ The temperature is 149°F.
Changing from Fahrenheit to Celsius temperature, p. 395.	1. To convert Fahrenheit to Celsius, we use the formula $$C = \dfrac{5 \times F - 160}{9}.$$ 2. We replace F by the Fahrenheit temperature. 3. We calculate to find the Celsius temperature.	Convert 50°F to Celsius. 1. $C = \dfrac{5 \times F - 160}{9}$ 2. $C = \dfrac{5 \times 50 - 160}{9}$ 3. $C = \dfrac{250 - 160}{9} = \dfrac{90}{9} = 10$ The temperature is 10°C.

Chapter 6 Review Problems

6.1 *Convert. When necessary, express your answer as a decimal. Round to the nearest hundredth.*

1. 33 ft = _____ yd $\qquad$ **2.** 27 ft = _____ yd $\qquad$ **3.** 5 mi = _____ yd $\quad$ **4.** 6 mi = _____ yd

5. 90 in. = _____ ft $\qquad$ **6.** 78 in. = _____ ft $\qquad$ **7.** 15,840 ft = _____ mi $\quad$ **8.** 10,560 ft = _____ mi

9. 7 tons = _____ lb $\quad$ **10.** 4 tons = _____ lb $\qquad$ **11.** 92 oz = _____ lb $\quad$ **12.** 100 oz = _____ lb

13. 15 gal = _____ qt $\qquad$ **14.** 21 gal = _____ qt $\qquad$ **15.** 31 pt = _____ qt $\quad$ **16.** 27 pt = _____ qt

6.2 *Convert. Do not round.*

17. 56 cm = _____ mm

18. 29 cm = _____ mm

19. 1763 mm = _____ cm

20. 2598 mm = _____ cm

21. 9.2 m = _____ cm

22. 7.4 m = _____ cm

23. 9 km = _____ m

24. 8 km = _____ m

Change all units to meters and add.

25. 6.2 m + 121 cm + 0.52 m

26. 9.8 m + 673 cm + 0.48 m

27. 0.024 km + 1.8 m + 983 cm

28. 0.078 km + 5.5 m + 609 cm

6.3 *Convert. Do not round.*

29. 17 kL = _____ L

30. 23 kL = _____ L

31. 196 kg = _____ g

32. 721 kg = _____ g

33. 778 mg = _____ g

34. 459 mg = _____ g

35. 76 kg = _____ g

36. 41 kg = _____ g

37. 765 cm^3 = _____ mL

38. 423 cm^3 = _____ mL

39. 2.43 L = _____ cm^3

40. 1.93 L = _____ cm^3

6.4 *Perform each conversion. Round to the nearest hundredth.*

41. 42 kg = _____ lb

42. 9 ft = _____ m

43. 13 oz = _____ g

44. 1.3 ft = _____ cm

45. 14 cm = _____ in.

46. 18 cm = _____ in.

47. 20 lb = _____ kg

48. 30 lb = _____ kg

49. 12 yd = _____ m

50. 14 yd = _____ m

51. 80 km/hr = _____ mi/hr

52. 70 km/hr = _____ mi/hr

53. 15°C = _____ °F

54. 25°C = _____ °F

55. 221°F = _____ °C

56. 185°F = _____ °C

57. 32°F = _____ °C

6.5 *Solve. Round to the nearest hundredth when necessary.*

58. A roof is layered with 1.76-cm plywood, 4.32-mm tar paper, and 0.93-cm shingles. How thick is the roof?

▲ **59.** Find the perimeter of the triangle.
(a) Express your answer in feet.
(b) Express your answer in inches.

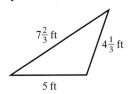

60. Find the perimeter of the rectangle.
(a) Express your answer in meters.
(b) Express your answer in kilometers

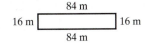

61. The unit price on a box of cereal was $0.14 per ounce. The net weight was 450 grams. How much did the cereal cost?

62. The width of a house in Mexico is 12.6 meters. Kim arrived with enough wooden trim for a house that is 43 feet wide. Did she have enough? How many feet extra or how short was this amount?

63. Keshia traveled at 90 km/hr for three hours. She needs to travel a total distance of 200 miles. How much farther does she need to travel?

64. A German cook wishes to bake a cake at 185° Celsius. The oven is set at 390° Fahrenheit. By how many degrees Fahrenheit is the oven temperature different from what is desired? Is the oven too hot or not hot enough?

65. A flagpole is 19 meters long. The bottom $\frac{1}{5}$ of it is coated with a special water seal before being placed in the ground. How many centimeters long is the portion that has the water seal?

66. A high school track member did a trial run of the 100 meter dash in 13.0 seconds. How many feet per second did he run?

67. Carlos was driving in Mexico City at 80 kilometers per hour. How many miles per hour was he driving?

68. Marcia and Melissa went up the Mt. Washington hiking trail each carrying a backpack tent that weighed 2.2 kg, a canteen with water that weighed 1.4 kg, and some supplies that weighed 3.8 kg. They were told to carry less than 16 pounds while hiking. How close were they to the limit? Did they succeed in carrying less than the weight limit?

69. When buying gas in Canada, Greg Salzman was told by the attendant that in U.S. currency he was paying $0.87 per liter. How much did the gas cost per gallon?

70. The Dodge Caravan that is made in Canada is built to accommodate a person who is 1.88 meters tall. A person who is taller than this will not be comfortable. Would a person who is 6 feet 2 inches tall be comfortable in this car?

▲ **71.** The driveway of Sir Arthur Jensen in London is 4 meters wide and 12 meters long. How many square feet of sealer does he need to cover his driveway?

72. While traveling through Berlin, Al Dundtreim purchased some powdered milk for $1.23 per kilogram. How much would it have cost him to buy 4 pounds of powdered milk?

Chapter 6 Test

Convert. Express your answer as a decimal rounded to the nearest hundredth when necessary.

1. 1.6 tons = _____ lb

2. 19 ft = _____ in.

3. 21 gal = _____ qt

4. 36,960 ft = _____ mi

5. 1800 sec = _____ min

6. 3 cups = _____ qt

Perform each conversion. Do not round.

7. 9.2 km = _____ m

8. 27.3 cm = _____ m

9. 9.88 cm = _____ m

10. 46 mm = _____ cm

11. 12.7 m = _____ cm

12. 0.936 cm = _____ mm

13. 46 L = _____ kL

14. 127 L = _____ mL

15. 28.9 mg = _____ g

16. 983 g = _____ kg

17. 0.92 L = _____ mL

18. 9.42 g = _____ mg

Perform each conversion. Round to the nearest hundredth when necessary.

19. 42 mi = _____ km

20. 1.78 yd = _____ m

1. _____

2. _____

3. _____

4. _____

5. _____

6. _____

7. _____

8. _____

9. _____

10. _____

11. _____

12. _____

13. _____

14. _____

15. _____

16. _____

17. _____

18. _____

19. _____

20. _____

Perform each conversion. Round to the nearest hundredth when necessary.

21. 9 cm = _____ in.

22. 38 L = _____ gal

23. 7.3 kg = _____ lb

24. 3 oz = _____ g

Solve. Round to the nearest hundredth when necessary.

▲**25.** A rectangular picture frame measures 3 m × 7 m.

```
      3 m
   ┌───────┐
   │       │
7 m│       │7 m
   │       │
   └───────┘
      3 m
```

(a) What is the perimeter of the picture frame in meters?
(b) What is the perimeter of the picture frame in yards?

26. The temperature is 80°F today. Kristen's computer has a warning not to operate above 35°C.
(a) How many degrees Fahrenheit are there between the two temperatures?
(b) Can she use her computer today?

27. A pump is running at 5.5 quarts per minute. How many gallons per hour is this?

28. The speed limit on a Canadian road is 100 km/hr.
(a) How far can Samuel travel at this speed limit in three hours?
(b) If Samuel has to travel 200 miles, how much farther will he need to go after three hours of driving at 100 km/hr?

29. The coldest day this winter in Boston was −16°F. What was the Celsius temperature? (Round to the nearest degree.)

30. The warmest day this year in Acapulco, Mexico, was 40°C. What was the Fahrenheit temperature?

Cumulative Test for Chapters 1–6

Approximately one-half of this test is based on Chapter 6 material. The remainder is based on material covered in Chapters 1–5.

Solve. Simplify your answer.

1. Subtract.   9824
           $-\ 3796$

2. Multiply.   608
             $\times\ 305$

3. Divide. $28\overline{)1932}$

4. Add. $\dfrac{1}{7} + \dfrac{3}{14} + \dfrac{2}{21}$

5. Subtract. $3\dfrac{1}{8} - 1\dfrac{3}{4}$

6. Is this equation a proportion? $\dfrac{21}{35} = \dfrac{12}{20}$

7. Solve the proportion. $\dfrac{0.4}{n} = \dfrac{2}{30}$

8. A piece of wire 6.5 centimeters long weighs 68 grams. What will a 20-centimeter length of the same wire weigh? (Round to the nearest hundredth.)

9. What percent of 66 is 165?

10. Find 26% of 7500.

11. 0.5% of what number is 100?

Convert. Express your answer as a decimal rounded to the nearest hundredth when necessary.

12. 38 qt = _____ gal

13. 2.5 tons = _____ lb

14. 7 pt = _____ qt

15. 25 feet = _____ in.

Perform each conversion. Do not round.

16. 3.7 km = _____ m

17. 62.8 g = _____ kg

1. _____

2. _____

3. _____

4. _____

5. _____

6. _____

7. _____

8. _____

9. _____

10. _____

11. _____

12. _____

13. _____

14. _____

15. _____

16. _____

17. _____

18. _____

19. _____

20. _____

21. _____

22. _____

23. _____

24. _____

25. _____

26. _____

27. _____

28. _____

Perform each conversion. Do not round answers to 18–20.

18. 0.79 L = _____ mL   **19.** 5 cm = _____ m

20. 42 lb = _____ oz

Perform each conversion. Round to the nearest hundredth when necessary.

21. 28 gal = _____ L   **22.** 96 lb = _____ kg

23. 7.87 m = _____ ft   **24.** 9 mi = _____ km

▲ **25.** Find the perimeter in meters of this triangle.

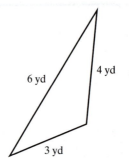

6 yd

4 yd

3 yd

26. Change 15°C to Fahrenheit temperature. Now find the difference between 15°C and 15°F. Which figure represents the higher temperature?

27. Ricardo traveled on a Mexican highway at 100 km/hr for $1\frac{1}{2}$ hours. He needs to travel a total distance of 100 miles. How far does he still need to travel? (Express your answer in miles.)

28. Two metal sheets are 0.72 centimeter thick and 0.98 centimeter thick, respectively. An insulating foil that is 0.38 millimeter thick is placed between them. When the three layers are placed together, what is the total thickness? Write the exact answer.

Geometry

Suppose a major city newspaper were to reduce the width of the paper by two inches. Do you think that this would save a lot of paper? Or would it be an insignificant amount of savings? Would such a change have any real impact on the environment? Turn to the Putting Your Skills to Work problems on page 480 and see if you can answer these questions.

Pretest Chapter 7

Section 7.1

1. Find the complement of an angle that is 72°.

2. Find the supplement of an angle that is 117°.

3. Find the measure of angle *a*, angle *b*, and angle *c* in the sketch to the right, which shows two intersecting straight lines.

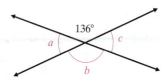

Section 7.2

Find the perimeter of each rectangle or square.

4. Length = 6.5 m, width = 2.5 m   **5.** Length = width = 3.5 m

Find the area of each square or rectangle.

6. Length = width = 4.8 cm   **7.** Length = 2.7 cm, width = 0.9 cm

Section 7.3

Find the perimeter.

8. A parallelogram with one side measuring 9.2 yd and another side measuring 3.6 yd.

9. A trapezoid with sides measuring 17 ft, 15 ft, 25 ft, and 21 ft.

Find the area.

10. A parallelogram with a base of 27 in. and a height of 13 in.

11. A trapezoid with a height of 9 in. and bases of 16 in. and 22 in.

12.

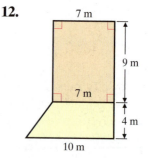

Section 7.4

13. Find the third angle in the triangle if two angles are 39° and 118°.

14. Find the perimeter of the triangle whose sides measure 7.2 m, 4.3 m, and 3.8 m.

15. Find the area of a triangle with a base of 16 m and a height of 9 m.

1. _____

2. _____

3. _____

4. _____

5. _____

6. _____

7. _____

8. _____

9. _____

10. _____

11. _____

12. _____

13. _____

14. _____

15. _____

Section 7.5

Evaluate exactly.

16. $\sqrt{64}$

17. $\sqrt{4} + \sqrt{100}$

18. Approximate $\sqrt{46}$ using a square root table or a calculator with a square root key. Round to the nearest thousandth.

Section 7.6

Find the unknown side of each right triangle.

19.

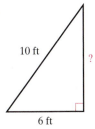

10 ft
?
6 ft

20.

13 ft
5 ft
?

Section 7.7

21. Find the diameter of a circle whose radius is 14 in.

22. Find the circumference of a circle whose diameter is 30 cm.

23. Find the area of a circle whose radius is 9 m.

24. Find the area of the shaded region.

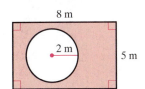

8 m
2 m
5 m

Section 7.8

Find the volume.

25. A rectangular solid with the dimensions of 6 yd by 5 yd by 8 yd.

26. A sphere of radius 3 ft.

27. A cylinder of height 12 in. and radius 7 in.

28. A pyramid of height 21 m with a square base measuring 25 m on a side.

29. A cone of height 30 m and with a radius of 6 m.

16. _____

17. _____

18. _____

19. _____

20. _____

21. _____

22. _____

23. _____

24. _____

25. _____

26. _____

27. _____

28. _____

29. _____

Section 7.9

Find n in each set of similar triangles.

30.

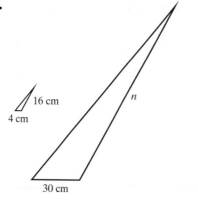

16 cm

4 cm

30 cm

n

31.

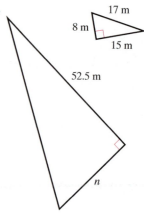

17 m

8 m

15 m

52.5 m

n

Section 7.10

32. A track field consists of two semicircles and a rectangle.
(a) Find the area of the field.
(b) It costs $0.22 per square yard to fertilize the field. What would it cost to complete this task?

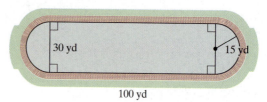

30 yd

15 yd

100 yd

7.1 Angles

1 Understanding and Using Angles

Geometry is a branch of mathematics that deals with the properties of and relationships between figures in space. One of the simplest figures is a *line*. A **line** ↔ extends indefinitely, but a portion of a line, called a **line segment**, has a beginning and an end. An **angle** is formed whenever two lines meet. The two line segments are called the **sides** of the angle. The point at which they meet is called the **vertex** of the angle.

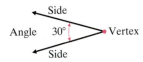

The "amount of opening" of an angle can be measured. Angles are commonly measured in **degrees**. In the preceding sketch the angle measures 30 degrees, or 30°. The symbol ° indicates degrees. If you fix one side of an angle and keep moving the other side, the angle measure will get larger and larger until eventually you have gone around in one complete revolution.

One complete revolution is 360°.

One-half revolution is 180°.

One-fourth revolution is 90°.

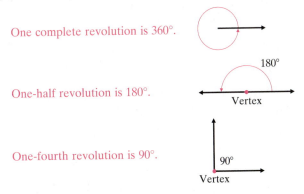

We call two lines **perpendicular** when they meet at an angle of 90°. A 90° angle is called a **right angle**. A 90° angle is often indicated by a small ☐ at the vertex. Thus when you see ⌐ you know that the angle is 90° and also that the sides are perpendicular to each other. The following three angles are right angles.

Often, to avoid confusion, angles are labeled with letters. Suppose we consider the angle with a vertex at point *B*. This angle can be called ∠ *ABC* or ∠ *CBA*. Notice that when three letters are used, the middle letter is the vertex. This angle can also be called ∠ *B* or ∠ *x*.

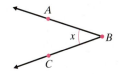

Now consider the following angles. We could label angle *y* as angle *DEF* or angle *FED*. However, we could not label it as ∠ *E* because this would be unclear. If we refer to ∠ *E*, people would not know for sure whether we mean

419

angle *DEF*, angle *FEG*, or angle *DEG*. In such cases where there might be some confusion, the use of the three-letter label is always preferred.

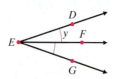

Certain types of angles are commonly encountered. It is important to learn their names. An angle that measures 180° is called a **straight angle**. Angle *ABC* in the following figure is a straight angle. As we mentioned previously, this is one-half of a revolution.

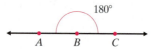

An angle whose measure is between 0° and 90° is called an **acute angle**. ∠ *DEF* and ∠ *GHJ* are both acute angles.

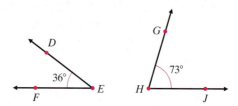

An angle whose measure is between 90° and 180° is called an **obtuse angle**. ∠ *ABC* and ∠ *JKL* are both obtuse angles.

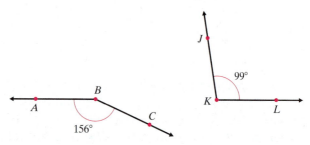

EXAMPLE 1   In the following sketch, determine which angles are acute, obtuse, right, or straight angles.

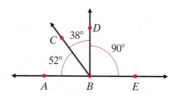

∠ *ABC* and ∠ *CBD* are acute angles, ∠ *CBE* is an obtuse angle, ∠ *ABD* and ∠ *DBE* are right angles, and ∠ *ABE* is a straight angle.

Practice Problem 1   In the following sketch, determine which angles are acute, obtuse, right, or straight angles.

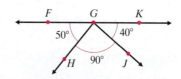

Two angles that have a sum of 90° are called **complementary angles**. We can therefore say that each angle is the **complement** of the other. Two angles that have a sum of 180° are called **supplementary angles**. In this case we say that each angle is the **supplement** of the other.

EXAMPLE 2 Angle A measures 39°.

(a) Find the complement of angle A.   **(b)** Find the supplement of angle A.

(a) Complementary angles have a sum of 90°. So the complement of angle A is $90° - 39° = 51°$.

(b) Supplementary angles have a sum of 180°. So the supplement of angle A is $180° - 39° = 141°$.

Practice Problem 2 Angle B measures 83°.

(a) Find the complement of angle B.   **(b)** Find the supplement of angle B.

Four angles are formed when two lines intersect. Think of how you have four angles if two straight streets intersect. The two angles that are opposite each other are called **vertical angles**. Vertical angles have the same measure. In the following sketch, angle x and angle z are vertical angles, and they have the same measure. Also, angle w and angle y are vertical angles, so they have the same measure.

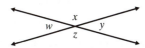

Now suppose we consider two angles that have a common side, such as angle w and angle x. Two angles that are formed by intersecting lines and share a common side are called **adjacent angles**. Adjacent angles of intersecting lines are supplementary. If we know that the measure of angle x is 120°, then we also know that the measure of angle w is 60°.

EXAMPLE 3 In the following sketch, two lines intersect forming four angles. The measure of angle a is 55°. Find the measure of all the other angles.

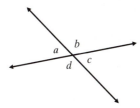

Since $\angle a$ and $\angle c$ are vertical angles, we know that they have the same measure. Thus we know that $\angle c$ measures 55°.

Since $\angle a$ and $\angle b$ are adjacent angles of intersecting lines, we know that they are supplementary angles. Thus we know that $\angle b$ measures $180° - 55° = 125°$.

Finally, $\angle b$ and $\angle d$ are vertical angles, so we know that they have the same measure. Thus we know that $\angle d$ measures 125°.

Practice Problem 3 In the following sketch, two lines intersect forming four angles. The measure of angle y is 133°. Find the measure of all the other angles.

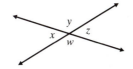

In mathematics there is a common notation for perpendicular lines. If line m is perpendicular to line n, we write $m \perp n$. **Parallel lines** never meet. If line p is parallel to line q, we write $p \parallel q$.

One more situation that is very important in geometry involves lines and angles. A line that intersects two or more lines at different points is called a **transversal**. In the following figure, line m is a transversal that intersects line n and line p. **Alternate interior angles** are two angles that are on opposite sides of the transversal and between the other two lines. In the figure, $\angle c$ and $\angle w$ are alternate interior angles.

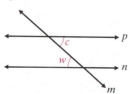

Corresponding angles are two angles that are on the same side of the transversal and are both above (or both below) the other two lines. In the following figure, angle a and angle b are corresponding angles.

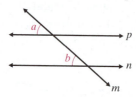

The most important case occurs when the two lines cut by the transversal are parallel. We will state this as follows:

Parallel Lines Cut by a Transversal

If two parallel lines are cut by a transversal, then the measures of **corresponding angles are equal** and the measures of **alternate interior angles** are equal.

EXAMPLE 4 In the following figure, $m \parallel n$ and the measure of $\angle a$ is $64°$. Find the measures of $\angle b$, $\angle c$, $\angle d$, and $\angle e$.

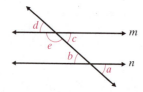

$\angle a = \angle b = 64°$.

$\angle b = \angle c = 64°$.

$\angle b = \angle d = 64°$.

$\angle e = 180° - 64° = 116°$.

Practice Problem 4 In the following figure, $p \parallel q$ and the measure of $\angle x$ is $105°$. Find the measures of $\angle w$, $\angle y$, $\angle z$, and $\angle v$.

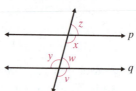

Verbal and Writing Skills

In your own words, give a definition for each term.

1. acute angle

2. obtuse angle

3. complementary angles

4. supplementary angles

5. vertical angles

6. adjacent angles

7. transversal

8. alternate interior angles

In exercises 9–14, two straight lines intersect at B, as shown in the following sketch.

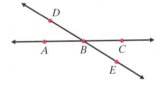

9. Name all the acute angles.

10. Name all the obtuse angles.

11. Name two pairs of angles that have the same measure.

12. Name two pairs of angles that are supplementary.

13. Name two pairs of angles that are complementary, if any exist.

14. Name two different straight angles.

In exercises 15–24, name the measure of each angle, as shown in the following sketch. Assume that angle PQV is a straight angle.

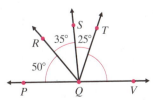

15. $\angle PQR$

16. $\angle PQS$

17. $\angle RQT$

18. $\angle TQP$

19. $\angle TQR$

20. $\angle SQT$

21. $\angle RQS$

22. $\angle RQV$

23. $\angle TQV$

24. $\angle SQV$

25. Find the complement of an angle that is 31°.

26. Find the complement of an angle that is 86°.

27. Find the supplement of an angle that is 127°.

28. Find the supplement of an angle that is 8°.

Find the measure of ∠ a.

29.

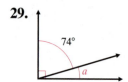

30.

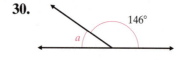

31.

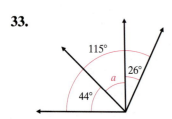

32.

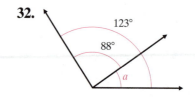

33.

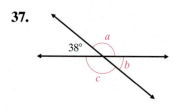

34.

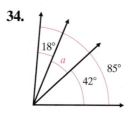

Find the measures of ∠ a, ∠ b, and ∠ c.

35.

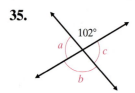

36.

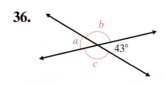

37.

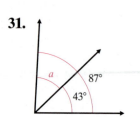

38.

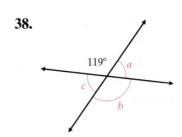

Find the measures of ∠ a, ∠ b, and ∠ c if we know that p ∥ q.

39.

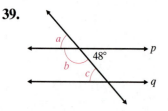

40.

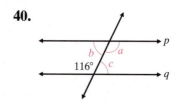

Find the measures of $\angle a$, $\angle b$, $\angle c$, $\angle d$, $\angle e$, $\angle f$, and $\angle g$ if we know that $p \parallel q$.

41.

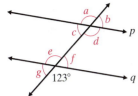

42.

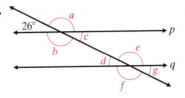

Applications

43. In Mexico the famous pyramids of Monte Albán are visited by thousands of tourists each month. The one most often climbed by tourists is steeper than the pyramids of Egypt and tourists find the climb very challenging. Find the angle x, which indicates the angle of inclination of the pyramid, based on the following sketch.

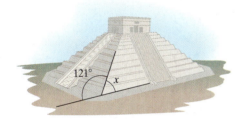

44. The famous Leaning Tower of Pisa has an angle of inclination of 84°. Scientists are working now to move the tower slightly so that it does not lean so much. Find the angle x, at which the tower deviates from the normal upright position.

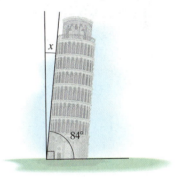

45. A jetliner is flying 56° north of east when it leaves the airport at Dallas–Fort Worth. The control tower orders the plane to change course by turning to the right 7°. Describe the new course in terms of how many degrees north of east the plane is flying.

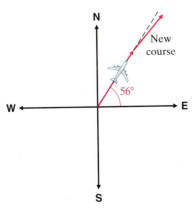

46. A cruise ship is leaving Bermuda and is heading on a course 72° north of west. The captain orders that the ship be turned to the left 9°. Describe the new course in terms of how many degrees north of west the ship is heading.

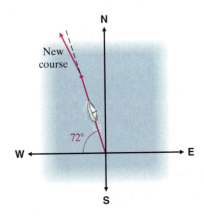

Cumulative Review Problems

47. Antonio jogged 3.2 miles on Monday, 5.8 miles on Tuesday, and 4.3 miles on Wednesday. He wants to jog a total of 23 miles over a five-day period, Monday to Friday. How many total miles will he need to jog on Thursday and Friday?

48. While driving in Mexico, Greg saw a sign that said MEXICO CITY 34 KILOMETERS AHEAD. He is driving an American car with an odometer that reads in miles. How many miles farther does he need to drive? Round to the nearest tenth of a mile.

49. Dr. Finkelstein owns 1000 shares of IBM, 500 shares of America Online, 300 shares of General Electric, and 400 shares of Microsoft. As he was reviewing his stocks over the last year, he observed that IBM had increased in value $4 per share, America Online had increased $3 per share, and General Electric had increased $1 per share. Unfortunately, Microsoft had lost value and was $5 less per share. In total, how much had the value of Dr. Finkelstein's stock increased over the last year?

50. In constructing a new suspension bridge over the Charles River in Boston, engineers have used 34 tons of cable on the first 800 feet of the bridge. How many additional tons of cable will be used to complete the remaining 1300 feet of the bridge? Round to the nearest tenth.

7.2 Rectangles and Squares

① Finding the Perimeters of Rectangles and Squares

Geometry has a visual aspect that numbers and abstract ideas do not have. We can take pen in hand and draw a picture of a rectangle that represents a room with certain dimensions. We can easily visualize problems such as "What is the distance around the outside edges of the room (perimeter)?" or "How much carpeting will be needed for the room (area)?"

A rectangle is a four-sided figure like those shown here.

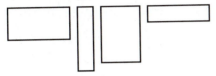

A rectangle has two interesting properties: (1) Any two adjoining sides are perpendicular and (2) the lengths of the opposite sides of a rectangle are equal. By "any two adjoining sides are perpendicular," we mean that any two sides that meet form an angle that measures 90°. When we say that "the lengths of the opposite sides of a rectangle are equal," we mean that the measure of one side is equal to the measure of the side opposite to it. If all four sides have the same length, then the rectangle is called a **square**.

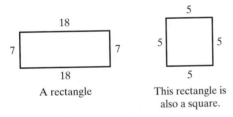

A rectangle	This rectangle is also a square.

A farmer owns some land in the Colorado mountains. It is in the shape of a rectangle. The **perimeter** of a rectangle is the sum of the lengths of all its sides. To find the perimeter of the rectangular field shown in the following figure, we add up the lengths of all the sides of the field.

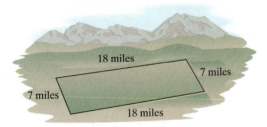

Perimeter = 7 miles + 18 miles + 7 miles + 18 miles
= 50 miles

Thus the perimeter of the field is 50 miles.

We could also use a formula to find the perimeter of a rectangle. In the formula we use letters to represent the measurements of the length and width of the rectangle. Let l represent the length, w represent the width, and P represent the perimeter. Note that the length is the longer side and the width is the shorter side. Since the perimeter is found by adding up the measurements all around the rectangle, we see that

$$P = w + l + w + l$$
$$= 2l + 2w.$$

Student Learning Objectives

After studying this section, you will be able to:

① Find the perimeters of rectangles and squares.

② Find the perimeters of shapes made up of rectangles and squares.

③ Find the areas of rectangles and squares.

④ Find the areas of shapes made up of rectangles and squares.

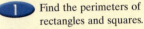

SSM   PH TUTOR   CD & VIDEO   MATH PRO   WEB
CENTER

427

When we write $2l$ and $2w$, we mean 2 times l and 2 times w. We can use the formula to find the perimeter of the rectangle.

$$P = 2l + 2w$$
$$= (2)(18 \text{ mi}) + (2)(7 \text{ mi})$$
$$= 36 \text{ mi} + 14 \text{ mi}$$
$$= 50 \text{ mi}$$

Thus the perimeter can be found quickly by using the following formula.

> The **perimeter (P) of a rectangle** is twice the length plus twice the width.
> $$P = 2l + 2w$$

EXAMPLE 1 A helicopter has a 3-cm by 5.5-cm insulation pad near the control panel that is rectangular. Find the perimeter of the rectangle.

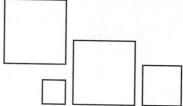

$$\text{Length} = l = 5.5 \text{ cm}$$

$$\text{Width} = w = 3 \text{ cm}$$

In the formula for the perimeter of a rectangle, we substitute 5.5 cm for l and 3 cm for w. Remember, $2l$ means 2 times l and $2w$ means 2 times w. Thus

$$P = 2l + 2w$$
$$= (2)(5.5 \text{ cm}) + (2)(3 \text{ cm})$$
$$= 11 \text{ cm} + 6 \text{ cm} = 17 \text{ cm}.$$

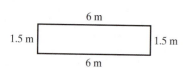

Practice Problem 1 Find the perimeter of the rectangle in the margin.

A square is a rectangle where all four sides have the same length. Some examples of squares are shown in the following figure.

A square, then, is only a special type of rectangle. We can find the perimeter of a square just as we found the perimeter of a rectangle—by adding the measurements of all the sides of the square. Because the lengths of all sides are the same, the formula for the perimeter of a square is very simple. Let s represent the length of one side and P represent the perimeter. So, to find the perimeter, we multiply the length of a side by 4.

> The **perimeter of a square** is four times the length of a side.
> $$P = 4s$$

● **EXAMPLE 2**   High Ridge Stables has a new sign at the highway entrance that is in the shape of a square, with each side measuring 8.6 yards. Find the perimeter of the sign.

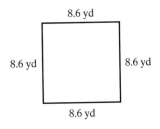

8.6 yd

8.6 yd   8.6 yd

8.6 yd

Side = s = 8.6 yd

$P = 4s$

$= (4)(8.6\text{ yd})$

$= 34.4\text{ yd}$

Practice Problem 2   Find the perimeter of the square in the margin.   ●

5.8 cm

5.8 cm   5.8 cm

5.8 cm

2   Finding the Perimeters of Shapes Made Up of Rectangles and Squares

Some figures are a combination of rectangles and squares. To find the perimeter of the total figure, look only at the outside edges.

We can apply our knowledge to everyday problems. For example, by knowing how to find the perimeter of a rectangle, we can find out how many feet of picture framing a painting will need or how many feet of weather stripping will be needed to seal a doorway. Consider the following problem.

● **EXAMPLE 3**   Find the cost of weather stripping needed to seal the edges of the hatch of a boat pictured at right. Weather stripping costs $0.12 per foot.

First we need to find the perimeter of the hatch. The perimeter is the sum of all the edges.

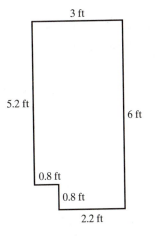

3 ft

5.2 ft

6 ft

0.8 ft

0.8 ft

2.2 ft

```
   3.0 ft
   6.0 ft
   2.2 ft
   0.8 ft
   0.8 ft
 + 5.2 ft
  18.0 ft
```

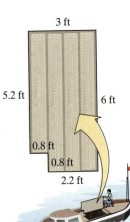

3 ft

5.2 ft   6 ft

0.8 ft

0.8 ft

2.2 ft

The perimeter is 18 ft. Now we calculate the cost.

$18.0\ \cancel{ft} \times \dfrac{0.12\text{ dollar}}{\cancel{ft}} = \$2.16$ for weather stripping materials

Practice Problem 3 Find the cost of weather stripping required to seal the edges of the hatch shown below. Weather seal stripping costs $0.16 per foot.

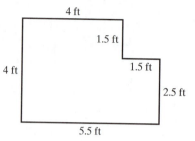

3 ▸ *Finding the Areas of Rectangles and Squares*

What do we mean by **area?** Area is the measure of the *surface inside* a geometric figure. For example, for a rectangular room, the area is the amount of floor in that room.

One *square meter* is the measure of a square that is 1 m long and 1 m wide.

We can abbreviate *square meter* as m^2. In fact, all areas are measured in square meters, square feet, square inches, and so on (written as m^2, ft^2, $in.^2$, and so on).

We can calculate the area of a rectangular region if we know its length and its width. To find the area, *multiply* the length by the width.

> The **area (A) of a rectangle** is the length times the width.
>
> $$A = lw$$

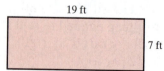

⬬ **EXAMPLE 4** Find the area of the rectangle shown at right.

Our answer must be in square feet because the measures of the length and width are in feet.

$$l = 19 \text{ ft} \quad w = 7 \text{ ft}$$

$$A = (l)(w) = (19 \text{ ft})(7 \text{ ft}) = 133 \text{ ft}^2$$

The area is 133 square feet.

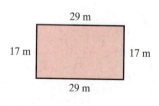

Practice Problem 4 Find the area of the rectangle in the margin.

To find the area of a square, we multiply the length of one side by itself.

> The **area of a square** is the square of the length of one side.
>
> $$A = s^2$$

EXAMPLE 5   A square measures 9.6 in. on each side. Find its area.

We know our answer will be measured in square inches. We will write this as in.^{2}.

$$A = s^2$$
$$= (9.6)^2$$
$$= (9.6 \text{ in.})(9.6 \text{ in.})$$
$$= 92.16 \text{ in.}^2$$

Practice Problem 5   Find the area of a square computer chip that measures 11.8 mm on each side.

4   **Finding the Areas of Shapes Made Up of Rectangles and Squares**

EXAMPLE 6   Consider the shape shown at right, which is made up of a rectangle and a square. Find the area of the shaded region.

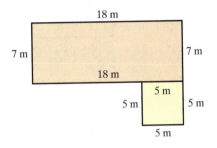

The shaded region is made up of two separate regions. You can think of each separately, and calculate the area of each one. The total area is just the sum of the two separate areas.

Area of rectangle = $(7)(18) = 126 \text{ m}^2$ Area of square = $5^2 = 25 \text{ m}^2$

$$
\begin{array}{ll}
\text{The area of the rectangle} & = 126 \text{ m}^2 \\
+ \text{ The area of the square} & = 25 \text{ m}^2 \\
\hline
\text{The total area is} & = 151 \text{ m}^2
\end{array}
$$

Practice Problem 6   Find the area of the shaded region shown in figure below.

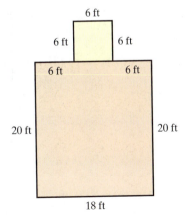

Verbal and Writing Skills

1. A rectangle has two properties: (1) any two adjoining sides are _____ and (2) the lengths of opposite sides are _____.

2. To find the perimeter of a figure, we _____ the lengths of all of the sides.

3. To find the area of a rectangle, we _____ the length by the width.

4. All area is measured in _____ units.

Find the perimeter of the rectangle or square.

5.

5.5 mi

2 mi   2 mi

5.5 mi

6.

9 cm

1.5 cm   1.5 cm

9 cm

7.

2.5 ft

9.3 ft   9.3 ft

2.5 ft

8.

11.3 ft

8.7 ft   8.7 ft

11.3 ft

9.

4.2 ft

12.8 ft   12.8 ft

4.2 ft

10.

15.6 ft

15.6 ft   15.6 ft

15.6 ft

11. Length = 0.84 mm, width = 0.12 mm

12. Length = 6.2 in., width = 1.5 in.

13. Length = width = 4.28 km

14. Length = width = 9.63 cm

15. Length = 3.2 ft, width = 48 in.
(*Hint*: Make the units of length the same.)

16. Length = 8.5 ft, width = 30 in.
(*Hint*: Make the units of length the same.)

Find the perimeter of the square. The length of the side is given.

17. 0.068 mm

18. 0.097 mm

19. 7.96 cm

20. 6.32 cm

Find the perimeter of each shape made up of rectangles and squares.

21.

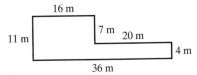

22.

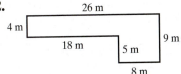

23.

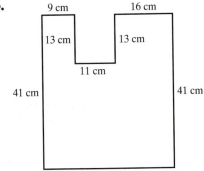

24.

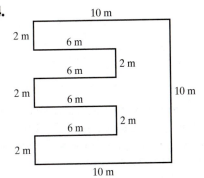

Find the area of the rectangle or square.

25. Length = 0.96 m, width = 0.3 m

26. Length = 0.132 m, width = 0.02 m

27. Length = 39 yd, width = 9 ft
(*Hint*: Make the units of length the same.)

28. Length = 57 yd, width = 15 ft
(*Hint*: Make the units of length the same.)

Find the shaded area.

29.

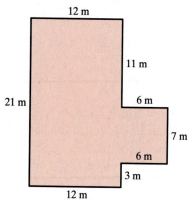

12 m

11 m

21 m

6 m

7 m

6 m

3 m

12 m

30.

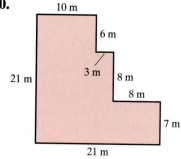

10 m

6 m

21 m

3 m

8 m

8 m

7 m

21 m

Applications

Some of the following exercises will require that you find a perimeter. Others will require that you find an area. Read each problem carefully to determine which you are to find.

31. A hotel conference center is building an indoor fitness area measuring 220 ft × 50 ft. The flooring to cover the space is made of a special three-layered cushioned tile and costs $12.00 per square foot. How much will the new flooring cost?

32. A beach volleyball area will feature four volleyball courts, which will take up an area of 500 ft × 90 ft. To put down an adequate amount of sifted sand, it will cost $0.27 per square foot. How much will it cost to sand the volleyball courts?

33. Arlene made a quilt that took first place at the county fair. The quilt measured 12.5 ft by 9.5 ft. She sewed a unique fringed border on each side of it. If the border material cost $1.35 per foot, how much did the border cost?

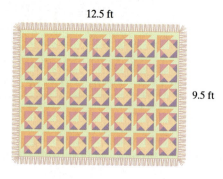

12.5 ft

9.5 ft

34. Sammy's Scuba Shop is installing a new sign measuring 5.4 ft × 8.1 ft. The sign will be framed in purple neon light, which will cost $32.50 per foot. How much will it cost to frame the sign in purple neon light?

A family decides to have custom carpeting installed. It will cost $14.50 per square yard. The binding, which runs along the outside edges of the carpet, will cost $1.50 per yard. Find the cost of carpeting and binding for each room. Note that dimensions are given in feet. (Remember, 1 square yard equals 9 square feet.)

35.

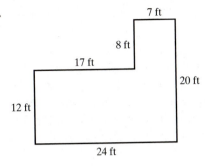

36.

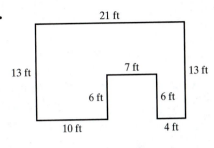

To Think About

37. Russ and Norma had a dining room rug that was rectangular. It was 11 feet long and 3 feet shorter in width than length. It was badly damaged by the pets in the house. When Russ and Norma replaced the rug, they could not find one the same size. They settled for a new rug that is 3 inches shorter in length and 2 inches shorter in width. How much smaller is the area of the new rug compared to the area of the old rug? Express your answer in square inches.

38. The perimeter of a rectangular school athletic field is 900 feet. The length of the field is twice the width. The school principal would like to cut down some trees and increase the length of the field by 50 feet and the width of the field by 70 feet. How much larger will the area of the new field be compared to the area of the existing field?

Cumulative Review Problems

39. Add.   156.8
              27.2
          + 39.3

40. Subtract.   200.57
              − 193.39

41. Multiply.   1076
              × 20.3

42. Divide. $12.3\overline{)19.384}$

43. On January 7, 2000, 1 share of Merck Pharmaceuticals stock was worth $72\frac{3}{4}$. Kendra bought 11 shares. The next day, the value of each share went down $2\frac{3}{16}$.
   (a) What was the total value of the 11 shares when Kendra first bought them?
   (b) What was the total value of the 11 shares the following day?

7.3 Parallelograms, Trapezoids, and Rhombuses

1 Finding the Perimeter and Area of a Parallelogram or a Rhombus

Parallelograms, rhombuses, and trapezoids are figures related to rectangles. Actually, they are in the same "family," the **quadrilaterals** (four-sided figures). For all these figures, the perimeter is the distance around the figure. But there is a different formula for finding the area of each.

A **parallelogram** is a four-sided figure in which both pairs of opposite sides are parallel. The opposite sides of a parallelogram are equal in length.

The following figures are parallelograms. Notice that the adjoining sides need not be perpendicular.

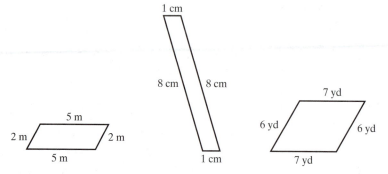

The **perimeter** of a parallelogram is the distance around the parallelogram. It is found by adding the lengths of all the sides of the figure.

⬤ EXAMPLE 1   Find the perimeter.

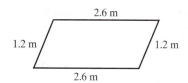

$$P = (2)(1.2 \text{ meters}) + (2)(2.6 \text{ meters})$$
$$= 2.4 \text{ meters} + 5.2 \text{ meters} = 7.6 \text{ meters}$$

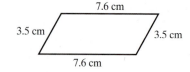

Practice Problem 1   Find the perimeter of the parallelogram in the margin.

To find the **area** of a parallelogram, we multiply the base times the height. Any side of a parallelogram can be considered the **base**. The **height** is the shortest distance between the base and the side opposite the base. The height is a line segment that is perpendicular to the base. When we write the formula for area, we use the lengths of the base (*b*) and the height (*h*).

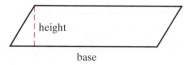

> The **area of a parallelogram** is the base (*b*) times the height (*h*).
>
> $$A = bh$$

Why is the area of a parallelogram equal to the base times the height? What reasoning leads us to that formula? Suppose that we cut off the triangular region on one side of the parallelogram and move it to the other side.

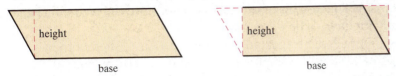

We now have a rectangle.

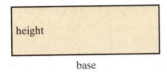

To find the area, we multiply the width by the length. In this case, $A = bh$. Thus finding the area of a parallelogram is like finding the area of a rectangle of length b and width h: $A = bh$.

EXAMPLE 2   Find the area of a parallelogram with base 7.5 m and height 3.2 m.

$$A = bh$$
$$= (7.5\ m)(3.2\ m)$$
$$= 24\ m^2$$

Practice Problem 2   Find the area of a parallelogram with base 10.3 km and height 1.5 km.

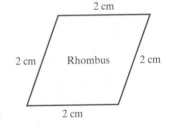

A **rhombus** is a parallelogram with all four sides equal. The figure to the right, with each side of length 2 centimeters, is a rhombus. We will solve a problem involving a rhombus in Example 3.

EXAMPLE 3   A truck is manufactured with an iron brace welded to the truck frame. The brace is shaped like a rhombus. The brace has a base of 9 inches and a height of 5 inches. Find the perimeter and the area of this iron brace.
    Since all four sides are equal, we merely multiply

$$P = 4(9\ in.) = 36\ in.$$

The perimeter of this brace is 36 inches.
    Since the rhombus is a special type of parallelogram, we can use the area formula for a parallelogram. In this case the base is 9 inches and the height is 5 inches.

$$A = bh = (9\ in.)(5\ in.) = 45\ in.^2$$

Thus the area of the brace is 45 square inches.

Practice Problem 3   An inlaid piece of cherry wood on the front of a hope chest is shaped like a rhombus. This piece has a base of 6 centimeters and a height of 4 centimeters. Find the perimeter and the area of this inlaid piece of cherry wood.

2   *Finding the Perimeter and Area of a Trapezoid*

A **trapezoid** is a four-sided figure with two parallel sides. The parallel sides are called **bases**. The lengths of the bases do not have to be equal. The adjoining sides do not have to be perpendicular.

Sometimes the trapezoid is sitting on a base. Then both bases are horizontal. But be careful. Sometimes the bases are vertical. You can recognize the bases because they are the two parallel sides. This becomes important when you use the formula for finding the area of a trapezoid.

Look at the following trapezoids. See if you can recognize the bases.

The perimeter of a trapezoid is the sum of the lengths of all of its sides.

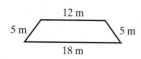

EXAMPLE 4   Find the perimeter of the trapezoid on the left.

$$P = 18 \text{ m} + 5 \text{ m} + 12 \text{ m} + 5 \text{ m}$$
$$= 40 \text{ m}$$

Practice Problem 4   Find the perimeter of a trapezoid with sides of 7 yd, 15 yd, 21 yd, and 13 yd.

Remember, we often use parentheses as a way to group numbers together. The numbers inside parentheses should be combined first.

$$(5)(7 + 2) = (5)(9) \quad \text{First we add numbers inside the parentheses.}$$
$$= 45 \quad \text{Then we multiply.}$$

The formula for the area of a trapezoid uses parentheses in this way.

The **height** of a trapezoid is the distance between the two parallel sides. The area of a trapezoid is one-half the height times the sum of the bases. (This means you add the bases *first*.)

The **area of a trapezoid** with a shorter base b, a longer base B, and height h is

$$A = \frac{h(b + B)}{2}.$$

base = b

height = h

base = B

● **EXAMPLE 5**   A roadside sign is in the shape of a trapezoid. It has a height of 30 ft, and the bases are 60 ft and 75 ft.

(a) What is the area of the sign?

(b) If 1 gallon of paint covers 200 ft^2, how many gallons of paint will be needed to paint the sign?

(a) We use the trapezoid formula with $h = 30$, $b = 60$, and $B = 75$.

$$A = \frac{h(b + B)}{2}$$

$$= \frac{(30 \text{ ft})(60 \text{ ft} + 75 \text{ ft})}{2}$$

$$= \frac{(30 \text{ ft})(135 \text{ ft})}{2} = \frac{4050}{2} \text{ ft}^2 = 2025 \text{ ft}^2$$

(b) Each gallon covers 200 ft^2, so we multiply the area by the fraction $\dfrac{1 \text{ gal}}{200 \text{ ft}^2}$.

This fraction is equivalent to 1.

$$2025 \text{ ft}^2 \times \frac{1 \text{ gal}}{200 \text{ ft}^2} = \frac{2025}{200} \text{ gal}$$

$$= 10.125 \text{ gal}$$

Thus 10.125 gallons of paint would be needed. In real life we would buy 11 gallons of paint.

Practice Problem 5   A corner parking lot is shaped like a trapezoid. The trapezoid has a height of 140 yd. The bases measure 180 yd and 130 yd.

(a) Find the area of the parking lot.

(b) If 1 gallon of sealant will cover 100 square yards of the parking lot, how many gallons are needed to cover the entire parking lot?   ●

Some area problems involve two or more separate regions. Remember, areas can be added or subtracted.

EXAMPLE 6 Find the area of the following piece for inlaid woodwork made by a master carpenter. Since this shape is hard to cut, it is made of one trapezoid and one rectangle laid together.

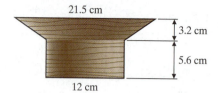

We separate the area into two portions and find the area of each portion separately.

The area of the trapezoid is

$$A = \frac{h(b + B)}{2}$$

$$= \frac{(3.2 \text{ cm})(12 \text{ cm} + 21.5 \text{ cm})}{2}$$

$$= \frac{(3.2 \text{ cm})(33.5 \text{ cm})}{2}$$

$$= \frac{107.2}{2} \text{ cm}^2$$

$$= 53.6 \text{ cm}^2.$$

The area of the rectangle is

$$A = lw$$
$$= (12 \text{ cm})(5.6 \text{ cm})$$
$$= 67.2 \text{ cm}^2.$$

We now add each area.

$$\begin{array}{r} 67.2 \text{ cm}^2 \\ + 53.6 \text{ cm}^2 \\ \hline 120.8 \text{ cm}^2 \end{array}$$

The total area of the piece for inlaid woodwork is 120.8 cm^2.

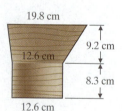

Practice Problem 6 Find the area of the piece for inlaid woodwork shown in the margin. The shape is made of one trapezoid and one rectangle.

7.3 Exercises

Verbal and Writing Skills

1. The perimeter of a parallelogram is found by _____ the lengths of all the sides of the figure.

2. To find the area of a parallelogram, multiply the base times the _____.

3. The height of a parallelogram is a line segment that is _____ to the base.

4. The area of a trapezoid is one-half the height times the _____ of the bases.

Find the perimeter of the parallelogram.

5. One side measures 2.8 m and a second side measures 17.3 m.

6. One side measures 4.6 m and a second side measures 20.5 m.

7.

8.

Find the area of the parallelogram.

9. The base is 14.2 m and the height is 21.25 m.

10. The base is 17.6 m and the height is 20.15 m.

11. A courtyard is shaped like a parallelogram. Its base is 126 yd and its height is 28 yd. Find its area.

12. The preferred seating area at the South Shore Music Theatre is in the shape of a parallelogram. Its base is 28 yd and its height is 21.5 yd. Find the area.

13. Find the perimeter and the area of a rhombus with height 6 meters and base 12 meters.

14. Find the perimeter and the area of a rhombus with height 9 yards and base 14 yards.

15. Walter made his son Daniel a kite in the shape of a rhombus. The height of the kite is 1.5 feet. The length of the base of the kite is 2.4 feet. Find the perimeter and the area of the kite.

16. The lawn in front of Bradley Palmer State Park is constructed in the shape of a rhombus. The height of the lawn region is 17 feet. The length of the base is 25 feet. Find the perimeter and the area of this lawn.

Find the perimeter of the trapezoid.

17.

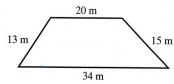

18.

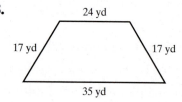

19. The four sides are 130 cm, 70 cm, 260 cm, and 40 cm.

20. The four sides are 90 cm, 110 cm, 150 cm, and 50 cm.

Find the area of the trapezoid.

21. The height is 12 yd and the bases are 9.6 yd and 10.2 yd.

22. The height is 18 yd and the bases are 8.4 yd and 17.8 yd.

23. A provincial park in Canada is laid out in the shape of a trapezoid. The trapezoid has a height of 20 km. The bases are 24 km and 31 km. Find the area of the park.

24. An underwater diving area for snorkelers and scuba divers in Key West, Florida, is designated by buoys and ropes, making the diving section into the shape of a trapezoid on the surface of the water. The trapezoid has a height of 265 meters. The bases are 300 meters and 280 meters. Find the area of the designated diving area.

Find the area of the entire shape made of trapezoids, parallelograms, squares, and rectangles.

25.

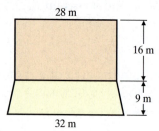

26.

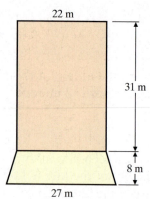

27.

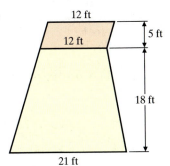

12 ft
12 ft
5 ft
18 ft
21 ft

28.

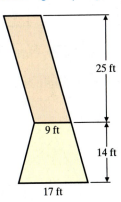

25 ft
9 ft
14 ft
17 ft

Each of the following shapes represents the lobby of a conference center. The lobby will be carpeted at a cost of $22 per square yard. How much will the carpeting cost?

29.

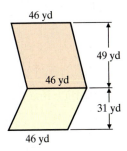

46 yd
49 yd
46 yd
31 yd
46 yd

30.

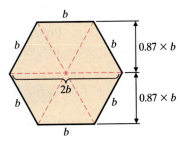

50 yd
72 yd   72 yd
50 yd
24 yd   30 yd
68 yd

To Think About

31. See if you can find a formula that would give the area of a regular octagon. (A regular octagon is an eight-sided figure with all sides of equal length.) The dimensions of the rectangles and trapezoids are labeled on the sketch.

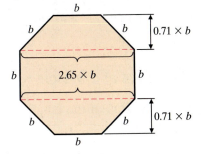

b
b b
$0.71 \times b$
b $2.65 \times b$ b
$0.71 \times b$
b b
b

32. See if you can find a formula for the area of a regular hexagon. Each side is b units long. (A regular hexagon is a six-sided figure with all sides of equal length.)

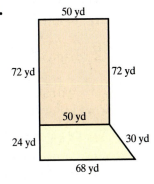

b
b b $0.87 \times b$
$2b$
b b $0.87 \times b$
b

Cumulative Review Problems   *Complete each conversion.*

33. 10 yd = _____ ft

34. 15,840 ft = _____ mi

35. 18 m = _____ cm

36. 26 mm = _____ cm

1 Finding the Measures of Angles in a Triangle

A **triangle** is a three-sided figure with three angles. The prefix *tri*-means "three." Some triangles are shown.

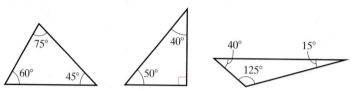

Although all triangles have three sides, not all triangles have the same shape. The shape of a triangle depends on the sizes of the angles and the lengths of the sides.

We will begin our study of triangles by looking at the angles. Although the sizes of the angles in triangles may be different, the sum of the angle measures of any triangle is always 180°.

> The sum of the measures of the angles in a triangle is 180°.

We can use this fact to find the measure of an unknown angle in a triangle if we know the measures of the other two angles.

EXAMPLE 1 In the triangle to the right, angle *A* measures 35° and angle *B* measures 95°. Find the measure of angle *C*.

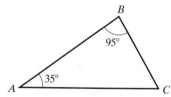

We will use the fact that the sum of the measures of the angles of a triangle is 180°.

$$35 + 95 + x = 180$$
$$130 + x = 180$$

What number *x* when added to 130 equals 180? Since $130 + 50 = 180$, *x* must equal 50.

Angle *C* must equal 50°.

Practice Problem 1 In a triangle, angle *B* measures 125° and angle *C* measures 15°. What is the measure of angle *A*?

2 Finding the Perimeter and the Area of a Triangle

Recall that the perimeter of any figure is the sum of the lengths of its sides. Thus the perimeter of a triangle is the sum of the lengths of its three sides.

EXAMPLE 2 Find the perimeter of a triangular sail whose sides are 12 ft, 14 ft, and 17 ft.

$$P = 12 \text{ ft} + 14 \text{ ft} + 17 \text{ ft} = 43 \text{ ft}$$

Practice Problem 2 Find the perimeter of a triangle whose sides are 10.5 m, 10.5 m, and 8.5 m.

Some triangles have special names. A triangle with two equal sides is called an **isosceles triangle**.

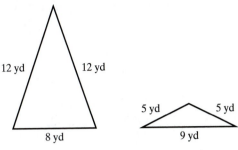

Isosceles triangles

A triangle with three equal sides is called an **equilateral triangle**. All angles in an equilateral triangle are exactly 60°.

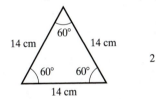

Equilateral triangles

A **scalene triangle** has no two sides of equal lengths and no two angles of equal measure.

A triangle with one 90° angle is called a **right triangle**.

The **height** of any triangle is the distance of a line drawn from a vertex perpendicular to the opposite side or an extension of the opposite side. The height may be one of the sides in a right triangle. The **base** of a triangle is perpendicular to the height.

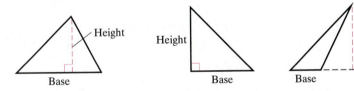

To find the area of a triangle, we need to be able to identify its height and base. The area of any triangle is half of the product of the base times the height of the triangle. The height is measured from the vertex above the base to that base.

The **area of a triangle** is the base times the height divided by 2.

$$A = \frac{bh}{2}$$

h = height

b = base

Where does the 2 come from in the formula $A = \dfrac{bh}{2}$? Why does this formula for the area of a triangle work? Suppose that we construct a triangle with base b and height h.

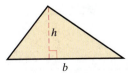

Now let us make an exact copy of the triangle and turn the copy around to the right exactly 180°. Carefully place the two triangles together. We now have a parallelogram of base b and height h. The area of a parallelogram is $A = bh$.

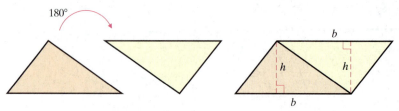

Because the parallelogram has area $A = bh$ and is made up of two triangles of identical shape and area, the area of one of the triangles is the area of the parallelogram divided by 2. Thus the area of a triangle is $A = \dfrac{bh}{2}$.

EXAMPLE 3 Find the area of the triangle.

$$A = \frac{bh}{2} = \frac{(23 \text{ m})(16 \text{ m})}{2} = \frac{368 \text{ m}^2}{2} = 184 \text{ m}^2$$

Practice Problem 3 Find the area of the triangle in the margin.

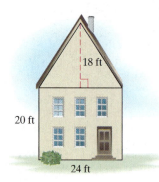

In some geometric shapes, a triangle is combined with rectangles, squares, parallelograms, and trapezoids.

EXAMPLE 4 Find the area of the side of the house shown in the margin.

Because the lengths of opposite sides of a rectangle are equal, the triangle has a base of 24 ft. Thus we can calculate its area.

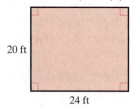

$$A = \frac{bh}{2} = \frac{(24 \text{ ft})(18 \text{ ft})}{2} = \frac{432 \text{ ft}^2}{2} = 216 \text{ ft}^2$$

The area of the rectangle is $A = lw = (24 \text{ ft})(20 \text{ ft}) = 480 \text{ ft}^2$.

20 ft

24 ft

Now we find the sum of the two areas.

$$\begin{array}{r} 216 \text{ ft}^2 \\ + 480 \text{ ft}^2 \\ \hline 696 \text{ ft}^2 \end{array}$$

Thus the area of the side of the house is 696 square feet.

Practice Problem 4 Find the area of the figure.

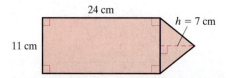

7.4 Exercises

Verbal and Writing Skills

1. A 90° angle is called a _____ angle.

2. The sum of the angle measures of a triangle is _____.

3. Explain in your own words how you would find the measure of an unknown angle in a triangle if you knew the measures of the other two angles.

4. If you were told that a triangle was an isosceles triangle, what could you conclude about the sides of that triangle?

5. If you were told that a triangle was an equilateral triangle, what could you conclude about the sides of the triangle?

6. How do you find the area of a triangle?

Write true or false for each statement.

7. Two lines that meet at a 90° angle are perpendicular.

8. A right triangle has two angles of 90°.

9. The sum of the angles of a triangle is 180°.

10. An isosceles triangle has two sides of equal length.

11. An equilateral triangle has one angle greater than 90°.

12. All equilateral triangles are the same size.

13. The measures of the angles of an equilateral triangle are all equal.

14. To find the perimeter of an equilateral triangle, you can multiply the length of one of the sides by 3.

Find the missing angle in the triangle.

15. Two angles are 36° and 74°.

16. Two angles are 23° and 95°.

17. Two angles are 44° and 8°.

18. Two angles are 136° and 18°.

Find the perimeter of the triangle.

19. A triangle whose sides are 36 m, 27 m, and 41 m.

20. A triangle whose sides are 71 m, 65 m, and 82 m.

21. An isosceles triangle whose sides are 45.25 in., 35.75 in., and 35.75 in.

22. An isosceles triangle whose sides are 36.2 in., 47.65 in., and 47.65 in.

447

23. An equilateral triangle whose side measures 3.5 mi.

24. An equilateral triangle whose side measures 4.6 mi.

Find the area of the triangle.

25.

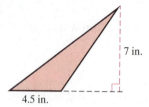

7 in.

4.5 in.

26.

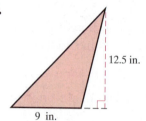

12.5 in.

9 in.

27. The base is 17.5 cm and the height is 9.5 cm.

28. The base is 3.6 cm and the height is 11.2 cm.

29. The base is 3.5 yd and the height is 30 ft.

30. The base is 5.6 yd and the height is 42 ft.

31.

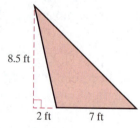

8.5 ft

2 ft   7 ft

32.

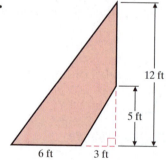

12 ft

5 ft

6 ft   3 ft

Find the area of the shaded region.

33.

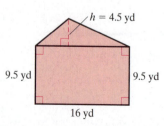

$h = 4.5$ yd

9.5 yd   9.5 yd

16 yd

34.

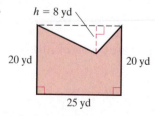

$h = 8$ yd

20 yd   20 yd

25 yd

Applications

Find the total area of all four vertical sides of the building.

35.

36.

The top surface of the wings of a test plane must be coated with a special lacquer that costs $90 per square yard. Find the cost to coat the shaded wing surface of the plane.

37.

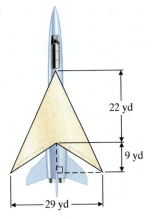

38.

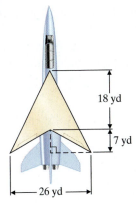

To Think About

An equilateral triangle has a base of 20 meters and a height of h meters. Inside that triangle is constructed a second equilateral triangle of base 10 meters and a height of 0.5h meters. Inside the second triangle is constructed a third equilateral triangle of base 5 meters and a height of 0.25h meters.

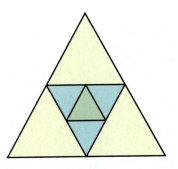

39. What percent of the area of the largest triangle is the area of the smallest triangle?

40. What percent of the perimeter of the largest triangle is the perimeter of the smallest triangle?

Cumulative Review Problems

Solve for n. Round to the nearest hundredth.

41. $\dfrac{5}{n} = \dfrac{7.5}{18}$

42. $\dfrac{n}{\frac{3}{4}} = \dfrac{7}{\frac{1}{8}}$

43. On the cruise ship *H.M.S. Salinora*, all restaurants on board must keep the ratio of waitstaff to patrons at 4 to 15. How many waitstaff should be serving if there are 2685 patrons dining at one time?

44. Recently, an airline found that after the transatlantic flight to Frankfurt, 68 people out of 300 passengers kept their in-flight magazines after being encouraged to take the magazines with them to read at their leisure. On a similar flight carrying 425 people, how many in-flight magazines would the airline expect to be taken? Round to the nearest whole number.

45. Ahmal wants to be able to take his laptop computer into any room of his apartment while he is online. The greatest distance from the living room phone jack to any point in the apartment is 52 feet. Ahmal plans to use three cords to span this distance. If he has attached a $6\frac{1}{2}$-foot cord to an existing cord that measures $14\frac{2}{3}$ feet, what must the minimum length of the third cord be?

7.5 Square Roots

1 Evaluating the Square Root of a Number That Is a Perfect Square

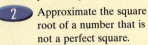
We know that by using the formula $A = s^2$ we can quickly find the area of a square with a side of 3 in. We simply square 3 in. That is, $A = (3 \text{ in.})(3 \text{ in.}) = 9 \text{ in.}^2$. Sometimes we want to ask another kind of question. If a square has an area of 64 in.², what is the length of its sides?

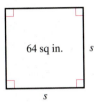

3 in.

3 in.

Area = 9 sq in.

64 sq in. s

s

The answer is 8 in. Why? The skill we need to find a number when we are given the square of that number is called *finding the square root*. The square root of 64 is 8.

If a number is a product of two identical factors, then either factor is called a **square root**.

> The square root of 64 is 8 because $(8)(8) = 64$.
>
> The square root of 9 is 3 because $(3)(3) = 9$.

The symbol for finding the square root of a number is $\sqrt{}$. To write the square root of 64, we write $\sqrt{64} = 8$. Sometimes we speak of finding the square root of a number as *taking* the square root of the number, or we can say that we will *evaluate* the square root of the number. Thus to take the square root of 9, we write $\sqrt{9} = 3$; to evaluate the square root of 9, we write $\sqrt{9} = 3$.

EXAMPLE 1 Find. **(a)** $\sqrt{25}$ **(b)** $\sqrt{121}$

(a) $\sqrt{25} = 5$ because $(5)(5) = 25$.
(b) $\sqrt{121} = 11$ because $(11)(11) = 121$.

Practice Problem 1 Find. **(a)** $\sqrt{49}$ **(b)** $\sqrt{169}$

If square roots are added or subtracted, they must be evaluated *first*, then added or subtracted.

EXAMPLE 2 Find. $\sqrt{25} + \sqrt{36}$

$$\sqrt{25} = 5 \text{ because } (5)(5) = 25.$$
$$\sqrt{36} = 6 \text{ because } (6)(6) = 36.$$

Thus $\sqrt{25} + \sqrt{36} = 5 + 6 = 11$.

Practice Problem 2 Find. $\sqrt{49} - \sqrt{4}$

When a whole number is multiplied by itself, the number that is obtained is called a **perfect square**.

> 36 is a perfect square because $(6)(6) = 36$.
> 49 is a perfect square because $(7)(7) = 49$.

The numbers 20 or 48 are *not* perfect squares. There is no *whole number* that when squared—multiplied by itself—yields 20 or 48. Consider 20. $4^2 = 16$, which is less than 20. $5^2 = 25$, which is more than 20. We realize, then, that the square root of 20 is between 4 and 5 because 20 is between 16 and 25. Since there is no whole number between 4 and 5, no whole number squared equals 20. Since the square root of a perfect square is a whole number, we can say that 20 is *not* a perfect square.

It is helpful to know the first 15 perfect squares. Take a minute to complete the following table.

Number, n	1	2	3	4	5	6	7	8	9	10	11	12	13	14	15
Number Squared, n^2	1	4	9	16										196	225

● EXAMPLE 3

(a) Is 81 a perfect square?　　　　**(b)** If so, find $\sqrt{81}$.

(a) Yes. 81 is a perfect square because $(9)(9) = 81$.　　　　**(b)** $\sqrt{81} = 9$

Practice Problem 3

(a) Is 144 a perfect square?　　　　**(b)** If so, find $\sqrt{144}$.　　●

　Approximating the Square Root of a Number That Is Not a Perfect Square

If a number is not a perfect square, we can only approximate its square root. This can be done by using a square root table such as the one that follows. Except for exact values such as $\sqrt{4} = 2.000$, all values are rounded to the nearest thousandth.

Number, n	Square Root of the Number, $\sqrt{n}$	Number, n	Square Root of the Number, $\sqrt{n}$
1	1.000	8	2.828
2	1.414	9	3.000
3	1.732	10	3.162
4	2.000	11	3.317
5	2.236	12	3.464
6	2.449	13	3.606
7	2.646	14	3.742

A square root table is located on page A-3. It gives you the square roots of whole numbers up to 200. Square roots can also be found with any calculator that has a square root key. Usually the key looks like this $\boxed{\sqrt{}}$ or this $\boxed{\sqrt{x}}$. To find the square root of 8 on a calculator, enter the number 8 and press $\boxed{\sqrt{}}$ or $\boxed{\sqrt{x}}$. You will see displayed 2.8284271. (Your calculator may display fewer or more digits.) Remember, no matter how many digits your calculator displays, when we find $\sqrt{8}$, we have only an **approximation**. It is not an exact answer. To emphasize this we use the $\approx$ notation to mean "is approximately equal to." Thus $\sqrt{8} \approx 2.828$.

EXAMPLE 4 Find approximate values using the square root table or a calculator. Round to the nearest thousandth.

(a) $\sqrt{2}$ **(b)** $\sqrt{12}$ **(c)** $\sqrt{7}$

(a) $\sqrt{2} \approx 1.414$ **(b)** $\sqrt{12} \approx 3.464$ **(c)** $\sqrt{7} \approx 2.646$

Practice Problem 4 Approximate to the nearest thousandth.

(a) $\sqrt{3}$ **(b)** $\sqrt{13}$ **(c)** $\sqrt{5}$

EXAMPLE 5 Approximate to the nearest thousandth of an inch the length of the side of a square that has an area of 6 in.^{2}.

$$\sqrt{6 \text{ in.}^2} \approx 2.449 \text{ in.}$$

6 sq in.   2.449 in.

2.449 in.

Thus, to the nearest thousandth of an inch, the side measures 2.449 in.

Practice Problem 5 Approximate to the nearest thousandth of a meter the length of the side of a square that has an area of 22 m^2.

Calculator

Square Roots

Locate the square root key $\boxed{\sqrt{}}$ on your calculator.

1. To find $\sqrt{289}$, enter

   289 $\boxed{\sqrt{}}$

   The display should read

   $\boxed{17}$

2. To find $\sqrt{194}$, enter

   194 $\boxed{\sqrt{}}$

   The display should read

   $\boxed{13.928388}$

   This is just an approximation of the actual square root. We will round the answer to the nearest thousandth.

   $\sqrt{194} \approx 13.928$

Verbal and Writing Skills

1. Why is $\sqrt{25} = 5$?

2. 25 is a perfect square because its square root, 5, is a _____ number.

3. Is 32 a perfect square? Why or why not?

4. How can you approximate the square root of a number that is not a perfect square?

Find each square root. Do not use a calculator. Do not refer to a table of square roots.

5. $\sqrt{25}$ **6.** $\sqrt{36}$ **7.** $\sqrt{49}$ **8.** $\sqrt{64}$ **9.** $\sqrt{121}$ **10.** $\sqrt{144}$ **11.** $\sqrt{0}$ **12.** $\sqrt{225}$

In exercises 13–20, evaluate the square roots first, then add or subtract the results. Do not use a calculator or a square root table.

13. $\sqrt{49} + \sqrt{9}$ **14.** $\sqrt{25} + \sqrt{64}$ **15.** $\sqrt{100} + \sqrt{1}$ **16.** $\sqrt{0} + \sqrt{121}$

17. $\sqrt{225} - \sqrt{144}$ **18.** $\sqrt{169} - \sqrt{64}$ **19.** $\sqrt{169} - \sqrt{121} + \sqrt{36}$

20. $\sqrt{225} - \sqrt{100} + \sqrt{64}$ **21.** $\sqrt{4} \times \sqrt{121}$ **22.** $\sqrt{225} \times \sqrt{9}$

23. (a) Is 256 a perfect square?
(b) If so, find $\sqrt{256}$.

24. (a) Is 289 a perfect square?
(b) If so, find $\sqrt{289}$.

Use a table of square roots or a calculator with a square root key to approximate to the nearest thousandth.

25. $\sqrt{18}$ **26.** $\sqrt{24}$ **27.** $\sqrt{76}$ **28.** $\sqrt{200}$ **29.** $\sqrt{194}$

Find the length of the side of the square. If the area is not a perfect square, approximate by using a square root table or a calculator with a square root key. Round to the nearest thousandth.

30. A square with area 34 m^2 **31.** A square with area 136 m^2 **32.** A square with area 180 m^2

Applications

High school basketball is played on a standard rectangular court that measures 92 feet in length and 50 feet in width. Some middle schools have smaller basketball courts that measure 80 feet in length and 42 feet in width.

33. The diagonal of a standard high school basketball court measures $\sqrt{10{,}964}$ feet in length. Find the length of this diagonal to the nearest tenth of a foot.

34. The diagonal of the smaller basketball court found in some middle schools measures $\sqrt{8164}$ feet in length. Find the length of this diagonal to the nearest tenth of a foot.

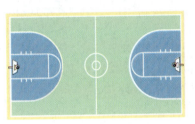

To Think About

35. Find each square root.
   (a) $\sqrt{4}$
   (b) $\sqrt{0.04}$
   (c) What pattern do you observe?

   (d) Can you find $\sqrt{0.004}$ exactly? Why?

36. Find each square root.
   (a) $\sqrt{25}$
   (b) $\sqrt{0.25}$
   (c) What pattern do you observe?

   (d) Can you find $\sqrt{0.025}$ exactly? Why?

Using a calculator with a square root key, evaluate and round to the nearest thousandth.

 37. $\sqrt{456} + \sqrt{322}$

 38. $\sqrt{578} + \sqrt{984}$

Cumulative Review Problems

39. The Taronga Zoo in Sydney, Australia, has a viewing tank for its platypuses. The tank is 60 in. high and 80 in. wide. What is the area of the front of the rectangular tank?

40. Ashrita Furman of Jamaica, New York, holds the world record in "joggling"—juggling and running at the same time—over 80.5 km. How many meters did he "joggle"?

41. The gate that leads from the street to the neighborhood basketball courts measures 92 centimeters wide. How many meters wide is the gate?

42. Goliath beetles in equatorial Africa can weigh up to 98.9 grams. How many kilograms can they weigh?

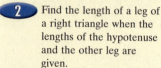
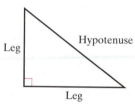 **1** *Finding the Hypotenuse of a Right Triangle When the Length of Each Leg Is Given*

The Pythagorean Theorem is a mathematical idea formulated long ago. It is as useful today as it was when it was discovered. The Pythagoreans lived in Italy about 2500 years ago. They studied various mathematical properties. They discovered that for any right triangle, the square of the hypotenuse equals the sum of the squares of the two legs of the triangle. This relationship is known as the **Pythagorean Theorem**. The side opposite the right angle is called the **hypotenuse**; the other two sides are called the **legs** of the right triangle.

Leg   Hypotenuse

Leg

$$(\text{hypotenuse})^2 = (\text{leg})^2 + (\text{leg})^2$$

Note how the Pythagorean Theorem applies to the right triangle shown here.

3   5

4

$$5^2 = 3^2 + 4^2$$
$$25 = 9 + 16$$
$$25 = 25 \ \checkmark$$

In a right triangle, the hypotenuse is the longest side. It is always opposite the largest angle, the right angle. The legs are the two shorter sides. When we know each leg of a right triangle, we use the following property.

$$\text{Hypotenuse} = \sqrt{(\text{leg})^2 + (\text{leg})^2}$$

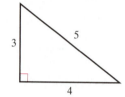 **EXAMPLE 1** Find the hypotenuse of a right triangle with legs of 5 in. and 12 in.

5   Hypotenuse

12

$$\text{Hypotenuse} = \sqrt{(5)^2 + (12)^2}$$
$$= \sqrt{25 + 144} \qquad \text{Square each value first.}$$
$$= \sqrt{169} \qquad \text{Add together the two values.}$$
$$= 13 \text{ in.} \qquad \text{Take the square root.}$$

Practice Problem 1 Find the hypotenuse of a right triangle with legs of 8 m and 6 m.

Sometimes we cannot find the hypotenuse exactly. In those cases, we often approximate the square root by using a calculator or a square root table.

🔴 **EXAMPLE 2**   Find the hypotenuse of a right triangle with legs of 4 m and 5 m. Round to the nearest thousandth.

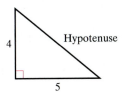

$$\text{Hypotenuse} = \sqrt{(4)^2 + (5)^2}$$
$$= \sqrt{16 + 25} \qquad \text{Square each value first.}$$
$$= \sqrt{41} \text{ m} \qquad \text{Add the two values together.}$$

Using the square root table or a calculator, we have the hypotenuse ≈ 6.403 m.

Practice Problem 2   Find to the nearest thousandth the hypotenuse of a right triangle with legs of 3 cm and 7 cm.   🔴

---

2 *Finding the Length of a Leg of a Right Triangle When the Lengths of the Hypotenuse and the Other Leg Are Given*

When we know the hypotenuse and one leg of a right triangle, we find the length of the other leg by using the following property.

$$\text{Leg} = \sqrt{(\text{hypotenuse})^2 - (\text{leg})^2}$$

🔴 **EXAMPLE 3**   A right triangle has a hypotenuse of 15 cm and a leg of 12 cm. Find the length of the other leg.

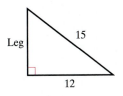

$$\text{Leg} = \sqrt{(15)^2 - (12)^2}$$
$$= \sqrt{225 - 144} \qquad \text{Square each value first.}$$
$$= \sqrt{81} \qquad \text{Subtract.}$$
$$= 9 \text{ cm} \qquad \text{Find the square root.}$$

Practice Problem 3   A right triangle has a hypotenuse of 17 m and a leg of 15 m. Find the length of the other leg.   🔴

🔴 **EXAMPLE 4**   A sail for a sailboat is in the shape of a right triangle. The right triangle has a hypotenuse of 14 feet and a leg of 8 feet. Find the length of the other leg. Round to the nearest thousandth.

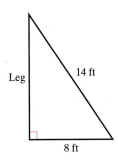

$$\text{Leg} = \sqrt{(14)^2 - (8)^2}$$
$$= \sqrt{196 - 64} \qquad \text{Square each value first.}$$
$$= \sqrt{132} \text{ feet} \qquad \text{Subtract the two numbers.}$$

Using a calculator or the square root table, we find that the leg ≈ 11.489 feet.

Practice Problem 4 A right triangle has a hypotenuse of 10 m and a leg of 5 m. Find the length of the other leg. Round to the nearest thousandth.

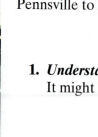

3 **Solving Applied Problems Using the Pythagorean Theorem**

Certain applied problems call for the use of the Pythagorean Theorem in the solution.

EXAMPLE 5 A pilot flies 13 mi east from Pennsville to Salem. She then flies 5 mi south from Salem to Elmer. What is the straight-line distance from Pennsville to Elmer? Round to the nearest tenth of a mile.

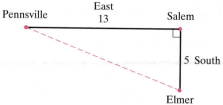

1. **Understand the problem.**
   It might help to draw a picture.

The distance from Pennsville to Elmer is the hypotenuse of the triangle.

2. **Solve and state the answer.**

$$\text{Hypotenuse} = \sqrt{(\text{leg})^2 + (\text{leg})^2}$$
$$= \sqrt{(13)^2 + (5)^2}$$
$$= \sqrt{169 + 25}$$
$$= \sqrt{194}$$
$$\sqrt{194} \approx 13.928 \text{ mi}$$

Rounded to the nearest tenth, the distance is 13.9 mi.

3. **Check.**
   Work backward to check. Use the Pythagorean Theorem.

$$13.9^2 \approx 13^2 + 5^2 \quad \text{(We use} \approx \text{because 13.9 is an approximate answer.)}$$
$$193.21 \approx 169 + 25$$
$$193.21 \approx 194 \quad \checkmark$$

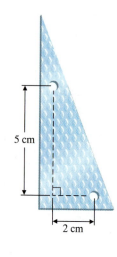

Practice Problem 5 Find the distance to the nearest thousandth between the centers of the holes in the triangular metal plate in the margin.

EXAMPLE 6 A 25-ft ladder is placed against a building at a point 22 ft from the ground. What is the distance of the base of the ladder from the building? Round to the nearest tenth.

1. **Understand the problem.**
   Draw a picture.

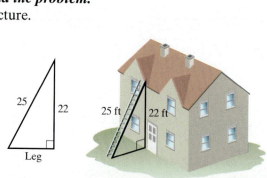

2. *Solve and state the answer.*

$$\text{Leg} = \sqrt{(\text{hypotenuse})^2 - (\text{leg})^2}$$
$$= \sqrt{(25)^2 - (22)^2}$$
$$= \sqrt{625 - 484}$$
$$= \sqrt{141}$$
$$\sqrt{141} \approx 11.874 \text{ ft}$$

If we round to the nearest tenth, the ladder is 11.9 ft from the base of the house.

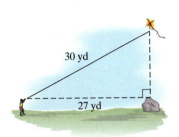

Practice Problem 6   A kite is out on 30 yd of string. The kite is directly above a rock. The rock is 27 yd from the boy flying the kite. How far above the rock is the kite? Round to the nearest tenth.

4   *Solving for the Missing Sides of Special Right Triangles*

If we use the Pythagorean Theorem and some other facts from geometry, we can find a relationship among the sides of two special right triangles. The first special right triangle is one that contains an angle that measures 30° and one that measures 60°. We call this the 30°–60°–90° right triangle.

> In a 30°–60°–90° triangle the length of the leg opposite the 30° angle is $\frac{1}{2}$ the length of the hypotenuse.

Notice that the hypotenuse of the first triangle is 10 m and the side opposite the 30° angle is exactly $\frac{1}{2}$ of that, or 5 m. The second triangle has a hypotenuse of 15 yd. The side opposite the 30° angle is exactly $\frac{1}{2}$ of that, or 7.5 yd.

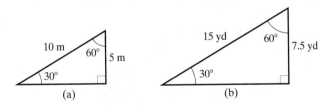

(a)   (b)

The second special right triangle is one that contains exactly two angles that each measure 45°. We call this the 45°–45°–90° right triangle.

> In a 45°–45°–90° triangle the lengths of the sides opposite the 45° angles are equal. The length of the hypotenuse is equal to $\sqrt{2} \times$ the length of either leg.

We will use the decimal approximation $\sqrt{2} \approx 1.414$ with this property.

$$\text{Hypotenuse} = \sqrt{2} \times 7$$
$$\approx 1.414 \times 7$$
$$\approx 9.898 \text{ cm}$$

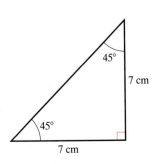

EXAMPLE 7 Find the requested sides of each special triangle. Round to the nearest tenth.

(a) Find the lengths of sides y and x.   **(b)** Find the length of hypotenuse z.

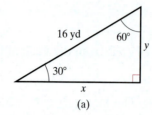

(a)

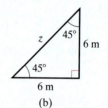

(b)

(a) In a 30°–60°–90° triangle the side opposite the 30° angle is $\frac{1}{2}$ of the hypotenuse.

$$\frac{1}{2} \times 16 = 8$$

Therefore, $y = 8$ yd.

When we know two sides of a right triangle, we find the third side using the Pythagorean Theorem.

$$\text{Leg} = \sqrt{(\text{hypotenuse})^2 - (\text{leg})^2}$$
$$= \sqrt{16^2 - 8^2} = \sqrt{256 - 64}$$
$$= \sqrt{192} \approx 13.856 \text{ yd}$$

Thus $x = 13.9$ yd rounded to the nearest tenth.

(b) In a 45°–45°–90° triangle we have the following.

$$\text{Hypotenuse} = \sqrt{2} \times \text{leg}$$
$$\approx 1.414(6)$$
$$= 8.484 \text{ m}$$

Rounded to the nearest tenth, the hypotenuse $= 8.5$ m.

Practice Problem 7 Find the requested sides of each special triangle. Round to the nearest tenth.

(a) Find the lengths of sides y and x.

(b) Find the length of hypotenuse z.

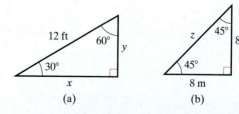

(a)   (b)

Verbal and Writing Skills

1. Explain in your own words how to obtain the length of the hypotenuse of a right triangle if you know the length of each of the legs of the triangle.

2. Explain in your own words how to obtain the length of one leg of a right triangle if you know the length of the hypotenuse and the length of the other leg.

Find the unknown side of the right triangle. Use a calculator or square root table when necessary and round to the nearest thousandth.

3.

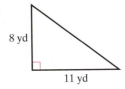

8 yd

11 yd

4.

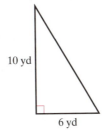

10 yd

6 yd

5.

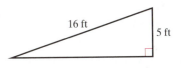

16 ft

5 ft

Find the unknown side of the right triangle using the information given to the nearest thousandth.

6. leg = 11 m, leg = 3 m

7. leg = 5 m, leg = 4 m

8. leg = 10 m, leg = 10 m

9. leg = 7 m, leg = 7 m

10. leg = 9 in., leg = 12 in.

11. hypotenuse = 13 yd, leg = 11 yd

12. hypotenuse = 14 yd, leg = 10 yd

Solve. Round to the nearest tenth.

13. Find the length of this ramp to a loading dock.

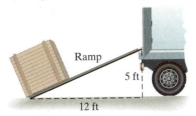

Ramp

5 ft

12 ft

14. Find the length of the guy wire supporting the telephone pole.

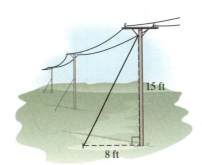

15 ft

8 ft

15. A construction project requires a stainless steel plate with holes drilled as shown. Find the distance between the centers of the holes in this triangular plate.

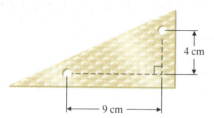

16. Juan runs out of gas in Los Lunas, New Mexico. He walks 4 mi west and then 3 mi south looking for a gas station. How far is he from his starting point?

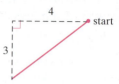

17. Barbara is flying her dragon kite on 32 yd of string. The kite is directly above the edge of a pond. The edge of the pond is 30 yd from where the kite is tied to the ground. How far is the kite above the pond?

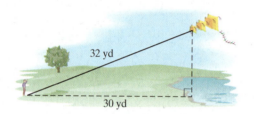

18. A 20-ft ladder is placed against a college classroom building at a point 18 ft above the ground. What is the distance from the base of the ladder to the building?

Using your knowledge of special right triangles, find the length of each leg. Round to the nearest tenth.

19.

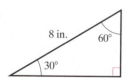

20.

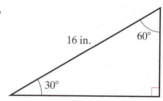

Using your knowledge of special right triangles, find the length of the hypotenuse. Round to the nearest tenth.

21.

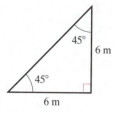

22.

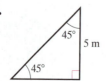

23.

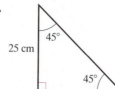

24.

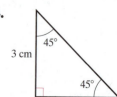

To Think About

25. A carpenter is going to use a wooden flagpole 10 in. in diameter, from which he will shape a rectangular base. The base will be 7 in. wide. The carpenter wishes to make the base as tall as possible, minimizing any waste. How tall will the rectangular base be? (Round to the nearest tenth.)

26. A 4-m shortwave antenna is placed on a garage roof that is 2 m above the lower part of the roof. The base of the garage roof is 16 m wide. How long is an antenna support from point *A* to point *B*?

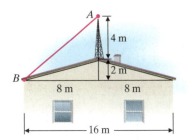

27. A woodworker who makes tables needs to make sure the rectangular leaves (the additions that make the table longer) will fit correctly. To do this, he measures the diagonals of each leaf to verify that its corners are right angles. What would be the diagonal length of a leaf with dimensions 2 ft by 5 ft? (Round to the nearest hundredth.)

28. Nancy makes pictures frames from strips of wood, plastic, and metal. She often measures the lengths of the frame's diagonals to be sure the corners are true right angles. If the dimensions of a frame are 7 in. by 9 in., how long would the diagonal measures be? (Round to the nearest hundredth.)

Find the approximate value of the unknown side of the right triangle. Round to the nearest thousandth.

29. The two legs are 7 yd and 11 yd.

30. The hypotenuse is 45 yd and one leg is 43 yd.

Cumulative Review Problems

31. Find the area of a triangular piece of land with altitude 22 m and base 31 m.

32. Find the area of a backyard rectangular vegetable garden with length 14 m and width 12 m.

33. Find the area of a square window that measures 21 inches on each side.

34. Find the area of the parallelogram-shaped roof of a building with a height of 48 yd and a base of 88 yd.

7.7 Circles

Student Learning Objectives

After studying this section, you will be able to:

1. Find the area and circumference of a circle.

2. Solve area problems containing circles and other geometric shapes.

SSM
PH TUTOR CENTER   CD & VIDEO   MATH PRO   WEB

1 Finding the Area and Circumference of a Circle

Every point on a circle is the same distance from the center of the circle, so the circle looks the same all around. In geometry we study the relationship between the parts of a circle and learn how to calculate the distance around a circle as well as the area of a circle.

A **circle** is a figure for which all points are at an equal distance from a given point. This given point is called the **center** of the circle.

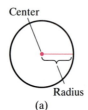

Center

Radius

(a)

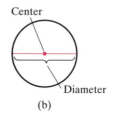

Center

Diameter

(b)

The **radius** is a line segment from the center to a point on the circle.

The **diameter** is a line segment across the circle that passes through the center.

We often use the words **radius** and **diameter** to mean the length of those segments. Note that the plural of radius is radii. Clearly, then,

$$\text{diameter} = 2 \times \text{radius} \quad \text{or} \quad d = 2r.$$

We could also say that

$$\text{radius} = \text{diameter} \div 2 \quad \text{or} \quad r = \frac{d}{2}.$$

Circumference

The distance around the circle is called the **circumference**. There is a special number called **pi**, which we denote by the symbol π. π is the number we get when we divide the circumference of a circle by the diameter $\frac{C}{d} = \pi$. The value of π is approximately 3.14159265359. We can approximate π to any number of digits. For all work in this book we will use the following.

> π is approximately 3.14, rounded to the nearest hundredth.

> We find the **circumference** C of a circle by multiplying the length of the diameter d times π.
>
> $$C = \pi d$$

EXAMPLE 1 Find the circumference of a quarter if we know the diameter is 2.4 cm. Use $\pi \approx 3.14$. Round to the nearest tenth.

$$C = \pi d = (3.14)(2.4 \text{ cm}) = 7.536 \text{ cm} \approx 7.5 \text{ cm} \text{ (rounded to the nearest tenth)}$$

Practice Problem 1 Find the circumference of a circle when the diameter is 9 m. Use $\pi \approx 3.14$. Round to the nearest tenth.

464

An alternative formula is $C = 2\pi r$. Remember, $d = 2r$. We can use this formula to find the circumference if we are given the length of the radius.

When solving word problems involving circles, be careful. Ask yourself, "Is the radius given, or is the diameter given?" Then do the calculations accordingly.

● **EXAMPLE 2**   A bicycle tire has a diameter of 24 in. How many feet does the bicycle travel if the wheel makes one revolution?

1. *Understand the problem*
   The distance the wheel travels when it makes 1 revolution is the circumference of the tire. Think of the tire unwinding.

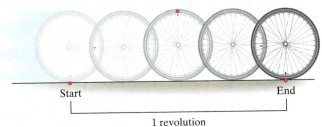

Start   End

1 revolution

   We are given the *diameter*. The diameter is given in *inches*. The answer should be in *feet*.

2. *Solve and state the answer.*
   Since we are given the diameter, we will use $C = \pi d$. We use 3.14 for π.

   $$C = \pi d$$
   $$= (3.14)(24 \text{ in.}) = 75.36 \text{ in.}$$

   We will change 75.36 inches to feet.

   $$75.36 \text{ in.} \times \frac{1 \text{ ft}}{12 \text{ in.}} = 6.28 \text{ ft}$$

   When the wheel makes 1 revolution, the bicycle travels 6.28 ft.

3. *Check.*
   We estimate to check. Since $\pi \approx 3.14$, we will use 3 for π.

   $$C \approx (3)(24 \text{ in.}) \times \frac{1 \text{ ft}}{12 \text{ in.}} \approx 6 \text{ ft} \quad \checkmark$$

Practice Problem 2   A bicycle tire has a diameter of 30 in. How many feet does the bicycle travel if the wheel makes two revolutions?   ●

> The **area of a circle** is the product of π times the radius squared.
>
> $$A = \pi r^2$$

● EXAMPLE 3

(a) Estimate the area of a circle whose radius is 6 cm.

(b) Find the exact area of a circle whose radius is 6 cm. Use $\pi \approx 3.14$. Round to the nearest tenth.

(a) Since π is approximately equal to 3.14, we will use 3 for π to estimate the area.

$$
\begin{aligned}
A &= \pi r^2 \\
&\approx (3)(6 \text{ cm})^2 \\
&\approx (3)(6 \text{ cm})(6 \text{ cm}) \\
&\approx (3)(36 \text{ cm}^2) \\
&\approx 108 \text{ cm}^2
\end{aligned}
$$

Thus our *estimated* area is 108 cm².

(b) Now let's compute the exact area.

$$
\begin{aligned}
A &= \pi r^2 \\
&= (3.14)(6 \text{ cm})^2 \\
&= 3.14(6 \text{ cm})(6 \text{ cm}) \\
&= (3.14)(36 \text{ cm}^2) \qquad \text{We } must \text{ square the radius first before multiplying by 3.14.} \\
&= 113.04 \text{ cm}^2 \\
&\approx 113.0 \text{ cm}^2 \text{ (rounded to the nearest tenth)}
\end{aligned}
$$

Our exact answer is close to the value 108 that we found in part (a).

Practice Problem 3   Find the area of a circle whose radius is 5 km. Use $\pi = 3.14$. Round to the nearest tenth. Estimate to check.   ●

The formula for the area of a circle uses the length of the radius. If we are given the diameter, we can use the property that $r = \frac{d}{2}$.

● EXAMPLE 4   Lexie Hatfield wants to buy a circular braided rug that is 8 ft in diameter. Find the cost of the rug at $35 a square yard.

1. **Understand the problem.**
   We are given the *diameter in feet*. We will need to find the radius. The cost of the rug is in *square yards*. We will need to change square feet to square yards.

2. **Solve and state the answer.**
   Find the radius.

   $$r = \frac{d}{2} = \frac{8 \text{ ft}}{2} = 4 \text{ ft}$$

Use 3.14 for π.

$$A = \pi r^2$$
$$= (3.14)(4 \text{ ft})^2$$
$$= (3.14)(4 \text{ ft})(4 \text{ ft})$$
$$= (3.14)(16 \text{ ft}^2)$$
$$= 50.24 \text{ ft}$$

Change square feet to square yards. Since 1 yd = 3 ft, $(1 \text{ yd})^2 = (3 \text{ ft})^2$. That is, 1 yd² = 9 ft².

$$50.24 \ \cancel{\text{ft}^2} \times \frac{1 \text{ yd}^2}{9 \ \cancel{\text{ft}^2}} \approx 5.58 \text{ yd}^2$$

Find the cost.

$$\frac{\$35}{1 \ \cancel{\text{yd}^2}} \times 5.58 \ \cancel{\text{yd}^2} = \$195.30$$

3. **Check.**
   You may use a calculator to check.

Practice Problem 4   Dorrington Little wants to buy a circular pool cover that is 10 ft in diameter. Find the cost of the pool cover at $12 a square yard. ⬭

2   *Solving Area Problems Containing Circles and Other Geometric Shapes*

Several applied area problems have a circular region combined with another region.

⬭ **EXAMPLE 5**   Find the area of the shaded region. Use $\pi \approx 3.14$. Round to the nearest tenth.

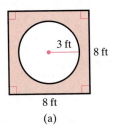

8 ft
(a)

We will subtract two areas to find the shaded region.

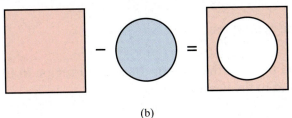

(b)

Area of the square − area of the circle = area of the shaded region

$$A = s^2 \qquad\qquad A = \pi r^2$$
$$= (8 \text{ ft})^2 \qquad\qquad = (3.14)(3 \text{ ft})^2$$
$$= 64 \text{ ft}^2 \qquad\qquad = (3.14)(9 \text{ ft}^2)$$
$$\qquad\qquad\qquad\qquad = 28.26 \text{ ft}^2$$

Area of the square	area of the circle	area of the shaded region
64 ft²	− 28.26 ft²	= 35.74 ft²

$$\approx 35.7 \text{ ft}^2$$
(rounded to nearest tenth)

Practice Problem 5   Find the area of the shaded region in the margin. Use $\pi \approx 3.14$. Round to the nearest tenth. ⬤

Many geometric shapes involve the semicircle. **A semicircle** is one half of a circle. The area of a semicircle is therefore one-half of the area of a circle.

⬤ **EXAMPLE 6**   Find the area of the shaded region. Use $\pi \approx 3.14$. Round to the nearest tenth.

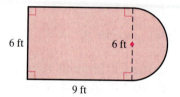

First we will find the area of the semicircle with diameter 6 ft.

$$r = \frac{d}{2} = \frac{6 \text{ ft}}{2} = 3 \text{ ft}$$

The radius is 3 ft. The area of a semicircle with radius 3 ft is

$$A_{\text{semicircle}} = \frac{\pi r^2}{2} = \frac{(3.14)(3 \text{ ft})^2}{2} = \frac{(3.14)(9 \text{ ft}^2)}{2}$$

$$= \frac{28.26 \text{ ft}^2}{2} = 14.13 \text{ ft}^2$$

Now we add the area of the rectangle.

$$A = lw = (9 \text{ ft})(6 \text{ ft}) = 54 \text{ ft}^2$$

$$
\begin{array}{ll}
54.00 \text{ ft}^2 & \text{area of rectangle} \\
+\ 14.13 \text{ ft}^2 & \text{area of semicircle} \\
\hline
68.13 \text{ ft}^2 & \text{total area}
\end{array}
$$

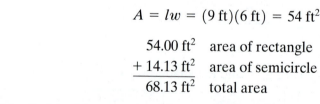

Rounded to the nearest tenth, area = 68.1 ft².

Practice Problem 6   Find the area of the shaded region in the margin. Use $\pi \approx 3.14$. Round to the nearest tenth. ⬤

7.7 Exercises

Verbal and Writing Skills

1. The distance around a circle is called the _____.

2. The radius is a line segment from the _____ to a point on the circle.

3. The diameter is two times the _____ of the circle.

4. Explain in your own words how to find the area of a circle if you are given the diameter.

5. Explain in your own words how to find the circumference of a circle if you are given the radius.

6. Explain in your own words how to find the area of a semicircle if you are given the radius.

In all exercises, use $\pi \approx 3.14$. Round to the nearest hundredth.

Find the diameter of a circle if the radius has the value given.

7. $r = 29$ in.

8. $r = 33$ in.

9. $r = 8.5$ mm

10. $r = 7.7$ m

Find the radius of a circle if the diameter has the value given.

11. $d = 45$ yd

12. $d = 65$ yd

13. $d = 19.84$ cm

14. $d = 27.06$ cm

Find the circumference of the circle.

15. diameter = 32 cm

16. diameter = 22 cm

17. radius = 18.5 in.

18. radius = 24.5 in.

A bicycle wheel makes five revolutions. Determine how far the bicycle travels in feet.

19. The diameter of the wheel is 32 in.

20. The diameter of the wheel is 24 in.

Find the area of each circle.

21. radius = 5 yd

22. radius = 7 yd

23. diameter = 32 cm

24. diameter = 44 cm

A water sprinkler sends water out in a circular pattern. Determine how large an area is watered.

25. The radius of watering is 8 ft.

26. The radius of watering is 12 ft.

A radio station sends out radio waves in all directions from a tower at the center of the circle of broadcast range. Determine how large an area is reached.

27. The diameter is 120 mi.

28. The diameter is 90 mi.

Find the area of the shaded region.

29.

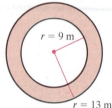

30.

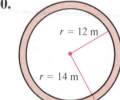

31.

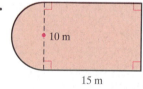

32.

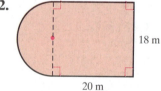

33.

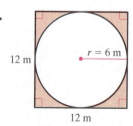

34.

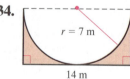

Find the cost of fertilizing a playing field at $0.20 per square yard for the conditions stated.

35. The rectangular part of the field is 120 yd long and the diameter of each semicircle is 40 yd.

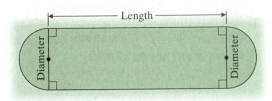

36. The rectangular part of the field is 110 yd long and the diameter of each semicircle is 50 yd.

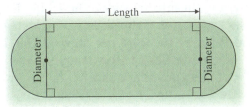

Applications

Use $\pi \approx 3.14$ *in exercises 37–44. Round to the nearest hundredth.*

37. A porthole window on a freighter ship has a diameter of 2 ft. What is the length of the insulating strip that encircles the window and keeps out wind and moisture?

38. A manhole cover has a diameter of 3 ft. What is the length of the brass grip-strip that encircles the cover, making it easier to manage?

39. Elena's car has tires with a radius of 14 in. How many feet does her car travel if the wheel makes 35 revolutions?

40. Jimmy's truck has tires with a radius of 30 inches. How many feet does his truck travel if the wheel makes nine revolutions?

 41. Lucy's car has tires with a radius of 16 in. How many complete revolutions do her wheels make in 1 mile? (*Hint*: First determine how many inches are in 1 mile.)

 42. Mickey's car has tires with a radius of 15 in. How many complete revolutions do his wheels make in 1 mile? (*Hint*: First determine how many inches are in 1 mile.)

43. Sarah bought a circular marble tabletop to use as her dining room table. The marble is 6 ft in diameter. Find the cost of the tabletop at $72 per square yard of marble.

44. An architect is designing a tranquil garden for the back of a business complex. She wants to put a bamboo enclosure around a pond. The pond is a perfect circle, and has a diameter of 40 feet. The bamboo enclosure will be installed, also in an exact circle, 14 feet from the edge of the pond. What will be the circumference of the circular bamboo enclosure?

To Think About

45. A 15-in.-diameter pizza costs $6.00. A 12-in.-diameter pizza costs $4.00. The 12-in.-diameter pizza is cut into six slices. The 15-in.-diameter pizza is cut into eight slices.

(a) What is the cost per slice of the 15-in.-diameter pizza? How many square inches of pizza are in one slice?

(b) What is the cost per slice of the 12-in.-diameter pizza? How many square inches of pizza are in one slice?

(c) If you want more value for your money, which slice of pizza should you buy?

46. A 14-in.-diameter pizza costs $5.50. It is cut into eight pieces. A 12.5 in. $\times$ 12.5 in. square pizza costs $6.00. It is cut into nine pieces.

(a) What is the cost of one piece of the 14-in.-diameter pizza? How many square inches of pizza are in one piece?

(b) What is the cost of one piece of the 12.5 in. $\times$ 12.5 in. square pizza? How many square inches of pizza are in one piece?

(c) If you want more value for your money, which piece of pizza should you buy?

Cumulative Review Problems

47. Find 16% of 87.

48. What is 0.5% of 60?

49. 12% of what number is 720?

50. 19% of what number is 570?

7.8 Volume

1 Finding the Volume of a Rectangular Solid (Box)

How much grain can that shed hold? How much water is in the polluted lake? How much air is inside a basketball? These are questions of **volume**. In this section we compute the volume of several three-dimensional geometric figures: the rectangular solid (box), cylinder, sphere, cone, and pyramid.

We can start with a box 1 in. $\times$ 1 in. $\times$ 1 in.

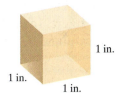

1 in.

1 in.

1 in.

This box has a volume of 1 cubic inch (written 1 in.^{3}). We can use this as a **unit of volume**. Volume is measured in cubic units such as cubic meters (abbreviated m^3) or cubic feet (abbreviated ft^3). When we measure volume, we are measuring the space inside an object.

Student Learning Objectives

After studying this section, you will be able to:

1 Find the volume of a rectangular solid (box).

2 Find the volume of a cylinder.

3 Find the volume of a sphere.

4 Find the volume of a cone.

5 Find the volume of a pyramid.

SSM   PH TUTOR   CD & VIDEO   MATH PRO   WEB
CENTER

The **volume of a rectangular solid** (box) is the length times the width times the height.

$$V = lwh$$

h

w

l

EXAMPLE 1   Find the volume of a box of width 2 ft, length 3 ft, and height 4 ft.

$$V = lwh = (3 \text{ ft})(2 \text{ ft})(4 \text{ ft}) = (6)(4) \text{ ft}^3 = 24 \text{ ft}^3$$

Practice Problem 1   Find the volume of a box of width 5 m, length 6 m, and height 2 m.

4

2

3

2 Finding the Volume of a Cylinder

Cylinders are the shape we observe when we see a tin can or a tube.

The **volume of a cylinder** is the area of its circular base, πr^2, times the height h.

$$V = \pi r^2 h$$

h

r

We will continue to use 3.14 as π, as we did in Section 7.7, in all volume problems requiring the use of π.

EXAMPLE 2   Find the volume of a cylinder of radius 3 in. and height 7 in. Round to the nearest tenth.

$$V = \pi r^2 h = (3.14)(3 \text{ in.})^2(7 \text{ in.}) = (3.14)(3 \text{ in.})(3 \text{ in.})(7 \text{ in.})$$

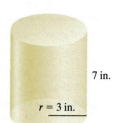

7 in.

r = 3 in.

Be sure to square the radius before doing any other multiplication.

$$V = (3.14)(9 \text{ in.}^2)(7 \text{ in.}) = (28.26 \text{ in.}^2)(7 \text{ in.})$$

$$= 197.82 \text{ in.}^3 \approx 197.8 \text{ in.}^3 \text{ rounded to the nearest tenth}$$

Practice Problem 2   Find the volume of a cylinder of radius 2 in. and height 5 in. Round to the nearest tenth.

To Think About   Take a minute to compare the formulas for the volumes of a rectangular solid and a cylinder. Do you see how they are similar? Consider the area of the base of each figure. In each case, what must you multiply the base area by to obtain the volume of the solid?

3   Finding the Volume of a Sphere

Have you ever considered how you would find the volume of the inside of a ball? How many cubic inches of air are inside a basketball? To answer these questions we need a volume formula for a *sphere*.

> The **volume of a sphere** is 4 times π times the radius cubed divided by 3.
>
> $$V = \frac{4\pi r^3}{3}$$
>
> r

r = 3 m

EXAMPLE 3   Find the volume of a sphere with radius 3 m. Round to the nearest tenth.

$$V = \frac{4\pi r^3}{3} = \frac{(4)(3.14)(3 \text{ m})^3}{3}$$

$$= \frac{(4)(3.14)(3)(3)\cancel{(3)} \text{ m}^3}{\cancel{3}}$$

$$= (12.56)(9) \text{ m}^3 = 113.04 \text{ m}^3$$

$$\approx 113.0 \text{ m}^3 \text{ rounded to the nearest tenth}$$

Practice Problem 3   Find the volume of a sphere with radius 6 m. Round to the nearest tenth.

4   Finding the Volume of a Cone

We see the shape of a cone when we look at the sharpened end of a pencil or at an ice cream cone. To find the volume of a cone we use the following formula.

> The **volume of a cone** is π times the radius of the base squared times the height divided by 3.
>
> $$V = \frac{\pi r^2 h}{3}$$
>
> h
> r

EXAMPLE 4   Find the volume of a cone of radius 7 m and height 9 m. Round to the nearest tenth.

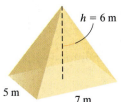

$$V = \frac{\pi r^2 h}{3}$$

$$= \frac{(3.14)(7 \text{ m})^2(9 \text{ m})}{3}$$

$$= \frac{(3.14)(7 \text{ m})(7 \text{ m})(9 \text{ m})}{3}$$

$$= (3.14)(49)(3) \text{ m}^3 = (153.86)(3) \text{ m}^3 = 461.58 \text{ m}^3$$

$$\approx 461.6 \text{ m}^3 \text{ rounded to the nearest tenth}$$

Practice Problem 4 Find the volume of a cone of radius 5 m and height 12 m. Round to the nearest tenth.

5 *Finding the Volume of a Pyramid*

You have seen pictures of the great pyramids of Egypt. These amazing stone structures are over 4000 years old.

The volume of a pyramid is obtained by multiplying the area B of the base of the pyramid by the height h and dividing by 3.

$$V = \frac{Bh}{3}$$

EXAMPLE 5 Find the volume of a pyramid with height 6 m, length of base 7 m, and width of base 5 m.

  The base is a rectangle.

$h = 6$ m

5 m   7 m

$$\text{Area of base} = (7 \text{ m})(5 \text{ m}) = 35 \text{ m}^2$$

Substituting the area of the base 35 m² and the height of 6 m, we have

$$V = \frac{Bh}{3} = \frac{(35 \text{ m}^2)(6 \text{ m})}{3}$$

$$= (35)(2) \text{ m}^3$$

$$= 70 \text{ m}^3$$

Practice Problem 5 Find the volume of a pyramid with the dimensions given.

(a) height 10 m, width 6 m, length 6 m

(b) height 15 m, width 7 m, length 8 m

7.8 Exercises

Verbal and Writing Skills

Match each formula for volume with the figure it belongs to on the right.

1. $V = lwh$

2. $V = \pi r^2 h$

3. $V = \dfrac{4\pi r^3}{3}$

4. $V = \dfrac{\pi r^2 h}{3}$

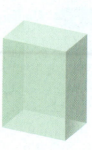

5. $V = \dfrac{Bh}{3}$

6. $V = s^3$

Find the volume. Use $\pi \approx 3.14$. Round to the nearest tenth when necessary.

7. a rectangular solid with width = 14 mm, length = 20 mm, height = 2.5 mm

8. a rectangular solid with width = 12 mm, length = 30 mm, height = 1.5 mm

9. a cylinder with radius 2 m and height 7 m

10. a cylinder with radius 3 m and height 8 m

11. a cylinder with diameter 30 m and height 9 m

12. a cylinder with diameter 22 m and height 17 m

13. a sphere with radius 9 yd

14. a sphere with radius 12 yd

Exercises 15 and 16 involve hemispheres. A hemisphere is exactly one half of a sphere.

15. Find the volume of a hemisphere with radius = 7 m.

16. Find the volume of a hemisphere with radius = 6 m.

Find the volume. Use $\pi \approx 3.14$. Round to the nearest tenth.

17. a cone with a height of 12 cm and a radius of 9 cm

18. a cone with a height of 14 cm and a radius of 8 cm

19. a cone with a height of 12 ft and a radius of 6 ft

20. a cone with a height of 10 ft and a radius of 5 ft

21. a pyramid with a height of 10 m and a square base of 7 m on a side

22. a pyramid with a height of 7 m and a square base of 3 m on a side

23. a pyramid with a height of 10 m and a rectangular base measuring 8 m by 14 m

24. a pyramid with a height of 5 m and a rectangular base measuring 6 m by 12 m

Applications

Use $\pi \approx 3.14$ when necessary.

25. Mr. Bledsoe wants to put down a crushed-stone driveway to his summer camp. The driveway is 7 yd wide and 120 yd long. The crushed stone is to be 4 in. thick. How many cubic yards of stone will he need?

26. The Essex Village neighborhood park has a great sandbox for young children that measures 20 yd × 18 yd. The parks committee will be adding 4 in. of new sand. How many cubic yards of sand will they need?

A collar of Styrofoam is made to insulate a pipe. Find the volume of the unshaded region (which represents the collar). The large radius R is to the outer rim. The small radius r is to the edge of the insulation.

27. $r = 3$ in.
$R = 5$ in.
$h = 20$ in.

28. $r = 4$ in.
$R = 6$ in.
$h = 25$ in.

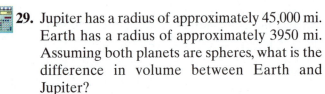

29. Jupiter has a radius of approximately 45,000 mi. Earth has a radius of approximately 3950 mi. Assuming both planets are spheres, what is the difference in volume between Earth and Jupiter?

30. A tennis ball has a diameter of 2.5 in. A baseball has a diameter of 2.9 in. What is the difference in volume between the baseball and the tennis ball? (Round your answer to the nearest tenth.)

31. A fragile glass box in the shape of a rectangular solid has width 6 in., length 18 in., and height 12 in. It is being shipped in a larger box of width 12 in., length 22 in., and height 16 in. All of the space between the glass box and shipping box will be packed with Styrofoam packing "peanuts." How many cubic inches of the shipping box will be Styrofoam peanuts?

32. A storage garage rents space for people who need to store items on a monthly basis. The rooms measure 6 ft by 6 ft by 8 ft and rent for $45 per month. If you rented storage space for 1 month, how much would you pay per cubic foot? (Round to the nearest cent.)

33. The nose cone of a passenger jet is used to receive and send radar. It is made of a special aluminum alloy that costs $4.00 per cm^3. The cone has a radius of 5 cm and a height of 9 cm. What is the cost of the aluminum needed to make this *solid* nose cone?

34. A special stainless steel cone sits on top of a cable television antenna. The cost of the stainless steel is $3.00 per cm^3. The cone has a radius of 6 cm and a height of 10 cm. What is the cost of the stainless steel needed to make this *solid* steel cone?

35. Suppose that a stone pyramid has a rectangular base that measures 87 yd by 130 yd. Also suppose that the pyramid has a height of 70 yd. Find the volume.

36. Suppose the pyramid in exercise 35 is made of solid stone. It is not hollow like the pyramids of Egypt. It is composed of layer after layer of cut stone. The stone weighs 422 lb per cubic yard. How many pounds does the pyramid weigh? How many tons does the pyramid weigh?

To Think About

37. You have seen that the volume of a rectangular solid or a cylinder is found by multiplying the area of the base of the solid by the height of the solid. Can you draw any other solids whose formulas might follow this same pattern?

38. What is the formula for the volume of the given solid?

Cumulative Review Problems

39. Add. $7\frac{1}{3} + 2\frac{1}{4}$

40. Subtract. $9\frac{1}{8} - 2\frac{3}{4}$

41. Multiply. $2\frac{1}{4} \times 3\frac{3}{4}$

42. Divide. $7\frac{1}{2} \div 4\frac{1}{5}$

Putting Your Skills to Work

Newspaper Math

In February 2000, the *Boston Globe* followed other newspapers in the United States and decided to reformat their paper. 2 inches has been taken off the width of the paper. The new size is easier to handle and more economical to produce.

According to researchers, 10,000 tons of newsprint and 54,000 gallons of ink are saved each year. You might wonder, does 2 inches really make a big difference? How much paper is saved each day? How much money will be saved by producing a smaller newspaper?

Two-page spread (old size)

Two-page spread (new size)

22.5 inches

22.5 inches

27 inches

25 inches

Problems for Individual Investigation and Calculation

1. (a) What was the perimeter and area of a two-page spread of the *Boston Globe* previously?

(b) What is the perimeter and area now that the paper has been reformatted?

(c) How many fewer square inches is the new size?

2. Previously, 2900 tons of newsprint were used each week to print the *Boston Globe*.

(a) How many pounds per week was this?

(b) How many pounds of newsprint were used each day? (Round to the nearest pound.)

3. Now that the paper has been reformatted, 10,000 tons of newsprint are being saved each year.

(a) How many pounds of newsprint are being saved each year?

(b) How many pounds of newsprint are being saved each day? (Round to the nearest pound.)

4. How many pounds of paper are used each day now that the *Boston Globe* has been reformatted? (Round to the nearest pound.)

480

Problems for Group Investigation and Cooperative Learning

5. One roll of newsprint is 6.5 miles long and weighs 1800 pounds. Use a proportion to find out how many miles long one ton of newsprint is. (*Hint:* First you'll need to convert one ton to pounds. Round to the nearest tenth of a mile.)

6. How many miles of newsprint were used each day at the *Boston Globe* previously? Recall that 2900 tons of paper were used each week. (Round to the nearest mile.)

7. (a) The price of newsprint averages about $515 per ton. How much money was the *Boston Globe* spending on newsprint each week?

(b) How much was this a day? (Round to nearest dollar.)

8. The *Boston Globe* prints 550,000 papers each day. Use your answer from question 7(b) to determine how much it previously cost to print each paper. (Round to nearest cent.)

9. Find out how much money was saved each day when the paper was downsized. (*Hint:* Determine how many tons of newsprint were saved each day. Round to the nearest tenth of a ton.)

Internet Connections

 Netsite: http://www.prenhall.com/tobey_basic

This site provides environmental information.

10. Use your answer from question 3 and this Web site to determine the average number of trees per year that are saved with this decrease in the number of pounds of newsprint used each year.

11. Use your answer from question 4 and this Web site to determine the average number of trees per year that would be saved if one-half of all users of the *Boston Globe* recycled their newspapers after reading them.

1 Finding the Corresponding Parts of Similar Triangles

In English, "similar" means that the things being compared are, in general, alike. But in mathematics, "similar" means that the things being compared are alike in a special way—they are *alike in shape*, even though they may be different in size. So photographs that are enlarged produce images *similar* to the original; a floor plan of a building is *similar* to the actual building; a model car is *similar* to the actual vehicle.

Two triangles with the same shape but not necessarily the same size are called **similar triangles**. Here are two pairs of similar triangles.

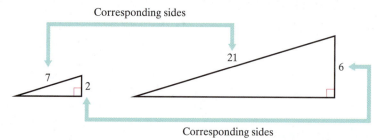

(a)   (b)   (c)   (d)

The two triangles on the right are similar. The smallest angle in the first triangle is angle *A*. The smallest angle in the second triangle is angle *D*. Both angles measure 36°. We say that angle *A* and angle *D* are **corresponding angles** in these similar triangles.

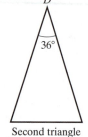

First triangle   Second triangle

> The **corresponding angles** of similar triangles are equal.

The following two triangles are similar. Notice the **corresponding sides**.

Corresponding sides

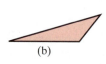

Corresponding sides

We see that the ratio of 7 to 21 is the same as the ratio of 2 to 6.

$$\frac{7}{21} = \frac{2}{6} \quad \text{is obviously true since} \quad \frac{1}{3} = \frac{1}{3}.$$

> The corresponding sides of similar triangles have the same ratio.

We can use the fact that corresponding sides of similar triangles have the same ratio to find missing sides of triangles.

● **EXAMPLE 1**   These two triangles are similar. Find the length of side *n*. Round to the nearest tenth.

The ratio of 12 to 19 is the same as the ratio of 5 to n.

$$\frac{12}{19} = \frac{5}{n}$$

$12n = (5)(19)$ Cross multiply.

$12n = 95$ Simplify.

$$\frac{12n}{12} = \frac{95}{12}$$ Divide each side by 12.

$n = 7.91\overline{6}$

≈ 7.9 Round to the nearest tenth.

Side n is of length 7.9.

Practice Problem 1   The two triangles in the margin are similar. Find the length of side n. Round to the nearest tenth.

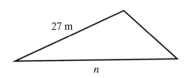

Similar triangles are not always oriented the same way. You may find it helpful to rotate one of the triangles so that the similarity is more apparent.

EXAMPLE 2   These two triangles are similar. Name the sides that correspond.

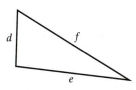

First we turn the second triangle so that the shortest side is on the top, the intermediate side is to the left, and the longest side is on the right.

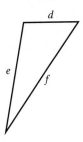

Now the shortest side of each triangle is on the top, the longest side of each triangle is on the right, and so on. We can see that

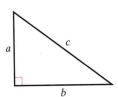

a corresponds to d.

b corresponds to e.

c corresponds to f.

Practice Problem 2   The two triangles in the margin are similar. Name the sides that correspond.

> The perimeters of similar triangles have the same ratio as the corresponding sides.

Similar triangles can be used to find distances or lengths that are difficult to measure.

EXAMPLE 3 A flagpole casts a shadow of 36 ft. At the same time, a nearby tree that is 3 ft tall has a shadow of 5 ft. How tall is the flagpole?

1. ***Understand the problem.***

   The shadows cast by the sun shining on vertical objects at the same time of day form similar triangles. We draw a picture.

   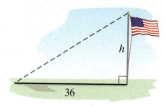

2. ***Solve and state the answer.***

   Let h = the height of the flagpole. Thus we can say h is to 3 as 36 is to 5.

$$\frac{h}{3} = \frac{36}{5}$$
$$5h = (3)(36)$$
$$5h = 108$$
$$\frac{5h}{5} = \frac{108}{5}$$
$$h = 21.6$$

   The flagpole is about 21.6 feet tall.

3. ***Check.***

   The check is up to you.

Practice Problem 3 What is the height (h) of the side wall of the building in the margin if the two triangles are similar?

2 · Finding the Corresponding Parts of Similar Geometric Figures

Geometric figures such as rectangles, trapezoids, and circles can be similar figures.

> The corresponding sides of similar geometric figures have the same ratio.

EXAMPLE 4 The two rectangles shown here are similar because the corresponding sides of the two rectangles have the same ratio. Find the width of the larger rectangle.

Let w = the width of the larger rectangle.

$$\frac{w}{1.6} = \frac{9}{2}$$

$$2w = (1.6)(9)$$

$$2w = 14.4$$

$$\frac{2w}{2} = \frac{14.4}{2}$$

$$w = 7.2$$

The width of the larger rectangle is 7.2 meters.

Practice Problem 4   The two rectangles in the margin are similar. Find the width of the larger rectangle.

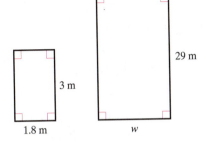

The perimeters of similar figures—whatever the figures—have the same ratio as their corresponding sides. Circles are special cases. *All* circles are similar. The circumferences of two circles have the same ratio as their radii.

To Think About   How would you find the relationship between the areas of two similar geometric figures? Consider the following two similar rectangles?

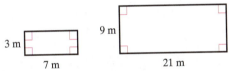

The area of the smaller rectangle is (3 m)(7 m) = 21 m². The area of the larger rectangle is (9 m)(21 m) = 189 m². How could you have predicted this result?

The ratio of small width to large width is $\frac{3}{9} = \frac{1}{3}$. The small rectangle has sides that are $\frac{1}{3}$ as large as the large rectangle. The ratio of the area of the small rectangle to the area of the large rectangle is $\frac{21}{189} = \frac{1}{9}$. Note that $\left(\frac{1}{3}\right)^2 = \frac{1}{9}$.

Thus we can develop the following principle: The areas of two similar figures are in the same ratio as the square of the ratio of two corresponding sides.

Verbal and Writing Skills

1. Similar figures may be different in _____ but they are alike in _____ .

2. The corresponding sides of similar triangles have the same _____ .

3. The perimeters of similar figures have the same ratio as their corresponding _____ .

4. You are given the lengths of the sides of a large triangle and the length of a corresponding side of a smaller, similar triangle. Explain in your own words how to find the perimeter of the smaller triangle.

For each pair of similar triangles, find the missing side n. Round to the nearest tenth when necessary.

5.

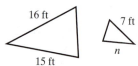

6.

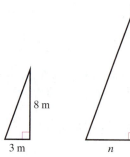

7.

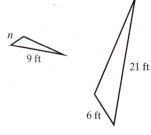

8.

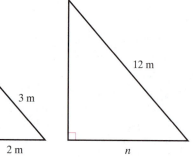

9.

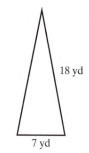

10.

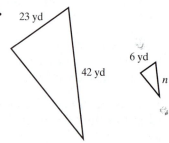

Each pair of triangles is similar. Determine the three pairs of corresponding sides in each case.

11.

12.

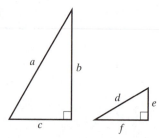

Applications

13. A sculptor is designing her new triangular masterpiece. In her scale drawing, the shortest side of the triangular piece to be made measures 8 cm. The longest side of the drawing measures 25 cm. The longest side of the actual triangular piece to be sculpted must be 10.5 m long. How long will the shortest side of this piece be? Round to the nearest tenth.

14. The zoo has hired a landscape architect to design the triangular lobby of the children's petting zoo. In his scale drawing, the longest side of the lobby is 9 cm. The shortest side of the lobby is 5 cm. The longest side of the actual lobby will be 30 m. How long will the shortest side of the actual lobby be? Round to the nearest tenth.

15. Janice took a great photo of the entire family at this year's reunion. She brought it to a professional photography studio and asked that the 3-in.-by-5-in. photo be blown up to poster size, which is 3.5 feet tall. What is the smaller dimension (width) of the poster?

16. Jeff and Shelley are planning a new kitchen. The old kitchen measured 9 ft by 12 ft. The new kitchen is similar in shape, but the longest dimension is 19 ft. What is the smaller dimension (width) of the new kitchen?

17. Marcy is making a banner to hang in the halls at her school to welcome students for the new school year. She first drew a smaller sample banner measuring 6 in. by 25 in. Marcy wants the actual banner to be similar in shape to the sample. Marcy wants the width of the actual banner to be $1\frac{1}{2}$ ft. How long will the banner be?

18. A theater company's prop designer sends drawings of the props to the person in charge of construction. An upcoming play will require a large brick wall to stretch across the stage floor. In the designer's drawing, the wall measures $\frac{1}{4}$ ft by $\frac{3}{4}$ ft. If the length of the stage is 36 ft, how tall will the wall be?

In exercises 19 and 20, a flagpole casts a shadow. At the same time, a nearby tree casts a shadow. Use the sketch to find the height n of each flagpole.

19.

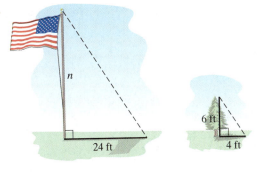

20.

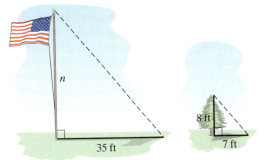

21. Lola is standing outside the shopping mall. She is 5.5 feet tall and her shadow measures 6.5 feet long. The outside of the department store casts a shadow of 96 feet. How tall is the store? Round to the nearest foot.

22. Thomas is rock climbing in Utah. He is 6 feet tall and his shadow measures 8 feet long. The rock he wants to climb casts a shadow of 610 feet. How tall is the rock he is about to climb?

Each pair of figures is similar. Find the missing side. Round to the nearest tenth when necessary.

23.

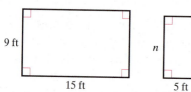

9 ft

15 ft

n

5 ft

24.

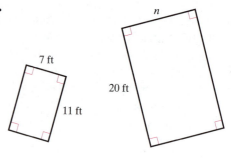

n

7 ft

20 ft

11 ft

25.

28 cm

12 cm

7 cm

n

26.

15 cm

9 cm

n

14 cm

To Think About

Each pair of geometric figures is similar. Find the unknown area.

27.

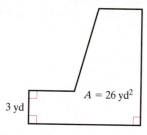

$A = 26 \text{ yd}^2$

3 yd

2 yd

$A = ?$

28.

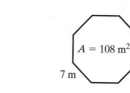

$A = 108 \text{ m}^2$

$A = ?$

7 m

10 m

Cumulative Review Problems

Calculate. Use the correct order of operations.

29. $2 \times 3^2 + 4 - 2 \times 5$

30. $300 \div (85 - 65) + 2^4$

31. $(5)(9) - (21 + 3) \div 8$

32. $108 \div 6 + 51 \div 3 + 3^3$

1 ▸ Solving Applied Problems Involving Geometric Shapes

We can solve many real-life problems with the geometric knowledge we now have. Our everyday world is filled with objects that are geometric in shape, so we can use our knowledge of geometry to find length, area, or volume. How far is the automobile trip? How much framing, edging, or fencing is required? How much paint, siding, or roofing is required? How much can we store? All of these questions can be answered with geometry.

If it is helpful to you, use the Mathematics Blueprint for Problem Solving to organize a plan to solve these applied problems.

🔴 **EXAMPLE 1**   A professional painter can paint 90 ft² of wall space in 20 minutes. How long will it take the painter to paint four walls with the following dimensions: 14 ft × 8 ft, 12 ft × 8 ft, 10 ft × 7 ft, and 8 ft × 7 ft?

1. Understand the problem.

Mathematics Blueprint For Problem Solving

Gather the Facts	What Am I Asked to Do?	How Do I Proceed?	Key Points to Remember
Painter paints four walls: 14 ft × 8 ft 12 ft × 8 ft 10 ft × 7 ft 8 ft × 7 ft Painter can paint 90 ft² in 20 minutes.	Find out how long it will take the painter to paint the four walls.	**(a)** Find the total area to be painted. **(b)** Then find how long it will take to paint the total area.	The area of each rectangular wall is obtained by multiplying the length by the width. To get the time, we set up a proportion.

2. Solve and state the answer.

(a) Find the total area of the four walls.

Each wall is a rectangle. The first one is 8 ft wide and 14 ft long.

$$A = lw$$

$$= (14 \text{ ft})(8 \text{ ft}) = 112 \text{ ft}^2$$

We find the areas of the other three walls.

$$(12 \text{ ft})(8 \text{ ft}) = 96 \text{ ft}^2 \quad (10 \text{ ft})(7 \text{ ft}) = 70 \text{ ft}^2 \quad (8 \text{ ft})(7 \text{ ft}) = 56 \text{ ft}^2$$

The total area is obtained by adding.

$$
\begin{array}{r}
112 \text{ ft}^2 \\
96 \text{ ft}^2 \\
70 \text{ ft}^2 \\
+ \ 56 \text{ ft}^2 \\
\hline
334 \text{ ft}^2
\end{array}
$$

(b) Determine how long it will take to paint the four walls.

489

Now we set up a proportion. If 90 ft^2 can be done in 20 minutes, then 334 ft^2 can be done in t minutes.

$$\frac{90 \text{ ft}^2}{20 \text{ minutes}} = \frac{334 \text{ ft}^2}{t \text{ minutes}}$$

$$\frac{90}{20} = \frac{334}{t}$$

$$90(t) = 334(20)$$

$$90t = 6680$$

$$\frac{90t}{90} = \frac{6680}{90}$$

$$t \approx 74 \qquad \text{We round our answer to the nearest minute.}$$

Thus the work can be done in approximately 74 minutes.

3. Check.

Estimate the answer.

$$14 \times 8 \approx 10 \times 8 = 80 \text{ ft}^2$$
$$12 \times 8 \approx 10 \times 8 = 80 \text{ ft}^2$$
$$10 \times 7 = 70 \text{ ft}^2$$
$$8 \times 7 = 56 \text{ ft}^2$$

If we estimate the sum of the number of square feet, we will have

$$80 + 80 + 70 + 56 = 286 \text{ ft}^2.$$

Now since 60 minutes = 1 hour, we know that if you can paint 90 ft^2 in 20 minutes, you can paint 270 ft^2 in one hour. Our estimate of 286 ft^2 is slightly more than 270 ft^2, so we would expect that the answer would be slightly more than one hour.

Thus our calculated value of 74 minutes (1 hour 14 minutes) seems reasonable. ✓

Practice Problem 1 Mike rented an electric floor sander. It will sand 80 ft^2 of hardwood floor in 15 minutes. He needs to sand the floors in three rooms. The floor dimensions are 24 ft $\times$ 13 ft, 12 ft $\times$ 9 ft, and 16 ft $\times$ 3 ft. How long will it take him to sand the floors in all three rooms?

● **EXAMPLE 2** Carlos and Rosetta want to put vinyl siding on the front of their home in West Chicago. The house dimensions are shown in the figure at right. The door dimensions are 6 ft $\times$ 3 ft. The windows measure 2 ft $\times$ 4 ft.

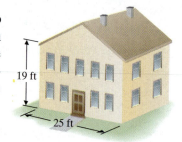

19 ft

25 ft

(a) Excluding windows and doors, how many square feet of siding will be needed?

(b) If the siding costs $2.25 per square foot, how much will it cost to side the front of the house?

1. *Understand the problem.*

Mathematics Blueprint For Problem Solving

Gather the Facts	What Am I Asked to Do?	How Do I Proceed?	Key Points to Remember
House measures 19 ft × 25 ft. Windows measure 2 ft × 4 ft. Door measures 6 ft × 3 ft. Siding costs $2.25 per square foot.	Find the cost to put siding on the front of the house.	**(a)** Find the area of the entire front of the house by multiplying 19 ft by 25 ft. Find the area of one window by multiplying 2 ft by 4 ft. Find the area of the door by multiplying 6 ft by 3 ft. **(b)** Multiply shaded area by $2.25.	**(a)** To obtain the shaded area we must subtract the area of nine windows and one door from the area of the entire front. **(b)** We must multiply the resulting area by the cost of the siding per foot.

2. *Solve and state the answer.*

(a) Find the area of the front of the house (shaded area).
We will find the area of the large rectangle representing the front of the house. Then we will subtract the area of the windows and the door.

$$\text{Area of each window} = (2 \text{ ft})(4 \text{ ft}) = 8 \text{ ft}^2$$
$$\text{Area of 9 windows} = (9)(8 \text{ ft})^2 = 72 \text{ ft}^2$$
$$\text{Area of 1 door} = (6 \text{ ft})(3 \text{ ft}) = 18 \text{ ft}^2$$
$$\text{Area of 9 windows} + 1 \text{ door} = 90 \text{ ft}^2$$
$$\text{Area of large rectangle} = (19 \text{ ft})(25 \text{ ft}) = 475 \text{ ft}^2$$

$$\begin{aligned}
\text{Total area of front of house} \quad & 475 \text{ ft}^2 \\
- \text{ Area of 9 windows and 1 door} \quad & -\ 90 \text{ ft}^2 \\
\hline
= \text{ Total area to be covered} \quad & 385 \text{ ft}^2
\end{aligned}$$

We see that 385 ft² of siding will be needed.

(b) Find the cost of the siding.

$$\text{Cost} = 385 \text{ ft}^2 \times \frac{\$2.25}{1 \text{ ft}^2} = \$866.25$$

The cost to put up siding on the front of the house will be $866.25.

3. *Check.*
We leave the check up to you.

Practice Problem 2   In the margin is a sketch of a roof of a commercial building.

(a) What is the area of the roof?

(b) How much would it cost to install new roofing on the roof area shown if the roofing costs $2.75 per square yard? (*Hint*: 9 square feet = 1 square yard.)

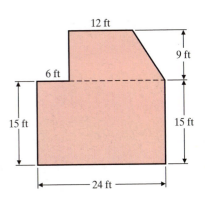

Applications

Round to the nearest tenth unless otherwise directed.

1. Monica drives to work each day from Bethel to Bridgeton. The following sketch shows the two possible routes.

(a) How many kilometers is the trip if she drives through Suffolk? What is her average speed if this trip takes her 0.4 hour?

(b) How many kilometers is the trip if she drives through Woodville and Palermo? What is her average speed if this trip takes her 0.5 hour?

(c) Over which route does she travel at a more rapid rate?

2. Robert drives from work to either a convenience store and then home or to a supermarket and then home. The sketch shows the distances.

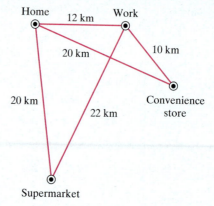

(a) How far does he travel if he goes from work to the supermarket and then home? How fast does he travel if the trip takes 0.6 hour?

(b) How far does he travel if he goes from work to the convenience store and then home? How fast does he travel if the trip takes 0.5 hour?

(c) Over which route does he travel at a more rapid rate?

3. A professional wallpaper hanger can wallpaper 120 ft² in 15 minutes. She will be papering four walls in a house. They measure 7 ft × 10 ft, 7 ft × 14 ft, 6 ft × 10 ft, and 6 ft × 8 ft. How many minutes will it take her to paper all four walls?

4. Dave and Linda McCormick are painting the walls of their apartment in Seattle. When they worked together at Linda's mother's house, they were able to paint 80 ft² in 25 minutes. The living room they wish to paint has one wall that measures 16 feet by 7 feet, one wall that measures 14 feet by 7 feet, and two walls that measure 12 feet by 7 feet. How long will it take Dave and Linda together to paint the living room of their apartment? Round to the nearest minute.

5. The floor area of the recreation room at Yvonne's house is shown in the following drawing. How much will it cost to carpet the room if the carpet costs $15 per square yard?

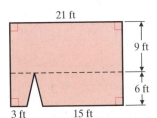

6. The side view of a barn is shown in the following diagram. The cost of aluminum siding is $18 per square yard. How much will it cost to put siding on this side of the barn?

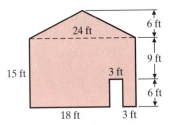

7. Find the volume of a concrete connector for the city sewer system. A diagram of the connector is shown. It is shaped like a box with a hole of diameter 2 m. If it is formed using concrete that costs $1.20 per cubic meter, how much will the necessary concrete cost?

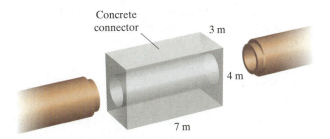

8. A dentist places a gold filling in the shape of a cylinder with a hemispherical top in a patient's tooth. The radius r of the filling is 1 mm. The height is 2 mm. Find the volume of the filling. If dental gold costs $95 per cubic millimeter, how much did the gold cost for the filling?

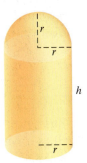

9. The Landstat satellite orbits the earth in an almost circular pattern. Assume that the radius of orbit (distance from center of Earth to the satellite) is 6500 km.

(a) How many kilometers long is one orbit of the satellite? (That is, what is the circumference of the orbit path?)

(b) If the satellite goes through one orbit around the Earth in two hours, what is its speed in kilometers per hour?

10. The North City Park is constructed in a shape that includes the region outside one-fourth of a circle. It is shaded in this sketch.

r = 140 m

140 m

←— 140 m —→

(a) What is the area of the park?

(b) How much will it cost to resod the park with new grass at $4.00 per square meter?

11. For Valentine's Day, a company makes decorative cylinder-shaped canisters and fills them with red hot candies. Each canister is 10 in. high and has a radius of 2 in. They want to make 400 canisters and need to determine how much candy to buy. What is the total number of cubic inches that will be filled with candy?

12. James is making cone-shaped candles. The mold he pours the wax into is 5 in. in diameter and 8 in. high. James needs to find the volume so he knows how much wax to buy. How many cubic inches of wax does he need to make 60 candles?

13. Dr. Dobson's ranch in Colorado has a rectangular field that measures 456 feet by 625 feet. The fence surrounding the field needs to be replaced. If the fencing costs $4.57 per yard, what will it cost for materials to replace the fence? Round to the nearest cent.

14. Boston Sunoco Dealerships owns a large gasoline storage tank in the shape of a cylinder in South Boston. The diameter of the tank is 122.9 feet. The height of the tank is 55.8 feet. How many gallons of gasoline will it hold if there are 7.5 gal in 1 ft^3? Round to the nearest tenth.

To Think About

15. Jonathan Wells has entered a bike race in Stowe, Vermont. He is traveling at a speed such that his bicycle wheel makes 200 revolutions per minute. The bicycle tire has a diameter of 28 inches. How fast is Jonathan traveling in feet per minute? (Use $\pi = 3.14$ and round to the nearest hundredth.)

16. A car in front of Jonathan is traveling at 26 miles per hour. If Jonathan increases his speed by 20% from the speed he was traveling in exercise 15, will he be able to pass the car?

Cumulative Review Problems

Divide.

17. $16\overline{)2048}$

18. $27\overline{)24,705}$

19. $1.3\overline{)0.325}$

20. $0.52\overline{)2.5324}$

Math in the Media

Just a Nonartist in the Art World, but Endlessly Seen and Cited

Roberta Smith
The New York Times,
January 21, 1998.

"The space-twisting, mind-bending images of this Dutch printmaker, *M. C. Escher*, rank among the 20th century's most popular.

They have been widely disseminated through unlimited editions, posters, T-shirts and album covers and are regularly found in college dormitories, bookstores, computer labs and mathematicians' offices."

In the early 1920s and again in 1936, Escher visited the Alhambra, where he studied and drew the pulsating patterns of the Islamic tilework and became fascinated with "the regular division of a plane," as he called it, a principle that would determine much of his later work. Divisions of the plane, called "tessellations," are arrangements of closed shapes that completely cover the plane without overlapping and without leaving gaps.

For example a 15′ by 20′ kitchen floor could be covered with 300 square tiles (also the area in square feet) with side length of 1 foot or 3600 one-inch square chips (3600 sq. in. of area). This pattern of square tiles forms a mosaic. We are assuming these tiles are not grouted. These would be tessellations easy to visualize and to create. The tiles themselves could have any kind of pattern or color desired. Escher went beyond simple tessellations and used shapes that would interlock in some way. Look up Escher in any encyclopedia for pictures of some of his tessellations and other interesting works.

EXERCISES

To create and explore a few tessellations, answer the questions below.

1. Draw with a ruler a rectangle 2 inches by 3 inches and cover it with 1 inch by 1 inch tiles. Drawing a diagonal for each square will cover the rectangle with right triangles. Work with a friend and you will note that you will agree as to the square tessellations, but your triangle tessellations might have a slightly different pattern.

2. Cover the 2 inch by 3 inch rectangle with rectangles. This is most easily done by bisecting the squares. For example, the width would be half of the length. Another easy tessellation would be using strips of the same width. Experiment with other rectangles and other triangles.

3. Cover the 2 inch by 3 inch rectangle with circles of 1/2 inch radius. What do you notice? Either there is space left in the rectangle which isn't covered or there are circles hanging over the edge of the rectangle. Putting circles in a rectangle is like packing baseballs in a box.

4. Finally, cover the rectangle with an irregular shape, for example, a fish, drawn to minimize gaps. You will probably overlap the edges of the rectangle.

Chapter 7 Organizer

Topic	Procedure	Examples
Perimeter of a rectangle, p. 428.	$P = 2l + 2w$	Find the perimeter of a rectangle with width = 3 m and length = 8 m. $P = (2)(8\text{ m}) + (2)(3\text{ m})$ $= 16\text{ m} + 6\text{ m} = 22\text{ m}$
Perimeter of a square, p. 428.	$P = 4s$	Find the perimeter of a square with side $s = 6$ m. $P = (4)(6\text{ m}) = 24\text{ m}$
Area of a rectangle, p. 430.	$A = lw$	Find the area of a rectangle with width = 2 m and length = 7 m. $A = (7\text{ m})(2\text{ m}) = 14\text{ m}^2$
Area of a square, p. 430.	$A = s^2$	Find the area of a square with a side of 4 m. $A = s^2 = (4\text{ m})^2 = 16\text{ m}^2$
Perimeter of parallelograms, trapezoids, and triangles, pp. 436, 437, 438.	Add up the lengths of all sides.	Find the perimeter of a triangle with sides of 3 m, 6 m, and 4 m. $3\text{ m} + 6\text{ m} + 4\text{ m} = 13\text{ m}$
Area of a parallelogram, p. 436.	$A = bh$ $b = $ length of base $h = $ height 	Find the area of a parallelogram with a base of 12 m and a height of 7 m. $A = bh = (12\text{ m})(7\text{ m}) = 84\text{ m}^2$
Area of a trapezoid, p. 438.	$A = \dfrac{h(b + B)}{2}$ $b = $ length of shorter base $B = $ length of longer base $h = $ heigth 	Find the area of a trapezoid whose height is 12 m and whose bases are 17 m and 25 m. $A = \dfrac{(12\text{ m})(17\text{ m} + 25\text{ m})}{2} = \dfrac{(12\text{ m})(42\text{ m})}{2}$ $= \dfrac{504\text{ m}^2}{2} = 252\text{ m}^2$
The sum of the measures of the three interior angles of a triangle is 180°, p. 444.	In a triangle, to find one missing angle if two are given: **1.** Add up the two known angles. **2.** Subtract the sum from 180°.	Find the missing angle if two known angles in a triangle are 60° and 70°. **1.** $60° + 70° = 130°$ **2.** $\begin{array}{r} 180° \\ -\ 130° \\ \hline 50° \end{array}$ The missing angle is 50°.
Area of a triangle, p. 445.	$A = \dfrac{bh}{2}$ $b = $ base $h = $ height 	Find the area of the triangle whose base is 1.5 m and whose height is 3 m. $A = \dfrac{bh}{2} = \dfrac{(1.5\text{ m})(3\text{ m})}{2} = \dfrac{4.5\text{ m}^2}{2} = 2.25\text{ m}^2$

Topic	Procedure	Examples
Evaluating square roots of numbers that are perfect squares, p. 452.	If a number is a product of two identical factors, then either factor is called a square root.	$\sqrt{0} = 0$ because $(0)(0) = 0$ $\sqrt{4} = 2$ because $(2)(2) = 4$ $\sqrt{100} = 10$ because $(10)(10) = 100$ $\sqrt{169} = 13$ because $(13)(13) = 169$
Approximating the square root of a number that is not a perfect square, p. 452.	1. If a calculator with a square root key is available, enter the number and then press the $\boxed{\sqrt{x}}$ or $\boxed{\sqrt{}}$ key. The approximate value will be displayed. 2. If using a square root table, find the number n, then look for the square root of that number. The approximate value will be correct to the nearest thousandth. | Number, n	Square Root of That Number, $\sqrt{n}$
---	---
31	5.568
32	5.657
33	5.745

34	5.831		1. Find on a calculator.
			   (a) $\sqrt{13}$ (b) $\sqrt{182}$
			   Round to the nearest thousandth.
			   (a) 13 $\boxed{\sqrt{x}}$ 3.60555127 rounds to 3.606.
			   (b) 182 $\boxed{\sqrt{x}}$ 13.49073756 rounds to 13.491.
			2. Find from a square root table.
			   (a) $\sqrt{31}$ (b) $\sqrt{33}$ (c) $\sqrt{34}$
			   To the nearest thousandth, the approximate values are as follows.
			   (a) $\sqrt{31} = 5.568$ (b) $\sqrt{33} = 5.745$
			   (c) $\sqrt{34} = 5.831$

Finding the hypotenuse of a right triangle when given the length of each leg, p. 456.	Hypotenuse $= \sqrt{(\text{leg})^2 + (\text{leg})^2}$ 	Find the hypotenuse of a triangle with legs of 9 m and 12 m. $\begin{aligned} \text{hypotenuse} &= \sqrt{(12)^2 + (9)^2} \\ &= \sqrt{144 + 81} = \sqrt{225} \\ &= 15 \text{ m} \end{aligned}$
Finding the leg of a right triangle when given the length of the other leg and the hypotenuse, p. 457.	Leg $= \sqrt{(\text{hypotenuse})^2 - (\text{leg})^2}$	Find the leg of a right triangle. The hypotenuse is 14 in. and the other leg is 12 in. Round to nearest thousandth. leg $= \sqrt{(14)^2 - (12)^2} = \sqrt{196 - 144} = \sqrt{52}$ Using a calculator or a square root table, the leg ≈ 7.211 in.

Solving applied problems involving the Pythagorean Theorem, p. 458.	1. Read the problem carefully.
	2. Draw a sketch.
	3. Label the two sides that are given.
	4. If the hypotenuse is unknown, use
	   hypotenuse $= \sqrt{(\text{leg})^2 + (\text{leg})^2}$.
	5. If one leg is unknown, use
	   leg $= \sqrt{(\text{hypotenuse})^2 - (\text{leg})^2}$.
		A boat travels 5 mi south and then 3 mi east. How far is it from the starting point? Round to the nearest tenth.
		$\begin{aligned} \text{hypotenuse} &= \sqrt{(5)^2 + (3)^2} \\ &= \sqrt{25 + 9} = \sqrt{34} \end{aligned}$
		Using a calculator or a square root table, the distance is approximately 5.8 mi.

The special 30°–60°–90° right triangle, p. 459.	The length of the leg opposite the 30° angle is $\frac{1}{2} \times$ the length of the hypotenuse.	Find y. $y = \frac{1}{2}(26) = 13$ m

Topic	Procedure	Examples
The special 45°–45°–90° right triangle, p. 459.	The sides opposite the 45° angles are equal. The hypotenuse is $\sqrt{2}\times$ the length of either leg.	Find z. $z = \sqrt{2}\,(13) \approx (1.414)(13) = 18.382$ m
Radius and diameter of a circle, p. 464.	$r = $ radius $d = $ diameter $r = \dfrac{d}{2}$ $d = 2r$	What is the radius of a circle with diameter 50 in. $r = \dfrac{50 \text{ in.}}{2} = 25$ in. What is the diameter of a circle with radius 16 in.? $d = (2)(16 \text{ in.}) = 32$ in.
Pi, p. 464.	Pi is a decimal that goes on forever. It can be approximated by as many decimal places as needed. $\pi = \dfrac{\text{circumference of a circle}}{\text{diameter of same circle}}$	Use $\pi \approx 3.14$ for all calculations. Unless otherwise directed, round your final answer to the nearest tenth when any calculation involves π.
Circumference of a circle, p. 464.	$C = \pi d$	Find the circumference of a circle with diameter 12 ft. $C = \pi d = (3.14)(12 \text{ ft}) = 37.68$ ≈ 37.7 ft (rounded to the nearest tenth)
Area of a circle, p. 466.	$A = \pi r^2$ **1.** Square the radius first. **2.** Then multiply the result by 3.14.	Find the area of a circle with radius 7 ft. $A = \pi r^2 = (3.14)(7 \text{ ft})^2 = (3.14)(49 \text{ ft}^2)$ $= 153.86 \text{ ft}^2$ $\approx 153.9 \text{ ft}^2$ (rounded to the nearest tenth)
Volume of a rectangular solid (box), p. 473.	$V = lwh$	Find the volume of a box whose dimensions are 5 m by 8 m by 2 m. $V = (5 \text{ m})(8 \text{ m})(2 \text{ m}) = (40)(2) \text{ m}^3 = 80 \text{ m}^3$
Volume of a cylinder, p. 473.	$r = $ radius $h = $ height $V = \pi r^2 h$ **1.** Square the radius first. **2.** Then multiply the result by 3.14. and by the height. 	Find the volume of a cylinder with radius 7 m and height 3 m. $V = \pi r^2 h = (3.14)(7 \text{ m})^2(3 \text{ m})$ $= (3.14)(49)(3) \text{ m}^3$ $= (153.86)(3) \text{ m}^3 = 461.58 \text{ m}^3$ $\approx 461.6 \text{ m}^3$ (rounded to the nearest tenth)
Volume of a sphere, p. 474.	$V = \dfrac{4\pi r^3}{3}$ $r = $ radius	Find the volume of a sphere of radius 3 m. $V = \dfrac{4\pi r^3}{3} = \dfrac{(4)(3.14)(3 \text{ m})^3}{3}$ $= \dfrac{(4)(3.14)\overset{9}{\cancel{(27)}} \text{ m}^3}{1}$ $= (12.56)(9) \text{ m}^3 = 113.04 \text{ m}^3$ $\approx 113.0 \text{ m}^3$ (rounded to the nearest tenth)

Topic	Procedure	Examples
Volume of a cone, **p. 474.**	$$V = \frac{\pi r^2 h}{3}$$ r = radius h = height	Find the volume of a cone of height 9 m and radius 7 m. $$V = \frac{\pi r^2 h}{3} = \frac{(3.14)(7 \text{ m})^2 (9 \text{ m})}{3}$$ $$= \frac{(3.14)(7^2) \overset{3}{\cancel{(9)}} \text{ m}^3}{\underset{1}{\cancel{3}}} = (3.14)(49)(3) \text{ m}^3$$ $$= (153.86)(3) \text{ m}^3 = 461.58 \text{ m}^3$$ $$\approx 461.6 \text{ m}^3 \quad \text{(rounded to the nearest tenth)}$$
Volume of a **pyramid, p. 475.**	$$V = \frac{Bh}{3}$$ B = area of the base h = height **1.** Find the area of the base. **2.** Multiply this area by the height and divide the result by 3.	Find the volume of a pyramid whose height is 6 m and whose rectangular base is 10 m by 12 m. **1.** $B = (12 \text{ m})(10 \text{ m}) = 120 \text{ m}^2$ **2.** $V = \dfrac{(120)\overset{2}{\cancel{(6)}} \text{ m}^3}{\underset{1}{\cancel{3}}} = (120)(2) \text{ m}^3 = 240 \text{ m}^3$
Similar figures, **corresponding sides,** **p. 482.**	The corresponding sides of similar figures have the same ratio.	Find n in the following similar figures. $$\frac{n}{4} = \frac{9}{3}$$ $$3n = 36$$ $$n = 12 \text{ m}$$
Similar figures, **corresponding** **perimeters, p. 483.**	The perimeters of similar figures have the same ratio as the corresponding sides. For reasons of space, the procedure for the areas of similar figures is not given here but may be found in the text (see p. 485).	These two figures are similar. Find the perimeter of the larger figure. $$\frac{6}{12} = \frac{29}{p}$$ $$6p = (12)(29)$$ $$6p = 348$$ $$\frac{6p}{6} = \frac{348}{6}$$ $$p = 58$$ The perimeter of the larger figure is 58 m.

Chapter 7 Review Problems

Round to the nearest tenth when necessary. Use $\pi \approx 3.14$ in all calculations requiring the use of π.

7.1

1. Find the complement of an angle of 76°.

2. Find the supplement of an angle of 12° angle.

3. Find the measures of $\angle a$, $\angle b$, and $\angle c$ in the following sketch.

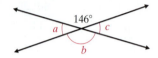

4. Find $\angle s$, $\angle t$, $\angle u$, $\angle w$, $\angle x$, $\angle y$, and $\angle z$ in the following sketch if we know that line p is parallel to line q.

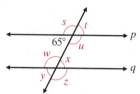

7.2 *Find the perimeter of the square or rectangle.*

5. length = 8.3 m, width = 1.6 m

6. length = width = 2.4 yd

Find the area of the square or rectangle.

7. length = 5.9 cm, width = 2.8 cm

8. length = width = 7.2 in.

Find the perimeter of each object made up of rectangles and squares.

9.

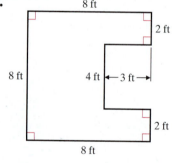

10.

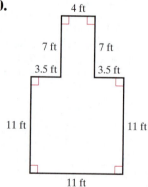

Find the area of each shaded region made up of rectangles and squares.

11.

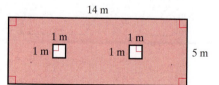

12.

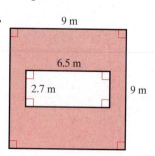

7.3 *Find the perimeter of the parallelogram or trapezoid.*

13. Two sides of the parallelogram are 52 m and 20.6 m.

14. The sides of the trapezoid are 5 mi, 22 mi, 5 mi, and 30 mi.

Find the area of the parallelogram or trapezoid.

15. The parallelogram has a base of 90 m and a height of 30 m.

16. The trapezoid has a height of 36 yd and bases of 17 yd and 23 yd.

Find the total area of each region made up of parallelograms, trapezoids, and rectangles

17.

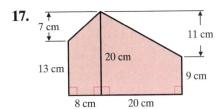

18.

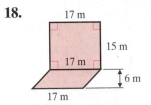

7.4 *Find the perimeter of the triangle.*

19. The sides are 10 ft, 5 ft, and 7 ft.

20. The sides are 5.5 ft, 3 ft, and 5.5 ft.

Find the measure of the third angle in the triangle.

21. Two known angles are 15° and 12°.

22. Two known angles are 37° and 96°.

Find the area of the triangle.

23. base = 8.5 m, height = 12.3 m

24. base = 12.5 m, height = 9.5 m

Find the total area of each region made up of triangles and rectangles.

25.

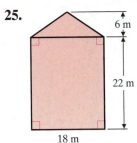

26.

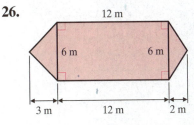

7.5

Evaluate exactly.

27. $\sqrt{81}$ **28.** $\sqrt{64}$ **29.** $\sqrt{121}$ **30.** $\sqrt{36} + \sqrt{0}$ **31.** $\sqrt{9} + \sqrt{4}$

Approximate using a square root table or a calculator with a square root key. Round to the nearest thousandth when necessary.

32. $\sqrt{35}$ **33.** $\sqrt{45}$ **34.** $\sqrt{165}$ **35.** $\sqrt{180}$

7.6 *Find the unknown side. If the answer cannot be obtained exactly, use a square root table or a calculator with a square root key. Round to the nearest hundredth when necessary.*

36.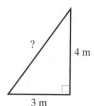

37.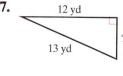

38.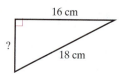

39.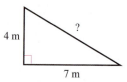

Round to the nearest tenth.

40. Find the distance between the centers of the holes of a metal plate with the dimensions labeled in the following sketch.

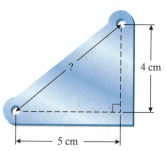

41. A building ramp has the following dimensions. Find the length of the ramp.

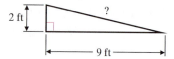

42. A shed is built with the following dimensions. Find the distance from the peak of the roof to the horizontal support brace.

43. Find the width of a door if it is 6 ft tall and the diagonal measures 7 ft.

7.7

44. What is the diameter of a circle whose radius is 53 cm?

45. What is the radius of a circle whose diameter is 126 cm?

46. Find the circumference of a circle with diameter 14 in.

47. Find the circumference of a circle with radius 9 in.

Find the area of each circle.

48. radius = 9 m

49. diameter = 16 ft

Find the area of each shaded region made up of circles, semicircles rectangles, trapezoids, and parallelograms. Round your answer to the nearest tenth.

50.
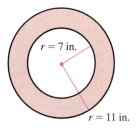
r = 7 in.
r = 11 in.

51.
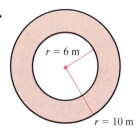
r = 6 m
r = 10 m

52.
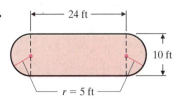
24 ft
10 ft
r = 5 ft

53.
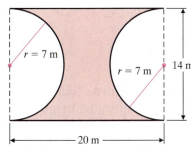
r = 7 m
r = 7 m
14 m
20 m

54.
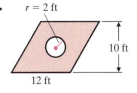
r = 2 ft
10 ft
12 ft

55.
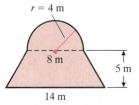
r = 4 m
8 m
5 m
14 m

7.8   *In exercises 56–62, find the volume.*

56. a rectangular box measuring 3 ft by 6 ft by 2.5 ft

57. a sphere with radius 1.2 ft

58. Find the volume of a storage can that is 2 m tall and has a radius of 7 m.

59. Find the volume of a pyramid that is 18 m high and whose rectangular base measures 16 m by 18 m.

60. Find the volume of a pyramid that is 15 m high and whose square base measures 7 m by 7 m.

61. Find the volume of a cone of sand 9 ft tall with a radius of 20 ft.

62. A chemical has polluted a volume of ground in a cone shape. The depth of the cone is 30 yd. The radius of the cone is 17 yd. Find the volume of polluted ground.

7.9   *Find n in each set of similar triangles.*

63.
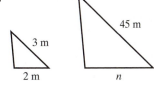
45 m
3 m
2 m
n

64.
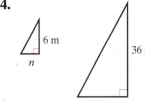
6 m
n
36 m
20 m

Determine the perimeter of the unlabeled figure.

65.

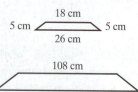

18 cm

5 cm   5 cm

26 cm

108 cm

66.

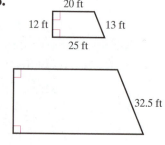

20 ft

12 ft   13 ft

25 ft

32.5 ft

67. Anastasio is in charge of decorations for the "International Cars of the Future" show. He has designed a rectangular banner that will hang in front of the 1ˢᵗ prize-winning car, so that all he has to do is push a button and the banner will fly up into the ceiling space to reveal the car. The model of the banner used 12 square yards of fabric. The dimensions of the actual banner will be $3\frac{1}{2}$ times the length and the width of the model. How much fabric will the finished banner need?

7.10

68. The Wilsons are carpeting a recreation room with the dimensions shown. Carpeting costs $8 per square yard. How much will the carpeting cost?

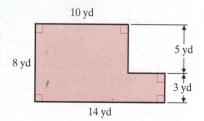

10 yd

5 yd

8 yd

3 yd

14 yd

69. A conical tank holds acid in a chemistry lab. The tank has a radius of 9 in. and a height of 24 in. How many cubic inches does the tank hold? The acid weighs 16 g per cubic inch. What is the weight of the acid if the tank is full?

70. A silo has a cylindrical shape with a hemispherical dome. It has dimensions as shown in the following figure.
   (a) What is its volume in cubic feet?

   (b) If 1 cubic foot ≈ 0.8 bushel, how many bushels of grain will it hold?

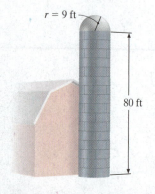

r = 9 ft

80 ft

71. (a) In the following diagram, how many kilometers is it from Homeville to Seaview if you drive through Ipswich? How fast do you travel if it takes 0.5 hour to travel that way?

   (b) How many kilometers is it from Homeville to Seaview if you drive through Acton and Westville? How fast do you travel if it takes 0.8 hour to travel that way?

   (c) Over which route do you travel at a more rapid rate?

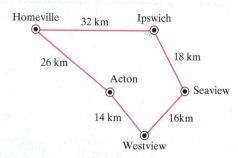

Homeville   32 km   Ipswich

26 km   18 km

Acton

                                   Seaview

14 km   16km

Westview

72. During 1998, U.S. farms produced 2.757 billion bushels of soybeans. (*Source*: U.S. Department of Agriculture) Each bushel takes up 1.244 cubic feet of storage. How many cubic feet of storage was needed for the 1998 soybean crop?

73. If all of the soybeans in exercise 72 were stored in a huge rectangular storage bin that is 10,000 feet wide and 20,000 feet long, how many feet high would the storage bin need to be?

74. A farmer needs to construct a holding pen for his cattle. It will be a rectangular pen that measures 21 feet by 30 feet. What will be the perimeter of the pen in feet? If the farmer needs to put up fencing around the pen that costs $7 per foot, how much will the fencing cost? (Assume that he will use the fencing for the gate as well as the sides.)

75. If the farmer puts up the fencing in exercise 74 with fence posts that are 3 feet apart, how many fence posts will he need? (Assume that he will use a gate that is 3 feet across with a fence post on either side.)

76. Marvin uses 8.5-inch-by-11-inch paper in his computer printer. When he examined a document printed by his printer, he found that the printed area measures 7 inches by 9 inches. How many square inches of each sheet of paper are not used for printing?

77. Frank and Dolly manage a roller rink in Springfield. The floor of the rink consists of a large rectangle with a semicircle at each end. The rectangular part measures 16 yards by 20 yards. Each semicircle has a diameter of 16 yards. What is the area of the floor?

78. Frank and Dolly have decided to put down a new hardwood floor for the roller rink in exercise 77. The hardwood flooring will cost $50 per square yard to install (including both labor and materials). How much will it cost them to put down a new hardwood floor?

79. A kite is flying exactly 30 feet above the edge of a pond. The person flying the kite is using exactly 33 feet of string. Assuming that the string is so tight that it forms a straight line, how far is the person standing from the edge of the pond? Round to the nearest tenth.

80. The Suburban Gas Company has a spherical gas tank. The diameter of the tank is 90 meters. Find the volume of the spherical tank.

81. Charlie and Ginny have a cylindrical hot water tank that is 5 feet high and has a diameter of 18 inches. How many cubic feet does the tank hold?

82. The tank in exercise 81 is filled with water. One cubic foot of water is about 7.5 gallons. How many gallons does the tank hold?

83. The front lawn at Central High School in Hanover is in the shape of a trapezoid. The bases of the trapezoid are 45 feet and 50 feet. The height of the trapezoid is 35 feet. What is the area of the front lawn?

84. The principal of the high school in exercise 83 has stated that the front lawn needs to be fertilized three times a year. The lawn company charges $0.50 per square foot to apply fertilizer. How much will it cost to have the front lawn fertilized three times a year?

1. _____

2. _____

3. _____

4. _____

5. _____

6. _____

7. _____

8. _____

9. _____

10. _____

11. _____

12. _____

13. _____

14. _____

15. _____

16. _____

17. _____

18. _____

Find the perimeter.

1. a rectangle that measures 9 yd × 11 yd

2. a square with side 6.3 ft

3. a parallelogram with sides measuring 6.5 m and 3.5 m

4. a trapezoid with sides measuring 22 m, 13 m, 32 m, and 13 m

5. a triangle with sides measuring 58.6 m, 32.9 m, and 45.5 m

Find the area. Round to the nearest tenth.

6. a rectangle that measures 10 yd × 18 yd

7. a square 10.2 m on a side

8. a parallelogram with a height of 6 m and a base of 13 m

9. a trapezoid with a height of 9 m and bases of 7 m and 25 m

10. a triangle with a base of 4 cm and a height of 6 cm

11. a triangle with a base of 15 m and a height of 7 m

Evaluate exactly.

12. $\sqrt{81}$ **13.** $\sqrt{121}$

14. Find the complement of an angle that measures 63°.

15. Find the supplement of an angle that measures 107°.

Approximate using a square root table or a calculator with a square root key. Round to the nearest thousandth when necessary.

16. $\sqrt{54}$ **17.** $\sqrt{135}$ **18.** $\sqrt{187}$

In exercises 19 and 20, find the unknown side. Use a calculator or a square root table to approximate square roots to the nearest thousandth.

19.

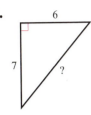

20.

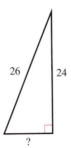

21. Find the distance between the centers of the holes drilled in a rectangular metal plate with the dimensions labeled in the following sketch.

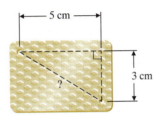

22. A 15-ft-tall ladder is placed so that it reaches 12 ft up on the wall of a house. How far is the base of the ladder from the wall of the house?

23. Find the circumference of a circle with radius 6 in.

24. Find the area of a circle with diameter 18 ft.

19.

20.

21.

22.

23.

24.

Find the shaded area of each region made up of circles, semicircles, rectangles, squares, trapezoids, and parallelograms.

25.

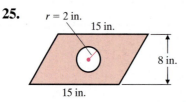

26.

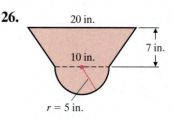

Find the volume.

27. a rectangular box measuring 7 m by 12 m by 10 m

28. a cone with height 12 m and radius 8 m

29. a sphere of radius 3 m

30. a cylinder of height 2 ft and radius 9 ft

31. a pyramid of height 14 m and whose rectangular base measures 4 m by 3 m

Each pair of triangles is similar. Find the missing side n.

32.

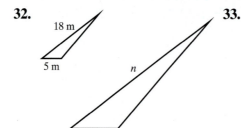

33.

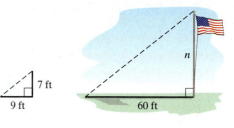

Solve.

An athletic field has the dimensions shown in the figure below. Assume you are considering only the darker green shaded area.

34. What is the area of the athletic field?

35. How much will it cost to fertilize it at $0.40 per square yard?

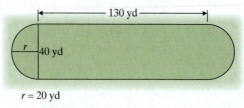

Approximately one-half of this test is based on Chapter 7 material. The remainder is based on material covered in Chapters 1–6.

Solve. Simplify.

1. Add.   126,350
             278,120
           + 531,290

2. Multiply.   163
                 $\times\,205$

3. Subtract. $\dfrac{17}{18} - \dfrac{11}{30}$

4. Divide. $\dfrac{3}{7} \div 2\dfrac{1}{4}$

5. Round to the nearest hundredth. 56.1279

6. Multiply.   9.034
                 $\times\quad 0.8$

7. Divide. $0.021\overline{)1.743}$

8. Find *n*. $\dfrac{3}{n} = \dfrac{2}{18}$

9. There are seven teachers for every 100 students. If there are 56 teachers at the university, how many students would you expect?

10. Michael scored 18 baskets out of 24 shots on the court. What percent of his shots went into the basket?

11. 0.8% of what number is 16?

12. What is 18.5% of 220?

13. Convert 586 cm to m.

14. Convert 42 yd to in.

15. Ben traveled 88 km. How many miles did he travel? (1 kilometer ≈ 0.62 mile; 1 mile ≈ 1.61 kilometers.)

In questions 16–30, round to the nearest tenth when necessary. Use π ≈ 3.14 in calculations involving π.

16. Find the perimeter of a rectangle of length 17 m and width 8 m

17. Find the perimeter of a trapezoid with sides of 86 cm, 13 cm, 96 cm, and 13 cm

18. Find the circumference of a circle with diameter 18 yd.

1. _____

2. _____

3. _____

4. _____

5. _____

6. _____

7. _____

8. _____

9. _____

10. _____

11. _____

12. _____

13. _____

14. _____

15. _____

16. _____

17. _____

18. _____

19.
20.
21.
22.
23.
24.
25.
26.
27.
28.
29.

Find the area.

19. a triangle with base 1.2 cm and height 2.4 cm

20. a trapezoid with height 18 m and bases of 26 m and 34 m

21.

12 m
12 m
12 m
12 m
4 m

22.

35 yd
20 yd
20 yd
6 yd
6 yd
5 yd
24 yd

23. a circle with radius 4 m

Find the volume.

24. a cylinder with height 12 m and radius 8 m

25. a sphere with radius 9 cm

26. a pyramid with height 32 cm and a rectangular base 14 cm by 21 cm

27. A cone with height 18 m and radius 12 m

Find the value of n in each pair of similar figures.

28.

26 m
n
7 m
9 m

29.

11 ft
4 ft
n
1.5 ft

Solve.

30. (a) _____

30. Mary Ann and Wong Twan have a recreation room with the dimensions shown in the figure below. They wish to carpet it at a cost of $8.00 per square yard.
   (a) How many square yards of carpet are needed? (Include the area of the rectangle, the triangle, and the square.)
   (b) How much will it cost?

(b) _____

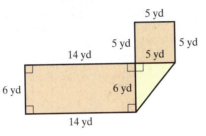

31. _____

31. Evaluate exactly. $\sqrt{36} + \sqrt{25}$

32. _____

32. Approximate to the nearest thousandth using a table or a calculator. $\sqrt{57}$

33. _____

In questions 33 and 34, find the unknown side. Use a calculator or square root table if necessary. Round to the nearest thousandth.

33.

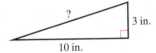

34.

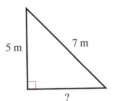

34. _____

35. _____

35. A boat travels 12 mi south and then 7 mi east. How far is it from its original starting point? Round to the nearest tenth of a mile.

36. Bo Sigvarson of Malmö, Sweden, constructed a giant paintbrush measuring 20 feet long and weighing 100 pounds. If the original paintbrush model measures $7\frac{1}{2}$ inches long, how many of them would you need to span the length of the giant paintbrush if you placed them end-to-end?

36. _____

Statistics

Older people are living longer and staying healthier and active longer these days. Plans for the future of the United States, especially in the critical area of health care, must take into account the growing number of senior citizens. Demand for nursing homes, hospital beds, doctors, nurses, and medical supplies will increase greatly as the number of older people increases in U.S. society. Do you think you could accurately predict future populations of elderly people in the United States? Turn to page 529 and try the Putting Your Skills to Work problems to find out.

Pretest Chapter 8

1. _____

2. _____

3. _____

4. _____

5. _____

If you are familiar with the topics in this chapter, take this test now. Check your answers with those in the back of the book. If an answer is wrong or you can't answer a question, study the appropriate section of the chapter.

If you are not familiar with the topics in this chapter, don't take this test now. Instead, study the examples, work the practice problems, and then take the test.

This test will help you identify which concepts you have mastered and which you need to study further.

Section 8.1

The ages of 5000 students on campus were recorded. The following circle graph depicts the distribution of ages.

Age Distribution of Students on Campus

Over age 27 7%
Age 25–27 10%
Under age 18 6%
Age 21–24 33%
Age 18–20 44%

1. What age group comprises the smallest percent of the student body?

2. What percent of the students are between 21 and 27?

3. What percent of the students are 25 or older?

4. If 5000 students are at the university, how many students are 21 to 24 years old?

5. How many students are over age 27?

Section 8.2

The following double-bar graph indicates the number of new housing starts in Manchester during each quarter of 2000 and 2001.

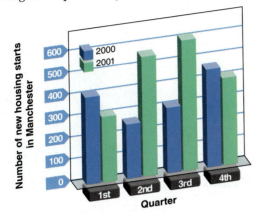

6. How many housing starts were there in Manchester in the first quarter of 2001?

7. How many housing starts were there in Manchester in the fourth quarter of 2000?

8. When were the smallest number of housing starts in Manchester?

9. When were the greatest number of housing starts in Manchester?

10. How many more housing starts were there in the second quarter of 2001 than in the second quarter of 2000?

11. How many fewer housing starts were there in the fourth quarter of 2001 than in the fourth quarter of 2000?

The line graph indicates sales and production of color television sets by a major manufacturer during the specified months.

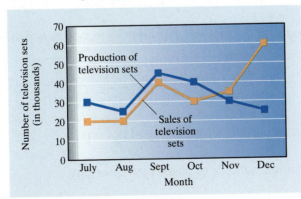

12. During what months was the production of television sets the lowest?

13. During what month was the sales of television sets the highest?

14. What was the first month in which the production of television sets was lower than the sales of television sets?

15. How many television sets were sold in August?

16. How many television sets were produced in November?

6. _____

7. _____

8. _____

9. _____

10. _____

11. _____

12. _____

13. _____

14. _____

15. _____

16. _____

17. _____

18. _____

19. _____

20. _____

21. _____

22. _____

23. _____

Section 8.3

The histogram tells us the number of miles a car was driven before the car was discarded or sold to a junk dealer.

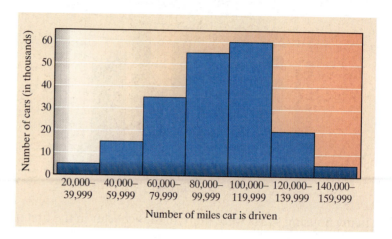

17. How many discarded or junked cars had been driven 80,000 to 99,999 mi?

18. How many discarded or junked cars had been driven 100,000 to 119,999 mi?

19. How many discarded or junked cars had been driven 120,000 mi or more?

20. How many discarded or junked cars had been driven less than 60,000 mi?

Section 8.4

A secretary produced the following number of pages on her computer.

Mon.	Tues.	Wed.	Thurs.	Fri.	Sat.
38	42	44	27	32	27

21. Find the mean number of pages typed per day.

22. Find the median number of pages typed per day.

23. Find the mode for the number of pages typed per day.

8.1 Circle Graphs

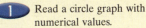

1 Reading a Circle Graph with Numerical Values

Statistics is that branch of mathematics that collects and studies data. Once the data are collected, they must be organized so that the information is easily readable. We use **graphs** to give a visual representation of the data that is easy to read. Graphs appeal to the eye. Their visual nature allows them to communicate information about the complicated relationships among statistical data. For this reason, newspapers often use graphs to help their readers quickly grasp information.

Circle graphs are especially helpful for showing the relationship of parts to a whole. The entire circle represents 100%; the pie-shaped pieces represent the subcategories. The following circle graph divides the 10,000 students at Westline College into five categories.

Distribution of Students at Westline College

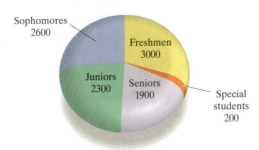

🔴 **EXAMPLE 1** What is the largest category of students?

The largest pie-shaped section of the circle is labeled "Freshmen." Thus the largest category is freshmen students.

Practice Problem 1 What is the smallest category of students? 🔴

🔴 **EXAMPLE 2**

(a) How many students are either sophomores or juniors?

(b) What percent of the students are sophomores or juniors?

(a) There are 2600 sophomores and 2300 juniors. If we add these two numbers, we have $2600 + 2300 = 4900$. Thus we see that there are 4900 students who are either sophomores or juniors.

(b) 4900 out of 10,000 are sophomores or juniors.

$$\frac{4900}{10,000} = 0.49 = 49\%$$

Practice Problem 2

(a) How many students are either freshmen or special students?

(b) What percent of the students are freshmen or special students? 🔴

🔴 **EXAMPLE 3** What is the ratio of freshmen to seniors?

Number of freshmen ⟶ 3000   Thus $\frac{3000}{1900} = \frac{30}{19}$.
Number of seniors ⟶ 1900

The ratio of freshmen to seniors is $\frac{30}{19}$.

Practice Problem 3 What is the ratio of freshmen to sophomores?

517

● **EXAMPLE 4**   What is the ratio of seniors to the total number of students?

There are 1900 seniors. We find the total of all the students by adding the number of students in each section of the graph. There are 10,000 students. The ratio of seniors to the total number of students is

$$\frac{1900}{10,000} = \frac{19}{100}.$$

Practice Problem 4   What is the ratio of freshmen to the total number of students?

② Reading a Circle Graph with Percentage Values

Together, the Great Lakes form the largest body of fresh water in the world. The total area of these five lakes is about 94,680 mi². The percentage of this total area taken up by each of the Great Lakes is shown in the circle graph below.

Percentage of Area Occupied by Each of the Great Lakes

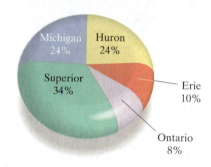

Source: U.S. Department of the Interior

● **EXAMPLE 5**   Which of the Great Lakes occupies the largest area in square miles?

The largest percent corresponds to the biggest area, which is occupied by Lake Superior. Lake Superior has the largest area in square miles.

Practice Problem 5   Which of the Great Lakes occupies the smallest area in square miles?

● **EXAMPLE 6**   What percent of the total area is occupied either by Lake Erie or Lake Ontario?

If we add 10% for Lake Erie and 8% for Lake Ontario, we get

$$10\% + 8\% = 18\%.$$

Thus 18% of the area is occupied by either Lake Erie or Lake Ontario.

Practice Problem 6   What percent of the total area is occupied by either Lake Superior or Lake Michigan?

● **EXAMPLE 7**   How many of the total 94,680 mi^2 are occupied by Lake Michigan? Round to the nearest whole number.

Remember that we multiply the percent times the base to obtain the amount. Here 24% of 94,680 is what is occupied by Lake Michigan.

$$(0.24)(94,680) = n$$
$$22,723.2 = n$$

Rounded to the nearest whole number, 22,723 mi^2 are occupied by Lake Michigan.

Practice Problem 7   How many of the total 94,680 mi^2 are occupied by Lake Superior? Round to the nearest whole number.   ●

Sometimes a circle graph is used to investigate the distribution of one part of a larger group. For example, approximately 736,000 bachelor's degrees were awarded in the United States in 1999 to students majoring in the six most popular subject areas: business, social sciences, education, engineering, health sciences, and information science and technology. The circle graph shows how the degrees in these six subject areas were distributed.

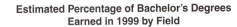

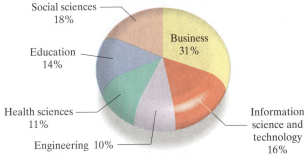

Estimated Percentage of Bachelor's Degrees Earned in 1999 by Field

Social sciences 18%
Education 14%
Health sciences 11%
Engineering 10%
Business 31%
Information science and technology 16%

Source: U.S. National Center for Educational Statistics

● **EXAMPLE 8**

(a) What percent of the bachelor's degrees represented in this circle graph are in the fields of information science and technology or business?

(b) Of the approximately 736,000 degrees awarded in these six fields, how many were awarded in the field of engineering?

(a) We add 16% to 31% to obtain 47%. Thus 47% of the bachelor's degrees represented by this circle are in the fields of information science and technology or business.

(b) We take 10% of the 736,000 people who obtained degrees in these six areas. Thus we have $(0.10)(736,000) = 73,600$. Approximately 73,600 degrees in engineering were awarded in 1999.

Practice Problem 8

(a) What percent of the bachelor's degrees represented in this circle graph are in the fields of health sciences or education?

(b) How many bachelor's degrees in social sciences were awarded in 1999?   ●

The following circle graph displays Bob and Linda McDonald's monthly $2700 family budget. Use the circle graph to answer exercises 1–10.

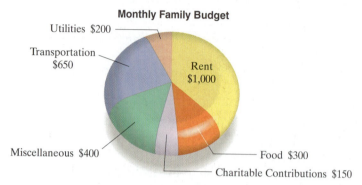

Monthly Family Budget

Utilities $200
Transportation $650
Rent $1,000
Miscellaneous $400
Food $300
Charitable Contributions $150

1. What category takes the largest amount of the budget?

2. What category takes the least amount of the budget?

3. How much money is allotted each month for utilities?

4. How much money is allotted each month for transportation (this includes car payments, insurance, and gas)?

5. How much money in total is allotted each month for transportation and charitable contributions?

6. How much money is allotted for either food or rent?

7. What is the ratio of money spent for transportation to money spent on utilities?

8. What is the ratio of money spent on rent to money spent on miscellaneous items?

9. What is the ratio of money spent on rent to the total amount of the monthly budget?

10. What is the ratio of money spent on food to the total amount of the monthly budget?

A major league pitcher has thrown 650 pitches during the first part of the baseball season. The following circle graph shows the results of his pitches. Use the circle graph to answer exercises 11–20.

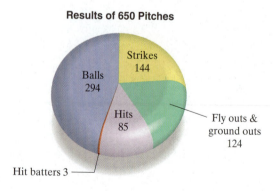

Results of 650 Pitches

Strikes 144
Balls 294
Hits 85
Fly outs & ground outs 124
Hit batters 3

11. What category had the least number of pitches?

12. What category had the second-highest number of pitches?

13. How many pitches were balls?

14. How many pitches were strikes?

15. How many pitches were either hits or balls?

16. How many pitches were either strikes or fly outs and ground outs?

17. What is the ratio of the number of strikes to the total number of pitches?

18. What is the ratio of the number of balls to the total number of pitches?

19. What is the ratio of the number of balls to the number of strikes?

20. What is the ratio of the number of fly outs and ground outs to the number of hits?

The following circle graph indicates the different primary sources from which Americans got their news in 1999. Use the circle graph to answer exercises 21–26. The adult population of the United States in 1999 was about 198,000,000.

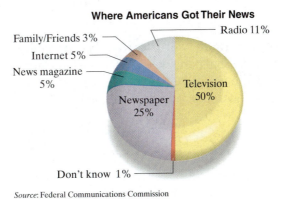

Where Americans Got Their News

Radio 11%
Family/Friends 3%
Internet 5%
News magazine 5%
Television 50%
Newspaper 25%
Don't know 1%

Source: Federal Communications Commission

21. What percent of Americans got their news from either radio or television?

22. What percent of Americans got their news from either newspapers or news magazines?

23. What percent of Americans got their news from sources other than newspapers or television?

24. How many Americans got their news from the Internet?

25. How many Americans turned on the television to get their news?

26. How many Americans indicated that they got their news from either newspapers or television?

Researchers estimate that the religious faith distribution of the 6,000,000,000 people in the world in 1999 was approximately that displayed in the following circle graph. Use the graph to answer exercises 27–32.

Distribution of Religious Faith in the World Population of 1999

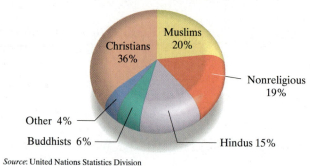

Source: United Nations Statistics Division

27. Approximately how many of the 6,000,000,000 people are Christians?

28. Approximately how many of the 6,000,000,000 people are Muslims?

29. What percent of the world's population is either Muslim or nonreligious?

30. What percent of the world's population is either Hindu or Buddhist?

31. What percent of the world's population is *not* Muslim?

32. What percent of the world's population is *not* Christian?

33. In 1999, it was estimated that there were 69,000,000 Anglicans in the world. What percent of the 6,000,000,000 people in the world were Anglican? Round to the nearest tenth of a percent.

34. The preceding circle graph contains a sector labeled Christians. What percent of the Christian sector represents Anglicans? Round to the nearest tenth of a percent.

Cumulative Review Problems

▲ **35.** Find the area of a triangle with base 6 in. and height 14 in.

▲ **36.** Find the area of a parallelogram with base of 17 in. and height 12 in.

▲ **37.** How many gallons of paint will it take to cover the four sides of a barn with two sides that measure 7 yd by 12 yd and two sides that measure 7 yd by 20 yd? Assume that a gallon of paint covers 28 square yards.

▲ **38.** A circular reflector has a radius of 8 cm. How many grams of silver will it take to cover the reflector if each gram will cover 64 sq cm? Assume that the reflector is covered on one side only. (Use $\pi \approx 3.14$.)

8.2 Bar Graphs and Line Graphs

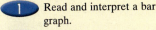
1 Reading and Interpreting a Bar Graph

Bar graphs are helpful for seeing changes over a period of time. Bar graphs or line graphs are especially helpful when the same type of data is repeatedly studied. The following bar graph shows the approximate population of California from 1940 to 2000.

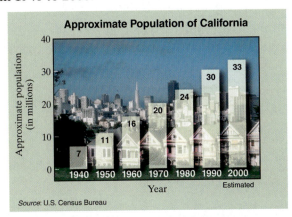

Approximate Population of California

Source: U.S. Census Bureau

EXAMPLE 1 What was the approximate population of California in 2000?

The bar for 2000 rises to 33. This represents 33 million; thus the approximate population was 33,000,000.

Practice Problem 1 What was the approximate population of California in 1980?

EXAMPLE 2 What was the increase in population from 1980 to 1990?

The bar for 1980 rises to 24. Thus the approximate population is 24,000,000. The bar for 1990 rises to 30. Thus the approximate population is 30,000,000. To find the increase in population from 1980 to 1990, we subtract.

$$30,000,000 - 24,000,000 = 6,000,000$$

Practice Problem 2 What was the increase in population from 1940 to 1960?

2 Reading and Interpreting a Double-Bar Graph

Double-bar graphs are useful for making comparisons. For example, when a company is analyzing its sales, it may want to compare different years or different quarters. The following double-bar graph illustrates the sales of new cars at a Ford dealership for two different years, 2000 and 2001. The sales are recorded for each quarter of the year.

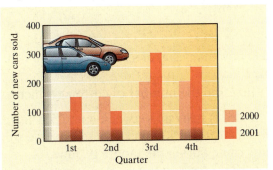

523

EXAMPLE 3   How many cars were sold in the second quarter of 2000?

The bar rises to 150 for the second quarter of 2000. Therefore, 150 cars were sold.

Practice Problem 3   How many cars were sold in the fourth quarter of 2001?

EXAMPLE 4   How many more cars were sold in the third quarter of 2001 than in the third quarter of 2000?

From the double-bar graph, we see that 300 cars were sold in the third quarter of 2001 and that 200 cars were sold in the third quarter of 2000.

$$\begin{array}{r} 300 \\ -\ 200 \\ \hline 100 \end{array}$$

Thus, 100 more cars were sold.

Practice Problem 4   How many fewer cars were sold in the second quarter of 2001 than in the second quarter of 2000?

3   Reading and Interpreting a Line Graph

A **line graph** is useful for showing trends over a period of time. In a line graph only a few points are actually plotted from measured values. The points are then connected by straight lines to show a trend. The intervening values between points may not lie exactly on the line. The following line graph shows the number of customers per month coming into a restaurant in a vacation community.

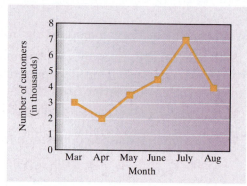

EXAMPLE 5   In which month did the smallest number of customers come into the restaurant?

The lowest point on the graph occurs for the month of April. Thus the fewest number of customers came in April.

Practice Problem 5   In which month did the greatest number of customers come into the restaurant?

EXAMPLE 6

(a) Approximately how many customers came into the restaurant during the month of June?

(b) From May to June, did the number of customers increase or decrease?

(a) Notice that the dot is halfway between 4 and 5. This represents a value halfway between 4000 and 5000 customers. Thus we would estimate that 4500 customers came during the month of June.

(b) From May to June the line goes up, so the number of customers increased.

Practice Problem 6

(a) Approximately how many customers came into the restaurant during the month of May?

(b) From March to April, did the number of customers increase or decrease?

 EXAMPLE 7 Between what two months was the increase in the number of customers the largest?

The line from June to July goes upward at the steepest angle. This represents the largest increase. (You can check this by reading the numbers from the left axis.) Thus the greatest increase in attendance was between June and July.

Practice Problem 7 Between what two months did the biggest decrease occur?

4 Reading and Interpreting a Comparison Line Graph

Two or more sets of data can be compared by using a **comparison line graph**. A comparison line graph shows two or more line graphs together. A different style for each line distinguishes them. Note that using a blue line and a red line in the following graph makes it easy to read.

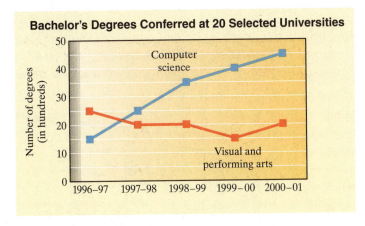

 EXAMPLE 8 How many bachelor's degrees in computer science were awarded in the academic year 1998–1999?

Because the dot corresponding to 1998–1999 is at 35 and the scale is in hundreds, we have $35 \times 100 = 3500$. Thus 3500 degrees were awarded in computer science in 1998–1999.

Practice Problem 8 How many bachelor's degrees in visual and performing arts were awarded in the academic year 1999–2000?

 EXAMPLE 9 In what academic year were more degrees awarded in the visual and performing arts than in computer science?

The only year when more bachelor's degrees were awarded in the visual and performing arts was the academic year 1996–1997.

Practice Problem 9 What was the first academic year in which more degrees were awarded in computer science than in the visual and performing arts?

The following bar graph shows the approximate population of Texas from 1940 to 2000. Use the graph to answer exercises 1–6.

1. What was the approximate population in 2000?

2. What was the approximate population in 1950?

3. What was the approximate population in 1980?

4. What was the approximate population in 1990?

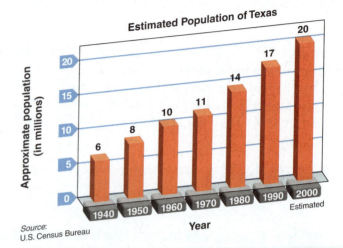

5. Between what years did the population of Texas increase by the smallest amount?

6. Between what years did the population of Texas increase by the largest amount?

The following double-bar graph displays the production levels of oil and coal in the United States. Because oil is measured in barrels and coal is measured in tons, energy officials compare the production using British thermal units (Btu), which is a measure of how much heat is produced from the oil or coal. Use the following bar graph to answer exercises 7–18.

7. How much coal was produced in 1990?

8. How much oil was produced in 1995?

9. What year had the highest production of oil?

10. In what years was the production of oil and coal the same in terms of Btu?

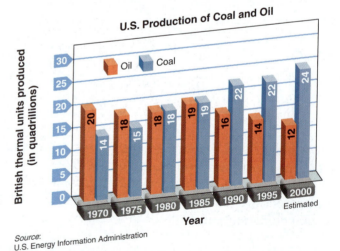

11. How much more oil was produced in 1975 than in 1995?

12. How much more coal was produced in 2000 than in 1980?

13. How much more coal was produced than oil in 1990?

14. How much more oil was produced than coal in 1975?

15. During what five-year period(s) did the biggest increase in coal production occur?

16. During what five year period did the biggest decrease in oil production occur?

17. If the production of coal increases at the same rate from 2000 to 2020 as it did from 1980 to 2000, how much coal will be produced in 2020?

18. If the production of oil decreases at the same rate from 2000 to 2015 as it did from 1985 to 2000, how much oil will be produced in 2015?

The following line graph shows Wentworth Construction Company's profits during the last six years. Use the graph to answer exercises 19–24.

19. What was the profit in 2000?

20. What was the profit in 1996?

21. What year had the lowest profit?

22. What year had the highest profit?

23. How much greater was the profit in 1998 than in 1997?

24. How much greater was the profit in 2001 than in 1996?

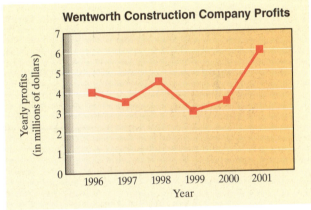

The following comparison line graph indicates the rainfall for the last six months of two different years in Springfield. Use the graph to answer exercises 25–30.

25. In September 2001, how many inches of rain were recorded?

26. In October 2000 how many inches of rain were recorded?

27. During what months was the rainfall of 2001 less than the rainfall of 2000?

28. During what months was the rainfall of 2001 greater than the rainfall of 2000?

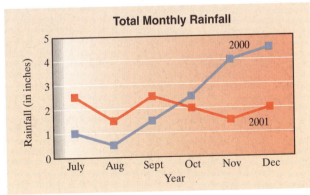

29. How many more inches of rain fell in November 2000 than in October 2000?

30. How many more inches of rain fell in September 2000 than in August 2000?

To Think About

The following table shows the number of pizzas sold at a pizza parlor near a college campus.

Number of Pizzas Sold by Alfredo's Pizza Parlor	300	400	100	200	600
Month of the Year	Jan.	Feb.	Mar.	Apr.	May

31. Use the graph paper below to construct a line graph of the information in the table. Let the vertical scale (height) represent the number of pizzas sold. Let the horizontal scale (width) represent the month.

32. Is the biggest change on the graph an increase or a decrease in the number of pizzas sold per month?

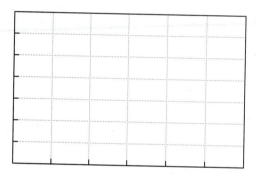

Cumulative Review Problems

Do each calculation in the proper order.

33. $7 \times 6 + 3 - 5 \times 2$

34. $(5 + 6)^2 - 18 \div 9 \times 3$

35. $\dfrac{1}{5} + \left(\dfrac{1}{5} - \dfrac{1}{6} \right) \times \dfrac{2}{3}$

Putting Your Skills to Work

Studying the Changes in America's Aging Population

Officials are studying trends in the changes in the population of America. Each year, more people live longer. One of the most important trends is the significant increase in the number of people who are age 85 or older. The following double-bar graph shows the projected increase over a 50-year period.

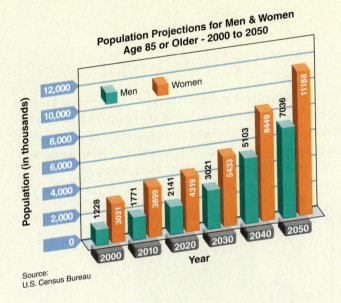

Population Projections for Men & Women Age 85 or Older - 2000 to 2050

Population (in thousands)

Men ■ Women ■

Men: 2000: 1228, 2010: 1771, 2020: 2141, 2030: 3021, 2040: 5103, 2050: 7036
Women: 2000: 3031, 2010: 3899, 2020: 4319, 2030: 5433, 2040: 8449, 2050: 11188

Year

Source:
U.S. Census Bureau

Problems for Individual Investigation

1. How many more men are expected to be age 85 or older in 2030 than in 2010?

2. How many more women are expected to be age 85 or older in 2050 than in 2020?

3. In what 10-year period is the greatest increase expected in the number of women age 85 or older?

4. In what 10-year period is the greatest increase expected in the number of men age 85 or older?

Problems for Group Investigation and Cooperative Study

5. What is the expected ratio of women to men in 2020 for people who are age 85 or older?

6. What is the expected ratio of women to men in 2050 for people who are age 85 or older?

7. For what year is the ratio of women to men the largest?

8. During this 50-year period, what is increasing faster, the number of men age 85 or older or the number of women age 85 or older?

Internet Connections

Netsite: http://www.prenhall.com/tobey_basic

This site contains data from the U.S. Census Bureau.

9. Determine the increase in the population of people age 75 to 84. By how much will the population of men in this age category increase during the period from 2000 to 2050? By how much will the population of women in this age category increase during the same period? Is the ratio of women to men in this age category increasing or decreasing during this period?

10. Determine the increase in the population of people age 25 to 34. By how much will the population of men in this age category increase during the period from 2010 to 2040? By how much will the population of women in this age category increase during the same period? Is the ratio of women to men in this age category increasing or decreasing during this period?

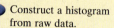

1 **Understanding and Interpreting a Histogram**

In business or in higher education you are often asked to take data and organize them in some way. This section shows you the technique for making a *histogram*—a type of bar graph.

Suppose that a mathematics professor announced the results of a class test. The 40 students in the class scored between 50 and 99 on the test. The results are displayed on the following chart.

Scores on the Test	Class Frequency
50–59	4
60–69	6
70–79	16
80–89	8
90–99	6

The results in the table can be organized in a special type of bar graph known as a **histogram**. In a histogram the width of each bar is the same. The width represents the range of scores on the test. This is called a **class interval**. The height of each bar gives the class frequency of each class interval. The **class frequency** is the number of times a score occurs in a particular class interval.

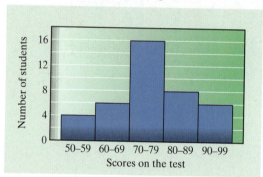

EXAMPLE 1 How many students scored a B on the test if the professor considers a test score of 80–89 a B?

Since the 80–89 bar rises to a height of 8, eight students scored a B on the test.

Practice Problem 1 How many students scored a D on the test if the professor considers a test score of 60–69 a D?

EXAMPLE 2 How many students scored less than 80 on the test?

From the histogram, we see that there are three different bar heights to be included. Four tests were scored 50–59, six tests were scored 60–69, and 16 tests were scored 70–79. When we combine $4 + 6 + 16 = 26$, we can see that 26 students scored less than 80 on the test.

Practice Problem 2 How many students scored greater than 69 on the test?

The following histogram tells us about the length of life of 110 new light bulbs tested at a research center. The number of hours the bulbs lasted is indicated on the horizontal scale. The frequency of bulbs lasting that long is indicated on the vertical scale.

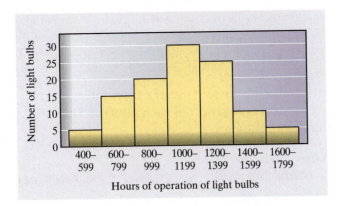

EXAMPLE 3   How many light bulbs lasted between 1400 and 1599 hours?

The bar with a range of 1400–1599 hours rises to 10. Thus 10 light bulbs lasted that long.

Practice Problem 3   How many light bulbs lasted between 800 and 999 hours?

EXAMPLE 4   How many light bulbs lasted less than 1000 hours?

We see that there are three different bar heights to be included. Five bulbs lasted 400–599 hours, 15 bulbs lasted 600–799 hours, and 20 bulbs lasted 800–999 hours. We add $5 + 15 + 20 = 40$. Thus 40 light bulbs lasted less than 1000 hours.

Practice Problem 4   How many light bulbs lasted more than 1199 hours?

2   *Constructing a Histogram from Raw Data*

To construct a histogram, we start with *raw data*, data that have not yet been organized or interpreted. We perform the following steps.

1. Select data class intervals of equal width for the data.
2. Make a table with class intervals and a *tally* (count) of how many numbers occur in each interval. Add up the tally to find the class frequency for each class interval.
3. Draw the histogram.

First we will practice making the table. Later we will use the table to draw the histogram.

EXAMPLE 5   Each number in the following chart represents the number of kilowatt-hours of electricity used in a home during a one-month period. Create a set of class intervals for this data and then determine the frequency of each class interval.

770	520	850	900	1100
1200	1150	730	680	900
1160	590	670	1230	980

1. We select class intervals of equal width for the data. We choose intervals of 200. We might have chosen smaller or larger intervals, but we choose 200 because it gives us a convenient number of intervals to work with, as we will see.

2. We make a table. We write down the class intervals, then count (tally) how many numbers occur within each interval. Then we write the total. This is the class frequency.

Kilowatt-Hours Used (Class Interval)	Tally	Frequency
500–699	||||	4
700–899	|||	3
900–1099	|||	3
1100–1299	⊬⊬	5

Practice Problem 5   Each number in the following chart represents the weight in pounds of a new car.

2250	1760	2000	2100	1900
1640	1820	2300	2210	2390
2150	1930	2060	2350	1890

Complete the following table to determine the frequency of each class interval for the preceding data.

Weight In Pounds (Class Interval)	Tally	Frequency
1600–1799		
1800–1999		
2000–2199		
2200–2399		

EXAMPLE 6   Draw a histogram from the table in Example 5.

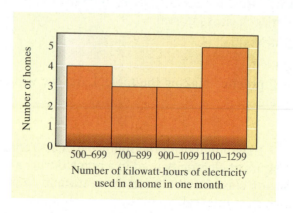

Number of kilowatt-hours of electricity used in a home in one month

Practice Problem 6 Draw a histogram using the data from the table in Practice Problem 5.

Note: Usually it is desirable for the class intervals to be of equal size. However, sometimes data is collected such that this is not the case. We will see this situation in Example 7. Here government data was collected with unequal class intervals.

EXAMPLE 7 Draw a histogram for the following table of data showing the number of people in the United States as of July 1, 1998, in each of five age categories.

Age Category	Number of People in the United States
17 or younger	69,872,000
18–34	64,244,000
35–54	79,105,000
55–74	41,071,000
75 or older	16,006,000

Source: U.S. Census Bureau

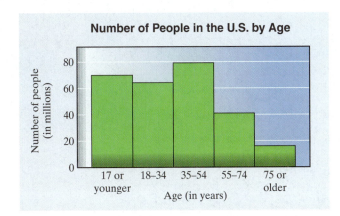

Practice Problem 7 Based on the preceding histogram, between what two age categories is there the greatest difference in population in the United States?

The number of miles per gallon achieved by 190 rental cars is depicted in the following histogram. Use the histogram to answer exercises 1–8.

1. How many cars achieved between 28 and 30.9 miles per gallon?

2. How many cars achieved between 19 and 21.9 miles per gallon?

3. How many cars achieved between 25 and 27.9 miles per gallon?

4. How many cars achieved between 16 and 18.9 miles per gallon?

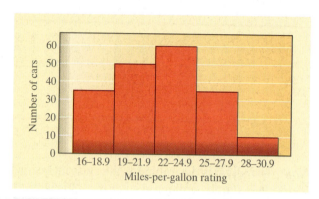

5. How many cars achieved more than 24.9 miles per gallon?

6. How many cars achieved less than 22 miles per gallon?

7. How many cars achieved between 19 and 27.9 miles per gallon?

8. How many cars achieved between 16 and 24.9 miles per gallon?

A large company comprising three bookstores studied its yearly report to find out its customers' spending habits. The company sold a total of 70,000 books. The following histogram indicates the number of books sold in certain price ranges. Use the histogram to answer exercises 9–18.

9. How many books priced at $3.00 to $4.99 were sold?

10. How many books priced at $25.00 or more were sold?

11. Which price category of books did the bookstore sell the most of?

12. What price category of books did the bookstore sell the least of?

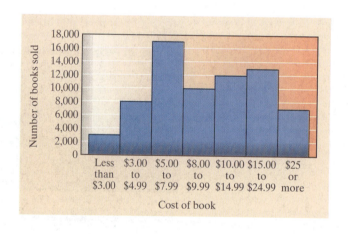

13. How many books priced at less than $8.00 were sold?

14. How many books priced at more than $9.99 were sold?

15. How many books priced between $5.00 and $24.99 were sold?

16. How many books priced between $3.00 and $9.99 were sold?

17. What percent of the 70,000 books sold were over $14.99?

18. What percent of the 70,000 books sold were under $8.00?

The numbers in the following chart are the daily high temperatures in degrees Fahrenheit in Boston during February. In exercises 19–26, determine the frequencies of the class intervals for this data.

23°	26°	30°	18°	42°	17°	19°
51°	42°	38°	36°	12°	18°	14°
20°	24°	26°	30°	18°	17°	16°
35°	38°	40°	33°	19°	22°	26°

	Temperature (Class Interval)	*Tally*	*Frequency*		*(Class Interval)*	*Tally*	*Temperature Frequency*
19.	12°–16°	_____	_____	**20.**	17°–21°	_____	_____
21.	22°–26°	_____	_____	**22.**	27°–31°	_____	_____
23.	32°–36°	_____	_____	**24.**	37°–41°	_____	_____
25.	42°–46°	_____	_____	**26.**	47°–51°	_____	_____

27. Construct a histogram using the table prepared in exercises 19–26.

28. How many days in February was the temperature in Boston less than 27°?

29. How many days in February was the temperature in Boston greater than 36°?

Each number in the following chart is the cost of a prescription purchased by the Lin family this year. In exercises 30–37, determine the frequencies of the class intervals for this data.

$28.50	$16.00	$32.90	$46.20	$ 9.85
$27.30	$16.00	$41.95	$36.00	$24.20
$ 7.65	$ 8.95	$ 4.50	$11.35	$ 7.75
$12.30	$21.85	$46.20	$15.50	$ 4.50

	Purchase Price (Class Interval)	Tally	Frequency		Purchase Price (Class Interval)	Tally	Frequency
30.	$ 4.00–$ 9.99	_____	_____	**31.**	$10.00–$15.99	_____	_____
32.	$16.00–$21.99	_____	_____	**33.**	$22.00–$27.99	_____	_____
34.	$28.00–$33.99	_____	_____	**35.**	$34.00–$39.99	_____	_____
36.	$40.00–$45.99	_____	_____	**37.**	$46.00–$51.99	_____	_____

38. Construct a histogram using the table constructed in exercises 30–37.

39. How many prescriptions cost less than $22.00?

40. How many prescriptions cost more than $33.99?

Cumulative Review Problems

41. Solve for *n*. $\dfrac{126}{n} = \dfrac{36}{17}$

42. Solve for *n*. $\dfrac{n}{18} = \dfrac{3.5}{9}$

43. Ben and Trish Hale are going to have a Texas barbeque next week. According to Trish's grandmother, the recipe calls for 3 pounds of chicken for every 5 people. The Hales are expecting 20 people. How much chicken should they buy?

44. Tim and Judy Newitt worked as scientists on Mount Washington last year. They found that every 23 in. of snow corresponded to 2 in. of water. During the month of January they measured 150 in. of snow at the mountain weather observatory. How many inches of water does this correspond to? Round to the nearest tenth of an inch.

8.4 Mean, Median, and Mode

1 Finding the Mean of a Set of Numbers

We often want to know the "middle value" of a group of numbers. In this section we learn that, in statistics, there is more than one way of describing this middle value: there is the *mean* of the group of numbers, there is the *median* of the group of numbers, and in most cases there is a *mode* of the group of numbers. In some situations it's more helpful to look at the mean; in others it's more helpful to look at the median, and in yet others the mode. We'll learn to tell which situations lend themselves to one or the other.

The **mean** of a set of values is the sum of the values divided by the number of values. The mean is often called the **average**.

The mean value is often rounded to a certain decimal-place accuracy.

EXAMPLE 1 Carl recorded the miles per gallon achieved by his car for the last two months. His results were as follows:

Week	1	2	3	4	5	6	7	8
Miles per Gallon	26	24	28	29	27	25	24	23

What is the mean miles-per-gallon figure for the last eight weeks? Round to the nearest whole number.

Sum of values $\longrightarrow$
Number of values $\longrightarrow$
$$\frac{26 + 24 + 28 + 29 + 27 + 25 + 24 + 23}{8}$$

$$= \frac{206}{8} = 25.75 \approx 26 \text{ rounded to the nearest whole number}$$

The mean miles-per-gallon rating is 26.

Practice Problem 1 Bob and Wally kept records of their phone bills for the last six months. Their bills were $39.20, $43.50, $81.90, $34.20, $51.70, and $48.10. Find the mean monthly bill. Round to the nearest cent.

2 Finding the Median of a Set of Numbers

If a set of numbers is arranged in order from smallest to largest, the **median** is that value that has the same number of values above it as below it.

If the numbers are not arranged in order, then the first step for finding the median is to put the numbers in order.

EXAMPLE 2 Find the median value of the following costs for a microwave oven: $100, $60, $120, $200, $190, $120, $320, $290, $180.

We must arrange the numbers in order from smallest to largest (or largest to smallest).

$\underbrace{\$60, \$100, \$120, \$120}_{\text{four numbers}}$ $\underset{\underset{\text{middle number}}{\uparrow}}{\$180}$ $\underbrace{\$190, \$200, \$290, \$320}_{\text{four numbers}}$

Thus $180 is the median cost.

Student Learning Objectives

After studying this section, you will be able to:

1. Find the mean of a set of numbers.
2. Find the median of a set of numbers.
3. Find the mode of a set of numbers.

SSM PH TUTOR CD & VIDEO MATH PRO WEB
CENTER

537

Practice Problem 2   Find the median value of the following weekly salaries: $320, $150, $400, $600, $290, $150, $450.

If a list of numbers contains an even number of items, then of course there is no one middle number. In this situation we obtain the median by taking the average of the two middle numbers.

EXAMPLE 3   Find the median of the following numbers: 13, 16, 18, 26, 31, 33, 38, 39.

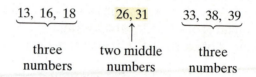

13, 16, 18	26, 31	33, 38, 39
three numbers	two middle numbers	three numbers

The average (mean) of 26 and 31 is

$$\frac{26 + 31}{2} = \frac{57}{2} = 28.5.$$

Thus the median value is 28.5.

Practice Problem 3   Find the median value of the following numbers: 88, 90, 100, 105, 118, 126.

Sidelight

When would someone want to use the mean, and when would someone want to use the median? Which is more helpful?

The mean, or average, is used more frequently. It is most helpful when the data are distributed fairly evenly, that is, when no one value is "much larger" or "much smaller" than the rest.

For example, suppose a company had employees with annual salaries of $9000, $11,000, $14,000, $15,000, $17,000, and $20,000. All the salaries fall within a fairly limited range. The mean salary

$$\frac{9000 + 11,000 + 14,000 + 15,000 + 17,000 + 20,000}{6} = \$14,333.33$$

gives us a reasonable idea of the typical salary.

However, suppose the company had six employees with salaries of $9000, $11,000, $14,000, $15,000, $17,000, and $90,000. Talking about the mean salary, which is $26,000, is deceptive. No one earns a salary very close to the mean salary. The typical worker in that company does not earn around $26,000. In this case, the median value is more appropriate. Here the median is $14,500. See exercises 31 and 32 in Exercises 8.4 for more on this.

3 **Finding the Mode of a Set of Numbers**

Another value that is sometimes used to describe a set of data is the mode. The **mode** of a set of data is the number or numbers that occur most often.

EXAMPLE 4   The following numbers are the weights of automobiles measured in pounds:

> 2345, 2567, 2785, 2967, 3105, 3105, 3245, 3546

Find the mode of these weights.

   The value 3105 occurs twice, whereas each of the other values occurs just once. Thus the mode is 3105 pounds.

Practice Problem 4   The following numbers are the heights in inches of 10 male students in Basic Mathematics: 64, 66, 67, 69, 70, 71, 71, 73, 75, 76.

   Find the mode of these heights.

   A set of numbers may have more than one mode.

EXAMPLE 5   The following numbers are finish times for 12 high school students who ran a distance of one mile. The finish times are measured in minutes.

> 290, 272, 268, 260, 290, 272, 330, 355, 368, 290, 370, 272

Find the mode of these finish times.

   First we need to arrange the numbers in order from smallest to largest and include all repeats.

> 260, 268, 272, 272, 272, 290, 290, 290, 330, 355, 368, 370

Now we can see that the value 272 occurs three times, as does the value 290. Thus the modes for these finish times are 272 minutes and 290 minutes.

Practice Problem 5   The following numbers are distances in miles that 16 students traveled to take classes at Massasoit Community College each day.

> 2, 5, 8, 3, 12, 15, 28, 8, 3, 14, 16, 31, 33, 27, 3, 28

   Find the mode of these distances.

A set of numbers may have **no mode** at all. For example, the set of numbers 50, 60, 70, 80, 90 has no mode because each number occurs just once. The set of numbers 33, 33, 44, 44, 55, 55 has no mode because each number occurs twice. If all numbers occur the same number of times, there is no mode.

Verbal and Writing Skills

1. Explain the difference between a median and a mean.

2. Explain why some sets of numbers have one mode, others two modes, and others no modes.

In exercises 3–12, find the mean. Round to the nearest tenth when necessary.

3. The numbers of customers who were served at Grinders Coffeehouse between 8:00 A.M. and 9:00 A.M. in the past seven days were as follows: 30, 29, 28, 35, 34, 37, 31.

4. The numbers of pizzas delivered by Papa John's over the last 7 days were as follows: 28, 17, 18, 21, 24, 30, 30.

5. Sam studied for the following number of hours during the past week:

S	M	T	W	Th	F	S
6	2	3	3.5	2.5	1	0

6. Steve's car got the following miles-per-gallon results during the last six months:

Jan.	Feb.	Mar.	Apr.	May	June
23	22	25	28	29	30

7. The population on the island of Guam has increased significantly over the last 40 years. Find an approximate value for the mean population for this 40-year period from the following population chart:

1960 Population	1970 Population	1980 Population	1990 Population	2000 Population
67,000	86,000	107,000	134,000	152,000

Source: U.S. Census Bureau

8. Desmond has taken four three-credit courses at college each of four semesters. His semester grade averages are as follows:

Semester 1	Semester 2	Semester 3	Semester 4
2.90	2.66	2.52	2.48

9. The captain of the college baseball team achieved the following results:

	Game 1	Game 2	Game 3	Game 4	Game 5
Hits	0	2	3	2	2
Times at Bat	5	4	6	5	4

Find his batting average by dividing his total number of hits by the total times at bat.

10. The captain of the college bowling team had the following results after practice:

	Practice 1	Practice 2	Practice 3	Practice 4
Score (Pins)	541	561	840	422
Number of Games	3	3	4	2

Find her bowling average by dividing the total number of pins scored by the total number of games.

11. Frank and Wally traveled to the West Coast during the summer. The number of miles they drove and the number of gallons of gas they used are recorded in the following chart.

	Day 1	Day 2	Day 3	Day 4
Miles Driven	276	350	391	336
Gallons of Gas	12	14	17	14

Find the average miles per gallon achieved by the car on the trip by dividing the total number of miles driven by the total number of gallons used.

12. Cindy and Andrea traveled to Boston this fall. The number of miles they drove and the number of gallons of gas they used are recorded in the following chart.

	Day 1	Day 2	Day 3	Day 4
Miles Driven	260	375	408	416
Gallons of Gas	10	15	17	16

Find the average miles per gallon achieved by the car on the trip by dividing the total number of miles driven by the total number of gallons used.

In exercises 13–24, find the median value.

13. 1052, 968, 1023, 999, 865, 1152

14. 1400, 1329, 1200, 1386, 1427, 1350

15. 0.52, 0.69, 0.71, 0.34, 0.58

16. 0.26, 0.12, 0.35, 0.43, 0.28

17. The annual salaries of the employees of a local cable television office are $17,000, $11,600, $23,500, $15,700, $26,700, and $31,500.

18. The annual incomes of six families are $24,000, $60,000, $32,000, $18,000, $29,000, and $35,000.

19. The ages of 10 people studying for their nursing degree are 19, 32, 21, 44, 24, 30, 33, 28, 35, and 20.

20. The numbers of minutes Carl spent searching for files of clients at work in the past 10 days were 20, 45, 42, 28, 108, 38, 10, 44, 84, and 64.

21. The phone bills for Dr. Price's cellular phone over the last seven months were as follows: $109, $207, $420, $218, $97, $330, and $185.

22. The prices of the same compact disc sold at several different music stores or by mail order were as follows: $15.99, $11.99, $5.99, $12.99, $14.99, $9.99, $13.99, $7.99, and $10.99.

23. The grade point averages (GPA) for eight students were 1.8, 1.9, 3.1, 3.7, 2.0, 3.1, 2.0, and 2.4.

24. The numbers of pounds of smoked turkey breast purchased at a deli by the last eight customers were 1.2, 2.0, 1.7, 2.5, 2.4, 1.6, 1.5, and 2.3.

Find the mean. Round to the nearest cent when necessary.

 25. The salaries of eight small business owners in Big Rapids are $30,000, $74,500, $47,890, $89,000, $57,645, $78,090, $110,370, and $65,800.

26. The prices of nine laptop computers with a Pentium III chip are $5679, $6902, $1530, $2738, $2999, $4105, $3655, $5980, and $4430.

In exercises 27 and 28, find the median.

 27. 2576, 8764, 3700, 5000, 7200, 4700, 9365, 1987

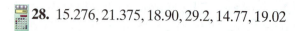 **28.** 15.276, 21.375, 18.90, 29.2, 14.77, 19.02

29. The numbers of computers bought in a store during five days in December were 18, 27, 101, 93, and 111. Find the mean and the median.

30. The total expenses for a salesperson over the last six months were $581, $902, $218, $446, $833, and $1112. Find the mean and the median.

To Think About

31. A local travel office has 10 employees. Their monthly salaries are $1500, $1700, $1650, $1300, $1440, $1580, $1820, $1380, $2900, and $6300.
   (a) Find the mean.
   (b) Find the median.
   (c) Which of these numbers best represents what the typical person earns? Why?

32. A college track star in California ran the 100-meter event in eight track meets. Her times were 11.7 seconds, 11.6 seconds, 12.0 seconds, 12.1 seconds, 11.9 seconds, 18 seconds, 11.5 seconds, and 12.4 seconds.
   (a) Find the mean.
   (b) Find the median.
   (c) Which of these numbers represents her typical running time? Why?

Find the mode.

33. 60, 65, 68, 60, 72, 59, 80

34. 86, 84, 82, 87, 84, 88, 90

35. 121, 150, 116, 150, 121, 181, 117, 123

36. 144, 143, 140, 141, 149, 144, 141, 150

37. The last six bicycles sold at the Skol Bike shop cost $249, $649, $439, $259, $269, and $249.

38. The last six color televisions sold at the local Circuit City cost $315, $430, $515, $330, $430, and $615.

39. An Internet shopping site received the following numbers of inquiries over the last seven days: 869, 992, 482, 791, 399, 855, and 869. Find the mean, the median, and the mode.

40. The numbers of passengers taking the Rockport to Boston train during the last seven days were 568, 388, 588, 688, 750, 900, and 388. Find the mean, the median, and the mode.

To Think About

Finding the grade point average or GPA.

At most universities and colleges, students are assigned grade point values for the grades they have earned. A student's grade point average is the average of all the grade point values for the credit hours taken. In most schools, the grade point values are $A = 4, B = 3, C = 2, D = 1$, and $F = 0$.

To find your GPA you need to multiply the number of credit hours of each course you have taken by the grade point value you received for the course. Add the results. Then divide by the total number of credit hours.

Suppose you took a six credit hour course and received a grade of B and a three credit hour course and receival a grade of A. Multiply $6(3) = 18$ as well as $3(4) = 12$. Add the results: $18 + 12 = 30$. Since you have a total of nine credit hours, you need to divide by 9. $30 \div 9 = 3.33333\ldots$. Rounded to the nearest tenth, your GPA is 3.3.

In exercises 41 and 42, find the GPA rounded to the nearest tenth.

41.

Number of Credit Hours	Grade
3	A
4	B
3	C
3	B

42.

Number of Credit Hours	Grade
3	B
5	A
3	C
4	C

Cumulative Review Problems

Round to the nearest tenth. Use $\pi \approx 3.14$.

▲ **43.** A triangular piece of insulation is located under the dash of a Ford Explorer. It has a base of 7 inches and a height of 5.5 inches. What is the area of this piece of insulation?

▲ **44.** The Canaan Family Farm has two fields that are irrigated with a rotating sprinkler system. This system waters a circular area. The system is designed to deliver 2 gallons per hour for each square foot of the field. If each circular area has a radius of 40 feet, how many gallons per hour are needed to water these fields?

▲ **45.** Collette Camp has made a huge advertising sign in the shape of a rhombus. The sign has a base of 5 feet and a height of 4 feet. The sign is made out of aluminum that costs $16.50 per square foot. How much did it cost for the aluminum used to make the sign?

▲ **46.** Liquid nitrogen is stored in a steel cylinder at Swenson Industries in Taunton. The base of the cylinder is a circle that has a radius of 2 feet. The height of the cylinder is 5 feet. How many cubic feet of nitrogen can be stored in the cylinder?

Math in the Media

Sex, Statistically Speaking

John Allen Paulos

A professor of mathematics at Temple University, John Allen Paulos is the author of *Innumeracy* and *A Mathematician Reads the Newspaper* and writes the "Who's Counting" column for ABCnews.com. Reprinted with Permission of John Allen Paulos.

"Conducting surveys can seem as straightforward as adding numbers. But when the surveys get personal, well, accuracy can get elusive."

Professor Paulos suggests, by using a simple coin flip, that it is mathematically possible to obtain sensitive information about a group of people without compromising anyone's privacy. If we want to discover what percentage of a large group of people has ever X-ed—betrayed a spouse, cheated in school, or whatever—we can use the following technique:

Have each member of the survey group flip a coin, but without anyone else seeing the results. If the coin lands tails, the person is told to answer the question honestly: Has he or she ever X-ed, yes or no? If the coin lands heads, the person is to answer yes, regardless of the correct answer. So a yes response could mean one of two things, one quite innocuous (obtaining a head), the other potentially embarrassing (X-ing). Since the experimenter can't know what the yes means, people presumably will be more honest. It's not difficult to infer from the number of yes responses what percentage of the population have X-ed—without learning anything about any particular individual.

For example, if 580 out of 1000 people answer "yes," what percentage of this group has ever X-ed? Since on a fair coin toss the expected number of heads in 1000 tosses would be 500, then 80 of those 580 yeses represent people who have X-ed which would be 80/500 or 16%.

This somewhat controversial approach to conducting a statistical survey has caused a variety of reactions from mathematicians. Assuming that this method is valid, try to answer the following questions.

EXERCISES

Apply the technique to answer questions 1–4.

1. If 500 people flipped one fair coin, what would be the expected number of heads?

2. Try an experiment with 30 flips of a fair coin. The expected number of heads is 15. Did you get 15? Do you have to get 15? Was it close to 15?

3. In an attempt to estimate the extent of cheating at a university, a random selection of 500 students were asked if they had ever seen cheating or had cheated themselves. If there were 270 yes responses, what is the estimated number and percentage of students who had cheated or had seen cheating?

4. From Problem 3, what conclusions are possible if 250 yeses were received? Less than 250 yeses?

Chapter 8 Organizer

Topic	Procedure	Examples
Circle graphs, p. 517.	The following circle graph describes the ages of the 200 men and women of the Glover City police force. **Age Distribution of Grover City Police Force** Over age 50 12% Under age 23 10% Age 32–50 48% Age 23–32 30%	1. What percent of the police force is between 23 and 32 years old? 30% 2. How many men and women in the police force are over 50 years old? 12% of 200 = (0.12)(200) = 24 people
Bar graphs and double-bar graphs, p. 523.	The following double-bar graph illustrates the sales of color television sets by a major store chain for 2000 and 2001 in three regions of the country. Number of televisions sold (in thousands) West Coast · Midwest · East Coast Region ■ 2000 ■ 2001	1. How many color television sets were sold by the chain on the East Coast in 2001? 6000 sets 2. How many *more* color television sets were sold in 2001 than in 2000 on the West Coast? 3000 sets were sold in 2001; 2000 sets were sold in 2000. 3000 -2000 $\overline{1000}$ sets more in 2001
Line graphs and comparison line graphs, pp. 524–525.	The following line graph indicates the number of visitors to Wetlands State Park during a four-month period in 2000 and 2001. Number of visitors (in thousands) 2000 2001 July · Aug · Sept · Oct Month	1. How many visitors came to the park in July 2000? 3000 visitors 2. In what months were there more visitors in 2000 than in 2001? September and October 3. The sharpest *decrease* in attendance took place between what two months? Between August 2001 and September 2001
Histograms, p. 530.	The following histogram indicates the number of students in a math class who scored within each interval on a 15-point quiz. Number of students 0–3 · 4–7 · 8–11 · 12–15 Score	1. How many students had a score between 8 and 11? 20 students 2. How many students had a score of less than 8? 12 + 6 = 18 students

Topic	Procedure	Examples
Finding the mean, p. 537.	The *mean* of a set of values is the sum of the values divided by the number of values. The mean is often called the *average*.	**1.** Find the mean of 19, 13, 15, 25, and 18. $$\frac{19 + 13 + 15 + 25 + 18}{5} = \frac{90}{5} = 18$$ The mean is 18.

Finding the median, p. 537.	**1.** Arrange the numbers in order from smallest to largest.
	2. If there is an odd number of values, the middle value is the median.
	3. If there is an even number of values, the average of the two middle values is the median.
		1. Find the median of 19, 29, 36, 15, and 20.
		   First we arrange in order from smallest to largest: 15, 19, 20, 29, 36.
		
		$$\underbrace{15, \quad 19}_{\substack{\text{two} \\ \text{numbers}}} \quad \underset{\substack{\uparrow \\ \text{middle} \\ \text{number}}}{20} \quad \underbrace{29, \quad 36}_{\substack{\text{two} \\ \text{numbers}}}$$
		
		The median is 20.
		2. Find the median of 67, 28, 92, 37, 81, and 75.
		   First we arrange in order from smallest to largest: 28, 37, 67, 75, 81, 92. There is an even number of values
		
		$$28, 37, \quad \underset{\substack{\uparrow \\ \text{two middle} \\ \text{numbers}}}{67, 75,} \quad 81, 92$$
		
		$$\frac{67 + 75}{2} = \frac{142}{2} = 71$$
		The median is 71.

Finding the mode, p. 539.	The *mode* of a set of values is the value that occurs most often. A set of values may have more than one mode or no mode.	**1.** Find the mode of 12, 15, 18, 26, 15, 9, 12, and 27. The modes are 12 and 15. **2.** Find the mode of 4, 8, 15, 21, and 23. There is no mode.

Chapter 8 Review Problems

A student found that there were a total of 140 personal computers owned by students in the dormitory. The following circle graph displays the distribution of manufacturers of these computers. Use the graph to answer exercises 1–8.

1. How many personal computers were manufactured by IBM?

2. How many personal computers were manufactured by Apple?

3. How many personal computers were manufactured by Dell or Compaq?

4. How many personal computers were manufactured by Gateway or Acer?

Distribution of Computers by Manufacturer in a Dormitory

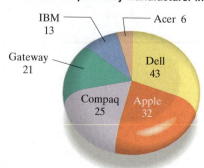

IBM 13 · Acer 6 · Gateway 21 · Dell 43 · Compaq 25 · Apple 32

5. What is the ratio of the number of computers manufactured by IBM to the number of computers manufactured by Gateway?

6. What is the ratio of the number of computers manufactured by Dell to the number of computers manufactured by Apple?

7. What percent of the 140 computers are manufactured by compaq?

8. What percent of the computers are manufactured by Apple?

The accompanying graph represents the favorite colors of a random poll of 500 students at John Tyler Community College. Use the graph to answer exercises 9–16.

9. What percent of the students chose colors other than red?

10. What percent of the students chose colors other than blue?

11. What is the percent of students who like the least favorite color?

12. What is the third favorite color?

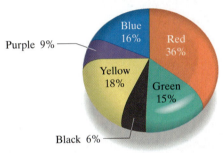

Distribution of Favorite Colors of Students at John Tyler Community College

Blue 16%
Purple 9%
Red 36%
Yellow 18%
Green 15%
Black 6%

13. What is the second-to-last favorite color?

14. How many students chose green or yellow?

15. How many students did not choose black?

16. How many students chose blue, purple, or black?

The following double-bar graph illustrates the numbers of glasses of milk consumed each week by children in various age categories for the years 1990 and 2000. Use the graph to answer exercises 17–24.

17. How many glasses of milk per week were consumed by children age 2–5 years in 1990?

18. How many glasses of milk per week were consumed by children age 6–9 years in 2000?

19. What age group saw the greatest decrease in milk consumption between 1990 and 2000?

20. What age group saw an increase in milk consumption between 1990 and 2000?

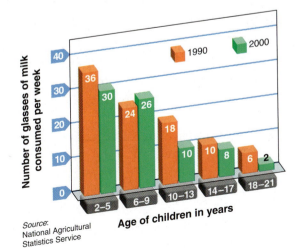

Number of glasses of milk consumed per week

1990   2000

36   30   24   26   18   10   10   8   6   2

2–5   6–9   10–13   14–17   18–21

Age of children in years

Source: National Agricultural Statistics Service

21. In 1990, how many fewer glasses of milk were consumed per week by children age 18–21 years than children age 10–13 years?

22. In 2000, how many more glasses of milk were consumed per week by children age 2–5 years than by children age 14–17 years?

23. What is the ratio of number of glasses of milk consumed each week by children age 10–13 years to that consumed by children age 18–21 years in 2000?

24. What is the ratio of number of glasses of milk consumed each week by children age 2–5 years to that consumed by children aged 10–13 years in 1990?

The following double-bar graph shows the average salaries paid to public school classroom teachers and principals in the U.S. for selected years from 1980 to 2000. Use the bar graph to answer exercises 25–36.

25. What was the average salary of a classroom teacher in 1990?

26. What was the average salary of a principal in 2000?

27. During what five-year period was there the greatest increase in salary for a principal?

28. During what five-year period was there the greatest increase in salary for a classroom teacher?

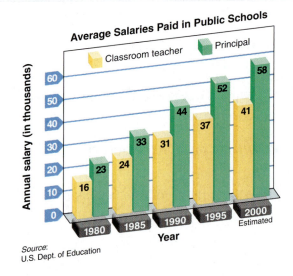

29. How much more did the average principal make per year than the average classroom teacher in 1985?

30. How much more did the average principal make per year than the average classroom teacher in 1995?

31. In what year was the difference between the average salary of a principal and the average salary of a classroom teacher become the greatest?

32. In what year was the difference between the average salary of a principal and the average salary of a classroom teacher the smallest?

33. What is the average increase in salary per five years for a principal?

34. What is the average increase in salary per five years for a classroom teacher?

35. If the same five-year increase occurs from 2000 to 2005 as from 1995 to 2000, what will be the average salary of a classroom teacher in 2005?

36. If the same ten-year increase occurs from 2000 to 2010 as from 1990 to 2000, what will be the average salary of a principal in 2010?

The following line graph shows the numbers of graduates of Williamston University during the last six years. Use the graph to answer exercises 37–44.

37. How many Williamston University students graduated in 2000?

38. How many Williamston University students graduated in 1999?

39. How many Williamston University students graduated in 2001?

40. How many Williamston University students graduated in 1998?

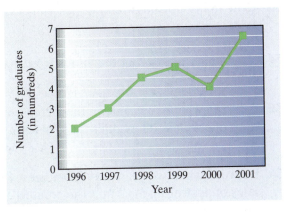

41. How many more Williamston University students graduated in 1997 than in 1996?

42. How many more Williamston University students graduated in 1999 than in 1998?

43. Between what two years did the number of graduates decline?

44. Between what two years did the number of graduates increase by the greatest amount?

The following comparison line graph shows the number of ice cream cones purchased at the Junction Ice Cream Stand during a five-month period in 2000 and in 2001. Use this graph to answer exercises 45–52.

45. How many ice cream cones were purchased in July 2001?

46. How many ice cream cones were purchased in August 2000?

47. How many more ice cream cones were purchased in May 2000 than in May 2001?

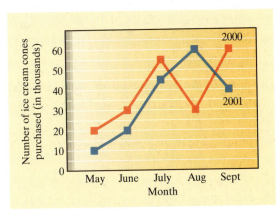

48. How many more ice cream cones were purchased in August 2001 than in August 2000?

49. How many more ice cream cones were purchased in July 2000 than in June 2000?

50. How many more ice cream cones were purchased in September 2000 than in August 2000?

51. At the location of the ice cream stand, July 2000 was warm and sunny, whereas August 2000 was cold and rainy. Describe how the weather might have played a role in the trend shown on the graph from July to August 2000.

52. At the location of the ice cream stand, June 2001 was cold and rainy, whereas July 2001 was warm and sunny. Describe how the weather might have played a role in the trend shown on the graph from June to July 2001.

The following comparison line graph shows the annual sales of cordless telephones and answering machines in the United States for selected years. Please use the graph to answer exercises 53–64.

53. How many cordless telephones were sold in 1990?

54. How many answering machines were sold in 1996?

55. Between what two years was the increase in the sales of cordless telephones the greatest?

56. Between what two years was the increase in the sales of answering machines the greatest?

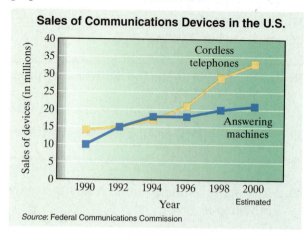

57. How many more answering machines than cordless telephones were sold in 1994?

58. How many more cordless telephones than answering machines were sold in 2000?

59. In what year were the sales of answering machines and cordless phones at the same level?

60. In what year were the sales of answering machines and cordless phones the most significantly different?

61. What was the average increase in sales per two-year period for cordless telephones?

62. What was the average increase in sales per two-year period for answering machines?

63. If the amount of increase from 1996 to 2000 continues in the four years from 2000 to 2004, what will be the sales of answering machines in 2004?

64. If the amount of increase from 1994 to 2000 continues in the next six years from 2000 to 2006, what will be the sales of cordless telephones in 2006?

The state highway department made a survey of the bridges over all of its county and state highways. The distribution of the ages of the bridges is displayed in the following histogram. Use the histogram to answer exercises 65–70.

65. How many bridges are between 40 and 59 years old?

66. How many bridges are between 60 and 79 years old?

67. The greatest number of bridges in the state are between _____ and _____ years old.

68. The highway commissioner has ordered an immediate inspection of all bridges older than 79 years old. How many bridges will be inspected?

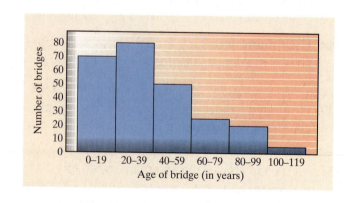

69. All bridges less than 40 years old have had a recent inspection by the highway department. How many bridges were inspected recently?

70. How many more bridges in the state are 60–79 years old than are 80–99 years old?

During the last 28 days, a major manufacturer produced 400 new color television sets each day. The manufacturer recorded the number of defective television sets produced each day. The results are shown in the following chart. In exercises 71–75, determine frequencies of the class intervals for this data.

	Mon.	Tues.	Wed.	Thurs.	Fri.
Week 1	3	5	8	2	0
Week 2	13	6	3	4	1
Week 3	0	2	16	5	7
Week 4	12	10	17	5	4
Week 5	1	7	8	12	13
Week 6	14	0	3	closed	

Mon.   Tues.   Wed.   Thurs.   Fri.

	Number of Defective Televisions Produced (Class Interval)	**Tally**	**Frequency**
71.	0–3	_____	_____
72.	4–7	_____	_____
73.	8–11	_____	_____
74.	12–15	_____	_____
75.	16–19	_____	_____

76. Construct a histogram using the table prepared in exercises 71–75.

77. Based on the data of exercises 71–75, how often were between 0 and 7 defective television sets identified in the production?

Find the mean.

78. The maximum temperature readings in Los Angeles for the last seven days in July were: 86°, 83°, 88°, 95°, 97°, 100°, and 81°.

79. The weekly amounts of groceries purchased by the Michael Stallard family for the last seven weeks were $87, $105, $89, $120, $139, $160, and $98.

80. The numbers of cars parked daily in the Central City garage for the last five days were 1327, 1561, 1429, 1307, and 1481.

81. The numbers of employees throughout the nation employed annually by Freedom Rent a Car for the last six years were 882, 913, 1017, 1592, 1778, and 1936.

Find the median.

82. The costs of eight trucks purchased by the highway department: $28,500, $29,300, $21,690, $35,000, $37,000, $43,600, $45,300, $38,600.

83. The costs of 10 houses recently purchased at Stillwater: $98,000, $150,000, $120,000, $139,000, $170,000, $156,000, $135,000, $144,000, $154,000, $126,000.

In exercises 84 and 85, find the median and the mode.

84. The numbers of cups of coffee consumed by the students in the 7:00 A.M. Biology III class during the last semester: 38, 19, 22, 4, 0, 1, 5, 9, 18, 36, 43, 27, 21, 19, 22, 20.

85. The daily numbers of deliveries made by the Northfield House of Pizza: 21, 16, 15, 3, 19, 24, 13, 18, 9, 31, 36, 25, 28, 14, 15, 26.

86. The scores on eight tests taken by Wong Yin in calculus last semester were 96, 98, 88, 100, 31, 89, 94, and 98. Which is a better measure of his usual score, the *mean* or the *median*? Why?

87. The ten sales people at People's Dodge sold the following numbers of cars last month: 13, 16, 8, 4, 5, 19, 15, 18, 39, 12. Which is a better measure of the usual sales of these salespersons, the *mean* or the *median*? Why?

Chapter 8 Test

A state highway safety commission recently reported the results of inspecting 300,000 automobiles. The following circle graph depicts the percent of automobiles that passed and the percent that had one or more safety violations. Use this graph to answer questions 1–5.

1. What percent of the automobiles passed inspection?

2. What percent of the automobiles had two safety violations?

3. What percent of the automobiles had more than two safety violations?

4. If 300,000 automobiles were inspected, how many of them had one safety violation?

5. If 300,000 automobiles were inspected, how many of them had either two violations or three violations?

The following double-bar graph indicates the number of cars sold at Danvers Ford during each quarter of 2000 and 2001. Use the graph to answer questions 6–11.

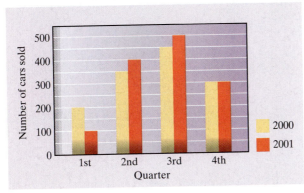

6. How many cars were sold in the second quarter of 2000?

7. How many cars were sold in the third quarter of 2001?

8. When was the greatest number of cars sold?

9. During which quarter were more cars sold in 2000 than in 2001?

10. How many more cars were sold in the second quarter of 2001 than in the second quarter of 2000?

11. How many more cars were sold in the third quarter of 2000 than in the fourth quarter of 2000?

1.

2.

3.

4.

5.

6.

7.

8.

9.

10.

11.

553

A research study by 10 midwestern universities produced the following line graph. Use the graph to answer questions 12–16.

12. Approximately how many more years is a 45-year-old American man expected to live if he smokes?

13. Approximately how many more years is a 55-year-old American man expected to live if he does not smoke?

14. According to this graph, approximately how much longer is a 25-year-old nonsmoker expected to live than a 25-year-old smoker?

15. According to this graph, at what age is the difference between the life expectancy of a smoker and a nonsmoker the greatest?

16. According to this graph, at what age is the difference between the life expectancy of a smoker and a nonsmoker the smallest?

The following histogram was prepared by a consumer research group. Use the histogram to answer questions 17–20.

17. How many color television sets lasted 6–8 years?

18. How many color television sets lasted 3–5 years?

19. How many color television sets lasted more than 11 years?

20. How many color television sets lasted 9–14 years?

A chemistry student had the following scores on ten quizzes in her chemistry class: 19, 16, 15, 12, 18, 17, 14, 10, 16, 20.

21. Find the mean quiz score.

22. Find the median quiz score.

23. Find the mode quiz score.

Approximately one-half of this test is based on Chapter 8 material. The remainder is based on material covered in Chapters 1–7.

1. Add. $1376 + 2804 + 9003 + 7642$

2. Multiply. 2008×37

3. Subtract. $7\frac{1}{5} - 3\frac{3}{8}$ **4.** Divide. $10\frac{4}{5} \div 3\frac{1}{2}$

5. Round to the nearest hundredth. 1796.4289

6. Subtract.
$$\begin{array}{r} 200.58 \\ -\ 127.93 \\ \hline \end{array}$$

7. Divide. $52.0056 \div 0.72$ **8.** Find *n*. $\dfrac{7}{n} = \dfrac{35}{3}$

9. Of every 2030 cars manufactured, 3 have major engine defects. If the total number of these cars manufactured was 26,390, approximately how many had major engine defects?

10. What is 1.3% of 25?   **11.** 72% of what number is 252?

12. Convert 198 cm to m.   **13.** Convert 18 yd to ft.

▲ **14.** Find the area of a circle with radius of 3 in. Round to the nearest tenth.

▲ **15.** Find the perimeter of a square with a side of 17 in.

The 12,000-member student body of Mason University consists of five groups: freshmen, sophomores, juniors, seniors, and graduate students. The distribution by category is displayed in the following graph.

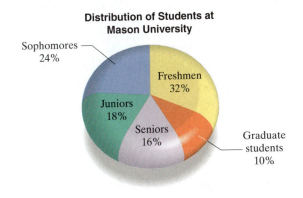

Distribution of Students at Mason University

Sophomores 24%
Freshmen 32%
Juniors 18%
Seniors 16%
Graduate students 10%

16. What percent are either juniors or seniors?

17. How many of the 12,000 students are freshmen?

1. _____

2. _____

3. _____

4. _____

5. _____

6. _____

7. _____

8. _____

9. _____

10. _____

11. _____

12. _____

13. _____

14. _____

15. _____

16. _____

17. _____

18.

19.

20.

21.

22.

23.

24.

25.

26.

The following double-bar graph indicates the quarterly profits for Dedalon Corporation for 2000 and 2001.

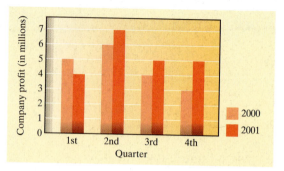

18. What was the quarterly profit for Dedalon Corporation in the fourth quarter of 2000?

19. How much greater was the profit of Dedalon Corporation in the second quarter of 2001 than in the second quarter of 2000?

The following comparison line graph depicts the annual rainfall in Dixville compared to the annual rainfall in Weston for five specific years.

20. How many inches of rain fell in Dixville in 1980?

21. In what years was the annual rainfall in Weston greater than the annual rainfall in Dixville?

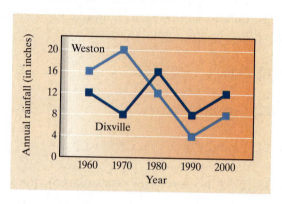

The following histogram depicts the number of students in the Basic Mathematics course who fall into various age groups.

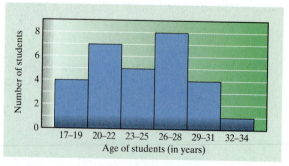

22. How many students are 26–28 years old?

23. How many students are under 26 years old?

The following are the hourly wages of eight employees of the Hamilton House of Pizza: $5.00, $4.50, $3.95, $4.90, $7.00, $12.15, $4.50, $6.00.

24. Find the mean hourly wage.   **25.** Find the median hourly wage.

26. Find the mode hourly wage.

Signed Numbers

Trying to contain and control radioactive waste from a nuclear power plant is one of the most important problems facing scientists today. An innovative new approach for controlling strontium 90 contamination in ground water uses zeolite, a material commonly used for cat litter. Can you make some of the calculations needed to decide how effectively the contamination can be controlled? Turn to the Putting Your Skills to Work problems on page 595 to find out.

Pretest Chapter 9

1. _____
2. _____
3. _____
4. _____
5. _____
6. _____
7. _____
8. _____
9. _____
10. _____
11. _____
12. _____
13. _____
14. _____
15. _____
16. _____
17. _____
18. _____
19. _____
20. _____

If you are familiar with the topics in this chapter, take this test now. Check your answers with those in the back of the book. If an answer is wrong or you can't answer a question, study the appropriate section of the chapter.

If you are not familiar with the topics in this chapter, don't take this test now. Instead, study the examples, work the practice problems, and then take the test.

This test will help you identify which concepts you have mastered and which you need to study further. Be sure to simplify or reduce all answers on this test.

Section 9.1

Add.

1. $-7 + (-12)$

2. $-23 + 19$

3. $7.6 + (-3.1)$

4. $8 + (-5) + 6 + (-9)$

5. $\dfrac{5}{12} + \left(-\dfrac{3}{4}\right)$

6. $-\dfrac{5}{6} + \left(-\dfrac{1}{3}\right)$

7. $-2.8 + (-4.2)$

8. $-3.7 + 5.4$

Section 9.2

Subtract.

9. $13 - 21$

10. $-26 - 15$

11. $\dfrac{5}{17} - \left(-\dfrac{9}{17}\right)$

12. $-19 - (-7)$

13. $-4.9 - (-6.3)$

14. $2.8 - 5.6$

15. $21 - (-21)$

16. $\dfrac{2}{3} - \left(-\dfrac{3}{5}\right)$

Section 9.3

Multiply or divide.

17. $(-3)(-8)$

18. $-48 \div (-12)$

19. $-72 \div 9$

20. $(5)(-4)(2)(-1)\left(-\dfrac{1}{4}\right)$

21. $\dfrac{72}{-3}$

22. $\dfrac{-\dfrac{3}{4}}{-\dfrac{4}{5}}$

23. $(-8)(-2)(-4)$

24. $120 \div (-12)$

Section 9.4

Perform the operations in the proper order.

25. $24 \div (-4) + 28 \div (-7)$

26. $18 \div 3(5) + (-5) \div (-5)$

27. $7 + (-9) + 2(-5)$

28. $8(-6) \div (-10)$

29. $5 - (-6) + 18 \div (-3)$

30. $9(-3) + 4(-2) - (-6)$

31. $\dfrac{12 + 8 - 4}{(-4)(3)(4)}$

32. $\dfrac{56 \div (-7) - 1}{3(-9) + 6(-3)}$

Section 9.5

Write in scientific notation.

33. $80{,}000$

34. 0.0005

35. $128{,}000{,}000$

Write in standard notation.

36. 6.7×10^{-3}

37. 1.32×10^{6}

38. 1.5678×10^{-8}

21.	
22.	
23.	
24.	
25.	
26.	
27.	
28.	
29.	
30.	
31.	
32.	
33.	
34.	
35.	
36.	
37.	
38.	

9.1 Addition of Signed Numbers

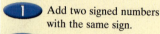
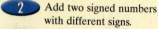
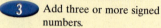

Student Learning Objectives

After studying this section, you will be able to:

1. Add two signed numbers with the same sign.

2. Add two signed numbers with different signs.

3. Add three or more signed numbers.

SSM

PH TUTOR CD & VIDEO MATH PRO WEB
CENTER

1 Adding Two Signed Numbers with the Same Sign

In Chapters 1–8 we worked with whole numbers, fractions, and decimals. In this chapter we enlarge the set of numbers we work with to include numbers that are less than zero. Many real-life situations require using numbers that are less than zero. A debt that is owed, a financial loss, temperatures that fall below zero, and elevations that are below sea level can be expressed only in numbers that are less than zero, or negative numbers.

The following is a graph of the financial reports of four small airlines for the year. It shows **positive numbers**—those numbers that rise above zero—and **negative numbers**—those numbers that fall below zero. The positive numbers represent money gained. The negative numbers represent money lost.

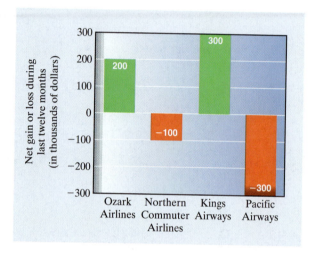

A value of −100,000 is shown for Northern Commuter Airlines. This means that Northern Commuter Airlines lost $100,000 during the year. A value of −300,000 is recorded for Pacific Airways. What does this mean?

Another way to picture positive and negative numbers is on a number line. A **number line** is a line on which each point is associated with a number. The numbers may be positive or negative, whole numbers, fractions, or decimals. Positive numbers are to the right of zero on the number line. Negative numbers are to the left of zero on the number line. Zero is neither positive nor negative.

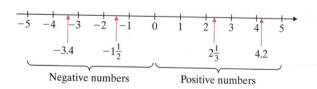

Positive numbers can be written with a plus sign—for example, +2—but this is not usually done. Positive 2 is usually written as 2. It is understood that the *sign* of the number is positive although it is not written. Negative numbers must always have the negative sign so that we know they are negative numbers. Negative 2 is written as −2. The *sign* of the number is negative. The set of positive numbers, negative numbers, and zero is called the set of **signed numbers**.

Order

Signed numbers are named in **order** on the number line. Smaller numbers are to the left. Larger numbers are to the right. For any two numbers on the number line, the number on the left is less than the number on the right.

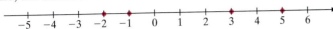

We use the symbol $<$ to mean "is less than." Thus the mathematical sentence $-2 < -1$ means "-2 is less than -1." We use the symbol $>$ to mean "is greater than." Thus the mathematical sentence $5 > 3$ means "5 is greater than 3."

● EXAMPLE 1   In each case, replace the ? with $<$ or $>$.

(a) $-8 \; ? \; -4$ **(b)** $7 \; ? \; 1$ **(c)** $-2 \; ? \; 0$

(d) $-6 \; ? \; 3$ **(e)** $2 \; ? \; -5$

(a) Since -8 lies to the left of -4, we know that $-8 < -4$.

(b) Since 7 lies to the right of 1, we know that $7 > 1$.

(c) Since -2 lies to the left of 0, we know that $-2 < 0$.

(d) Since -6 lies to the left of 3, we know that $-6 < 3$.

(e) Since 2 lies to the right of -5, we know that $2 > -5$.

Practice Problem 1   In each case replace the ? with $<$ or $>$.

(a) $4 \; ? \; 2$ **(b)** $-3 \; ? \; -5$ **(c)** $0 \; ? \; -6$

(d) $-2 \; ? \; 1$ **(e)** $5 \; ? \; -7$

Absolute Value

Sometimes we are only interested in the distance a number is from zero. For example, the distance from 0 to $+3$ is 3. The distance from 0 to -3 is also 3. Notice that distance is always a positive number, regardless of which direction we travel on the number line. This distance is called the *absolute value*.

> The **absolute value** of a number is the distance between that number and zero on the number line.

The symbol for absolute value is $|\;\;|$. When we write $|5|$, we are looking for the distance from 0 to 5 on the number line. Thus $|5| = 5$. This is read, "The absolute value of 5 is 5." $|-5|$ is the distance from 0 to -5 on the number line. Thus $|-5| = 5$. This is read, "The absolute value of -5 is 5."

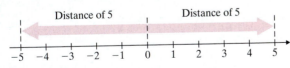

Other examples of absolute value are shown next.

$$|6| = 6 \qquad\qquad |-3| = 3$$

$$|7.2| = 7.2 \qquad\qquad \left|-\frac{1}{5}\right| = \frac{1}{5}$$

$$|0| = 0 \qquad\qquad |-26| = 26$$

When we find the absolute value of any nonzero number, we always get a positive value. We use the concept of absolute value to develop rules for adding signed numbers. We begin by looking at addition of numbers with the same sign. Although you are already familiar with the addition of positive numbers, we will look at an example.

Suppose that we earn $52 one day and earn $38 the next day. To learn what our two-day total is, we add the positive numbers. We earn

$$\$52 + \$38 = +\$90.$$

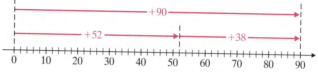

We can show this sum on a number line by drawing an arrow that starts at 0 and points 52 units to the right (because 52 is positive). At the end of this arrow we draw a second arrow that points 38 units to the right. Note that the second arrow ends at the sum, 90.

The money is coming in, and the plus sign records a gain. Notice that we added the numbers and that the sign of the sum is the same as the sign of the addends.

Now let's consider an example of addition of two negative numbers.

Suppose that we consider money spent as negative dollars. If we spend $52 one day (−$52) and we spend $38 the next day (−$38), we must add two negative numbers. What is our financial position?

$$-\$52 + (-\$38) = -\$90$$

We have spent $90. The negative sign tells us the direction of the money: out! Notice that we added the numbers and that the sign of the sum is the same as the sign of the addends.

We can illustrate this sum on a number line by drawing a line that starts at 0 and points 52 units to the left (because −52 is negative). At the end of this arrow we add a second arrow that points 38 units to the left. This second arrow ends at the sum, −90.

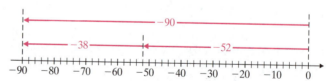

These examples suggest the addition rule for two numbers with the same sign.

> **Addition Rule for Two Numbers with the Same Sign**
>
> To add two numbers with the same sign:
> **1.** Add the absolute value of the numbers.
> **2.** Use the common sign in the answer.

● **EXAMPLE 2**   Add.   **(a)** 7 + 5   **(b)** −3.2 + (−5.6)

(a) 7   We add the absolute value of the numbers 7 and 5. The positive
       + 5   sign, although not written, is common to both numbers. The
       ———   answer is a positive 12. (The + sign is not written.)
        12

(b) -3.2
      $+ \ -5.6$
      $\overline{\ -8.8\ }$ ┌ We add the absolute value of the numbers 3.2 and 5.6.

We use a negative sign in our answer because we added two negative numbers.

Practice Problem 2 Add. **(a)** $9 + 14$ **(b)** $-4.5 + (-1.9)$

These rules can be applied to fractions as well.

EXAMPLE 3 Add. **(a)** $\dfrac{5}{18} + \dfrac{1}{3}$ **(b)** $-\dfrac{1}{7} + \left(-\dfrac{3}{5}\right)$

(a) The LCD = 18. The first fraction already has the LCD.

$$\dfrac{5}{18} = \dfrac{5}{18}$$

$$+ \ \dfrac{1}{3} \cdot \dfrac{6}{6} = + \dfrac{6}{18}$$

We add two positive numbers, so the answer is positive.

$$\dfrac{11}{18}$$

(b) The LCD = 35.

$$\dfrac{1}{7} \cdot \dfrac{5}{5} = \dfrac{5}{35}$$

Because $\dfrac{1}{7} = \dfrac{5}{35}$ it follows that $-\dfrac{1}{7} = -\dfrac{5}{35}$.

$$\dfrac{3}{5} \cdot \dfrac{7}{7} = \dfrac{21}{35}$$

Because $\dfrac{3}{5} = \dfrac{21}{35}$ it follows that $-\dfrac{3}{5} = -\dfrac{21}{35}$. Thus

$$-\dfrac{1}{7}$$
$$+ \ -\dfrac{3}{5}$$

is equivalent to

$$-\dfrac{5}{35}$$
$$+ \ -\dfrac{21}{35}$$
$$-\dfrac{26}{35}$$

We add two negative numbers, so the answer is negative.

Practice Problem 3 Add. **(a)** $\dfrac{5}{12} + \dfrac{1}{4}$ **(b)** $-\dfrac{1}{6} + \left(-\dfrac{2}{7}\right)$

It is interesting to see how often negative numbers appear in statements of the Federal Budget. Observe the data in the following bar graph.

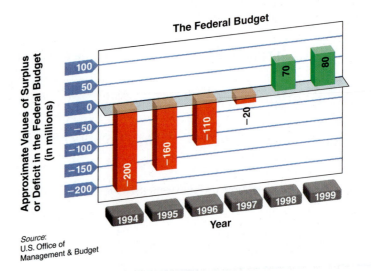

The Federal Budget

Source:
U.S. Office of
Management & Budget

🔴 **EXAMPLE 4**   Find the total value of surplus or deficit for the two years 1995 and 1996.

We add (−$160 million) + (−$110 million) to obtain −$270 million. The total debt for these two years is $270,000,000.

Practice Problem 4   Find the total value of surplus or deficit for the two years 1994 and 1997.   🔴

2   Adding Two Signed Numbers with Different Signs

Let's look at some real-life situations involving addition of signed numbers with different signs. Suppose that we earn $52 one day and we spend $38 the next day. If we combine (add) the two transactions, it would look like

$$\$52 + (-\$38) = +\$14.$$

On a number line we draw an arrow that starts at zero and points 52 units to the right. From the end of this arrow we draw an arrow that points 38 units to the left. (Remember, the arrow points to the left for a negative number.)

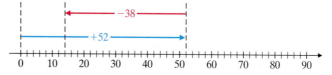

This is a situation with which we are familiar. What we actually do is subtract. That is, we take the difference between $52 and $38. Notice that the sign of the larger number is positive and that the sign of the answer is also positive.

Let's look at another situation. Suppose we earn $38 and we spend $52. The situation would look like this.

$$(-\$52) + \$38 = -\$14$$

On a number line we draw an arrow that starts at zero and points 52 units to the left. Again the arrow must point to the left to represent a negative number.

From the end of this arrow we draw an arrow that points 38 units to the right.

On our number line we end up at -14.

In our real-life situation we end up owing \$14, which is represented by a negative number. To find the sum, we actually find the difference between \$52 and \$38. Notice that if we do not account for sign, the larger number is 52. The sign of that number is negative and the sign of the answer is also negative. This suggests the addition rule for two numbers with different signs.

Addition Rule for Two Numbers with Different Signs

To add two numbers with different signs:

1. Subtract the absolute value of the numbers.

2. Use the sign of the number with the larger absolute value.

EXAMPLE 5 Add. **(a)** $8 + (-10)$ **(b)** $-16.6 + 12.3$ **(c)** $\dfrac{3}{4} + \left(-\dfrac{2}{3}\right)$

(a)
$$\begin{array}{r} 8 \\ +\ -10 \\ \hline -2 \end{array}$$
← The signs are different, so we find the difference: $10 - 8 = 2$.

The sign of the number with the larger absolute value is negative, so the answer is negative.

(b)
$$\begin{array}{r} -16.6 \\ +\ \ 12.3 \\ \hline -4.3 \end{array}$$
← The signs are different, so we find the difference.

The sign of the number with the larger absolute value is negative, so the answer is negative.

(c) $\dfrac{3}{4} + \left(-\dfrac{2}{3}\right) = \dfrac{9}{12} + \left(-\dfrac{8}{12}\right) = \dfrac{9 + (-8)}{12} = \dfrac{1}{12}$

The signs are different, so we find the difference. The sign of the number with the larger absolute value is positive, so the answer is positive.

Practice Problem 5 Add.

(a) $7 + (-12)$ **(b)** $-20.8 + 15.2$ **(c)** $\dfrac{5}{6} + \left(-\dfrac{3}{4}\right)$

Notice that in **(b)** of Example 5, the number with the larger absolute value is on top. This makes the numbers easier to subtract. Because addition is commutative, we could have written **(a)** as

$$\begin{array}{r} -10 \\ +\ \ 8 \\ \hline \end{array}$$

This makes the computation easier. If you are adding two numbers with different signs, place the number with the larger absolute value on top so that you can find the difference easily. As noted, the commutative property of addition holds for signed numbers.

Commutative Property of Addition

For any real numbers a and b,

$$a + b = b + a.$$

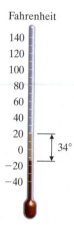

Fahrenheit

140
120
100
80
60
40
20
0
−20
−40

34°

EXAMPLE 6 Last night the temperature dropped to −14°F. From that low, today the temperature rose 34°F. What was the highest temperature today?

We want to add −14°F and 34°F. Because addition is commutative, it does not matter whether we add −14 + 34 or 34 + (−14).

$$
\begin{array}{r}
34°\text{F} \\
+\ -14°\text{F} \\
\hline
20°\text{F} \longleftarrow
\end{array}
$$

The 34 is larger than 14. The difference between 34 and 14 is 20. The number with the larger absolute value is positive, so the answer is positive.

Practice Problem 6 Last night the temperature dropped to −19°F. From that low, today the temperature rose 28°F. What was the highest temperature today?

3 Adding Three or More Signed Numbers

We can add three or more numbers using these rules. Since addition is associative, we may group the numbers to be added in any way. That is, it does not matter which two numbers are added first.

EXAMPLE 7 Add. 24 + (−16) + (−10)

We can go from left to right and start with 24, or we can start with −16.

Step 1
$$
\begin{array}{r}
24 \\
+\ -16 \\
\hline
8
\end{array}
$$
or
Step 1
$$
\begin{array}{r}
-16 \\
+\ -10 \\
\hline
-26
\end{array}
$$

Step 2
$$
\begin{array}{r}
8 \\
+\ -10 \\
\hline
-2
\end{array}
$$
Step 2
$$
\begin{array}{r}
-26 \\
+\ \ \ 24 \\
\hline
-2
\end{array}
$$

Practice Problem 7 Add. 36 + (−21) + (−18)

Associative Property of Addition

For any three real numbers a, b, and c,

$$(a + b) + c = a + (b + c).$$

If there are many numbers to add, it may be easier to add the positive numbers and the negative numbers separately and then combine the results.

● **EXAMPLE 8**   The results of a new company's operations over five months are listed in the following table. What is the company's overall profit or loss over the five-month period?

NET OPERATIONS
Profit/Loss Statement in Dollars

Month	Profit	Loss
January	30,000	
February		−50,000
March		−10,000
April	20,000	
May	15,000	

First we will add separately the positive numbers and the negative numbers.

$$
\begin{array}{r}
30,000 \\
20,000 \\
+\ 15,000 \\
\hline
65,000
\end{array}
\qquad
\begin{array}{r}
-\ 50,000 \\
+\ -\ 10,000 \\
\hline
-\ 60,000
\end{array}
$$

Now we add the positive number 65,000 and the negative number −60,000.

$$
\begin{array}{r}
65,000 \\
+\ -\ 60,000 \\
\hline
5,000
\end{array}
$$

The company had an overall profit of $5000 for the five-month period.

Practice Problem 8   The results of the next five months of operations for the same company are listed in the following table. What is the overall profit or loss over this five-month period?

NET OPERATIONS
Profit/Loss Statement in Dollars

Month	Profit	Loss
June		−20,000
July	30,000	
August	40,000	
September		−5,000
October		−35,000

Verbal and Writing Skills

1. Explain in your own words how to add two signed numbers if the signs are the same.

2. Explain in your own words how to add two signed numbers if one number is positive and one number is negative.

Add each pair of signed numbers that have the same sign.

3. $-6 + (-11)$ **4.** $-5 + (-13)$ **5.** $-4.9 + (-2.1)$ **6.** $-8.3 + (-3.7)$

7. $8.9 + 7.6$ **8.** $5.8 + 2.7$ **9.** $\dfrac{1}{5} + \dfrac{2}{7}$ **10.** $\dfrac{2}{3} + \dfrac{1}{4}$

11. $-\dfrac{1}{12} + \left(-\dfrac{5}{6}\right)$ **12.** $-\dfrac{1}{16} + \left(-\dfrac{3}{4}\right)$

Add each pair of signed numbers that have different signs.

13. $14 + (-5)$ **14.** $15 + (-6)$ **15.** $-17 + 12$ **16.** $-21 + 15$

17. $-36 + 58$ **18.** $-42 + 57$ **19.** $-9.3 + 6.5$ **20.** $-7.2 + 4.4$

21. $\dfrac{1}{12} + \left(-\dfrac{3}{4}\right)$ **22.** $\dfrac{1}{15} + \left(-\dfrac{3}{5}\right)$

Mixed Practice

Add.

23. $\dfrac{7}{9} + \left(-\dfrac{2}{9}\right)$ **24.** $\dfrac{5}{12} + \left(-\dfrac{7}{12}\right)$ **25.** $-18 + (-4)$ **26.** $-34 + (-2)$

27. $1.5 + (-2.2)$ **28.** $3.6 + (-4.1)$ **29.** $-14.6 + (-5.7)$ **30.** $-11.8 + (-4.6)$

31. $13 + (-9)$ **32.** $-18 + 7$ **33.** $-\dfrac{1}{3} + \left(-\dfrac{2}{7}\right)$ **34.** $-\dfrac{3}{4} + \left(-\dfrac{1}{5}\right)$

35. $-7.3 + 13.8$ **36.** $-6.4 + 15.9$ **37.** $-5 + \left(-\dfrac{1}{2}\right)$ **38.** $\dfrac{5}{6} + (-3)$

Add.

39. $6 + (-14) + 4$

40. $15 + (-2) + 4$

41. $11 + (-9) + (-10) + 8$

42. $(-13) + 8 + (-12) + 17$

43. $-7 + 6 + (-2) + 5 + (-3) + (-5)$

44. $-2 + 1 + (-12) + 7 + (-4) + (-1)$

45. $\left(-\dfrac{1}{5}\right) + \left(-\dfrac{2}{3}\right) + \left(\dfrac{4}{25}\right)$

46. $\left(-\dfrac{1}{7}\right) + \left(-\dfrac{5}{21}\right) + \left(\dfrac{3}{14}\right)$

Applications

Use signed numbers to represent the total profit or loss for a company after the following reports.

47. A \$43,000 loss in February followed by a \$51,000 loss in March.

48. A \$16,000 loss in May followed by a \$25,000 loss in June.

49. A \$28,000 profit in July followed by a \$19,000 loss in August.

50. An \$89,000 profit in April followed by a \$33,000 loss in May.

51. A \$35,000 loss in April, a \$17,000 profit in May, and a \$20,000 loss in June.

52. A \$34,000 loss in October, a \$12,000 loss in November, and a \$15,500 profit in December.

Solve.

53. Last night, the temperature was −1°F. In the morning, the temperature dropped 17°F. What was the new temperature?

54. This morning the temperature was −12°F. This afternoon, the temperature rose 19°F. What was the new temperature?

55. This morning, the temperature was −5°F. This evening, the temperature rose 4°F. What was the new temperature?

56. Last night the temperature was −7°F. The temperature dropped 15°F this afternoon. What was the new temperature?

57. On a January morning the temperature was 15°. After this reading, the temperature was measured each hour for four hours. It dropped 6°, rose 2°, rose 1°, and dropped 3°. What was the final temperature? (Assume all temperature readings are degrees Fahrenheit.)

58. Last night at midnight the temperature was −2°. The temperature was measured each hour for the next four hours. It rose 2°, dropped 3°, rose 3°, and rose 1°. What was the final temperature? (Assume all temperature readings are degrees Fahrenheit.)

59. In three plays of a football game, the quarterback threw passes that lost 8 yards, gained 13 yards, and lost 6 yards. What was the total gain or loss of the three plays?

60. In three plays of a football game, the quarterback threw passes that gained 20 yards, lost 13 yards, and gained 5 yards. What was the total gain or loss of the three plays?

To Think About

61. Jeffrey had a credit card balance of −$28. He immediately made a payment of $30. The next day the credit card company notified him that they had charged him $15 because his account balance was negative. What was the credit card balance in Jeffrey's account after these actions took place?

62. Susan had a credit card balance of −$39. She immediately made a payment of $100. Before the payment was credited, the credit card bank charged her $0.79 in interest. What was the credit card balance in Susan's account after these actions took place?

63. Bob examined his checking account register. He thought his balance was $89.50. However, he forgot to subtract an ATM withdrawal of $50.00. Since the ATM that he used was at a different bank, he was also charged $2.50 for using the ATM. What was the actual balance in his checking account?

64. Nancy examined her checking account register. She thought her balance was $97.40. However, she forgot to subtract a check of $95.00 that she had made out the previous week. She also forgot to subtract the monthly $4.50 fee charged by her bank for having a checking account. What was the actual balance in her checking account?

Cumulative Review Problems

▲ **65.** Use $V = \dfrac{4\pi r^3}{3}$ to find the volume of a sphere of radius 6 feet. Use $\pi \approx 3.14$ and round to the nearest tenth.

▲ **66.** Use $V = \dfrac{Bh}{3}$ to find the volume of a pyramid whose rectangular base measures 9 meters by 7 meters and whose height is 10 meters.

9.2 Subtraction of Signed Numbers

1 Subtracting One Signed Number from Another

We begin our discussion by defining the word **opposite**. The opposite of a positive number is a negative number with the same absolute value. For example, the opposite of 7 is −7.

The opposite of a negative number is a positive number with the same absolute value. For example, the opposite of −9 is 9. If a number is the opposite of another number, these two numbers are at an equal distance from zero on the number line.

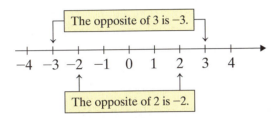

The opposite of 3 is −3.

The opposite of 2 is −2.

The sum of a number and its opposite is zero.

$$-3 + 3 = 0 \qquad 2 + (-2) = 0$$

We will use the concept of opposite to develop a way to subtract integers.

Let's think about how a checking account works. Suppose that you deposit $25 and the bank adds a service charge of $5 for a new checkbook. Your account looks like this:

$$\$25 + (-\$5) = \$20$$

Suppose instead that you deposit $25 and the bank adds no charge. The next day, you write a check for $5. The result of these two transactions is

$$\$25 - \$5 = \$20$$

Note that your account has the same amount of money ($20) in both cases.

We see that adding a negative 5 to 25 is the same as subtracting a positive 5 from 25. That is, $25 + (-5) = 20$ and $25 - 5 = 20$.

Subtracting is equivalent to adding the opposite.

Subtracting	*Adding the Opposite*
$25 - 5 = 20$	$25 + (-5) = 20$
$19 - 6 = 13$	$19 + (-6) = 13$
$7 - 3 = 4$	$7 + (-3) = 4$
$15 - 5 = 10$	$15 + (-5) = 10$

Student Learning Objectives

After studying this section, you will be able to:

1 Subtract one signed number from another.

2 Solve problems involving both addition and subtraction of signed numbers.

3 Solve simple applied problems that involve the subtraction of signed numbers.

SSM

PH TUTOR CENTER   CD & VIDEO   MATH PRO   WEB

We define a rule for the subtraction of signed numbers.

> ### Subtraction of Signed Numbers
>
> To subtract signed numbers, add the opposite of the second number to the first number.

Thus, to do a subtraction problem, we first change it to an equivalent addition problem in which the first number does not change but the second number is replaced by its opposite. Then we follow the rules of *addition* for signed numbers.

EXAMPLE 1   Subtract. $-8 - (-2)$

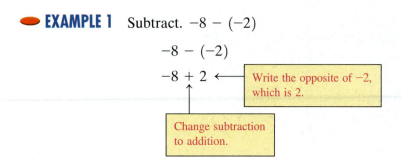

Practice Problem 1   Subtract. $-10 - (-5)$

EXAMPLE 2   Subtract.   **(a)** $7 - 8$ **(b)** $-12 - 16$

(a)

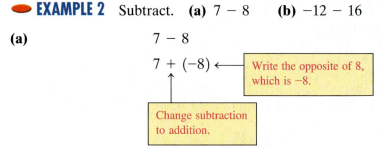

Now we use the rules of addition for two numbers with opposite signs.

$$7 + (-8) = -1$$

(b) $-12 - 16$

$$-12 + (-16) \longleftarrow \boxed{\text{Write the opposite of 16, which is } -16.}$$

Change subtraction to addition.

Now we follow the rules of addition of two numbers with the same sign.

$$-12 + (-16) = -28$$

Practice Problem 2   Subtract.   **(a)** $5 - 12$ **(b)** $-11 - 17$

Sometimes the numbers we subtract are fractions or decimals.

EXAMPLE 3 Subtract. **(a)** $5.6 - (-8.1)$ **(b)** $-\dfrac{6}{11} - \left(-\dfrac{1}{22}\right)$

(a) Change the subtraction to adding the opposite. Then add.

$$5.6 - (-8.1) = 5.6 + 8.1$$
$$= 13.7$$

(b) Change the subtraction to adding the opposite. Then add.

$$-\frac{6}{11} - \left(-\frac{1}{22}\right) = -\frac{6}{11} + \frac{1}{22}$$

$$= -\frac{6}{11} \cdot \frac{2}{2} + \frac{1}{22} \qquad \text{We see that the LCD = 22. We change } \tfrac{6}{11}$$
$$\text{to a fraction with a denominator of 22.}$$

$$= -\frac{12}{22} + \frac{1}{22} \qquad \text{Add.}$$

$$= -\frac{11}{22} \quad \text{or} \quad -\frac{1}{2}$$

Practice Problem 3 Subtract. **(a)** $3.6 - (-9.5)$ **(b)** $-\dfrac{5}{8} - \left(-\dfrac{5}{24}\right)$

Remember that in performing subtraction of two signed numbers:

1. The first number does not change.

2. The subtraction sign is changed to addition.

3. We write the opposite of the second number.

4. We find the result of this addition problem.

Think of each subtraction problem as a problem of adding the opposite.

If you see $7 - 10$, think $7 + (-10)$.

If you see $-3 - 19$, think $-3 + (-19)$.

If you see $6 - (-3)$, think $6 + (+3)$.

EXAMPLE 4 Subtract.

(a) $6 - (+3)$ **(b)** $-\dfrac{1}{2} - \left(-\dfrac{1}{3}\right)$ **(c)** $2.7 - (-5.2)$

(a) $6 - (+3) = 6 + (-3) = 3$

(b) $-\dfrac{1}{2} - \left(-\dfrac{1}{3}\right) = -\dfrac{1}{2} + \dfrac{1}{3} = -\dfrac{3}{6} + \dfrac{2}{6} = -\dfrac{1}{6}$

(c) $2.7 - (-5.2) = 2.7 + 5.2 = 7.9$

Practice Problem 4 Subtract.

(a) $20 - (-5)$ **(b)** $-\dfrac{1}{5} - \left(-\dfrac{1}{2}\right)$ **(c)** $3.6 - (-5.5)$

Calculator

Negative Numbers

To enter a negative number on most scientific calculators, find the key marked $\boxed{+/-}$. To enter the number -3, press the key 3 and then the key $\boxed{+/-}$. The display should read

$$\boxed{-3}$$

To find $(-32) + (-46)$, enter

$$32 \boxed{+/-} \boxed{+} 46 \boxed{+/-}$$
$$\boxed{=}$$

The display should read

$$\boxed{-78}$$

Try the following.

(a) $-756 + 184$

(b) $92 + (-51)$

(c) $-618 - (-824)$

(d) $-36 + (-10) - (-15)$

Note: The $\boxed{+/-}$ key changes the sign of a number from $+$ to $-$ or $-$ to $+$.

2 Solving Problems Involving Both Addition and Subtraction of Signed Numbers

EXAMPLE 5 Perform the following set of operations, working from left to right.

$$-8 - (-3) + (-5) = -8 + 3 + (-5)$$

First we change subtracting a -3 to adding a 3.

$$= -5 + (-5) = -10$$

Practice Problem 5 Perform the following set of operations.

$$-5 - (-9) + (-14)$$

3 Solving Simple Applied Problems That Involve the Subtraction of Signed Numbers

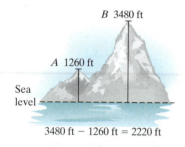

B 3480 ft

A 1260 ft

Sea level

3480 ft − 1260 ft = 2220 ft

When we want to find the difference in altitude between two mountains, we subtract. We subtract the lower altitude from the higher altitude. Look at the illustration at the left. The difference in altitude between A and B is 3480 feet − 1260 feet = 2220 feet.

Land that is below sea level is considered to have a negative altitude. The Dead Sea is 1312 feet below sea level. Look at the following illustration. The difference in altitude between C and D is 2590 feet − (−1312 feet) = 3902 feet.

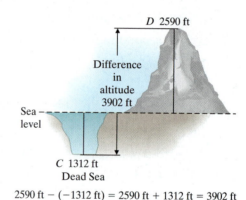

D 2590 ft

Difference in altitude 3902 ft

Sea level

C 1312 ft
Dead Sea

2590 ft − (−1312 ft) = 2590 ft + 1312 ft = 3902 ft

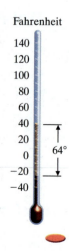

Fahrenheit

140
120
100
80
60
40
20
0
−20
−40

64°

EXAMPLE 6 Find the difference in temperature between 38°F during the day in Anchorage, Alaska, and −26°F at night.

We subtract the lower temperature from the higher temperature.

$$38 - (-26) = 38 + 26 = 64$$

The difference is 64°F.

Practice Problem 6 Find the difference in temperature between 31°F during the day in Fairbanks, Alaska, and −37° at night.

Subtract the signed numbers by adding the opposite of the second number to the first number.

1. $3 - 9$

2. $5 - 12$

3. $-14 - 3$

4. $-6 - 18$

5. $-12 - (-10)$

6. $-27 - (-12)$

7. $3 - (-21)$

8. $10 - (-23)$

9. $12 - 30$

10. $10 - 14$

11. $-12 - (-15)$

12. $-17 - (-30)$

13. $150 - 210$

14. $500 - 150$

15. $300 - (-256)$

16. $420 - (-300)$

17. $-2.5 - 4.2$

18. $-4.1 - 3.9$

19. $4.2 - 10.7$

20. $3.8 - 12.3$

21. $-10.9 - (-2.3)$

22. $-6.8 - (-2.9)$

23. $20.23 - (-12.71)$

24. $13.92 - (-14.86)$

25. $\dfrac{1}{4} - \left(-\dfrac{3}{4}\right)$

26. $\dfrac{5}{7} - \left(-\dfrac{6}{7}\right)$

27. $-\dfrac{5}{6} - \dfrac{1}{3}$

28. $-\dfrac{2}{8} - \dfrac{1}{4}$

29. $-\dfrac{5}{12} - \left(-\dfrac{1}{4}\right)$

30. $-\dfrac{7}{14} - \left(-\dfrac{1}{7}\right)$

31. $\dfrac{5}{9} - \dfrac{2}{7}$

32. $\dfrac{5}{11} - \dfrac{1}{5}$

Perform each set of operations, working from left to right.

33. $2 - (-8) + 5$

34. $7 - (-3) + 9$

35. $-5 - 6 - (-11)$

36. $-3 - 12 - (-5)$

37. $7 - (-2) - (-8)$

38. $6 - (-3) - (-5)$

39. $-16 - (-6) - 12$

40. $-13 - (-4) - 15$

41. $9 - 3 - 2 - 6$

42. $12 - 5 - 4 - 8$

43. $-16 + 9 - (-2) - 8$

44. $-12 + 8 - (-16) - 9$

Applications

Use your knowledge of signed numbers to answer exercises 45–50.

45. Find the difference in altitude between a mountain 5277 feet high and a canyon 844 feet below sea level.

46. Find the difference in altitude between a mountain 4128 feet high and a gorge 312 feet below sea level.

47. Find the difference in temperature in Alta, Utah, between 23°F during the day and −19°F at night.

48. Find the difference in temperature in Fairbanks, Alaska, between 27°F during the day and −33°F at night.

49. In Thule, Greenland, yesterday, the temperature was −29°F. Today the temperature rose 16°F. What is the new temperature?

50. Find the difference in height between the top of a hill 642 feet high and a crack caused by an earthquake 57 feet below sea level.

A company's profit and loss statement in dollars for the last five months is shown in the following table.

Month	Profit	Loss
January	18,700	
February		−32,800
March		−6,300
April	43,500	
May		−12,400

51. What is the change in the profit/loss status of the company from the first of January to the end of February?

52. What is the change in the profit/loss status of the company from the first of February to the end of March?

53. What is the change in the profit/loss status of the company from the first of March to the end of April?

54. What is the change in the profit/loss status of the company from the first of January to the end of May?

55. In January 2000, the value of one share of a certain stock was $15\frac{1}{2}$. During the next three days, the value fell $1\frac{1}{2}$, rose $2\frac{3}{4}$, and fell $3\frac{1}{4}$. What was the value of one share at the end of the three days?

56. In February 2000, the value of one share of a certain stock was $28. During the next three days, the value rose $2\frac{1}{4}$, fell $5\frac{1}{2}$, and fell $1\frac{1}{2}$. What was the value of one share at the end of those three days?

To Think About

57. Write a word problem about a bank and the calculation $50 - (-80) = 50 + 80 = 130$.

58. Write a word problem about a bank and the calculation $-100 - 50 = -100 + (-50) = -150$.

Cumulative Review Problems

In exercises 59–60, perform the operations in the proper order.

59. $20 \times 2 \div 10 + 4 - 3$

60. $5 \times 7 + 6 \times 3 - 8$

▲ **61.** A metalworker is making a copper sign. Find the area of the largest possible copper circle that can be made from a square piece of copper that measures 6 inches on each side. Use $\pi \approx 3.14$.

▲ **62.** A triangular wooden sign is mounted in the laundry room of Dori and Elizabeth Little. The sign reads "Blessed are they who wash the clothes." The piece of wood is 7 inches high and has a base of 11 inches. Dori needs to replace the wood with a larger triangle with the same height but a new base of 14 inches. How much does the area of the piece of wood increase with this change?

9.3 Multiplication and Division of Signed Numbers

① Multiplying and Dividing Two Signed Numbers

Recall the different ways we can indicate multiplication.

$$3 \times 5 \quad 3 \cdot 5 \quad (3)(5) \quad 3(5)$$

It is common to use parentheses to mean multiplication.

EXAMPLE 1 Evaluate.

(a) $(7)(8)$

(b) $3(12)$

(a) $(7)(8) = 56$

(b) $3(12) = 36$

Practice Problem 1 Evaluate.

(a) $(6)(9)$

(b) $7(12)$

Now suppose we multiply a positive number times a negative number. What will happen? Let us look for a pattern.

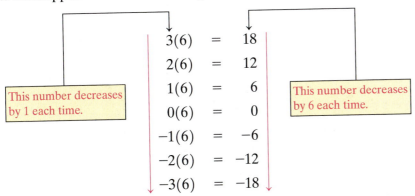

$$
\begin{array}{rcl}
3(6) & = & 18 \\
2(6) & = & 12 \\
1(6) & = & 6 \\
0(6) & = & 0 \\
-1(6) & = & -6 \\
-2(6) & = & -12 \\
-3(6) & = & -18
\end{array}
$$

This number decreases by 1 each time.

This number decreases by 6 each time.

Our pattern suggests that when we multiply a positive number times a negative number, we get a negative number. Thus we will state the following rule.

> **Multiplication Rule for Two Numbers with Different Signs**
>
> To multiply two numbers with different signs, multiply the absolute values. The result is negative.

EXAMPLE 2 Multiply.

(a) $2(-8)$

(b) $(-3)(25)$

In each case, we are multiplying two signed numbers with different signs. We will always get a negative number for an answer.

(a) $2(-8) = -16$

(b) $(-3)(25) = -75$

Practice Problem 2 Multiply.

(a) $(-8)(5)$

(b) $3(-60)$

Student Learning Objectives

After studying this section, you will be able to:

① Multiply and divide two signed numbers.

② Multiply three or more signed numbers.

SSM   PH TUTOR CD & VIDEO MATH PRO   WEB
        CENTER

A similar rule applies to division of two numbers when the signs are not the same.

> ### Division Rule for Two Numbers with Different Signs
>
> To divide two numbers with different signs, divide the absolute values. The result is negative.

⬤ EXAMPLE 3 Divide. **(a)** $-20 \div 5$ **(b)** $36 \div (-18)$

(a) $-20 \div 5 = -4$ **(b)** $36 \div (-18) = -2$

Practice Problem 3 Divide. **(a)** $-50 \div 25$ **(b)** $49 \div (-7)$ ⬤

What happens if we multiply $(-2)(-6)$? What sign will we obtain? Let us once again look for a pattern.

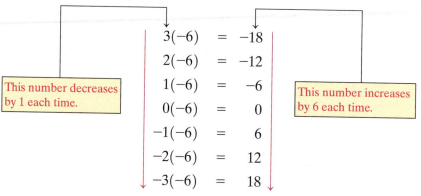

$$
\begin{aligned}
3(-6) &= -18 \\
2(-6) &= -12 \\
1(-6) &= -6 \\
0(-6) &= 0 \\
-1(-6) &= 6 \\
-2(-6) &= 12 \\
-3(-6) &= 18
\end{aligned}
$$

This number decreases by 1 each time.

This number increases by 6 each time.

Our pattern suggests that when we multiply a negative number by a negative number, we get a positive number. A similar pattern occurs for division. Thus we are ready to state the following rule.

> ### Multiplication and Division Rule for Two Numbers with the Same Sign
>
> To multiply or divide two numbers with the same sign, multiply or divide the absolute values. The sign of the result is positive.

⬤ EXAMPLE 4 Multiply.

(a) $-5(-6)$ **(b)** $\left(-\dfrac{1}{2}\right)\left(-\dfrac{3}{5}\right)$

In each case, we are multiplying two numbers with the same sign. We will always obtain a positive number.

(a) $-5(-6) = 30$ **(b)** $\left(-\dfrac{1}{2}\right)\left(-\dfrac{3}{5}\right) = \dfrac{3}{10}$

Practice Problem 4 Multiply.

(a) $-10(-6)$ **(b)** $\left(-\dfrac{1}{3}\right)\left(-\dfrac{2}{7}\right)$ ⬤

Because division is related to multiplication, we find, just as in multiplication, that whenever we divide two numbers with the same sign, the result is a positive number.

● **EXAMPLE 5** Divide.   **(a)** $(-50) \div (-2)$ **(b)** $(-9.9) \div (-3.0)$

(a) $(-50) \div (-2) = 25$ **(b)** $(-9.9) \div (-3.0) = 3.3$

Practice Problem 5 Divide.   **(a)** $-78 \div (-2)$ **(b)** $(-1.2) \div (-0.5)$ ●

---

2 *Multiplying Three or More Signed Numbers*

When multiplying more than two numbers, multiply any two numbers first, then multiply the result by another number. Continue until each factor has been used.

● **EXAMPLE 6**   Multiply. $5(-2)(-3)$

$$5(-2)(-3) = -10(-3) \qquad \text{First multiply } 5(-2) = -10.$$
$$= 30 \qquad \qquad \text{Then multiply } -10(-3) = 30.$$

Practice Problem 6   Multiply. $(-6)(3)(-4)$ ●

● **EXAMPLE 7**   Chemists have determined that a phosphate ion has an electrical charge of -3. If 10 phosphate ions are removed from a substance, what is the change in the charge of the remaining substance?

Removing 10 ions from a substance can be represented by the number -10. We can use multiplication to determine the result of removing 10 ions with an electrical charge of -3.

$$(-3)(-10) = 30$$

Thus the change in charge would be $+30$.

Practice Problem 7   An oxide ion has an electrical charge of -2. If 6 oxide ions are added to a substance, what is the change in the charge of the new substance? ●

● **EXAMPLE 8**   Travis Tobey went outside his house to measure the temperature at 4:00 P.M. for seven days in October in Copper Center, Alaska. His temperature readings in degrees Fahrenheit were $-11°, -8°, -15°, -3°, 5°, 12°$, and $-1°$. Find the average temperature for this seven-day period.

To find the average, we take the sum of the seven days of temperature readings and divide by seven.

$$\frac{-11 + (-8) + (-15) + (-3) + 5 + 12 + (-1)}{7} = \frac{-21}{7} = -3$$

The average temperature was $-3°$F.

Practice Problem 8   In March Tony Pitkin measured the morning temperature at his farm in North Dakota at 7:00 A.M. each day. For a six-day period the temperatures in degrees Fahrenheit were $17°, 19°, 2°, -4°, -3°$, and $-13°$. What was the average temperature at his farm over this six-day period? ●

Verbal and Writing Skills

1. In your own words, state the rule for multiplying two signed numbers if the signs are the same.

2. In your own words, state the rule for multiplying two signed numbers if the signs are different.

Multiply.

3. $(12)(3)$

4. $(15)(5)$

5. $(-20)(-3)$

6. $(-30)(-6)$

7. $(-20)(8)$

8. $(-15)(12)$

9. $(2.5)(-0.6)$

10. $(8.5)(-0.3)$

11. $(-12.5)(-2.25)$

12. $(-7.35)(-10.5)$

13. $\left(-\dfrac{2}{5}\right)\left(\dfrac{3}{7}\right)$

14. $\left(-\dfrac{5}{12}\right)\left(-\dfrac{2}{3}\right)$

15. $\left(-\dfrac{4}{12}\right)\left(-\dfrac{3}{23}\right)$

16. $\left(-\dfrac{7}{9}\right)\left(\dfrac{3}{28}\right)$

Divide.

17. $-64 \div 8$

18. $-63 \div 7$

19. $\dfrac{48}{-6}$

20. $\dfrac{52}{-13}$

21. $\dfrac{-120}{-20}$

22. $\dfrac{-180}{-10}$

23. $\dfrac{3}{10} \div \left(-\dfrac{9}{20}\right)$

24. $\dfrac{12}{25} \div \left(-\dfrac{4}{5}\right)$

25. $\dfrac{-\dfrac{4}{5}}{-\dfrac{7}{10}}$

26. $\dfrac{-\dfrac{26}{15}}{-\dfrac{13}{7}}$

27. $50.28 \div (-6)$

28. $30.45 \div (-5)$

29. $\dfrac{-55.8}{-9}$

30. $\dfrac{-27.3}{-7}$

Multiply.

31. $3(-6)(-4)$

32. $3(-5)(-9)$

33. $(-8)(4)(-6)$

34. $(-7)(-3)(6)$

35. $2(-8)(3)\left(-\dfrac{1}{3}\right)$ **36.** $7(-2)(-5)\left(\dfrac{1}{7}\right)$ **37.** $(-5)(2)(-1)(-3)$ **38.** $(-8)(-2)(3)(-1)$

39. $8(-3)(-5)(0)(-2)$ **40.** $9(-6)(-4)(-3)(0)$

Perform each set of operations. Simplify your answer. Work from left to right.

41. $\left(-\dfrac{2}{3}\right)\left(-\dfrac{3}{4}\right)\left(-\dfrac{5}{6}\right)$ **42.** $\left(-\dfrac{2}{3}\right) \div \left(-\dfrac{2}{3}\right)\left(\dfrac{3}{5}\right)$

Applications

43. Shirley has a checking account. She made six ATM withdrawals last month. Four of those withdrawals were from a bank that charged her $1.25 for each withdrawal. What impact will that have on her checking account?

44. The New England Patriots recently played a game in which they lost three yards on each consecutive play for four plays. What did this do to the position of the football?

45. In Missoula, Montana, Bob recorded the following temperature readings in degrees Fahrenheit at 7 A.M. each morning for eight days in a row: $-12°, -14°, -3°, 5°, 8°, -1°, -10°,$ and $-23°$. What was the average temperature?

46. In Rhinelander, Wisconsin, Nancy recorded the following temperature readings in degrees Fahrenheit at 9 A.M. each morning for seven days in a row: $-8°, -5°, -18°, -22°, -6°, 3°,$ and $7°$. What was the average temperature?

47. An underwater photographer made seven sets of underwater shots. He started at the surface, dropped down ten feet, and shot some pictures. Then he dropped ten more feet and shot more pictures. He continued this pattern until he had dropped seven times. How far beneath the surface was he at that point?

48. A new high-tech company reported losses of $3 million for five quarters in a row. What was the amount of total losses after five quarters?

Ions are atoms or groups of atoms with positive or negative electrical charges. The charges of some ions are given in the following box.

| aluminum +3 | chloride −1 | magnesium +2 |
| oxide −2 | phosphate −3 | silver +1 |

In exercises 49–52, find the total charge.

49. 17 magnesium ions

50. 8 phosphate ions

51. 4 silver ions and 2 chloride ions (*Hint:* multiply first, and then add.)

52. 3 oxide ions and 9 chloride ions (*Hint:* multiply first, and then add.)

53. Eight chloride ions are removed from a substance. What is the change in the charge of the remaining substance?

54. Six oxide ions are removed from a substance. What is the change in the charge of the remaining substance?

A round of golf is nine holes. Pine Hills Golf Course is a par 3 course. This means that the expected number of strokes it takes to complete each hole is 3. The following table indicates points above or below par for each hole. For example, if someone shoots a 5 on one of the holes (that is, takes 5 strokes to complete the hole), this is called a double bogey and is 2 points over par.

Birdie −1	Bogey +1
Eagle −2	Double Bogey +2

55. During a round of golf with his friend, Louis got two birdies, one eagle, and two double bogeys. He made par on the rest of the holes. How many points above or below par for the entire nine-hole course is this? (*Hint*: Multiply first, and then add.)

56. Judy shot three birdies, two bogeys, and one double bogey. On the other three holes, she made par. How many points above or below par for the entire nine-hole course is this? (*Hint*: Multiply first, and then add.)

To Think About

57. Bob multiplied −12 times a mystery number that was not negative and he did not get a negative answer. What was the mystery number?

58. Fred divided a mystery number that was not negative by −3 and did not get a negative answer. What was the mystery number?

In exercises 59 and 60, the letters a, b, c, and d represent numbers that may be either positive or negative. The first three numbers are multiplied to obtain the fourth number. This can be stated by the equation $(a)(b)(c) = d$.

59. In the case where a, c, and d are all negative numbers, is the number b positive or negative?

60. In the case where a is positive but b and d are negative, is the number c positive or negative?

Cumulative Review Problems

▲ **61.** Find the area of a parallelogram with height of 6 inches and base of 15 inches.

▲ **62.** Find the area of a trapezoid with height of 12 meters and bases of 18 meters and 26 meters.

1 **Calculating with Signed Numbers Using More Than One Operation**

Student Learning Objectives

After studying this section, you will be able to:

1 Calculate with signed numbers using more than one operation.

SSM   PH TUTOR   CD & VIDEO   MATH PRO   WEB
        CENTER

The order of operations we discussed for whole numbers applies to signed numbers as well. The rules that we will use is this chapter are listed in the following box.

> **Order of Operations for Signed Numbers**
>
> With grouping symbols:
>
> Do first   **1.** Perform operations inside the parentheses.
>
> **2.** Simplify any expressions with exponents, and find any square roots.
>
> **3.** Multiply or divide from left to right.
>
> Do last   **4.** Add or subtract from left to right.

⬭ EXAMPLE 1 Perform the indicated operations in the proper order.

$$-6 \div (-2)(5)$$

Multiplication and division are of equal priority. So we work from left to right and divide first.

$$\underbrace{-6 \div (-2)}_{3}(5)$$
$$(5) = 15$$

Practice Problem 1 Perform the indicated operations in the proper order.

$$20 \div (-5)(-3)$$

⬭ EXAMPLE 2 Perform the indicated operations in the proper order.

(a) $7 + 6(-2)$ **(b)** $9 \div 3 - 16 \div (-2)$

(a) Multiplication and division must be done first. We begin with $6(-2)$.

$$7 + \underbrace{6(-2)}_{}$$
$$7 + (-12) = -5$$

(b) There is no multiplication, but there is division, and we do that first.

$$\underbrace{9 \div 3}_{3} - \underbrace{16 \div (-2)}_{(-8)} = 3 + 8$$ Transform subtraction to adding
$$= 11$$ the opposite.

Practice Problem 2 Perform the indicated operations in the proper order.

(a) $25 \div (-5) + 16 \div (-8)$ **(b)** $9 + 20 \div (-4)$

583

If a fraction has operations written in the numerator, in the denominator, or both, these operations must be done first. Then the fraction may be simplified or the division carried out.

● **EXAMPLE 3** Perform the indicated operations in the proper order.

$$\frac{7(-2) + 4}{8 \div (-2)(5)}$$

$$\frac{7(-2) + 4}{8 \div (-2)(5)} = \frac{-14 + 4}{(-4)(5)}$$

We perform the multiplication and division first in the numerator and the denominator, respectively.

$$= \frac{-10}{-20}$$

Simplify the fraction.

$$= \frac{1}{2}$$

Note that the answer is positive since a negative number divided by a negative number gives a positive result.

Practice Problem 3 Perform the indicated operations in the proper order.

$$\frac{9(-3) - 5}{2(-4) \div (-2)}$$

●

Some problems involve parentheses and exponents and will require additional steps.

● **EXAMPLE 4** Perform the indicated operations in the proper order.

$$4(6 - 9) + (-2)^3 + 3(-5)$$

$$4(-3) + (-2)^3 + 3(-5)$$

First we combine the numbers inside the parentheses.

$$= 4(-3) + (-8) + 3(-5)$$

Next we simplify the expression with exponents. We obtain $(-2)^3 = (-2)(-2)(-2) = -8$.

$$= -12 + (-8) + (-15)$$

Next we perform each multiplication.

$$= -35$$

Now we add three numbers.

Practice Problem 4 Perform the indicated operations in the proper order.

$$-2(-12 + 15) + (-3)^4 + 2(-6)$$

●

Be sure to use extra caution with problems involving fractions or decimals. It is easy to make an error in finding a common denominator or in placing a decimal point.

● **EXAMPLE 5**   Perform the indicated operations in the proper order.

$$\left(\frac{1}{2}\right)^3 + 2\left(\frac{3}{4} - \frac{3}{8}\right) \div \left(-\frac{3}{5}\right)$$

$$\left(\frac{1}{2}\right)^3 + 2\left(\frac{6}{8} - \frac{3}{8}\right) \div \left(-\frac{3}{5}\right)$$ First find the LCD and write $\dfrac{3}{4} = \dfrac{6}{8}$.

$$= \left(\frac{1}{2}\right)^3 + 2\left(\frac{3}{8}\right) \div \left(-\frac{3}{5}\right)$$ Next we combine the two fractions inside the parentheses.

$$= \frac{1}{8} + 2\left(\frac{3}{8}\right) \div \left(-\frac{3}{5}\right)$$ Next we simplify $\left(\dfrac{1}{2}\right)^3 = \left(\dfrac{1}{2}\right)\left(\dfrac{1}{2}\right)\left(\dfrac{1}{2}\right) = \dfrac{1}{8}$.

$$= \frac{1}{8} + \frac{3}{4} \div \left(-\frac{3}{5}\right)$$ Next multiply: $\left(\dfrac{2}{1}\right)\left(\dfrac{3}{8}\right) = \dfrac{3}{4}$.

$$= \frac{1}{8} + \frac{3}{4} \cdot \left(-\frac{5}{3}\right)$$ To divide two fractions, invert and multiply the second fraction.

$$= \frac{1}{8} + \left(-\frac{5}{4}\right)$$ Multiply $\left(\dfrac{3}{4}\right)\left(-\dfrac{5}{3}\right) = -\dfrac{5}{4}$.

$$= \frac{1}{8} + \left(-\frac{10}{8}\right)$$ Change $-\dfrac{5}{4}$ to the equivalent $-\dfrac{10}{8}$.

$$= -\frac{9}{8}$$ Add the two fractions.

Practice Problem 5   Perform the indicated operations in the proper order.

$$\left(\frac{1}{5}\right)^2 + 4\left(\frac{1}{5} - \frac{3}{10}\right) \div \frac{2}{3}$$

●

Perform the indicated operations in the proper order.

1. $16 + 32 \div (-4)$

2. $15 - (-18) \div 3$

3. $24 \div (-3) + 16 \div (-4)$

4. $(-56) \div (-7) + 30 \div (-15)$

5. $3(-4) + 5(-2) - (-3)$

6. $6(-3) + 8(-1) - (-2)$

7. $-54 \div 6 + 9$

8. $-72 \div 9 + 8$

9. $5 - 30 \div 3$

10. $8 - 70 \div 5$

11. $36 \div 12(-2)$

12. $9(-6) + 6$

13. $3(-4) + 6(-2) - 3$

14. $-6(7) + 8(-3) + 5$

15. $11(-6) - 3(12)$

16. $10(-5) - 4(12)$

17. $16 - 4(8) + 18 \div (-9)$

18. $20 - 3(-2) + (-20) \div (-5)$

In exercises 19–28, simplify the numerator and denominator first, using the proper order of operations. Then reduce the fraction if possible.

19. $\dfrac{8 + 6 - 12}{3 - 6 + 5}$

20. $\dfrac{10 - 5 - 7}{8 + 3 - 9}$

21. $\dfrac{6(-2) + 4}{6 - 3 - 5}$

22. $\dfrac{8 + 2 - 6}{3(-5) + 13}$

23. $\dfrac{-16 \div (-2)}{3(-4) - 4}$

24. $\dfrac{2(-3) - 5 + 1}{10 \div (-5)}$

25. $\dfrac{4(-5) \div 4 + 7}{20 \div (-10)}$

26. $\dfrac{6(-3) \div (-2) + 5}{32 \div (-8)}$

27. $\dfrac{12 \div 3 + (-2)(2)}{9 - 9 \div (-3)}$

28. $\dfrac{16 \div 2 + (4)(-2)}{2 - 2 \div (-2)}$

Perform the operations in the proper order.

29. $3(2 - 6) + 4^2$

30. $-5(8 - 3) + 6^2$

31. $12 \div (-6) + (7 - 2)^3$

32. $-20 \div 2 + (6 - 3)^4$

33. $\dfrac{3}{16} - \dfrac{3}{8} + \dfrac{1}{2}\left(-\dfrac{1}{2}\right)$

34. $-\dfrac{2}{15} - \dfrac{1}{5} + \dfrac{1}{3}\left(\dfrac{4}{5}\right)$

35. $\left(\dfrac{3}{5}\right)^2 - \dfrac{3}{5}\left(\dfrac{1}{3}\right)$

36. $\left(\dfrac{1}{3}\right)^2 - \dfrac{1}{4}\left(\dfrac{3}{5}\right)$

37. $(1.2)^2 - 3.6(-1.5)$

38. $(0.6)^2 - 5.2(-3.4)$

Applications

The following chart gives the average monthly temperatures in Fahrenheit for two cities in Alaska.

	Jan.	Feb.	Mar.	Apr.	May	June	July	Aug.	Sept.	Oct.	Nov.	Dec.
Barrow	−14	−20	−16	−2	19	33	39	38	31	14	−1	−13
Fairbanks	−13	−4	9	30	48	59	62	57	45	25	4	−10

Use the chart to solve. Round to the nearest tenth of a degree.

39. What is the average temperature in Barrow during December, January, and February?

40. What is the average temperature in Fairbanks during December, January, and February?

41. What is the average temperature in Barrow from October through March?

42. What is the average temperature in Fairbanks from October through March?

43. What is the average yearly temperature in Barrow?

44. What is the average yearly temperature in Fairbanks?

45. In Hveragerdi, Iceland, yesterday the average temperature was −12°C. Today the average temperature rose 3°C. What is the new average temperature? If the next four days experience the same rise, what will the temperature be in four days?

46. In January 2000, Jeff Slater was at the research weather station on the top of Mount Washington. One day the low temperature was −15°F. For the next four days, the low temperature dropped 6 degrees each day. What was the low temperature at the end of those four days?

Cumulative Review Problems

47. A telephone wire that is 3840 meters long is how long measured in kilometers?

48. A container with 36.8 grams of protein contains how many milligrams of protein?

▲ **49.** Find the area of a trapezoid with a height of 14 inches and bases of 23 inches and 37 inches.

▲ **50.** Janice receives a lot of mail-order catalogs and junk mail. In June she received 28 catalogs and 48 pieces of junk mail. In July she received exactly twice as many of each. In August she received exactly half of what she received in June. Over the three months, how many catalogs did she receive? In July and August, how many pieces of junk mail did she receive?

9.5 Scientific Notation

Student Learning Objectives

After studying this section, you will be able to:

1. Change numbers in standard notation to scientific notation.

2. Change numbers in scientific notation to standard notation.

3. Add and subtract numbers in scientific notation.

SSM
PH TUTOR CENTER   CD & VIDEO   MATH PRO   WEB

1. Changing Numbers in Standard Notation to Scientific Notation

Scientists who frequently work with very large or very small measurements use a certain way to write numbers, called *scientific notation*. In our usual way of writing a number, which we call "standard notation" or "ordinary form," we would express the distance to the nearest star, Proxima Centauri, as 24,800,000,000,000 miles. In scientific notation, we more conveniently write this as

$$2.48 \times 10^{13} \text{ miles.}$$

For a very small number, like two millionths, the standard notation is

$$0.000002.$$

The same quantity in scientific notation is

$$2 \times 10^{-6}.$$

Notice that each number in scientific notation has two parts: (1) a number that is 1 or greater but less than 10, which is multiplied by (2) a power of 10. That power is either a whole number or the negative of a whole number. In this section we learn how to go back and forth between standard and scientific notation.

> A positive number is in **scientific notation** if it is in the form $a \times 10^n$, where a is a number greater than (or equal to) 1 and less than 10, and n is an integer.

We begin our investigation of scientific notation by looking at large numbers. We recall from Section 1.6 that

$$10 = 10^1$$
$$100 = 10^2$$
$$1000 = 10^3$$

The number of zeros tells us the number for the exponent. To write a number in scientific notation, we want the first number to be greater than or equal to 1 and less than 10, and the second number to be a power of 10.

Let's see how this can be done. Consider the following.

$$6700 = \underbrace{6.7}_{\substack{\text{greater than 1} \\ \text{and less than 10}}} \times 1000 = 6.7 \times \overset{\uparrow}{\underset{\substack{\text{a power} \\ \text{of 10}}}{10^3}}$$

Let us look at two more cases.

$$530 = 5.3 \times 100 \qquad\qquad 156{,}000 = 1.56 \times 100{,}000$$
$$= \underline{5.3 \times 10^2} \qquad\qquad\qquad = \underline{1.56 \times 10^5}$$

These numbers are in scientific notation.

Now that we have seen some examples of numbers in scientific notation, let us think through the steps in the next example.

It is important to remember that all numbers *greater than* 10 always have a positive exponent when expressed in scientific notation.

EXAMPLE 1   Write in scientific notation.

(a) 9826   **(b)** 163,457

(a)

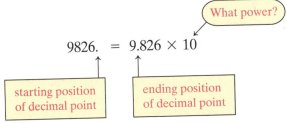

The decimal point moves **3** places to the left. We therefore use **3** for the power of 10.

$$9826 = 9.826 \times 10^3$$

(b)

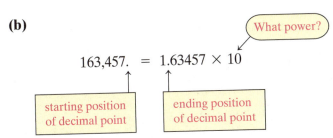

The decimal point moves **5** places to the left. We therefore use **5** for the power of 10.

$$163,457 = 1.63457 \times 10^5$$

Practice Problem 1   Write in scientific notation.

(a) 3729   **(b)** 506,936

When changing to scientific notation, all zeros to the *right* of the final nonzero digit may be eliminated. This does not change the value of your answer.

Now we will look at numbers that are *less than* 1. In the introduction to this section we saw that 0.000002 can be written as 2×10^{-6}. How is it possible to have a negative exponent? Let's take another look at the powers of 10.

$$10^3 = 1000$$
$$10^2 = 100$$
$$10^1 = 10$$
$$10^0 = 1 \qquad \text{Recall that any number to the zero power is 1.}$$

Now, following this pattern, what would you expect the next number on the left of the equal sign to be? What would be the next number on the right of the equal sign? Each number on the right is one-tenth the number above it. We continue the pattern.

$$10^{-1} = 0.1$$
$$10^{-2} = 0.01$$
$$10^{-3} = 0.001$$

Thus you can see how it is possible to have negative exponents. We use negative exponents to write numbers that are less than 1 in scientific notation.

Calculator

Standard to Scientific
Many calculators have a setting that will display numbers in scientific notation. Consult your manual on how to do this. To convert 154.32 into scientific notation, first change your setting to display in scientific notation. Often this is done by pressing [SCI] or [2nd] [SCI]. Often SCI is then displayed on the calculator. Then enter:

154.32 [=]

Display:

| 1.5432   02 |

1.5432   02 means 1.5432×10^2.

Note that your calculator display may show the power of 10 in a different manner. Be sure to change your setting back to the regular display when you are done.

Let's look at 0.76. This number is less than 1. Recall that in our definition for scientific notation the first number must be greater than or equal to 1 and less than 10. To change 0.76 to such a number, we will have to move the decimal point.

$$0.76 = 7.6 \times 10^{-1}$$

The decimal point moves 1 place to the right. We therefore put a -1 for the power of 10. We use a negative exponent because the original number is less than 1. Let's look at two more cases.

$$0.0025 = \underbrace{2.5 \times 10^{-3}} \qquad\qquad 0.00088 = \underbrace{8.8 \times 10^{-4}}$$

These numbers are in scientific notation.

When we start with a number that is less than 1 and write it in scientific notation, we will get a result with 10 to a negative power. Think carefully through the steps of the following example.

EXAMPLE 2 Write in scientific notation.

(a) 0.036 **(b)** 0.72

(a) We change the given number to a number greater than or equal to 1 and less than 10. Thus we change 0.036 to 3.6.

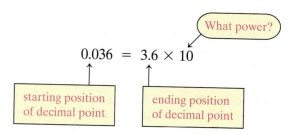

$$0.036 = 3.6 \times 10$$

What power?

starting position of decimal point

ending position of decimal point

The decimal point moved **2** places to the right. Since the original number is less than 1, we use **-2** for the power of 10.

$$0.036 = 3.6 \times 10^{-2}$$

(b)

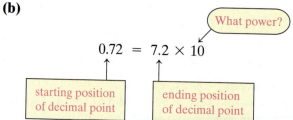

$$0.72 = 7.2 \times 10$$

What power?

starting position of decimal point

ending position of decimal point

The decimal point moved **1** place to the right. Because the original number is less than 1, we use **-1** for the power of 10.

$$0.72 = 7.2 \times 10^{-1}$$

Practice Problem 2 Write in scientific notation.

(a) 0.076 **(b)** 0.982

It is very helpful when writing numbers in scientific notation to remember these two concepts.

> **1.** A number that is larger than 10 and written in scientific notation will always have a positive exponent as the power of 10.
>
> **2.** A number that is smaller than 1 and written in scientific notation will always have a negative exponent as the power of 10.

2 Changing Numbers in Scientific Notation to Standard Notation

Often we are given a number in scientific notation and want to write it in standard notation—that is, we want to write it in what we consider "ordinary form." To do this, we reverse the process we've just used. If the number in scientific notation has a positive power of 10, we know the number is greater than 10. Therefore, we move the decimal point to the right to convert to standard notation.

EXAMPLE 3 Write in standard notation.

5.8671×10^4

$5.8671 \times 10^4 = 58,671$ Move the decimal point four places to the right.

Practice Problem 3 Write in standard notation.

$$6.543 \times 10^3$$

In some cases, we will need to add zeros as we move the decimal point to the right.

EXAMPLE 4 Write in standard notation.

(a) 9.8×10^5 **(b)** 3×10^3

(a) $9.8 \times 10^5 = 980,000$ Move the decimal point **five** places to the right.
                              Add four zeros.

(b) $3 \times 10^3 = 3000$ Move the decimal point **three** places to the right.
                          Add three zeros.

Practice Problem 4 Write in standard notation.

(a) 4.3×10^5 **(b)** 6×10^4

If the number in scientific notation has a negative power of 10, we know the number is less than 1. Therefore, we move the decimal point to the left to convert to standard notation. In some cases, we need to add zeros as we move the decimal point to the left.

EXAMPLE 5 Write in standard notation.

(a) 2.48×10^{-3} **(b)** 1.2×10^{-4}

(a) $2.48 \times 10^{-3} = 0.00248$ Move the decimal point **three** places to the left.
                                 Add two zeros between the decimal point and 2.

(b) $1.2 \times 10^{-4} = 0.00012$ Move the decimal point **four** places to the left.
                                Add three zeros between the decimal point and 1.

Practice Problem 5 Write in standard notation.

(a) 7.72×10^{-3} **(b)** 2.6×10^{-5}

Adding and Subtracting Numbers in Scientific Notation

Scientists calculate with numbers in scientific notation when they study galaxies (the huge) and when they study microbes (the tiny). To add or subtract numbers in scientific notation, the numbers must have the same power of 10.

> Numbers in scientific notation may be added or subtracted if they have the same power of 10. We add or subtract the decimal part and leave the power of 10 unchanged.

EXAMPLE 6   Add. 5.89×10^{-20} meters + 3.04×10^{-20} meters

$$\begin{array}{r} 5.89 \times 10^{-20} \text{ meters} \\ + 3.04 \times 10^{-20} \text{ meters} \\ \hline 8.93 \times 10^{-20} \text{ meters} \end{array}$$

Practice Problem 6   Add. 6.85×10^{22} kilograms + 2.09×10^{22} kilograms

EXAMPLE 7   Add. $7.2 \times 10^6 + 5.2 \times 10^5$

Note that these two numbers have different powers of ten, so we cannot add them as written. We will change one number from scientific notation to another form.

Now $5.2 \times 10^5 = 520,000$ can be rewritten as 0.52×10^6. Now we can add.

$$\begin{array}{r} 7.2 \times 10^6 \\ + 0.52 \times 10^6 \\ \hline 7.72 \times 10^6 \end{array}$$

Practice Problem 7   Subtract. $4.36 \times 10^5 - 3.1 \times 10^4$

Write in scientific notation.

1. 120   **2.** 340   **3.** 10,000   **4.** 100,000   **5.** 26,300   **6.** 78,100

7. 288,000   **8.** 238,000   **9.** 4,632,000   **10.** 2,034,000   **11.** 12,000,000   **12.** 28,000,000

13. 0.0931   **14.** 0.0242   **15.** 0.00279   **16.** 0.00613   **17.** 0.4   **18.** 0.9

19. 0.000016   **20.** 0.000072   **21.** 0.00000531   **22.** 0.00000198

23. 0.00000018   **24.** 0.000000027

Write in standard notation.

25. 5.36×10^4 **26.** 2.19×10^4 **27.** 6.2×10^{-2} **28.** 3.5×10^{-2} **29.** 3.71×10^{-1}

30. 7.44×10^{-4} **31.** 9×10^{11} **32.** 2×10^{12} **33.** 3.862×10^{-8} **34.** 8.139×10^{-9}

35. 4.6×10^{12} **36.** 3.8×10^{11} **37.** 6.721×10^{10} **38.** 4.039×10^9

Applications

Write in scientific notation.

39. In 1 year, light will travel 5,878,000,000,000 miles.

40. The world's forests total 2,700,000,000 acres of wooded area.

41. On July 19, 1999, the world population reached 6,000,000,000.

42. The flowing duckweed of Australia has a small fruit that weighs 0.0000024 ounce.

Write in standard notation.

43. During the period August 15–25, 1875, a huge swarm of Rocky Mountain locusts invaded Nebraska. The swarm contained an estimated 1.25×10^{13} insects.

44. The gross national product (GNP) of the United States in 1999 was 6.8681×10^{12} dollars.

▲ 45. The diameter of a red corpuscle of human blood is about 7.5×10^{-5} centimeter.

▲ 46. Recently a tiny hole with an approximate diameter of 3.16×10^{-10} meter was produced by Doctors Heckl and Maddocks using a chemical method involving a mercury drill.

47. During the eruption of the Taupo volcano in New Zealand in 130 A.D., approximately 1.4×10^{10} tons of pumice were carried up in the air.

48. The number of grains of sand it would take to fill a sphere the size of the Earth is 1×10^{23}

Add.

49. 3.38×10^7 dollars $+ 5.63 \times 10^7$ dollars

50. 8.17×10^9 atoms $+ 2.76 \times 10^9$ atoms

51. The mass of the Earth is 5.87×10^{21} tons, and the mass of Venus is 4.81×10^{21} tons. What is the total mass of the two planets?

52. The masses of Mars, Pluto, and Mercury are 6.34×10^{20} tons, 1.76×10^{20} tons, and 3.21×10^{20} tons, respectively. What is the total mass of the three planets?

Subtract.

53. 4×10^8 feet $- 3.76 \times 10^7$ feet

54. 9×10^{10} meters $- 1.26 \times 10^9$ meters

▲ **55.** The two largest continents are Asia and Africa with areas of 1.76×10^7 square miles and 1.16×10^7 square miles, respectively. How much larger is Asia than Africa?

▲ **56.** The area of Canada is 3.852×10^6 square miles and the area of the United States is 3.619×10^6 square miles. How much larger is Canada than the United States?

To Think About

57. Recently astronomers have discovered evidence of a planet named Epsilon Eridani b. This planet is in a different solar system and is 10.5 light-years from Earth. A light-year is approximately 5.88 trillion miles. Find in scientific notation how many miles Epsilon Eridani b is from Earth.

58. Recently astronomers have discovered evidence of a planet named Cancri c. This planet is in a different solar system and is 41 light-years from Earth. A light-year is approximately 5.88 trillion miles. Find in scientific notation how many miles Cancri c is from Earth.

Cumulative Review Problems

Calculate.

59. 7.63×2.18

60. $0.53\overline{)0.13674}$

▲ **61.** A radar mounting plate is constructed in the shape of a parallelogram. Two sides of the parallelogram are 9 feet and 13 feet. A rubber piece of safety bumper must be placed around all sides of the parallelogram. It costs $4 per foot. How much will it cost to place the safety bumper around the plate?

62. Michael is mountain climbing in the Rocky Mountains. He is 70 inches tall. At 2 P.M. his shadow measures 95 inches. He wants to climb to the top of a cliff that casts a shadow of 800 feet at 2 P.M. How tall is the cliff? Round to the nearest whole number.

Putting Your Skills to Work

Using Cat Litter to Clean Up Nuclear Waste

The West Valley nuclear power plant in New York was closed in 1972 after only six years of operation. The site of the nuclear power plant is contaminated. Over 30 million curies of radioactive strontium 90 remain at the site. (For comparison, the accident at the Three Mile Island nuclear plant released about 50 curies into the air.) About one-third of this radioactive waste is contained in a 700-foot-long amoeba-shaped underground plume of water. This slow-moving underground stream of radioactive liquid is flowing downhill from the old nuclear power plant.

Now it appears that help is available from the use of zeolite. Zeolite is a compound containing 48 minerals and is commonly used as cat litter as well as in a number of industrial applications. Zeolite contains clinoptilolite, a substance that can attract and retain strontium 90. After small tests with this material were successful at nuclear power sites in Washington State as well as Chalk River, Ontario, Canada, scientists began a large scale effort to use zeolite to combat nuclear contamination at the West Valley site in New York.

Scientists are now designing a porous mineral wall containing zeolite. The wall will be 30 feet across and 26 feet deep. It will act as a giant molecular sieve, letting water flow downhill but capturing any radioactive strontium 90 molecules.

Problems for Individual Investigation and Analysis

1. Assume that a specially constructed wall three feet thick made up of zeolite is able to absorb 1×10^{-6} curies per liter of water that passes through it. How many feet thick would the wall need to be to absorb 1×10^{-4} curies per liter?

2. Again, assume that the wall is three feet thick and absorbs 1×10^{-6} curies per liter of water. If an area of the wall measuring 1 foot by 1 foot can filter one liter of water each minute, how many liters of water could be filtered each minute through the entire wall measuring 30 feet across and 26 feet deep?

Problems for Group Investigation and Cooperative Problem Solving

3. Suppose a 1-foot-by-1-foot section of a 3-foot-thick wall filters 1×10^{-6} curies per liter of water per minute. How many liters of water would pass through this section of wall in one year? How many curies of strontium 90 would be filtered out in one year by this 1-foot-by-1-foot section?

4. Using the answer from question 3, determine how many curies of strontium 90 would be filtered out in five years by the entire wall measuring 30 feet across and 26 feet deep? How much would the entire wall filter out in five years if the wall were 300 feet thick instead of 3 feet thick?

Internet Connections

 Netsite: http://www.prenhall.com/tobey_basic

This site contains information about strontium 90.

5. Determine the half-life of strontium 90. (Round to the nearest year.) If 1300 kilograms of strontium 90 were left undisturbed at a nuclear power plant, how much of it would still be there after 112 years?

6. If 2500 kilograms of strontium 90 were left undisturbed at a nuclear power plant, how much of it would still be there after 196 years?

Math in the Media

Wind Chill: A Rough Guide to Danger

USA TODAY, February 08, 2000

Copyright 2000, USA TODAY. Reprinted with permission.

"As temperatures fall and the wind howls, we begin hearing about the danger of 'wind chill.' The wind chill index, shown below, combines the temperature and wind speed to tell you how cold the wind makes it 'feel.' Even though the chill is given as a temperature, it's not really a different kind of temperature. it's based on polar research going back to the early 1940s and is meant to be a guide to how much you should bundle up when going outside."

In the table below the wind chill for a wind speed of 10 miles/hour and 20 degrees Fahrenheit is the cell which is the intersect of the row containing 10 miles/hour and the column containing 20 degrees Fahrenheit which is 3 degrees. All other wind chill factor are intersections of a row and a column.

Wind speed [mpH]	AIR TEMPERATURE [Degrees Fahrenheit]										
		35	30	25	20	15	10	5	0	−5	−10	−15
4	35	30	25	20	15	10	5	0	−5	−10	−15
5	32	27	22	16	11	6	0	−5	−10	−15	−21
10	22	16	10	3	−3	−9	−15	−22	−27	−34	−41
15	16	9	2	−5	−11	−18	−25	−31	−38	−45	−51
20	12	4	−3	−10	−17	−24	−31	−39	−46	−53	−60
25	8	1	−7	−15	−22	−29	−36	−44	−51	−59	−66

EXERCISES

Refer to the table to answer questions 1–3.

1. Find the wind chill for a wind speed of 20 miles/hour and a Fahrenheit temperature of 0 degrees.

2. Find the wind chill for a wind speed of 5 miles/hour and a Fahrenheit temperature of −10 degrees.

3. Fahrenheit freezing is at 32 degrees while Celsius freezing is at 0 degrees. If the table above was to be rewritten using the Celsius scale for temperature, what changes would you notice?

Chapter 9 Organizer

Topic	Procedure	Examples
Absolute value, p. 561.	The absolute value of a number is the distance between the number and zero on the number line.	$\|-6\| = 6 \qquad \|3\| = 3 \qquad \|0\| = 0$
Adding signed numbers with the same sign, p. 562.	To add two numbers with the same sign: 1. Add the absolute value of the numbers. 2. Use the common sign in the answer.	$12 + 5 = 17$ $-6 + (-8) = -14$ $-5.2 + (-3.5) = -8.7$ $-\dfrac{1}{7} + \left(-\dfrac{3}{7}\right) = -\dfrac{4}{7}$
Adding signed numbers with different signs, p. 565.	To add two numbers with different signs: 1. Subtract the absolute value of the numbers. 2. Use the sign of the number with the larger absolute value.	$14 + (-8) = 6$ $-4 + 8 = 4$ $-3.2 + 7.1 = 3.9$ $\dfrac{5}{13} + \left(-\dfrac{8}{13}\right) = -\dfrac{3}{13}$
Subtracting signed numbers, p. 572.	To subtract signed numbers, add the opposite of the second number to the first number.	$-9 - (-3) = -9 + 3 = -6$ $5 - (-7) = 5 + 7 = 12$ $8 - 12 = 8 + (-12) = -4$ $-4 - 13 = -4 + (-13) = -17$ $-\dfrac{1}{12} - \left(-\dfrac{5}{12}\right) = -\dfrac{1}{12} + \dfrac{5}{12} = \dfrac{4}{12} = \dfrac{1}{3}$
Multiplying or dividing signed numbers with the same sign, p. 578.	To multiply or divide two numbers with the same sign, multiply or divide the absolute values. The sign of the result is positive.	$(0.5)(0.3) = 0.15$ $(-6)(-2) = 12$ $\dfrac{-20}{-2} = 10$ $\dfrac{-\dfrac{1}{3}}{-\dfrac{1}{7}} = \left(-\dfrac{1}{3}\right)\left(-\dfrac{7}{1}\right) = \dfrac{7}{3}$
Multiplying or dividing signed numbers with different signs, p. 577.	To multiply or divide two numbers with different signs, multiply or divide the absolute values. The result is negative.	$7(-3) = -21$ $(-6)(4) = -24$ $(-36) \div (2) = -18$ $\dfrac{41.6}{-8} = -5.2$
Order of operations for signed numbers, p. 583.	With grouping symbols: Do first 1. Perform operations inside the parentheses. 2. Simplify any expressions with exponents, and find any square roots. 3. Multiply or divide from left to right. Do last 4. Add or subtract from left to right.	Perform the operations in the proper order. $-2(12 - 8) + (-3)^3 + 4(-6)$ $= -2(4) + (-3)^3 + 4(-6)$ $= -2(4) + (-27) + 4(-6)$ $= -8 + (-27) + (-24)$ $= -59$

Topic	Procedure	Examples
Simplifying fractions with combined operations in numerator and denominator, p. 584.	1. Perform the operations in the numerator. 2. Perform the operations in the denominator. 3. Simplify the fraction.	Perform the operations in the proper order. $$\frac{7(-4) - (-2)}{8 - (-5)}$$ The numerator is $$7(-4) - (-2) = -28 - (-2) = -26.$$ The denominator is $$8 - (-5) = 8 + 5 = 13.$$ Thus the fraction becomes $-\dfrac{26}{13} = -2.$
Writing a number in scientific notation, p. 588.	1. Move the decimal point to the position immediately to the right of the first nonzero digit. 2. Count the number of decimal places you moved the decimal point. 3. Multiply the result by a power of 10 equal to the number of places moved. If the number you started with is larger than 10, use a positive exponent. If the number you started with is less than 1, use a negative exponent.	Write in scientific notation. **(a)** 178 **(b)** 25,000,000 **(c)** 0.006 **(d)** 0.00001732 **(a)** $178 = 1.78 \times 10^2$ **(b)** $25{,}000{,}000 = 2.5 \times 10^7$ **(c)** $0.006 = 6 \times 10^{-3}$ **(d)** $0.00001732 = 1.732 \times 10^{-5}$
Changing from scientific notation to standard notation, p. 591.	1. If the exponent of 10 is *positive*, move the decimal point to the *right* as many places as the exponent shows. Insert extra zeros as necessary. 2. If the exponent of 10 is *negative*, move the decimal point to the *left* as many places as the exponent shows. *Note:* Remember that numbers in scientific notation that have positive exponents are always greater than 10. Numbers in scientific notation that have negative exponents are always less than 1.	Write in standard notation. **(a)** 8×10^6 **(b)** 1.23×10^4 **(c)** 7×10^{-2} **(d)** 8.45×10^{-5} **(a)** $8 \times 10^6 = 8{,}000{,}000$ **(b)** $1.23 \times 10^4 = 12{,}300$ **(c)** $7 \times 10^{-2} = 0.07$ **(d)** $8.45 \times 10^{-5} = 0.0000845$

Chapter 9 Review Problems

9.1 *Add.*

1. $-20 + 5$

2. $-18 + 4$

3. $-3.6 + (-5.2)$

4. $-7.6 + (-1.2)$

5. $-\dfrac{1}{5} + \left(-\dfrac{1}{3}\right)$

6. $-\dfrac{2}{7} + \dfrac{5}{14}$

7. $20 + (-14)$

8. $-80 + 60$

9. $7 + (-2) + 9 + (-3)$

10. $6 + (-3) + 8 + (-9)$

9.2   *Subtract.*

11. $-36 - (-21)$ **12.** $-21 - (-28)$ **13.** $12 - (-7)$ **14.** $14 - (-3)$

15. $1.6 - 3.2$ **16.** $-5.2 - 7.1$ **17.** $-\dfrac{2}{5} - \left(-\dfrac{1}{3}\right)$ **18.** $\dfrac{1}{6} - \left(-\dfrac{5}{12}\right)$

Perform the indicated operations from left to right.

19. $5 - (-2) - (-6)$ **20.** $-15 - (-3) + 9$ **21.** $9 - 8 - 6 - 4$ **22.** $-7 - 8 - (-3)$

9.3   *Multiply or divide.*

23. $\left(-\dfrac{2}{7}\right)\left(-\dfrac{1}{5}\right)$ **24.** $\left(-\dfrac{6}{15}\right)\left(\dfrac{5}{12}\right)$ **25.** $(5.2)(-1.5)$ **26.** $(-3.6)(-1.2)$ **27.** $-60 \div (-20)$

28. $-18 \div (-3)$ **29.** $\dfrac{-36}{4}$ **30.** $\dfrac{-60}{12}$ **31.** $\dfrac{-13.2}{-2.2}$ **32.** $\dfrac{48}{-3.2}$

33. $\dfrac{-\dfrac{2}{5}}{\dfrac{4}{7}}$ **34.** $\dfrac{-\dfrac{1}{3}}{-\dfrac{7}{9}}$ **35.** $3(-5)(-2)$ **36.** $6(-8)(-1)$

9.4   *Perform the indicated operations in the proper order.*

37. $8 - (-30) \div 6$ **38.** $26 + (-28) \div 4$ **39.** $2(-6) + 3(-4) - (-13)$

40. $-49 \div (-7) + 3(-2)$ **41.** $36 \div (-12) + 50 \div (-25)$ **42.** $21 - (-30) \div 15$

43. $50 \div 25(-4)$ **44.** $-80 \div 20(-3)$ **45.** $9(-9) + 9$

In exercises 46–49, simplify the numerator and denominator first, using the proper order of operations. Then reduce the fraction if possible.

46. $\dfrac{5 - 9 + 2}{3 - 5}$ **47.** $\dfrac{4(-6) + 8 - 2}{15 - 7 + 2}$ **48.** $\dfrac{20 \div (-5) - (-6)}{(2)(-2)(-5)}$ **49.** $\dfrac{6 - (-3) - 2}{5 - 2 \div (-1)}$

Perform the operations in the proper order.

50. $-3 + 4(2 - 6)^2 \div (-2)$ 　　　**51.** $-6(12 - 15) + 2^3$ 　　　**52.** $-50 \div (-10) + (5 - 3)^4$

53. $\dfrac{1}{2} - \dfrac{1}{6} + \dfrac{2}{3}\left(\dfrac{3}{7}\right)$ 　　　**54.** $\left(\dfrac{2}{3}\right)^2 - \dfrac{3}{8}\left(\dfrac{8}{5}\right)$ 　　　**55.** $(0.8)^2 - 3.2(-1.6)$

56. $1.4(4.7 - 4.9) - 12.8 \div (-0.2)$

9.5 *Write in scientific notation.*

57. 4160 　　　**58.** 3,700,000 　　　**59.** 218,000

60. 0.007 　　　**61.** 0.0000218 　　　**62.** 0.00000763

Write in standard notation.

63. 1.89×10^4 　　　**64.** 3.76×10^3 　　　**65.** 7.52×10^{-2} 　　　**66.** 6.61×10^{-3}

67. 9×10^{-7} 　　　**68.** 8×10^{-8} 　　　**69.** 5.36×10^{-4} 　　　**70.** 1.98×10^{-5}

Add or subtract. Express your answer in scientific notation.

71. $5.26 \times 10^{11} + 3.18 \times 10^{11}$ 　　　**72.** $7.79 \times 10^{15} + 1.93 \times 10^{15}$

73. $3.42 \times 10^{14} - 1.98 \times 10^{14}$ 　　　**74.** $1.76 \times 10^{26} - 1.08 \times 10^{26}$

75. There are 123,120,000,000,000 drops of water in a river 12 kilometers long, 270 meters wide, and 38 meters deep. Write this number in scientific notation.

76. The mass of the Earth is approximately 5,983,000,000,000,000,000,000,000 kilograms. Since writing down all of these zeros is not fun, write this number in scientific notation. This can also be characterized as 5983 Yg (yottagrams).

77. How many feet are in 2,500,000,000 miles? Write this number in scientific notation.

78. We know that the prefix *kilo-* means 1000, or 10^3. The prefix *zepto-* means 0.000000000000000000001. Write this number in scientific notation.

79. The mass of a proton is about 1.67 yg (yoctograms), and the mass of an electron is about 0.00091 yg. If *yocto-* means 0.0000000000000000000000001, write the mass of a proton and that of an electron in grams in scientific notation.

80. The prefix *femto-* means 10^{-15}. Write this number in standard form.

81. The average distance to the moon is 384.4 Mm (megameters). If a megameter is equal to 10^6 meters, write out the number in meters.

82. In three plays of a football game, the quarterback threw passes that lost 5 yards, gained 6 yards, and lost 7 yards. What was the total gain or loss of the three plays?

83. Find the difference in height between the top of a hill 785 feet high and a crack caused by an earthquake 98 feet below sea level.

84. Max has overdrawn his checking account by $18. His bank charged him $20 for an overdraft fee. He quickly deposited $40. What is his current balance?

85. In Duluth, Minnesota, the high temperature in degrees Fahrenheit for five days during January was $-16°$, $-18°$, $-5°$, $3°$, and $-12°$. What was the average high temperature for these five days?

86. Frank played golf with his friend Samuel. They played nine holes of golf on a special practice course. The expected number of strokes for each hole is 3. A birdie is 1 below par. An eagle is 2 below par. A bogie is 1 above par. A double bogey is 2 above par. Frank played on par for 1 hole and got two birdies, one eagle, four bogies, and one double bogey on the rest of the course. How many points above or below par was Frank on this nine-hole course?

Chapter 9 Test

1. _____

2. _____

3. _____

4. _____

5. _____

6. _____

7. _____

8. _____

9. _____

10. _____

11. _____

12. _____

13. _____

14. _____

15. _____

16. _____

17. _____

18. _____

19. _____

20. _____

21. _____

22. _____

Add.

1. $-26 + 15$

2. $-31 + (-12)$

3. $12.8 + (-8.9)$

4. $-3 + (-6) + 7 + (-4)$

5. $10 + (-7) + 3 + (-9)$

6. $-\dfrac{1}{4} + \left(-\dfrac{5}{8}\right)$

Subtract.

7. $-32 - 6$

8. $23 - 18$

9. $\dfrac{4}{5} - \left(-\dfrac{1}{3}\right)$

10. $-50 - (-7)$

11. $-2.5 - (-6.5)$

12. $4.8 - 2.7$

13. $\dfrac{1}{12} - \left(-\dfrac{5}{6}\right)$

14. $23 - (-23)$

Multiply or divide.

15. $(-20)(-6)$

16. $27 \div \left(-\dfrac{3}{4}\right)$

17. $-40 \div (-4)$

Multiply or divide.

18. $(-9)(-1)(-2)(4)\left(\dfrac{1}{4}\right)$

19. $\dfrac{-39}{-13}$

20. $\dfrac{-\dfrac{3}{5}}{\dfrac{6}{7}}$

21. $(-7)(-2)(4)$

22. $96 \div (-3)$

Perform the indicated operations in the proper order.

23. $7 - 2(-5)$

24. $(-42) \div (-7) + 8$

25. $18 \div (-3) + 24 \div (-12)$

26. $8(-5) + 6(-8 - 6 + 12)$

27. $1.3 - 9.5 - (-2.5) + 3(-0.5)$

28. $-48 \div (-6) - 7(-2)^2$

29. $\dfrac{3 + 8 - 5}{(-4)(6) + (-6)(3)}$

30. $\dfrac{5 + 28 \div (-4)}{7 - (-5)}$

Write in scientific notation.

31. $80{,}540$

32. 0.000007

Write in standard notation.

33. 9.36×10^{-5}

34. 7.2×10^4

35. In Chicago the high temperatures in degrees Fahrenheit for five days during February were $-14°, -8°, -5°, 7°$, and $-11°$. What was the average high temperature for these five days?

▲ **36.** A rectangular computer chip is 5.8×10^{-5} meter wide and 7.8×10^{-5} meter long. Find the perimeter of the chip and express your answer in scientific notation.

23. _____

24. _____

25. _____

26. _____

27. _____

28. _____

29. _____

30. _____

31. _____

32. _____

33. _____

34. _____

35. _____

36. _____

Approximately one-half of this test is based on Chapter 9 material. The remainder is based on material covered in Chapters 1–8.

Solve. Simplify your answers.

1. Subtract.

$$28{,}981 \\ -\,16{,}598$$

2. Divide. $36\overline{)4572}$

3. Add. $3\dfrac{1}{4} + 8\dfrac{2}{3}$

4. Multiply. $1\dfrac{5}{6} \times 2\dfrac{1}{2}$

5. Round to the nearest thousandth. 9.812456

6. Add. $5.82 + 38.964 + 0.571 + 9.305 + 8.8$

7. Multiply. 12.89×5.12

8. Find n. $\dfrac{n}{8} = \dfrac{56}{7}$

9. For every 156 parts manufactured, there are seven defects. If 2808 parts are manufactured, how many defects would you expect?

10. What is 0.8% of 38?

11. 12% of what number is 480?

12. Convert 94 kilometers to meters.

13. Convert 180 inches to yards.

▲ **14.** Find the area of a circle with radius 5 meters. Round to the nearest tenth.

15. The following histogram depicts the ages of students at Wolfville College.
 (a) How many students are between ages 23 and 25?
 (b) How many students are older than 19 years?

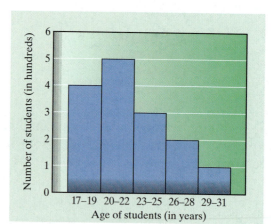

16. Evaluate. $\sqrt{36} + \sqrt{49}$

1.

2.

3.

4.

5.

6.

7.

8.

9.

10.

11.

12.

13.

14.

15. (a)

(b)

16.

Add.

17. $-1.2 + (-3.5)$

18. $-\dfrac{1}{4} + \dfrac{2}{3}$

Subtract.

19. $7 - 18$

20. $-8 - (-3)$

Multiply or divide.

21. $(5)(-3)(-1)(-2)(2)$

22. $\dfrac{-\dfrac{3}{7}}{-\dfrac{5}{14}}$

Perform the indicated operations in the proper order.

23. $6 - 3(-4)$

24. $(-20) \div (-2) + (-6)$

25. $\dfrac{(-2)(-1) + (-4)(-3)}{1 + (-4)(2)}$

26. $\dfrac{(-11)(-8) \div 22}{1 - 7(-2)}$

Write in scientific notation.

27. $579{,}863$

28. 0.00078

Write in standard notation.

29. 3.85×10^{7}

30. 7×10^{-5}

17. _____

18. _____

19. _____

20. _____

21. _____

22. _____

23. _____

24. _____

25. _____

26. _____

27. _____

28. _____

29. _____

30. _____

Introduction to Algebra

Students in college often wonder if taking classes and studying hard is worth all the effort. Will I be more successful in my career? Will I be more likely to earn a greater income if I obtain a bachelor's degree? If I am a woman, am I less likely to experience sex discrimination in my job if I have a bachelor's degree? The U.S. Census Bureau has produced some information on these topics that you may find quite interesting. Turn to the Putting Your Skills To Work problems on page 637 and see if you can answer some thought-provoking questions on this subject.

Pretest Chapter 10

1. _____

2. _____

3. _____

4. _____

5. _____

6. _____

7. _____

8. _____

9. _____

10. _____

11. _____

12. _____

13. _____

14. _____

15. _____

16. _____

17. _____

18. _____

19. _____

20. _____

If you are familiar with the topics in this chapter, take this test now. Check your answers with those in the back of the book. If an answer is wrong or you can't answer a question, study the appropriate section of the chapter.

If you are not familiar with the topics in this chapter, don't take this test now. Instead, study the examples, work the practice problems, and then take the test.

This test will help you identify which concepts you have mastered and which you need to study further.

Section 10.1

Combine like terms.

1. $23x - 40x$

2. $-8y + 12y - 3y$

3. $6a - 5b - 9a + 7b$

4. $5x - y + 2 - 17x - 3y + 8$

5. $7x - 14 + 5y + 8 - 7y + 9x$

6. $4a - 7b + 3c - 5b$

Section 10.2

Simplify.

7. $(6)(7x - 3y)$

8. $(-3)(a + 5b - 1)$

9. $(-2)(1.5a + 3b - 6c - 5)$

10. $(5)(2x - y) - (3)(3x + y)$

Section 10.3–Section 10.5

Solve for the variable.

11. $5 + x = 42$

12. $x + 2.5 = 6$

13. $x - \dfrac{5}{8} = \dfrac{1}{4}$

14. $7x = -56$

15. $5.4x = 27$

16. $\dfrac{3}{5}x = \dfrac{9}{10}$

17. $5x - 9 = 26$

18. $12 - 3x = 7x - 4$

19. $5(x - 1) = 7 - 3(x - 4)$

20. $3x + 7 = 5(5 - x)$

608

Section 10.6

Translate the English sentence into an equation.

21. The computer weighs 9 kilograms more than the printer. Use c to represent the weight of the computer and p to represent the weight of the printer.

▲ **22.** The length of the rectangle is 5 meters longer than double the width. Use l to represent the length and w to represent the width of the rectangle.

Write algebraic expressions for each of the specified quantities using the same variable.

23. Mount Hood is 1758 meters shorter than Mount Ararat in Turkey. Use the variable a to describe the height of each mountain in meters.

Section 10.7

Solve using an equation.

▲ **24.** A rectangular field has a perimeter of 108 meters. The length is 3 meters less than double the width. Find the dimensions of the field.

25. An 18-foot board is cut into two pieces. One piece is 2.5 foot longer than the other. Find the length of each piece.

10.1 Variables and Like Terms

Student Learning Objectives

After studying this section, you will be able to:

1. Recognize the variable in an equation or a formula.

2. Combine like terms containing a variable.

SSM PH TUTOR   CD & VIDEO   MATH PRO   WEB
CENTER

1 Recognizing the Variable in an Equation or a Formula

In algebra we reason and solve problems by means of symbols. A **variable** is a symbol, usually a letter of the alphabet, that stands for a number. We can use the variable even though we may not know what number the variable stands for. We can find that number by following a logical order of steps. These are the rules of algebra.

We begin by taking a closer look at variables. In the formula for the area of a circle, the equation $A = \pi r^2$ contains two variables. r represents the value of the radius. A represents the value of the area. π is a known value. We often use the decimal approximation 3.14 for π.

▲ ● **EXAMPLE 1** Name the variables in each equation.

(a) $A = lw$ **(b)** $V = \dfrac{4\pi r^3}{3}$

(a) $A = lw$ The variables are A, l, and w.

(b) $V = \dfrac{4\pi r^3}{3}$ The variables are V and r.

▲ **Practice Problem 1** Name the variables.

(a) $A = \dfrac{bh}{2}$ **(b)** $V = lwh$ ●

We have seen various ways to write multiplication. For example, three times n can be written as $3 \times n$. In algebra we usually do not write the multiplication symbol. We can simply write $3n$ to mean three times n. A number just to the left of a variable indicates that the number is multiplied by the variable. Thus $4ab$ means 4 times a times b.

A number just to the left of a set of parentheses also means multiplication. Thus $3(n + 8)$ means $3 \times (n + 8)$. So a product can be written with or without a multiplication sign.

▲ ● **EXAMPLE 2** Write the formula without a multiplication sign.

(a) $V = \dfrac{B \times h}{3}$ **(b)** $A = \dfrac{h \times (B + b)}{2}$

(a) $V = \dfrac{Bh}{3}$ **(b)** $A = \dfrac{h(B + b)}{2}$

▲ **Practice Problem 2** Write the formula without a multiplication sign.

(a) $p = 2 \times w + 2 \times l$ **(b)** $A = \pi \times r^2$ ●

2   *Combining Like Terms Containing a Variable*

Recall that when we work with measurements we combine like quantities. A carpenter, for example, might perform the following calculations.

$$20 \text{ m} - 3 \text{ m} = 17 \text{ m}$$

$$7 \text{ in.} + 9 \text{ in.} = 16 \text{ in.}$$

We cannot combine quantities that are not the same. We cannot add 5 yd + 7 gal. We cannot subtract 12 lb − 3 in.

Similarly, when using variables, we can add or subtract only when the same variable is used. For example, we can add $4a + 5a = 9a$, but we cannot add $4a + 5b$.

A **term** is a number, a variable, or a product of a number and one or more variables separated from other terms in an expression by a + sign or a − sign. In the expression $2x + 4y + (-1)$ there are three terms: $2x$, $4y$, and −1. **Like terms** have identical variables and identical exponents, so in the expression $3x + 4y + (-2x)$, the two terms $3x$ and $(-2x)$ are called *like terms*. The terms $2x^2y$ and $5xy$ are not like terms, since the exponents for the variable x are not the same. To combine like terms, you combine the numbers, called the **numerical coefficients**, that are directly in front of the terms by using the rules for adding signed numbers. Then you use this new number as the coefficient of the variable.

EXAMPLE 3   Add. $5x + 7x$

We add $5 + 7 = 12$. Thus $5x + 7x = 12x$.

Practice Problem 3   Add. $9x + 2x$

When we combine signed numbers, we try to combine them mentally. Thus to combine $7 - 9$, we think $7 + (-9)$ and we write −2. In a similar way, to combine $7x - 9x$, we think $7x + (-9x)$ and we write −2x. Your instructor may ask you to write out this "think" step as part of your work. In the following example we show this extra step inside a ⬚ box. You should determine from your instructor whether he or she feels this step is necessary.

EXAMPLE 4   Combine.

(a) $3x + 7x - 15x$ **(b)** $9x - 12x - 11x$

(a) $3x + 7x - 15x = 3x + 7x + (-15x) = 10x + (-15x) = -5x$

(b) $9x - 12x - 11x = 9x + (-12x) + (-11x) = -3x + (-11x) = -14x$

Practice Problem 4   Combine.

(a) $8x - 22x + 5x$ **(b)** $19x - 7x - 12x$

A variable without a numerical coefficient is understood to have a coefficient of 1.

$$7x + y \text{ means } 7x + 1y. \qquad 5a - b \text{ means } 5a - 1b.$$

EXAMPLE 5 Combine.

(a) $3x - 8x + x$ **(b)** $12x - x - 20.5x$

(a) $3x - 8x + x = 3x - 8x + 1x = -5x + 1x = -4x$

(b) $12x - x - 20.5x = 12x - 1x - 20.5x = 11.0x - 20.5x = -9.5x$

Practice Problem 5 Combine.

(a) $9x - 12x + x$ **(b)** $5.6x - 8x - x$

> Numbers cannot be combined with variable terms.

EXAMPLE 6 Combine.

$$7.8 - 2.3x + 9.6x - 10.8$$

In each case, we combine the numbers separately and the variable terms separately. It may help to use the commutative and associative properties first.

$$7.8 - 2.3x + 9.6x - 10.8$$
$$= 7.8 - 10.8 - 2.3x + 9.6x$$
$$= -3 + 7.3x$$

Practice Problem 6 Combine.

$$17.5 - 6.3x - 8.2x + 10.5$$

There may be more than one variable in a problem. Keep in mind, however, that only like terms may be combined.

EXAMPLE 7 Combine.

(a) $5x + 2y + 8 - 6x + 3y - 4$ **(b)** $\dfrac{3}{4}x - 12 + \dfrac{1}{6}x + \dfrac{2}{3}$

For convenience, we will rearrange the problem to place like terms next to each other. This is an optional step; you do not need to do this.

(a) $5x - 6x + 2y + 3y + 8 - 4 = -1x + 5y + 4 = -x + 5y + 4$

(b) $\dfrac{3}{4}x - 12 + \dfrac{1}{6}x + \dfrac{2}{3} = \dfrac{9}{12}x + \dfrac{2}{12}x - \dfrac{36}{3} + \dfrac{2}{3}$

$$= \dfrac{11}{12}x - \dfrac{34}{3}$$

The order of the terms in an answer is not important in this type of problem. The answer to Example 7(a) could have been $5y + 4 - x$ or $4 + 5y - x$. Often we give the answer with the letters in alphabetical order.

Practice Problem 7 Combine.

(a) $2w + 3z - 12 - 5w - z - 16$ **(b)** $\dfrac{3}{5}x + 5 - \dfrac{7}{15}x - \dfrac{1}{3}$

Verbal and Writing Skills

1. In your own words, write a definition for the word *variable*.

2. In your own words, define *like terms*.

Name the variables in each equation.

3. $G = 5xy$

4. $S = 3\pi r^3$

5. $p = \dfrac{4ab}{3}$

6. $p = \dfrac{7ab}{4}$

Write each equation without multiplication signs.

7. $r = 3 \times m + 5 \times n$

▲ 8. $p = 2 \times w + 2 \times l$

9. $H = 2 \times a - 3 \times b$

10. $F = 4 \times p - 2 \times q$

Combine like terms.

11. $-16x + 26x$

12. $-12x + 40x$

13. $2x - 8x + 5x$

14. $4x - 10x + 3x$

15. $-\dfrac{1}{2}x + \dfrac{3}{4}x + \dfrac{1}{12}x$

16. $\dfrac{2}{5}x - \dfrac{2}{3}x + \dfrac{7}{15}x$

17. $x + 3x + 8 - 7$

18. $5x - x + 9 + 16$

19. $1.3x + 10 - 2.4x - 3.6$

20. $3.8x + 2 - 1.9x - 3.5$

21. $-13x + 7 - 19x - 10$

22. $12 - 28x - 15 + 13x$

23. $16x + 9y - 11 + 21x - 17y + 30$

24. $22x - 13y - 23 - 8x - 5y + 8$

25. $5a - 3b - c + 8a - 2b - 6c$

26. $10a + 5b - 7c - a - 8b - 3c$

27. $\dfrac{1}{2}x + \dfrac{1}{7}y - \dfrac{3}{4}x + \dfrac{5}{21}y - \dfrac{5}{6}x$

28. $\dfrac{1}{4}x + \dfrac{1}{3}y - \dfrac{7}{12}x - \dfrac{1}{2}y + \dfrac{1}{6}x$

29. $7.3x + 1.7x + 4 - 6.4x - 5.6x - 10$

30. $3.1x + 2.9x - 8 - 12.8x - 3.2x + 3$

31. $-7.6n + 1.2 + 11.2m - 3.5n - 8.1m$

32. $4.5n - 5.9m + 3.9 - 7.2n + 9m$

Find the perimeter of each geometric figure.

 33.

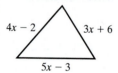

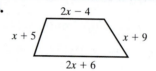 **34.**

Cumulative Review Problems

Solve for n.

35. $\dfrac{n}{6} = \dfrac{12}{15}$

36. $\dfrac{n}{9} = \dfrac{36}{40}$

37. $6n = 18$

38. $5.5n = 46.75$

1 *Removing Parentheses Using the Distributive Property*

What do we mean by the word *property* in mathematics? A **property** is an essential characteristic. A property of addition is an essential characteristic of addition. In this section we learn about the distributive property and how to use this property to simplify expressions.

Sometimes we encounter expressions like $4(x + 3)$. We'd like to be able to simplify the expression so that no parentheses appear. Notice that this expression contains two operations, multiplication and addition. We will use the distributive property to distribute the 4 to both terms in the addition statement. That is,

$$4(x + 3) \quad \text{is equal to} \quad 4(x) + 4(3).$$

The numerical coefficient 4 can be "distributed over" the expression $x + 3$ by multiplying the 4 by each of the terms in the parentheses. The expression $4(x) + 4(3)$ can be simplified to $4x + 12$, which has no parentheses. Thus we can use the distributive property to remove the parentheses.

Using variables, we can write the distributive property two ways.

> **Distributive Properties of Multiplication over Addition**
>
> If a, b, and c are signed numbers, then
>
> $$a(b + c) = ab + ac \quad \text{and} \quad (b + c)a = ba + ca.$$

A numerical example shows that the distributive property works.

$$(7)(4 + 6) = (7)(4) + (7)(6)$$
$$(7)(10) = 28 + 42$$
$$70 = 70$$

When the number is to the left of the parentheses, we use

$$a(b + c) = ab + ac.$$

There is also a distributive property of multiplication over subtraction: $a(b - c) = ab - ac$.

EXAMPLE 1 Simplify.

(a) $(8)(x + 5)$ **(b)** $(-3)(x + 3y)$ **(c)** $6(3a - 7b)$

(a) $(8)(x + 5) = 8x + (8)(5) = 8x + 40$
(b) $(-3)(x + 3y) = -3x + (-3)(3)(y) = -3x + (-9y) = -3x - 9y$
(c) $6(3a - 7b) = 6(3a) - (6)(7b) = 18a - 42b$

Practice Problem 1 Simplify.

(a) $(7)(x + 5)$ **(b)** $(-4)(x + 2y)$ **(c)** $5(6a - 2b)$

Sometimes the number is to the right of the parentheses, so we use

$$(b + c)a = ba + ca.$$

EXAMPLE 2 Simplify. $(5x + y)(2)$

$$(5x + y)(2) = (5x)(2) + (y)(2)$$
$$= 10x + 2y$$

Notice in Example 2 that we write our final answer with the numerical coefficient to the left of the variable. We would not leave $(y)(2)$ as an answer but would write $2y$.

Practice Problem 2 Simplify. $(x + 3y)(8)$

Sometimes the distributive property is used with three terms within the parentheses. When parentheses are used inside parentheses, the outside ( ) is often changed to a bracket [ ] notation.

EXAMPLE 3 Simplify.

(a) $(-5)(x + 2y - 8)$ **(b)** $(1.5x + 3y + 7)(2)$

(a) $(-5)(x + 2y - 8) = (-5)[x + 2y + (-8)]$
$$= -5(x) + (-5)(2y) + (-5)(-8)$$
$$= -5x + (-10y) + 40$$
$$= -5x - 10y + 40$$

(b) $(1.5x + 3y + 7)(2) = (1.5x)(2) + (3y)(2) + (7)(2) = 3x + 6y + 14$

In this example every step is shown in detail. You may find that you do not need to write so many steps.

Practice Problem 3 Simplify.

(a) $(-5)(x + 4y + 5)$ **(b)** $3(2.2x + 5.5y + 6)$

The parentheses may contain four terms. The coefficients may be decimals or fractions.

EXAMPLE 4 Simplify. $\dfrac{2}{3}\left(x + \dfrac{1}{2}y - \dfrac{1}{4}z + \dfrac{1}{5}\right)$

$$\dfrac{2}{3}\left(x + \dfrac{1}{2}y - \dfrac{1}{4}z + \dfrac{1}{5}\right)$$

$$= \left(\dfrac{2}{3}\right)(x) + \left(\dfrac{2}{3}\right)\left(\dfrac{1}{2}y\right) + \left(\dfrac{2}{3}\right)\left(-\dfrac{1}{4}z\right) + \left(\dfrac{2}{3}\right)\left(\dfrac{1}{5}\right)$$

$$= \dfrac{2}{3}x + \dfrac{1}{3}y - \dfrac{1}{6}z + \dfrac{2}{15}$$

Practice Problem 4 Simplify. $\dfrac{3}{2}\left(\dfrac{1}{2}x - \dfrac{1}{3}y + 4z - \dfrac{1}{2}\right)$

2   *Simplifying Expressions by Removing Parentheses and Combining Like Terms*

After removing parentheses we may have a chance to combine like terms. The direction "simplify" means remove parentheses, combine like terms, and leave the answer in as simple and correct a form as possible.

EXAMPLE 5   Simplify. $2(x + 3y) + 3(4x + 2y)$

$2(x + 3y) + 3(4x + 2y) = 2x + 6y + 12x + 6y$ Use the distributive property.

$\qquad\qquad\qquad\qquad = 14x + 12y$ Combine like terms.

Practice Problem 5   Simplify. $3(2x + 4y) + 2(5x + y)$

EXAMPLE 6   Simplify. $2(x - 3y) - 5(2x + 6)$

$2(x - 3y) - 5(2x + 6) = 2x - 6y - 10x - 30$ Use the distributive property.

$\qquad\qquad\qquad\qquad = -8x - 6y - 30$ Combine like terms.

Notice that in the final step of Example 6 only the x terms could be combined. There are no other like terms.

Practice Problem 6   Simplify. $-4(x - 5) + 3(-1 + 2x)$

Verbal and Writing Skills

1. A _____ is a symbol, usually a letter of the alphabet, that stands for a number.

2. What is the variable in the expression $5x + 9$?

3. Identify the like terms in the expression $3x + 2y - 1 + x - 3y$.

4. Explain the distributive property in your own words. Give an example.

Simplify.

5. $9(3x - 2)$

6. $8(4x - 5)$

7. $(-2)(x + y)$

8. $(-5)(x + y)$

9. $(-7)(1.5x - 3y)$

10. $(-4)(3.5x - 6y)$

11. $(-3x + 7y)(-10)$

12. $(-2x + 8y)(-12)$

13. $4(p + 9q - 10)$

14. $6(2p - 7q + 11)$

15. $3\left(\dfrac{1}{5}x + \dfrac{2}{3}y - \dfrac{1}{4}\right)$

16. $4\left(\dfrac{2}{3}x + \dfrac{1}{4}y - \dfrac{3}{8}\right)$

17. $15(-12a + 2.2b + 6.7)$

18. $14(-10a + 3.2b + 4.5)$

19. $(8a + 12b - 9c - 5)(4)$

20. $(-7a + 11b - 10c - 9)(7)$

21. $(-2)(1.3x - 8.5y - 5z + 12)$

22. $(-3)(1.4x - 7.6y - 9z - 4)$

23. $\dfrac{1}{2}\left(2x - 3y + 4z - \dfrac{1}{2}\right)$

24. $\dfrac{1}{3}\left(-3x + \dfrac{1}{2}y + 2z - 3\right)$

Applications

▲ **25.** The perimeter of a rectangle is $p = 2(l + w)$. Write this formula without parentheses and without multiplication signs.

▲ **26.** The surface area of a rectangular solid is $S = 2(lw + lh + wh)$. Write this formula without parentheses and without multiplication signs.

▲ **27.** The area of a trapezoid is $A = \dfrac{h(B + b)}{2}$. Write this formula without parentheses and without multiplication signs.

▲ **28.** The surface area of a cylinder is $S = 2\pi r(h + r)$. Write this formula without parentheses and without multiplication signs.

Simplify. Be sure to combine like terms.

29. $4(5x - 1) + 7(x - 5)$

30. $6(5x - 1) + 3(x - 4)$

31. $6(3x + 2y) - 4(x + 7)$

32. $5(4x + 3y) - 3(x + 8)$

33. $1.5(x + 2.2y) + 3(2.2x + 1.6y)$

34. $2.4(x + 3.5y) + 2(1.4x + 1.9y)$

35. $2(3b + c - 2a) - 5(a - 2c + 5b)$

36. $3(-4a + c + 4b) - 4(2c + b - 6a)$

To Think About

▲ **37.** Illustrate the distributive property by using the area of two rectangles.

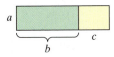

▲ **38.** Show that multiplication is distributive over subtraction by using the area of two rectangles.

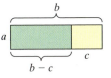

▲ **39.** Illustrate how the distributive property for three terms, $a(b + c + d) = ab + ac + ad$, could be used to discuss the area of three rectangles in the following figure.

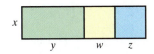

▲ **40.** Illustrate how the distributive property for three terms, $a(b + c + d) = ab + ac + ad$, could be used to discuss the area of three parallelograms in the following figure.

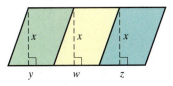

Cumulative Review Problems

Round to the nearest tenth. Use $\pi \approx 3.14$ when necessary.

▲ **41.** Find the circumference of a circular fan with a diameter of 12 inches.

▲ **42.** Find the area of a circular security mirror in a convenience store with a radius of 8 inches.

▲ **43.** Find the area of the shaded figure.

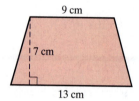

9 cm

7 cm

13 cm

▲ **44.** Find the volume of a cylinder with radius 4 inches and height 6 inches.

45. One afternoon, two Icelandic fishing vessels catch a total of 16,440 pounds of fish. The *Nereid* catches 1,110 pounds less than the *Auriole*. How much did each boat catch?

10.3 Solving Equations Using the Addition Property

1 Solving Equations Using the Addition Property

One of the most important skills in algebra is that of **solving an equation**. Starting from an equation with a variable whose value is unknown, we transform the equation into a simpler, equivalent equation by choosing a logical step. In the following sections we'll learn some of the logical steps for solving an equation successfully.

We begin with the concept of an equation. An **equation** is a mathematical statement that says that two expressions are equal. The statement $x + 5 = 13$ means that some number (x) plus 5 equals 13. Experience with the basic addition facts tells us that $x = 8$ since $8 + 5 = 13$. Therefore, 8 is the solution to the equation. The **solution** of an equation is that number which makes the equation true. What if we did not know that $8 + 5 = 13$? How could we find the value of x? Let's look at the first logical step in solving this equation.

The important thing to remember is that an equation is like a balance scale. Whatever you do to one side of the equation, you must do to the other side of the equation to maintain the balance. Our goal is to do this in such a way that we isolate the variable.

> **Addition Property of Equations**
>
> You may add the same number to each side of an equation to obtain an equivalent equation.

We want to make $x + 5 = 13$ into a simpler equation, $x =$ some number. What can we add to each side of the equation so that $x + 5$ becomes simply x? We can add the opposite of $+5$. Let's see what happens when we do this addition.

$$x + 5 = 13$$
$$x + 5 + (-5) = 13 + (-5) \qquad \text{Remember to add } -5 \text{ to both sides of the equation.}$$
$$x + 0 = 8$$
$$x = 8$$

We found the solution to the equation.

Notice that in the second step, the number 5 was removed from the left side of the equation. We know that we may add a number to both sides of the equation. But what number should we choose to add? We always add the opposite of the number we want to remove from one side of the equation.

EXAMPLE 1 Solve. $x - 9 = 3$

We want to isolate the variable x.

$$x - 9 = 3 \qquad \text{Think: "Add the opposite of } -9 \text{ to both sides of the equation."}$$
$$x - 9 + 9 = 3 + 9$$
$$x + 0 = 12$$
$$x = 12$$

To check the solution, we substitute 12 for x in the original equation.

$$x - 9 = 3$$
$$12 - 9 \overset{?}{=} 3$$
$$3 = 3 \quad ✓$$

It is always a good idea to check your solution, especially in more complicated equations.

Practice Problem 1   Solve. $x + 7 = -8$

When we solve an equation we want to isolate the variable. We do this by performing the same operation on both sides of the equation.

Equations can contain integers, decimals, or fractions.

EXAMPLE 2   Solve. Check your solution.

(a) $x + 1.5 = 4$ **(b)** $\dfrac{3}{8} = x - \dfrac{3}{4}$

We use the addition property to solve these equations since each equation involves only addition or subtraction. In each case we want to isolate the variable.

(a) $x + 1.5 = 4$ Think: "Add the opposite of 1.5 to both sides of the equation."

$$x + 1.5 + (-1.5) = 4 + (-1.5)$$

$$x = 2.5$$

Check.

$$x + 1.5 = 4 \quad \text{Substitute 2.5 for } x \text{ in the original equation.}$$
$$2.5 + 1.5 \overset{?}{=} 4$$
$$4 = 4$$

(b) $\dfrac{3}{8} = x - \dfrac{3}{4}$ Note that here the variable is on the right-hand side.

$\dfrac{3}{8} + \dfrac{3}{4} = x - \dfrac{3}{4} + \dfrac{3}{4}$ Think: "Add the opposite of $-\dfrac{3}{4}$ to both sides of the equation."

$\dfrac{3}{8} + \dfrac{6}{8} = x + 0$ We need to change $\dfrac{3}{4}$ to $\dfrac{6}{8}$.

$\dfrac{9}{8}$ or $1\dfrac{1}{8} = x$

Check.

$\dfrac{3}{8} = x - \dfrac{3}{4}$ Substitute $\dfrac{9}{8}$ for x in the original equation.

$\dfrac{3}{8} \overset{?}{=} \dfrac{9}{8} - \dfrac{3}{4}$

$\dfrac{3}{8} \overset{?}{=} \dfrac{9}{8} - \dfrac{6}{8}$

$\dfrac{3}{8} = \dfrac{3}{8} \quad ✓$

Practice Problem 2   Solve. Check your solution.

(a) $y - 3.2 = 9$ **(b)** $\dfrac{2}{3} = x + \dfrac{1}{6}$

Sometimes you will need to use the addition property twice. If variables and numbers appear on both sides of the equation, you will want to get all of the variables on one side of the equation and all of the numbers on the other side of the equation.

EXAMPLE 3   Solve $2x + 7 = x + 9$. Check your solution.

We want to remove the +7 on the left-hand side of the equation.

$$2x + 7 = x + 9$$

$2x + 7 - 7 = x + 9 - 7$ Add −7 to both sides of the equation.

$2x = x + 2$ Now we need to remove the x on the right-hand side of the equation.

$2x - x = x - x + 2$ Add $-x$ to both sides of the equation.

$x = 2$

We would certainly want to check this solution. Solving the equation took several steps and we might have made a mistake along the way. To check, we substitute 2 for x in the original equation.

Check.

$$2x + 7 = x + 9$$
$$2(2) + 7 \overset{?}{=} 2 + 9$$
$$4 + 7 \overset{?}{=} 2 + 9$$
$$11 = 11 \ \checkmark$$

Practice Problem 3   Solve. $3x - 5 = 2x + 1$. Check your solution.

10.3 Exercises

Verbal and Writing Skills

1. An _____ is a mathematical statement that says that two expressions are equal.

2. The _____ of an equation is that number which makes the equation true.

3. To use the addition property, we add to both sides of the equation the _____ of the number we want to remove from one side of the equation.

4. To use the addition property to solve the equation $x - 8 = 9$, we add _____ to both sides of the equation.

Solve for the variable.

5. $y - 5 = 9$

6. $y - 7 = 6$

7. $x + 6 = 15$

8. $x + 8 = 12$

9. $x + 16 = -2$

10. $y + 12 = -8$

11. $14 + x = -11$

12. $10 + y = -9$

13. $-12 + x = 7$

14. $-20 + y = 10$

15. $5.2 = x - 4.6$

16. $3.1 = y - 5.4$

17. $x + 3.7 = -5$

18. $y + 6.2 = -3.1$

19. $x - 25.2 = -12$

20. $y - 29.8 = -15$

21. $x + \dfrac{1}{4} = \dfrac{3}{4}$

22. $y + \dfrac{3}{8} = \dfrac{5}{8}$

23. $x - \dfrac{3}{5} = \dfrac{2}{5}$

24. $y - \dfrac{2}{7} = \dfrac{6}{7}$

25. $x + \dfrac{2}{3} = -\dfrac{5}{6}$

26. $y + \dfrac{1}{2} = -\dfrac{3}{4}$

27. $\dfrac{1}{5} + y = -\dfrac{2}{3}$

28. $\dfrac{5}{9} + y = -\dfrac{3}{18}$

Solve for the variable. You may need to use the addition property twice.

29. $3x - 5 = 2x + 9$

30. $5x + 1 = 4x - 3$

31. $2x - 7 = x - 19$

32. $4x + 6 = 3x + 5$

33. $7x - 9 = 6x - 7$

34. $4x - 8 = 3x + 12$

35. $18x + 28 = 17x + 19$

36. $15x + 41 = 14x + 27$

Mixed Practice

Solve for the variable.

37. $y - \dfrac{1}{2} = 6$

38. $x + 1.2 = -3.8$

39. $5 = z + 13$

40. $-15 = -6 + x$

41. $-5.9 + y = -4.7$

42. $z + \dfrac{2}{3} = \dfrac{7}{12}$

43. $2x - 1 = x + 5$

44. $3x - 8 = 2x - 15$

To Think About

45. In the equation $x + 9 = 12$, we add -9 to both sides of the equation to solve. What would you do to solve the equation $3x = 12$? Why?

Cumulative Review

46. Combine like terms. $5x - y + 3 - 2x + 4y$

47. Simplify. $7(2x + 3y) - 3(5x - 1)$

48. In 1999 there were 139 million people in the total labor force of the United States. Of this number, 6.7 million people were unemployed. (*Source:* Bureau of Labor Statistics) What percent of the total labor force was unemployed in 1999? Round to the nearest tenth of a percent.

▲ 49. In the front of the Digitech Company building, there is a stone pyramid with a rectangular base. The base measures 3 feet wide and 4 feet long. The pyramid is 8 feet tall. What is the volume of this pyramid?

10.4 Solving Equations Using the Division or Multiplication Property

Student Learning Objectives

After studying this section, you will be able to:

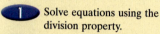

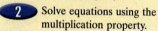

1 Solve equations using the division property.

2 Solve equations using the multiplication property.

SSM PH TUTOR CD & VIDEO MATH PRO WEB
CENTER

1 Solving Equations Using the Division Property

Recall that when we solve an equation using the addition property, we transform the equation to a simpler one where $x =$ some number. We use the same idea to solve the equation $3n = 75$. Think: "What can we do to the left side of the equation so that n stands alone?" If we divide the left side of the equation by 3, n will stand alone. Remember, however, whatever you do to one side of the equation, you must do to the other side of the equation. Our goal once again is to isolate the variable.

$$3n = 75$$
$$\frac{3n}{3} = \frac{75}{3}$$
$$n = 25$$

This is another important procedure used to solve equations.

> ### Division Property of Equations
>
> You may divide each side of an equation by the same nonzero number to obtain an equivalent equation.

EXAMPLE 1 Solve for n. $6n = 72$

$6n = 72$ The variable n is multiplied by 6.

$\dfrac{6n}{6} = \dfrac{72}{6}$ Divide each side by 6.

$n = 12$ We have $72 \div 6 = 12$.

Practice Problem 1 Solve for n. $8n = 104$

Sometimes the coefficient of the variable is a negative number. Therefore, in solving problems of this type, we need to divide each side of the equation by that negative number.

EXAMPLE 2 Solve for the variable. $-3n = 20$

$-3n = 20$ The coefficient of n is -3.

$\dfrac{-3n}{-3} = \dfrac{20}{-3}$ Divide each side of the equation by -3.

$n = -\dfrac{20}{3}$ Watch your signs!

Practice Problem 2 Solve for the variable. $-7n = 30$

Sometimes the coefficient of the variable is a decimal. In solving problems of this type, we need to divide each side of the equation by that decimal number.

EXAMPLE 3   Solve for the variable. $2.5y = 20$

$2.5y = 20$ The coefficient of y is 2.5.

$\dfrac{2.5y}{2.5} = \dfrac{20}{2.5}$ Divide each side of the equation by 2.5.

$$2.5_\wedge\overline{)20.0_\wedge}\ \ ^{8}$$

$y = 8$

To check, substitute 8 for y in the original equation.

$$2.5(8) \overset{?}{=} 20$$

$$20 = 20\ \checkmark$$

It is always best to check the solution to equations involving decimals or fractions.

Practice Problem 3   Solve for the variable. Check your solution.

$$3.2x = 16$$

2 Solving Equations Using the Multiplication Property

Sometimes the coefficient of the variable is a fraction, as in the equation $\frac{3}{4}x = 6$. Think: "What can we do to the left side of the equation so that x will stand alone?" Recall that when you multiply a fraction by its reciprocal, the product is 1. That is, $\frac{4}{3} \cdot \frac{3}{4} = 1$. We will use this idea to solve the equation. But remember, whatever you do to one side of the equation, you must do to the other side of the equation.

$$\frac{3}{4}x = 6$$

$$\frac{4}{3} \cdot \frac{3}{4}x = 6 \cdot \frac{4}{3}$$

$$1x = \frac{\overset{2}{\cancel{6}}}{1} \cdot \frac{4}{\cancel{3}}$$

$$x = 8$$

> **Multiplication Property of Equations**
>
> You may multiply each side of an equation by the same nonzero number to obtain an equivalent equation.

EXAMPLE 4   Solve for the variable and check your solution.

(a) $\dfrac{5}{8}y = 1\dfrac{1}{4}$

(b) $1\dfrac{1}{2}z = 3$

(a) $\dfrac{5}{8}y = 1\dfrac{1}{4}$

$\dfrac{5}{8}y = \dfrac{5}{4}$ Change the mixed number to a fraction. It will be easier to work with.

$\dfrac{8}{5} \cdot \dfrac{5}{8}y = \dfrac{5}{4} \cdot \dfrac{8}{5}$ Multiply both sides of the equation by $\dfrac{8}{5}$ because $\dfrac{8}{5} \cdot \dfrac{5}{8} = 1$.

$y = 2$

Check.

$\dfrac{5}{8}y = 1\dfrac{1}{4}$

$\dfrac{5}{8}(2) \stackrel{?}{=} 1\dfrac{1}{4}$ Substitute 2 for y in the original equation.

$\dfrac{10}{8} \stackrel{?}{=} 1\dfrac{1}{4}$

$1\dfrac{2}{8} \stackrel{?}{=} 1\dfrac{1}{4}$

$1\dfrac{1}{4} = 1\dfrac{1}{4}$ ✓

(b) $1\dfrac{1}{2}z = 3$

$\dfrac{3}{2}z = 3$ Change the mixed number to a fraction.

$\dfrac{2}{3} \cdot \dfrac{3}{2}z = 3 \cdot \dfrac{2}{3}$ Multiply both sides of the equation by $\dfrac{2}{3}$. Why?

$z = 2$

It is always a good idea to check the solution to an equation involving fractions. We leave the check for this solution up to you.

Practice Problem 4   Solve for the variable and check your solution.

(a) $\dfrac{1}{6}y = 2\dfrac{2}{3}$ **(b)** $3\dfrac{1}{5}z = 4$ ●

Hint: Remember to write all mixed numbers as improper fractions before solving linear equations. An alternate method that may also be used is to convert the mixed number to decimal form. Thus equations like $4\dfrac{1}{8}x = 12$ can first be written as $4.125x = 12$.

10.4 Exercises

Verbal and Writing Skills

1. How is an equation similar to a balance scale?

2. The division property states that we may divide each side of an equation by _____ to obtain an equivalent equation.

3. To change $\frac{3}{4}x = 5$ to a simpler equation, multiply both sides of the equation by ____.

4. Given the equation $1\frac{3}{5}y = 2$, we multiply both sides of the equation by $\frac{5}{8}$ to solve for y. Why?

Solve for the variable. Use the division or multiplication property of equations.

5. $4x = 36$

6. $8x = 56$

7. $7y = -28$

8. $5y = -45$

9. $-9x = 16$

10. $-7y = 22$

11. $-5x = -40$

12. $-4y = -28$

13. $-64 = -4m$

14. $-88 = -8m$

15. $0.6x = 6$

16. $0.5y = 50$

17. $5.5z = 9.9$

18. $3.5n = 7.7$

19. $-0.5x = 6.75$

20. $-0.4y = 6.88$

21. $\frac{2}{3}x = 6$

22. $\frac{3}{4}x = 3$

23. $\frac{2}{5}y = 4$

24. $\frac{5}{6}y = 10$

25. $\frac{3}{5}n = \frac{3}{4}$

26. $\frac{2}{3}z = \frac{1}{3}$

27. $\frac{3}{8}x = -\frac{3}{5}$

28. $\frac{4}{9}y = -\frac{4}{7}$

29. $\frac{1}{2}x = -2\frac{1}{4}$

30. $\frac{3}{4}y = -3\frac{3}{8}$

31. $1\frac{1}{4}z = 10$

32. $3\frac{1}{2}n = 21$

To Think About

33. Mical solves an equation by multiplying both sides of the equation by 0.5 when he should have divided. He gets an answer of 4.5. What should the correct answer be?

34. Alexis solves an equation by multiplying both sides of the equation by $\frac{2}{3}$ when she should have multiplied both sides by the reciprocal of $\frac{2}{3}$. She gets an answer of 4. What should the correct answer be?

Cumulative Review

Combine like terms.

35. $6 - 3x + 5y + 7x - 12y$

36. It is estimated that the amount of electricity that could be produced for $26.10 in 1970 cost $140.50 to produce in 2000. (*Source:* U.S. Bureau of Labor Statistics) What is the percent of increase in the cost to produce electricity from 1970 to 2000? Round to the nearest tenth.

37. It is estimated that in 2000, a total of 68.4 million barrels of oil were produced in the world. In that year the United Stated produced 6.5 million barrels of oil. (*Source:* U.S. Energy Administration) What percent of the world's oil production was produced in the United States in 2000? Round to the nearest tenth.

38. The local food warehouse club sells all kinds of paper products at wholesale prices. New England Camp Cherith needs a 30-day supply of paper plates, paper napkins, and paper towels for its campers. If the camp purchaser allows 5 plates, 7 napkins, and 4 paper towels per camper per day for 118 campers, and the same amount for each of its 32 staff members, how many of each paper product does the purchaser need to buy?

10.5 Solving Equations Using Two Properties

1 Using Two Properties to Solve an Equation

Student Learning Objectives

After studying this section, you will be able to:

1 Use two properties to solve an equation.

2 Solve equations where the variable is on both sides of the equal sign.

3 Solve equations with parentheses.

SSM   PH TUTOR   CD & VIDEO   MATH PRO   WEB
CENTER

To solve an equation, we take logical steps to change the equation to a simpler equivalent equation. The simpler equivalent equation is $x =$ some number. To do this, we use the addition property, the division property, or the multiplication property. In this section you will use more than one property to solve complex equations. Each time you use a property, you take a step toward solving the equation. At each step you try to isolate the variable. That is, you try to get the variable to stand alone.

EXAMPLE 1   Solve $3x + 18 = 27$. Check your solution.

We want only x terms on the left and only numbers on the right. We begin by removing 18 from the left side of the equation.

$$3x + 18 + (-18) = 27 + (-18)$$ Add the opposite of 18 to both sides of the equation so that $3x$ stands alone.

$$3x = 9$$

$$\frac{3x}{3} = \frac{9}{3}$$ Divide both sides of the equation by 3 so that x stands alone.

$$x = 3$$ The solution to the equation is $x = 3$.

Check.

$$3(3) + 18 \overset{?}{=} 27$$ Substitute 3 for x in the original equation.

$$9 + 18 \overset{?}{=} 27$$

$$27 = 27 \quad \checkmark$$

Practice Problem 1   Solve $5x + 13 = 33$. Check your solution.

Note that the variable is on the right-hand side in the next example.

EXAMPLE 2   Solve $-41 = 9x - 5$. Check your solution.

We begin by removing -5 from the right-hand side of the equal sign.

$$-41 + 5 = 9x - 5 + 5$$ Add the opposite of -5 to both sides of the equation.

$$-36 = 9x$$

$$\frac{-36}{9} = \frac{9x}{9}$$ Divide both sides of the equation by 9.

$$-4 = x$$

The check is left up to you.

Practice Problem 2   Solve $-50 = 7x - 8$. Check your solution.

2 Solving Equations Where the Variable Is on Both Sides of the Equal Sign

Sometimes variables appear on both sides of the equation. When this occurs, we need to isolate the variable. This requires us to add a variable term to each side of the equation.

EXAMPLE 3 Solve. $8x = 5x - 21$

We want to remove the $5x$ from the right-hand side of the equation so that all of the variables are on one side of the equation and all of the numbers are on the other side of the equation.

$$8x + (-5x) = 5x + (-5x) - 21 \quad \text{Add the opposite of } 5x \text{ to both sides of the equation.}$$

$$3x = -21$$

$$\frac{3x}{3} = \frac{-21}{3} \quad \text{Divide both sides of the equation by 3.}$$

$$x = -7$$

The check is left up to you.

Practice Problem 3 Solve. $4x = -8x + 42$

Suppose there is a variable term and a numerical term on each side of an equation. We want to collect all the variables on one side of the equation and all the numerical terms on the other side of the equation. To do this, we will have to use the addition property twice.

EXAMPLE 4 Solve. $2x + 9 = 5x - 3$

We begin by collecting the numerical terms on the right-hand side of the equation.

$$2x + 9 + (-9) = 5x - 3 + (-9) \quad \text{Add } -9 \text{ to both sides of the equation.}$$

$$2x = 5x - 12$$

Now we want to collect all the variable terms on the left-hand side of the equation.

$$2x + (-5x) = 5x + (-5x) - 12 \quad \text{Add } -5x \text{ to both sides of the equation.}$$

$$-3x = -12$$

$$\frac{-3x}{-3} = \frac{-12}{-3} \quad \text{Divide both sides of the equation by } -3.$$

$$x = 4$$

Check:

$$2(4) + 9 \overset{?}{=} 5(4) - 3 \quad \text{Substitute 4 for } x \text{ in the original equation.}$$

$$8 + 9 \overset{?}{=} 20 - 3$$

$$17 = 17 \quad \checkmark$$

Practice Problem 4 Solve. $4x - 7 = 9x + 13$

If there are like terms on one side of the equation, these should be combined first. Then proceed as before.

EXAMPLE 5 Solve for the variable. $-5 + 2y + 8 = 7y + 23$

We begin by combining the like terms on the left-hand side of the equation.

$$-5 + 2y + 8 = 7y + 23$$

$$2y + 3 = 7y + 23 \qquad \text{Combine } -5 + 8.$$

$$2y + (-7y) + 3 = 7y + (-7y) + 23 \qquad \text{Remove the variables on the right-hand side of the equation.}$$

$$-5y + 3 = 23$$

$$-5y + 3 + (-3) = 23 + (-3) \qquad \text{Remove the 3 on the left-hand side of the equation.}$$

$$-5y = 20$$

$$\frac{-5y}{-5} = \frac{20}{-5} \qquad \text{Divide both sides of the equation by } -5.$$

$$y = -4$$

Check:

$$-5 + (2)(-4) + 8 \overset{?}{=} (7)(-4) + 23$$

$$-5 + (-8) + 8 \overset{?}{=} -28 + 23$$

$$-5 = -5 \ \checkmark$$

Practice Problem 5 Solve for the variable. $4x - 23 = 3x + 7 - 2x$

It is wise to check your answer when solving this type of linear equation. The chance of making a simple error with signs is quite high. Checking gives you a chance to detect this type of error.

3 *Solving Equations with Parentheses*

If a problem contains one or more sets of parentheses, remove them using the distributive property. Then collect like terms on each side of the equation. Then solve.

EXAMPLE 6 Isolate the variable on the right-hand side. Then solve for x.

$$7x - 3(x - 4) = 9(x + 2)$$

$7x - 3x + 12 = 9x + 18$	Remove parentheses by using the distributive property.
$4x + 12 = 9x + 18$	Add like terms.
$4x + 12 + (-18) = 9x + 18 + (-18)$	Add -18 to each side.
$4x + (-6) = 9x$	Simplify.
$4x + (-4x) - 6 = 9x + (-4x)$	Add $-4x$ to each side. This isolates the variable on the right-hand side.
$-6 = 5x$	Simplify.
$\dfrac{-6}{5} = \dfrac{5x}{5}$	Divide each side of the equation by 5.
$-\dfrac{6}{5} = x$	We obtain a solution that is a fraction.

Practice Problem 6 Isolate the variable on the right-hand side. Then solve for x. $8(x - 3) + 5x = 15(x - 2)$

We now list a procedure that may be used to help you remember all the steps we are using to solve equations.

Procedure to Solve Equations

1. Remove any parentheses by using the distributive property.
2. Collect like terms on each side of the equation.
3. Add the appropriate value to both sides of the equation to get all numbers on one side.
4. Add the appropriate term to both sides of the equation to get all variable terms on the other side.
5. Divide both sides of the equation by the numerical coefficient of the variable term.
6. Check by substituting the solution back into the original equation.

You have probably noticed that steps 3 and 4 are interchangeable. You can do step 3 and then step 4, or you can do step 4 and then step 3.

Verbal and Writing Skills

1. Explain how you would decide what to add to each side of the equation as you begin to solve $-5x - 6 = 29$ for x.

2. Explain how you would decide what to add to each side of the equation as you begin to solve $-11x = 4x - 45$ for x.

Check to see whether the given answer is a solution to the equation.

3. Is 2 a solution to $3 - 4x = 5 - 3x$?

4. Is $x = 5$ a solution to $3x + 2 = -2x - 23$?

5. Is $x = \dfrac{1}{2}$ a solution to $8x - 2 = 10 - 16x$?

6. Is $x = \dfrac{1}{3}$ a solution to $12x - 7 = 3 - 18x$?

Solve.

7. $12x - 30 = 6$

8. $15x - 10 = 35$

9. $9x - 3 = -7$

10. $6x - 9 = -12$

11. $-9x = 3x - 10$

12. $-3x = 7x + 14$

13. $18 - 2x = 4x + 6$

14. $3x + 4 = 7x - 12$

15. $8 + x = 3x - 6$

16. $9 - 8x = 3 - 2x$

17. $7 + 3x = 6x - 8$

18. $2x - 7 = 3x + 9$

19. $5 + 2y = 7 + 5y$

20. $12 + 5y = 9 - 3y$

21. $2x + 6 = -8 - 12x$

22. $4x + 5 = -19 - 8x$

23. $-10 + 6y + 2 = 3y - 26$

24. $6 - 5x + 2 = 4x + 35$

25. $12 + 4y - 7 = 6y - 9$

26. $\dfrac{2}{3}x - 5 = 17$

27. $\dfrac{3}{4}x + 2 = -10$

28. $8x - 6 + 3x = 5 + 7x - 19$

29. $-30 - 12y + 18 = -24y + 13 + 7y$

30. $15y - 22 - 29y = 15 - 18y - 21$

31. $5(x + 4) = 4x + 15$

32. $2x + 3(x - 4) = 3$

33. $7(y - 1) = 4(y + 2) + 18$

34. $13 + 7(2y - 1) = 5(y + 6)$

35. $7x - 3(x - 6) = 2(x - 3) + 8$

36. $5x - 6(x + 1) = 2x - 6$

37. $7x - 16 = 3(x + 2)$

38. $5x - 18 = 2(x + 3)$

39. $7x - 6(x + 3) = -2(x - 4)$

40. $8x - 5(x + 2) = -3(x - 5)$

To Think About

41. (a) Solve $7 + 3x = 6x - 8$ by collecting x-terms on the left.

(b) Solve by collecting x-terms on the right.

(c) Which method is easier? Why?

42. (a) Solve $8 + 5x = 2x - 6 + x$ by collecting x-terms on the left.

(b) Solve by collecting x-terms on the right.

(c) Which method is easier? Why?

Cumulative Review Problems

Use $\pi \approx 3.14$. Round to the nearest tenth.

▲ **43.** A topographic globe has a radius of 46 centimeters. Find the volume of this sphere.

▲ **44.** Find the area of the shaded region in the given figure.

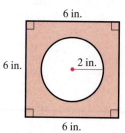

6 in.

6 in. 2 in.

6 in.

45. At the annual "I Love My Pet" supply convention of 640 dealers, each dealer sold pet food or pet toys or both. If 300 dealers sold both pet food and toys, and 150 dealers sold only toys, how many dealers sold only pet food?

▲ **46.** Challenge Problem: What is the greatest number of pieces that can result from slicing a spherical watermelon with five straight cuts? (*Hint:* Do not make all your cuts in the same way.)

Putting Your Skills to Work

Examining the Relationship of Gender and Level of Education to Income

In 1999 the U.S. Census Bureau released figures showing the number of people in the United States who earned an annual income of $50,000 or more. The report was broken down by gender and by level of education. Four levels of education were specified: not a high school graduate, high school graduate, some college, bachelor's degree or higher. The numbers from the report are listed in the following double-bar graph:

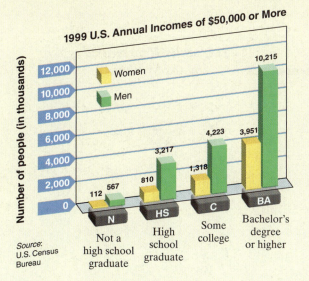

1999 U.S. Annual Incomes of $50,000 or More

Number of people (in thousands)

Women
Men

N — Not a high school graduate: 112, 567
HS — High school graduate: 810, 3,217
C — Some college: 1,318, 4,223
BA — Bachelor's degree or higher: 3,951, 10,215

Source: U.S. Census Bureau

Study the bar graph to answer the following questions.

Problems for Individual Investigation and Analysis

1. How many women earned $50,000 or more? What percent of the women who earned $50,000 or more had at least a bachelor's degree? Round to the nearest tenth.

2. How many men earned $50,000 or more? What percent of the men who earned $50,000 or more had at least a bachelor's degree? Round to the nearest tenth.

Problems for Group Investigation and Cooperative Problem Solving

3. Of the people who earned $50,000 or more, was the ratio of women to men higher for people who had earned a bachelor's degree or for those who had not earned a bachelor's degree?

4. Of the people who earned $50,000 or more, what percent of these people were women? Round to the nearest tenth.

Internet Connections

 WWW | Netsite: http://www.prenhall.com/tobey_basic

This site contains information from the U.S. Census Bureau.

Find the table that contains the following title: Money Income of Persons—Selected Characteristics, by Income Level.

5. For 1999, what was the median income of female full-time year-round workers as a percentage of the equivalent figure for men?

6. Was this figure obtained in exercise 5 higher or lower as compared with 1998? By how much?

Translating English into Mathematical Equations Using Two Given Variables

In the preceding section you learned how to solve an equation. We can use equations to solve applied problems, but before we can do that, we need to know how to write an equation that will represent the situation in a word problem.

In this section we practice *translating* to help you to write your own mathematical equations. That is, we translate English expressions into algebraic expressions. In the next section we'll apply this translation skill to a variety of word problems.

The following chart presents the mathematical symbols generally used in translating English phrases into equations.

The English Phrase:	Is Usually Represented by the Symbol:
greater than increased by more than added to sum of	+
less than decreased by smaller than fewer than shorter than difference of	−
multiplied by of product of times	×
double	2 ×
triple	3 ×
divided by ratio of quotient of	÷
is was has costs equals represents amounts to	=

An English sentence describing the relationship between two or more quantities can often be translated into a short equation using variables. For example, if we say in English "Bob's salary is $1000 greater than Fred's salary," we can express the mathematical relationship by the equation

$$b = 1000 + f$$

where b represents Bob's salary and f represents Fred's salary.

EXAMPLE 1 Translate the English sentence into an equation using variables. Use r to represent Roberto's weight and j to represent Juan's weight.

Roberto's weight is 42 pounds more than Juan's weight.

Roberto's weight is 42 pounds more than Juan's weight.
↓   ↓   ↓   ↓   ↓
r $=$ 42 $+$ j

The equation $r = 42 + j$ could also be written as

$$r = j + 42.$$

Both are correct translations because addition is commutative.

Practice Problem 1 Translate the English sentence into an equation using variables. Use t to represent Tom's height and a to represent Abdul's height.

Tom's height is 7 inches more than Abdul's height.

When translating the phrase "less than" or "fewer than," be sure that the number that appears before the phrase is the value that is subtracted. Words like "costs," "weighs," or "has the value of" are translated into an equal sign $(=)$.

EXAMPLE 2 Translate the English sentence into an equation using variables. Use c to represent the cost of the chair in dollars and s to represent the cost of the sofa in dollars.

The chair costs $200 less than the sofa.

The chair costs $ 200 *less than* the sofa.
↓   ↓   ↓   ↓
c $=$ s $-$ 200

Note the order of the equation. We subtract 200 from s, so we have $s - 200$. It would be incorrect to write $200 - s$. Do you see why?

Practice Problem 2 Translate the English sentence into an equation using variables. Use n to represent the number of students in the noon class and m to represent the number of students in the morning class.

The noon class has 24 fewer students than the morning class.

In a similar way we can translate the phrases "more than" or "greater than," but since addition is commutative, we find it easier to write the mathematical symbols in the same order as the words in the English sentence.

EXAMPLE 3 Translate the English sentence into an equation using variables. Use f for the cost of a 14-foot truck and e for the cost of an 11-foot truck.

The daily cost of a 14-foot truck is 20 dollars more than the daily cost of an 11-foot truck.

The 14-foot truck cost is 20 more than the 11-foot truck cost.

$$f = 20 + e$$

Practice Problem 3 Translate the English sentence into an equation with variables. Use t to represent the number of boxes carried on Thursday and f to represent the number of boxes carried on Friday.

On Thursday Adrianne carried five more boxes into the dorm than she did on Friday.

2 *Writing Algebraic Expressions for Several Quantities Using One Given Variable*

In each of the examples so far we have used two *different variables*. Now we'll learn how to write algebraic expressions for several quantities using the *same variable*. In the next section we'll use this skill to write and solve equations.

A mathematical expression that contains a variable is often called an **algebraic expression.**

EXAMPLE 4 Write algebraic expressions for Bob's salary and Fred's salary. Fred's salary is $150 more than Bob's salary. Use the letter b.

Let $b =$ Bob's salary.

Let $\underbrace{b + 150}_{\$150 \text{ more than Bob's salary}} =$ Fred's salary.

Notice that Fred's salary is described in terms of Bob's salary. Thus it is logical to let Bob's salary be b and then to express Fred's salary as $150 more than Bob's.

Practice Problem 4 Write algebraic expressions for Sally's trip and Melinda's trip. Melinda's trip is 380 miles longer than Sally's trip. Use the letter s.

Taking the time to master problems like Example 4 will enable you to develop skills that are absolutely essential in Section 10.7.

▲ ⬭ **EXAMPLE 5**   Write algebraic expressions for the size of each of two angles of a triangle. Angle *B* of the triangle is 34° less than angle *A*. Use the letter *A*.

Let $A =$ the number of degrees in angle *A*.

Let $A - 34 =$ the number of degrees in angle *B*.

$\underbrace{\hspace{3cm}}$

34° less than angle *A*

Practice Problem 5   Write algebraic expressions for the height of each of two buildings. Larson Center is 126 feet shorter than McCormick Hall. Use the letter *m*. ⬭

Often in algebra when we write expressions for one or two unknown quantities, we use the letter *x*.

▲ ⬭ **EXAMPLE 6**   Write algebraic expressions for the length of each of three sides of a triangle. The second side is 4 inches longer than the first. The third side is 7 inches shorter than triple the length of the first side. Use the letter *x*.

Since the other two sides are described in terms of the first side, we start by writing an expression for the first side.

Let $x =$ the length of the first side.

Let $x + 4 =$ the length of the second side.

Let $3x - 7 =$ the length of the third side.

▲ **Practice Problem 6**   Write an algebraic expression for the length of each of three sides of a triangle. The second side is double the length of the first side. The third side is 6 inches longer than the first side. Use the letter *x*. ⬭

10.6 Exercises

Verbal and Writing Skills

Translate the English sentence into an equation using the variables indicated.

1. Harry weighs 34 pounds more than Rita. Use h for Harry's weight and r for Rita's weight.

2. The large cereal box contains 7 ounces more than the small box of cereal. Use l for the number of ounces in the large cereal box and s for the number of ounces in the small cereal box.

3. The bracelet costs $107 less than the necklace. Use b for the cost of the bracelet and n for the cost of the necklace.

4. There were 42 fewer students taking algebra in the fall semester than the spring semester. Use s to represent the number of students registered in spring and f to represent the number of students registered in fall.

5. Dwight threw the javelin 5 meters less than Evan did. Use d for the distance Dwight threw the javelin and e for the distance that Evan threw the javelin.

6. In a long-jump competition, Raul's best jump was 7 inches longer than Neil's best jump. Use r to represent Raul's best jump and n to represent Neil's best jump.

▲ 7. The length of the rectangle is 7 meters longer than double the width. Use l for the length and w for the width of the rectangle.

▲ 8. The length of the rectangle is 5 meters longer than triple the width. Use l for the length and w for the width of the rectangle.

▲ 9. The length of the second side of a triangle is 2 inches shorter than triple the length of the first side of the triangle. Use s to represent the length of the second side and f to represent the length of the first side.

▲ 10. The length of the second side of a triangle is 4 inches shorter than double the length of the first side of the triangle. Use s to represent the length of the second side and f to represent the length of the first side.

11. The cost of the average home in 2000 was $3000 more than triple the cost of the average home in the 1960s. Use s to represent the cost of the average home in the 1960s and n to represent the cost of the average home in 2000.

12. The cost of buying an average bag of groceries in 2000 was $40 more than double the cost of an average bag of groceries in the 1950s. Use e to represent the cost of an average bag of groceries in 2000 and f to represent the cost of an average bag of groceries in the 1950s.

13. The combined number of hours Jenny and Sue watch television per week is 26 hours. Use j for the number of hours Jenny watches television and s for the number of hours Sue watches television.

14. The attendance at the Giants game was greater than the attendance at the Jets game. The difference in attendance between the two games was 7234 fans. Use g for the number of fans at the Giants game and j for the number of fans at the Jets game.

15. The product of your hourly wage and the amount of time worked is $500. Let h = the hourly wage and t = the number of hours worked.

16. The ratio of men to women at Central College is 5 to 3. Let m = the number of men and w = the number of women.

Write algebraic expressions for each quantity using the given variable.

17. Barbara's car trip was 386 miles longer than Julie's car trip. Use the letter j.

18. The cost of the television set was $212 more than the cost of the compact disc player. Use the letter p.

▲ **19.** Angle A of the triangle is 46° less than angle B. Use the letter b.

20. The top of the box is 38 centimeters shorter than the side of the box. Use the letter s.

21. Mount Everest is 4430 meters taller than Mount Whitney. Use the letter w.

22. Mount McKinley is 1802 meters taller than Mount Rainier. Use the letter r.

23. Sam made $12 more in tips than Lisa one Friday night. Brenda made $6 less than Lisa. Use the letter l.

24. During the summer, Nina read twice as many books as Aaron. Molly read five more books than Aaron. Use the letter a.

▲ **25.** The height of a box is 7 inches longer than the width. The length is 1 inch shorter than double the width. Use the letter w.

▲ **26.** The length of a box is 5 inches longer than its height. The width is triple the height. Use the letter h.

▲ **27.** The second angle of a triangle is double the first. The third angle of the triangle is 14° smaller than the first. Use the letter x.

▲ **28.** The second angle of a triangle is triple the first. The third angle of the triangle is 36° larger than the first. Use the letter x.

To Think About

29. A Caravan is traveling down the Massachusetts Turnpike at a speed s. A Lexus sedan is traveling 10 miles per hour faster and passes the Caravan. Both vehicles come to the top of a steep hill. As they go down the hill, both vehicles increase in speed by 8%. Write an expression for the speed of the Caravan going down the hill. Write an expression for the speed of the Lexus sedan going down the hill.

30. Recently astronomers discovered evidence of two planets that are not part of our solar system. The first planet is 73 light-years from Earth. It was given the designation HD 19994b. The second planet is directly behind the first one. It is 105 light-years from Earth. The second planet has the designation HD 92788b. Suppose a space probe is sent out from Earth in the direction of these two distant planets and it has traveled x light-years from Earth. Write an expression for the distance from the probe to each planet.

Cumulative Review Problems

Perform the indicated operations in the proper order.

31. $-6 - (-7)(2)$

32. $5 - 5 + 8 - (-4) + 2 - 15$

33. The following are recent statistics for some NBA players. Fill in the blanks to complete the season totals for three-point shots. (Round to three significant digits.)

Player	Team	3-Point Attempts	3-Points Made Made	3-Point %
Dee Brown	Toronto	349	135	
Tim Hardaway	Miami		112	0.360
Reggie Miller	Indiana	275		0.385

10.7 Applied Problems

1 Solving Problems Involving Comparisons

Student Learning Objectives

After studying this section, you will be able to:

1 Solve problems involving comparisons.

2 Solve problems involving geometric formulas.

3 Solve problems involving rates and percents.

SSM PH TUTOR CD & VIDEO MATH PRO WEB
CENTER

To solve the following problem, we use the three steps for problem solving with which you are familiar, plus another step: *Write an equation.*

🔴 **EXAMPLE 1** A 12-foot board is cut into two pieces. The longer piece is 3.5 feet longer than the shorter piece. What is the length of each piece?

You may find it helpful to use the Mathematics Blueprint for Problem Solving to organize the data and make a plan for solving.

Mathematics Blueprint For Problem Solving

Gather the Facts	What Am I Asked to Do?	How Do I Proceed?	Key Points to Remember
The board is 12 feet long. It is cut into two pieces. One piece is 3.5 feet longer than the other.	I need to find the length of each piece.	I need to use algebra to describe the length of each piece.	I will make an equation by adding up the length of each piece to get 12 feet.

1. **Understand the problem.**
   Draw a diagram.

   Shorter piece   Longer piece

   |← 12 ft →|

   Since the longer piece is described in terms of the shorter piece, we let the variable represent the shorter piece. Let x = the length of the shorter piece. The longer piece is 3.5 feet longer than the shorter piece. Let $x + 3.5$ = the length of the longer piece. The sum of the two pieces is 12 feet. We write an equation.

2. **Write an equation.**

$$x + (x + 3.5) = 12$$

3. **Solve and state the answer.**

$$x + x + 3.5 = 12$$
$$2x + 3.5 = 12 \qquad \text{Collect like terms.}$$
$$2x + 3.5 + (-3.5) = 12 + (-3.5) \qquad \text{Add } -3.5 \text{ to each side.}$$
$$2x = 8.5 \qquad \text{Collect like terms.}$$
$$\frac{2x}{2} = \frac{8.5}{2} \qquad \text{Divide each side by 2.}$$
$$x = 4.25$$

The shorter piece is 4.25 feet long.

$$x + 3.5 = \text{the longer piece}$$
$$4.25 + 3.5 = 7.75$$

The longer piece is 7.75 feet long.

4. *Check.*

We verify solutions to word problems by making sure that all the calculated values satisfy the original conditions. Do the two pieces add up to 12 feet?

$$4.25 + 7.75 \overset{?}{=} 12$$

$$12 = 12 \checkmark$$

Is one piece 3.5 feet longer than the other?

$$7.75 \overset{?}{=} 3.5 + 4.25$$

$$7.75 = 7.75 \checkmark$$

Practice Problem 1   An 18-foot board is cut into two pieces. The longer piece is 4.5 feet longer than the shorter piece. What is the length of each piece?

Sometimes three items are compared. Let a variable represent the quantity to which the other two quantities are compared. Then write an expression for the other two quantities.

EXAMPLE 2   Professor Jones is teaching 332 students in three sections of general psychology this semester. His noon class has 23 more students than his 8:00 A.M. class. His 2:00 P.M. class has 36 fewer students than his 8:00 A.M. class. How many students are in each class?

1. *Understand the problem.*

Each class enrollment is described in terms of the enrollment in the 8:00 A.M. class.

Let x = the number of students in the 8:00 A.M. class.

The noon class has 23 more students than the 8:00 A.M. class.

Let $x + 23$ = the number of students in the noon class.

The 2:00 P.M. class has 36 fewer students than the 8:00 A.M. class.

Let $x - 36$ = the number of students in the 2:00 P.M. class.

The total enrollment for the three sections is 332.
You can draw a diagram.

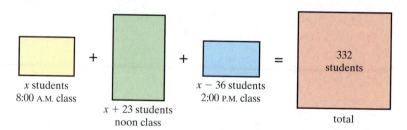

x students
8:00 A.M. class

$x + 23$ students
noon class

$x - 36$ students
2:00 P.M. class

332
students

total

2. *Write an equation.*

$$x + (x + 23) + (x - 36) = 332$$

3. *Solve and state the answer.*

$$x + x + 23 + x - 36 = 332$$

$3x - 13 = 332$	Collect like terms.
$3x + (-13) + 13 = 332 + 13$	Add 13 to each side.
$3x = 345$	Simplify.
$\dfrac{3x}{3} = \dfrac{345}{3}$	Divide each side by 3.
$x = 115$	8:00 A.M. class
$x + 23 = 115 + 23 = 138$	noon class
$x - 36 = 115 - 36 = 79$	2:00 P.M. class

Thus there are 115 students in the 8:00 A.M. class, 138 students in the noon class, and 79 students in the 2:00 P.M. class.

4. *Check.*

Do the numbers of students in the classes total 332?

$$115 + 138 + 79 \stackrel{?}{=} 332$$
$$332 = 332 \checkmark$$

Does the noon class have 23 more students than the 8:00 A.M. class?

$$138 \stackrel{?}{=} 23 + 115$$
$$138 = 138 \checkmark$$

Does the 2:00 P.M. class have 36 fewer students than the 8:00 A.M. class?

$$79 \stackrel{?}{=} 115 - 36$$
$$79 = 79 \checkmark$$

Practice Problem 2   The city airport had a total of 349 departures on Monday, Tuesday, and Wednesday. There were 29 more departures on Tuesday than on Monday. There were 16 fewer departures on Wednesday than on Monday. How many departures occurred on each day?

② Solving Problems Involving Geometric Formulas

The following applied problems concern the geometric properties of two-dimensional figures. The problems involve perimeter or the measure of the angles in a triangle.

Recall that when we double something, we are multiplying by 2. That is, if something is x units, then double that value is $2x$. Triple that value is $3x$.

▲ ● **EXAMPLE 3**   A farmer wishes to fence in a rectangular field with 804 feet of fence. The length is to be 3 feet longer than *double the width*. How long and how wide is the field?

1. *Understand the problem.*

The perimeter of a rectangle is given by $P = 2w + 2l$.

Let w = the width.

The length is 3 feet longer than double the width.

$$\text{Length} = 3 + 2w$$

Thus $2w + 3$ = the length.

You may wish to draw a diagram and label the figures with the given facts.

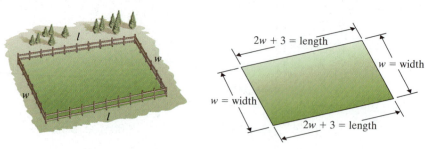

2. Write an equation.

Substitute the given facts into the perimeter formula.

$$2w + 2l = P$$
$$2w + 2(2w + 3) = 804$$

3. Solve and state the answer.

$$2w + 2(2w + 3) = 804$$
$$2w + 4w + 6 = 804 \qquad \text{Use the distributive property.}$$
$$6w + 6 = 804 \qquad \text{Collect like terms.}$$
$$6w + 6 + (-6) = 804 + (-6) \qquad \text{Add } -6 \text{ to each side.}$$
$$6w = 798 \qquad \text{Simplify.}$$
$$\frac{6w}{6} = \frac{798}{6} \qquad \text{Divide each side by 6.}$$
$$w = 133$$

The width is 133 feet.
The length = $2w + 3$. When $w = 133$, we have

$$(2)(133) + 3 = 266 + 3 = 269.$$

Thus the length is 269 feet.

4. Check.

Is the length 3 feet longer than double the width?

$$269 \overset{?}{=} 3 + (2)(133)$$
$$269 \overset{?}{=} 3 + 266$$
$$269 = 269 \quad ✓$$

Is the perimeter 804 feet?

$$(2)(133) + (2)(269) \overset{?}{=} 804$$
$$266 + 538 \overset{?}{=} 804$$
$$804 = 804 \quad ✓$$

```
        269
   ┌──────────┐
133 │          │ 133
   └──────────┘
        269
```

▲ **Practice Problem 3**   What are the length and width of a rectangular field that has a perimeter of 772 feet and a length that is 8 feet longer than double the width?

▲ ⬤ **EXAMPLE 4**   The perimeter of a triangular rug section is 21 feet. The second side is double the length of the first side. The third side is 3 feet longer than the first side. Find the length of the three sides of the rug.

> Let x = the length of the first side.
>
> Let $2x$ = the length of the second side.
>
> Let $x + 3$ = the length of the third side.

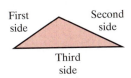 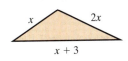

The distance around the three sides totals 21 feet.

Thus

$$x + 2x + (x + 3) = 21 \qquad \text{Use the perimeter formula.}$$
$$4x + 3 = 21 \qquad \text{Collect like terms.}$$
$$4x + 3 + (-3) = 21 + (-3) \qquad \text{Add } -3 \text{ to each side.}$$
$$4x = 18 \qquad \text{Simplify.}$$
$$\frac{4x}{4} = \frac{18}{4} \qquad \text{Divide each side by 4.}$$
$$x = 4.5$$

The first side is 4.5 feet long.

$$2x = (2)(4.5) = 9 \text{ feet}$$

The second side is 9 feet long.

$$x + 3 = 4.5 + 3 = 7.5 \text{ feet}$$

The third side is 7.5 feet long.

Check.
Do the three sides add up to a perimeter of 21 feet?

$$4.5 + 9 + 7.5 \stackrel{?}{=} 21$$
$$21 = 21 \;\; ✓$$

Is the second side double the length of the first side?

$$9 \stackrel{?}{=} (2)(4.5)$$
$$9 = 9 \;\; ✓$$

Is the third side 3 feet longer than the first side?

$$7.5 \stackrel{?}{=} 3 + 4.5$$
$$7.5 = 7.5 \;\; ✓$$

▲ **Practice Problem 4**   The perimeter of a triangle is 36 meters. The second side is double the first side. The third side is 10 meters longer than the first side. Find the length of each side. Check your solutions. ⬤

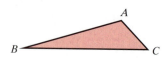

▲ ● **EXAMPLE 5** A triangle has three angles, A, B, and C. Angle C is triple angle B. Angle A is 105° larger than angle B. Find the measure of each angle. Check your answer.

Let x = the number of degrees in angle B.

Let $3x$ = the number of degrees in angle C.

Let $x + 105$ = the number of degrees in angle A.

The sum of the interior angles of a triangle is 180°. Thus we can write the following.

$$x + 3x + (x + 105) = 180$$
$$5x + 105 = 180$$
$$5x + 105 + (-105) = 180 + (-105)$$
$$5x = 75$$
$$\frac{5x}{5} = \frac{75}{5}$$
$$x = 15$$

Angle B measures 15°.

$$3x = (3)(15) = 45$$

Angle C measures 45°.

$$x + 105 = 15 + 105 = 120$$

Angle A measures 120°.

Check.
Do the angles total 180°?

$$15 + 45 + 120 \overset{?}{=} 180$$
$$180 = 180 \quad \checkmark$$

Is angle C triple angle B?

$$45 \overset{?}{=} (3)(15)$$
$$45 = 45 \quad \checkmark$$

Is angle A 105° larger than angle B?

$$120 \overset{?}{=} 105 + 15$$
$$120 = 120 \quad \checkmark$$

▲ **Practice Problem 5** Angle C of a triangle is triple angle A. Angle B is 30° less than angle A. Find the measure of each angle. ●

3 *Solving Problems Involving Rates and Percents*

You can use equations to solve problems that involve rates and percents. Recall that the commission a salesperson earns is based on the total sales made. For example, a saleswoman earns $40 if she gets a 4% commission and she sells $1000 worth of products. That is, 4% of $1000 = $40. Sometimes a salesperson earns a base salary. The commission will then be added to the base salary to determine the total salary. You can find the total salary if you know the amount of the sales. How would you find the amount of sales if the salary were known? We will use an equation.

EXAMPLE 6 This month's salary for an encyclopedia saleswoman was $3000. This includes her base monthly salary of $1800 plus a 5% commission on total sales. Find the total sales for the month.

$$\boxed{\begin{array}{c}\text{total salary}\\\text{of \$3000}\end{array}} = \boxed{\begin{array}{c}\text{base salary}\\\text{of \$1800}\end{array}} + \boxed{\begin{array}{c}\text{5\% commission}\\\text{of total sales}\end{array}}$$

Let $s =$ the amount of total sales.
Then $0.05s =$ the amount of commission earned from the sales.

$$3000 = 1800 + 0.05s$$
$$1200 = 0.05s$$
$$\frac{1200}{0.05} = \frac{0.05s}{0.05}$$
$$24{,}000 = s$$

She sold $24,000 worth of encylopedias.

Check.
Does 5% of $24,000 added to $1800 yield a salary of $3000?

$$(0.05)(24{,}000) + 1800 \stackrel{?}{=} 3000$$
$$1200 + 1800 \stackrel{?}{=} 3000$$
$$3000 = 3000 \ \checkmark$$

Practice Problem 6 A salesperson at a boat dealership earns $1000 a month plus a 3% commission on the total sales of the boats he sells. Last month he earned $3250. What were the total sales of the boats he sold?

Applications

Solve using an equation. Show what you let the variable equal.

1. A 16-foot board is cut into two pieces. The longer piece is 5.5 feet longer than the shorter piece. What is the length of each piece?

2. A 20-foot board is cut into two pieces. The longer piece is 4.5 feet longer than the shorter piece. What is the length of each piece?

3. At the gym, Jodie can leg-press 140 pounds less than Jack. Together they are able to leg-press 670 pounds. How much can each leg-press separately?

4. Darcy and Jonah each bought a new car last month. Jonah spent $840 less than Darcy. Together they spent $22,160. How much did each spend on his or her new car?

5. The Business Club's thrice-yearly car wash serviced 398 cars this year. A total of 84 more cars participated in May than in November. A total of 43 fewer cars were washed in July than in November. How many cars were washed during each month?

6. In the game Scrabble, wooden tiles with letters on them are placed on a board to spell words. There are three times as many A tiles as there are G tiles. The number of O tiles is two more than twice the number of G tiles. The total number of A, O, and G tiles is 20. How many tiles of each are there?

Solve using an equation. Show what you let the variable equal. Check your answers.

7. A 12-foot solid cherry wood tabletop is cut into two pieces to allow for an insert later on. Of the two original pieces, the shorter piece is 4.7 feet shorter than the longer piece. What is the length of each piece?

8. An artist has created a huge painting 18 feet long. Her goal is to cut the canvas and have two pieces of the same painting. The longer piece of canvas will be 6.5 feet longer than the shorter piece. What will the length of each piece be?

▲ 9. The playing board of a new game has a perimeter of 76 inches. It was designed so that the length is 4 inches shorter than double the width. What are the dimensions of the playing board?

▲ 10. The perimeter of a rectangle is 64 centimeters. The length is 4 centimeters less than triple the width. What are the dimensions of the rectangle?

▲ **11.** An unusual triangular wall flag at the United Nations has a perimeter of 199 millimeters. The second side is 20 millimeters longer than the first side. The third side is 4 millimeters shorter than the first side. Find the length of each side.

▲ **12.** There is a triangular piece of land adjoining an oil field in Texas, with a perimeter of 271 meters. The length of the second side is double the first side. The length of the third side is 15 meters longer than the first side. Find the length of each side.

▲ **13.** A geometric puzzle has a triangular playing piece with a perimeter of 44 centimeters. The length of the second side is double the first side. The length of the third side is 12 centimeters longer than the first side. Find the length of each side.

▲ **14.** An unusual triangular pennant has a perimeter of 63 inches. The length of the first side is twice the length of the second side. The third side is 3 inches longer than twice the second side. Find the length of each side.

▲ **15.** A triangle has three angles, A, B, and C. Angle B is triple angle A. Angle C is 40° larger than angle A. Find the measure of each angle.

▲ **16.** A triangle has three angles, A, B, and C. Angle B is triple angle A. Angle C is 15° smaller than angle A. Find the measure of each angle.

17. A saleswoman at a car dealership earns $1200 per month plus a 5% commission on her total sales. If she earned $5000 last month, what was the amount of her sales?

18. A salesman at a jewelry store earns $1000 per month plus a 6% commission on his total sales. If he earned $2200 last month, what was the amount of his sales?

19. A real estate agent charges $100 to place a rental listing plus 12% of the yearly rent. An apartment in Central City was rented by the agent for one year. She charged the landowner $820. How much did the apartment cost to rent for one year?

20. A real estate agent charges $50 to place a rental listing plus 9% of the yearly rent. An apartment in the town of West Longmeadow was rented by the agent for one year. He charged the landowner $482. How much did the apartment cost to rent for one year?

21. A community in South Florida has decided to paint a mural. Adults and children each have one section, but since there are more children interested in participating than adults, the children are awarded the larger piece of wall. If the wall is 32 feet long and the children's section is 6.2 feet longer than the adult section, what is the length of each section of wall?

▲ **22.** A triangular steel support for a computer case has a perimeter of 266 millimeters. The second side is 48 millimeters longer than the first side. The third side is 4 millimeters shorter than the first side. Find the length of each side.

To Think About

23. The cost of three computer programs at a computer discount store is $570.33. The cost of the second program is $20 less than double the cost of the first program. The cost of the third program is $17 more than triple the cost of the second program. How much does each program cost?

24. The cost of three vintage cars at an antique car club auction is $45,000. The cost of the Model T is $6000 less than double the cost of an Edsel. The cost of a GTO is $3000 less than triple the cost of the Model T. How much does each vehicle cost?

Cumulative Review Problems

25. What percent of 20 is 12?

26. 38% of what number is 190?

▲ **27.** The largest pizza ever cooked was in Norwood, South Africa, and was 37.4 meters in diameter. Find the circumference of the pizza. Round to the nearest tenth.

Math in the Media

Use a Vehicle? Think about Taxes.

Sandra Block
USA TODAY, 03/07/2000

Copyright 2000, USA TODAY. Reprinted with permission.

"With gas prices topping $1.50 a gallon in many areas, it's tough to fill your car without emptying your wallet.

Fortunately, transportation costs, like other costs of running your business, are tax deductible. But you can't take full credit for the miles you spend in traffic unless you keep good records. And depending on your circumstances, your might want to rethink the way you deduct your driving costs.

The IRS offers two options for deducting expenses for business use of an automobile:

- **Actual expenses.** This method lets you claim all the expenses associated with operating a car for business, such as gas, maintenance, and depreciation. If you use your car for business and non-business purposes, you can only deduct part of those costs.

- **The standard mileage allowance.** Here, you simply deduct a specific amount for every business-related mile you drove during the year. The mileage allowance for 2000 is 32.5 cents a mile. So, if you drove 5,000 miles for business in 2000, you can deduct $1625 when figuring your taxes."

EXERCISES

Answer questions 1–4. Use actual expenses for question 1. For questions 2–4 refer to the standard mileage.

1. If Jane drives a $12,000 car, uses it 90% for business, and records 7200 miles in business travel at an average price per gallon of gas at $1.50, she claims actual expenses of $400 for gas, $200 for car washes, $96 for maintenance, $599 for insurance, and $2160 for depreciation. What is Jane's total deduction?

2. Write a formula for determining amount of deduction in dollars, D, from business-related miles driven, M.

3. Compute, using this formula, the amount of deduction, D, for the following business-related miles driven: 3500, 5000, and 7200.

4. Notice at 7200 miles, the deduction using actual expenses was larger than the standard mileage allowance. Under what easily observed situation might the standard allowance result in a larger deduction?

Chapter 10 Organizer

Topic	Procedure	Examples
Combining like terms, p. 611.	If the terms are like terms, combine the numerical coefficients directly in front of the variables.	Combine like terms. **(a)** $7x - 8x + 2x = -1x + 2x = x$ **(b)** $3a - 2b - 6a - 5b = -3a - 7b$ **(c)** $a - 2b + 3 - 5a = -4a - 2b + 3$
The distributive properties, p. 615.	$a(b + c) = ab + ac$ and $(b + c)a = ba + ca$	$5(x - 4y) = 5x - 20y$ $3(a + 2b - 6) = 3a + 6b - 18$ $(-2x + y)(7) = -14x + 7y$
Problems involving parentheses and like terms, p. 617.	1. Remove the parentheses using the distributive property. 2. Combine like terms.	Simplify. $2(4x - y) - 3(-2x + y) = 8x - 2y + 6x - 3y$ $\qquad\qquad\qquad\qquad\qquad = 14x - 5y$
Solving equations using the addition property, p. 621.	1. Add the appropriate value to both sides of the equation so that the variable is on one side and a number is on the other side of the equal sign. 2. Check by substituting your answer back into the original equation.	Solve for x. $$x - 2.5 = 7$$ $$x - 2.5 + 2.5 = 7 + 2.5$$ $$x + 0 = 9.5$$ $$x = 9.5$$ Check. $$x - 2.5 = 7$$ $$9.5 - 2.5 \overset{?}{=} 7$$ $$7 = 7 \checkmark$$
Solving equations using the division property, p. 626.	1. Divide both sides of the equation by the numerical coefficient of the variable. 2. Check by substituting your answer back into the original equation.	Solve for x. $$-12x = 60$$ $$\frac{-12x}{-12} = \frac{60}{-12}$$ $$x = -5$$ Check. $$-12x = 60$$ $$(-12)(-5) \overset{?}{=} 60$$ $$60 = 60 \checkmark$$
Solving equations using the multiplication property, p. 627.	1. Multiply both sides of the equation by the reciprocal of the numerical coefficient of the variable. 2. Check by substituting your answer back into the original equation.	Solve for x. $$\frac{3}{4}x = \frac{5}{8}$$ $$\frac{4}{3} \cdot \frac{3}{4}x = \frac{5}{8} \cdot \frac{4}{3}$$ $$x = \frac{5}{6}$$ Check. $$\frac{3}{4}x = \frac{5}{8}$$ $$\left(\frac{3}{4}\right)\left(\frac{5}{6}\right) \overset{?}{=} \frac{5}{8}$$ $$\frac{5}{8} = \frac{5}{8} \checkmark$$

Solving equations using more than one step, p. 634.

1. Remove any parantheses by using the distributive property.
2. Collect like terms on each side of the equation.
3. Add the appropriate value to both sides of the equation to get all numbers on one side.
4. Add the appropriate term to both sides of the equation to get all variable terms on the other side.
5. Divide both sides of the equation by the numerical coefficient of the variable term.
6. Check by substituting back into the original equation.

Solve for x.

$$5x - 2(6x - 1) = 3(1 + 2x) + 12$$
$$5x - 12x + 2 = 3 + 6x + 12$$
$$-7x + 2 = 15 + 6x$$
$$-7x + 2 + (-2) = 15 + (-2) + 6x$$
$$-7x = 13 + 6x$$
$$-7x + (-6x) = 13 + 6x + (-6x)$$
$$-13x = 13$$
$$\frac{-13x}{-13} = \frac{13}{-13}$$
$$x = -1$$

Check.

$$5x - 2(6x - 1) = 3(1 + 2x) + 12$$
$$5(-1) - 2[6(-1) - 1] \overset{?}{=} 3[1 + 2(-1)] + 12$$
$$-5 - 2[-6 - 1] \overset{?}{=} 3[1 - 2] + 12$$
$$-5 - 2[-7] \overset{?}{=} 3[-1] + 12$$
$$-5 + 14 \overset{?}{=} -3 + 12$$
$$9 = 9 \checkmark$$

Translating an English sentence into an equation, p. 639.

When translating English into an equation:

The English Phrase:	Is Usually Represented by the Symbol:
greater than increased by more than added to taller than total of sum of	+
less than smaller than fewer than shorter than difference of	−
multiplied by of product of times	×
double	2 ×
triple	3 ×
divided by ratio of quotient of	÷
is costs has the value of weighs has was equals represents amounts to	=

Translate a comparison in English into an equation using two given variables. Use t to represent Thursday's temperature and w to represent Wednesday's temperature. The temperature Thursday was 12 degrees higher than the temperature on Wednesday.

Temperature on Thursday	was	12°	higher than	Temperature on Wednesday
t	$=$	12	$+$	w

657

Topic	Procedure	Examples
Writing algebraic expressions for several quantities, p. 640.	1. Use a variable to describe the quantity that other quantities are described in terms of. 2. Write an expression in terms of that variable for each of the other quantities.	▲Write algebraic expressions for the size of each angle of a triangle. The second angle of a triangle is 7° less than the first angle. The third angle of a triangle is double the first angle. Use the letter x. Since two angles are described in terms of the first angle, we let the variable x represent that angle. Let x = the number of degrees in the first angle. Let $x - 7$ = the number of degrees in the second angle. Let $2x$ = the number of degrees in the third angle.
Solving applied problems using equations, p. 645.	1. *Understand the problem.* **(a)** Draw a sketch. **(b)** Choose a variable. **(c)** Represent other variables in terms of the first variable. 2. *Write an equation.* 3. *Solve the equation and state the answer.* 4. *Check.*	▲The perimeter of a field is 128 meters. The length of this rectangular field is 4 meters less than triple the width. Find the dimensions of the field. 1. *Understand the problem.* We need to find the dimensions of a rectangular field. We label the sides of our sketch. Let w = the width of the rectangle in meters. Let $3w - 4$ = the length of the rectangle in meters. 2. *Write an equation.* Perimeter = 2(width) + 2(length) $128 = 2(w) + 2(3w - 4)$ 3. *Solve and state the answer.* $128 = 2w + 6w - 8$ $128 = 8w - 8$ $136 = 8w$ $17 = w$ The width is 17 meters. $3w - 4 = 3(17) - 4 = 47$ The length is 47 meters. 4. *Check.* Is the perimeter 128 meters? Does $17 + 47 + 17 + 47 = 128$? $128 = 128$ ✓ Is the length 4 less than triple the width? Is $47 = 3(17) - 4$? $47 = 47$ ✓

Chapter 10 Review Problems

10.1 *Combine like terms.*

1. $-8a + 6 - 5a - 3$
2. $\dfrac{3}{4}x + \dfrac{2}{3} + \dfrac{1}{8}x + \dfrac{1}{4}$
3. $5x + 2y - 7x - 9y$

4. $3x - 7y + 8x + 2y$
5. $5x - 9y - 12 - 6x - 3y + 18$
6. $7x - 2y - 20 - 5x - 8y + 13$

10.2 *Simplify.*

7. $-3(5x + y)$
8. $-4(2x + 3y)$
9. $2(x - 3y + 4)$
10. $3(2x - 6y - 1)$

11. $-8(3a - 5b - c)$
12. $-9(2a - 8b - c)$
13. $5(1.2x + 3y - 5.5)$
14. $6(1.4x - 2y + 3.4)$

Simplify.

15. $2(x + 3y) - 4(x - 2y)$
16. $2(5x - y) - 3(x + 2y)$

17. $-2(a + b) - 3(2a + 8)$
18. $-4(a - 2b) + 3(5 - a)$

10.3 *Solve for the variable.*

19. $x - 3 = 9$
20. $-8 = x - 12$
21. $x + 8.3 = 20$
22. $2.4 = x - 5$

23. $3.1 + x = -9$
24. $x - 7 = 5.8$
25. $x - \dfrac{3}{4} = 2$
26. $x + \dfrac{1}{2} = 3\dfrac{3}{4}$

27. $x + \dfrac{3}{8} = \dfrac{1}{2}$
28. $x - \dfrac{5}{6} = \dfrac{2}{3}$
29. $2x + 20 = 25 + x$
30. $5x - 3 = 4x - 15$

10.4 *Solve for the variable.*

31. $8x = -20$
32. $-12y = 60$
33. $1.5x = 9$
34. $1.8y = 12.6$

35. $-7.2x = 36$
36. $6x = 1.5$
37. $\dfrac{3}{4}x = 6$
38. $\dfrac{2}{3}x = \dfrac{5}{9}$

10.5 *Solve for the variable.*

39. $3 - 2x = 9 - 8x$ **40.** $8 - 6x = -7 - 3x$ **41.** $10 + x = 3x - 6$

42. $8x - 7 = 5x + 8$ **43.** $9x - 3x + 18 = 36$ **44.** $4 + 3x - 8 = 12 + 5x + 4$

45. $5x - 2 = 27$ **46.** $2(3x - 4) = 7 - 2x + 5x$

47. $5 + 2y + 5(y - 3) = 6(y + 1)$ **48.** $3 + 5(y + 4) = 4(y - 2) + 3$

10.6 *Translate the English sentence into an equation using the variables indicated.*

49. The weight of the truck is 3000 pounds more than the weight of the car. Use w for the weight of the truck and c for the weight of the car.

50. The afternoon class had 18 fewer students than the morning class. Use a for the number of students in the afternoon class and m for the number of students in the morning class.

▲ **51.** The number of degrees in angle A is triple the number of degrees in angle B. Use A for the number of degrees in angle A and B for the number of degrees in angle B.

▲ **52.** The length of a rectangle is 3 inches shorter than double the width of the rectangle. Use w for the width of the rectangle in inches and l for the length of the rectangle in inches.

Write an algebraic expression for each quantity using the given variable.

53. Michael's salary is $2050 more than Roberto's salary. Use the letter r.

▲ **54.** The length of the second side of a triangle is double the length of the first side of the triangle. Use the letter x.

55. Nancy has completed six fewer graduate courses than Connie has. Use the letter c.

56. The number of books in the new library is 450 more than double the number of books in the old library. Use the letter b.

10.7   *Solve using an equation. Show what you let the variable equal.*

57. A 60-ft length of pipe is divided into two pieces. One piece is 6.5 ft longer than the other. Find the length of each piece.

58. Two clerks work in a store. The new employee earns $28 less per week than a person hired six months ago. Together they earn $412 per week. What is the weekly salary of each person?

59. A local fast-food restaurant had twice as many customers in March as in February. It had 3000 more customers in April than in February. Over the three months 45,200 customers came to the restaurant. How many came each month?

60. Alfredo drove 856 mi during three days of travel. He drove 106 more miles on Friday than on Thursday. He drove 39 fewer miles on Saturday than on Thursday. How many miles did he drive each day?

▲ **61.** A rectangle has a perimeter of 72 in. The length is 3 in. less than double the width. Find the dimensions of the rectangle.

▲ **62.** A rectangle has a perimeter of 180 m. The length is 2 m more than triple the width. Find the dimensions of the rectangle.

▲ **63.** A triangle has a perimeter of 99 m. The second side is 7 m longer than the first side. The third side is 4 m shorter than the first side. How long is each side?

▲ **64.** A triangle has three angles labeled *A*, *B*, and *C*. Angle *C* is triple the measure of angle *B*. Angle *A* is 74 degrees larger than angle *B*. Find the measure of each angle.

65. Dick and Anne Wright went to the furniture discount warehouse to purchase a new dining room table and chairs. They could purchase the desired table and four chairs for $275. They could purchase the same table and six chairs for $345. What is the cost of each chair? What is the cost of the table?

▲ **66.** A regulation NBA basketball court is in the shape of a rectangle. The width of the court is 44 feet shorter than the length. The perimeter of the court is 288 feet. Find the width and length of the court.

67. Victor and Samuel drove 760 miles to visit their mother over Christmas vacation. It took two days for them to make the trip. On the first day they drove 88 more miles than they did on the second day. How many miles did they drive each day?

68. During the second week of July, the North Lake Community College admissions office received 156 more applications than it did during the first week of July. During the third week of July, it received 142 fewer applications than it did during the first week of July. During these three weeks it received 800 applications. How many were received each week?

69. The Amtrak train from Boston to Philadelphia usually has four times as many coach passengers as first-class passengers. A recent trip had a total of 760 passengers. How many coach passengers were there? How many first-class passengers?

70. Megan receives an 8% commission on the furniture that she sells. Last month her total salary was $3050. Her base salary for the month was $1500. What was the cost of the furniture she sold last month?

Chapter 10 Test

1. _____

2. _____

3. _____

4. _____

5. _____

6. _____

7. _____

8. _____

9. _____

10. _____

11. _____

12. _____

13. _____

14. _____

15. _____

16. _____

17. _____

18. _____

Combine like terms.

1. $5a - 11a$

2. $\dfrac{1}{3}x + \dfrac{5}{8}y - \dfrac{1}{5}x + \dfrac{1}{2}y$

3. $-3x + 7y - 8x - 5y$

4. $6a - 5b + 7 - 5a - 3b + 4$

5. $7x - 8y + 2z - 9z + 8y$

6. $x + 5y - 6 - 5x - 7y + 11$

Simplify.

7. $5(12x - 5y)$

8. $-4(2x - 3y + 7)$

9. $-1.5(3a - 2b + c - 8)$

10. $2(-3a + 2b) - 5(a - 2b)$

Solve for the variable.

11. $-5 - 3x = 19$

12. $-3(4 - x) = 5(6 + 2x)$

13. $-5x + 9 = -4x - 6$

14. $8x - 2 - x = 3x - 9 - 10x$

15. $2x - 5 + 7x = 4x - 1 + x$

16. $3 - (x + 2) = 5 + 3(x + 2)$

Translate the English sentence into an equation using the variables indicated.

17. The second floor of Trabor Laboratory has 15 more classrooms than the first floor. Use s to represent the number of classrooms on the second floor and f to represent the number of classrooms on the first floor.

18. The north field yields 15,000 fewer bushels of wheat than the south field. Use n to represent the number of bushels of wheat in the north field and s to represent the number of bushels of wheat in the south field.

Write an algebraic expression for each quantity using the given variable.

▲ **19.** The first angle of a triangle is half the second angle. The third angle of the triangle is twice the second angle. Use the variable s.

▲ **20.** The length of a rectangle is 5 inches shorter than double the width. Use the letter w.

Solve using an equation.

21. The number of acres of land in the old Smithfield farm is three times the number of acres of land in the Prentice farm. Together the two farms have 348 acres. How many acres of land are there on each farm?

22. Sam earns $1500 less per year than Marcia does. The combined income of the two people is $46,500 per year. How much does each person earn?

23. Gina drove 975 miles in three days. She drove 56 more miles on Tuesday than on Monday. She drove 14 fewer miles on Wednesday than on Monday. How many miles did she drive each day?

▲ **24.** A rectangular field has a perimeter of 118 feet. The width is 8 feet longer than half the length. Find the dimensions of the rectangle.

19. _____

20. _____

21. _____

22. _____

23. _____

24. _____

Approximately one-half of this test is based on Chapter 10 material. The remainder is based on material covered in Chapter 1–9.

Solve. Simplify your answers.

1. Add. $456 + 89 + 123 + 79$

2. Multiply. $\begin{array}{r} 309 \\ \times\ 35 \\ \hline \end{array}$

3. Round to the nearest hundred. 45,678,934

4. Divide. $\dfrac{5}{12} \div \dfrac{1}{6}$

5. Multiply. $3\dfrac{1}{4} \times 2\dfrac{1}{2}$

6. Multiply. 9.3×0.0078

7. Subtract. $34{,}007.090 - 3456.789$

8. Find n. $\dfrac{9}{n} = \dfrac{40.5}{72}$

9. What is 28.5% of $5600?

10. 34% of what number is 1870?

11. Convert 345 millimeters to meters.

12. Convert 10 feet to inches.

▲ 13. Find the circumference of a circle with diameter 12 yards. Round to the nearest tenth.

▲ 14. Find the area of a triangle that has a base of 13 meters and a height of 22 meters.

1. _____

2. _____

3. _____

4. _____

5. _____

6. _____

7. _____

8. _____

9. _____

10. _____

11. _____

12. _____

13. _____

14. _____

Perform the indicated operations.

15. $4 - 8 + 12 - 32 - 7$

16. $(5)(-2)(3)(-1)$

Combine like terms.

17. $\dfrac{1}{2}a + \dfrac{1}{7}b + \dfrac{1}{4}a - \dfrac{3}{14}b$

18. $-4x + 5y - 9 - 2x - 3y + 12$

Simplify.

19. $-7(-3x + y - 8)$

20. $2(3x - 4y) - 8(x + 2y)$

Solve for the variable.

21. $5x - 5 = 7x - 13$

22. $7 - 9y - 12 = 3y + 5 - 8y$

23. $x - 2 + 5x + 3 = 183 - x$

24. $9(2x + 8) = 20 - (x + 5)$

Write an algebraic expression for each quantity, using the given variable.

25. The weight of the computer was 322 pounds more than the weight of the printer. Use the letter p.

26. The summer enrollment in algebra was 87 students less than the fall enrollment. Use the letter f.

Solve using an equation.

27. Barbara drove 1081 miles in three days. She drove 48 more miles on Friday than on Thursday. She drove 95 fewer miles on Saturday than on Thursday. How many miles did she drive each day?

▲ **28.** A rectangle has a perimeter of 98 feet. The length is 8 feet longer than double the width. Find each dimension.

15. _____

16. _____

17. _____

18. _____

19. _____

20. _____

21. _____

22. _____

23. _____

24. _____

25. _____

26. _____

27. _____

28. _____

Practice Final Examination

This examination is based on Chapters 1–10 of the book. There are 10 questions covering the content of each chapter.

Chapter 1

1. Write in words. 82,367

2. Add.   13,428
  + 16,905

3. Add.   19
  23
  16
  45
  + 70

4. Subtract.   89,071
  − 54,968

Multiply.

5.   78
  × 54

6.   2035
  × 107

In questions 7 and 8, divide. (Be sure to indicate the remainder if one exists.)

7. $7\overline{)1106}$

8. $26\overline{)15,756}$

9. Evaluate. Perform operations in the proper order. $3^4 + 20 \div 4 \times 2 + 5^2$

10. Melinda traveled 512 miles in her car. The car used 16 gallons of gas on the entire trip. How many miles per gallon did the car achieve?

Chapter 2

11. Reduce the fraction. $\dfrac{14}{30}$

12. Change to an improper fraction. $3\dfrac{9}{11}$

13. Add. $\dfrac{1}{10} + \dfrac{3}{4} + \dfrac{4}{5}$

14. Add. $2\dfrac{1}{3} + 3\dfrac{3}{5}$

15. Subtract. $4\dfrac{5}{7} - 2\dfrac{1}{2}$

16. Multiply. $1\dfrac{1}{4} \times 3\dfrac{1}{5}$

1. _____

2. _____

3. _____

4. _____

5. _____

6. _____

7. _____

8. _____

9. _____

10. _____

11. _____

12. _____

13. _____

14. _____

15. _____

16. _____

17. Divide. $\dfrac{7}{9} \div \dfrac{5}{18}$

18. Divide. $\dfrac{5\frac{1}{2}}{3\frac{1}{4}}$

19. Lucinda jogged $1\frac{1}{2}$ miles on Monday, $3\frac{1}{4}$ miles on Tuesday, and $2\frac{1}{10}$ miles on Wednesday. How many miles in all did she jog over the three-day period?

20. A butcher has $11\frac{2}{3}$ pounds of steak. She wishes to place them in several equal-sized packages. Each package will hold $2\frac{1}{3}$ pounds of steak. How many packages can be made?

Chapter 3

21. Express as a decimal. $\dfrac{719}{1000}$

22. Write in reduced fractional notation. 0.86

23. Fill in the blank with $<$, $=$, or $>$ 0.315 _____ 0.309

24. Round to the nearest hundredth. 506.3782

25. Add.
$$\begin{array}{r} 9.6 \\ 3.82 \\ 1.05 \\ + \ 7.3 \\ \hline \end{array}$$

26. Subtract.
$$\begin{array}{r} 3.61 \\ - \ 2.853 \\ \hline \end{array}$$

27. Multiply.
$$\begin{array}{r} 1.23 \\ \times \ 0.4 \\ \hline \end{array}$$

28. Divide. $0.24\overline{)0.8856}$

29. Write as a decimal. $\dfrac{13}{16}$

30. Evaluate by performing operations in proper order.
$0.7 + (0.2)^3 - 0.08(0.03)$

Chapter 4

31. Write a rate in simplest form to compare 7000 students to 215 faculty.

32. Is this a proportion? $\dfrac{12}{15} = \dfrac{17}{21}$

17. _____

18. _____

19. _____

20. _____

21. _____

22. _____

23. _____

24. _____

25. _____

26. _____

27. _____

28. _____

29. _____

30. _____

31. _____

32. _____

33.

34.

35.

36.

37.

38.

39.

40.

41.

42.

43.

44.

45.

46.

47.

48.

49.

50.

Solve the proportion. Round to the nearest tenth when necessary.

33. $\dfrac{5}{9} = \dfrac{n}{17}$

34. $\dfrac{3}{n} = \dfrac{7}{18}$

35. $\dfrac{n}{12} = \dfrac{5}{4}$

36. $\dfrac{n}{7} = \dfrac{36}{28}$

Solve using a proportion. Round to the nearest hundredth when necessary.

37. Bob earned $2000 for painting three houses. How much would he earn for painting five houses?

38. Two cities that are actually 200 miles apart appear 6 inches apart on the map. Two other cities are 325 miles apart. How far apart will they appear on the same map?

39. Roberta earned $68 last week on her part-time job. She had $5 withheld for federal income tax. Last year she earned $4000 on her part-time job. Assuming the same rate, how much was withheld for federal income tax last year?

40. Malaga's recipe feeds 18 people and calls for 1.2 pounds of butter. If she wants to feed 24 people, how many pounds of butter does she need?

Chapter 5

41. Write as a percent. 0.0063

42. Change $\dfrac{17}{80}$ to a percent.

Round to the nearest tenth when necessary.

43. Write as a decimal. 164%

44. What percent of 300 is 52?

45. Find 6.3% of 4800.

46. 145 is 58% of what number?

47. 126% of 3400 is what number?

48. Pauline bought a new car. She got an 8% discount. The car listed for $11,800. How much did she pay for the car?

49. A total of 1260 freshmen were admitted to Central College. This is 28% of the student body. How big is the student body?

50. There are 11.28 centimeters of water in the rain gauge this week. Last week the rain gauge held 8.40 centimeters of water. What is the percent of increase from last week to this week?

Chapter 6

Convert. Express your answers as a decimal rounded to the nearest hundredth when necessary.

51. 17 quarts = _____ gallons

52. 3.25 tons = _____ pounds

53. 16 feet = _____ inches

54. 5.6 kilometers = _____ meters

55. 69.8 grams = _____ kilogram

56. 2.48 milliliters = _____ liter

57. 12 miles = _____ kilometers

In questions 58 and 59, write in scientific notation.

58. 0.00063182

59. 126,400,000,000

60. Two metal sheets are 0.623 centimeter and 0.74 centimeter thick, respectively. An insulating foil is 0.0428 millimeter thick. When all three layers are placed tightly together, what is the total thickness? Express your answer in centimeters.

Chapter 7

Round to the nearest hundredth when necessary. Use $\pi = 3.14$ when necessary.

▲ **61.** Find the perimeter of a rectangle that is 6 meters long and 1.2 meters wide.

▲ **62.** Find the perimeter of a trapezoid with sides of 82 centimeters, 13 centimeters, 98 centimeters, and 13 centimeters.

▲ **63.** Find the area of a triangle with base 6 feet and height 1.8 feet.

▲ **64.** Find the area of a trapezoid with bases of 12 meters and 8 meters and a height of 7.5 meters.

▲ **65.** Find the area of a circle with radius 6 meters.

▲ **66.** Find the circumference of a circle with diameter 18 meters.

▲ **67.** Find the volume of a cone with a radius of 4 centimeters and a height of 10 centimeters.

▲ **68.** Find the volume of a rectangular pyramid with a base of 12 feet by 19 feet and a height of 2.7 feet.

51. _____

52. _____

53. _____

54. _____

55. _____

56. _____

57. _____

58. _____

59. _____

60. _____

61. _____

62. _____

63. _____

64. _____

65. _____

66. _____

67. _____

68. _____

69.

▲ **69.** Find the area of this object, consisting of a square and a triangle.

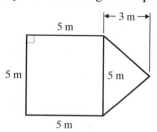

▲ **70.** In the following pair of similar triangles, find *n*.

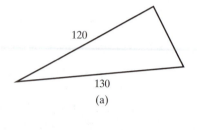

120

130

(a)

30

n

(b)

70.

Chapter 8

The following double-bar graph indicates the quarterly profits for Westar Corporation in 2000 and 2001.

71.

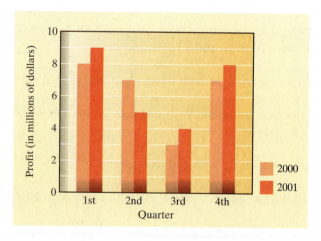

71. What were the profits in the fourth quarter of 2001?

72. How much greater were the profits in the first quarter of 2001 than the profits in the first quarter of 2000?

72.

The following line graph depicts the average annual temperature at West Valley for the years 1960, 1970, 1980, 1990, and 2000.

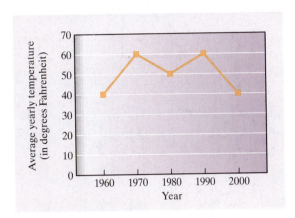

73. What was the average temperature in 1980?

74. In what 10-year period did the average temperature show the greatest decline?

The following histogram shows the number of students in each age category at Center City College.

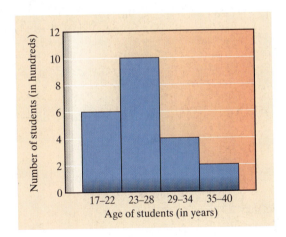

75. How many students are between 17 and 22 years old?

76. How many students are between 23 and 34 years old?

77. Find the *mean* and the *median* of the following. 8, 12, 16, 17, 20, 22

78. Evaluate exactly. $\sqrt{49} + \sqrt{81}$

79. Approximate to the nearest thousandth using a calculator or the square root table. $\sqrt{123}$

73.

74.

75.

76.

77.

78.

79.

80. _____

81. _____

82. _____

83. _____

84. _____

85. _____

86. _____

87. _____

88. _____

89. _____

90. _____

91. _____

92. _____

93. _____

94. _____

95. _____

96. _____

97. _____

98. _____

99. _____

100. _____

▲ **80.** Find the unknown side of the right triangle.

? / 12 ft / 9 ft

Chapter 9

Add.

81. $-8 + (-2) + (-3)$

82. $-\dfrac{1}{4} + \dfrac{3}{8}$

Subtract.

83. $9 - 12$

84. $-20 - (-3)$

85. Multiply. $(2)(-3)(4)(-1)$

86. Divide. $-\dfrac{2}{3} \div \dfrac{1}{4}$

Perform the indicated operations in the proper order.

87. $(-16) \div (-2) + (-4)$

88. $12 - 3(-5)$

89. $7 - (-3) + 12 \div (-6)$

90. $\dfrac{(-3)(-1) + (-4)(2)}{(0)(6) + (-5)(2)}$

Chapter 10

Combine like terms.

91. $5x - 3y - 8x - 4y$

92. $5 + 2a - 8b - 12 - 6a - 9b$

Simplify.

93. $-2(x - 3y - 5)$

94. $-2(4x + 2) - 3(x + 3y)$

Solve for the variable.

95. $5 - 4x = -3$

96. $5 - 2(x - 3) = 15$

97. $7 - 2x = 10 + 4x$

98. $-3(x + 4) = 2(x - 5)$

Solve using an equation.

99. There are 12 more students taking history than math. There are twice as many students taking psychology as there are students taking math. There are 452 students taking these three subjects. How many are taking history? How many are taking math?

▲ **100.** A rectangle has a perimeter of 106 feet. The length is 5 meters longer than double the width. Find the length and width of the rectangle.

Basic College Mathematics Glossary

Absolute value of a number (9.1) The absolute value of a number is the distance between that number and zero on the number line. When we find the absolute value of a number, we use the | | notation. To illustrate, $|-4| = 4$, $|6| = 6$, $|-20 - 3| = |-23| = 23$, $|0| = 0$.

Addends (1.2) When two or more numbers are added, the numbers being added are called addends. In the problem $3 + 4 = 7$, the numbers 3 and 4 are both addends.

Algebraic expression (10.6) An algebraic expression consists of variables, numerals, and operation signs.

Altitude of a triangle (7.4) The height of a triangle.

Amount of a percent equation (5.3A) The product we obtain when we multiply a percent times a number. In the equation $75 = 50\% \times 150$, the amount is 75.

Angle (7.1) An angle is formed whenever two lines meet.

Area (7.1) The measure of the surface inside a geometric figure. Area is measured in square units, such as square feet.

Associative property of addition (1.2) The property that tells us that when three numbers are added, it does not matter which two numbers are added first. An example of the associative property is $5 + (1 + 2) = (5 + 1) + 2$. Whether we add $1 + 2$ first and then add 5 to that, or add $5 + 1$ first and then add that result to 2, we will obtain the same result.

Associative property of multiplication (1.4) The property that tells us that when we multiply three numbers, it does not matter which two numbers we group together first to multiply; the result will be the same. An example of the associative property of multiplication is $2 \times (5 \times 3) = (2 \times 5) \times 3$.

Base (1.6) The number that is to be repeatedly multiplied in exponent form. When we write $16 = 2^4$, the number 2 is the base.

Base of a percent equation (5.3A) The quantity we take a percent of. In the equation $80 = 20\% \times 400$, the base is 400.

Billion (1.1) The number 1,000,000,000.

Borrowing (1.3) The renaming of a number in order to facilitate subtraction. When we subtract $42 - 28$, we rename 42 as 3 tens plus 12. This represents 3 tens and 12 ones. This renaming is called borrowing.

Box (7.8) A three-dimensional object whose every side is a rectangle. Another name for a box is a *rectangular solid*.

Building fraction property (2.6) For whole numbers a, b, and c, where neither b nor c equals zero,

$$\frac{a}{b} = \frac{a}{b} \times 1 = \frac{a}{b} \times \frac{c}{c} = \frac{a \times c}{b \times c}.$$

Building up a fraction (2.6) To make one fraction into an equivalent fraction by making the denominator and numerator larger numbers. For example, the fraction $\frac{3}{4}$ can be built up to the fraction $\frac{30}{40}$.

Caret (3.5) A symbol $_\wedge$ used to indicate the new location of a decimal point when performing division of decimal fractions.

Celsius temperature (6.4) A temperature scale in which water boils at 100 degrees ($100°C$) and freezes at 0 degrees ($0°C$). To convert Celsius temperature to Fahrenheit, we use the helpful formula $F = 1.8 \times C + 32$.

Center of a circle (7.7) The point in the middle of a circle from which all points on the circle are an equal distance.

Centimeter (6.2) A unit of length commonly used in the metric system to measure small distances. 1 centimeter = 0.01 meter.

Circle (7.7) A figure for which all points are at an equal distance from a given point.

Circumference of a circle (7.7) The distance around the rim of a circle.

Commission (5.5) The amount of money a salesperson is paid that is a percentage of the value of the sales made by that salesperson. The commission is obtained by multiplying the commission rate times the value of the sales. If a salesman sells $120,000 of insurance and his commission rate is 0.5%, then his commission is $0.5\% \times 120,000 = \$600.00$

Common denominator (2.7) Two fractions have a common denominator if the same number appears in the denominator of each fraction. $\frac{3}{7}$ and $\frac{1}{7}$ have a common denominator of 7.

Commutative property of addition (1.2) The property that tells us that the order in which two numbers are added does not change the sum. An example of the commutative property of addition is $3 + 6 = 6 + 3$.

Commutative property of multiplication (1.4) The property that tells us that the order in which two numbers are multiplied does not change the value of the answer. An example of the commutative property of multiplication is $7 \times 3 = 3 \times 7$.

Composite number (2.2) A composite number is a whole number greater than 1 that can be divided by whole numbers other than itself. The number 6 is a composite number since it can be divided exactly by 2 and 3 (as well as by 1 and 6).

Cone (7.8) A three-dimensional object shaped like an ice-cream cone or the sharpened end of a pencil.

Cross-multiplying (4.3) If you have a proportion such as $\frac{n}{5} = \frac{12}{15}$, then to cross-multiply, you form products to obtain $n \times 15 = 5 \times 12$.

Cubic centimeter (6.3) A metric measurement of volume equal to 1 milliliter.

Cup (6.1) One of the smallest units of volume in the American system. 2 cups = 1 pint.

Cylinder (7.8) A three-dimensional object shaped like a tin can.

Debit (1.2) A debit in banking is the removing of money from an account. If you had a savings account and took $300 out of it on Wednesday, we would say that you had a debit of $300 from your account. Often a bank will add a service charge to your account and use the word *debit* to mean that it has removed money from your account to cover the charge.

Decimal fraction (3.1) A fraction whose denominator is a power of 10.

Decimal places (3.4) The number of digits to the right of the decimal point in a decimal fraction. The number 1.234 has three decimal places, while the number 0.129845 has six decimal places. A whole number such as 42 is considered to have zero decimal places.

Decimal point (3.1) The period that is used when writing a decimal fraction. In the number 5.346, the period between the 5 and the 3 is the decimal point. It separates the whole number from the fractional part that is less than 1.

Decimal system (1.1) Our number system is called the decimal system or base 10 system because the value of numbers written in our system is based on tens and ones.

Decimeter (6.2) A unit of length not commonly used in the metric system. 1 decimeter = 0.1 meter.

Degree (7.1) A unit used to measure an angle. A degree is $\frac{1}{360}$ of a complete revolution. An angle of 32 degrees is written as 32°.

Dekameter (6.2) A unit of length not commonly used in the metric system. 1 dekameter = 10 meters.

Denominator (2.1) The number on the bottom of a fraction. In the fraction $\frac{2}{9}$ the denominator is 9.

Deposit (1.2) A deposit in banking is the placing of money in an account. If you had a checking account and on Tuesday you placed $124 into that account, we would say that you made a deposit of $124.

Diameter of a circle (7.7) A line segment across the circle that passes through the center of the circle. The diameter of a circle is equal to twice the radius of the circle.

Difference (1.3) The result of performing a subtraction. In the problem $9 - 2 = 7$ the number 7 is the difference.

Digits (1.1) The symbols 0, 1, 2, 3, 4, 5, 6, 7, 8, and 9 are called digits.

Discount (5.4) The amount of reduction in a price. The discount is a product of the discount rate times the list price. If the list price of a television is $430.00 and it has a discount rate of 35%, then the amount of discount is $35\% \times \$430.00 = \$150.50$. The price would be reduced by $150.50.

Distributive property of multiplication over addition (1.4) The property illustrated by $5 \times (4 + 3) = (5 \times 4) + (5 \times 3)$. In general, for any numbers a, b, and c, it is true that $a(b + c) = a \times b + a \times c$.

Dividend (1.5) The number that is being divided by another. In the problem $14 \div 7 = 2$, the number 14 is the dividend.

Divisor (1.5) The number that you divide into another number. In the problem $30 \div 5 = 6$, the number 5 is the divisor.

Earned run average (4.4) A ratio formed by finding the number of runs a pitcher would give up in a nine-inning game. If a pitcher has an earned run average of 2, it means that, on the average, he gives up two runs for every nine innings he pitches.

Equal fractions (2.2) Fractions that represent the same number. The fractions $\frac{3}{4}$ and $\frac{6}{8}$ are equal fractions.

Equality test of fractions (2.2) Two fractions $\frac{a}{b}$ and $\frac{c}{d}$ are equal if the product $a \times d = b \times c$. In this case, a, b, c, and d are whole numbers and b and $d \neq 0$.

Equations (10.3) Mathematical statements with variables that say that two expressions are equal, such as $x + 3 = -8$ and $2s + 5s = 34 - 4s$.

Equilateral triangle (7.4) A triangle with three equal sides.

Equivalent equations (10.3) Equations that have the same solution.

Equivalent fractions (2.2) Two fractions that are equal.

Expanded notation for a number (1.1) A number is written in expanded notation if it is written as a sum of hundreds, tens, ones, etc. The expanded notation for 763 is $700 + 60 + 3$.

Exponent (1.6) The number that indicates the number of times a factor occurs. When we write $8 = 2^3$, the number 3 is the exponent.

Factors (1.4) Each of the numbers that are multiplied. In the problem $8 \times 9 = 72$, the numbers 8 and 9 are factors.

Fahrenheit temperature (6.4) A temperature scale in which water boils at 212 degrees (212°F) and freezes at 32 degrees (32°F). To convert Fahrenheit temperature to Celsius, we use the formula $C = \frac{5 \times F - 160}{9}$.

Foot (6.1) American system unit of length. 3 feet = 1 yard. 12 inches = 1 foot.

Fundamental theorem of arithmetic (2.2) Every composite number has a unique product of prime numbers.

Gallon (6.1) A unit of volume in the American system. 4 quarts = 1 gallon.

Gigameter (6.2) A metric unit of length equal to 1,000,000,000 meters.

Gram (6.3) The basic unit of weight in the metric system. A gram is defined as the weight of the water in a box that is 1 centimeter on each side. 1 gram = 1000 milligrams. 1 gram = 0.001 kilogram.

Hectometer (6.2) A unit of length not commonly used in the metric system. 1 hectometer = 100 meters.

Height (7.3) The distance between two parallel sides in a four-sided figure such as a parallelogram or a trapezoid.

Height of a cone (7.8) The distance from the vertex of a cone to the base of the cone.

Height of a pyramid (7.8) The distance from the point on a pyramid to the base of the pyramid.

Height of a triangle (7.4) The distance of a line drawn from a vertex perpendicular to the other side, or an extension of the other side, of the triangle. This is sometimes called the *altitude of a triangle*.

Hexagon (7.3) A six-sided figure.

Hypotenuse (7.6) The side opposite the right angle in a right triangle. The hypotenuse is always the longest side of a right triangle.

Improper fraction (2.3) A fraction in which the numerator is greater than or equal to the denominator. The fractions $\frac{34}{29}, \frac{8}{7}$, and $\frac{6}{6}$ are all improper fractions.

Inch (6.1) The smallest unit of length in the American system. 12 inches = 1 foot.

Inequality symbol (3.2) The symbol that is used to indicate whether a number is greater than another number or less than another number. Since 5 is greater than 3, we would write this with a "greater than" symbol as follows: $5 > 3$. The statement "7 is less than 12" would be written as follows: $7 < 12$.

Interest (5.4) The money that is paid for the use of money. If you deposit money in a bank, the bank uses that money and pays you interest. If you borrow money, you pay the bank interest for the use of that money. Simple interest is determined by the formula $I = P \times R \times T$. Compound interest is usually determined by a table, a calculator, or a computer.

Invert a fraction (2.5) To invert a fraction is to interchange the numerator and the denominator. If we invert $\frac{5}{9}$, we obtain the fraction $\frac{9}{5}$. To invert a fraction is sometimes referred to as *to take the reciprocal of a fraction*.

Irreducible (2.2) A fraction that cannot be reduced (simplified) is called irreducible.

Isosceles triangle (7.4) A triangle with two sides equal.

Kilogram (6.3) The most commonly used metric unit of weight. 1 kilogram = 1000 grams.

Kiloliter (6.3) The metric unit of volume normally used to measure large volumes. 1 kiloliter = 1000 liters.

Kilometer (6.2) The unit of length commonly used in the metric system to measure large distances. 1 kilometer = 1000 meters.

Least common denominator (LCD) (2.6) The least common denominator (LCD) of two or more fractions is the smallest number that can be divided without remainder by each fraction's denominator. The LCD of $\frac{1}{3}$ and $\frac{1}{4}$ is 12. The LCD of $\frac{5}{6}$ and $\frac{4}{15}$ is 30.

Legs of a right triangle (7.6) The two shortest sides of a right triangle.

Length of a rectangle (7.2) Each of the longer sides of a rectangle.

Like terms (10.1) Like terms have identical variables with identical exponents. $-5x$ and $3x$ are like terms. $-7xyz$ and $-12xyz$ are like terms.

Line segment (7.3) A portion of a straight line that has a beginning and an end.

Liter (6.3) The standard metric measurement of volume. 1 liter = 1000 milliliters. 1 liter = 0.001 kiloliter.

Mean (8.4) The mean of a set of values is the sum of the values divided by the number of values. The mean of the numbers 10, 11, 14, 15 is 12.5. In everyday language, when people use the word *average*, they are usually referring to the mean.

Median (8.4) If a set of numbers is arranged in order from smallest to largest, the median is that value that has the same number of values above it as below it. The median of the numbers 3, 7, and 8 is 7. If the list contains an even number of items, we obtain the median by finding the mean of the two middle numbers. The median of the numbers 5, 6, 10, and 11 is 8.

Megameter (6.2) A metric unit of length equal to 1,000,000 meters.

Meter (6.2) The basic unit of length in the metric system. 1 meter = 1000 millimeters. 1 meter = 0.001 kilometer.

Metric ton (6.3) A metric unit of measurement for very heavy weights. 1 metric ton = 1,000,000 grams.

Microgram (6.3) A unit of weight equal to 0.000001 gram.

Micrometer (6.2) A metric unit of length equal to 0.000001 meter.

Mile (6.1) Largest unit of length in the American system. 5280 feet = 1 mile, 1760 yards = 1 mile.

Milligram (6.3) A metric unit of weight used for very, very small objects. 1 milligram = 0.001 gram.

Milliliter (6.3) The metric unit of volume normally used to measure small volumes. 1 milliliter = 0.001 liter.

Millimeter (6.2) A unit of length commonly used in the metric system to measure very small distances. 1 millimeter = 0.001 meter.

Million (1.1) The number 1,000,000.

Minuend (1.3) The number being subtracted from in a subtraction problem. In the problem $8 - 5 = 3$, the number 8 is the minuend.

Mixed number (2.3) A number created by the sum of a whole number greater than 1 and a proper fraction. The numbers $4\frac{5}{6}$ and $1\frac{1}{8}$ are both mixed numbers. Mixed numbers are sometimes referred to as *mixed fractions*.

Mode (8.4) The mode of a set of data is the number or numbers that occur most often.

Multiplicand (1.4) The first factor in a multiplication problem. In the problem $7 \times 2 = 14$, the number 7 is the multiplicand.

Multiplier (1.4) The second factor in a multiplication problem. In the problem $6 \times 3 = 18$, the number 3 is the multiplier.

Nanogram (6.3) A unit of weight equal to 0.000000001 gram.

Nanometer (6.2) A metric unit of length equal to 0.000000001 meter.

Negative numbers (9.1) All of the numbers to the left of zero on the number line. The numbers $-1.5, -16, -200.5, -4500$ are all negative numbers. All negative numbers are written with a negative sign in front of the digits.

Number line (1.7) A line on which numbers are placed in order from smallest to largest.

Numerator (2.1) The number on the top of a fraction. In the fraction $\frac{3}{7}$ the numerator is 3.

Numerical coefficients (10.1) The numbers in front of the variables in one or more terms. If we look at $-3xy + 12w$, we find that the numerical coefficient of the xy term is -3 while the numerical coefficient of the w term is 12.

Octagon (7.3) An eight-sided figure.

Odometer (1.8) A device on an automobile that displays how many miles the car has been driven since it was first put into operation.

Opposite of a signed number (9.2) The opposite of a signed number is a number that has the same absolute value. The opposite of -5 is 5. The opposite of 7 is -7.

Order of operations (1.6) An agreed-upon procedure to do a problem with several arithmetic operations in the proper order.

Ounce (6.1) Smallest unit of weight in the American system. 16 ounces = 1 pound.

Overtime (2.9) The pay earned by a person if he or she works more than a certain number of hours per week. In most jobs that pay by the hour, a person will earn $1\frac{1}{2}$ times as much per hour for every hour beyond 40 hours worked in one workweek. For example, Carlos earns $6.00 per hour for the first 40 hours in a week and overtime for each additional hour. He would earn $9.00 per hour for all hours he worked in that week beyond 40 hours.

Parallel lines (7.3) Two straight lines that are always the same distance apart.

Parallelogram (7.3) A four-sided figure with both pairs of opposite sides parallel.

Parentheses (1.4) One of several symbols used in mathematics to indicate multiplication. For example, (3)(5) means 3 multiplied by 5. Parentheses are also used as a grouping symbol.

Percent (5.1) The word *percent* means per one hundred. For example, 14 percent means $\frac{14}{100}$.

Percent of decrease (5.5) The percent that something decreases is determined by dividing the amount of decrease by the original amount. If a tape deck sold for $300 and its price was decreased by $60, the percent of decrease would be $\frac{60}{300} = 0.20 = 20\%$.

Percent of increase (5.5) The percent that something increases is determined by dividing the amount of increase by the original amount. If the population of a town was 5000 people and the population increased by 500 people, the percent of increase would be $\frac{500}{5000} = 0.10 = 10\%$.

Percent proportion (5.3B) The percent proportion is the equation $\frac{a}{b} = \frac{p}{100}$ where a is the amount, b is the base, and p is the percent number.

Percent symbol (5.1) A symbol that is used to indicate percent. To indicate 23 percent, we write 23%.

Perfect square (7.4) When a whole number is multiplied by itself, the number that is obtained is a perfect square. The numbers 1, 4, 9, 16, 25, 36, 49, 64, 81, and 100 are all perfect squares.

Perimeter (7.2) The distance around a figure.

Perpendicular lines (7.1) Lines that meet at an angle of 90 degrees.

Pi (7.7) Pi is an irrational number that we obtain if we divide the circumference of a circle by the diameter of a circle. It is represented by the symbol π. Accurate to eleven decimal places, the value of pi is given by 3.14159265359. For most work in this textbook, the value of 3.14 is used to approximate the value of pi.

Picogram (6.3) A unit of weight equal to 0.000000000001 gram.

Pint (6.1) Unit of volume in the American system. 2 pints = 1 quart.

Placeholder (1.1) The use of a digit to indicate a place. Zero is a placeholder in our number system. It holds a position and shows that there is no other digit in that place.

Place-value system (1.1) Our number system is called a place-value system because the placement of the digits tells the value of the number. If we use the digits 5 and 4 to write the number 54, the result is different than if we placed them in opposite order and wrote 45.

Positive numbers (9.1) All of the numbers to the right of zero on the number line. The numbers 5, 6.2, 124.186, 5000 are all positive numbers. A positive number such as +5 is usually written without the positive sign.

Pound (6.1) Basic unit of weight in the American system. 2000 pounds = 1 ton. 16 ounces = 1 pound.

Power of 10 (1.4) Whole numbers that begin with 1 and end in one or more zeros are called powers of 10. The numbers 10, 100, 1000, etc., are all powers of 10.

Prime factors (2.2) Factors that are prime numbers. If we write 15 as a product of prime factors, we have $15 = 5 \times 3$.

Prime number (2.2) A prime number is a whole number greater than 1 that can only be divided by 1 and itself. The first fifteen prime numbers are 2, 3, 5, 7, 11, 13, 17, 19, 23, 29, 31, 37, 41, 43, and 47. The list of prime numbers goes on indefinitely.

Principal (5.4) The amount of money deposited or borrowed on which interest is computed. In the simple interest formula $I = P \times R \times T$, the P stands for the principal. (The other letters are I = interest, R = interest rate, and T = amount of time.)

Product (1.4) The answer in a multiplication problem. In the problem $3 \times 4 = 12$ the number 12 is the product.

Proper fraction (2.3) A fraction in which the numerator is less than the denominator. The fractions $\frac{3}{4}$ and $\frac{15}{16}$ are proper fractions.

Proportion (4.2) A statement that two ratios or two rates are equal. The statement $\frac{3}{4} = \frac{15}{20}$ is a proportion. The statement $\frac{5}{7} = \frac{7}{9}$ is false, and is therefore not a proportion.

Pyramid (7.8) A three-dimensional object made up of a geometric figure for a base and triangular sides that meet at a point. Some pyramids are shaped like the great pyramids of Egypt.

Pythagorean Theorem (7.6) A statement that for any right triangle the square of the hypotenuse equals the sum of the squares of the two legs of the triangle.

Quadrilateral (7.3) A four-sided geometric figure.

Quadrillion (1.1) The number 1,000,000,000,000,000.

Quart (6.1) Unit of volume in the American system. 4 quarts = 1 gallon.

Quotient (1.5) The answer after performing a division problem. In the problem $60 \div 6 = 10$ the number 10 is the quotient.

Radius of a circle (7.7) A line segment from the center of a circle to any point on the circle. The radius of a circle is equal to one-half the diameter of the circle.

Rate (4.1) A rate compares two quantities that have different units. Examples of rates are $5.00 an hour and 13 pounds for every 2 inches. In fraction form, these two rates would be written as $\frac{\$5.00}{1 \text{ hour}}$ and $\frac{13 \text{ pounds}}{2 \text{ inches}}$.

Ratio (4.1) A ratio is a comparison of two quantities that have the same units. To compare 2 to 3, we can express the ratio in three ways: the ratio of 2 to 3; 2:3; or the fraction $\frac{2}{3}$.

Ratio in simplest form (4.1) A ratio is in simplest form when the two numbers do not have a common factor.

Rectangle (7.2) A four-sided figure that has four right angles.

Reduced fraction (2.2) A fraction for which the numerator and denominator have no common factor other than 1. The fraction $\frac{5}{7}$ is a reduced fraction. The fraction $\frac{15}{21}$ is not a reduced fraction because both numerator and denominator have a common factor of 3.

Regular hexagon (7.3) A six-sided figure with all sides equal.

Regular octagon (7.3) An eight-sided figure with all sides equal.

Remainder (1.5) When two numbers do not divide exactly, a part is left over. This part is called the remainder. For example, $13 \div 2 = 6$ with 1 left over; the 1 is the remainder.

Repeating decimals (3.6) Decimals that have a digit or a group of digits that repeat. The decimals $0.33333333333 \ldots$ and $1.234234234234 \ldots$ are repeating decimals. The pattern of repeating continues indefinitely. Repeating decimals can be written in a form with a bar over the repeating digit(s). Thus the preceding decimals could be written as $0.\overline{3}$ and $1.\overline{234}$.

Right angle (7.1) and (7.4) An angle that measures 90 degrees.

Right triangle (7.4) A triangle with one 90-degree angle.

Rounding (1.7) The process of writing a number in an approximate form for convenience. The number 9756 rounded to the nearest hundred is 9800.

Sales tax (5.4) The amount of tax on a purchase. The sales tax for any item is a product of the sales tax rate times the purchase price. If an item is purchased for $12.00 and the sales tax rate is 5%, the sales tax is $5\% \times 12.00 = \$0.60$.

Scientific notation (9.5) A positive number is written in scientific notation if it is in the form $a \times 10^n$ where a is a number greater than or equal to 1, but less than 10, and n is an integer. If we write 5678 in scientific notation, we have 5.678×10^3. If we write 0.00825 in scientific notation, we have 8.25×10^{-3}.

Semicircle (7.7) One-half of a circle. The semicircle usually includes the diameter of a circle connected to one-half the circumference of the circle.

Sides of an angle (7.1) The two line segments that meet to form an angle.

Signed numbers (9.1) All of the numbers on a number line. Numbers like -33, 2, 5, -4.2, 18.678, -8.432 are all signed numbers. A negative number always has a negative sign in front of the digits. A positive number such as $+3$ is usually written without the positive sign in front of it.

Similar triangles (7.9) Two triangles that have the same shape but are not necessarily the same size. The corresponding angles of similar triangles are equal. The corresponding sides of similar triangles have the same ratio.

Simple interest (5.4) The interest determined by the formula $I = P \times R \times T$ where I = the interest obtained, P = the principal or the amount borrowed or invested, R = the interest rate (usually on an annual basis), and T = the number of time periods (usually years).

Solution of an equation (10.3) A number is a solution of an equation if replacing the variable by the number makes the equation always true. The solution of $x - 5 = -20$ is the number -15.

Sphere (7.8) A three-dimensional object shaped like a perfectly round ball.

Square (7.2) A rectangle with all four sides equal.

Square root (7.5) The square root of a number is one of two identical factors of that number. The square root of 9 is 3. The square root of 121 is 11.

Square root sign (7.5) The symbol $\sqrt{}$. When we want to find the square root of 25, we write $\sqrt{25}$. The answer is 5.

Standard notation for a number (1.1) A number written in ordinary terms. For example, $70 + 2$ in standard notation is 72.

Subtrahend (1.3) The number being subtracted. In the problem $7 - 1 = 6$, the number 1 is the subtrahend.

Sum (1.2) The result of an addition of two or more numbers. In the problem $7 + 3 + 5 = 15$, the number 15 is the sum.

Term (10.1) A number, a variable, or a product of a number and one or more variables. $5x$, $2ab$, $-43cdef$ are three examples of terms, separated in an expression by a $+$ sign or a $-$ sign.

Terminating decimals (3.6) Every fraction can be written as a decimal. If the division process of dividing denominator into numerator ends with a remainder of zero, the decimal is a terminating decimal. Decimals such as 1.28, 0.007856, and 5.123 are terminating decimals.

Trapezoid (7.3) A four-sided figure with at least two parallel sides.

Triangle (7.4) A three-sided figure.

Trillion (1.1) The number 1,000,000,000,000.

Unit fraction (6.1) A fraction used to change one unit to another. For example, to change 180 inches to feet, we multiply by the unit fraction $\dfrac{1 \text{ foot}}{12 \text{ inches}}$. Thus we have

$$180 \text{ inches} \times \frac{1 \text{ foot}}{12 \text{ inches}} = 15 \text{ feet.}$$

Variable (10.1) A letter that is used to represent a number.

Vertex of a cone (7.8) The sharp point of a cone.

Vertex of an angle (7.1) The point at which two line segments meet to form an angle.

Volume (7.8) The measure of the space inside a three-dimensional object. Volume is measured in cubic units such as cubic feet.

Whole numbers (1.1) The whole numbers are the set of numbers 0, 1, 2, 3, 4, 5, 6, 7, 8, 9, 10, 11, 12, ... The set goes on forever. There is no largest whole number.

Width of a rectangle (7.2) Each of the shorter sides of a rectangle.

Word names for whole numbers (1.1) The notation for a number in which each digit is expressed by a word. To write 389 with a word name, we would write three hundred eighty-nine.

Zero (1.1) The smallest whole number. It is normally written 0.

Appendix A
Tables

Table of Basic Addition Facts

+	0	1	2	3	4	5	6	7	8	9
0	0	1	2	3	4	5	6	7	8	9
1	1	2	3	4	5	6	7	8	9	10
2	2	3	4	5	6	7	8	9	10	11
3	3	4	5	6	7	8	9	10	11	12
4	4	5	6	7	8	9	10	11	12	13
5	5	6	7	8	9	10	11	12	13	14
6	6	7	8	9	10	11	12	13	14	15
7	7	8	9	10	11	12	13	14	15	16
8	8	9	10	11	12	13	14	15	16	17
9	9	10	11	12	13	14	15	16	17	18

Table of Basic Multiplication Facts

×	0	1	2	3	4	5	6	7	8	9	10	11	12
0	0	0	0	0	0	0	0	0	0	0	0	0	0
1	0	1	2	3	4	5	6	7	8	9	10	11	12
2	0	2	4	6	8	10	12	14	16	18	20	22	24
3	0	3	6	9	12	15	18	21	24	27	30	33	36
4	0	4	8	12	16	20	24	28	32	36	40	44	48
5	0	5	10	15	20	25	30	35	40	45	50	55	60
6	0	6	12	18	24	30	36	42	48	54	60	66	72
7	0	7	14	21	28	35	42	49	56	63	70	77	84
8	0	8	16	24	32	40	48	56	64	72	80	88	96
9	0	9	18	27	36	45	54	63	72	81	90	99	108
10	0	10	20	30	40	50	60	70	80	90	100	110	120
11	0	11	22	33	44	55	66	77	88	99	110	121	132
12	0	12	24	36	48	60	72	84	96	108	120	132	144

Table of Prime Factors

Number	Prime Factors	Number	Prime Factors	Number	Prime Factors	Number	Prime Factors
2	prime	52	$2^2 \times 13$	102	$2 \times 3 \times 17$	152	$2^3 \times 19$
3	prime	53	prime	103	prime	153	$3^2 \times 17$
4	2^2	54	2×3^3	104	$2^3 \times 13$	154	$2 \times 7 \times 11$
5	prime	55	5×11	105	$3 \times 5 \times 7$	155	5×31
6	2×3	56	$2^3 \times 7$	106	2×53	156	$2^2 \times 3 \times 13$
7	prime	57	3×19	107	prime	157	prime
8	2^3	58	2×29	108	$2^2 \times 3^3$	158	2×79
9	3^2	59	prime	109	prime	159	3×53
10	2×5	60	$2^2 \times 3 \times 5$	110	$2 \times 5 \times 11$	160	$2^5 \times 5$
11	prime	61	prime	111	3×37	161	7×23
12	$2^2 \times 3$	62	2×31	112	$2^4 \times 7$	162	2×3^4
13	prime	63	$3^2 \times 7$	113	prime	163	prime
14	2×7	64	2^6	114	$2 \times 3 \times 19$	164	$2^2 \times 41$
15	3×5	65	5×13	115	5×23	165	$3 \times 5 \times 11$
16	2^4	66	$2 \times 3 \times 11$	116	$2^2 \times 29$	166	2×83
17	prime	67	prime	117	$3^2 \times 13$	167	prime
18	2×3^2	68	$2^2 \times 17$	118	2×59	168	$2^3 \times 3 \times 7$
19	prime	69	3×23	119	7×17	169	13^2
20	$2^2 \times 5$	70	$2 \times 5 \times 7$	120	$2^3 \times 3 \times 5$	170	$2 \times 5 \times 17$
21	3×7	71	prime	121	11^2	171	$3^2 \times 19$
22	2×11	72	$2^3 \times 3^2$	122	2×61	172	$2^2 \times 43$
23	prime	73	prime	123	3×41	173	prime
24	$2^3 \times 3$	74	2×37	124	$2^2 \times 31$	174	$2 \times 3 \times 29$
25	5^2	75	3×5^2	125	5^3	175	$5^2 \times 7$
26	2×13	76	$2^2 \times 19$	126	$2 \times 3^2 \times 7$	176	$2^4 \times 11$
27	3^3	77	7×11	127	prime	177	3×59
28	$2^2 \times 7$	78	$2 \times 3 \times 13$	128	2^7	178	2×89
29	prime	79	prime	129	3×43	179	prime
30	$2 \times 3 \times 5$	80	$2^4 \times 5$	130	$2 \times 5 \times 13$	180	$2^2 \times 3^2 \times 5$
31	prime	81	3^4	131	prime	181	prime
32	2^5	82	2×41	132	$2^2 \times 3 \times 11$	182	$2 \times 7 \times 13$
33	3×11	83	prime	133	7×19	183	3×61
34	2×17	84	$2^2 \times 3 \times 7$	134	2×67	184	$2^3 \times 23$
35	5×7	85	5×17	135	$3^3 \times 5$	185	5×37
36	$2^2 \times 3^2$	86	2×43	136	$2^3 \times 17$	186	$2 \times 3 \times 31$
37	prime	87	3×29	137	prime	187	11×17
38	2×19	88	$2^3 \times 11$	138	$2 \times 3 \times 23$	188	$2^2 \times 47$
39	3×13	89	prime	139	prime	189	$3^3 \times 7$
40	$2^3 \times 5$	90	$2 \times 3^2 \times 5$	140	$2^2 \times 5 \times 7$	190	$2 \times 5 \times 19$
41	prime	91	7×13	141	3×47	191	prime
42	$2 \times 3 \times 7$	92	$2^2 \times 23$	142	2×71	192	$2^6 \times 3$
43	prime	93	3×31	143	11×13	193	prime
44	$2^2 \times 11$	94	2×47	144	$2^4 \times 3^2$	194	2×97
45	$3^2 \times 5$	95	5×19	145	5×29	195	$3 \times 5 \times 13$
46	2×23	96	$2^5 \times 3$	146	2×73	196	$2^2 \times 7^2$
47	prime	97	prime	147	3×7^2	197	prime
48	$2^4 \times 3$	98	2×7^2	148	$2^2 \times 37$	198	$2 \times 3^2 \times 11$
49	7^2	99	$3^2 \times 11$	149	prime	199	prime
50	2×5^2	100	$2^2 \times 5^2$	150	$2 \times 3 \times 5^2$	200	$2^3 \times 5^2$
51	3×17	101	prime	151	prime		

Table of Square Roots

Square Root Values Are Rounded to the Nearest Thousandth Unless the Answer Ends in .000

n	$\sqrt{n}$	n	$\sqrt{n}$	n	$\sqrt{n}$	n	$\sqrt{n}$	n	$\sqrt{n}$
1	1.000	41	6.403	81	9.000	121	11.000	161	12.689
2	1.414	42	6.481	82	9.055	122	11.045	162	12.728
3	1.732	43	6.557	83	9.110	123	11.091	163	12.767
4	2.000	44	6.633	84	9.165	124	11.136	164	12.806
5	2.236	45	6.708	85	9.220	125	11.180	165	12.845
6	2.449	46	6.782	86	9.274	126	11.225	166	12.884
7	2.646	47	6.856	87	9.327	127	11.269	167	12.923
8	2.828	48	6.928	88	9.381	128	11.314	168	12.961
9	3.000	49	7.000	89	9.434	129	11.358	169	13.000
10	3.162	50	7.071	90	9.487	130	11.402	170	13.038
11	3.317	51	7.141	91	9.539	131	11.446	171	13.077
12	3.464	52	7.211	92	9.592	132	11.489	172	13.115
13	3.606	53	7.280	93	9.644	133	11.533	173	13.153
14	3.742	54	7.348	94	9.695	134	11.576	174	13.191
15	3.873	55	7.416	95	9.747	135	11.619	175	13.229
16	4.000	56	7.483	96	9.798	136	11.662	176	13.266
17	4.123	57	7.550	97	9.849	137	11.705	177	13.304
18	4.243	58	7.616	98	9.899	138	11.747	178	13.342
19	4.359	59	7.681	99	9.950	139	11.790	179	13.379
20	4.472	60	7.746	100	10.000	140	11.832	180	13.416
21	4.583	61	7.810	101	10.050	141	11.874	181	13.454
22	4.690	62	7.874	102	10.100	142	11.916	182	13.491
23	4.796	63	7.937	103	10.149	143	11.958	183	13.528
24	4.899	64	8.000	104	10.198	144	12.000	184	13.565
25	5.000	65	8.062	105	10.247	145	12.042	185	13.601
26	5.099	66	8.124	106	10.296	146	12.083	186	13.638
27	5.196	67	8.185	107	10.344	147	12.124	187	13.675
28	5.292	68	8.246	108	10.392	148	12.166	188	13.711
29	5.385	69	8.307	109	10.440	149	12.207	189	13.748
30	5.477	70	8.367	110	10.488	150	12.247	190	13.784
31	5.568	71	8.426	111	10.536	151	12.288	191	13.820
32	5.657	72	8.485	112	10.583	152	12.329	192	13.856
33	5.745	73	8.544	113	10.630	153	12.369	193	13.892
34	5.831	74	8.602	114	10.677	154	12.410	194	13.928
35	5.916	75	8.660	115	10.724	155	12.450	195	13.964
36	6.000	76	8.718	116	10.770	156	12.490	196	14.000
37	6.083	77	8.775	117	10.817	157	12.530	197	14.036
38	6.164	78	8.832	118	10.863	158	12.570	198	14.071
39	6.245	79	8.888	119	10.909	159	12.610	199	14.107
40	6.325	80	8.944	120	10.954	160	12.649	200	14.142

Appendix B
Scientific Calculators

This text *does not require* the use of a calculator. However, you may want to consider the purchase of an inexpensive scientific calculator. It is wise to ask your instructor for advice before you purchase any calculator for this course. It should be stressed that students are asked to avoid using a calculator for any of the exercises in which the calculations can be readily done by hand. The only problems in the text that really demand the use of a scientific calculator are marked with the ▦ symbol. Dependence on the use of the scientific calculator for regular exercises in the text will only hurt the student in the long run.

The Two Types of Logic Used in Scientific Calculators

Two major types of scientific calculators are popular today. The most common type employs a type of logic known as **algebraic** logic. The calculators manufactured by Casio, Sharp, and Texas Instruments as well as many other companies employ this type of logic. An example of calculation on such a calculator would be the following. To add $14 + 26$ on an algebraic logic calculator, the sequence of buttons would be:

$$14 \boxed{+} 26 \boxed{=}$$

The second type of scientific calculator requires the entry of data in **Reverse Polish Notation (RPN)**. Calculators manufactured by Hewlett-Packard and a few other specialized calculators use RPN. To add $14 + 26$ on an RPN calculator, the sequence of buttons would be:

$$14 \boxed{\text{enter}} 26 \boxed{+}$$

Graphing scientific calculators such as the TI-82 and TI-83 have a large liquid display for viewing graphs. To perform the calculation on most graphing calculators, the sequence of buttons would be:

$$14 \boxed{+} 26 \boxed{\text{enter}}$$

Mathematicians and scientists do not agree on which type of scientific calculator is superior. However, the clear majority of college students own calculators that employ *algebraic* logic. Therefore this section of the text is explained with reference to the sequence of steps employed by an *algebraic* logic calculator. If you already own or intend to purchase a scientific calculator that uses RPN or a graphing calculator, you are encouraged to study the instruction booklet that comes with the calculator and practice the problems shown in the booklet. After this practice you will be able to solve the calculator problems discussed in this section.

Performing Simple Calculations

The following example will illustrate the use of the scientific calculator in doing basic arithmetic calculations.

EXAMPLE 1   Add. 156 + 298

We first enter the number 156, then press the $\boxed{+}$ key, then enter the number 298, and finally press the $\boxed{=}$ key.

$$156 \boxed{+} 298 \boxed{=} 454$$

Practice Problem 1   Add. 3792 + 5896

EXAMPLE 2   Subtract. 1508 − 963

We first enter the number 1508, then press the $\boxed{-}$ key, then enter the number 963, and finally press the $\boxed{=}$ key.

$$1508 \boxed{-} 963 \boxed{=} 545$$

Practice Problem 2   Subtract. 7930 − 5096

EXAMPLE 3   Multiply. 196 × 358

$$196 \boxed{\times} 358 \boxed{=} 70168$$

Practice Problem 3   Multiply. 896 × 273

EXAMPLE 4   Divide. 2054 ÷ 13

$$2054 \boxed{\div} 13 \boxed{=} 158$$

Practice Problem 4   Divide. 2352 ÷ 16

Decimal Problems

Problems involving decimals can be readily done on the calculator. Entering numbers with a decimal point is done by pressing the decimal point key, the $\boxed{\cdot}$ key, at the appropriate time.

EXAMPLE 5   Calculate. 4.56 × 283

To enter 4.56, we press the $\boxed{4}$ key, the decimal point key, then the $\boxed{5}$ key, and finally the $\boxed{6}$ key.

$$4.56 \boxed{\times} 283 \boxed{=} 1290.48$$

The answer is 1290.48. Observe how your calculator displays the decimal point.

Practice Problem 5   Calculate. 72.8 × 197

EXAMPLE 6   Add. 128.6 + 343.7 + 103.4 + 207.5

$$128.6 \boxed{+} 343.7 \boxed{+} 103.4 \boxed{+} 207.5 \boxed{=} 783.2$$

The answer is 783.2. Observe how your calculator displays the answer.

Practice Problem 6   Add. 52.98 + 31.74 + 40.37 + 99.82

Combined Operations

You must use extra caution concerning the order of mathematical operations when you are doing a problem on the calculator that involves two or more different operations.

Any scientific calculator with algebraic logic uses a priority system that has a clearly defined order of operations. It is the same order we use in performing arithmetic operations by hand. In either situation, calculations are performed in the following order:

1. First calculations within parentheses are completed.
2. Then numbers are raised to a power or a square root is calculated.
3. Then multiplication and division operations are performed from left to right.
4. Then addition and subtraction operations are performed from left to right.

This order is carefully followed on *scientific calculators* and *graphing calculators*. Small inexpensive calculators that do not have scientific functions often do not follow this order of operations.

The number of digits displayed in the answer varies from calculator to calculator. In the following examples, your calculator may display more or fewer digits than the answer we have listed.

EXAMPLE 7   Evaluate. $5.3 \times 1.62 + 1.78 \div 3.51$

This problem requires that we multiply 5.3 by 1.62 and divide 1.78 by 3.51 first and then add the two results. If the numbers are entered directly into the calculator exactly as the problem is written, the calculator will perform the calculations in the correct order.

$$5.3 \;\boxed{\times}\; 1.62 \;\boxed{+}\; 1.78 \;\boxed{\div}\; 3.51 \;\boxed{=}\; 9.09312251$$

Practice Problem 7   Evaluate. $0.0618 \times 19.22 - 59.38 \div 166.3$

The Use of Parentheses

In order to perform some calculations on the calculator the use of parentheses is helpful. These parentheses may or may not appear in the original problem.

EXAMPLE 8   Evaluate. $5 \times (2.123 + 5.786 - 12.063)$

The problem requires that the numbers in the parentheses be combined first. By entering the parentheses on the calculator this will be accomplished.

$$5 \;\boxed{\times}\; \boxed{(}\; 2.123 \;\boxed{+}\; 5.786 \;\boxed{-}\; 12.063 \;\boxed{)}\; \boxed{=}\; -20.77$$

Note: The result is a negative number.

Practice Problem 8   Evaluate. $3.152 \times (0.1628 + 3.715 - 4.985)$

Negative Numbers

To enter a negative number, enter the number followed by the $\boxed{+/-}$ button.

● **EXAMPLE 9**   Evaluate. $(-8.634)(5.821) + (1.634)(-16.082)$

The products will be evaluated first by the calculator. Therefore parentheses are not needed as we enter the data.

8.634 $\boxed{+/-}$ $\boxed{\times}$ 5.821 $\boxed{+}$ 1.634 $\boxed{\times}$ 16.082 $\boxed{+/-}$ $\boxed{=}$ -76.536502

Note: The result is negative.

Practice Problem 9   Evaluate. $(0.5618)(-98.3) - (76.31)(-2.98)$ ●

Scientific Notation

If you wish to enter a number in scientific notation, you should use the special scientific notation button. On most calculators it is denoted as $\boxed{\text{EXP}}$ or $\boxed{\text{EE}}$.

● **EXAMPLE 10**   Multiply. $(9.32 \times 10^6)(3.52 \times 10^8)$

9.32 $\boxed{\text{EXP}}$ 6 $\boxed{\times}$ 3.52 $\boxed{\text{EXP}}$ 8 $\boxed{=}$ 3.28064   15

This notation means the answer is 3.28064×10^{15}.

Practice Problem 10   Divide. $(3.76 \times 10^{15}) \div (7.76 \times 10^7)$ ●

Raising a Number to a Power

All scientific calculators have a key for finding powers of numbers. It is usually labeled $\boxed{y^x}$. (On a few calculators the notation is $\boxed{x^y}$ or sometimes $\boxed{\wedge}$.) To raise a number to a power on most scientific calculators, first you enter the base, then push the $\boxed{y^x}$ key. Then you enter the exponent, then finally the $\boxed{=}$ button.

● **EXAMPLE 11**   Evaluate. $(2.16)^9$

2.16 $\boxed{y^x}$ 9 $\boxed{=}$ 1023.490369

Practice Problem 11   Evaluate. $(6.238)^6$ ●

There is a special key to square a number. It is usually labeled $\boxed{x^2}$.

● **EXAMPLE 12**   Evaluate. $(76.04)^2$

76.04 $\boxed{x^2}$ 5782.0816

Practice Problem 12   Evaluate. $(132.56)^2$ ●

Finding Square Roots of Numbers

There is usually a key to approximate square roots on a scientific calculator labeled $\boxed{\sqrt{}}$. In this example we will need to use parentheses.

● **EXAMPLE 13**   Evaluate. $\sqrt{5618 + 2734 + 3913}$

$\boxed{(}$ 5618 $\boxed{+}$ 2734 $\boxed{+}$ 3913 $\boxed{)}$ $\boxed{\sqrt{}}$ 110.7474605

Practice Problem 13   Evaluate. $\sqrt{0.0782 - 0.0132 + 0.1364}$ ●

Appendix B Scientific Calculator Exercises

Use your calculator to complete each of the following. Your answers may vary slightly because of the characteristics of individual calculators.

Complete the table.

To Do This Operation	Use These Keystrokes	Record Your Answer Here
1. 8963 + 2784	8963 $\boxed{+}$ 2784 $\boxed{=}$	
2. 15,308 − 7980	15308 $\boxed{-}$ 7980 $\boxed{=}$	
3. 2631 × 134	2631 $\boxed{\times}$ 134 $\boxed{=}$	
4. 70,221 ÷ 89	70221 $\boxed{\div}$ 89 $\boxed{=}$	
5. 5.325 − 4.031	5.325 $\boxed{-}$ 4.031 $\boxed{=}$	
6. 184.68 + 73.98	184.68 $\boxed{+}$ 73.98 $\boxed{=}$	
7. 2004.06 ÷ 7.89	2004.06 $\boxed{\div}$ 7.89 $\boxed{=}$	
8. 1.34 × 0.763	1.34 $\boxed{\times}$ 0.763 $\boxed{=}$	

Write down the answer and then show what problem you have solved.

9. 123.45 $\boxed{+}$ 45.9876 $\boxed{+}$ 8765.3 $\boxed{=}$

10. 0.0897 $\boxed{\times}$ 234.56 $\boxed{\times}$ 2.5428 $\boxed{=}$

11. 34 $\boxed{\div}$ 8 $\boxed{+}$ 12.56 $\boxed{=}$

12. 458 $\boxed{\div}$ 4 $\boxed{-}$ 16.897 $\boxed{=}$

Perform each calculation using your calculator.

13. 9.467 + 0.563

14. 0.347 + 23.457

15. 34.89 + 39.6 + 214.897

16. 12.567 + 48.31 + 189.38

17. 412,899 − 34,675

18. 87,456 − 2876

19. 3,567,089 − 2,876,805

20. 8,345,802 − 4,985,004

21. 234 × 4.567

22. 1.9876 × 347

23. 0.456 × 3.48

24. 67,876 × 0.0946

25. 3458 ÷ 2.5

26. 9764 ÷ 8

27. 12.107524 ÷ 15.86

28. 16.06513 ÷ 17.98

Perform each calculation using your calculator.

29.   1.98
    6.34
  + 7.71

30.   8.92
    9.31
  + 7.79

31. $103.91
    $2653.82
  + $9804.61

32. $3986.21
    $4502.89
  + $989.30

33.   368,781.5
  − 283,617.8

34.   571,809.6
  − 539,376.8

35. $1,393,271.86
  − $1,289,663.21

36. $8,571,300.76
  − $4,098,789.39

37.   345.34
  ×   45.7

38.   8954.34
  ×   425.4

39.   0.6314
  ×   3.96

40.   0.0789
  ×   12.38

41. $40.36 \overline{)36202.92}$

42. $52.98 \overline{)172,608.84}$

43. $0.7613 \overline{)17.12925}$

44. $0.9854 \overline{)3.59671}$

Perform the following operations in the proper order using your calculator.

45. $4.567 + 87.89 - 2.45 \times 3.3$

46. $4.891 + 234.5 - 0.98 \times 23.4$

47. $7 \div 8 + 3.56$

48. $9 \div 4.5 + 0.6754$

49. $(9.34)(0.345) + 98.345$

50. $(0.628)(398) + 34.4581$

51. $\dfrac{(95.34)(0.9874)}{381.36}$

52. $\dfrac{(0.8759)(45.87)}{183.48}$

53. $2.56 + 8.98 \times 3.14$

54. $1.62 + 3.81 - 5.23 \times 6.18$

55. $(-4.23)(1.863) - 5.998$

56. $12.34 - (26.314)(-1.856)$

57. $5.62(5 \times 3.16 - 18.12)$

58. $9.356(4.8 - 7.2 - 15.94)$

59. $(3.42 \times 10^8)(0.97 \times 10^{10})$

60. $(6.27 \times 10^{20})(1.35 \times 10^3)$

61. $\dfrac{(2.16 \times 10^3)(1.37 \times 10^{14})}{6.39 \times 10^5}$

62. $\dfrac{(3.84 \times 10^{12})(1.62 \times 10^5)}{7.78 \times 10^8}$

63. $\dfrac{2.3 + 5.8 - 2.6 - 3.9}{5.3 - 8.2}$

64. $\dfrac{(2.6)(-3.2) + (5.8)(-0.9)}{2.614 + 5.832}$

65. $\sqrt{253.12}$

66. $\sqrt{0.0713}$

67. $\sqrt{5.6213 - 3.7214}$

68. $\sqrt{3417.2 - 2216.3}$

69. $(1.78)^3 + 6.342$

70. $(2.26)^8 - 3.1413$

71. $\sqrt{(6.13)^2 + (5.28)^2}$

72. $\sqrt{(0.3614)^2 + (0.9217)^2}$

73. $\sqrt{56 + 83} - \sqrt{12}$

74. $\sqrt{98 + 33} - \sqrt{17}$

Find an approximate value. Round to five decimal places.

75. $\dfrac{7}{18} + \dfrac{9}{13}$

76. $\dfrac{5}{22} + \dfrac{1}{31}$

77. $\dfrac{7}{8} + \dfrac{3}{11}$

78. $\dfrac{9}{14} + \dfrac{5}{19}$

Solutions to Practice Problems

Chapter 1

1.1 Practice Problems

1. (a) $3182 = 3000 + 100 + 80 + 2$
   (b) $520,890 = 500,000 + 20,000 + 800 + 90$
   (c) $709,680,059 = 700,000,000 + 9,000,000 + 600,000$
   $+ 80,000 + 50 + 9$

2. (a) 492   (b) 80,427
3. (a) 7   (b) 9   (c) 4,000
   (d) 900,000 for the first 9; 9 for the last 9
4. two hundred sixty-seven million, three hundred fifty-eight thousand, nine hundred eighty-one
5. (a) two thousand, seven hundred thirty-six
   (b) nine hundred eighty thousand, three hundred six
   (c) twelve million, twenty-one
6. The world population on July 6, 2000 was six billion, one hundred fifty-three million, five hundred eighty-four thousand, eight hundred five.
7. (a) 803   (b) 30,229
8. (a) 13,000   (b) 88,000   (c) 10,000

1.2 Practice Problems

1. (a) $\begin{array}{r} 7 \\ + 5 \\ \hline 12 \end{array}$ (b) $\begin{array}{r} 9 \\ + 4 \\ \hline 13 \end{array}$ (c) $\begin{array}{r} 3 \\ + 0 \\ \hline 3 \end{array}$

2. $\begin{array}{r} 7 \quad 7 \\ 6 \\ 5 \end{array} \Big] 11 \\ \begin{array}{r} 8 \\ +2 \end{array} \Big] 10 \\ \hline 28$

3. $\begin{array}{r} 1 \\ 7 \\ 2 \quad \Big] 10 \\ 9 \quad \Big] 10 \\ +3 \\ \hline 22 \end{array}$

4. $\begin{array}{r} 8246 \\ + 1702 \\ \hline 9948 \end{array}$

5. $\begin{array}{r} 1 \\ 56 \\ + 36 \\ \hline 92 \end{array}$

6. $\begin{array}{r} 21 \\ 789 \\ 63 \\ + 297 \\ \hline 1149 \end{array}$

7. (a) $\begin{array}{r} 111 \\ 127 \\ 9876 \\ + 342 \\ \hline 10,345 \end{array}$ (b) Check by adding in opposite order. $\begin{array}{r} 111 \\ 342 \\ 9876 \\ + 127 \\ \hline 10,345 \end{array}$ same

8. $\begin{array}{r} 12 \; 12 \\ 18,316 \\ 24,789 \\ + 22,965 \\ \hline 66,070 \text{ total women} \end{array}$

9. $\begin{array}{r} 1000 \\ 2000 \\ 1000 \\ + 2000 \\ \hline 6000 \text{ ft} \end{array}$

1.3 Practice Problems

1. (a) $\begin{array}{r} 9 \\ - 6 \\ \hline 3 \end{array}$ (b) $\begin{array}{r} 12 \\ - 5 \\ \hline 7 \end{array}$ (c) $\begin{array}{r} 17 \\ - 8 \\ \hline 9 \end{array}$ (d) $\begin{array}{r} 14 \\ - 0 \\ \hline 14 \end{array}$ (e) $\begin{array}{r} 18 \\ - 9 \\ \hline 9 \end{array}$

2. $\begin{array}{r} 7695 \\ - 3481 \\ \hline 4214 \end{array}$

3. $\begin{array}{r} {}^{2\;14} \\ \cancel{3}\cancel{4} \\ - 1 \; 6 \\ \hline 1 \; 8 \end{array}$

4. $\begin{array}{r} {}^{8\;13} \\ 6 \cancel{9} \cancel{3} \\ - 4 2 \; 6 \\ \hline 2 6 \; 7 \end{array}$

5. $\begin{array}{r} {}^{\quad 16} \\ {}^{8\;9\;6\;10} \\ \cancel{9}\cancel{0}\cancel{7}\cancel{0} \\ - 5 \; 8 8 6 \\ \hline 3 \; 1 8 4 \end{array}$

6. (a) $\begin{array}{r} 8964 \\ - 985 \\ \hline 7979 \end{array}$ (b) $\begin{array}{r} 50,000 \\ - 32,508 \\ \hline 17,492 \end{array}$

7. Subtraction
   $\begin{array}{r} 9763 \\ - 5732 \\ \hline 4031 \end{array}$ | IT CHECKS | Checking by addition $\begin{array}{r} 5732 \\ + 4031 \\ \hline 9763 \end{array}$

8. (a) $\begin{array}{r} 284,000 \\ - 96,327 \\ \hline 187,673 \end{array}$ | IT CHECKS | Checking by addition $\begin{array}{r} 96,327 \\ + 187,673 \\ \hline 284,000 \end{array}$

   (b) $\begin{array}{r} 8,526,024 \\ - 6,397,518 \\ \hline 2,128,506 \end{array}$ | IT CHECKS | Checking by addition $\begin{array}{r} 6,397,518 \\ + 2,128,506 \\ \hline 8,526,024 \end{array}$

9. (a) $x = 5$ (b) $x = 12$
10. (a) $\begin{array}{r} 23,667,947 \\ - 14,227,799 \\ \hline 9,440,148 \end{array}$ (b) $\begin{array}{r} 11,198,655 \\ - 9,579,677 \\ \hline 1,618,978 \end{array}$

11. (a) From the bar graph:
   $\begin{array}{lr} 2000 \text{ sales} & 114 \\ 1999 \text{ sales} & - 78 \\ \hline \text{Sales increase} & 36 \end{array}$
   (b) From the bar graph:
   $\begin{array}{lr} \text{Springfield} & 91 \\ \text{Riverside} & - 78 \\ \hline & 13 \text{ more homes} \end{array}$
   (c) $\begin{array}{lr} 2000 \text{ sales} & 271 \\ 1999 \text{ sales} & - 240 \\ \hline & 31 \end{array}$
   $\begin{array}{lr} 2001 \text{ sales} & 284 \\ 2000 \text{ sales} & - 271 \\ \hline & 13 \end{array}$
   Therefore, the greatest increase in sales occurred from 1999 to 2000.

1.4 Practice Problems

1. (a) 64   (b) 42   (c) 40   (d) 63   (e) 81

2.
$$\begin{array}{r} 3021 \\ \times \quad 3 \\ \hline 9063 \end{array}$$

3.
$$\begin{array}{r} \overset{2}{4}3 \\ \times \quad 8 \\ \hline 344 \end{array}$$

4.
$$\begin{array}{r} \overset{5\,6}{5}79 \\ \times \quad 7 \\ \hline 4053 \end{array}$$

5. (a) $1267 \times 10 = 12{,}670$

(b) $1267 \times 1000 = 1{,}267{,}000$

(c) $1267 \times 10{,}000 = 12{,}670{,}000$

(d) $1267 \times 1{,}000{,}000 = 1{,}267{,}000{,}000$

6. (a) $9 \times 6 \times 10{,}000 = 54 \times 10{,}000 = 540{,}000$

(b) $15 \times 4 \times 100 = 60 \times 100 = 6{,}000$

(c) $270 \times 800 = 27 \times 8 \times 10 \times 100 = 216 \times 1000 = 216{,}000$

7.
$$\begin{array}{r} 323 \\ \times \quad 32 \\ \hline 646 \\ 969 \quad \\ \hline 10{,}336 \end{array}$$

8.
$$\begin{array}{r} 385 \\ \times \quad 69 \\ \hline 3465 \\ 2310 \quad \\ \hline 26{,}565 \end{array}$$

9.
$$\begin{array}{r} 34 \\ \times \quad 20 \\ \hline 680 \end{array}$$

10.
$$\begin{array}{r} 130 \\ \times \quad 50 \\ \hline 6500 \end{array}$$

11.
$$\begin{array}{r} 923 \\ \times \quad 675 \\ \hline 4615 \\ 6461 \quad \\ 5538 \quad\quad \\ \hline 623{,}025 \end{array}$$

12. $25 \times 4 \times 17 = (25 \times 4) \times 17 = 100 \times 17 = 1700$

13. $8 \times 4 \times 3 \times 25 = 8 \times 3 \times (4 \times 25)$

$$= 24 \times 100$$
$$= 2400$$

14.
$$\begin{array}{r} 17348 \\ \times \quad 378 \\ \hline 138784 \\ 121436 \quad \\ 52044 \quad\quad \\ \hline 6{,}557{,}544 \end{array}$$

The total sales of cars is $6,557,544.

15. Area $= 5$ yards $\times 7$ yards $= 35$ square yards.

1.5 Practice Problems

1. (a) $4\overline{)36}$ → 9 (b) $5\overline{)25}$ → 5 (c) $9\overline{)72}$ → 8 (d) $6\overline{)30}$ → 5

2. (a) $\dfrac{7}{1} = 7$ (b) $\dfrac{9}{9} = 1$ (c) $\dfrac{0}{5} = 0$

(d) $\dfrac{12}{0}$ cannot be done   (e) $\dfrac{7}{0}$ cannot be done

3.
$$\begin{array}{r} 7 \text{ R3} \\ 6\overline{)45} \\ \underline{42} \\ 3 \end{array}$$

Check
$$\begin{array}{r} 6 \\ \times \quad 7 \\ \hline 42 \\ + \quad 3 \\ \hline 45 \end{array}$$

4.
$$\begin{array}{r} 21 \text{ R } 3 \\ 6\overline{)129} \\ \underline{12} \\ 9 \\ \underline{6} \\ 3 \end{array}$$

Check
$$\begin{array}{r} 21 \\ \times \quad 6 \\ \hline 126 \\ + \quad 3 \text{ Remainder} \\ \hline 129 \end{array}$$

5.
$$\begin{array}{r} 529 \text{ R } 5 \\ 8\overline{)4237} \\ \underline{40} \\ 23 \\ \underline{16} \\ 77 \\ \underline{72} \\ 5 \end{array}$$

6.
$$\begin{array}{r} 7 \text{ R } 5 \\ 32\overline{)229} \\ \underline{224} \\ 5 \end{array}$$

7.
$$\begin{array}{r} 1278 \text{ R } 9 \\ 33\overline{)42183} \\ \underline{33} \\ 91 \\ \underline{66} \\ 258 \\ \underline{231} \\ 273 \\ \underline{264} \\ 9 \end{array}$$

8.
$$\begin{array}{r} 25 \text{ R } 27 \\ 128\overline{)3227} \\ \underline{256} \\ 667 \\ \underline{640} \\ 27 \end{array}$$

9.
$$\begin{array}{r} 16{,}852 \\ 7\overline{)117{,}964} \end{array}$$

Check
$$\begin{array}{r} 16{,}852 \\ \times \quad 7 \\ \hline 117{,}964 \end{array}$$

The cost of one car is $16,852.

10.
$$\begin{array}{r} 367 \\ 14\overline{)5138} \end{array}$$

The average speed was 367 mph.

1.6 Practice Problems

1. (a) $12 \times 12 \times 12 \times 12 = 12^4$. 12 is the base, 4 is the exponent.

(b) $2 \times 2 \times 2 \times 2 \times 2 \times 2 = 2^6$. 2 is the base, 6 is the exponent.

2. (a) $12^2 = 12 \times 12 = 144$

(b) $6^3 = 6 \times 6 \times 6 = 216$

(c) $2^6 = 2 \times 2 \times 2 \times 2 \times 2 \times 2 = 64$

(d) $1^{10} = 1 \times 1 \times 1 \times 1 \times 1 \times 1 \times 1 \times 1 \times 1 \times 1 = 1$

3. (a) $(7)(7)(7) + (8)(8) = 343 + 64 = 407$

(b) $(9)(9) + 1 = 81 + 1 = 82$

(c) $5^4 + 5 = (5)(5)(5)(5) + 5 = 625 + 5 = 630$

4. $7 + 4^3 \times 3 = 7 + 64 \times 3$ Exponents

$$= 7 + 192 \quad \text{Multiply}$$
$$= 199 \quad \text{Add}$$

5. $37 - 20 \div 5 + 2 - 3 \times 4$

$= 37 - 4 + 2 - 3 \times 4$	Divide
$= 37 - 4 + 2 - 12$	Multiply
$= 33 + 2 - 12$	Subtract
$= 35 - 12$	Add
$= 23$	Subtract

6. $4^3 - 2 + 3^2$

$= 4 \times 4 \times 4 - 2 + 3 \times 3$	Evaluate exponents.
$= 64 - 2 + 9$	$4^3 = 64$ and $3^2 = 9$.
$= 62 + 9$	Subtract
$= 71$	Add

7. $(17 + 7) \div 6 \times 2 + 7 \times 3 - 4$

$= 24 \div 6 \times 2 + 7 \times 3 - 4$	Combine inside parentheses
$= 4 \times 2 + 7 \times 3 - 4$	Divide
$= 8 + 7 \times 3 - 4$	Multiply
$= 8 + 21 - 4$	Multiply
$= 29 - 4$	Add
$= 25$	Subtract

8. $5^2 - 6 \div 2 + 3^4 + 7 \times (12 - 10)$

$= 5^2 - 6 \div 2 + 3^4 + 7 \times 2$	Combine inside parentheses
$= 25 - 6 \div 2 + 81 + 7 \times 2$	Exponents
$= 25 - 3 + 81 + 7 \times 2$	Divide
$= 25 - 3 + 81 + 14$	Multiply
$= 22 + 81 + 14$	Subtract
$= 117$	Add

1.7 Practice Problems

1. $6\,5\,,\!5\,2\,8$ Locate the thousands round-off place.

$6\,5\,,\!5\,2\,8$ The first digit to the right is 5 or more. We will increase the thousands digit by 1.

$6\,6\,,\!0\,0\,0$ All digits to the right of thousands are replaced by zero.

2. $1\,7\,2\,,\!9\,6\,3 = 170,000$ to the nearest ten thousand.

3. (a) $5\,3\,,\!2\,8\,2 = 53,280$ to the nearest ten. The digit to the right of tens was less than 5.

(b) $1\,6\,4\,,\!4\,8\,5 = 164,000$ to the nearest thousand. The digit to the right of thousands was less than 5.

(c) $1\,,\!3\,6\,5\,,\!2\,7\,3 = 1,400,000$ to the nearest hundred thousand. The digit to the right of hundred thousands was greater than 5.

4. (a) $9\,3\,5\,,\!6\,8\,2 = 936,000$. The digit to the right of thousands was greater than 5.

(b) $9\,3\,5\,,\!6\,8\,2 = 900,000$. The digit to the right of hundred thousands was less than 5.

(c) $9\,3\,5\,,\!6\,8\,2 = 1,000,000$. The digit to the right of millions was greater than 5.

5. $9,460,000,000,000,000 = 9,500,000,000,000,000$ meters to the nearest hundred trillion.

6.

Actual Sum	Estimated Sum	
3456	3000	
9876	10000	
5421	5000	
+ 1278	+ 1000	
20,031	19,000	Close to the actual sum

7.

$697	$700	
35	40	
+ 19	+ 20	
	$760	

We estimate that the total cost is $760. (The exact answer is $751, so we can see that our answer is quite close.)

8. $10,000 + 10,000 + 20,000 + 60,000 = 100,000$
This is significantly different from 81,358, so we would suspect that an error has been made. In fact, Ming did make an error. The exact sum is actually 101,358!

9. $30,000,000 - 10,000,000 = 20,000,000$
We estimate that 20,000,000 more people lived in California than in Florida.

10. $9000 \times 7000 = 63,000,000$
We estimate the product to be 63,000,000.

11. $40\overline{)80,000}$ $\dfrac{2,000}{}$ Our estimate is 2,000.

12. $60\overline{)2,000,000}$ $33,333$ R20   Our estimate is $33,333 for one truck.

1.8 Practice Problems

Practice Problem 1

1. Understand the problem.

MATHEMATICS BLUEPRINT FOR PROBLEM SOLVING

Gather the Facts	What Am I Asked to Do?	How Do I Proceed?	Key Points to Remember
The deductions are $135, $28, $13, and $34.	Find out the total amount of deductions.	I must add the four deductions to obtain the total.	Watch out! Gross pay of $1352 is not needed to solve the problem.

2. Solve and state the answer:
$135 + 28 + 13 + 34 = 210$
The total amount taken out of Diane's paycheck is $210.

3. Check. Estimate the answer to see if it is reasonable.

Practice Problem 2

1. Understand the problem.

MATHEMATICS BLUEPRINT FOR PROBLEM SOLVING			
Gather the Facts	What Am I Asked to Do?	How Do I Proceed?	Key Points to Remember
Clinton had 47,402,357 votes. Dole had 39,198,755 votes.	Find out by how many votes Clinton beat Dole.	I must subtract the amounts.	Clinton is a Democrat and Dole is a Republican.

2. Solve and state the answer:

$$\begin{array}{r} 47{,}402{,}357 \\ -\ 39{,}198{,}755 \\ \hline 8{,}203{,}602 \end{array}$$

Clinton beat Dole by 8,203,602 votes.

3. Check. Estimate the answer to see if it is reasonable.

Practice Problem 3

1. Understand the problem.

MATHEMATICS BLUEPRINT FOR PROBLEM SOLVING			
Gather the Facts	What Am I Asked to Do?	How Do I Proceed?	Key Points to Remember
1 gallon is 1,024 fluid drams.	Find out how many fluid drams in 9 gallons.	I need to multiply 1024 by 9.	I must use fluid drams as my measure in my answer.

2. Solve and state the answer:

$$\frac{1024 \text{ fluid drams}}{1 \text{ \sout{gal}}} \times 9 \text{ \sout{gal}} = 9{,}216 \text{ fluid drams.}$$

There would be 9,216 fluid drams in 9 gallons.

3. Check. Estimate the answer to see if it is reasonable.

Practice Problem 4

1. Understand the problem.

MATHEMATICS BLUEPRINT FOR PROBLEM SOLVING			
Gather the Facts	What Am I Asked to Do?	How Do I Proceed?	Key Points to Remember
Donna bought 45 shares of stock. She paid $1620 for them.	Find out the cost per share of stock.	I need to divide $1620 by 45.	Use dollars as the unit in the answer.

2. Solve and state the answer:

$$\begin{array}{r} 36 \\ 45\overline{)1620} \\ \underline{135} \\ 270 \\ \underline{270} \\ 0 \end{array}$$

Donna paid $36 per share for the stock.

3. Check. Estimate the answer to see if it is reasonable.

Practice Problem 5

1. Understand the problem.

MATHEMATICS BLUEPRINT FOR PROBLEM SOLVING			
Gather the Facts	What Am I Asked to Do?	How Do I Proceed?	Key Points to Remember
50 tables at $200 each. 180 chairs at $40 each. 6 carts at $65 each.	Find the total cost.	1. Multiply the number of purchases by each price. 2. Add the total costs of each of the items.	There are three types of items involved in the total purchase.

2. Solve and state the answer:

$$50 \text{ tables at } \$200 \text{ each} = 50 \times 200 = \$10,000 \quad \text{cost of tables}$$
$$180 \text{ chairs at } \$40 \text{ each} = 180 \times 40 = 7,200 \quad \text{cost of chairs}$$
$$6 \text{ carts at } \$65 \text{ each} = 6 \times 65 = \underline{390} \quad \text{cost of carts}$$
$$\$17,590 \quad \text{total cost}$$

The total purchase was $17,590.

3. Check. Estimate the answer to see if it is reasonable.

Practice Problem 6

1. Understand the problem.

MATHEMATICS BLUEPRINT FOR PROBLEM SOLVING			
Gather the Facts	What Am I Asked to Do?	How Do I Proceed?	Key Points to Remember
Old balance: $498 New deposits: $607 $163 Interest: $36 Withdrawals: $ 19 $158 $582 $ 74	Find her new balance after the transactions.	(a) Add the new deposits and interest to the old balance. (b) Add the withdrawals. (c) Subtract the results from steps (a) and (b).	Deposits and interest are added and withdrawals are subtracted from savings accounts.

2. Solve and state the answer:

(a)
$$\begin{array}{r} \$498 \\ 607 \\ 163 \\ +36 \\ \hline \$1304 \end{array}$$

(b)
$$\begin{array}{r} \$19 \\ 158 \\ 582 \\ +74 \\ \hline \$833 \end{array}$$

(c)
$$\begin{array}{r} \$1304 \\ -833 \\ \hline \$471 \end{array}$$

Her balance this month is $471.

3. Check. Estimate the answer to see if it is reasonable.

Practice Problem 7

1. Understand the problem.

MATHEMATICS BLUEPRINT FOR PROBLEM SOLVING			
Gather the Facts	What Am I Asked to Do?	How Do I Proceed?	Key Points to Remember
Odometer reading at end of trip: 51,118 miles Odometer reading at start of trip: 50,698 miles Used on trip: 12 gallons of gas	Find the number of miles per gallon that the car obtained on the trip.	(a) Subtract the two odometer readings. (b) Divide that number by 12.	The gas tank was full at the beginning of the trip. 12 gallons fill the tank at the end of the trip.

2. Solve and state the answer:

$$\begin{array}{l} 51{,}118 \quad \text{odometer at end of trip} \\ \underline{-\,50{,}698} \quad \text{odometer at start of trip} \\ 420 \quad \text{miles traveled on trip} \end{array}$$

$$\frac{420 \text{ miles}}{12 \text{ gallons of gas used}} = 12\overline{)420}$$

$$\begin{array}{r} 35 \\ 12\overline{)420} \\ \underline{36} \\ 60 \\ \underline{60} \end{array} = 35 \text{ miles per gallon on the trip}$$

3. Check. Estimate the answer to see if it is reasonable.

Chapter 2

2.1 Practice Problems

1. (a) Four parts of twelve are shaded. The fraction is $\frac{4}{12}$.

(b) Three parts out of six are shaded. The fraction is $\frac{3}{6}$.

(c) Two parts of three are shaded. The fraction is $\frac{2}{3}$.

2. (a) $\frac{4}{5}$ of the object is shaded.

(b) $\frac{3}{7}$ of the group is shaded.

3. (a) $\frac{9}{17}$ represents 9 players out of 17.

(b) The total class is $382 + 351 = 733$.

The fractional part that is men is $\frac{382}{733}$.

(c) $\frac{7}{8}$ of a yard of material.

4. Total number of defective items $1 + 2 = 3$. Total number of items $7 + 9 = 16$. A fraction that represents the portion of the items that were defective is $\frac{3}{16}$.

2.2 Practice Problems

1. (a) $18 = 9 \times 2$

$ = 3 \times 3 \times 2$

$ = 3^2 \times 2$

(b) $72 = 9 \times 8$

$ = 3 \times 3 \times 2 \times 2 \times 2$

$ = 3^2 \times 2^3$

(c) $400 = 10 \times 40$

$ = 5 \times 2 \times 5 \times 8$

$ = 5 \times 2 \times 5 \times 2 \times 2 \times 2$

$ = 5^3 \times 2^4$

2. (a) $\dfrac{30 \div 6}{42 \div 6} = \dfrac{5}{7}$

(b) $\dfrac{60 \div 12}{132 \div 12} = \dfrac{5}{11}$

3. (a) $\dfrac{120}{135} = \dfrac{2 \times 2 \times 2 \times \cancel{3} \times \cancel{5}}{3 \times 3 \times \cancel{3} \times \cancel{5}} = \dfrac{8}{9}$

(b) $\dfrac{715}{880} = \dfrac{\cancel{5} \times \cancel{11} \times 13}{\cancel{5} \times \cancel{11} \times 2 \times 2 \times 2 \times 2} = \dfrac{13}{16}$

4. (a) $\dfrac{84}{108} \overset{?}{=} \dfrac{7}{9}$

$84 \times 9 \overset{?}{=} 108 \times 7$

$756 = 756 \text{ Yes}$

(b) $\dfrac{3}{7} \overset{?}{=} \dfrac{79}{182}$

$3 \times 182 \overset{?}{=} 7 \times 79$

$546 \neq 553 \text{ No}$

2.3 Practice Problems

1. (a) $4\dfrac{3}{7} = \dfrac{7 \times 4 + 3}{7} = \dfrac{28 + 3}{7} = \dfrac{31}{7}$

(b) $6\dfrac{2}{3} = \dfrac{3 \times 6 + 2}{3} = \dfrac{18 + 2}{3} = \dfrac{20}{3}$

(c) $19\dfrac{4}{7} = \dfrac{7 \times 19 + 4}{7} = \dfrac{133 + 4}{7} = \dfrac{137}{7}$

2. (a) $4\overline{)17}$ so $\dfrac{17}{4} = 4\dfrac{1}{4}$

$\begin{array}{r} 4 \\ 4\overline{)17} \\ \underline{16} \\ 1 \end{array}$

(b) $5\overline{)36}$ so $\dfrac{36}{5} = 7\dfrac{1}{5}$

$\begin{array}{r} 7 \\ 5\overline{)36} \\ \underline{35} \\ 1 \end{array}$

(c) $27\overline{)116}$ so $\dfrac{116}{27} = 4\dfrac{8}{27}$

$\begin{array}{r} 4 \\ 27\overline{)116} \\ \underline{108} \\ 8 \end{array}$

(d) $13\overline{)91}$ so $\dfrac{91}{13} = 7$

$\begin{array}{r} 7 \\ 13\overline{)91} \\ \underline{91} \\ 0 \end{array}$

3. $\dfrac{51}{15} = \dfrac{3 \times 17}{3 \times 5} = \dfrac{\overset{1}{\cancel{3}} \times 17}{\underset{1}{\cancel{3}} \times 5} = \dfrac{17}{5}$

4. $\dfrac{16}{80} = \dfrac{1}{5}$

Therefore $3\dfrac{16}{80} = 3\dfrac{1}{5}$.

5. $\dfrac{1001}{572} = 1\dfrac{429}{572}$

Now the fraction $\dfrac{429}{572} = \dfrac{3 \times \overset{1}{\cancel{11}} \times \overset{1}{\cancel{13}}}{2 \times 2 \times \underset{1}{\cancel{11}} \times \underset{1}{\cancel{13}}} = \dfrac{3}{4}$.

Thus $\dfrac{1001}{572} = 1\dfrac{429}{572} = 1\dfrac{3}{4}$.

2.4 Practice Problems

1. (a) $\dfrac{6}{7} \times \dfrac{3}{13} = \dfrac{18}{91}$ **(b)** $\dfrac{1}{5} \times \dfrac{11}{12} = \dfrac{11}{60}$

2. $\dfrac{\overset{5}{\cancel{55}}}{\underset{9}{\cancel{72}}} \times \dfrac{\overset{2}{\cancel{16}}}{\underset{3}{\cancel{33}}} = \dfrac{5}{9} \times \dfrac{2}{3} = \dfrac{10}{27}$

3. (a) $7 \times \dfrac{5}{13} = \dfrac{7}{1} \times \dfrac{5}{13} = \dfrac{35}{13}$ or $2\dfrac{9}{13}$

(b) $\dfrac{13}{16} \times 8 = \dfrac{13}{\underset{2}{\cancel{16}}} \times \dfrac{\overset{1}{\cancel{8}}}{1} = \dfrac{13}{2}$ or $6\dfrac{1}{2}$

4. $\dfrac{3}{\underset{1}{\cancel{8}}} \times \overset{12{,}300}{\cancel{98{,}400}} = \dfrac{3}{1} \times 12{,}300 = 36{,}900$

There are $36{,}900 \text{ ft}^2$ in the wetland area.

5. (a) $2\dfrac{1}{6} \times \dfrac{4}{7} = \dfrac{13}{\underset{3}{\cancel{6}}} \times \dfrac{\overset{2}{\cancel{4}}}{7} = \dfrac{26}{21}$ or $1\dfrac{5}{21}$

(b) $10\dfrac{2}{3} \times 13\dfrac{1}{2} = \dfrac{\overset{16}{\cancel{32}}}{\underset{1}{\cancel{3}}} \times \dfrac{\overset{9}{\cancel{27}}}{\underset{1}{\cancel{2}}} = \dfrac{144}{1} = 144$

(c) $\dfrac{3}{5} \times 1\dfrac{1}{3} \times \dfrac{5}{8} = \dfrac{\overset{1}{\cancel{3}}}{\underset{1}{\cancel{5}}} \times \dfrac{\overset{1}{\cancel{4}}}{\underset{1}{\cancel{3}}} \times \dfrac{\overset{1}{\cancel{5}}}{\underset{2}{\cancel{8}}} = \dfrac{1}{2}$

(d) $3\dfrac{1}{5} \times 2\dfrac{1}{2} = \dfrac{\overset{8}{\cancel{16}}}{\underset{1}{\cancel{5}}} \times \dfrac{\overset{1}{\cancel{5}}}{\underset{1}{\cancel{2}}} = \dfrac{8}{1} = 8$

6. Area $= lw = 1\frac{1}{5} \times 4\frac{5}{6} = \frac{\overset{1}{\cancel{6}}}{5} \times \frac{29}{\underset{1}{\cancel{6}}} = \frac{29}{5} = 5\frac{4}{5}\,\text{m}^2$

7. Since $8 \cdot 10 = 80$ and $9 \cdot 9 = 81$,

we know that $\frac{8}{9} \cdot \frac{10}{9} = \frac{80}{81}$.

Therefore $x = \frac{10}{9}$.

2.5 Practice Problems

1. (a) $\frac{7}{13} \div \frac{3}{4} = \frac{7}{13} \times \frac{4}{3} = \frac{28}{39}$

(b) $\frac{16}{35} \div \frac{24}{25} = \frac{\overset{2}{\cancel{16}}}{\underset{7}{\cancel{35}}} \times \frac{\overset{5}{\cancel{25}}}{\underset{3}{\cancel{24}}} = \frac{10}{21}$

2. (a) $\frac{3}{17} \div \frac{6}{1} = \frac{\overset{1}{\cancel{3}}}{17} \times \frac{1}{\underset{2}{\cancel{6}}} = \frac{1}{34}$

(b) $\frac{14}{1} \div \frac{7}{15} = \frac{\overset{2}{\cancel{14}}}{1} \times \frac{15}{\underset{1}{\cancel{7}}} = 30$

3. (a) $1 \div \frac{11}{13} = \frac{1}{1} \times \frac{13}{11} = \frac{13}{11}$ or $1\frac{2}{11}$

(b) $\frac{14}{17} \div 1 = \frac{14}{17} \times \frac{1}{1} = \frac{14}{17}$

(c) $\frac{3}{11} \div 0$ Division by zero cannot be done. This problem cannot be done.

(d) $0 \div \frac{9}{16} = \frac{0}{1} \times \frac{16}{9} = \frac{0}{9} = 0$

4. (a) $1\frac{1}{5} \div \frac{7}{10} = \frac{6}{\underset{1}{\cancel{5}}} \times \frac{\overset{2}{\cancel{10}}}{7} = \frac{12}{7}$ or $1\frac{5}{7}$

(b) $2\frac{1}{4} \div 1\frac{7}{8} = \frac{9}{4} \div \frac{15}{8} = \frac{\overset{3}{\cancel{9}}}{\underset{1}{\cancel{4}}} \times \frac{\overset{2}{\cancel{8}}}{\underset{5}{\cancel{15}}} = \frac{6}{5}$ or $1\frac{1}{5}$

5. (a) $\frac{5\frac{2}{3}}{7} = 5\frac{2}{3} \div 7 = \frac{17}{3} \times \frac{1}{7} = \frac{17}{21}$

(b) $\frac{1\frac{2}{5}}{2\frac{1}{3}} = 1\frac{2}{5} \div 2\frac{1}{3} = \frac{7}{5} \div \frac{7}{3} = \frac{\overset{1}{\cancel{7}}}{5} \times \frac{3}{\underset{1}{\cancel{7}}} = \frac{3}{5}$

6. $x \div \frac{3}{2} = \frac{22}{36}$

$x \cdot \frac{2}{3} = \frac{22}{36}$

$\frac{11}{12} \cdot \frac{2}{3} = \frac{22}{36}$ Thus $x = \frac{11}{12}$.

7. $19\frac{1}{4} \div 14 = \frac{\overset{11}{\cancel{77}}}{4} \times \frac{1}{\underset{2}{\cancel{14}}} = \frac{11}{8}$ or $1\frac{3}{8}$

Each piece will be $1\frac{3}{8}$ feet long.

2.6 Practice Problems

1. The multiples of 14 are $14, 28, \boxed{42}, 56, 70, 84, \ldots$
The multiples of 21 are $21, \boxed{42}, 63, 84, 105, 126, \ldots$
42 is the least common multiple of 14 and 21.
2. The multiples of 15 are $15, 30, 45, 60, \boxed{75}, 90, \ldots$
The multiples of 25 are $25, 50, \boxed{75}, 100, 125, 150, \ldots$
75 is the least common multiple of 15 and 25.
3. 54 is a multiple of 6. We know that $6 \times 9 = 54$.
The least common multiple of 6 and 54 is 54.

4. (a) The LCD of $\frac{3}{4}$ and $\frac{11}{12}$ is 12.
12 can be divided by 4 and 12.
(b) The LCD of $\frac{1}{7}$ and $\frac{8}{35}$ is 35.
35 can be divided by 7 and 35.
5. The LCD of $\frac{3}{7}$ and $\frac{5}{6}$ is 42.
42 can be divided by 7 and 6.
6. (a) $14 = 2 \times 7$
$10 = 2 \times 5$
The LCD $= 2 \times 5 \times 7 = 70$.
(b) $15 = 3 \times 5$
$50 = 2 \times 5 \times 5$
The LCD $= 2 \times 3 \times 5 \times 5 = 150$.
(c) $16 = 2 \times 2 \times 2 \times 2$
$12 = 2 \times 2 \times 3$
The LCD $= 2 \times 2 \times 2 \times 2 \times 3 = 48$.
7. $49 = 7 \times 7$
$21 = 7 \times 3$
$7 = 7 \times 1$
LCD $= 7 \times 7 \times 3 = 147$.

8. (a) $\frac{3}{5} \times \frac{8}{8} = \frac{24}{40}$ **(c)** $\frac{2}{7} = \frac{2}{7} \times \frac{4}{4} = \frac{8}{28}$

(b) $\frac{7}{11} \times \frac{4}{4} = \frac{28}{44}$ $\frac{3}{4} = \frac{3}{4} \times \frac{7}{7} = \frac{21}{28}$

9. (a) The LCD is 60 because it can be divided by 20 and 15.

(b) $\frac{3}{20} \times \frac{3}{3} = \frac{9}{60}$

$\frac{11}{15} \times \frac{4}{4} = \frac{44}{60}$

2.7 Practice Problems

1. $\frac{3}{17} + \frac{12}{17} = \frac{15}{17}$

2. (a) $\frac{1}{12} + \frac{5}{12} = \frac{6}{12} = \frac{1}{2}$

(b) $\frac{13}{15} + \frac{7}{15} = \frac{20}{15} = \frac{4}{3}$ or $1\frac{1}{3}$

3. (a) $\frac{5}{19} - \frac{2}{19} = \frac{3}{19}$ **(b)** $\frac{21}{25} - \frac{6}{25} = \frac{15}{25} = \frac{3}{5}$

4. $\frac{2}{15} = \frac{2}{15}$

$\underline{\frac{2}{5} \times \frac{3}{3} = \frac{6}{15}}$

$\frac{8}{15}$

5. LCD $= 48$ $\frac{5}{12} \times \frac{4}{4} = \frac{20}{48}$ $\frac{5}{16} \times \frac{3}{3} = \frac{15}{48}$

$\frac{5}{12} + \frac{5}{16} = \frac{20}{48} + \frac{15}{48} = \frac{35}{48}$

6. LCD $= 48$

$\frac{3}{16} \times \frac{3}{3} = \frac{9}{48}$

$\frac{1}{8} \times \frac{6}{6} = \frac{6}{48}$

$\underline{\frac{1}{12} \times \frac{4}{4} = \frac{4}{48}}$

$\frac{19}{48}$

7. LCD $= 96$ $\frac{9}{48} \times \frac{2}{2} = \frac{18}{96}$

$\frac{5}{32} \times \frac{3}{3} = \frac{15}{96}$ $\frac{9}{48} - \frac{5}{32} = \frac{18}{96} - \frac{15}{96} = \frac{3}{96} = \frac{1}{32}$

8.
$$\frac{9}{10} \times \frac{2}{2} = \frac{18}{20}$$
$$-\frac{1}{4} \times \frac{5}{5} = -\frac{5}{20}$$
$$\frac{13}{20}$$

There is $\frac{13}{20}$ gallon left.

9. The LCD of $\frac{3}{10}$ and $\frac{23}{25}$ is 50.

$$\frac{3}{10} \times \frac{5}{5} = \frac{15}{50}$$
$$\frac{23}{25} \times \frac{2}{2} = \frac{46}{50}$$

Now rewriting: $x + \frac{15}{50} = \frac{46}{50}$

$$\frac{31}{50} + \frac{15}{50} = \frac{46}{50}$$

So, $x = \frac{31}{50}$

10. $\frac{15}{16} + \frac{3}{40}$

$$\frac{15}{16} \times \frac{40}{40} = \frac{600}{640} \qquad \frac{3}{40} \times \frac{16}{16} = \frac{48}{640}$$

Thus $\frac{15}{16} + \frac{3}{40} = \frac{600}{640} + \frac{48}{640} = \frac{648}{640} = \frac{81}{80}$ or $1\frac{1}{80}$

2.8 Practice Problems

1.
$$5\frac{1}{12}$$
$$+9\frac{5}{12}$$
$$14\frac{6}{12} = 14\frac{1}{2}$$

2.
$$6\frac{1}{4} = 6\frac{5}{20}$$
$$+2\frac{2}{5} = +2\frac{8}{20}$$
$$8\frac{13}{20}$$

3. LCD = 12
$$7\boxed{\frac{1}{4} \times \frac{3}{3}} = 7\frac{3}{12}$$
$$+3\boxed{\frac{5}{6} \times \frac{2}{2}} = +3\frac{10}{12}$$
$$10\frac{13}{12} = 10 + 1\frac{1}{12} = 11\frac{1}{12}$$

4. LCD = 12
$$12\frac{5}{6} = 12\frac{10}{12}$$
$$-7\frac{5}{12} = -7\frac{5}{12}$$
$$5\frac{5}{12}$$

5. (a) LCD = 24
$$9\boxed{\frac{1}{8} \times \frac{3}{3}} = 9\frac{3}{24} = 8\frac{27}{24}$$
$$-3\boxed{\frac{2}{3} \times \frac{8}{8}} = -3\frac{16}{24} = -3\frac{16}{24}$$
$$5\frac{11}{24}$$

Borrow 1 from 9:
$$9\frac{3}{24} = 8 + 1 + \frac{3}{24} = 8\frac{27}{24}$$

(b)
$$18 = 17\frac{18}{18}$$
$$-6\frac{7}{18} = -6\frac{7}{18}$$
$$11\frac{11}{18}$$

6.
$$6\frac{1}{4} = 6\frac{3}{12} = 5\frac{15}{12}$$
$$-4\frac{2}{3} = -4\frac{8}{12} = -4\frac{8}{12}$$
$$1\frac{7}{12}$$

They had $1\frac{7}{12}$ gallons left over.

7. $\frac{3}{5} - \frac{1}{15} \times \frac{10}{13}$

$$= \frac{3}{5} - \frac{2}{39} \qquad LCD = 5 \cdot 39 = 195$$
$$= \frac{117}{195} - \frac{10}{195}$$
$$= \frac{107}{195}$$

8. $\frac{1}{7} \times \frac{5}{6} + \frac{5}{3} \div \frac{7}{6} = \frac{1}{7} \times \frac{5}{6} + \frac{5}{3} \times \frac{6}{7}$

$$= \frac{5}{42} + \frac{10}{7} \qquad LCD = 42$$
$$= \frac{5}{42} + \frac{60}{42} = \frac{65}{42} \text{ or } 1\frac{23}{42}$$

2.9 Practice Problems

Practice Problem 1

1. Understand the problem.

MATHEMATICS BLUEPRINT FOR PROBLEM SOLVING			
Gather The Facts	What Am I Asked To Do?	How Do I Proceed?	Key Points To Remember
Gas amounts: $18\frac{7}{10}$ gal $15\frac{2}{5}$ gal $14\frac{1}{2}$ gal	Find out how many gallons of gas she bought altogether.	Add the three amounts.	When adding mixed numbers, the LCD is needed for the fractions.

2. Solve and state the answer:

LCD = 10

$$18\frac{7}{10} = 18\frac{7}{10}$$

$$15\frac{2}{5} = 15\frac{4}{10}$$

$$14\frac{1}{2} = 14\frac{5}{10}$$

$$47\frac{16}{10} = 48\frac{6}{10}$$

$$= 48\frac{3}{5} \text{ gallons}$$

Total is $48\frac{3}{5}$ gallons.

3. Check. Estimate the answer to see if it is reasonable.

Practice Problem 2

1. Understand the problem.

MATHEMATICS BLUEPRINT FOR PROBLEM SOLVING			
Gather The Facts	What Am I Asked To Do?	How Do I Proceed?	Key Points To Remember
Poster: $12\frac{1}{4}$ in. Top border: $1\frac{3}{8}$ in. Bottom border: 2 in.	Find the length of the inside portion of the poster.	(a) Add the two border lengths. (b) Subtract this total from the poster length.	When adding mixed numbers, the LCD is needed for the fractions.

2. Solve and state the answer:

(a)
$$1\frac{3}{8}$$
$$+ 2$$
$$3\frac{3}{8}$$

(b)
$$12\frac{1}{4} = \quad 12\frac{2}{8} = \quad 11\frac{10}{8}$$
$$- 3\frac{3}{8} = \quad - 3\frac{3}{8} = \quad - 3\frac{3}{8}$$
$$\qquad\qquad\qquad\qquad\qquad 8\frac{7}{8}$$

The inside portion is $8\frac{7}{8}$ inches.

3. Check. Estimate the answer to see if it is reasonable.

Practice Problem 3

1. Understand the problem.

MATHEMATICS BLUEPRINT FOR PROBLEM SOLVING			
Gather The Facts	What Am I Asked To Do?	How Do I Proceed?	Key Points To Remember
Regular tent uses $8\frac{1}{4}$ yards. Large tent uses $1\frac{1}{2}$ times the regular. She makes 6 regular and 16 large tents.	Find out how many yards of cloth will be needed to make the tents.	Find the amount used for regular tents, and the amount used for large tents. Then add the two.	Large tents use $1\frac{1}{2}$ times the regular amount.

2. Solve and state the answer:

We multiply $6 \times 8\frac{1}{4}$ for regular tents and $16 \times 1\frac{1}{2} \times 8\frac{1}{4}$ for large tents. Then add total yardage.

$$6 \times 8\frac{1}{4} \text{ (regular tents)} = \overset{3}{\cancel{6}} \times \frac{33}{\underset{2}{\cancel{4}}} = \frac{99}{2} = 49\frac{1}{2} \text{ yards.}$$

(Large tents)

$$16 \times 1\frac{1}{2} \times 8\frac{1}{4} = \overset{2}{\cancel{16}} \times \frac{3}{\underset{1}{\cancel{2}}} \times \frac{33}{\underset{1}{\cancel{4}}} = \frac{198}{1} = 198 \text{ yards.}$$

Total yards for all tents $198 + 49\frac{1}{2} = 247\frac{1}{2}$ yards.

3. Check. Estimate the answer to see if it is reasonable.

Practice Problem 4

1. Understand the problem.

MATHEMATICS BLUEPRINT FOR PROBLEM SOLVING			
Gather The Facts	What Am I Asked To Do?	How Do I Proceed?	Key Points To Remember
He purchases 12-foot boards. Each shelf is $2\frac{3}{4}$ ft. He needs four shelves for each bookcase and he is making two bookcases.	(a) Find out how many boards he needs to buy. (b) Find out how many feet of shelving are actually needed. (c) Find out how many feet will be left over.	Find out how many $2\frac{3}{4}$-ft. shelves he can get from one board. Then see how many boards he needs to make all eight shelves.	There will be three answers to this problem. Don't forget to calculate the leftover wood.

2. Solve and state answer:

We want to know how many $2\frac{3}{4}$-ft. shelves are in a 12-ft. board.

$$12 \div 2\frac{3}{4} = \frac{12}{1} \div \frac{11}{4} = \frac{12}{1} \times \frac{4}{11} = \frac{48}{11} = 4\frac{4}{11} \text{ shelves}$$

He will get 4 shelves from each board with some left over.

(a) For two bookcases, he needs eight shelves. He gets four shelves out of each board. $8 \div 4 = 2$. He will need two 12-ft. boards.

(b) He needs 8 shelves at $2\frac{3}{4}$ feet.

$$8 \times 2\frac{3}{4} = 8 \times \frac{11}{4} = 22 \text{ feet.}$$

He actually needs 22 feet of shelving.

(c)
$$\begin{array}{r} 24 \text{ feet of shelving bought} \\ - 22 \text{ feet of shelving used} \\ \hline 2 \text{ feet of shelving left over.} \end{array}$$

3. Check. Estimate the answer to see if it is reasonable.

Practice Problem 5

1. Understand the problem.

MATHEMATICS BLUEPRINT FOR PROBLEM SOLVING			
Gather The Facts	What Am I Asked To Do?	How Do I Proceed?	Key Points To Remember
Distance is $199\frac{3}{4}$ miles. He uses $8\frac{1}{2}$ gallons of gas	Find out how many miles per gallon he gets.	Divide the distance by the number of gallons.	Change mixed numbers to improper fractions before dividing.

2. Solve and state the answer:

$$199\frac{3}{4} \div 8\frac{1}{2} = \frac{799}{4} \div \frac{17}{2}$$

$$= \frac{\overset{47}{\cancel{799}}}{\underset{2}{\cancel{4}}} \times \frac{\overset{1}{\cancel{2}}}{\underset{1}{\cancel{17}}}$$

$$= \frac{47}{2} = 23\frac{1}{2}$$

He gets $23\frac{1}{2}$ miles per gallon.

3. Check. Estimate the answer to see if it is reasonable.

Chapter 3

3.1 Practice Problems

1. (a) 0.073   seventy-three thousandths
(b) 4.68   four and sixty-eight hundredths
(c) 0.0017   seventeen ten-thousandths
(d) 561.78   five hundred sixty-one and seventy-eight hundredths

2. seven thousand eight hundred sixty-three and $\frac{4}{100}$ dollars

3. (a) $\frac{9}{10} = 0.9$ **(b)** $\frac{136}{1000} = 0.136$ **(c)** $2\frac{56}{100} = 2.56$

    (d) $34\frac{86}{1000} = 34.086$

4. (a) $0.37 = \frac{37}{100}$ **(b)** $182.3 = 182\frac{3}{10}$

    (c) $0.7131 = \frac{7131}{10,000}$ **(d)** $42.019 = 42\frac{19}{1000}$

5. (a) $8.5 = 8\frac{5}{10} = 8\frac{1}{2}$ **(b)** $0.58 = \frac{58}{100} = \frac{29}{50}$

    (c) $36.25 = 36\frac{25}{100} = 36\frac{1}{4}$ **(d)** $106.013 = 106\frac{13}{1000}$

6. $\frac{2}{1,000,000,000} = \frac{1}{500,000,000}$

The concentration of PCBs is $\frac{1}{500,000,000}$.

3.2 Practice Problems

1. Since $4 < 5$,

5.7 4   5.7 5 ;   therefore $5.74 < 5.75$.

2. $0.894 > 0.890$, so $0.894 > 0.89$

3. 2.45, 2.543, 2.46, 2.54, 2.5
It is helpful to add extra zeros and to place the decimals that begin with 2.4 in a group and the decimals that begin with 2.5 in the other.

     2.450, 2.460, 2.543, 2.540, 2.500

In order, we have from smallest to largest

     2.450, 2.460, 2.500, 2.540, 2.543.

It is OK to leave the extra terminal zeros in our answer.

4. 723.88
723.9   Since the digit to right of thousandths is greater than 5, we round up.

5. (a) 12.92 6 47
12.926   Since the digit to right of thousandths is less than 5, we drop the digits 4 and 7.

    (b) 0.00 7 892
0.008   Since the digit to right of thousandths is greater than 5, we round up.

6. 15,699.953
15,700.0   Since the digit to right of tenths is five, we round up.

7.

		Rounded to Nearest Dollar
Medical bills	375.50	376
Taxes	981.39	981
Retirement	980.49	980
Charity	817.65	818

3.3 Practice Problems

1. (a) 9.8
3.6
+ 5.4
18.8

(b) 300.72
163.75
+ 291.08
755.55

(c) 8.9000
37.0560
0.0023
+ 945.0000
990.9583

2. 93,521.8
+ 1,634.8 miles
95,156.6 miles

3. $ 80.95
133.91
256.47
53.08
+ 381.32
$905.73

4. (a) 3 8 . 8
− 2 6 . 9
1 1 . 9

(b) 2034.908
− 1986.325
48.583

5. (a) 19.000
− 12.579
6.421

(b) 283.076
− 96.380
186.696

6. 87,160.1
− 82,370.9
4,789.2 miles

7. 15.3 x is 4.5.
− 10.8
4.5

3.4 Practice Problems

1. 0.09   2 decimal places
× 0.6   + 1 decimal place
0.054   3 decimal places in product

2. (a) 0.47   2 decimal places
× 0.28   + 2 decimal places
376
94
0.1316   4 decimal places in product

    (b) 0.436   3 decimal places
× 18.39   + 2 decimal places
8.01804   5 decimal places in product

3. 0.4264   4 decimal places
× 38   + 0 decimal place
16.2032   4 decimal places in product

4. Area = length × width
1.26
× 2.3
2.898 square millimeters

5. (a) $0.0561 \times 1\underline{0} = 0.561$ Decimal point moved one place to the right.

    (b) $1462.37 \times 1\underline{00} = 146,237$ Decimal point moved two places to the right.

6. (a) $0.26 \times 1\underline{000} = 260$ Decimal point moved three places to the right. One extra zero needed.

    (b) $5862.89 \times 1\underline{0,000} = 58,628,900$ Decimal point moved four places to the right. Two extra zeros needed.

7. (a) $7.684 \times 10^4 = 76,840$ Decimal point moved four places to the right. One extra zero needed.

8. $156.2 \times 1000 = 156,200$ meters

3.5 Practice Problems

1. (a) $\begin{array}{r} 0.258 \\ 7\overline{)1.806} \\ \underline{1\ 4} \\ 40 \\ \underline{35} \\ 56 \\ \underline{56} \\ 0 \end{array}$

    (b) $\begin{array}{r} 0.0058 \\ 16\overline{)0.0928} \\ \underline{80} \\ 128 \\ \underline{128} \\ 0 \end{array}$

2.

$0.517 = 0.52$ to the nearest hundredth

$$46\overline{)23.820}$$
$$\underline{2\ 30}$$
$$82$$
$$\underline{46}$$
$$360$$
$$\underline{322}$$
$$38$$

3. $186.25 each month

$$19\overline{)\$3538.75}$$
$$\underline{19}$$
$$163$$
$$\underline{152}$$
$$118$$
$$\underline{114}$$
$$4\ 7$$
$$\underline{3\ 8}$$
$$95$$
$$\underline{95}$$
$$0$$

4. (a)

$$0.09_\wedge\overline{)0.10_\wedge08}\quad 1.12$$
$$\underline{9}$$
$$10$$
$$\underline{9}$$
$$18$$
$$\underline{18}$$
$$0$$

(b)

$$0.037_\wedge\overline{)1.702_\wedge}\quad 46.$$
$$\underline{1\ 48}$$
$$222$$
$$\underline{222}$$
$$0$$

5. (a)

$$1.8_\wedge\overline{)0.0_\wedge414}\quad 0.023$$
$$\underline{36}$$
$$54$$
$$\underline{54}$$
$$0$$

(b)

$$0.0036_\wedge\overline{)8.3160_\wedge}\quad 2310.$$
$$\underline{72}$$
$$111$$
$$\underline{108}$$
$$36$$
$$\underline{36}$$
$$0$$

6. (a)

$$3.8_\wedge\overline{)521.6_\wedge00}\quad 137.26$$
$$\underline{38}$$
$$141$$
$$\underline{114}$$
$$27\ 6$$
$$\underline{26\ 6}$$
$$1\ 00$$
$$\underline{76}$$
$$24\ 0$$
$$\underline{22\ 8}$$
$$1\ 2$$

The answer rounded to the nearest tenth is 137.3.

(b)

$$8.05_\wedge\overline{)0.17_\wedge0000}\quad 0.0211$$
$$\underline{16\ 10}$$
$$900$$
$$\underline{805}$$
$$950$$
$$\underline{805}$$
$$145$$

The answer rounded to the nearest thousandth is 0.021.

7.

$$28.5_\wedge\overline{)454.4_\wedge0}\quad 15\ .\ 9$$
$$\underline{285}$$
$$169\ 4$$
$$\underline{142\ 5}$$
$$269\ 0$$
$$\underline{256\ 5}$$
$$12\ 5$$

The truck got approximately 16 miles per gallon.

8.

$$0.12_\wedge\overline{)0.69_\wedge6}\quad 5.8 \qquad n \text{ is } 5.8.$$
$$\underline{60}$$
$$9\ 6$$
$$\underline{9\ 6}$$
$$0$$

9. Find the sum of levels for the years 1985, 1990, and 1995.

$$\begin{array}{r}9.30\\8.68\\+\ 7.37\\\hline25.35\end{array}$$

Then divide by three to obtain the average.

$$3\overline{)25.35}\quad 8.45$$
$$\underline{24}$$
$$13$$
$$\underline{12}$$
$$15$$
$$\underline{15}$$
$$0$$

The three-year average is 8.45 million tons. The five-year average was found to be 8.156 in Example 9. Find the difference between the averages.

$$\begin{array}{r}8.450\\-\ 8.156\\\hline0.294\end{array}$$

The three-year average differs from the five-year average by 0.294 million tons.

3.6 Practice Problems

1. (a)

$$16\overline{)5.0000}\quad 0.3125$$
$$\underline{48}$$
$$20$$
$$\underline{16}$$
$$40$$
$$\underline{32}$$
$$80$$
$$\underline{80}$$
$$0$$

(b)

$$80\overline{)11.0000}\quad 0.1375$$
$$\underline{80}$$
$$300$$
$$\underline{240}$$
$$600$$
$$\underline{560}$$
$$400$$
$$\underline{400}$$
$$0$$

2. (a)

$$11\overline{)7.0000}\quad 0.6363 = 0.\overline{63}$$
$$\underline{66}$$
$$40$$
$$\underline{33}$$
$$70$$
$$\underline{66}$$
$$40$$
$$\underline{33}$$
$$7$$

(b)

$$15\overline{)8.000}\quad 0.533 = 0.5\overline{3}$$
$$\underline{75}$$
$$50$$
$$\underline{45}$$
$$50$$
$$\underline{45}$$
$$5$$

(c)

$$44\overline{)13.000000}\quad 0.295454 = 0.29\overline{54}$$
$$\underline{88}$$
$$420$$
$$\underline{396}$$
$$240$$
$$\underline{220}$$
$$200$$
$$\underline{176}$$
$$240$$
$$\underline{220}$$
$$200$$
$$\underline{176}$$
$$24$$

3. (a) $2\dfrac{11}{18} = 2 + \dfrac{11}{18}$

$= 2 + 0.6\overline{1}$

$= 2.6\overline{1}$

$$\begin{array}{r} 0.611 = 0.6\overline{1} \\ 18\overline{)11.000} \\ \underline{108} \\ 20 \\ \underline{18} \\ 20 \\ \underline{18} \\ 2 \end{array}$$

(b)

$$\begin{array}{r} 1.03703 = 1.\overline{037} \\ 27\overline{)28.00000} \\ \underline{27} \\ 100 \\ \underline{81} \\ 190 \\ \underline{189} \\ 100 \\ \underline{81} \\ 19 \end{array}$$

4. $0.7916 = 0.792$ rounded to the nearest thousandth

$$\begin{array}{r} 24\overline{)19.0000} \\ \underline{168} \\ 220 \\ \underline{216} \\ 40 \\ \underline{24} \\ 160 \\ \underline{144} \\ 16 \end{array}$$

5. Divide to find the decimal equivalent of $\dfrac{5}{8}$.

$$\begin{array}{r} 0.625 \\ 8\overline{)5.000} \\ \underline{48} \\ 20 \\ \underline{16} \\ 40 \\ \underline{40} \\ 0 \end{array}$$

In the hundredths place $2 < 3$, so we know

$$0.6\underline{2}5 < 0.6\underline{3}0.$$

Therefore, $\dfrac{5}{8} < 0.63$.

6. $0.3 \times 0.5 + (0.4)^3 - 0.036 = 0.3 \times 0.5 + 0.064 - 0.036$

$= 0.15 + 0.064 - 0.036$

$= 0.214 - 0.036$

$= 0.178$

7. $6.56 \div (2 - 0.36) + (8.5 - 8.3)^2$

$= 6.56 \div (1.64) + (0.2)^2$	Parentheses
$= 6.56 \div 1.64 + 0.04$	Exponents
$= 4 + 0.04$	Divide
$= 4.04$	Add

3.7 Practice Problems

Practice Problem 1

(a) $385.98 + 875.34 \approx 400 + 900 = 1300$

(b) $623{,}999 - 212{,}876 \approx 600{,}000 - 200{,}000 = 400{,}000$

(c) $5876.34 \times 0.087 \approx$

$$\begin{array}{r} 6000 \\ \times\ \ 0.09 \\ \hline 540.00 \end{array}$$

(d) $46{,}873 \div 8.456 \approx$

$$\begin{array}{r} 6250 \\ 8\overline{)50{,}000} \\ \underline{48} \\ 20 \\ \underline{16} \\ 40 \\ \underline{40} \\ 0 \end{array}$$

Practice Problem 2

1. Understand the problem.

MATHEMATICS BLUEPRINT FOR PROBLEM SOLVING			
Gather the Facts	What Am I Asked to Do?	How Do I Proceed?	Key Points to Remember
She worked 51 hours. She gets paid $9.36 per hour for 40 hours. She gets paid time-and-a-half for 11 hours.	Find the amount Melinda earned working 51 hours last week.	Add the earnings of 40 hours at $9.36 per hour to the earnings of 11 hours at overtime pay.	Overtime pay is time-and-a-half, which is $1.5 \times \$9.36$.

2. Solve and state the answer:

(a) Calculate regular earnings for 40 hours.

$$\begin{array}{r} \$9.36 \\ \times\ \ 40 \\ \hline \$374.40 \end{array}$$

(b) Calculate overtime pay rate.

$$
\begin{array}{r}
\$9.36 \\
\times\ 1.5 \\
\hline
\$14.04
\end{array}
$$

(c) Calculate overtime earnings for 11 hours.

$$
\begin{array}{r}
\$14.04 \\
\times\ \ \ 11 \\
\hline
\$154.44
\end{array}
$$

(d) Add the two amounts.

$$
\begin{array}{rl}
\$374.40 & \text{Regular earnings} \\
+\ \ 154.44 & \text{Overtime earnings} \\
\hline
\$528.84 & \text{Total earnings}
\end{array}
$$

Melinda earned $528.84 last week.

3. Check. $40 \times \$9 = \$360$ $\begin{array}{r}\$360 \\ 10 \times \$14 = \$140 \quad +\ \ 140 \\ \hline \$500 \quad \text{The answer is reasonable.}\end{array}$

Practice Problem 3

1. Understand the problem.

MATHEMATICS BLUEPRINT FOR PROBLEM SOLVING			
Gather the Facts	What Am I Asked to Do?	How Do I Proceed?	Key Points to Remember
The total amount of steak is 17.4 pounds. Each package contains 1.45 pounds. Each package costs $4.60 per pound.	**(a)** Find out how many packages of steak the butcher will have. **(b)** Find the cost of each package.	**(a)** Divide the total, 17.4, by the amount in each package, 1.45, to find the number of packages. **(b)** Multiply the cost of one pound, $4.60, by the amount in one package, 1.45.	There will be two answers to this problem.

2. Solve and state the answer.

(a)
$$
1.45_\wedge\overline{)17.40_\wedge} \quad \text{The butcher will have}
$$
$$
\begin{array}{r}
12. \\
\underline{14\,5} \\
2\,90 \\
\underline{2\,90}
\end{array}
$$
12 packages of steak.

(b)
$$
\begin{array}{r}
\$4.60 \\
\times\ 1.45 \\
\hline
\$6.67
\end{array}
$$
Each package will cost $6.67.

3. Check.

(a)
$$
1.5_\wedge\overline{)17.0_\wedge}
$$
$$
\begin{array}{r}
1\,1. \\
\underline{15} \\
2\,0
\end{array}
$$

(b) $\$5 \times 1 = \$5$

The answers are reasonable.

Chapter 4

4.1 Practice Problems

1. (a) $\dfrac{36}{40} = \dfrac{9}{10}$ **(b)** $\dfrac{18}{15} = \dfrac{6}{5}$ **(c)** $\dfrac{220}{270} = \dfrac{22}{27}$

2. (a) $\dfrac{200}{450} = \dfrac{4}{9}$ **(b)** $\dfrac{300}{1200} = \dfrac{1}{4}$

3. $\dfrac{44 \text{ dollars}}{900 \text{ tons}} = \dfrac{11 \text{ dollars}}{225 \text{ tons}}$

4. $\dfrac{212 \text{ miles}}{4 \text{ hours}} = \dfrac{53 \text{ miles}}{1 \text{ hour}}$ 53 miles/hour

5.

selling price	$170.40
− purchase price	− 129.60
= profit	$ 40.80

She made a profit of $40.80 on 120 batteries.

$$\begin{array}{r} .34 \\ 120\overline{)40.80} \\ \underline{360} \\ 480 \\ \underline{480} \\ 0 \end{array}$$

Her profit was $0.34 per battery.

6. (a) $\dfrac{\$2.04}{12 \text{ ounces}} = \$0.17/\text{ounce}$

$\dfrac{\$2.80}{20 \text{ ounces}} = \$0.14/\text{ounce}$

(b) Fred saves $0.03 per ounce by buying the larger size.

4.2 Practice Problems

1. 6 is to 8 as 9 is to 12.

$\dfrac{6}{8} = \dfrac{9}{12}$

2. $\dfrac{2 \text{ hours}}{72 \text{ miles}} = \dfrac{3 \text{ hours}}{108 \text{ miles}}$

3. (a) $\dfrac{10}{18} \stackrel{?}{=} \dfrac{25}{45}$

$18 \times 25 = 450$

$\dfrac{10}{18} \bowtie \dfrac{25}{45}$ The cross products are equal.

$10 \times 45 = 450$

Thus $\dfrac{10}{18} = \dfrac{25}{45}$. This is a proportion.

(b) $\dfrac{42}{100} \stackrel{?}{=} \dfrac{22}{55}$

$100 \times 22 = 2200$

$\dfrac{42}{100} \bowtie \dfrac{22}{55}$ The cross products are not equal.

$42 \times 55 = 2310$

Thus $\dfrac{42}{100} \neq \dfrac{22}{55}$. This is a not proportion.

4. (a) $\dfrac{2.4}{3} \stackrel{?}{=} \dfrac{12}{15}$

$3 \times 12 = 36$

$\dfrac{2.4}{3} \bowtie \dfrac{12}{15}$ The cross products are equal.

$2.4 \times 15 = 36$

Thus $\dfrac{2.4}{3} = \dfrac{12}{15}$. This is a proportion.

(b) $\dfrac{2\frac{1}{3}}{6} \stackrel{?}{=} \dfrac{14}{38}$

$2\dfrac{1}{3} \times 38 = \dfrac{7}{3} \times \dfrac{38}{1} = \dfrac{266}{3} = 88\dfrac{2}{3}$

$6 \times 14 = 84$

The cross products are not equal.

$\dfrac{2\frac{1}{3}}{6} \bowtie \dfrac{14}{38}$ $2\dfrac{1}{3} \times 38 = 88\dfrac{2}{3}$

Thus $\dfrac{2\frac{1}{3}}{6} \neq \dfrac{14}{38}$. This is not a proportion.

5. (a) $\dfrac{1260}{7} \stackrel{?}{=} \dfrac{3530}{20}$

$7 \times 3530 = 24{,}710$

$\dfrac{1260}{7} \bowtie \dfrac{3530}{20}$ The cross products are not equal.

$1260 \times 20 = 25{,}200$

The rates are not equal.

(b) $\dfrac{2}{11} \stackrel{?}{=} \dfrac{16}{88}$

$11 \times 16 = 176$

$\dfrac{2}{11} \bowtie \dfrac{16}{88}$ The cross products are equal.

$2 \times 88 = 176$

The rates are equal.

4.3 Practice Problems

1. (a) $5 \times n = 45$

$\dfrac{5 \times n}{5} = \dfrac{45}{5}$

$n = 9$

(b) $7 \times n = 84$

$\dfrac{7 \times n}{7} = \dfrac{84}{7}$

$n = 12$

2. (a) $108 = 9 \times n$

$\dfrac{108}{9} = \dfrac{9 \times n}{9}$

$12 = n$

(b) $210 = 14 \times n$

$\dfrac{210}{14} = \dfrac{14 \times n}{14}$

$15 = n$

3. (a) $15 \times n = 63$

$\dfrac{15 \times n}{15} = \dfrac{63}{15}$

$n = 4.2$

$$\begin{array}{r} 4.2 \\ 15\overline{)63.0} \\ \underline{60} \\ 30 \\ \underline{30} \\ 0 \end{array}$$

(b) $39.2 = 5.6 \times n$

$\dfrac{39.2}{5.6} = \dfrac{5.6 \times n}{5.6}$

$7 = n$

$$\begin{array}{r} 7. \\ 5.6_\wedge\overline{)39.2_\wedge} \\ \underline{39\,2} \\ 0 \end{array}$$

4. $\dfrac{24}{n} = \dfrac{3}{7}$

$24 \times 7 = n \times 3$

$168 = n \times 3$

$\dfrac{168}{3} = \dfrac{n \times 3}{3}$

$56 = n$

5. $\dfrac{176}{4} = \dfrac{286}{n}$

$176 \times n = 286 \times 4$

$176 \times n = 1144$

$\dfrac{176 \times n}{176} = \dfrac{1144}{176}$

$n = 6.5$

$$\begin{array}{r} 6.5 \\ 176\overline{)1144.0} \\ \underline{1056} \\ 88\,0 \\ \underline{88\,0} \\ 0 \end{array}$$

6. $80 \times 6 = 5 \times n$

$480 = 5 \times n$

$\dfrac{480}{5} = \dfrac{5 \times n}{5}$

$96 = n$

The answer is 96 dollars.

7. $264.25 \times 2 = 3.5 \times n$

$528.5 = 3.5 \times n$

$\dfrac{528.5}{3.5} = \dfrac{3.5 \times n}{3.5}$

$151 = n$

The answer is 151 meters.

4.4 Practice Problems

1. $\dfrac{27 \text{ defective engines}}{243 \text{ engines produced}} = \dfrac{n \text{ defective engines}}{4131 \text{ engines produced}}$

$27 \times 4131 = 243 \times n$

$111{,}537 = 243 \times n$

$\dfrac{111{,}537}{243} = \dfrac{243 \times n}{243}$

$459 = n$

Thus we estimate that 459 engines are defective.

2. $\dfrac{9 \text{ gallons of gas}}{234 \text{ miles traveled}} = \dfrac{n \text{ gallons of gas}}{312 \text{ miles traveled}}$

$9 \times 312 = 234 \times n$

$2808 = 234 \times n$

$\dfrac{2808}{234} = \dfrac{234 \times n}{234}$

$12 = n$

She will need 12 gallons of gas.

3. $\dfrac{80 \text{ revolutions per minute}}{16 \text{ miles per hour}} = \dfrac{90 \text{ revolutions per minute}}{n \text{ miles per hour}}$

$80 \times n = 16 \times 90$

$80 \times n = 1440$

$\dfrac{80 \times n}{80} = \dfrac{1440}{80}$

$n = 18$

Alicia will be riding 18 miles per hour.

4. $\dfrac{4050 \text{ walk-in}}{729 \text{ purchase}} = \dfrac{5500 \text{ walk-in}}{n \text{ purchase}}$

$4050 \times n = 729 \times 5500$

$4050 \times n = 4{,}009{,}500$

$\dfrac{4050 \times n}{4050} = \dfrac{4{,}009{,}500}{4050}$

$n = 990$

Tom will expect 990 people to make a purchase in his store.

5. $\dfrac{50 \text{ bears tagged in 1st sample}}{n \text{ bears in forest}} = \dfrac{4 \text{ bears tagged in 2nd sample}}{50 \text{ bears caught in 2nd sample}}$

$50 \times 50 = n \times 4$

$2500 = n \times 4$

$\dfrac{2500}{4} = \dfrac{n \times 4}{4}$

$625 = n$

We estimate that there are 625 bears in the forest.

Chapter 5

5.1 Practice Problems

1. (a) $\dfrac{51}{100} = 51\%$ **(b)** $\dfrac{68}{100} = 68\%$

(c) $\dfrac{7}{100} = 7\%$ **(d)** $\dfrac{26}{100} = 26\%$

2. (a) $\dfrac{238}{100} = 238\%$ **(b)** $\dfrac{121}{100} = 121\%$

3. (a) $\dfrac{0.5}{100} = 0.5\%$ **(b)** $\dfrac{0.06}{100} = 0.06\%$ **(c)** $\dfrac{0.003}{100} = 0.003\%$

4. (a) $47\% = 0.47$ **(b)** $2\% = 0.02$

5. (a) $80.6\% = 0.806$ **(b)** $2.5\% = 0.025$

(c) $0.29\% = 0.0029$ **(d)** $231\% = 2.31$

6. (a) $0.78 = 78\%$ **(b)** $0.02 = 2\%$

(c) $5.07 = 507\%$ **(d)** $0.029 = 2.9\%$

(e) $0.006 = 0.6\%$

5.2 Practice Problems

1. (a) $71\% = \dfrac{71}{100}$ **(b)** $25\% = \dfrac{25}{100} = \dfrac{1}{4}$ **(c)** $8\% = \dfrac{8}{100} = \dfrac{2}{25}$

2. (a) $8.4\% = 0.084 = \dfrac{84}{1000} = \dfrac{21}{250}$

(b) $28.5\% = 0.285 = \dfrac{285}{1000} = \dfrac{57}{200}$

3. (a) $170\% = 1.70 = 1\dfrac{7}{10}$

(b) $288\% = 2.88 = 2\dfrac{88}{100} = 2\dfrac{22}{25}$

4. $7\dfrac{5}{8}\% = 7\dfrac{5}{8} \div 100$

$= \dfrac{61}{8} \times \dfrac{1}{100}$

$= \dfrac{61}{800}$

5. $23\dfrac{1}{6}\% = 23\dfrac{1}{6} \div 100 = 23\dfrac{1}{6} \times \dfrac{1}{100} = \dfrac{139}{6} \times \dfrac{1}{100} = \dfrac{139}{600}$

6. $\dfrac{5}{8}$ $\quad 8)\overline{5.000}^{\,0.625}$ $\quad 62.5\%$

7. (a) $\dfrac{21}{25} = 0.84 = 84\%$

(b) $\dfrac{7}{16} = 0.4375 = 43.75\%$

8. (a) $\dfrac{7}{9} = 0.77777 \approx 0.7778 = 77.78\%$

(b) $\dfrac{19}{30} = 0.63333 \approx 0.6333 = 63.33\%$

9. $\dfrac{7}{12}$ If we divide

$12)\overline{7.00}^{\,0.58}$

$\dfrac{60}{100}$

$\dfrac{96}{4}$ Thus $\dfrac{7}{12} = 58\dfrac{1}{3}\%$.

10.

Fraction	Decimal	Percent
$\dfrac{23}{99}$	0.2323	23.23%
$\dfrac{129}{250}$	0.516	51.6%
$\dfrac{97}{250}$	0.388	$38\dfrac{4}{5}\%$

5.3A Practice Problems

1. What is 26% of 35?

$\downarrow \quad \downarrow \; \downarrow \quad \downarrow \; \downarrow$

$n \quad = 26\% \times 35$

2. Find 0.08% of 350.

$\downarrow \qquad \downarrow \quad \downarrow \; \downarrow$

$n = 0.08\% \times 350$

3. (a) $58\% \times n = 400$ **(b)** $9.1 = 135\% \times n$

4. What percent of 250 is 36?

$$n \times 250 = 36$$

5. (a) $50 = n \times 20$ **(b)** $n \times 2000 = 4.5$

6. What is 82% of 350?

$$n = 82\% \times 350$$
$$n = 0.82(350)$$
$$n = 287$$

7. (a) $n = 230\% \times 400$
$$n = (2.30)(400)$$
$$n = 920$$

8. The problem asks: What is 8% of $350?
$$n = 8\% \times 350$$
$$n = \$28 \text{ tax}$$

9. $\quad 32 = 0.4\% \times n$
$$32 = 0.004n$$
$$\frac{32}{0.004} = \frac{0.004n}{0.004}$$
$$8000 = n$$

10. The problem asks: 30% of what is 6?
$$30\% \times n = 6$$
$$0.30n = 6$$
$$\frac{0.30n}{0.30} = \frac{6}{0.30}$$
$$n = 20$$

11. What percent of 9000 is 4.5?

$$n \times 9000 = 4.5$$
$$9000n = 4.5$$
$$\frac{9000n}{9000} = \frac{4.5}{9000}$$
$$n = 0.0005$$
$$n = 0.05\%$$

12. $\quad 198 = n \times 33$
$$\frac{198}{33} = \frac{n \times 33}{33}$$
$$6 = n$$
Now express n as a percent: 600%

13. The problem asks: 5 is what percent of 16?
$$5 = n \times 16$$
$$\frac{5}{16} = \frac{n \times 16}{16}$$
$$0.3125 = n$$
Now express n as a percent rounded to tenths: 31.3%

5.3B Practice Problems

1. (a) Find 83% of 460.
   $$p = 83$$
   (b) 18% of what number is 90?
   $$p = 18$$
   (c) What percent of 64 is 8?
   The percent is unknown. Use the variable p.
2. (a) $b = 52, a = 15.6$ **(b)** Base $= b, a = 170$
3. (a) What is 18% of 240?
   Percent $p = 18$
   Base $b = 240$
   Amount is unknown; use the variable a.
   (b) What percent of 64 is 4?
   Percent is unknown; use the variable p.
   Base $b = 64$
   Amount $a = 4$

4. Percent $p = 340$
Base $b = 70$
Variable a

$$\frac{a}{70} = \frac{340}{100}$$
$$100a = (70)(340)$$
$$a = 238$$
Thus 340% of 70 is 238.

5. 68% of what is 476?
Percent $p = 68$
Base is unknown; use base $= b$.
Amount $a = 476$

$$\frac{a}{b} = \frac{p}{100} \quad \text{becomes} \quad \frac{476}{b} = \frac{68}{100}$$

If we reduce the right-hand fraction, we have

$$\frac{476}{b} = \frac{17}{25}$$

$(476)(25) = 17b$ Using cross multiplication.
$11,900 = 17b$ Simplify.

$$\frac{11,900}{17} = \frac{17b}{17}$$ Divide each side by 17.

$700 = b$ Result of $11,900 \div 17$.
Thus 68% of 700 is 476.

6. Percent $p = 0.3$
Variable b
Amount $a = 216$

$$\frac{216}{b} = \frac{0.3}{100}$$
$$0.3b = (216)(100)$$
$$b = 72,000$$
Thus 216 is 0.3% of 72,000.
Therefore $72,000 was exchanged.

7. Variable p
Base $b = 3500$
Amount $a = 105$

$$\frac{105}{3500} = \frac{p}{100}$$
$$\frac{3}{100} = \frac{p}{100}$$
$$100p = (3)(100)$$
$$p = 3$$
Thus 3% of 3500 is 105.

5.4 Practice Problems

1. Let $n =$ number of people with reserved airline tickets.
$$12\% \text{ of } n = 4800$$
$$0.12 \times n = 4800$$
$$\frac{0.12 \times n}{0.12} = \frac{4800}{0.12}$$
$$n = 40,000$$
40,000 people held airline tickets that month.

2. The problem asks: What is 8% of $62.30?
$$n = 0.08 \times 62.30$$
$$n = 4.984$$
The tax is $4.98.

3. The problem asks: 105 is what percent of 130?
$$105 = n \times 130$$
$$\frac{105}{130} = n$$
$$0.8077 \approx n$$

Now express n as a percent rounded to tenths: 80.8%.
Thus 80.8% of the flights were on time.

4. 100% Cost of meal + tip of 15% = $46.00
Let $n =$ Cost of meal
$$100\% \text{ of } n + 15\% \text{ of } n = \$46.00$$
$$115\% \text{ of } n = 46.00$$
$$1.15 \times n = 46.00$$
$$n = 40.00$$
They can spend $40.00 on the meal itself.

5. (a) 7% of $13,600 is the discount.
$0.07 \times 13,600 =$ the discount
$952 is the discount.

(b) $\begin{array}{rl} \$13,600 & \text{list price} \\ -\quad 952 & \text{discount} \\ \hline \$12,648 & \text{Amount Betty paid for the car.} \end{array}$

5.5 Practice Problems

1. Commission = commission rate $\times$ value of sales
Commission = 6% $\times$ $156,000
$= 0.06 \times 156,000$
$= 9360$

His commission is $9360.

2. $\begin{array}{r} 15,000 \\ -\ 10,500 \\ \hline 4500 \end{array}$ the amount of decrease

Percent of decrease $= \dfrac{\text{amount of decrease}}{\text{original amount}} = \dfrac{\$4500}{\$15,000}$
$= 0.30 = 30\%$

The percent of decrease is 30%.

3. $I = P \times R \times T$
$P = \$5600 \qquad R = 12\% \qquad T = 1 \text{ year}$
$I = 5600 \times 12\% \times 1$
$\quad = 5600 \times 0.12$
$\quad = 672$

The interest is $672.

4. (a) $I = P \times R \times T$
$\quad = 1800 \times 0.11 \times 4$
$\quad = \$792$

(b) $I = P \times R \times T$
$\quad = 1800 \times 0.11 \times \dfrac{1}{2}$
$\quad = \$99$

Chapter 6

6.1 Practice Problems

1. (a) 3 **(b)** 5280 **(c)** 60 **(d)** 7 **(e)** 16 **(f)** 2 **(g)** 4

2. $15,840 \text{ ft} \times \dfrac{1 \text{ mile}}{5280 \text{ ft}} = \dfrac{15,840}{5280} \text{ miles} = 3 \text{ miles}$

3. (a) $18.93 \text{ miles} \times \dfrac{5280 \text{ feet}}{1 \text{ mile}} = 99,950.4 \text{ feet}$

(b) $16\dfrac{1}{2} \text{ inches} \times \dfrac{1 \text{ yard}}{36 \text{ inches}} = 16\dfrac{1}{2} \times \dfrac{1}{36} \text{ yard}$

$= \dfrac{\overset{11}{\cancel{33}}}{2} \times \dfrac{1}{\underset{12}{\cancel{36}}} \text{ yard} = \dfrac{11}{24} \text{ yard}$

4. $760.5 \text{ lb} \times \dfrac{16 \text{ oz}}{1 \text{ lb}} = 760.5 \times 16 \text{ oz} = 12,168 \text{ oz}$

5. $19 \text{ pints} \times \dfrac{1 \text{ quart}}{2 \text{ pints}} = \dfrac{19}{2} \text{ quarts} = 9.5 \text{ quarts}$

6. Step 1: $7 \text{ lbs} \times \dfrac{16 \text{ oz}}{1 \text{ lb}} = 7 \times 16 \text{ oz} = 112 \text{ oz}$

Step 2: $112 \text{ oz} + 12 \text{ oz} = 124$ ounces
The potatoes weigh 124 ounces.

7. Step 1: $1\dfrac{3}{4} \text{ days} \times \dfrac{24 \text{ hours}}{1 \text{ day}} = \dfrac{7}{4} \times \dfrac{24}{1} \text{ hours} = 42 \text{ hr parked}$

Step 2: $42 \text{ hr} \times \dfrac{1.50 \text{ dollars}}{1 \text{ hour}} = 63.00 \text{ dollars}$

She paid $63.00.

6.2 Practice Problems

1. (a) deka- **(b)** milli-
2. (a) 4 meters = $4.00_\wedge$ centimeters = 400 cm
(b) 30 centimeters = $30.0_\wedge$ millimeters = 300 mm
3. (a) 3 mm = $_\wedge003.$ m = 0.003 m
(b) 47 cm = $_\wedge00047.$ km = 0.00047 km

4. The car length would logically be choice (b) 3.8 meters. (A meter is close to a yard and 3.8 yards seems reasonable)
5. (a) 3.75 m **(b)** 46,000 mm
6. (a) 389 mm = 0.0389 dam (four places to left)
(b) 0.48 hm = 4800 cm (four places to right)
7. $\begin{array}{rl} 782 \text{ cm} = & 7.82 \text{ m} \\ 2 \text{ m} = & 2.00 \text{ m} \\ 537 \text{ m} = & \underline{537.00 \text{ m}} \\ & 546.82 \text{ m} \end{array}$

6.3 Practice Problems

1. (a) 5 L = 5000 mL **(b)** 84 kL = 84,000 L
(c) 0.732 L = 732 mL
2. (a) 15.8 mL = 0.0158 L **(b)** 12,340 mL = 12.34 L
(c) 86.3 L = 0.0863 kL
3. (a) 396 mL = 396 cm^3
(because 1 milliliter = 1 cubic centimeter)
(b) 0.096 L = 96 cm^3
4. (a) 3.2 t = 3200 kg **(b)** 7.08 kg = 7080 g
5. (a) 59 kg = 0.059 t **(b)** 28.3 mg = 0.0283 g
6. A gram is $\frac{1}{1000}$ of a kilogram. If the coffee costs $10.00 per kilogram, then 1 gram would be $\frac{1}{1000}$ of $10.

$\dfrac{1}{1000} \times \$10 = \dfrac{\$10.00}{1000} = \$0.01$

The coffee costs 1¢ per gram.
7. (a) 120 kg (A kg is slightly more than 2 lb.)

6.4 Practice Problems

1. (a) $7 \text{ ft} \times \dfrac{0.305 \text{ m}}{1 \text{ ft}} = 2.135 \text{ m}$

2. (a) $17 \text{ m} \times \dfrac{1.09 \text{ yd}}{1 \text{ m}} = 18.53 \text{ yard}$

(b) $29.6 \text{ km} \times \dfrac{0.62 \text{ mi}}{1 \text{ km}} = 18.352 \text{ mi}$

(c) $26 \text{ gal} \times \dfrac{3.79 \text{ L}}{1 \text{ gal}} = 98.54 \text{ L}$

(d) $6.2 \text{ L} \times \dfrac{1.06 \text{ qt}}{1 \text{ L}} = 6.572 \text{ qt}$

3. $180 \text{ cm} \times \dfrac{0.394 \text{ in.}}{1 \text{ cm}} \times \dfrac{1 \text{ ft}}{12 \text{ in.}} = 5.91 \text{ ft}$

4. $\dfrac{88 \text{ km}}{\text{hr}} \times \dfrac{0.62 \text{ mi}}{1 \text{ km}} = 54.56 \text{ mi/hr}$

5. $\dfrac{900 \text{ miles}}{\text{hr}} \times \dfrac{5280 \text{ ft}}{1 \text{ mile}} \times \dfrac{1 \text{ hr}}{60 \text{ min}} \times \dfrac{1 \text{ min}}{60 \text{ sec}}$

$= \dfrac{900 \times 5280 \text{ ft}}{60 \times 60 \text{ sec}} = \dfrac{4,752,000 \text{ ft}}{3600 \text{ sec}}$

$= 1320 \text{ ft/sec}$

The jet is traveling at 1320 feet per second.
6. $F = 1.8 \times C + 32$
$\quad = 1.8 \times 20 + 32$
$\quad = 36 + 32$
$\quad = 68$

The temperature is 68°F.

6.5 Practice Problems

Practice Problem 1

Step 1: $\begin{array}{r} 2\dfrac{2}{3} \text{ yd} \\[4pt] 8\dfrac{1}{3} \text{ yd} \\[4pt] 2\dfrac{2}{3} \text{ yd} \\[4pt] +\ 8\dfrac{1}{3} \text{ yd} \\ \hline 22 \text{ yd} \end{array}$

Step 2: $22 \text{ yd} \times \dfrac{3 \text{ ft}}{1 \text{ yd}} = 66 \text{ ft}$

The perimeter is 66 ft.

Practice Problem 2

1. Understand the problem.

MATHEMATICS BLUEPRINT FOR PROBLEM SOLVING			
Gather the Facts	What Am I Asked to Do?	How Do I Proceed?	Key Points to Remember
He must use 18.06 liters of solution. He has 42 jars to fill.	Find out how many milliliters of solution will go into each jar.	We need to convert 18.06 liters to milliliters, and then divide that result by 42.	To convert 18.06 liters to milliliters, we move the decimal point three places to the right.

2. Solve and state the answer:
   (a) 18.06 L = 18,060 mL
   (b) $\dfrac{18{,}060 \text{ mL}}{42 \text{ jars}} = 430$ mL/jar
   Thus, 430 mL of a solution will go into each jar.
3. Check. 18.06 L is approximately 18 L, or 18,000 mL.
   $\dfrac{18{,}000}{40} = 450$ and is reasonable.

Chapter 7

7.1 Practice Problems

1. $\angle FGH$ and $\angle KGJ$ are acute angles, $\angle HGK$ and $\angle FGJ$ are obtuse angles, $\angle HGJ$ is a right angle, and $\angle FGK$ is a straight angle.
2. The complement of angle B is $90° - 83° = 7°$.
   The supplement of angle B is $180° - 83° = 97°$.
3. $\angle y$ and $\angle w$ are vertical angles and so have the same measure. Thus $\angle w = 133°$. $\angle y$ and $\angle z$ are adjacent angles, so we know they are supplementary. Thus $\angle z$ measures $180° - 133° = 47°$. Finally, $\angle x$ and $\angle z$ are vertical angles, so we know they have the same measure. Thus $\angle x$ measures $47°$.
4. $\angle z = 180° - 105° = 75°$ ($\angle x$ and $\angle z$ are adjacent angles).
   $\angle x = \angle y = 105°$ ($\angle x$ and $\angle y$ are alternate interior angles).
   $\angle v = \angle x = 105°$ ($\angle v$ and $\angle x$ are corresponding angles).
   $\angle w = 180° - 105° = 75°$ ($\angle w$ and $\angle v$ are adjacent angles).

7.2 Practice Problems

1. $P = 2l + 2w$
   $= 2(6) + 2(1.5)$
   $= 12 + 3 = 15$ m
2. $P = 4S$
   $= 4(5.8) = 23.2$ cm
3. $P = 4 + 4 + 5.5 + 2.5 + 1.5 + 1.5 = 19$ ft
   $\text{Cost} = 19 \text{ ft} \times \dfrac{0.16 \text{ dollar}}{1 \text{ ft}} = \$3.04$
4. $A = lw = (29)(17) = 493$ m^2
5. $A = (s)^2$
   $= (11.8)^2$
   $= 139.24$ mm^2
6. Area of rectangle $= (18)(20) = 360$ ft^2
   Area of square $= 6^2 = 36$ ft^2
   Total area $= 396$ ft^2

7.3 Practice Problems

1. $P = (2)(7.6) + (2)(3.5)$
   $= 15.2 + 7.0 = 22.2$ cm
2. $A = bh$
   $= (10.3)(1.5)$
   $= 15.45$ km^2

3. $P = 4(6 \text{ cm}) = 24$ cm
   $A = bh$
   $= (4 \text{ cm})(6 \text{ cm}) = 24$ cm^2
4. $P = 7 + 15 + 21 + 13 = 56$ yd
5. (a) $A = \dfrac{h(b + B)}{2} = \dfrac{140(180 + 130)}{2} = 21{,}700$ yd^2
   (b) $21{,}700 \text{ yd}^2 \times \dfrac{1 \text{ gallon}}{100 \text{ yd}^2} = 217$ gallons
   Thus 217 gallons of sealer are needed.
6. The area of the trapezoid is
   $A = \dfrac{(9.2)(12.6 + 19.8)}{2}$
   $= \dfrac{(9.2)(32.4)}{2} = \dfrac{298.08}{2}$
   $= 149.04$ cm^2.
   The area of the rectangle is
   $A = (8.3)(12.6) = 104.58$ cm^2
   Total area $= 253.62$ cm^2

7.4 Practice Problems

1. The sum of the angles in a triangle is $180°$. The two given angles total $125° + 15° = 140°$. Thus $180° - 140° = 40°$. Angle A must be $40°$.
2. $P = 10.5 + 10.5 + 8.5 = 29.5$ m
3. $A = \dfrac{bh}{2} = \dfrac{(38)(13)}{2} = \dfrac{494}{2} = 247$ m^2
4. Area of rectangle $= (11)(24) = 264$ cm^2
   Area of triangle $= \dfrac{(11)(7)}{2} = \dfrac{77}{2} = 38.5$ cm^2
   Total area $= 302.5$ cm^2

7.5 Practice Problems

1. (a) $\sqrt{49} = 7$ because $(7)(7) = 49$.
   (b) $\sqrt{169} = 13$ because $(13)(13) = 169$.
2. $\sqrt{49} = 7$ because $(7)(7) = 49$.
   $\sqrt{4} = 2$ because $(2)(2) = 4$.
   Thus $\sqrt{49} - \sqrt{4} = 7 - 2 = 5$.
3. (a) Yes. 144 is a perfect square because $(12)(12) = 144$.
   (b) $\sqrt{144} = 12$
4. (a) $\sqrt{3} \approx 1.732$ **(b)** $\sqrt{13} \approx 3.606$ **(c)** $\sqrt{5} \approx 2.236$
5. $\sqrt{22 \text{ m}^2} \approx 4.690$ m
   Thus, to the nearest thousandth of an inch, the side measures 4.690 m.

7.6 Practice Problems

1. Hypotenuse $= \sqrt{(8)^2 + (6)^2}$
   $= \sqrt{64 + 36}$ — Square each value first.
   $= \sqrt{100}$ — Add together the two values.
   $= 10$ m — Take the square root.

2. Hypotenuse $= \sqrt{(3)^2 + (7)^2}$
$\qquad = \sqrt{9 + 49}$ Square each value first.
$\qquad = \sqrt{58}$ cm   Add the two values together
Using the square root table or a calculator, we have the hypotenuse ≈ 7.616 m.

3. Leg $= \sqrt{(17)^2 - (15)^2}$
$\qquad = \sqrt{289 - 225}$ Square each value first.
$\qquad = \sqrt{64}$ Subtract.
$\qquad = 8$ m   Find the square root.

4. Leg $= = \sqrt{(10)^2 - (5)^2}$
$\qquad = \sqrt{100 - 25}$ Square each value first.
$\qquad = \sqrt{75}$ m   Subtract the two numbers.
Using a calculator or the square root table, we find that the leg ≈ 8.660 m.

5. **1.** Understand the problem.
We are given a picture.
The distance between the centers of the holes is the hypotenuse of the triangle.
   2. Solve and state the answer.
$\qquad$ Hypotenuse $= \sqrt{(\text{leg})^2 + (\text{leg})^2}$
$\qquad\qquad = \sqrt{(2)^2 + (5)^2}$
$\qquad\qquad = \sqrt{4 + 25}$
$\qquad\qquad = \sqrt{29}$
$\qquad\qquad \sqrt{29} \approx 5.385$ cm
Rounded to the nearest thousandth, the distance is 5.385 cm.
   3. Check.
Work backward to check. Use the Pythagorean Theorem.
$5.385^2 \approx 2^2 + 5^2$?   (We use $\approx$ because 5.385 is an approximate answer.)
$28.998 \approx 4 + 25$
$28.998 \approx 29$ ✓

6. **1.** Understand the problem.
We are given a picture.
   2. Solve and state the answer.
$\qquad$ Leg $= \sqrt{(\text{hypotenuse})^2 - (\text{leg})^2}$
$\qquad\qquad = \sqrt{(30)^2 - (27)^2}$
$\qquad\qquad = \sqrt{900 - 729}$
$\qquad\qquad = \sqrt{171}$
$\qquad\qquad \sqrt{171} \approx 13.1$ yd
If we round to the nearest tenth, the kite is 13.1 yd above the rock.

7. **(a)** In a 30°–60°–90° triangle the side opposite the 30° angle is $\frac{1}{2}$ of the hypotenuse.
$$\frac{1}{2} \times 12 = 6$$
Therefore, $y = 6$ ft.
When we know two sides of a right triangle, we find the third side using the Pythagorean Theorem.
$\qquad$ Leg $= \sqrt{(\text{hypotenuse})^2 - (\text{leg})^2}$
$\qquad\qquad = \sqrt{(12)^2 - (6)^2} = \sqrt{144 - 36}$
$\qquad\qquad = \sqrt{108} \approx 10.4$ ft
$x = 10.4$ ft rounded to the nearest tenth.
   (b) In a 45°–45°–90° triangle we have the following:
$\qquad$ Hypotenuse $= \sqrt{2} \times$ leg
$\qquad\qquad \approx 1.414(8)$
$\qquad\qquad = 11.312$ m
Rounded to the nearest tenth, the hypotenuse = 11.3 m.

7.7 Practice Problems

1. $C = \pi d$
$\quad = (3.14)(9)$
$\quad = 28.26$
$\quad \approx 28.3$ m to nearest tenth

2. (a) $C = \pi d$
$\qquad = (3.14)(30)$
$\qquad = 94.2$ in.
   (b) Change 94.2 in. to ft.
$\qquad 94.2 \text{ in.} \times \dfrac{1 \text{ ft}}{12 \text{ in.}} = 7.85 \text{ ft}$
   (c) When the wheel makes 2 revolutions, the bicycle travels $7.85 \times 2 = 15.7$ ft.

3. $A = \pi r^2$
$\quad = (3.14)(5^2)$
$\quad = (3.14)(25)$
$\quad = 78.5$ km^2

4. $r = \dfrac{d}{2} = \dfrac{10}{2} = 5$ ft
$\quad A = \pi r^2$
$\qquad = (3.14)(5)^2$
$\qquad = 78.5$ ft^2
Change 78.5 ft^2 to yd^2
$\quad 78.5 \text{ ft}^2 \times \dfrac{1 \text{ yd}^2}{9 \text{ ft}^2} \approx 8.7222 \text{ yd}^2$
Find the cost: $\dfrac{\$12}{1 \text{ ft}^2} \times 8.7222 \text{ yd}^2 = \$104.67.$
The cost of the pool cover is $104.67.

5. Area of square $-$ area of circle $=$ shaded area
$\qquad s^2 \qquad\qquad - \qquad \pi r^2 \qquad = $ shaded area
$\qquad 5^2 \qquad\qquad - \quad (3.14)(2^2)$
$\qquad 25 \qquad\qquad - \quad (3.14)(4)$
$\qquad 25.00 \qquad\quad - \quad 12.56 \qquad \approx 12.4$ ft^2
$\qquad\qquad\qquad\qquad\qquad\qquad$ (rounded to nearest tenth)

6. $r = \dfrac{d}{2} = \dfrac{8}{2} = 4$ ft
$\quad A \text{ semicircle} = \dfrac{\pi r^2}{2}$
$\qquad\qquad\qquad = \dfrac{(3.14)(4)^2}{2}$
$\qquad\qquad\qquad = 25.12$ ft^2
$\quad A \text{ rectangle} = lw = (12)(8) = 96$ ft^2.
$\qquad 25.12$ ft^2
$\qquad + 96.00$ ft^2
$\qquad \overline{121.12 \text{ ft}^2}$ The total area is approximately 121.1 ft^2.

7.8 Practice Problems

1. $V = lwh$
$\quad = (6)(5)(2)$
$\quad = (30)(2)$
$\quad = 60$ m^3

2. $V = \pi r^2 h$
$\quad = (3.14)(2)^2(5)$
$\quad = (3.14)(4)(5)$
$\quad = 62.8$ in.^{3}

3. $V = \dfrac{4\pi r^3}{3} = \dfrac{(4)(3.14)(6^3)}{3} = \dfrac{(4)(3.14)(6)(6)(6)^2}{3}$
$\quad = (12.56)(36)(2) = 904.32$
$\quad \approx 904.3$ m^3 rounded to nearest tenth

4. $V = \dfrac{\pi r^2 h}{3}$
$\quad = \dfrac{(3.14)(5)^2(12)}{3}$
$\quad = 314$ m^3

5. (a) $V = \dfrac{Bh}{3}$
$\qquad = \dfrac{(6)(6)(10)}{3} = \dfrac{(36)(10)}{3}$
$\qquad = \dfrac{(36)^{12}(10)}{3} = 120$ m^3

(b) $V = \dfrac{Bh}{3}$

$$= \dfrac{(7)(8)(15)}{3} = \dfrac{(7)(8)\overset{5}{\cancel{(15)}}}{\underset{1}{\cancel{3}}}$$

$$= (56)(5) = 280 \text{ m}^3$$

7.9 Practice Problems

1. $\dfrac{11}{27} = \dfrac{15}{n}$

$11n = (27)(15)$

$11n = 405$

$\dfrac{11n}{11} = \dfrac{405}{11}$

$n = 36.\overline{81}$

$= 36.8$ meters measured to the nearest tenth

2. a corresponds to p, b corresponds to m, c corresponds to n

3. $\dfrac{h}{5} = \dfrac{20}{2}$

$2h = 100$

$h = 50$ ft

4. $\dfrac{3}{29} = \dfrac{1.8}{w}$

$3w = (1.8)(29)$

$3w = 52.2$

$\dfrac{3w}{3} = \dfrac{52.2}{3}$

$w = 17.4$ The width is 17.4 meters.

7.10 Practice Problems

Practice Problem 1

1. Understand the problem.

MATHEMATICS BLUEPRINT FOR PROBLEM SOLVING			
Gather the Facts	What Am I Asked to Do?	How Do I Proceed?	Key Points to Remember
Mike needs to sand three rooms: 24 ft × 13 ft 12 ft × 9 ft 16 ft × 3 ft He can sand 80 ft² in 15 min.	Find out how long it will take him to sand all three rooms.	**(a)** Find the total area to be sanded. **(b)** Then find out how long it will take him to sand the total area.	Area = length × width To get the total time, multiply the total area by the fraction: $\dfrac{15 \text{ min.}}{80 \text{ ft}^2}$

2. Solve and state the answer:

$24 \times 13 = 312 \text{ ft}^2$ room 1

$12 \times 9 = 108 \text{ ft}^2$ room 2

$16 \times 3 = 48 \text{ ft}^3$ room 3

Total area $= 468 \text{ ft}^2$

$468 \text{ ft}^2 \times \dfrac{15 \text{ min}}{80 \text{ ft}^2} = 87.75 \text{ min.}$ It will take Mike 87.75 min. to sand the rooms.

3. Check. Estimate the answer to see if it is reasonable.

Practice Problem 2

1. Understand the problem.

MATHEMATICS BLUEPRINT FOR PROBLEM SOLVING			
Gather the Facts	What Am I Asked to Do?	How Do I Proceed?	Key Points to Remember
The trapezoid has a height of 9 ft. The bases are 18 ft and 12 ft. The rectangular portion measures 24 ft × 15 ft. Roofing costs $2.75 per square yard.	**(a)** Find the area of the roof. **(b)** Find the cost to install new roofing.	**(a)** Find the area of the entire roof. Change square feet to square yards. **(b)** Multiply by $2.75.	9 square feet = 1 square yard

2. Solve and state the answer:

(a) Area of trapezoid = $\dfrac{1}{2}h(b + B)$

$$= \dfrac{1}{2}(9)(12 + 18)$$

$$= 135 \text{ ft}^2$$

Area of rectangle = lw

$$= (15)(24)$$

$$= 360 \text{ ft}^2$$

Total Area = $135 \text{ ft}^2 + 360 \text{ ft}^2 = 495 \text{ ft}^2$

Change square feet to square yards.

$$495 \text{ ft}^2 \times \dfrac{1 \text{ yd}^2}{9 \text{ ft}^2} = 55 \text{ yd}^2$$

The area of the roof is 55 yd^2.

(b) Cost = $55 \text{ yd}^2 \times \dfrac{\$2.75}{1 \text{ yd}^2} = \$151.25$

The cost to install new roofing would be \$151.25.

3. Check. Estimate the answers to see if they seem reasonable.

Chapter 8

8.1 Practice Problems

1. The smallest category of students is special students.
2. (a) 3000 freshmen + 200 special students = 3200
There are 3200 students who are either freshmen or special students.
(b) 3200 out of 10,000 are freshmen or special students.
$$\dfrac{3200}{10,000} = 0.32 = 32\%$$
3. There are 3000 freshmen but only 2600 sophomores. The ratio of freshmen to sophomores is $\dfrac{3000}{2600} = \dfrac{15}{13}$.
4. The ratio of freshmen to the total number of students is
$$\dfrac{3000}{10,000} = \dfrac{3}{10}.$$
5. Lake Ontario occupies the smallest area with 8%.
6. The percent of the total area occupied by either Lake Superior or Lake Michigan is: $34\% + 24\% = 58\%$.
7. Lake Superior has 34% of the area. 34% of $94,680 \text{ mi}^2$ is $(10.34)(94,680) = 32,191 \text{ mi}^2$.
8. (a) $11\% + 14\% = 25\%$
(b) 18% of $736,000 = (0.18)(736,000) = 132,480$

8.2 Practice Problems

1. The bar rises to 24. The approximate population was 24,000,000.
2. $16,000,000 - 7,000,000 = 9,000,000$. The population increased by 9,000,000.
3. The bar rises to 250. The number of new cars sold in the fourth quarter of 2001 was 250.
4. $150 - 100 = 50$. Thus, 50 fewer cars were sold.
5. The greatest number of customers came in July, since the highest point of the graph occurs in July.
6. (a) 3500 **(b)** The number decreased.
7. The sharpest line of decrease is from July to August. The line slopes downward most steeply between those two months. Thus the biggest decrease in customers is between July and August.
8. Because the dot corresponding to 1999–2000 is at 15 and the scale is in hundreds, we have $15 \times 100 = 1500$. Thus, 1500 degrees in visual and performing arts were awarded.
9. The computer science line goes above the visual and performing arts line first in 1997–1998. Thus the first academic year with more degrees in computer science was 1997–1998.

8.3 Practice Problems

1. The 60–69 bar rises to a height of 6. Thus six tests would have a D grade.
2. From the histogram, 16 tests were 70–79, 8 tests were 80–89, and 6 tests were 90–99. When we combine $16 + 8 + 6 = 30$, we can see that 30 students scored greater than 69 on the test.

3. The 800–999 bar rises to a height of 20. Thus 20 light bulbs lasted between 800 and 999 hours.
4. From the histogram, 25 bulbs lasted 1200–1399 hours, 10 lasted 1400–1599 hours, and 5 lasted 1600–1799 hours. When we combine $25 + 10 + 5 = 40$, we can see that 40 light bulbs lasted more than 1199 hours.
5. Weight in Pounds

(Class Interval)	Tally	Frequency
1600–1799	||	2
1800–1999	||||	4
2000–2199	||||	4
2200–2399	⊬||	5

6.

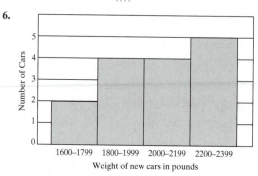

Weight of new cars in pounds

7. The greatest difference occurs between the 35–54 age category and the 75 or older category.

8.4 Practice Problems

1. $\dfrac{\$39.20 + \$43.50 + \$81.90 + \$34.20 + \$51.70 + \$48.10}{6} = \$49.77$

The mean monthly phone bill is \$49.77.

2. $\underbrace{\$150, \$150, \$290}_{\text{three numbers}}$ $\underset{\substack{\uparrow \\ \text{middle} \\ \text{number}}}{\$320}$ $\underbrace{\$400, \$450, \$600}_{\text{three numbers}}$

Thus, \$320 is the median salary.

3. $\underbrace{88, 90}_{\text{two numbers}}$ $\underset{\substack{\sqcup \\ \text{two middle} \\ \text{numbers}}}{100, 105}$ $\underbrace{118, 126}_{\text{three numbers}}$

$\dfrac{100 + 105}{2} = \dfrac{205}{2} = 102.5$

The median is 102.5.
4. The value 71 occurs twice. The mode is 71 inches.
5. The value 3 occurs three times. The mode is 3.

Chapter 9

9.1 Practice Problems

1. (a) 4 lies to the right of 2, so $4 > 2$.
(b) -3 lies to the right of -5, so $-3 > -5$.
(c) 0 lies to the right of -6, so $0 > -6$.
(d) -2 lies to the left of 1, so $-2 < 1$.
(e) 5 lies to the right of -7, so $5 > -7$.

2. (a) $\begin{array}{r} 9 \\ + 14 \\ \hline 23 \end{array}$ **(b)** $\begin{array}{r} -4.5 \\ + -1.9 \\ \hline -6.4 \end{array}$

3. (a) $\dfrac{5}{12} = \dfrac{5}{12}$

$\begin{array}{r} + \dfrac{1}{4} \times \dfrac{3}{3} = + \dfrac{3}{12} \\ \hline \dfrac{8}{12} = \dfrac{2}{3} \end{array}$

(b) The LCD = 42.

$$\frac{1}{6} \times \frac{7}{7} = \frac{7}{42}$$

Because $\frac{1}{6} = \frac{7}{42}$ it follows that $-\frac{1}{6} = -\frac{7}{42}$.

$$\frac{2}{7} \times \frac{6}{6} = \frac{12}{42}$$

Because $\frac{2}{7} = \frac{12}{42}$ it follows that $-\frac{2}{7} = -\frac{12}{42}$.

Thus

$$\begin{array}{c} -\dfrac{1}{6} \\ + \; -\dfrac{2}{7} \end{array} \text{ is equivalent to } \begin{array}{c} -\dfrac{7}{42} \\ + \; -\dfrac{12}{42} \\ \hline -\dfrac{19}{42} \end{array}$$

4. Add $(-\$200 \text{ million}) + (-\$20 \text{ million})$ to obtain $-\$220$ million. The total debt for these two years is $\$220,000,000$.

5. (a)
$$\begin{array}{r} 7 \\ + -12 \\ \hline -5 \end{array}$$

(b)
$$\begin{array}{r} -20.8 \\ + \;\; 15.2 \\ \hline -5.6 \end{array}$$

(c) $\dfrac{5}{6} + \left(-\dfrac{3}{4}\right) = \dfrac{10}{12} + \left(-\dfrac{9}{12}\right) = \dfrac{10 + (-9)}{12} = \dfrac{1}{12}$

6.
$$\begin{array}{r} 28°\text{F} \\ + -19°\text{F} \\ \hline 9°\text{F} \end{array}$$

7.
$$\begin{array}{r} 36 \\ + -21 \\ \hline 15 \end{array} \text{ Then we add } \begin{array}{r} 15 \\ + -18 \\ \hline -3 \end{array}$$

8.
$$\begin{array}{r} \$30,000 \\ + \$40,000 \\ \hline \$70,000 \end{array} \quad \begin{array}{r} -\$20,000 \\ - \$5,000 \\ \hline + -\$35,000 \\ - \$60,000 \end{array} \quad \begin{array}{r} \$70,000 \\ + -\$60,000 \\ \hline \$10,000 \end{array}$$

The company had an overall profit of $\$10,000$ in the five-month period.

9.2 Practice Problems

1. $-10 - (-5) = -10 + 5 = -5$

2. (a) $5 - 12 = 5 + (-12) = -7$

 (b) $-11 - 17 = -11 + (-17) = -28$

3. (a) $3.6 - (-9.5) = 3.6 + 9.5 = 13.1$

 (b) $-\dfrac{5}{8} - \left(-\dfrac{5}{24}\right) = -\dfrac{5}{8} + \dfrac{5}{24}$

$$= -\dfrac{5}{8} \times \dfrac{3}{3} + \dfrac{5}{24}$$

$$= -\dfrac{15}{24} + \dfrac{5}{24}$$

$$= -\dfrac{10}{24} \text{ or } -\dfrac{5}{12}$$

4. (a) $20 - (-5) = 20 + 5 = 25$

 (b) $-\dfrac{1}{5} - \left(-\dfrac{1}{2}\right) = -\dfrac{1}{5} + \dfrac{1}{2} = -\dfrac{2}{10} + \dfrac{5}{10} = \dfrac{3}{10}$

 (c) $3.6 - (-5.5) = 3.6 + 5.5 = 9.1$

5. $-5 - (-9) + (-14) = -5 + 9 + (-14) = 4 + (-14) = -10$

6. $31 - (-37) = 31 + 37 = 68$ The difference is $68°$F.

9.3 Practice Problems

1. (a) $(6)(9) = 54$ **(b)** $(7)(12) = 84$

2. (a) $(-8)(5) = -40$ **(b)** $3(-60) = -180$

3. (a) $-50 \div 25 = -2$

 (b) $49 \div (-7) = -7$

4. (a) $-10(-6) = 60$ **(b)** $\left(-\dfrac{1}{3}\right)\left(-\dfrac{2}{7}\right) = \dfrac{2}{21}$

5. (a) $-78 \div (-2) = 39$ **(b)** $(-1.2) \div (-0.5) = 2.4$

6. $(-6)(3)(-4) = (-18)(-4) = 72$

7. $(-2)(6) = -12$

Thus the change in charge would be -12.

8. $\dfrac{17 + 19 + 2 + (-4) + (-3) + (-13)}{6} = \dfrac{18}{6} = 3$

The average temperature was $3°$.

9.4 Practice Problems

1. $20 \div (-5)(-3)$

$= (-4) \;\; (-3)$

$= 12$

2. (a) $25 \div (-5) + 16 \div (-8)$

$\quad = (-5) \;\; + \;\; (-2)$

$\quad = -7$

 (b) $9 + 20 \div (-4)$

$\quad = 9 + (-5)$

$\quad = 4$

3. $\dfrac{9(-3) - 5}{2(-4) \div (-2)} = \dfrac{-27 - 5}{-8 \div (-2)} = \dfrac{-32}{+4} = -8$

4. $-2(-12 + 15) + (-3)^4 + 2(-6)$

$= -2(3) + (-3)^4 + 2(-6)$

$= -2(3) + 81 + 2(-6)$

$= -6 + 81 + (-12)$

$= 75 + (-12)$

$= 63$

5. $\left(\dfrac{1}{5}\right)^2 + 4\left(\dfrac{1}{5} - \dfrac{3}{10}\right) \div \dfrac{2}{3}$

$= \left(\dfrac{1}{5}\right)^2 + 4\left(\dfrac{2}{10} - \dfrac{3}{10}\right) \div \dfrac{2}{3}$

$= \left(\dfrac{1}{5}\right)^2 + 4\left(-\dfrac{1}{10}\right) \div \dfrac{2}{3}$

$= \dfrac{1}{25} + 4\left(-\dfrac{1}{10}\right) \div \dfrac{2}{3}$

$= \dfrac{1}{25} + \left(-\dfrac{2}{5}\right) \div \dfrac{2}{3}$

$= \dfrac{1}{25} + \left(-\dfrac{3}{5}\right)$

$= \dfrac{1}{25} + \left(-\dfrac{15}{25}\right)$

$= -\dfrac{14}{25}$

9.5 Practice Problems

1. (a) 3.729×10^3 **(b)** 5.06936×10^5

2. (a) 7.6×10^{-2} **(b)** 9.82×10^{-1}

3. $6.543 \times 10^3 = 6543$

4. (a) $430,000$ **(b)** $60,000$

5. (a) $7.72 \times 10^{-3} = 0.00772$ **(b)** $2.6 \times 10^{-5} = 0.000026$

6.
$$\begin{array}{l} 6.85 \times 10^{22} \text{ kilograms} \\ + 2.09 \times 10^{22} \text{ kilograms} \\ \hline 8.94 \times 10^{22} \text{ kilograms} \end{array}$$

7. $3.1 \times 10^4 = 31,000$

But $31,000 = 0.31 \times 10^5$

$$\begin{array}{r} 4.36 \times 10^5 \\ - 0.31 \times 10^5 \\ \hline 4.05 \times 10^5 \end{array}$$

Chapter 10

10.1 Practice Problems

1. (a) The variables are A, b and h.

 (b) The variables are V, l, w, and h.

2. (a) $p = 2w + 2l$ **(b)** $A = \pi r^2$

3. We add $9 + 2 = 11$, therefore $9x + 2x = 11x$.

4. (a) $8x - 22x + 5x = 8x + (-22x) + 5x = (-14x) + 5x = -9x$

 (b) $19x - 7x - 12x = 19x + (-7x) + (-12x)$

$\quad = 12x + (-12x) = 0$

5. (a) $9x - 12x + x = 9x + (-12x) + 1x = -3x + 1x = -2x$
   (b) $5.6x - 8x - x = 5.6x - 8x - 1x = -2.4x - 1x = -3.4x$
6. $17.5 - 6.3x - 8.2x + 10.5 =$
   $17.5 + 10.5 + (-6.3x) + (-8.2x) = 28 - 14.5x$
7. (a) $2w + 3z - 12 - 5w - z - 16 =$
   $2w + (-5w) + 3z + (-1z) + (-12) + (-16) = -3w + 2z - 28$
   (b) $\dfrac{3}{5}x + 5 - \dfrac{7}{15}x - \dfrac{1}{3} = \dfrac{9}{15}x - \dfrac{7}{15}x + \dfrac{15}{3} - \dfrac{1}{3} = \dfrac{2}{15}x + \dfrac{14}{3}$

10.2 Practice Problems

1. (a) $(7)(x + 5) = 7(x) + 7(5) = 7x + 35$
   (b) $(-4)(x + 2y) = -4(x) + (-4)(2y) = -4x - 8y$
   (c) $5(6a - 2b) = 5(6a) - 5(2b) = 30a - 10b$
2. $(x + 3y)(8) = (x)(8) + (3y)(8) = 8x + 24y$
3. (a) $(-5)(x + 4y + 5) = (-5)(1x) + (-5)(4y) + (-5)(5)$
   $= -5x - 20y - 25$
   (b) $(3)(2.2x + 5.5y + 6) = (3)(2.2x) + (3)(5.5y) + (3)(6)$
   $= 6.6x + 16.5y + 18$
4. $\left(\dfrac{3}{2}\right)\left(\dfrac{1}{2}x - \dfrac{1}{3}y + 4z - \dfrac{1}{2}\right)$

   $= \left(\dfrac{3}{2}\right)\left(\dfrac{1}{2}\right)(x) + \left(\dfrac{3}{2}\right)\left(-\dfrac{1}{3}\right)(y) + \left(\dfrac{3}{2}\right)(4)(z) + \left(\dfrac{3}{2}\right)\left(-\dfrac{1}{2}\right)$

   $= \dfrac{3}{4}x - \dfrac{1}{2}y + 6z - \dfrac{3}{4}$
5. $3(2x + 4y) + 2(5x + y) = 6x + 12y + 10x + 2y = 16x + 14y$
6. $-4(x - 5) + 3(-1 + 2x)$
   $= -4x + 20 - 3 + 6x$
   $= 2x + 17$

10.3 Practice Problems

1. $x + 7 = -8$
   $x + 7 + (-7) = -8 + (-7)$
   $x + 0 = -15$
   $x = -15$

2. (a) $y - 3.2 = 9$
   $y - 3.2 + 3.2 = 9.0 + 3.2$
   $y = 12.2$

   Check.
   $y - 3.2 = 9$
   $(12.2) - 3.2 \overset{?}{=} 9$
   $9 = 9$ ✓

   (b) $\dfrac{2}{3} = x + \dfrac{1}{6}$

   $\dfrac{4}{6} + \left(-\dfrac{1}{6}\right) = x + \dfrac{1}{6} + \left(-\dfrac{1}{6}\right)$

   $\dfrac{3}{6} = x$

   $\dfrac{1}{2} = x$

   Check.
   $\dfrac{2}{3} = x + \dfrac{1}{6}$

   $\dfrac{2}{3} \overset{?}{=} \left(\dfrac{1}{2}\right) + \dfrac{1}{6}$

   $\dfrac{2}{3} \overset{?}{=} \dfrac{3}{6} + \dfrac{1}{6}$

   $\dfrac{2}{3} = \dfrac{4}{6}$ ✓

3. $3x - 5 = 2x + 1$
   $3x - 5 + 5 = 2x + 1 + 5$
   $3x = 2x + 6$
   $3x + (-2x) = 2x + (-2x) + 6$
   $x = 6$

   Check.
   $3x - 5 = 2x + 1$
   $3(6) - 5 \overset{?}{=} 2(6) + 1$
   $18 - 5 \overset{?}{=} 12 + 1$
   $13 = 13$ ✓

10.4 Practice Problems

1. $8n = 104$
   $\dfrac{8n}{8} = \dfrac{104}{8}$
   $n = 13$
2. $-7n = 30$
   $\dfrac{-7n}{-7} = \dfrac{30}{-7}$
   $n = -\dfrac{30}{7}$
3. $3.2x = 16$
   $\dfrac{3.2x}{3.2} = \dfrac{16}{3.2}$
   $x = 5$

   Check.
   $3.2(5) \overset{?}{=} 16$
   $16 = 16$ ✓

4. $\dfrac{1}{6}y = 2\dfrac{2}{3}$

   $\left(\dfrac{1}{6}\right)y = \dfrac{8}{3}$

   $\left(\dfrac{6}{1}\right)\left(\dfrac{1}{6}\right)y = \left(\dfrac{8}{3}\right)\left(\dfrac{6}{1}\right)$

   $y = 16$

   Check.
   $\dfrac{1}{6}y = 2\dfrac{2}{3}$

   $\left(\dfrac{1}{6}\right)(16) \overset{?}{=} 2\dfrac{2}{3}$

   $\dfrac{16}{6} \overset{?}{=} \dfrac{8}{3}$

   $\dfrac{8}{3} = \dfrac{8}{3}$ ✓

   (b) $\left(3\dfrac{1}{5}\right)z = 4$

   $\dfrac{16}{5}z = 4$

   $\left(\dfrac{5}{16}\right)\left(\dfrac{16}{5}\right)z = (4)\left(\dfrac{5}{16}\right)$

   $z = \dfrac{5}{4}$

   Check.
   $3\dfrac{1}{5}z = 4$

   $\left(3\dfrac{1}{5}\right)\left(\dfrac{5}{4}\right) \overset{?}{=} 4$

   $\left(\dfrac{16}{5}\right)\left(\dfrac{5}{4}\right) \overset{?}{=} 4$

   $4 = 4$ ✓

10.5 Practice Problems

1. $5x + 13 = 33$
   $5x + 13 + (-13) = 33 + (-13)$
   $5x = 20$
   $\dfrac{5x}{5} = \dfrac{20}{5}$
   $x = 4$

   Check.
   $5(4) + 13 \overset{?}{=} 33$
   $20 + 13 \overset{?}{=} 33$
   $33 = 33$ ✓

2. $-50 = 7x - 8$
   $-50 + 8 = 7x - 8 + 8$
   $-42 = 7x$
   $\dfrac{-42}{7} = \dfrac{7x}{7}$
   $-6 = x$

   Check.
   $-50 \overset{?}{=} 7(-6) - 8$
   $-50 \overset{?}{=} -42 - 8$
   $-50 = -50$ ✓

3. $4x = -8x + 42$
   $4x + 8x = -8x + 8x + 42$
   $12x = 42$
   $\dfrac{12x}{12} = \dfrac{42}{12}$
   $x = \dfrac{7}{2}$

4. $4x - 7 = 9x + 13$
   $4x - 7 + 7 = 9x + 13 + 7$
   $4x = 9x + 20$
   $4x + (-9x) = 9x + (-9x) + 20$
   $-5x = 20$
   $\dfrac{-5x}{-5} = \dfrac{20}{-5}$
   $x = -4$

5. $4x - 23 = 3x + 7 - 2x$
   $4x - 23 = x + 7$
   $4x + (-1x) - 23 = (-1x) + x + 7$
   $3x - 23 = 7$
   $3x - 23 + 23 = 7 + 23$
   $3x = 30$
   $\dfrac{3x}{3} = \dfrac{30}{3}$
   $x = 10$

6. $8(x - 3) + 5x = 15(x - 2)$
   $8x - 24 + 5x = 15x - 30$
   $13x - 24 = 15x - 30$
   $13x - 24 + (30) = 15x - 30 + (30)$
   $13x + 6 = 15x$
   $13x + (-13x) + 6 = 15x + (-13x)$
   $6 = 2x$
   $\dfrac{6}{2} = \dfrac{2x}{2}$
   $3 = x$

10.6 Practice Problems

1. Tom's height is 7 inches more than Abdul's height
 ↓ ↓ ↓ ↓ ↓
 t = 7 + a

2. The noon class has 24 fewer students than the morning class
 ↓ ↓ ↓ ↓
 n = m − 24

3. On Thursday carried 5 more than Friday.
$t = 5 + f$

4. Let s = length in miles of Sally's trip
$\underline{s + 380}$ = length in miles of Melinda's trip
380 miles longer then Sally's trip.

5. Let m = height in feet of McCormick Hall
$\underline{m - 126}$ = height in feet of Larson Center
126 feet shorter than McCormick Hall.

6. Let x = length in inches of the first side of the triangle
$2x$ = length in inches of the second side of the triangle
$x + 6$ = length in inches of the third side of the triangle

10.7 Practice Problems

1. Let x = length in feet of shorter piece of board
$x + 4.5$ = length in feet of longer piece of board
$$x + x + 4.5 = 18$$
$$2x + 4.5 = 18$$
$$2x + 4.5 + (-4.5) = 18 + (-4.5)$$
$$2x = 13.5$$
$$\frac{2x}{2} = \frac{13.5}{2}$$
$$x = 6.75$$
The shorter piece is 6.75 feet long.
$x + 4.5 = 6.75 + 4.5 = 11.25$
The longer piece is 11.25 feet long.
Check.
$6.75 + 11.25 \overset{?}{=} 18$
$18 = 18$ ✓

2. Let x = the number of departures on Monday
$x + 29$ = the number of departures on Tuesday
$x - 16$ = the number of departures on Wednesday
$$x + x + 29 + x - 16 = 349$$
$$3x + 13 = 349$$
$$3x + 13 + (-13) = 349 + (-13)$$
$$3x = 336$$
$$\frac{3x}{3} = \frac{336}{3}$$
$$x = 112$$
$$x + 29 = 112 + 29 = 141$$
$$x - 16 = 112 - 16 = 96$$
There were 112 departures on Monday, 141 departures on Tuesday, and 96 departures on Wednesday.
Check.
$112 + 141 + 96 \overset{?}{=} 349$
$349 = 349$ ✓
$141 \overset{?}{=} 112 + 29$
$141 = 141$ ✓
$96 \overset{?}{=} 112 - 16$
$96 = 96$ ✓

3. Let w = the width of the field measured in feet
$2w + 8$ = the length of the field measured in feet
$2(\text{width}) + 2(\text{length}) = \text{perimeter}$
$$2(w) + 2(2w + 8) = 772$$
$$2w + 4w + 16 = 772$$
$$6w + 16 = 772$$
$$6w + 16 + (-16) = 772 + (-16)$$
$$6w = 756$$
$$\frac{6w}{6} = \frac{756}{6}$$
$$w = 126$$

The width of the field is 126 feet.
$2w + 8 = 2(126) + 8 = 252 + 8 = 260$
The length of the field is 260 feet.
Check.
$260 \overset{?}{=} 2(126) + 8$
$260 \overset{?}{=} 252 + 8$
$260 = 260$ ✓
$2(126) + 2(260) \overset{?}{=} 772$
$252 + 520 \overset{?}{=} 772$
$772 = 772$ ✓

4. Let x = the length in meters of the first side of the triangle
$2x$ = the length in meters of the second side of the triangle
$x + 10$ = the length in meters of the third side of the triangle
$$x + 2x + x + 10 = 36$$
$$4x + 10 = 36$$
$$4x + 10 + (-10) = 36 + (-10)$$
$$4x = 26$$
$$\frac{4x}{4} = \frac{26}{4}$$
$$x = 6.5$$
The first side of the triangle is 6.5 meters long.
$2x = 2(6.5) = 13$
The second side of the triangle is 13 meters long.
$x + 10 = 6.5 + 10 = 16.5$
The third side of the triangle is 16.5 meters long.
Check.
$6.5 + 13 + 16.5 \overset{?}{=} 36$
$36 = 36$ ✓
$13 \overset{?}{=} 2(6.5)$
$13 = 13$ ✓
$16.5 \overset{?}{=} 10 + 6.5$
$16.5 = 16.5$

5. Let x = the measure of angle A in degrees
$3x$ = the measure of angle C in degrees
$x - 30$ = the measure of angle B in degrees
$$x + 3x + x - 30 = 180$$
$$5x - 30 = 180$$
$$5x - 30 + 30 = 180 + 30$$
$$5x = 210$$
$$\frac{5x}{5} = \frac{210}{5}$$
$$x = 42$$
Angle A measures 42 degrees.
$3x = 3(42) = 126$
Angle C measures 126 degrees.
$x - 30 = 42 - 30 = 12$
Angle B measures 12 degrees.
Check.
$42 + 126 + 12 \overset{?}{=} 180$
$180 = 180$ ✓
$126 \overset{?}{=} 3(42)$
$126 = 126$ ✓
$12 \overset{?}{=} 42 - 30$
$12 = 12$ ✓

6. Let s = the total amount of sales of the boats in dollars
$0.03s$ = the amount of the commission earned on sales of s dollars
$$3250 = 1000 + 0.03s$$
$$2250 = 0.03s$$
$$\frac{2250}{0.03} = \frac{0.03s}{0.03}$$
$$75,000 = s$$
Therefore he sold \$75,000 worth of boats for the month.
Check.
$(0.03)(75,000) + 1000 \overset{?}{=} 3250$
$2250 + 1000 \overset{?}{=} 3250$
$3250 = 3250$ ✓

Selected Answers

Chapter 1

Pretest Chapter 1

1. seventy-eight million, three hundred ten thousand, four hundred thirty-six **2.** $30,000 + 8,000 + 200 + 40 + 7$ **3.** 5,064,122 **4.** 2,747,000 **5.** 2,583,000 **6.** 244 **7.** 50,570 **8.** 14,729 **9.** 2111 **10.** 76,311 **11.** 1,981,652 **12.** 108 **13.** 18,606 **14.** 6734 **15.** 331,420 **16.** 10,605 **17.** 7376 R 1 **18.** 139 **19.** 6^4 **20.** 81 **21.** 29 **22.** 52 **23.** 23 **24.** 271,000 **25.** 26,500 **26.** 60,000,000 **27.** 2600 **28.** 30,000,000,000 **29.** 1166 miles **30.** $22.00 **31.** $299 **32. (a)** 148,192 **(b)** 787,890

1.1 Exercises

1. $6000 + 700 + 30 + 1$ **3.** $100,000 + 8,000 + 200 + 70 + 6$ **5.** $20,000,000 + 3,000,000 + 700,000 + 60,000 + 1000 + 300 + 40 + 5$ **7.** $100,000,000 + 3,000,000 + 200,000 + 60,000 + 700 + 60 + 8$ **9.** 671 **11.** 9863 **13.** 39,733 **15.** 706,200 **17.** 7; 50,000 **19.** 4; 40,000 **21.** fifty-three **23.** eight thousand, nine hundred thirty-six **25.** thirty-six thousand, one hundred eighteen **27.** one hundred five thousand, two hundred sixty-one **29.** fourteen million, two hundred three thousand, three hundred twenty-six **31.** Four billion, three hundred two million, one hundred fifty-six thousand, two hundred **33.** 375 **35.** 27,382 **37.** 100,079,826 **39.** one thousand, nine hundred sixty-five **41.** 9 million or 9,000,000 **43.** 33 million or 33,000,000 **45.** $281 billion or $281,000,000,000 **47.** $574 billion or $574,000,000,000 **49. (a)** 5 **(b)** 2 **51. (a)** 7 **(b)** 8 **52. (a)** 9 **(b)** 7 **53.** 613,001,033,208,003 **54.** three quintillion, six hundred eighty-two quadrillion, nine hundred sixty-eight trillion nine billion, nine hundred thirty-one million, nine hundred sixty thousand, seven hundred forty-seven **55.** 60,000,000,000,000,000,000,000 **56.** You would obtain 2E20. This is 20,000,000,000,000,000,000 in standard form. **57.** There are 195 numbers between 200 and 800 that contain the digit 6.

1.2 Exercises

1. Answers may vary. A sample is: You can change the order of the addends without changing the sum.

3.

+	3	5	4	8	0	6	7	2	9	1
2	5	7	6	10	2	8	9	4	11	3
7	10	12	11	15	7	13	14	9	16	8
5	8	10	9	13	5	11	12	7	14	6
3	6	8	7	11	3	9	10	5	12	4
0	3	5	4	8	0	6	7	2	9	1
4	7	9	8	12	4	10	11	6	13	5
1	4	6	5	9	1	7	8	3	10	2
8	11	13	12	16	8	14	15	10	17	9
6	9	11	10	14	6	12	13	8	15	7
9	12	14	13	17	9	15	16	11	18	10

5. 23 **7.** 26 **9.** 49 **11.** 99 **13.** 4579 **15.** 12,356 **17.** 13,861 **19.** 42,479 **21.** 121 **23.** 1132 **25.** 7378 **27.** 11,579,426 **29.** 1,135,280,240 **31.** 2,303,820 **33.** 248 **35.** 20,909 **37.** $794 **39.** $4211 **41.** 468 feet **43.** 121,100,000 square miles **45.** 2,922,339 votes **47. (a)** 1134 students **(b)** 1392 students **49.** 202 miles **51.** 434 feet **53. (a)** $9553 **(b)** $7319 **(c)** $13,047 **55.** 1161 **57.** Answers may vary. A sample is: You could not group the addends in groups that sum to 10s to make column addition easier. **59.** one hundred twenty-one million, three hundred seventy-four **60.** 8,724,396 **61.** 9,051,719

1.3 Exercises

1. In subtraction the minuend minus the subtrahead equals the difference. To check the problem we add the subtrahead and the difference to see if we get the minuend. If we do, the answer is correct. **3.** We know that $1683 + 1592 = 32?5$. Therefore if we add 8 tens and 9 tens we get 17 tens, which is 1 hundred and 7 tens. Thus the ?, should be replaced by 7. **5.** 1 **7.** 9 **9.** 16 **11.** 9 **13.** 7 **15.** 6 **17.** 3 **19.** 9

21.
$$\begin{array}{r} 21 \\ + 21 \\ \hline 47 \end{array} \quad 26$$

23.
$$\begin{array}{r} 12 \\ + 12 \\ \hline 85 \end{array} \quad 73$$

25.
$$\begin{array}{r} 31 \\ + 31 \\ \hline 126 \end{array} \quad 95$$

27.
$$\begin{array}{r} 625 \\ + 625 \\ \hline 768 \end{array} \quad 143$$

29.
$$\begin{array}{r} 2321 \\ + 2321 \\ \hline 2893 \end{array} \quad 572$$

31.
$$\begin{array}{r} 11,191 \\ + 11,191 \\ \hline 24,396 \end{array} \quad 13,205$$

33. 553,101
$$\begin{array}{r} 433,201 \\ + 553,101 \\ \hline 986,302 \end{array}$$

35.
$$\begin{array}{r} 19 \\ + 110 \\ \hline 129 \end{array}$$
Correct

37. 1213
$$\begin{array}{r} 1213 \\ + 6134 \\ \hline 7347 \end{array}$$
Correct

39.
$$\begin{array}{r} 6030 \\ - 5020 \\ \hline 1010 \end{array}$$
Incorrect

41. 41,181
$$\begin{array}{r} 41,181 \\ + 58,402 \\ \hline 99,583 \end{array}$$
Correct

43. 46 **45.** 37 **47.** 75 **49.** 581 **51.** 10,692 **53.** 34,092 **55.** 7447 **57.** 223,116 **59.** $x = 5$ **61.** $x = 21$ **63.** $x = 27$ **65.** 182 votes **67.** 6,285,096 **69.** $248 **71.** 1,416,920 people **73.** 3,686,434 people **75.** 320,317 people **77.** 93,154 people **79.** 29 homes **81.** 80 homes **83.** between 1999 and 2000 **85.** Willow Creek and Irving **87.** It is true if a and b represent the same number, for example, if $a = 10$ and $b = 10$. **88.** It is true for all a and b if $c = 0$, for example, if $a = 5$, $b = 3$, and $c = 0$. **89.** $550 **90.** $720 **91.** 8,466,084 **92.** two hundred ninety-six thousand, three hundred eight **93.** 168 **94.** 1,119,534

1.4 Exercises

1. (a) You can change the order of the factors without changing the product. **(b)** You can group the factors in any way without changing the product.

3.

x	6	2	3	8	0	5	7	9	12	4
5	30	10	15	40	0	25	35	45	60	20
7	42	14	21	56	0	35	49	63	84	28
1	6	2	3	8	0	5	7	9	12	4
0	0	0	0	0	0	0	0	0	0	0
6	36	12	18	48	0	30	42	54	72	24
2	12	4	6	16	0	10	14	18	24	8
3	18	6	9	24	0	15	21	27	36	12
8	48	16	24	64	0	40	56	72	96	32
4	24	8	12	32	0	20	28	36	48	16
9	54	18	27	72	0	45	63	81	108	36

5. 96   **7.** 609   **9.** 18,306   **11.** 80,412   **13.** 70   **15.** 522   **17.** 1630   **19.** 12,216   **21.** 100,208   **23.** 942,808   **25.** 1560   **27.** 2,715,800   **29.** 482,000   **31.** 372,560,000   **33.** 8460   **35.** 63,600   **37.** 56,000,000   **39.** 67,296   **41.** 7884   **43.** 5696   **45.** 15,175   **47.** 41,537   **49.** 69,312   **51.** 148,567   **53.** 823,823   **55.** 1,240,064   **57.** 89,496   **59.** 217,980   **61.** 1,435,083   **63.** 30,211,210   **65.** 10,400   **67.** 80,600   **69.** 70   **71.** 308   **73.** 700   **75.** 770   **77.** $x = 0$ **79.** 3216 square feet   **81.** 195 square feet   **83.** $1200   **85.** $3192   **87.** 612 miles   **89.** 2375 books   **91.** $7,372,300,000   **93.** 198   **94.** 194   **95.** 62   **96.** 134   **97.** $x = 8$ **98.** $x = 8$ **99.** $x = 9$ **100.** $x = 7$ **101.** No, it would not always be true. In our number system $62 = 60 + 2$. But in roman numerals $IV \neq I + V$. The digit system in roman numerals involves subtraction. Thus $(XII) \times (IV) \neq (XII \times I) + (XII \times V)$.   **102.** yes, $5 \times (8 - 3) = 5 \times 8 - 5 \times 3, a \times (b - c) = (a \times b) - (a \times c)$ **103.** 6756   **104.** 1249   **105.** $139   **106.** $59

1.5 Exercises

1. (a) When you divide a nonzero number by itself, the result is one. **(b)** When you divide a number by 1, the result is that number. **(c)** When you divide zero by a nonzero number, the result is zero. **(d)** You cannot divide a number by zero. Division by zero is undefined.   **3.** 7   **5.** 3   **7.** 5   **9.** 4   **11.** 3   **13.** 7   **15.** 9   **17.** 9   **19.** 7   **21.** 4   **23.** 7   **25.** 0   **27.** 9   **29.** 1   **31.** 4 R 5   **33.** 9 R 4   **35.** 25 R 3   **37.** 21 R 7   **39.** 32   **41.** 37   **43.** 322 R 1   **45.** 127 R 1   **47.** 753   **49.** 1122 R 1   **51.** 1757 R 5   **53.** 2562 R 3   **55.** 5 R 5   **57.** 5 R 7   **59.** 7   **61.** 160 R 10   **63.** 48 R 12   **65.** 314 R 10   **67.** 210 R 8   **69.** 202 R 7   **71.** 4 R 4   **73.** 7 R 26   **75.** 37   **77.** 61,693 runs per day   **79.** 24 pairs of earrings per hour   **81.** $21,053   **83.** $171 for each monthly payment   **85.** 165 sandwiches   **87.** 78 feet   **88.** 82 feet   **89.** a and b must represent the same number. For example, if $a = 12$, then $b = 12$.   **91.** 5504   **92.** 1,038,490   **93.** 406,195   **94.** 66,844

Putting Your Skills to Work

1. $23   **2.** $93

1.6 Exercises

1. 5^3 means $5 \times 5 \times 5. 5^3 = 125.$ **3.** base
5. To ensure consistency we
   1. perform operations inside parentheses
   2. simplify any expressions with exponents
   3. multiply or divide from left to right
   4. add or subtract from left to right
7. 6^4 **9.** 4^5 **11.** 8^4 **13.** 9^1 **15.** 16   **17.** 16   **19.** 216   **21.** 10,000   **23.** 1   **25.** 64   **27.** 243   **29.** 225   **31.** 343   **33.** 256   **35.** 1   **37.** 625   **39.** 1,000,000   **41.** 1024   **43.** 9   **45.** 2401   **47.** 17   **49.** 225   **51.** 520   **53.** $30 - 3 + 10; 27 + 10 = 37$ **55.** $27 - 5 = 22$ **57.** $48 \div 8 + 4; 6 + 4 = 10$ **59.** $7 \times 9 + 4 - 8; 63 + 4 - 8 = 59$ **61.** $4 \times 81 - 4 \times 5; 324 - 20 = 304$:   **63.** $20 \div 20 = 1$ **65.** $950 \div 5 = 190$ **67.** $60 - 17 = 43$ **69.** $9 + 16 \div 4 = 9 + 4 = 13$ **71.** $42 - 4 \div 4 = 42 - 1 = 41$ **73.** $100 - 9 \times 4 = 100 - 36 = 64$ **75.** $25 + 4 + 27 = 56$ **77.** $8 \times 3 \times 1 \div 2; 24 \div 2 = 12$ **79.** $144 - 0; 144$ **81.** $16 \times 6 \div 3; 96 \div 3 = 32$ **83.** $5 + 3 - 1; 7$ **85.** $3 + 9 \times 6 + 4; 3 + 54 + 4 = 61$ **87.** $12 \div 2 \times 3 - 16; 18 - 16 = 2$ **89.** $9 \times 6 \div 9 + 4 \times 3; 6 + 12 = 18$ **91.** $4(4)^3 \div 8 = 4(64) \div 8 = 32$ **93.** $1200 - 8(3) \div 6 = 1200 - 4 = 1196$ **95.** $50 + 20 - 9 = 61$ **97.** $15 - 6 + 10; 19$ **99.** $6 + 16 - 20 = 2$ **101.** 86,164 seconds   **102.** 612 minutes; 36,720 seconds   **103.** $100,000 + 50,000 + 6000 + 300 + 10 + 2$ **104.** 200,765,909   **105.** two hundred sixty-one million, seven hundred sixty-three thousand, two   **106.** 1460 feet of fencing will be needed. 120,000 square feet of grass must be planted

1.7 Exercises

1. Locate the rounding place. If the digit to the right of the rounding place is 5 or greater than 5, round up. If the digit to the right of the rounding place is less than 5, round down.   **3.** 80   **5.** 70   **7.** 90   **9.** 530   **11.** 1670   **13.** 200   **15.** 2800   **17.** 7700   **19.** 8000   **21.** 2000   **23.** 28,000   **25.** 800,000   **27.** 15,000,000 stars   **29. (a)** 163,000   **(b)** 163,300   **31. (a)** 3,700,000 square miles; 9,600,000 square kilometers   **(b)** 3,710,000 square miles; 9,600,000 square kilometers

33.
$$\begin{array}{r} 600 \\ 300 \\ + 100 \\ \hline 1000 \end{array}$$

35.
$$\begin{array}{r} 30 \\ 80 \\ 60 \\ + 30 \\ \hline 200 \end{array}$$

37.
$$\begin{array}{r} 100,000 \\ 50,000 \\ + 100,000 \\ \hline 250,000 \end{array}$$

39.
$$\begin{array}{r} 600,000 \\ - 100,000 \\ \hline 500,000 \end{array}$$

41.
$$\begin{array}{r} 80,000,000 \\ - 60,000,000 \\ \hline 20,000,000 \end{array}$$

43.
$$\begin{array}{r} 30,000,000 \\ - 20,000,000 \\ \hline 10,000,000 \end{array}$$

45.
$$\begin{array}{r} 60 \\ \times 50 \\ \hline 3000 \end{array}$$

47.
$$\begin{array}{r} 1000 \\ \times 8 \\ \hline 8000 \end{array}$$

49.
$$\begin{array}{r} 600,000 \\ \times\ \ \ \ 300 \\ \hline 180,000,000 \end{array}$$
51. $20\overline{)20,000}$ → $1,000$
53. $40\overline{)200,000}$ → $5,000$
55. $800\overline{)4,000,000}$ → $5,000$
57. Incorrect
$$\begin{array}{r} 400 \\ 500 \\ 900 \\ +\ 200 \\ \hline 2000 \end{array}$$
59. Incorrect
$$\begin{array}{r} 100,000 \\ 50,000 \\ +\ 40,000 \\ \hline 190,000 \end{array}$$

61. Correct
$$\begin{array}{r} 300,000 \\ -\ 90,000 \\ \hline 210,000 \end{array}$$
63. Incorrect
$$\begin{array}{r} 80,000,000 \\ -\ 50,000,000 \\ \hline 30,000,000 \end{array}$$
65. Incorrect
$$\begin{array}{r} 200 \\ \times\ 20 \\ \hline 4000 \end{array}$$
67. Correct
$$\begin{array}{r} 6000 \\ \times\ 70 \\ \hline 420,000 \end{array}$$
69. $80\overline{)400,000}$ → 5000 Correct
71. $400\overline{)200,000}$ → 500 Correct

73. 2,400 square feet   **75.** $140,000   **77.** 30,000 pizzas   **79.** 200,000,000 passengers   **81.** $590,000 - 270,000 = 320,000$ square miles
83. (a) 400,000 hours   **(b)** 20,000 days   **84. (a)** 300,000 hours   **(b)** 15,000 days   **85.** 83   **86.** 27   **87.** 28   **88.** 66

1.8 Exercises

1. 25,231; 466   **3.** $4668   **5.** 3750 hors d'oeuvres   **7.** 6¢ per ounce   **9.** $23   **11.** 5 hours; 300 minutes   **13.** 800,000 people   **15.** $20,382
17. $786   **19.** $1482   **21.** $16,405   **23.** 25 miles per gallon   **25.** There are 54 oak trees, 108 maple trees, and 756 pine trees. In total there are 936 trees.   **27.** 118   **29.** 185   **31.** $14,734,000,000   **33.** $60,909,000,000   **35.** 343   **36.** 21   **37.** 4788   **38.** 258

Chapter 1 Review Problems

1. three hundred seventy-six   **2.** five thousand, eighty-two   **3.** one hundred nine thousand, two hundred seventy-six   **4.** four hundred twenty-three million, five hundred seventy-six thousand, fifty-five   **5.** $4000 + 300 + 60 + 4$ **6.** $20,000 + 7000 + 900 + 80 + 6$ **7.** $40,000,000 + 2,000,000 + 100,000 + 60,000 + 6000 + 30 + 7$ **8.** $1,000,000 + 300,000 + 5000 + 100 + 20 + 8$ **9.** 924   **10.** 6095
11. 1,328,828   **12.** 45,092,651   **13.** 115   **14.** 130   **15.** 690   **16.** 150   **17.** 400   **18.** 1598   **19.** 1469   **20.** 125,423   **21.** 14,703
22. 10,582   **23.** 17   **24.** 6   **25.** 27   **26.** 171   **27.** 159   **28.** 4828   **29.** 63,146   **30.** 3026   **31.** 224,757   **32.** 1,230,691   **33.** 114
34. 36   **35.** 0   **36.** 24   **37.** 18   **38.** 64   **39.** 105   **40.** 120   **41.** 144   **42.** 0   **43.** 240   **44.** 168   **45.** 2,612,100   **46.** 84,312,000
47. 83,200,000   **48.** 563,000,000   **49.** 1856   **50.** 864   **51.** 4050   **52.** 13,680   **53.** 25,524   **54.** 24,096   **55.** 87,822   **56.** 268,513
57. 543,510   **58.** 255,068   **59.** 150,000   **60.** 240,000   **61.** 7,200,000   **62.** 7,500,000   **63.** 2,000,000,000   **64.** 12,000,000,000   **65.** 2
66. 5   **67.** 14   **68.** 4   **69.** 0   **70.** 12   **71.** 7   **72.** 0   **73.** 7   **74.** 7   **75.** undefined   **76.** 4   **77.** 7   **78.** 6   **79.** 8   **80.** undefined   **81.** 125   **82.** 125   **83.** 258   **84.** 369   **85.** 25,874   **86.** 3692   **87.** 36,958   **88.** 36,921   **89.** 15,986 R 2   **90.** 35,783 R 4
91. 7 R 21   **92.** 4 R 37   **93.** 31 R 15   **94.** 14 R 11   **95.** 38 R 30   **96.** 54 R 38   **97.** 95   **98.** 258   **99.** 54   **100.** 19   **101.** 13^2
102. 24^2 **103.** 8^5 **104.** 9^5 **105.** 64   **106.** 81   **107.** 128   **108.** 125   **109.** 49   **110.** 81   **111.** 216   **112.** 256   **113.** 8   **114.** 11
115. 22   **116.** 66   **117.** 107   **118.** 76   **119.** 17   **120.** 48   **121.** 44   **122.** 93   **123.** 1280   **124.** 5670   **125.** 15,310   **126.** 42,640
127. 12,000   **128.** 23,000   **129.** 676,000   **130.** 202,000   **131.** 5,700,000   **132.** 10,000,000

133.
$$\begin{array}{r} 600 \\ 600 \\ 900 \\ +\ 900 \\ \hline 3000 \end{array}$$
134.
$$\begin{array}{r} 30,000 \\ 40,000 \\ +\ 70,000 \\ \hline 140,000 \end{array}$$
135.
$$\begin{array}{r} 4,000,000 \\ -\ 3,000,000 \\ \hline 1,000,000 \end{array}$$
136.
$$\begin{array}{r} 30,000 \\ -\ 20,000 \\ \hline 10,000 \end{array}$$
137.
$$\begin{array}{r} 2000 \\ \times\ 6000 \\ \hline 12,000,000 \end{array}$$
138.
$$\begin{array}{r} 3,000,000 \\ \times\ 900 \\ \hline 2,700,000,000 \end{array}$$
139. $20\overline{)80,000}$ → $4,000$

140. $400\overline{)8,000,000}$ → $20,000$
141. Correct
$$\begin{array}{r} 90 \\ 40 \\ 90 \\ +\ 60 \\ \hline 280 \end{array}$$
142. Correct
$$\begin{array}{r} 900,000 \\ -\ 400,000 \\ \hline 500,000 \end{array}$$
143. Incorrect
$$\begin{array}{r} 200,000 \\ \times\ 5000 \\ \hline 1,000,000,000 \end{array}$$
144. Correct $30\overline{)900,000}$ → $30,000$

145. 204 cans   **146.** 175 words   **147.** 2397 km   **148.** $59,470   **149.** 10,301 feet   **150.** $7028   **151.** $1356   **152.** $64 per share
153. $278   **154.** $334   **155.** 25 miles per gallon   **156.** 24 miles per gallon   **157.** $5041   **158.** $3456   **159.** 40,500,000 tons
160. 21,400,000 tons, from 1990 to 1995   **161.** 72,000,000 tons   **162.** 1,700,000 tons per year   **163.** $19,648,200,000   **164.** 501 red pieces

Chapter 1 Test

1. forty-four million, seven thousand, six hundred thirty-five   **2.** $20,000 + 6000 + 800 + 50 + 9$ **3.** 3,581,076   **4.** 977   **5.** 1045   **6.** 318,977
7. 8067   **8.** 38,341   **9.** 5,225,768   **10.** 378   **11.** 4320   **12.** 91,875   **13.** 129,437   **14.** 3014 R 1   **15.** 2283   **16.** 352   **17.** 11^3
18. 64   **19.** 23   **20.** 78   **21.** 79   **22.** 88,070   **23.** 6,460,000   **24.** 4,800,000   **25.** 150,000,000,000   **26.** 18,000   **27.** $2148   **28.** 467 feet   **29.** $127   **30.** $292   **31.** 748,000 square feet   **32.** 46 feet

Chapter 2

Pretest Chapter 2

1. $\dfrac{3}{8}$ **2.** Answers will vary.   **3.** $\dfrac{5}{124}$ **4.** $\dfrac{1}{6}$ **5.** $\dfrac{1}{3}$ **6.** $\dfrac{1}{7}$ **7.** $\dfrac{7}{8}$ **8.** $\dfrac{4}{11}$ **9.** $\dfrac{11}{3}$ **10.** $\dfrac{55}{9}$ **11.** $24\dfrac{1}{4}$ **12.** $5\dfrac{4}{5}$ **13.** $2\dfrac{2}{17}$
14. $\dfrac{5}{44}$ **15.** $\dfrac{2}{3}$ **16.** $67\dfrac{5}{6}$ **17.** 1   **18.** $\dfrac{1}{2}$ **19.** $5\dfrac{4}{13}$ **20.** $4\dfrac{2}{3}$ **21.** 8   **22.** 45   **23.** 55   **24.** 72   **25.** $\dfrac{19}{72}$ **26.** $\dfrac{49}{72}$
27. $\dfrac{13}{3}$ or $4\dfrac{1}{3}$ **28.** $3\dfrac{41}{42}$ **29.** $\dfrac{5}{3}$ or $1\dfrac{2}{3}$ **30.** $7\dfrac{1}{6}$ miles   **31.** $7\dfrac{11}{36}$ tons   **32.** 18 students

2.1 Exercises

1. fraction   **3.** denominator   **5.** N: 3; D: 5   **7.** N: 2; D: 3   **9.** N: 1; D: 17   **11.** $\frac{1}{3}$ **13.** $\frac{5}{6}$ **15.** $\frac{3}{4}$ **17.** $\frac{3}{8}$ **19.** $\frac{1}{4}$ **21.** $\frac{7}{10}$ **23.** $\frac{5}{8}$ **25.** $\frac{4}{7}$ **27.** $\frac{7}{8}$ **29.** $\frac{3}{5}$

31.   **33.**   **35.**

37. $\frac{31}{95}$ **39.** $\frac{209}{750}$ **41.** $\frac{89}{211}$ **43.** $\frac{7}{29}$ **45.** $\frac{9}{14}$ **47. (a)** $\frac{90}{195}$ **(b)** $\frac{22}{195}$

49. The amount of money each of six business owners gets if the business has a profit of $0.   **50.** We cannot do it. Division by zero is undefined.
51. 241   **52.** 11,106   **53.** 119,944   **54.** 177 R 6   **55.** 181

2.2 Exercises

1. 11, 19, 41, 5   **3.** composite number   **5.** $56 = 2 \times 2 \times 2 \times 7$ **7.** 3×5 **9.** 2×3 **11.** 7^2 **13.** 2^6 **15.** 5×11 **17.** $3^2 \times 7$
19. 3×5^2 **21.** 2×3^3 **23.** $2^2 \times 3 \times 7$ **25.** 2×7^2 **27.** prime   **29.** 3×19 **31.** prime   **33.** 2×31 **35.** prime   **37.** prime
39. 11×11 **41.** prime   **43.** $\frac{18 \div 9}{27 \div 9} = \frac{2}{3}$ **45.** $\frac{32 \div 16}{48 \div 16} = \frac{2}{3}$ **47.** $\frac{30 \div 6}{48 \div 6} = \frac{5}{8}$ **49.** $\frac{35 \div 5}{60 \div 5} = \frac{7}{12}$ **51.** $\frac{3 \times 1}{3 \times 5} = \frac{1}{5}$
53. $\frac{2 \times 3 \times 11}{2 \times 2 \times 2 \times 11} = \frac{3}{4}$ **55.** $\frac{2 \times 3 \times 3}{2 \times 2 \times 2 \times 3} = \frac{3}{4}$ **57.** $\frac{3 \times 3 \times 3}{3 \times 3 \times 5} = \frac{3}{5}$ **59.** $\frac{3 \times 11}{3 \times 12} = \frac{11}{12}$ **61.** $\frac{9 \times 7}{9 \times 12} = \frac{7}{12}$ **63.** $\frac{11 \times 8}{11 \times 11} = \frac{8}{11}$
65. $\frac{3 \times 50}{4 \times 50} = \frac{3}{4}$ **67.** $\frac{11 \times 20}{13 \times 20} = \frac{11}{13}$ **69.** yes   **71.** no   **73.** no   **75.** yes   **77.** yes   **79.** $\frac{3}{5}$ **81.** $\frac{16}{19}$ **83.** $\frac{1}{9}$ **85.** $\frac{17}{45}$ **87.** $\frac{8}{45}$
89. 164,050   **90.** 1296   **91.** 6,120,000

2.3 Exercises

1. (a) Multiply the whole number by the denominator of the fraction. **(b)** Add the numerator of the fraction to the product formed in step (a). **(c)** Write the sum found in step (b) over the denominator of the fraction.   **2. (a)** Divide the numerator by the denominator. **(b)** Write the quotient followed by the fraction with the remainder over the denominator.

3. $\frac{14}{3}$ **5.** $\frac{17}{7}$ **7.** $\frac{43}{7}$ **9.** $\frac{32}{3}$ **11.** $\frac{65}{3}$ **13.** $\frac{55}{6}$ **15.** $\frac{169}{6}$ **17.** $\frac{131}{12}$ **19.** $\frac{79}{10}$ **21.** $\frac{201}{25}$ **23.** $\frac{211}{2}$ **25.** $\frac{494}{3}$ **27.** $\frac{131}{15}$ **29.** $\frac{138}{25}$
31. $1\frac{1}{3}$ **33.** $2\frac{3}{4}$ **35.** $2\frac{1}{2}$ **37.** $3\frac{3}{8}$ **39.** 5   **41.** $9\frac{5}{9}$ **43.** $2\frac{2}{13}$ **45.** $3\frac{3}{16}$ **47.** $9\frac{1}{3}$ **49.** $17\frac{1}{2}$ **51.** 13   **53.** 14   **55.** 6
57. $36\frac{7}{11}$ **59.** $2\frac{3}{4}$ **61.** $4\frac{1}{6}$ **63.** $12\frac{3}{8}$ **65.** 4   **67.** $\frac{12}{5}$ **69.** $\frac{26}{3}$ **71.** $2\frac{88}{126} = 2\frac{44}{63}$ **73.** $2\frac{138}{424} = 2\frac{69}{212}$ **75.** $1\frac{212}{296} = 1\frac{53}{74}$
77. $\frac{1082}{3}$ yards   **79.** $50\frac{1}{3}$ acres   **81.** no; 101 is prime and is not a factor of 5687   **82.** no; 157 is prime and not a factor of 9810   **83.** 260,247
84. 16,000,000,000   **85.** 3000   **86.** 37; 5

2.4 Exercises

1. $\frac{21}{55}$ **3.** $\frac{15}{52}$ **5.** 1   **7.** $\frac{35}{54}$ **9.** $\frac{5}{12}$ **11.** $\frac{21}{8}$ or $2\frac{5}{8}$ **13.** $\frac{24}{7}$ or $3\frac{3}{7}$ **15.** $\frac{5}{2}$ or $2\frac{1}{2}$ **17.** $\frac{1}{6}$ **19.** $\frac{1}{2}$ **21.** $\frac{22}{9}$ or $2\frac{4}{9}$ **23.** $\frac{55}{12}$ or $4\frac{7}{12}$
25. 15   **27.** $\frac{69}{50}$ or $1\frac{19}{50}$ **29.** 0   **31.** $\frac{154}{3}$ or $51\frac{1}{3}$ **33.** $\frac{8}{5}$ or $1\frac{3}{5}$ **35.** $11\frac{5}{7}$ **37.** $x = \frac{9}{5}$ **39.** $x = \frac{8}{9}$ **41.** $37\frac{11}{12}$ square miles
43. 1560 miles   **45.** 1629 grams   **47.** $9\frac{3}{16}$ ounces   **49.** 377 companies   **51.** $2\frac{3}{16}$ miles   **53. (a)** $\frac{5}{8}$ of the students **(b)** 5375 students
55. The step of dividing the numerator and denominator by the same number allows us to work with smaller numbers when we do the multiplication. Also, this allows us to avoid the step of having to simplify the fraction in the answer.   **56.** There are an infinite number of answers. Any fraction that can be simplified to $\frac{3}{7}$ would be a correct answer. Thus three possible answers to this problem are $\frac{6}{14}, \frac{9}{21}$, or $\frac{12}{28}$ **57.** 529 cars   **58.** 368 calls
59. 1752 lines   **60.** 173,040 gallons

2.5 Exercises

1. Think of a simple problem like $3 \div \frac{1}{2}$. One way to think of it is how many $\frac{1}{2}$'s can be placed in 3? For example, how many $\frac{1}{2}$-pound rocks could be put in a bag that holds 3 pounds of rocks? The answer is 6. If we inverted the first fraction by mistake, we would have $\frac{1}{3} \times \frac{1}{2} = \frac{1}{6}$. We know this is wrong since there are obviously several $\frac{1}{2}$-pound rocks in a bag that holds 3 pounds of rocks. The answer $\frac{1}{6}$ would make no sense.   **2.** One way to think of it is to imagine how many $\frac{1}{3}$-pound rocks could be put in a bag that holds 2 pounds of rocks and then imagine how many $\frac{1}{2}$-pound rocks could be put in the same bag. The number of $\frac{1}{3}$-pound rocks would be larger. Therefore $2 \div \frac{1}{3}$ is a larger number.
3. $\frac{21}{16}$ or $1\frac{5}{16}$ **5.** $\frac{9}{2}$ or $4\frac{1}{2}$ **7.** $\frac{3}{2}$ or $1\frac{1}{2}$ **9.** $\frac{4}{5}$ **11.** $\frac{3}{44}$ **13.** $\frac{4}{3}$ or $1\frac{1}{3}$ **15.** 1   **17.** $\frac{9}{49}$ **19.** $\frac{36}{7}$ or $5\frac{1}{7}$ **21.** 0   **23.** undefined
25. $\frac{3}{4}$ **27.** $\frac{4}{5}$ **29.** 1   **31.** 10   **33.** $\frac{7}{32}$ **35.** 5000   **37.** $\frac{1}{10}$ **39.** $\frac{7}{40}$ **41.** $\frac{32}{3}$ or $10\frac{2}{3}$ **43.** $\frac{7}{18}$ **45.** 2   **47.** 4   **49.** $\frac{15}{7}$ or $2\frac{1}{7}$
51. 3   **53.** $\frac{63}{32}$ or $1\frac{31}{32}$ **55.** $\frac{5}{12}$ **57.** 14   **59.** $\frac{2}{3}$ **61.** $\frac{7}{44}$ **63.** $\frac{32}{15}$ or $2\frac{2}{15}$ **65.** $x = \frac{7}{5}$ **67.** $x = \frac{4}{7}$ **69.** $2\frac{1}{4}$ gallons
71. $37\frac{1}{2}$ miles per hour   **73.** 28 flags   **75.** 6 dresses   **76.** 36 test tubes   **77.** six drill attempts

79. We estimate by dividing $15 \div 5$, which is 3. The exact value is $2\frac{26}{31}$. Our estimate is very close. It is off by only $\frac{5}{31}$. **80.** We estimate by multiplying 18×28 to obtain 504. The exact value is $501\frac{7}{8}$. Our estimate is very close. It is off by only $2\frac{1}{8}$. **81.** thirty-nine million, five hundred seventy-six thousand, three hundred four **82.** $400{,}000 + 50{,}000 + 9000 + 200 + 70 + 3$ **83.** 1099 **84.** $87{,}595{,}631$

Test on Sections 2.1 to 2.5

1. $\frac{23}{32}$ **2.** $\frac{85}{113}$ **3.** $\frac{1}{2}$ **4.** $\frac{4}{5}$ **5.** $\frac{4}{11}$ **6.** $\frac{25}{31}$ **7.** $\frac{3}{4}$ **8.** $\frac{3}{2}$ or $1\frac{1}{2}$ **9.** $\frac{43}{12}$ **10.** $\frac{33}{8}$ **11.** $6\frac{3}{7}$ **12.** $8\frac{1}{4}$ **13.** $\frac{8}{35}$ **14.** $\frac{9}{7}$ or $1\frac{2}{7}$

15. 15 **16.** $16\frac{1}{2}$ or $\frac{33}{2}$ **17.** $13\frac{5}{12}$ **18.** $4\frac{16}{21}$ **19.** $\frac{16}{21}$ **20.** $5\frac{1}{3}$ or $\frac{16}{3}$ **21.** 7 **22.** $11\frac{1}{5}$ **23.** $7\frac{7}{8}$ **24.** 14 **25.** $2\frac{2}{3}$ **26.** $2\frac{7}{8}$

27. $\frac{13}{16}$ **28.** $\frac{1}{14}$ **29.** $\frac{9}{32}$ **30.** $\frac{13}{15}$ **31.** $45\frac{15}{16}$ square feet **32.** 4 cups **33.** $1\frac{1}{2}$ miles **34.** 16 full packages; $\frac{3}{8}$ lb. left over

35. 51 computers **36.** 8000 homes **37.** 16 hours **38.** 6 tents; $\frac{28}{33}$ of a yard left over **39.** 160 days

2.6 Exercises

1. 56 **3.** 100 **5.** 60 **7.** 36 **9.** 147 **11.** 10 **13.** 28 **15.** 35 **17.** 18 **19.** 60 **21.** 16 **23.** 90 **25.** 60 **27.** 105 **29.** 36

31. 6 **33.** 120 **35.** 132 **37.** 84 **39.** 120 **41.** 3 **43.** 45 **45.** 20 **47.** 14 **49.** 96 **51.** 63 **53.** $\frac{21}{36}$ and $\frac{20}{36}$ **55.** $\frac{24}{200}$ and $\frac{35}{200}$

57. $\frac{144}{160}$ and $\frac{152}{160}$ **59.** LCD $= 35$; $\frac{14}{35}$ and $\frac{9}{35}$ **61.** LCD $= 48$; $\frac{20}{48}$ and $\frac{3}{48}$ **63.** LCD $= 80$; $\frac{52}{80}$ and $\frac{55}{80}$ **65.** LCD $= 60$; $\frac{16}{60}$ and $\frac{25}{60}$

67. LCD $= 72$; $\frac{15}{72}, \frac{22}{72}, \frac{3}{72}$ **69.** LCD $= 56$; $\frac{3}{56}, \frac{49}{56}, \frac{40}{56}$ **71.** LCD $= 63$; $\frac{5}{63}, \frac{12}{63}, \frac{56}{63}$ **73.** (a) LCD $= 16$ (b) $\frac{3}{16}, \frac{12}{16}, \frac{6}{16}$

75. 178R3 **76.** 3624 **77.** 9963 **78.** between 1980 and 1985 **79.** between 1985 and 1990 **80.** more than $7469 a year
81. more than $8640 a year **82.** $21,111 **83.** $23,321

2.7 Exercises

1. $\frac{7}{9}$ **3.** $\frac{2}{3}$ **5.** $\frac{1}{12}$ **7.** $\frac{17}{44}$ **9.** $\frac{11}{14}$ **11.** $\frac{9}{20}$ **13.** $\frac{7}{8}$ **15.** $\frac{23}{20}$ or $1\frac{3}{20}$ **17.** $\frac{37}{100}$ **19.** $\frac{26}{175}$ **21.** $\frac{31}{24}$ or $1\frac{7}{24}$ **23.** $\frac{27}{40}$ **25.** $\frac{1}{4}$

27. $\frac{8}{35}$ **29.** $\frac{19}{27}$ **31.** $\frac{11}{60}$ **33.** $\frac{1}{4}$ **35.** $\frac{6}{25}$ **37.** $\frac{5}{48}$ **39.** 0 **41.** $\frac{5}{12}$ **43.** $\frac{8}{5}$ or $1\frac{3}{5}$ **45.** $\frac{11}{30}$ **47.** $1\frac{7}{15}$ **49.** $\frac{3}{14}$ **51.** $\frac{5}{33}$ **53.** $\frac{8}{15}$

55. $1\frac{5}{12}$ cups **57.** $1\frac{1}{2}$ pounds **59.** $\frac{19}{60}$ of the book report **61.** 16 chocolates **63.** $\frac{4}{21}$ of the membership **65.** $\frac{3}{17}$ **66.** $\frac{2}{3}$ **67.** $7\frac{11}{16}$

68. $\frac{101}{7}$ **69.** $9\frac{3}{8}$ **70.** $2\frac{8}{9}$

2.8 Exercises

1. $9\frac{3}{4}$ **3.** $4\frac{1}{7}$ **5.** $17\frac{1}{2}$ **7.** $16\frac{1}{10}$ **9.** $\frac{4}{7}$ **11.** $2\frac{17}{24}$ **13.** $16\frac{1}{4}$ **15.** $2\frac{13}{24}$ **17.** $4\frac{14}{15}$ **19.** $14\frac{4}{7}$ **21.** $41\frac{4}{5}$ **23.** $6\frac{7}{12}$ **25.** $8\frac{1}{6}$

27. $73\frac{37}{40}$ **29.** $5\frac{1}{2}$ **31.** 0 **33.** $4\frac{41}{60}$ **35.** $8\frac{8}{15}$ **37.** $102\frac{5}{8}$ **39.** $13\frac{5}{24}$ **41.** $43\frac{1}{8}$ miles **43.** $2\frac{1}{24}$ pounds **45.** $2\frac{3}{4}$ inches

47. (a) $9\frac{19}{24}$ pounds (b) $6\frac{5}{24}$ pounds **49.** $\frac{2607}{40}$ or $65\frac{7}{40}$ **50.** $\frac{277}{42}$ or $6\frac{25}{42}$

51. We estimate by adding $35 + 24$ to obtain 59. The exact answer is $59\frac{7}{12}$. Our estimate is very close. We are off by only $\frac{7}{12}$.
52. We estimate by subtracting $103 - 87$ to obtain 16. The exact answer is $16\frac{1}{21}$. Our estimate is very close. We are off by only $\frac{1}{21}$.

53. $\frac{2}{3}$ **55.** 1 **57.** $\frac{5}{3}$ or $1\frac{2}{3}$ **59.** $\frac{3}{5}$ **61.** $\frac{9}{25}$ **63.** $\frac{32}{21}$ or $1\frac{11}{21}$ **65.** $\frac{1}{9}$ **67.** $\frac{16}{11}$ or $1\frac{5}{11}$ **69.** 512,012 **70.** 8,529,300 **71.** $1893

Putting Your Skills to Work

1. 10 eggs, $2\frac{1}{2}$ cups milk, $2\frac{1}{6}$ cups shortening, 3 tablespoons ground orange peel, $4\frac{3}{4}$ teaspoons ground lemon peel, $8\frac{1}{6}$ cups sifted flour, $7\frac{7}{8}$ teaspoons baking powder, $2\frac{1}{4}$ teaspoons salt, 6 tablespoons lemon juice, $3\frac{5}{8}$ cups sugar. **2.** $6\frac{2}{3}$ eggs, $1\frac{2}{3}$ cups milk, $1\frac{4}{9}$ cups shortening, 2 tablespoons ground orange peel, $3\frac{1}{6}$ teaspoons ground lemon peel, $5\frac{4}{9}$ cups sifted flour, $5\frac{1}{4}$ teaspoons baking powder, $1\frac{1}{2}$ teaspoons salt, 4 tablespoons lemon juice, $2\frac{5}{12}$ cups sugar

2.9 Exercises

1. $16\frac{2}{3}$ feet **3.** $25\frac{11}{24}$ tons **5.** $1\frac{9}{16}$ inches **7.** $9\frac{19}{20}$ miles **9.** $\frac{7}{15}$ of a foot **11.** $275\frac{5}{8}$ gallons **13.** $106\frac{7}{8}$ nautical miles

15. $451 per week **17.** (a) 42 bags (b) $147 (c) $145 **19.** (a) $14\frac{1}{8}$ ounces of bread (b) $\frac{5}{8}$ ounce **21.** (a) $30\frac{1}{2}$ knots (b) 7 hours

23. (a) 5485 bushels (b) $11{,}998\frac{7}{16}$ cubic feet (c) $9598\frac{3}{4}$ bushels **25.** $14\frac{71}{72}$ lbs **26.** $15\frac{5}{24}$ lbs **27.** 44,245 **28.** 27,814 **29.** 45,441

30. 278

Chapter 2 Review Problems

1. $\frac{5}{12}$ **2.** $\frac{3}{8}$ **3.** Answers will vary.   **4.** Answers will vary.   **5.** $\frac{87}{100}$ **6.** $\frac{6}{31}$ **7.** 2×3^3 **8.** $2 \times 3 \times 7$ **9.** $2^3 \times 3 \times 7$ **10.** prime

11. $2 \times 3 \times 13$ **12.** prime   **13.** $\frac{2}{7}$ **14.** $\frac{1}{4}$ **15.** $\frac{7}{12}$ **16.** $\frac{13}{17}$ **17.** $\frac{7}{8}$ **18.** $\frac{17}{35}$ **19.** $\frac{35}{8}$ **20.** $\frac{65}{23}$ **21.** $5\frac{5}{8}$ **22.** $4\frac{11}{13}$ **23.** $3\frac{3}{11}$

24. $\frac{117}{8}$ **25.** $1\frac{29}{48}$ **26.** $\frac{20}{77}$ **27.** $\frac{7}{15}$ **28.** 0   **29.** $\frac{4}{63}$ **30.** $\frac{492}{5}$ or $98\frac{2}{5}$ **31.** $\frac{51}{2}$ or $25\frac{1}{2}$ **32.** $\frac{817}{40}$ or $20\frac{17}{40}$ **33.** $24\frac{1}{2}$ **34.** $\$677\frac{1}{4}$

35. $486\frac{17}{20}$ square inches   **36.** $\frac{15}{14}$ or $1\frac{1}{14}$ **37.** $\frac{5}{34}$ **38.** 1920   **39.** $\frac{27}{40}$ **40.** $\frac{25}{6}$ or $4\frac{1}{6}$ **41.** $\frac{17}{164}$ **42.** 0   **43.** $\frac{46}{33}$ or $1\frac{13}{33}$

44. 7 dresses   **45.** $186\frac{2}{3}$ calories   **46.** 98   **47.** 120   **48.** 90   **49.** $\frac{24}{56}$ **50.** $\frac{33}{72}$ **51.** $\frac{36}{172}$ **52.** $\frac{187}{198}$ **53.** $\frac{1}{14}$ **54.** $\frac{13}{12}$ or $1\frac{1}{12}$

55. $\frac{85}{63}$ or $1\frac{22}{63}$ **56.** $\frac{11}{40}$ **57.** $\frac{23}{70}$ **58.** $\frac{25}{36}$ **59.** $\frac{19}{48}$ **60.** $\frac{61}{75}$ **61.** $\frac{6}{23}$ **62.** $5\frac{4}{9}$ **63.** $8\frac{2}{3}$ **64.** $20\frac{5}{7}$ **65.** $\frac{39}{8}$ or $4\frac{7}{8}$ **66.** $\frac{209}{48}$ or $4\frac{17}{48}$

67. $\frac{9}{10}$ **68.** $\frac{20}{63}$ **69.** $8\frac{29}{40}$ miles   **70.** $283\frac{1}{12}$ miles   **71.** $1\frac{2}{3}$ cups sugar, $2\frac{1}{8}$ cups flour   **72.** $206\frac{1}{8}$ miles   **73.** 15 lengths   **74.** $9\frac{5}{8}$ liters

75. 30 words per minute   **76.** $\$56\frac{1}{4}$ **77.** $\$8\frac{3}{4}$ **78.** $1\frac{1}{16}$ inch   **79.** \$242   **80. (a)** 25 miles per gallon   **(b)** $\$22\frac{2}{25}$

Chapter 2 Test

1. $\frac{3}{5}$ **2.** $\frac{311}{388}$ **3.** $\frac{3}{7}$ **4.** $\frac{3}{14}$ **5.** $\frac{9}{2}$ **6.** $\frac{34}{5}$ **7.** $8\frac{1}{7}$ **8.** 12   **9.** $\frac{14}{45}$ **10.** $31\frac{1}{5}$ **11.** $\frac{77}{40}$ or $1\frac{37}{40}$ **12.** $\frac{39}{62}$ **13.** $6\frac{12}{13}$ **14.** $\frac{12}{7}$ or $1\frac{5}{7}$

15. 72   **16.** 48   **17.** 24   **18.** $\frac{30}{72}$ **19.** $\frac{13}{36}$ **20.** $\frac{11}{20}$ **21.** $\frac{25}{28}$ **22.** $14\frac{6}{35}$ **23.** $4\frac{13}{14}$ **24.** $\frac{19}{28}$ **25.** $\frac{7}{6}$ or $1\frac{1}{6}$ **26.** 42 square yards

27. 8 packages   **28.** $\frac{7}{10}$ mile   **29.** $14\frac{1}{24}$ miles   **30. (a)** 40 oranges   **(b)** $\$9\frac{3}{5}$ **31. (a)** 77 candles   **(b)** $\$1\frac{1}{4}$ **(c)** $\$827\frac{3}{4}$

Cumulative Test for Chapter 1–2

1. eighty-four million, three hundred sixty-one thousand, two hundred eight   **2.** 869   **3.** 719,220   **4.** 2075   **5.** 17,216   **6.** 4788   **7.** 202,896

8. 4658   **9.** 369   **10.** 49   **11.** 6,037,000   **12.** 38   **13.** \$174   **14.** \$306   **15.** $\frac{55}{84}$ **16.** $\frac{7}{13}$ **17.** $\frac{75}{4}$ **18.** $14\frac{2}{7}$ **19.** $\frac{527}{48}$ or $10\frac{47}{48}$

20. $\frac{12}{35}$ **21.** 39   **22.** $\frac{31}{54}$ **23.** $\frac{71}{8}$ or $8\frac{7}{8}$ **24.** $\frac{113}{15}$ or $7\frac{8}{15}$ **25.** $\frac{5}{6}$ **26.** $23\frac{1}{8}$ tons   **27.** $24\frac{3}{5}$ miles per gallon   **28.** $8\frac{1}{8}$ cups of sugar;

$5\frac{5}{6}$ cups of flour   **29.** 60,000,000 miles

Chapter 3

Pretest Chapter 3

1. forty-seven and eight hundred thirteen thousandths   **2.** 0.0567   **3.** $2\frac{11}{100}$ **4.** $\frac{21}{40}$ **5.** 1.59, 1.6, 1.601, 1.61   **6.** 123.5   **7.** 1.053   **8.** 19.45
9. 27.191   **10.** 10.59   **11.** 7.671   **12.** 0.3501   **13.** 4780.5   **14.** 37.96   **15.** 0.354   **16.** 0.128   **17.** 0.6875   **18.** $0.2\overline{27}$ or $0.22727 \ldots$
19. 0.97   **20.** 17.9 miles per gallon   **21.** \$489.51   **22.** \$8.53

3.1 Exercises

1. A decimal fraction is a fraction whose denominator is a power of 10. $\frac{23}{100}$ and $\frac{563}{1000}$ are decimal fractions.   **3.** hundred-thousandths   **5.** fifty-seven
hundredths   **7.** three and eight tenths   **9.** five and two hundred eighty-three thousandths   **11.** twenty-eight and thirty-seven ten-thousandths
13. one hundred twenty-four and $\frac{20}{100}$ dollars   **15.** one thousand, two hundred thirty-six and $\frac{8}{100}$ dollars   **17.** ten thousand and $\frac{76}{100}$ dollars   **19.** 0.7
21. 0.45   **23.** 0.022   **25.** 0.000286   **27.** 0.7   **29.** 0.76   **31.** 0.01   **33.** 0.053   **35.** 0.2403   **37.** 8.3   **39.** 84.13   **41.** 1.019
43. 126.0571   **45.** $\frac{1}{50}$ **47.** $3\frac{3}{5}$ **49.** $8\frac{6}{25}$ **51.** $12\frac{5}{8}$ **53.** $7\frac{3}{2000}$ **55.** $235\frac{627}{5000}$ **57.** $\frac{187}{10,000}$ **59.** $8\frac{27}{2500}$ **61.** $\frac{1}{250,000}$ **63.** 818
64. 938   **65.** 56,800   **66.** 8,069,000

3.2 Exercises

1. >   **3.** <   **5.** <   **7.** >   **9.** <   **11.** >   **13.** =   **15.** >   **17.** 12.6, 12.65, 12.8   **19.** 0.007, 0.0071, 0.05   **21.** 5.12, 5.2, 5.23, 5.3
23. 26.003, 26.033, 26.034, 26.04   **25.** 18.006, 18.060, 18.065, 18.066, 18.606   **27.** 5.7   **29.** 29.4   **31.** 578.1   **33.** 2176.8   **35.** 26.03   **37.** 5.77
39. 156.17   **41.** 2786.71   **43.** 1.061   **45.** 0.0474   **47.** 5.00761   **49.** 136   **51.** \$2537   **53.** \$10,098   **55.** \$56.98   **57.** \$5783.72

59. 0.10 kilogram; 0.07 kilogram   **61.** 365.24   **62.** 3.14; 2.72   **63.** 0.0059, 0.006, 0.0519, $\frac{6}{100}$, 0.0601, 0.0612, 0.062, $\frac{6}{10}$, 0.61

65. You should consider only one digit to the right of the decimal place that you wish to round to. 86.23498 is closer to 86.23 than to 86.24.

67. $12\frac{1}{8}$ **68.** $10\frac{9}{20}$ **69.** 692 miles   **70.** \$9658

3.3 Exercises

1. 76.8 **3.** 1215.55 **5.** 296.2 **7.** 23.00 **9.** 36.7287 **11.** 67.42 **13.** 1112.16 **15.** 19.86 m **17.** 8.6 pounds **19.** $78.12
21. $1411.97 **23.** 3.9 **25.** 26.66 **27.** 115.53 **29.** 508.313 **31.** 2.19901 **33.** 4.6465 **35.** 6.737 **37.** 1189.07 **39.** 1.4635
41. 1.464 meters **43.** $36,947.16 **45.** $45.30 **47.** 11.64 centimeters **49.** 2.95 liters **51.** 0.0061 milligram; yes **53.** $6.2 billion;
$6,200,000,000 **55.** $13.3 billion; $13,300,000,000 **57.** $8.40; yes; $8.34; very close, the estimate was off by 6¢ **59.** $x = 8.4$ **61.** $x = 43.7$

63. $x = 2.109$ **65.** 20,288 **66.** 18,213 **67.** $\frac{77}{25}$ or $3\frac{2}{25}$ **68.** $\frac{35}{4}$ or $8\frac{3}{4}$

3.4 Exercises

1. 0.12 **3.** 0.06 **5.** 0.00288 **7.** 0.00711 **9.** 0.000516 **11.** 0.6582 **13.** 981.73 **15.** 0.017304 **17.** 768.1517 **19.** 8460
21. 53.926 **23.** 6.5237 **25.** $9324 **27.** $382.00 **29.** 297.6 square feet **31.** $664.20 **33.** 514.8 miles **35.** 28.6 **37.** 430
39. 128,650 **41.** 56,098.2 **43.** 28,056,020 **45.** 816,320 **47.** 593.2 centimeters **49.** 2980 meters **51.** $248.00 **53.** $62,279.00
55. To multiply by numbers such as 0.1, 0.01, 0.001, and 0.0001, count the number of decimal places in this first number. Then, in the other number, move the decimal point to the left from its present position the same number of decimal places as was in the first number. **57.** 98 **58.** 86
59. 125 R4 **60.** 129 R4 **61.** $0.11 billion or $110,000,000 **63.** $3.386 billion

3.5 Exercises

1. 2.1 **3.** 17.83 **5.** 18.31 **7.** 136.5 **9.** 118.2 **11.** 11.59 **13.** 18 **15.** 130 **17.** 5.3 **19.** 1.2 **21.** 49.3 **23.** 94.21 **25.** 26.778
27. 12.246 **29.** 116 **31.** 13 **33.** 6.2 ounces **35.** (a) 25.3 ounces (b) 20.24 ounces **37.** $376.50 **39.** nine payments
41. (a) 8.65 lb (b) 0.325 lb **43.** 0.123 **45.** 69 **47.** 45 **49.** 74 **51.** $\frac{71}{40}$ or $1\frac{31}{40}$ **52.** $\frac{15}{16}$ **53.** $\frac{91}{12}$ or $7\frac{7}{12}$ **54.** $\frac{5}{3}$ or $1\frac{2}{3}$

Putting Your Skills to Work

1. $3\frac{1}{3}$ feet **2.** 2 feet

3.6 Exercises

1. same quantity **3.** The digits 8942 repeat. **5.** 0.25 **7.** 0.8 **9.** 0.4375 **11.** 0.35 **13.** 0.62 **15.** 2.25 **17.** 2.125 **19.** 1.4375
21. $0.9\overline{3}$ **23.** $0.\overline{45}$ **25.** $3.58\overline{3}$ **27.** $2.2\overline{7}$ **29.** 0.571 **31.** 0.905 **33.** 0.146 **35.** 1.296 **37.** 0.404 **39.** 0.944 **41.** 3.143
43. 0.474 **45.** 0.125 inch **47.** too small, by 0.125 inch **49.** too thick; 0.00125 inch **51.** 2.3 **53.** 114 **55.** 0 **57.** 21.414 **59.** 0.0072
61. 0.325 **63.** 28.6 **65.** 20.836 **67.** 0.586930 **69.** (a) 0.16 (b) $0.1449\overline{49}$ (c) (b) is repeating and (a) is a non-repeating decimal.
71. 312 square feet **72.** 29,273 votes **73.** $377 **74.** $83

3.7 Exercises

1. 700,000,000 **3.** 30,000 **5.** 320,000 **7.** 150 **9.** $10 **11.** 4875 kroners **13.** 2748.27 square feet **15.** 96 molds **17.** 11.59 meters
19. 12 days **21.** $1263.09 **23.** $359.60 **25.** $426.55 **27.** $17,319; $5819 **29.** yes; by 0.149 milligram per liter **31.** 137 minutes
33. 22.9 quadrillion Btu **35.** approximately 47.8 quadrillion Btu; 47,800,000,000,000,000 Btu **37.** $\frac{29}{70}$ **38.** $\frac{11}{34}$ **39.** $\frac{5}{9}$ **40.** 22

Chapter 3 Review Problems

1. thirteen and six hundred seventy-two thousandths **2.** eighty-four hundred-thousandths **3.** 0.7 **4.** 0.81 **5.** 1.523 **6.** 0.0079 **7.** $\frac{17}{100}$
8. $\frac{73}{200}$ **9.** $34\frac{6}{25}$ **10.** $1\frac{1}{4000}$ **11.** > **12.** > **13.** 0.901, 0.918, 0.98, 0.981 **14.** 5.2, 5.26, 5.59, 5.6, 5.62 **15.** 0.6 **16.** 19.21 **17.** 1.100
18. $156 **19.** 77.6 **20.** 6.639 **21.** 3.894 **22.** 156.814 **23.** 0.003136 **24.** 887.81 **25.** 405.6 **26.** 2398.02 **27.** 0.613
28. 123,540 **29.** $1.26 **30.** 0.00258 **31.** 36.8 **32.** 232.9 **33.** 574.4 **34.** 0.059 **35.** $0.2\overline{7}$ **36.** 0.175 **37.** 1.83 **38.** 1.1875
39. 0.786 **40.** 2.294 **41.** 3.6 **42.** 1.152 **43.** 1.301208 **44.** 87.13 **45.** 439.19 **46.** 0.402216 **47.** 0.8403 **48.** 69.2 **49.** 20.004
50. 1.25 **51.** 112 people **52.** 24.8 miles per gallon **53.** $2170.30 **54.** ABC Company **55.** no; by 0.0005 milligram per liter
56. seven test tubes **57.** 55.8 feet **58.** 1396.75 square feet **59.** 48.6 miles **60.** 259.9 feet **61.** $12,750.00; $12,255.00; they should change
to the new loan **62.** $262 **63.** $207 **64.** $15.97 **65.** $24.00 **66.** $30.00 **67.** $33.90

Chapter 3 Test

1. one hundred fifty-seven thousandths **2.** 0.3977 **3.** $7\frac{3}{20}$ **4.** $\frac{261}{1000}$ **5.** 2.19, 2.9, 2.907, 2.91 **6.** 78.66 **7.** 0.0342 **8.** 99.698
9. 37.53 **10.** 0.0979 **11.** 71.155 **12.** 0.5817 **13.** 218.9 **14.** 25.7 **15.** 47 **16.** $1.\overline{2}$ **17.** 0.5625 **18.** 1.487 **19.** 6.1952
20. $33.15 **21.** 18.8 miles per gallon **22.** 3.43 centimeters **23.** $390.55

Cumulative Test for Chapters 1–3

1. thirty-eight million, fifty-six thousand, nine hundred fifty-four **2.** 479,587 **3.** 54,480 **4.** 39,463 **5.** 258 **6.** 16 **7.** $\frac{3}{8}$ **8.** $7\frac{1}{2}$
9. $\frac{9}{35}$ **10.** $\frac{23}{24}$ **11.** 16 **12.** $\frac{33}{10}$ or $3\frac{3}{10}$ **13.** 24,000,000,000 **14.** 0.571 **15.** 2.01, 2.1, 2.11, 2.12, 20.1 **16.** 26.080 **17.** 19.54
18. 8.639 **19.** 1.136 **20.** 36,512.3 **21.** 1.058 **22.** 0.8125 **23.** 13.597 **24.** 456 miles **25.** $195.57 **26.** 60 months

Chapter 4

Pretest Chapter 4

1. $\frac{13}{18}$ **2.** $\frac{1}{5}$ **3.** $\frac{9}{2}$ **4.** $\frac{11}{12}$ **5. (a)** $\frac{7}{24}$ **(b)** $\frac{11}{120}$ **6.** $\frac{3 \text{ flight attendants}}{100 \text{ passengers}}$ **7.** $\frac{31 \text{ gallons}}{42 \text{ square feet}}$ **8.** $\frac{30.5 \text{ miles}}{1 \text{ hour}}$ or 30.5 miles/hour   **9.** $\frac{\$29}{1 \text{ CD player}}$ or \$29/CD player   **10.** $\frac{13}{40} = \frac{39}{120}$ **11.** $\frac{116}{158} = \frac{29}{37}$ **12.** It is a proportion.   **13.** It is not a proportion.   **14.** $n = 17$ **15.** $n = 18$ **16.** $n = 4$ **17.** $n = 7$ **18.** $n = 600$ **19.** $n = 2$ **20.** 3.5 cups   **21.** 364.5 miles   **22.** 146 miles   **23.** 54 defective bulbs   **24.** 2.5 runs   **25.** 13,680 people

4.1 Exercises

1. ratio   **3.** 5 to 8   **5.** $\frac{1}{3}$ **7.** $\frac{7}{6}$ **9.** $\frac{3}{11}$ **11.** $\frac{2}{3}$ **13.** $\frac{15}{16}$ **15.** $\frac{2}{3}$ **17.** $\frac{8}{5}$ **19.** $\frac{2}{3}$ **21.** $\frac{3}{2}$ **23.** $\frac{6}{7}$ **25.** $\frac{13}{1}$ **27.** $\frac{10}{17}$ **29.** $\frac{165}{285} = \frac{11}{19}$ **31.** $\frac{35}{165} = \frac{7}{33}$ **33.** $\frac{205}{1225} = \frac{41}{245}$ **35.** $\frac{450}{205} = \frac{90}{41}$ **37.** $\frac{1}{16}$ **39.** $\frac{\$5}{2 \text{ magazines}}$ **41.** $\frac{\$85}{6 \text{ bushes}}$ **43.** $\frac{62 \text{ gallons}}{125 \text{ square feet}}$ **45.** $\frac{410 \text{ revolutions}}{1 \text{ mile}}$ or 410 revolutions/mile   **47.** $\frac{\$27,500}{1 \text{ employee}}$ or \$27,500/employee   **49.** \$13/hour   **51.** 16 miles/gallon   **53.** 155 gallons/hour   **55.** 125 pencils/box   **57.** 69 miles/hour   **59.** \$30/share   **61.** \$5.50/calendar   **63. (a)** \$0.08/ounce small box; \$0.07/ounce large box   **(b)** \$0.01/ounce   **(c)** The consumer saves \$0.48.   **65. (a)** 13 moose   **(b)** 12 moose   **(c)** North Slope   **67.** \$24.53   **69.** increased by Mach 0.2   **71.** $2\frac{5}{8}$ **72.** 5   **73.** $\frac{5}{24}$ **74.** $1\frac{1}{48}$ **75.** \$12.25/square yard   **76.** \$24,150; \$16,800

4.2 Exercises

1. equal   **3.** $\frac{18}{9} = \frac{2}{1}$ **5.** $\frac{20}{36} = \frac{5}{9}$ **7.** $\frac{84}{105} = \frac{12}{15}$ **9.** $\frac{5\frac{1}{2}}{16} = \frac{7\frac{2}{3}}{23}$ **11.** $\frac{6.5}{14} = \frac{13}{28}$ **13.** $\frac{3 \text{ inches}}{40 \text{ miles}} = \frac{27 \text{ inches}}{360 \text{ miles}}$ **15.** $\frac{10 \text{ runs}}{45 \text{ games}} = \frac{36 \text{ runs}}{162 \text{ games}}$ **17.** $\frac{3 \text{ hours}}{\$525} = \frac{7 \text{ hours}}{\$1225}$ **19.** $\frac{3 \text{ teaching assistants}}{40 \text{ children}} = \frac{21 \text{ teaching assistants}}{280 \text{ children}}$ **21.** $\frac{4800 \text{ people}}{3 \text{ restaurants}} = \frac{11,200 \text{ people}}{7 \text{ restaurants}}$ **23.** It is a proportion.   **25.** It is not a proportion.   **27.** It is not a proportion.   **29.** It is a proportion.   **31.** It is a proportion.   **33.** It is not a proportion.   **35.** It is not a proportion.   **37.** It is a proportion.   **39.** It is a proportion.   **41.** It is a proportion.   **43.** It is not a proportion.   **45.** yes   **47.** no   **49.** yes   **51. (a)** yes   **(b)** yes   **(c)** the equality test for fractions   **52.** 23.1405   **53.** 14.15566   **54.** 402.408   **55.** 25.8   **56.** $4\frac{5}{24}$ gallons

4.3 Exercises

1. Divide each side of the equation by the number a. Calculate $\frac{b}{a}$. The value of n is $\frac{b}{a}$.   **3.** $n = 11$ **5.** $n = 5.6$ **7.** $n = 25$ **9.** $n = 7$ **11.** $n = 34\frac{2}{3}$ **13.** $n = 15$ **15.** $n = 16$ **17.** $n = 7.5$ **19.** $n = 32$ **21.** $n = 75$ **23.** $n = 22.5$ **25.** $n = 66$ **27.** $n = 31.5$ **29.** $n = 18$ **31.** $n = 8$ **33.** $n = 162$ **35.** $n = 40$ **37.** $n \approx 3.7$ **39.** $n \approx 5.5$ **41.** $n = 9$ **43.** $n \approx 3.82$ **45.** $n = 2.5$ **47.** $n = 13.65$ **49.** $n \approx 6.45$ **51.** $n = 80$ **53.** $n = 4\frac{7}{8}$ **55.** $8\frac{1}{3}$ inches   **57.** $n = 3\frac{13}{15}$ **59.** $n = 4\frac{33}{34}$ **60.** 76   **61.** 16.86   **62.** five hundred sixty-three thousandths   **63.** 0.0034   **64.** 56 games

4.4 Exercises

1. 380 desserts   **3.** 3 cups   **5.** 645 Francs   **7.** 197.6 feet   **9.** 356 miles   **11.** $11\frac{2}{3}$ cups   **13.** 102 free throws   **15.** 4898 people   **17.** 78 giraffes   **19.** \$3570   **21.** 270 chips   **23.** 1 cup of water and $\frac{3}{8}$ cup of milk   **25.** 2 cups of water and 1 cup of milk   **27.** Ken Griffey, Jr., approximately \$177,303/home run; Jay Bell, approximately \$157,895/home run   **29.** Roger Clemens, \$275,000/game; Pedro Martinez, approximately \$354,839/game   **31.** 56,200   **32.** 196,380,000   **33.** 56.1   **34.** 1.96   **35. (a)** $\frac{19}{20}$ square foot   **(b)** 1425 square feet

Putting Your Skills to Work

1. 88,236   **2.** 193,400

Chapter 4 Review Problems

1. $\frac{11}{5}$ **2.** $\frac{5}{3}$ **3.** $\frac{4}{5}$ **4.** $\frac{10}{19}$ **5.** $\frac{28}{51}$ **6.** $\frac{1}{3}$ **7.** $\frac{40}{93}$ **8.** $\frac{52}{147}$ **9.** $\frac{2}{5}$ **10.** $\frac{7}{43}$ **11.** $\frac{4}{43}$ **12.** $\frac{5 \text{ gallons}}{9 \text{ people}}$ **13.** $\frac{4 \text{ revolutions}}{11 \text{ minutes}}$ **14.** $\frac{47 \text{ vibrations}}{4 \text{ seconds}}$ **15.** $\frac{6 \text{ cups}}{19 \text{ people}}$ **16.** \$17.00/share   **17.** \$17.00/chair   **18.** \$13.50/square yard   **19.** \$12.40/ticket

20. (a) $0.22 **(b)** $0.25 **(c)** $0.03   **21. (a)** $0.74 **(b)** $0.58 **(c)** $0.16   **22.** $\dfrac{12}{48} = \dfrac{7}{28}$ **23.** $\dfrac{1\frac{1}{2}}{5} = \dfrac{4}{13\frac{1}{3}}$ **24.** $\dfrac{7.5}{45} = \dfrac{22.5}{135}$

25. $\dfrac{\$4.50}{15\ \text{pounds}} = \dfrac{\$8.10}{27\ \text{pounds}}$ **26.** $\dfrac{138\ \text{passengers}}{3\ \text{buses}} = \dfrac{230\ \text{passengers}}{5\ \text{buses}}$ **27.** It is not a proportion.   **28.** It is a proportion.
29. It is a proportion.   **30.** It is a proportion.   **31.** It is not a proportion.   **32.** It is not a proportion.   **33.** It is a proportion.
34. It is not a proportion.   **35.** $n = 23$ **36.** $n = 5\frac{1}{4}$ or 5.25   **37.** $n = 31$ **38.** $n = 17$ **39.** $n = 33$ **40.** $n = 42$ **41.** $n = 7$
42. $n = 24$ **43.** $n = 19$ **44.** $n = 5\frac{3}{5}$ or 5.6   **45.** $n = 3$ **46.** $n = 1$ **47.** $n = 87$ **48.** $n = 324$ **49.** $n = 16.8$ **50.** $n \approx 4.7$
51. $n = 12$ **52.** $n = 550$ **53.** 15 gallons   **54.** 1691 employees   **55.** 2016 francs   **56.** 50 pounds   **57.** 600 miles   **58.** 16 miles/hour
59. 120 feet   **60. (a)** 7.65 gallons   **(b)** $8.80   **61.** 5.71 centimeters tall   **62.** 7.5 grams   **63.** 182 bricks   **64.** 9600 people
65. 9 gallons   **66.** 85 shares   **67.** approximately 113.93 feet   **68.** approximately 27.38 minutes   **69.** 86 goals   **70.** 552 calories

Chapter 4 Test

1. $\dfrac{9}{26}$ **2.** $\dfrac{14}{37}$ **3.** $\dfrac{98\ \text{miles}}{3\ \text{gallons}}$ **4.** $\dfrac{140\ \text{square feet}}{3\ \text{pounds}}$ **5.** 3.8 tons/day   **6.** $8.28/hour   **7.** 245.45 feet/pole   **8.** $85.21/share   **9.** $\dfrac{17}{29} = \dfrac{51}{87}$

10. $\dfrac{3\frac{1}{2}}{5} = \dfrac{18}{25\frac{5}{7}}$ **11.** $\dfrac{490\ \text{miles}}{21\ \text{gallons}} = \dfrac{280\ \text{miles}}{12\ \text{gallons}}$ **12.** $\dfrac{5\ \text{tablespoons}}{18\ \text{people}} = \dfrac{15\ \text{tablespoons}}{54\ \text{people}}$ **13.** It is not a proportion.   **14.** It is a proportion.

15. It is a proportion.   **16.** It is not a proportion.   **17.** $n = 16$ **18.** $n = 14$ **19.** $n = 19$ **20.** $n = 29.4$ **21.** $n = 120$ **22.** $n = 8.4$
23. $n = 120$ **24.** $n = 5.13$ **25.** 6 eggs   **26.** 80.95 pounds   **27.** 19 miles   **28.** $360   **29.** 4 quarts   **30.** 696.67 kilometers
31. 13.8 gallons   **32.** 32 hits

Cumulative Test for Chapters 1–4

1. twenty-six million, five hundred ninety-seven thousand, eighty-nine   **2.** 68   **3.** $\dfrac{11}{32}$ **4.** $\dfrac{27}{35}$ **5.** $\dfrac{117}{8}$ or $14\frac{5}{8}$ **6.** 8.2584   **7.** 0.179586
8. $\dfrac{\$0.14}{1\ \text{banana}}$ or $0.14/banana   **9.** $\dfrac{4\ \text{yen}}{1\ \text{peso}}$ or 4 yen/peso   **10.** It is a proportion.   **11.** It is a proportion.   **12.** $n = 3$ **13.** $n = 2$
14. $n = 128$ **15.** $n = 9$ **16.** $n \approx 3.4$ **17.** $n = 21$ **18.** 8.33 inches   **19.** $750.00   **20.** 5 pounds   **21.** 214.5 gallons

Chapter 5

Pretest Chapter 5

1. 17%   **2.** 38.7%   **3.** 134%   **4.** 894%   **5.** 0.6%   **6.** 0.4%   **7.** 17%   **8.** 27%   **9.** 13.4   **10.** 19.8%   **11.** $6\frac{1}{2}\%$ **12.** $1\frac{3}{8}\%$
13. 80%   **14.** 2.5%   **15.** 260%   **16.** 106.25%   **17.** 71.43%   **18.** 28.57%   **19.** 95.65%   **20.** 68.42%   **21.** 440%   **22.** 275%
23. 0.33%   **24.** 0.25%   **25.** $\dfrac{11}{50}$ **26.** $\dfrac{53}{100}$ **27.** $\dfrac{3}{2}$ or $1\frac{1}{2}$ **28.** $\dfrac{8}{5}$ or $1\frac{3}{5}$ **29.** $\dfrac{19}{300}$ **30.** $\dfrac{7}{150}$ **31.** $\dfrac{41}{80}$ **32.** $\dfrac{7}{16}$ **33.** 55.2   **34.** 48.36
35. 94.44%   **36.** 44.74%   **37.** 3000   **38.** 885   **39.** 63.16%   **40.** $14.40   **41.** 30%   **42.** $5400   **43.** $408

5.1 Exercises

1. hundred   **3.** two; left; Drop   **5.** 48%   **7.** 9%   **9.** 80%   **11.** 245%   **13.** 5.3%   **15.** 0.07%   **17.** 13%   **19.** 28%   **21.** 0.51
23. 0.07   **25.** 0.2   **27.** 0.436   **29.** 0.0003   **31.** 0.0072   **33.** 1.26   **35.** 3.66   **37.** 74%   **39.** 50%   **41.** 8%   **43.** 56.3%   **45.** 0.2%
47. 0.57%   **49.** 135%   **51.** 272%   **53.** 27%   **55.** 9%   **57.** 94%   **59.** 231%   **61.** 10%   **63.** 0.9%   **65.** 0.62   **67.** 138   **69.** 0.003
71. 0.8   **73.** 59%

75. $36\% = 36$ percent $= 36$ "per one hundred" $= 36 \times \dfrac{1}{100} = \dfrac{36}{100} = 0.36$. The rule is using the fact that 36% means 36 per one hundred.

77. (a) 555.62   **(b)** $\dfrac{55,562}{100}$ **(c)** $\dfrac{27,781}{50}$ **79.** $\dfrac{14}{25}$ **80.** $\dfrac{39}{50}$ **81.** 0.6875   **82.** 0.875   **83.** 5336 vases

5.2 Exercises

1. Write the number in front of the percent symbol as the numerator of a fraction. Write the number 100 as the denominator of the fraction. Reduce the fraction if possible.   **3.** $\dfrac{1}{10}$ **5.** $\dfrac{33}{100}$ **7.** $\dfrac{11}{20}$ **9.** $\dfrac{3}{4}$ **11.** $\dfrac{1}{5}$ **13.** $\dfrac{19}{200}$ **15.** $\dfrac{7}{40}$ **17.** $\dfrac{81}{125}$ **19.** $\dfrac{57}{80}$ **21.** $1\frac{19}{25}$ **23.** $3\frac{2}{5}$ **25.** 12
27. $\dfrac{13}{600}$ **29.** $\dfrac{1}{8}$ **31.** $\dfrac{11}{125}$ **33.** $\dfrac{7}{45}$ **35.** $\dfrac{13}{550}$ **37.** 75%   **39.** 33.33%   **41.** 35%   **43.** 28%   **45.** 27.5%   **47.** 41.67%   **49.** 360%
51. 283.33%   **53.** 412.5%   **55.** 42.86%   **57.** 193.75%   **59.** 52%   **61.** 2.5%   **63.** 1.17%   **65.** $83\frac{1}{3}\%$ **67.** $35\frac{5}{7}\%$ **69.** $37\frac{1}{2}\%$

71. $7\frac{1}{2}\%$

	Fraction	Decimal	Percent
73.	$\dfrac{11}{12}$	0.9167	91.67%
75.	$\dfrac{14}{25}$	0.56	56%
77.	$\dfrac{3}{200}$	0.015	1.5%
79.	$\dfrac{5}{9}$	0.5556	55.56%
81.	$\dfrac{1}{32}$	0.0313	$3\dfrac{1}{8}\%$

83. $\dfrac{463}{1600}$ **85.** 15.375%

87. It will have at least two zeros to the right of the decimal point.

89. $n = 5.625$ **90.** $n = 4$ **91.** 549,165 documents

5.3A Exercises

1. $n = 5\% \times 90$ **3.** $70\% \times n = 2$ **5.** $17 = n \times 85$ **7.** 28 **9.** 912 **11.** $6.39 **13.** 56 **15.** 1300 **17.** $150 **19.** 50% **21.** 28%
23. 65% **25.** 31 **27.** 85 **29.** 12% **31.** 3.28 **33.** 64% **35.** 445 **37.** 35% **39.** 39.6 **41.** 80% **43.** 85% **45.** 554 students
47. 40 years **49.** 57.60 **51.** width 120 feet; length 400 feet **53.** 2.448 **54.** 14.892 **55.** 2834 **56.** 2917

5.3B Exercises

	p	b	a
1.	75	660	495
3.	22	60	a
5.	49	b	2450
7.	p	50	30

9. 36 **11.** 196 **13.** 56 **15.** 80 **17.** 80 **19.** 600,000 **21.** 20 **23.** 4 **25.** 27 **27.** 80 **29.** 16.4% **31.** 3.64 **33.** 25%
35. 170 **37.** $2280 **39.** 15% **41.** 20% **43.** $4000 per month **45.** 45.3% **47.** 9.7% **49.** $1\dfrac{31}{45}$ **50.** $\dfrac{1}{26}$ **51.** $4\dfrac{1}{5}$ **52.** $1\dfrac{13}{15}$

5.4 Exercises

1. 180,000 pencils **3.** $70 **5.** 9% **7.** 333 people **9.** $240 **11.** 25% **13.** $9,600,000 **15.** 12.62% **17.** 216 babies **19.** $55
21. $150,000 **23.** 2200 pounds **25.** $12,210,000 for personnel, food and decorations; $20,790,000 for security, facility rental, and all other
expenses **27.** $123.50 **29.** $30 **31. (a)** $1280 **(b)** $14,720 **33. (a)** $333 **(b)** $777 **35.** 1,698,000 **36** 2,452,400 **37.** 1.63
38. 0.793 **39.** 0.0556 **40.** 0.0792

5.5 Exercises

1. $3400 **3.** $4140 **5.** 52% **7.** 60% **9.** $140 **11.** $7.50 **13.** $480 **15.** 0.6% **17.** $1,600,000 **19.** $39.75 **21.** 39 boxes
23. 800 parts **25. (a)** $118.40 **(b)** $3818.40 **27. (a)** $6.96 **(b)** $122.96 **29.** $849.01 **31.** 2 **32.** 120 **33.** $\dfrac{11}{24}$ **34.** 3.64

Putting Your Skills to Work

1. 13.4% of potassium; 13.3% of total carbohydrates **2.** 21.1% of potassium; 19.6% of total carbohydrates

Chapter 5 Review Problems

1. 62% **2.** 43% **3.** 37.2% **4.** 52.9% **5.** 8.28% **6.** 7.19% **7.** 252% **8.** 437% **9.** 103.6% **10.** 105.2% **11.** 0.6% **12.** 0.2%
13. 0.52% **14.** 0.67% **15.** $4\dfrac{1}{12}\%$ **16.** $3\dfrac{5}{12}\%$ **17.** 317% **18.** 225% **19.** 76% **20.** 52% **21.** 55% **22.** 22.5% **23.** 45.45%
24. 44.44% **25.** 225% **26.** 375% **27.** 442.86% **28.** 555.56% **29.** 190% **30.** 183.33% **31.** 0.38% **32.** 0.63% **33.** 0.008
34. 0.006 **35.** 0.827 **36.** 0.596 **37.** 2.36 **38.** 1.77 **39.** 0.32125 **40.** 0.26375 **41.** $\dfrac{18}{25}$ **42.** $\dfrac{23}{25}$ **43.** $\dfrac{37}{20}$ **44.** $\dfrac{9}{4}$ **45.** $\dfrac{41}{250}$
46. $\dfrac{61}{200}$ **47.** $\dfrac{5}{16}$ **48.** $\dfrac{7}{16}$ **49.** $\dfrac{1}{2000}$ **50.** $\dfrac{3}{5000}$

	Fraction	Decimal	Percent
51.		0.6	60%
52.		0.875	87.5%
53.	$\frac{3}{8}$	0.375	
54.	$\frac{9}{16}$	0.5625	
55.	$\frac{1}{125}$		0.8%
56.	$\frac{9}{20}$		45%

57. 332   **58.** 405   **59.** 90   **60.** 175   **61.** 40%   **62.** 40%   **63.** 97.2
64. 99.2   **65.** 160   **66.** 140   **67.** 20%   **68.** 58%   **69.** 51 students
70. 96 trucks   **71.** $11,200   **72.** $80,200   **73.** 60%   **74.** 7.5%
75. 4.09%   **76.** $558   **77.** $1800   **78.** 4%   **79.** 6%   **80.** $1200
81. (a) $319   **(b)** $1276   **82. (a)** $255   **(b)** $1870
83. (a) $3360   **(b)** $20,640   **84. (a)** $330   **(b)** $1320   **85. (a)** $60   **(b)** $720

Chapter 5 Test

1. 57%   **2.** 1%   **3.** 0.8%   **4.** 13.9%   **5.** 356%   **6.** 71%   **7.** 1.8%   **8.** $3\frac{1}{7}$%   **9.** 47.5%   **10.** 40%   **11.** 300%   **12.** 175%
13. 17.13%   **14.** 302.4%   **15.** $1\frac{13}{25}$ **16.** $\frac{31}{400}$ **17.** 26.69   **18.** 130   **19.** 55.56%   **20.** 200   **21.** 5000   **22.** 46%   **23.** 699.6
24. 2.29%   **25.** $6092   **26. (a)** $150.81   **(b)** $306.19   **27.** 89.29%   **28.** 8.93%   **29.** 12,000 registered voters   **30. (a)** $240   **(b)** $960

Cumulative Test for Chapters 1–5

1. 2241   **2.** 8444   **3.** 5292   **4.** 89   **5.** $\frac{67}{12}$ or $5\frac{7}{12}$ **6.** $\frac{1}{12}$ **7.** $\frac{35}{12}$ or $2\frac{11}{12}$ **8.** $\frac{5}{21}$ **9.** 5731.7   **10.** 34.118   **11.** 1.686   **12.** 0.368
13. $\frac{3 \text{ pounds}}{5 \text{ square feet}}$ **14.** yes   **15.** $n = 24$ **16.** 673 faculty members   **17.** 2.3%   **18.** 46.8%   **19.** 198%   **20.** 3.75%   **21.** 2.43
22. 0.0675   **23.** 17.76%   **24.** 114.58   **25.** 300   **26.** 718.2   **27.** $8370   **28.** 3200 students   **29.** 11.31%   **30.** $352

Chapter 6

Pretest Chapter 6

1. 228   **2.** 40   **3.** 5280   **4.** 6400   **5.** 1320   **6.** 24   **7.** 6750   **8.** 7390   **9.** 98.6   **10.** 0.027   **11.** 529.6   **12.** 0.482
13. 2376 m   **14.** 94.262 m   **15.** 5660   **16.** 7.835   **17.** 0.0563   **18.** 4800   **19.** 0.568   **20.** 8900   **21.** 5.52   **22.** 1.28   **23.** 59.52
24. 1826.78   **25.** 39.69   **26.** 118.8   **27.** 55.8 feet   **28. (a)** 95°F   **(b)** no   **29.** 12.2 miles   **30.** 22.5 gal/hr

6.1 Exercises

1. We know that each mile is 5280 feet. Each foot is 12 inches. So we know that one mile is $5280 \times 12 = 63,360$ inches. The unit fraction we want is $\frac{63,360 \text{ inches}}{1 \text{ mile}}$. So we multiply 23 miles $\times \frac{63,360 \text{ inches}}{1 \text{ mile}}$. The mile unit divides out. We obtain 1,457,280 inches. Thus 23 miles = 1,457,280 inches.
3. 1760   **5.** 2000   **7.** 4   **9.** 2   **11.** 7   **13.** 9   **15.** 2   **17.** 12,320   **19.** 144   **21.** 128   **23.** 6.25   **25.** 12   **27.** 6   **29.** 36
31. 28   **33.** 12   **35.** 62   **37.** 64   **39.** 84   **41.** 64,800   **43.** $9.75   **45.** 138,336 feet   **47.** 2665 seconds   **49. (a)** 144 inches
(b) $122.40   **51.** 28,800 cups   **53.** 14,739 land miles   **55.** $10,800   **56.** 22%   **57.** 161 miles   **58.** 104 students

6.2 Exercises

1. hecto-   **3.** deci-   **5.** kilo-   **7.** 460   **9.** 5200   **11.** 1.67   **13.** 0.0732   **15.** 200,000   **17.** 0.078   **19.** 3.5; 0.035   **21.** 3.582; 0.003582
23. a   **25.** c   **27.** b   **29.** 39   **31.** 1.98   **33.** 0.482   **35.** 3255 m   **37.** 463 cm   **39.** 63.5 cm
41. 2.5464 cm *or* 25.464 mm   **43.** 695.2 cm   **45.** 939.86 m   **47.** 0.964   **49.** false   **51.** true   **53.** true   **55.** true
57. (a) 481,800 centimeters   **(b)** 4.818 kilometers   **59.** 0.00000000254   **61.** 5,040,000 metric tons   **63.** 18,560,000,000 kilograms
65. approximately 113,020,000 metric tons   **66.** 5000   **67.** 1.77

6.3 Exercises

1. 1 kL   **3.** 1 mg   **5.** 1 g   **7.** 58,000   **9.** 2600   **11.** 0.0189   **13.** 0.752   **15.** 2,430,000   **17.** 82   **19.** 0.005261   **21.** 74,000
23. 1.62   **25.** 0.035   **27.** 6.328   **29.** 2920   **31.** 17,000   **33.** 0.007; 0.000007   **35.** 0.128; 0.000128   **37.** 33; 33,000   **39.** 2580; 2,580,000
41. (b)   **43. (a)**   **45.** 113.922 L   **47.** 1520.052 g   **49.** true   **51.** false   **53.** false   **55.** true   **57.** $108.75   **59.** $340,000
61. $10,102,500   **63.** 0.005632   **65.** 2050 kg   **67.** approximately $9195.40; $9.20   **69.** 1997   **71.** 20%   **72.** 57.5   **73.** $321.30
74. $716.80

6.4 Exercises

1. 2.14 m **3.** 22.86 cm **5.** 15.26 yd **7.** 28.15 m **9.** 132.02 km **11.** 4.88 m **13.** 6.90 in. **15.** 218 yd **17.** 165.1 cm **19.** 1061.2 L **21.** 21.76 L **23.** 5.02 gal **25.** 4.77 qt **27.** 59.02 kg **29.** 4.45 oz **31.** 41.37 ft **33.** 1173 ft/sec **35.** 273 mi/hr **37.** 0.51 in. **39.** 185°F **41.** 53.6°F **43.** 60°C **45.** 4.44°C **47.** 143.87 km **49.** 18.85 liters **51.** 1397 pounds **53.** 66.2°F at 4:00 A.M.; 113°F after 7:00 A.M. **55.** 59,861 miles **57.** 180.6448 sq cm

59. $896 for the American carpet; $802 for the German carpet; the German carpet is $94 cheaper **61.** 47 **62.** 49 **63.** $\frac{13}{40}$ **64.** $\frac{7}{12}$

Putting Your Skills to Work

1. 37.4 miles per hour **2.** 32.2 miles per hour

6.5 Exercises

1. 3 ft **3.** 11 yd **5.** $33.46 **7.** 310 cm per piece **9.** 131.2 feet **11.** 2.38 g **13.** The discrepancy is 4.4°F. **15.** The difference is 6°F. The temperature reading of 180°C is hotter. **17.** (a) about 105 kilometers per hour (b) Probably not. We cannot be sure, but we have no evidence to indicate that they broke the speed limit. **19.** 15 gallons **21.** $102 **23.** 311.85 g **25.** (a) 4.34 quarts extra (b) $33.70 **27.** (a) $5.46 (b) about 132 miles per gallon **29.** yes; 240 gal/hr is equivalent to $533\frac{1}{3}$ pt/sec **31.** $n = 0.64$ **32.** (a) 26,538 meters (b) 87,045 feet **33.** 15.5 miles **34.** 41.25 yards

Chapter 6 Review Problems

1. 11 **2.** 9 **3.** 8800 **4.** 10,560 **5.** 7.5 **6.** 6.5 **7.** 3 **8.** 2 **9.** 14,000 **10.** 8000 **11.** 5.75 **12.** 6.25 **13.** 60 **14.** 84 **15.** 15.5 **16.** 13.5 **17.** 560 **18.** 290 **19.** 176.3 **20.** 259.8 **21.** 920 **22.** 740 **23.** 9000 **24.** 8000 **25.** 7.93 m **26.** 17.01 m **27.** 35.63 m **28.** 89.59 m **29.** 17,000 **30.** 23,000 **31.** 196,000 **32.** 721,000 **33.** 0.778 **34.** 0.459 **35.** 76,000 **36.** 41,000 **37.** 765 **38.** 423 **39.** 2430 **40.** 1930 **41.** 92.4 **42.** 2.75 **43.** 368.55 **44.** 39.62 **45.** 5.52 **46.** 7.09 **47.** 9.08 **48.** 13.62 **49.** 10.97 **50.** 12.80 **51.** 49.6 **52.** 43.4 **53.** 59 **54.** 77 **55.** 105 **56.** 85 **57.** 0 **58.** 3.12 cm **59.** (a) 17 ft **(b)** 204 in. **60.** (a) 200 m (b) 0.2 km **61.** $2.22 **62.** yes; 1.672 feet extra **63.** 32.6 miles **64.** 25°F too hot **65.** 380 centimeters **66.** approximately 25.2 ft/sec **67.** 49.6 mi/hr **68.** they are carrying 16.28 pounds; they are slightly over the weight limit **69.** about $3.30 **70.** yes **71.** approximately 516.4 square feet **72.** approximately $2.24

Chapter 6 Test

1. 3200 **2.** 228 **3.** 84 **4.** 7 **5.** 30 **6.** 0.75 **7.** 9200 **8.** 0.273 **9.** 0.0988 **10.** 4.6 **11.** 1270 **12.** 9.36 **13.** 0.046 **14.** 127,000 **15.** 0.0289 **16.** 0.983 **17.** 920 **18.** 9420 **19.** 67.62 **20.** 1.63 **21.** 3.55 **22.** 10.03 **23.** 16.06 **24.** 85.05 **25.** (a) 20 meters (b) 21.8 yards **26.** (a) 15°F (b) yes **27.** 82.5 gal/hr **28.** (a) 300 km (b) 14 miles **29.** −27°C **30.** 104°F

Cumulative Test for Chapters 1–6

1. 6028 **2.** 185,440 **3.** 69 **4.** $\frac{19}{42}$ **5.** $1\frac{3}{8}$ **6.** yes **7.** $n = 6$ **8.** 209.23 grams **9.** 250% **10.** 1950 **11.** 20,000 **12.** 9.5 **13.** 5000 **14.** 3.5 **15.** 300 **16.** 3700 **17.** 0.0628 **18.** 790 **19.** 0.05 **20.** 672 **21.** 106.12 **22.** 43.58 **23.** 25.81 **24.** 14.49 **25.** 11.88 meters **26.** 59°F; the difference is 44°F; the 15°C temperature is higher **27.** 7 miles **28.** 1.738 centimeters

Chapter 7

Pretest Chapter 7

1. 18° **2.** 63° **3.** $\angle b = 136°$, $\angle a = \angle c = 44°$ **4.** 18 m **5.** 14 m **6.** 23 sq cm **7.** 2.4 sq cm **8.** 25.6 yd **9.** 78 ft **10.** 351 sq in. **11.** 171 sq in. **12.** 97 sq m **13.** 23° **14.** 15.3 m **15.** 72 sq m **16.** 8 **17.** 12 **18.** 6.782 **19.** 8 ft **20.** 12 ft **21.** 28 in. **22.** 94.2 cm **23.** 254.3 sq m **24.** 27.4 sq m **25.** 240 cu yd **26.** 113 cu ft **27.** 1846.3 cu in. **28.** 4375 cu m **29.** 1130.4 cu m **30.** $n = 120$ cm **31.** $n = 28$ m **32.** (a) 3706.5 sq yd (b) $815.43

7.1 Exercises

1. An acute angle is an angle whose measure is between 0° and 90°. **3.** Complementary angles are two angles that have a sum of 90°. **5.** When two lines intersect, the two angles that are opposite each other are called vertical angles. **7.** A transversal is a line that intersects two or more other lines at different points. **9.** $\angle ABD, \angle CBE$ **11.** $\angle ABD$ and $\angle CBE$; $\angle DBC$ and $\angle ABE$ **13.** There are no complementary angles. **15.** 50° **17.** 60° **19.** 60° **21.** 35° **23.** 70° **25.** 59° **27.** 53° **29.** 16° **31.** 44° **33.** 45° **35.** $\angle b = 102°$, $\angle a = \angle c = 78°$ **37.** $\angle b = 38°$, $\angle a = \angle c = 142°$ **39.** $\angle a = \angle c = 48°$, $\angle b = 132°$ **41.** $\angle e = \angle d = \angle a = 123°$, $\angle b = \angle c = \angle f = \angle g = 57°$ **43.** 59° **45.** 49° north of east **47.** 9.7 miles **48.** 21.1 miles **49.** $3800 **50.** 55.3 tons

7.2 Exercises

1. perpendicular; equal **3.** multiply **5.** 15 mi **7.** 23.6 ft **9.** 34 ft **11.** 1.92 mm **13.** 17.12 km **15.** 14.4 ft or 172.8 in. **17.** 0.272 mm **19.** 31.84 cm **21.** 94 m **23.** 180 cm **25.** 0.288 m² **27.** 117 yd² or 1053 ft² **29.** 294 m² **31.** $132,000 **33.** $59.40 **35.** $598.22 **37.** 546 in.² **39.** 223.3 **40.** 7.18 **41.** 21,842.8 **42.** approximately 1.5759 **43.** (a) 800\frac{1}{4}$ (b) 776\frac{3}{16}$

7.3 Exercises

1. adding **3.** perpendicular **5.** 40.2 m **7.** 29.8 in. **9.** 301.75 m^2 **11.** 3528 yd^2 **13.** $P = 48$ m; $A = 72$ m^2 **15.** $P = 9.6$ ft; $A = 3.6$ ft^2
17. 82 m **19.** 500 cm **21.** 118.8 yd^2 **23.** 550 km^2 **25.** 718 m^2 **27.** 357 ft^2 **29.** $80,960 **31.** 5.2415 $\times$ b^2 sq units **33.** 30 **34.** 3
35. 1800 **36.** 2.6

7.4 Exercises

1. right **3.** Add the measures of the two known angles and subtract that value from 180°. **5.** You could conclude that the lengths of all three
sides of the triangle are equal. **7.** true **9.** true **11.** false **13.** true **15.** 70° **17.** 128° **19.** 104 m **21.** 116.75 in.
23. 10.5 mi **25.** 15.75 in.^{2} **27.** 83.125 cm^2 **29.** 17.5 yd^2 or 157.5 ft^2 **31.** 29.75 ft^2 **33.** 188 yd^2 **35.** 3375 ft^2 **37.** $28,710 **39.** 6.25%

41. $n = 12$ **42.** $n = 42$ **43.** 716 **44.** 96 magazines **45.** 30$\frac{5}{6}$ feet

7.5 Exercises

1. $\sqrt{25} = 5$ because $(5)(5) = 25$. **3.** 32 is not a perfect square because no whole number when multiplied by itself equals 32. **5.** 5 **7.** 7
9. 11 **11.** 0 **13.** 10 **15.** 11 **17.** 3 **19.** 8 **21.** 22 **23. (a)** yes **(b)** 16 **25.** $\approx$ 4.243 **27.** $\approx$ 8.718 **29.** $\approx$ 13.928
31. $\approx$ 11.662 m **33.** 104.7 ft **35. (a)** 2 **(b)** 0.2 **(c)** Each answer is obtained from the previous answer by dividing by 10. **(d)** no; because
0.004 isn't a perfect square **37.** 39.299 **39.** 4800 sq in. **40.** 80,500 meters **41.** 0.92 m **42.** 0.0989 kg

7.6 Exercises

1. Square the length of each leg and add those two results. Then take a square root of the remaining number. **3.** 13.601 yd **5.** 15.199 ft
7. $\approx$ 6.403 m **9.** $\approx$ 9.899 m **11.** 6.928 yd **13.** 13 ft **15.** 9.8 cm **17.** 11.1 yd **19.** 4 in.; 6.9 in. **21.** 8.5 m **23.** 35.4 cm **25.** 7.1 in.
27. 5.39 ft **29.** 13.038 yd **31.** 341 m^2 **32.** 168 m^2 **33.** 441 in.^{2} **34.** 4224 yd^2

7.7 Exercises

1. circumference **3.** radius **5.** You need to multiply the radius by 2, and then use the formula $C = \pi d$. **7.** 58 in. **9.** 17 mm **11.** 22.5 yd
13. 9.92 cm **15.** 100.48 cm **17.** 116.18 in. **19.** 41.87 ft **21.** 78.5 yd^2 **23.** 803.84 cm^2 **25.** 200.96 ft^2 **27.** 11,304 mi^2 **29.** 276.32 m^2
31. 189.25 m^2 **33.** 30.96 m^2 **35.** $1211.20 **37.** 6.28 ft **39.** 256.43 ft **41.** 630.57 revolutions **43.** $226.08
45. (a) $0.75 per slice, 22.1 in.^{2} **(b)** $0.67 per slice, 18.84 in.^{2} **(c)** For 12-in. pizza, it is $0.035 per in.^{2}; for 15-in. pizza, it is $0.034 per in.^{2}; the 15-inch
pizza is a better value **47.** 13.92 **48.** 0.3 **49.** 6000 **50.** 3000

7.8 Exercises

1. box **3.** sphere **5.** pyramid **7.** 700 mm^3 **9.** 87.9 m^3 **11.** 6358.5 m^3 **13.** 3052.1 yd^3 **15.** $\approx$ 718.0 m^3 **17.** 1017.4 cm^3
19. 452.2 ft^3 **21.** 163.3 m^3 **23.** 373.3 m^3 **25.** $\approx$ 93.3 yd^3 **27.** 1004.8 in.^{3} **29.** 381,251,976,256,667 mi^3 **31.** 2928 in.^{3} **33.** $942

35. 263,900 yd^3 **37.** Many possibilities. **39.** 9$\frac{7}{12}$ **40.** 6$\frac{3}{8}$ **41.** $\frac{135}{16}$ or 8$\frac{7}{16}$ **42.** $\frac{25}{14}$ or 1$\frac{11}{14}$

Putting Your Skills to Work

1. (a) 99 in., 607.5 in.^{2} **(b)** 95 in., 562.5 in.^{2} **(c)** 45 in.^{2} **2. (a)** 5,800,000 pounds **(b)** 828,571 pounds **3. (a)** 20,000,000 pounds
(b) 54,795 pounds **4.** 773,776 pounds

7.9 Exercises

1. size; shape **3.** sides **5.** 8 m **7.** 2.6 ft **9.** 1.9 yd **11.** a corresponds to f, b corresponds to e, c corresponds to d **13.** 3.4 m
15. 2.1 ft **17.** 75 in. **19.** 36 ft **21.** 81 ft **23.** 8.3 ft **25.** 16.3 cm **27.** 11.6 yd^2 **29.** 12 **30.** 31 **31.** 42 **32.** 62

7.10 Exercises

1. (a) 75 km/hr **(b)** 76 km/hr **(c)** through Woodville and Palermo **3.** 34.5 min **5.** $510 **7.** $74.42 **9. (a)** 40,820 km **(b)** 20,410 km/hr
11. 50,240 in.^{3} **13.** $3293.45 **15.** 1465.33 feet per minute **17.** 128 **18.** 915 **19.** 0.25 **20.** 4.87

Chapter 7 Review Problems

1. 14° **2.** 168° **3.** $\angle b = 146°, \angle a = \angle c = 34°$ **4.** $\angle t = \angle x = \angle y = 65°, \angle s = \angle u = \angle w = \angle z = 115°$ **5.** 19.8 m **6.** 9.6 yd
7. 16.5 cm^2 **8.** 51.8 in.^{2} **9.** 38 ft **10.** 58 ft **11.** 68 m^2 **12.** 63.5 m^2 **13.** 145.2 m **14.** 62 mi **15.** 2700 m^2 **16.** 720 yd^2 **27.** 9
17. 422 cm^2 **18.** 357 m^2 **19.** 22 ft **20.** 14 ft **21.** 153° **22.** 47° **23.** 52.3 m^2 **24.** 59.4 m^2 **25.** 450 m^2 **26.** 87 m^2 **27.** 9
28. 8 **29.** 11 **30.** 6 **31.** 5 **32.** 5.916 **33.** 6.708 **34.** 12.845 **35.** 13.416 **36.** 5 m **37.** 5 yd **38.** 8.25 cm **39.** 8.06 m
40. 6.4 cm **41.** 9.2 ft **42.** 6.3 ft **43.** 3.6 ft **44.** 106 cm **45.** 63 cm **46.** 44 in. **47.** 56.5 in. **48.** 254.3 m^2 **49.** 201.0 ft^2
50. 226.1 in.^{2} **51.** 201.0 m^2 **52.** 318.5 ft^2 **53.** 126.1 m^2 **54.** 107.4 ft^2 **55.** 80.1 m^2 **56.** 45 ft^3 **57.** 7.2 ft^3 **58.** 307.7 m^3
59. 1728 m^3 **60.** 245 m^3 **61.** 3768 ft^3 **62.** 9074.6 yd^3 **63.** 30 m **64.** 3.3 m **65.** 324 cm **66.** 175 ft **67.** 147 sq yd **68.** $736
69. $V \approx$ 2034.7 in.^{3}; $W = 32,555.2$ g **70. (a)** 21,873.2 ft^3 **(b)** $\approx$ 17,498.6 bushels **71. (a)** 50 km; 100 km/hr **(b)** 56 km; 70 km/hr
(c) through Ipswich **72.** 3,429,708,000 ft^3 **73.** 17.1 ft **74.** 102 ft; $714 **75.** 34 fence posts **76.** 30.5 in.^{2} **77.** $\approx$ 521.0 yd^2 **78.** 26,048
79. 13.7 ft **80.** 381,510 m^3 **81.** $\approx$ 8.8 ft^3 **82.** $\approx$ 66 gallons **83.** 1662.5 ft^2 **84.** $2493.75

Chapter 7 Test

1. 40 yd **2.** 25.2 ft **3.** 20 m **4.** 80 m **5.** 137 m **6.** 180 yd^2 **7.** 104.0 m^2 **8.** 78 m^2 **9.** 144 m^2 **10.** 12 cm^2 **11.** 52.5 m^2
12. 9 **13.** 11 **14.** 27° **15.** 73° **16.** $\approx$ 7.348 **17.** $\approx$ 11.619 **18.** $\approx$ 13.675 **19.** 9.220 **20.** 10 **21.** 5.83 cm **22.** 9 ft
23. 37.7 in. **24.** 254.3 ft^2 **25.** 107.4 in.^{2} **26.** 144.3 in.^{2} **27.** 840 m^3 **28.** 803.8 m^3 **29.** 113.04 m^3 **30.** 508.7 ft^3 **31.** 56 m^3
32. 43.2 m **33.** 46.7 ft **34.** 6456 yd^2 **35.** $2582.40

Cumulative Test for Chapters 1–7

1. 935,760 **2.** 33,415 **3.** $\frac{26}{45}$ **4.** $\frac{4}{21}$ **5.** 56.13 **6.** 7.2272 **7.** 83 **8.** $n = 27$ **9.** 800 students **10.** 75% **11.** 2000 **12.** 40.7 **13.** 5.86 m **14.** 1512 in. **15.** 54.56 mi **16.** 50 m **17.** 208 cm **18.** 56.5 yd **19.** 1.4 cm^2 **20.** 540 m^2 **21.** 192 m^2 **22.** 664 yd^2 **23.** 50.2 m^2 **24.** 2411.5 m^3 **25.** 3052.1 cm^3 **26.** 3136 cm^3 **27.** 2713.0 m^3 **28.** 33.4 m **29.** 4.1 ft **30. (a)** 124 yd^2 **(b)** \$992.00 **31.** 11 **32.** ≈ 7.550 **33.** 10.440 in. **34.** 4.899 m **35.** 13.9 mi **36.** 32 paintbrushes

Chapter 8

Pretest Chapter 8

1. under age 18 **2.** 43% **3.** 17% **4.** 1650 students **5.** 350 students **6.** 300 housing starts **7.** 450 housing starts **8.** 2nd quarter of 2000 **9.** 3rd and 4th quarters of 2001 **10.** 300 **11.** 150 **12.** August and December **13.** December **14.** November **15.** 20,000 sets **16.** 30,000 sets **17.** 55,000 cars **18.** 60,000 cars **19.** 25,000 cars **20.** 20,000 cars **21.** 35 pages **22.** 35 pages **23.** 27 pages

8.1 Exercises

1. rent **3.** \$200 **5.** \$800 **7.** $\frac{13}{4}$ **9.** $\frac{10}{27}$ **11.** hit batters **13.** 294 **15.** 379 pitches **17.** $\frac{72}{325}$ **19.** $\frac{49}{24}$ **21.** 61% **23.** 25% **25.** 99,000,000 people **27.** 2,160,000,000 **29.** 39% **31.** 80% **33.** 1.2% **35.** 42 in.^{2} **36.** 204 in.^{2} **37.** 16 gal **38.** about 3 g

8.2 Exercises

1. 20 million people **3.** 14 million people **5.** 1960–1970 **7.** 22 quadrillion Btu **9.** 1970 **11.** 4 quadrillion Btu **13.** 6 quadrillion Btu **15.** from 1975 to 1980 and from 1985 to 1990 **17.** 30 quadrillion Btu **19.** \$3.5 million **21.** 1999 **23.** \$1 million **25.** 2.5 in. **27.** October, November, and December **29.** 1.5 in.

31.

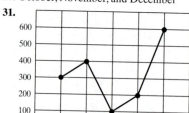

33. 35 **34.** 115 **35.** $\frac{2}{9}$

Putting Your Skills to Work

1. 1,250,000 **2.** 6,869,000 **3.** 2030 to 2040 **4.** 2030 to 2040

8.3 Exercises

1. 10 cars **3.** 35 cars **5.** 45 cars **7.** 145 cars **9.** 8,000 **11.** books costing \$5.00–\$7.99 **13.** 28,000 books **15.** 52,000 books **17.** 28.6%

	Tally	Frequency
19.	|||	3
21.	ℍℍ		6
23.	|||	3
25.	||	2

27.

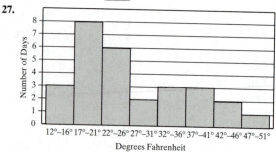

29. 6 days

	Tally	Frequency
31.	|||	3
33.	||	2
35.			1
37.	||	2

39. 12 prescriptions **41.** $n = 59.5$ **42.** $n = 7$ **43.** 12 pounds **44.** 13 in.

8.4 Exercises

1. The median of a set of numbers when they are arranged in order from smallest to largest is that value that has the same number of values above it as below it. The mean is most likely to be not typical of the values you would expect if there are many extremely low values or many extremely high values. The median is more likely to be typical of the value you would expect.   **3.** 32   **5.** 2.6 hours   **7.** 109,200   **9.** 0.375   **11.** 23.7 mi/gal   **13.** 1011   **15.** 0.58   **17.** $20,250   **19.** 29 years old   **21.** $207   **23.** 2.2   **25.** $69,161.88   **27.** 4850   **29.** mean = 70, median = 93   **31. (a)** $2157   **(b)** $1615   **(c)** The median, because the mean is affected by the high amount $6300.   **33.** 60   **35.** 121 and 150   **37.** $249   **39.** mean = 751, median = 855, mode = 869   **41.** 3.0   **43.** 19.3 in.^{2} **44.** 20,096 gallons per hour   **45.** $330   **46.** 62.8 ft^3

Chapter 8 Review Problems

1. 13 computers   **2.** 32 computers   **3.** 68 computers   **4.** 27 computers   **5.** $\frac{13}{21}$ **6.** $\frac{43}{32}$ **7.** ≈17.9%   **8.** ≈22.9%   **9.** 64%   **10.** 84%   **11.** 6%   **12.** blue   **13.** purple   **14.** 165 students   **15.** 470 students   **16.** 155 students   **17.** 36   **18.** 26   **19.** 10–13 years   **20.** 6–9 years   **21.** 12 glasses   **22.** 22 glasses   **23.** $\frac{5}{1}$ **24.** $\frac{2}{1}$ **25.** $31,000   **26.** $58,000   **27.** between 1985 and 1990   **28.** between 1980 and 1985   **29.** $9000   **30.** $15,000   **31.** 2000   **32.** 1980   **33.** $8750   **34.** $6250   **35.** $45,000   **36.** $72,000   **37.** 400 students   **38.** 500 students   **39.** 650 students   **40.** 450 students   **41.** 100 students   **42.** 50 students   **43.** 1999–2000 students   **44.** 2000–2001 students   **45.** 45,000 cones   **46.** 30,000 cones   **47.** 10,000 cones   **48.** 30,000 cones   **49.** 25,000 cones   **50.** 30,000 cones   **51.** The sharp drop in the number of ice cream cones purchased from July 2000 to August 2000 is probably directly related to the weather. Since August was cold and rainy significantly fewer people wanted ice cream during August.   **52.** The sharp increase in the number of ice cream cones purchased from June 2001 to July 2001 is probably directly related to the weather. Since June was cold and rainy and July was warm and sunny, significantly more people wanted ice cream during July.   **53.** 14,000,000   **54.** 18,000,000   **55.** between 1996 and 1998   **56.** between 1990 and 1992   **57.** 1,000,000   **58.** 12,000,000   **59.** 1992   **60.** 2000   **61.** 3,800,000   **62.** 2,200,000   **63.** 24,000,000   **64.** 49,000,000   **65.** 50 bridges   **66.** 25 bridges   **67.** 20, 39   **68.** 25 bridges   **69.** 150 bridges   **70.** 5 bridges

	Number of Defective Televisions (Class Intervals)	Tally	Frequency
71.	0–3	⫟⫟⫟⫟⫟ ⫟⫟⫟⫟⫟	10
72.	4–7	⫟⫟⫟⫟⫟ ⫟⫟⫟	8
73.	8–11	⫟⫟⫟	3
74.	12–15	⫟⫟⫟⫟⫟	5
75.	16–19	⫟⫟	2

76.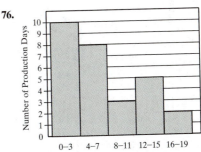

77. 18 times   **78.** 90°   **79.** $114   **80.** 1421 cars   **81.** 1353 employees   **82.** $36,000   **83.** $141,500   **84.** median = 19.5 cups, modes = 19 cups and 22 cups   **85.** median = 18.5 deliveries, mode = 15 deliveries   **86.** The median is better because the mean is skewed by the one low score, 31.   **87.** The median is better because the mean is skewed by the one high data item, 39.

Chapter 8 Test

1. 37%   **2.** 21%   **3.** 12%   **4.** 90,000 automobiles   **5.** 81,000 automobiles   **6.** 350 cars   **7.** 500 cars   **8.** 3rd quarter 2001   **9.** 1st quarter   **10.** 50 cars   **11.** 150 cars   **12.** 20 yr   **13.** 26 yr   **14.** 12 yr   **15.** age 35   **16.** age 65   **17.** 60,000 televisions   **18.** 25,000 televisions   **19.** 20,000 televisions   **20.** 60,000 televisions   **21.** 15.7   **22.** 16   **23.** 16

Cumulative Test for Chapters 1–8

1. 20,825   **2.** 74,296   **3.** $\frac{153}{40}$ or $3\frac{33}{40}$ **4.** $\frac{108}{35}$ or $3\frac{3}{35}$ **5.** 1796.43   **6.** 72.65   **7.** 72.23   **8.** $n = 0.6$ **9.** 39 cars   **10.** 0.325   **11.** 350   **12.** 1.98 m   **13.** 54 ft   **14.** 28.3 in.^{2} **15.** 68 in.   **16.** 34%   **17.** 3840 students   **18.** $3 million   **19.** $1 million   **20.** 16 in.   **21.** 1960 and 1970   **22.** 8 students   **23.** 16 students   **24.** $6.00   **25.** $4.95   **26.** $4.50

Chapter 9

Pretest Chapter 9

1. −19   **2.** −4   **3.** 4.5   **4.** 0   **5.** $-\frac{1}{3}$ **6.** $-\frac{7}{6}$ or $-1\frac{1}{6}$ **7.** −7   **8.** 1.7   **9.** −8   **10.** −41   **11.** $\frac{14}{17}$ **12.** −12   **13.** 1.4   **14.** −2.8   **15.** 42   **16.** $\frac{19}{15}$ or $1\frac{4}{15}$ **17.** 24   **18.** 4   **19.** −8   **20.** −10   **21.** −24   **22.** $\frac{15}{16}$ **23.** −64   **24.** −10   **25.** −10   **26.** 31   **27.** −12   **28.** 4.8   **29.** 5   **30.** −29   **31.** $-\frac{1}{3}$ **32.** $\frac{1}{5}$ **33.** 8×10^4 **34.** 5×10^{-4} **35.** 1.28×10^8 **36.** 0.0067   **37.** 1,320,000   **38.** 0.000000015678

9.1 Exercises

1. First find the absolute value of each number. Then add those two absolute values. Use the common sign in the answer. **3.** -17 **5.** -7
7. 16.5 **9.** $\frac{17}{35}$ **11.** $-\frac{11}{12}$ **13.** 9 **15.** -5 **17.** 22 **19.** -2.8 **21.** $-\frac{2}{3}$ **23.** $\frac{5}{9}$ **25.** -22 **27.** -0.7 **29.** -20.3 **31.** 4
33. $-\frac{13}{21}$ **35.** 6.5 **37.** $-\frac{11}{2}$ or $-5\frac{1}{2}$ **39.** -4 **41.** 0 **43.** -6 **45.** $-\frac{53}{75}$ **47.** $-\$94,000$ **49.** $\$9000$ **51.** $-\$38,000$ **53.** $-18°F$
55. $-1°F$ **57.** $9°F$ **59.** loss of 1 yard (-1 yard) **61.** $-\$13$ **63.** $\$37.00$ **65.** 904.3 cubic feet **66.** 210 cubic meters

9.2 Exercises

1. -6 **3.** -17 **5.** -2 **7.** 24 **9.** -18 **11.** 3 **13.** -60 **15.** 556 **17.** -6.7 **19.** -6.5 **21.** -8.6 **23.** 32.94 **25.** 1
27. $-\frac{7}{6}$ or $-1\frac{1}{6}$ **29.** $-\frac{1}{6}$ **31.** $\frac{17}{63}$ **33.** 15 **35.** 0 **37.** 17 **39.** -22 **41.** -2 **43.** -13 **45.** 6121 feet **47.** $42°F$ **49.** $-13°F$
51. $-\$14,100$ **53.** $+\$37,200$ **55.** $\$13\frac{1}{2}$ **57.** The bank finds that a customer has $50 in his checking account. However, the bank must remove an erroneous debit of $80 from the customer's account. When the bank makes the correction, what will the new balance be? **59.** 5 **60.** 45
61. 28.26 square inches **62.** 10.5 square inches

9.3 Exercises

1. To multiply two numbers with the same sign, multiply the absolute values. The sign of the result is positive. **3.** 36 **5.** 60 **7.** -160 **9.** -1.5
11. 28.125 **13.** $-\frac{6}{35}$ **15.** $\frac{1}{23}$ **17.** -8 **19.** -8 **21.** 6 **23.** $-\frac{2}{3}$ **25.** $\frac{8}{7}$ or $1\frac{1}{7}$ **27.** -8.38 **29.** 6.2 **31.** 72 **33.** 192 **35.** 16
37. -30 **39.** 0 **41.** $-\frac{5}{12}$ **43.** The account will be debited $5.00. **45.** $-6.25°F$ **47.** 70 feet **49.** 34 **51.** 2 **53.** $+8$
55. 0; at par **57.** 0 **59.** b is negative **61.** 90 square inches **62.** 264 square meters

9.4 Exercises

1. 8 **3.** -12 **5.** -19 **7.** 0 **9.** -5 **11.** -6 **13.** -27 **15.** -102 **17.** -18 **19.** 1 **21.** 4 **23.** $-\frac{1}{2}$ **25.** -1 **27.** 0 **29.** 4
31. 123 **33.** $-\frac{7}{16}$ **35.** $\frac{4}{25}$ **37.** 6.84 **39.** $-15.7°F$ **41.** $-8.3°F$ **43.** $9°F$ **45.** $-9°C; 3°C$ **47.** 3.84 kilometers
48. $36,800$ milligrams **49.** 420 square inches **50.** 98 catalogs; 120 pieces of junk mail

9.5 Exercises

1. 1.2×10^2 **3.** 1×10^4 **5.** 2.63×10^4 **7.** 2.88×10^5 **9.** 4.632×10^6 **11.** 1.2×10^7 **13.** 9.31×10^{-2} **15.** 2.79×10^{-3}
17. 4×10^{-1} **19.** 1.6×10^{-5} **21.** 5.31×10^{-6} **23.** 1.8×10^{-7} **25.** $53,600$ **27.** 0.062 **29.** 0.371 **31.** $900,000,000,000$
33. 0.00000003862 **35.** $4,600,000,000,000$ **37.** $67,210,000,000$ **39.** 5.878×10^{12} miles **41.** 6×10^9 **43.** $12,500,000,000,000$ insects
45. 0.000075 centimeter **47.** $14,000,000,000$ tons **49.** 9.01×10^7 dollars **51.** 1.068×10^{22} tons **53.** 3.624×10^8 feet
55. 6.0×10^6 square miles **57.** 6.174×10^{13} miles **59.** 16.6334 **60.** 0.258 **61.** $\$176$ **62.** 589 feet

Putting Your Skills to Work

1. 300 feet **2.** 780 liters

Chapter 9 Review Problems

1. -15 **2.** -14 **3.** -8.8 **4.** -8.8 **5.** $-\frac{8}{15}$ **6.** $\frac{1}{14}$ **7.** 6 **8.** -20 **9.** 11 **10.** 2 **11.** -15 **22.** 7 **13.** 19 **14.** 17
15. -1.6 **16.** -12.3 **17.** $-\frac{1}{15}$ **18.** $\frac{7}{12}$ **19.** 13 **20.** -3 **21.** -9 **22.** -12 **23.** $\frac{2}{35}$ **24.** $-\frac{1}{6}$ **25.** -7.8 **26.** 4.32 **27.** 3
28. 6 **29.** -9 **30.** -5 **31.** 6 **32.** -15 **33.** $-\frac{7}{10}$ **34.** $\frac{3}{7}$ **35.** 30 **36.** 48 **37.** 13 **38.** 19 **39.** -11 **40.** 1 **41.** -5
42. 23 **43.** -8 **44.** 12 **45.** -72 **46.** 1 **47.** $-\frac{9}{5}$ or $-1\frac{4}{5}$ **48.** $\frac{1}{10}$ **49.** 1 **50.** -35 **51.** 26 **52.** 21 **53.** $\frac{13}{21}$ **54.** $-\frac{7}{45}$
55. 5.76 **56.** 63.72 **57.** 4.16×10^3 **58.** 3.7×10^6 **59.** 2.18×10^5 **60.** 7.0×10^{-3} **61.** 2.18×10^{-5} **62.** 7.63×10^{-6} **63.** $18,900$
64. 3760 **65.** 0.0752 **66.** 0.00661 **67.** 0.0000009 **68.** 0.00000008 **69.** 0.000536 **70.** 0.0000198 **71.** 8.44×10^{11} **72.** 9.72×10^{15}
73. 1.44×10^{14} **74.** 6.8×10^{25} **75.** 1.2312×10^{14} drops **76.** 5.983×10^{24} kilograms **77.** 1.32×10^{13} feet **78.** 1.0×10^{-21}
79. 1.67×10^{-24} grams; 9.1×10^{-28} grams **80.** 0.000000000000001 **81.** $384,400,000$ meters **82.** total loss 6 yards **83.** 883 feet **84.** $\$2$
85. $-9.6°F$ **86.** 2 points above par

Chapter 9 Test

1. -11 **2.** -43 **3.** 3.9 **4.** -6 **5.** -3 **6.** $-\frac{7}{8}$ **7.** -38 **8.** 5 **9.** $\frac{17}{15}$ or $1\frac{2}{15}$ **10.** -43 **11.** 4 **12.** 2.1 **13.** $\frac{11}{12}$ **14.** 46
15. 120 **16.** -36 **17.** 10 **18.** -18 **19.** 3 **20.** $-\frac{7}{10}$ **21.** 56 **22.** -32 **23.** 17 **24.** 14 **25.** -8 **26.** -52 **27.** -7.2
28. -20 **29.** $-\frac{1}{7}$ **30.** $-\frac{1}{6}$ **31.** 8.054×10^4 **32.** 7×10^{-6} **33.** 0.0000936 **34.** $72,000$ **35.** $-6.2°F$ **36.** 2.72×10^{-4} meters

Cumulative Test for Chapters 1–9

1. 12,383   **2.** 127   **3.** $\frac{143}{12}$ or $11\frac{11}{12}$ **4.** $\frac{55}{12}$ or $4\frac{7}{12}$ **5.** 9.812   **6.** 63.46   **7.** 65.9968   **8.** $n = 64$ **9.** 126 defects   **10.** 0.304

11. 4000   **12.** 94,000 meters   **13.** 5 yards   **14.** 78.5 square meters   **15. (a)** 300 students   **(b)** 1100 students   **16.** 13   **17.** -4.7 **18.** $\frac{5}{12}$

19. -11 **20.** -5 **21.** -60 **22.** $\frac{6}{5}$ or $1\frac{1}{5}$ **23.** 18   **24.** 4   **25.** -2 **26.** $\frac{4}{15}$ **27.** 5.79863×10^5 **28.** 7.8×10^{-4} **29.** 38,500,000

30. 0.00007

Chapter 10

Pretest Chapter 10

1. $-17x$ **2.** y **3.** $-3a + 2b$ **4.** $-12x - 4y + 10$ **5.** $16x - 2y - 6$ **6.** $4a - 12b + 3c$ **7.** $42x - 18y$ **8.** $-3a - 15b + 3$

9. $-3a - 6b + 12c + 10$ **10.** $x - 8y$ **11.** $x = 37$ **12.** $x = 3.5$ **13.** $x = \frac{7}{8}$ **14.** $x = -8$ **15.** $x = 5$ **16.** $x = \frac{3}{2}$ or $1\frac{1}{2}$

17. $x = 7$ **18.** $x = \frac{8}{5}$ or $1\frac{3}{5}$ **19.** $x = 3$ **20.** $x = \frac{9}{4}$ or $2\frac{1}{4}$ **21.** $c = 9 + p$ **22.** $l = 2w + 5$

23. a = height of Mt. Ararat; $a - 1758$ = height of Mt. Hood   **24.** width = 19 meters; length = 35 meters   **25.** 7.75 feet; 10.25 feet

10.1 Exercises

1. A variable is a symbol, usually a letter of the alphabet, that stands for a number.   **3.** G, x, y **5.** p, a, b **7.** $r = 3m + 5n$ **9.** $H = 2a - 3b$

11. $10x$ **13.** $-x$ **15.** $\frac{1}{3}x$ **17.** $4x + 1$ **19.** $-1.1x + 6.4$ **21.** $-32x - 3$ **23.** $37x - 8y + 19$ **25.** $13a - 5b - 7c$

27. $-\frac{13}{12}x + \frac{8}{21}y$ **29.** $-3x - 6$ **31.** $-11.1n + 3.1m + 1.2$ **33.** $12x + 1$ **35.** $n = 4.8$ **36.** $n = 8.1$ **37.** $n = 3$ **38.** $n = 8.5$

10.2 Exercises

1. variable   **3.** $3x$ and x; $2y$ and $-3y$ **5.** $27x - 18$ **7.** $-2x - 2y$ **9.** $-10.5x + 21y$ **11.** $30x - 70y$ **13.** $4p + 36q - 40$

15. $\frac{3}{5}x + 2y - \frac{3}{4}$ **17.** $-180a + 33b + 100.5$ **19.** $32a + 48b - 36c - 20$ **21.** $-2.6x + 17y + 10z - 24$ **23.** $x - \frac{3}{2}y + 2z - \frac{1}{4}$

25. $p = 2l + 2w$ **27.** $A = \frac{hB + hb}{2}$ **29.** $27x - 39$ **31.** $14x + 12y - 28$ **33.** $8.1x + 8.1y$ **35.** $-9a - 19b + 12c$

37. $A = ab + ac$, $A = a(b + c)$ **39.** $A = xy + xw + xz$, $A = x(y + w + z)$ **41.** 37.7 inches   **42.** 201.0 square inches

43. 77 square centimeters   **44.** 301.4 cubic inches   **45.** *Nereid*, 7665 pounds; *Auriole*, 8775 pounds

10.3 Exercises

1. equation   **3.** opposite   **5.** $y = 14$ **7.** $x = 9$ **9.** $x = -18$ **11.** $x = -25$ **13.** $x = 19$ **15.** $9.8 = x$ **17.** $x = -8.7$

19. $x = 13.2$ **21.** $x = \frac{1}{2}$ **23.** $x = 1$ **25.** $x = -\frac{3}{2}$ or $-1\frac{1}{2}$ **27.** $y = -\frac{13}{15}$ **29.** $x = 14$ **31.** $x = -12$ **33.** $x = 2$

35. $x = -9$ **37.** $y = \frac{13}{2}$ or $6\frac{1}{2}$ **39.** $-8 = z$ **41.** $y = 1.2$ **43.** $x = 6$ **45.** To solve the equation $3x = 12$, divide both sides of the

equation by 3 so that x stands alone.   **46.** $3x + 3y + 3$ **47.** $-x + 21y + 3$ **48.** 4.8%   **49.** 32 cubic feet

10.4 Exercises

1. A sample answer is: To maintain the balance, whatever you do to one side of the scale, you need to do the exact same thing to the other side of

the scale.   **3.** $\frac{4}{3}$ **5.** $x = 9$ **7.** $y = -4$ **9.** $x = -\frac{16}{9}$ **11.** $x = 8$ **13.** $16 = m$ **15.** $x = 10$ **17.** $z = 1.8$ **19.** $x = -13.5$

21. $x = 9$ **23.** $y = 10$ **25.** $n = \frac{5}{4}$ or $1\frac{1}{4}$ **27.** $x = -\frac{8}{5}$ or $-1\frac{3}{5}$ **29.** $x = -\frac{9}{2}$ or $-4\frac{1}{2}$ **31.** $z = 8$ **33.** 18

35. $4x - 7y + 6$ **36.** 438.3%   **37.** 9.5%   **38.** 22,500 plates; 31,500 napkins; 18,000 towels

10.5 Exercises

1. You want to obtain the x-term all by itself on one side of the equation. So you want to remove the -6 from the left side of the equation. There-fore you would add the opposite of -6. This means you would add 6 to each side.

3. no   **5.** yes   **7.** $x = 3$ **9.** $x = -\frac{4}{9}$ **11.** $x = \frac{5}{6}$ **13.** $x = 2$ **15.** $x = 7$ **17.** $x = 5$ **19.** $y = -\frac{2}{3}$ **21.** $x = -1$ **23.** $y = -6$

25. $y = 7$ **27.** $x = -16$ **29.** $y = 5$ **31.** $x = -5$ **33.** $y = 11$ **35.** $x = -8$ **37.** $x = \frac{11}{2}$ **39.** $x = \frac{26}{3}$

41. (a) $x = 5$ **(b)** $5 = x$ **(c)** For most students, collecting x-terms on the left is easier, since we usually write answers with x on the left. But either

method is OK.   **43.** 407,513.4 cubic centimeters   **44.** 23.4 square inches   **45.** 190 dealers   **46.** 22 pieces

Putting Your Skills to Work

1. 6,191,000; 63.8%   **2.** 18,222,000; 56.1%

10.6 Exercises

1. $h = 34 + r$ **3.** $b = n - 107$ **5.** $d = e - 5$ **7.** $l = 2w + 7$ **9.** $s = 3f - 2$ **11.** $n = 3000 + 3s$ **13.** $j + s = 26$ **15.** $ht = 500$ **17.** j = Julie's mileage; $j + 386$ = Barbara's mileage **19.** b = number of degrees in angle B; $b - 46$ = number of degrees in angle A **21.** w = height of Mt. Whitney in meters; $w + 4430$ = height of Mt. Everest in meters **23.** l = Lisa's tips; $l + 12$ = Sam's tips; $l - 6$ = Brenda's tips **25.** w = width; $w + 7$ = height; $2w - 1$ = length **27.** x = first angle; $2x$ = second angle; $x - 14$ = third angle **29.** $1.08s$; $1.08(s + 10)$ **31.** 8 **32.** −1 **33.** from top to bottom in the table: 0.387; 311; 106

10.7 Exercises

1. x = length of shorter piece; $x + 5.5$ = length of longer piece; 5.25 feet; 10.75 feet **3.** x = amount Jack can leg-press; $x - 140$ = amount Jodie can leg-press 265 pounds; Jack can leg-press 405 pounds **5.** x = the number of cars in November; $x + 84$ = the number of cars in May; $x - 43$ = the number of cars in July; 119 cars in November; 203 cars in May; 76 cars in July **7.** x = length of shorter piece; $x + 4.7$ = length of the longer piece; the shorter piece is 3.65 feet long; the longer piece is 8.35 feet long **9.** x = width; $2x - 4$ = length; width is 14 inches; length is 24 inches **11.** x = length of the first side; $x + 20$ = length of the second side; $x - 4$ = length of the third side; 61 millimeters; 81 millimeters; 57 millimeters **13.** x = length of the first side; $2x$ = length of the second side; $x + 12$ = length of the third side; 8 centimeters; 16 centimeters; 20 centimeters **15.** x = number of degrees in angle A; $3x$ = number of degrees in angle B; $x + 40$ = number of degrees in angle C; angle A measures 28°; angle B measures 84°; angle C measures 68° **17.** x = total sales; \$76,000 **19.** x = yearly rent; \$6000 **21.** x = length of the adult section; $x + 6.2$ = length of the children section; the adult section is 12.9 feet; the children section is 19.1 feet **23.** first program = \$70.37; second program = \$120.74; third program = \$379.22 **25.** 60% **26.** 500 **27.** 117.4 meters

Chapter 10 Review Problems

1. $-13a + 3$ **2.** $\frac{7}{8}x + \frac{11}{12}$ **3.** $-2x - 7y$ **4.** $11x - 5y$ **5.** $-x - 12y + 6$ **6.** $2x - 10y - 7$ **7.** $-15x - 3y$ **8.** $-8x - 12y$ **9.** $2x - 6y + 8$ **10.** $6x - 18y - 3$ **11.** $-24a + 40b + 8c$ **12.** $-18a + 72b + 9c$ **13.** $6x + 15y - 27.5$ **14.** $8.4x - 12y + 20.4$ **15.** $-2x + 14y$ **16.** $7x - 8y$ **17.** $-8a - 2b - 24$ **18.** $-7a + 8b + 15$ **19.** $x = 12$ **20.** $x = 4$ **21.** $x = 11.7$ **22.** $x = 7.4$ **23.** $x = -12.1$ **24.** $x = 12.8$ **25.** $x = \frac{11}{4}$ or $2\frac{3}{4}$ **26.** $x = \frac{13}{4}$ or $3\frac{1}{4}$ **27.** $x = \frac{1}{8}$ **28.** $x = \frac{3}{2}$ or $1\frac{1}{2}$ **29.** $x = 5$ **30.** $x = -12$ **31.** $x = -\frac{5}{2}$ or $-2\frac{1}{2}$ **32.** $y = -5$ **33.** $x = 6$ **34.** $y = 7$ **35.** $x = -5$ **36.** $x = 0.25$ **37.** $x = 8$ **38.** $x = \frac{5}{6}$ **39.** $x = 1$ **40.** $x = 5$ **41.** $x = 8$ **42.** $x = 5$ **43.** $x = 3$ **44.** $x = -10$ **45.** $x = \frac{29}{5}$ **46.** $x = 5$ **47.** $y = 16$ **48.** $y = -28$ **49.** $w = c + 3000$ **50.** $a = m - 18$ **51.** $A = 3B$ **52.** $l = 2w - 3$ **53.** r = Roberto's salary; $r + 2050$ = Michael's salary **54.** x = length of first side; $2x$ = length second side **55.** c = number of Connie's courses; $c - 6$ = number of Nancy's courses **56.** b = number of books in old library; $2b + 450$ = number of books in new library **57.** x = length of the shorter piece; $x + 6.5$ = length of the longer piece; 26.75 feet; 33.25 feet **58.** x = the old employee's salary; $x - 28$ = the new employee's salary; \$192; \$220 **59.** x = number of customers in February; $2x$ = number of customers in March; $x + 3000$ = number of customers in April; Feb = 10,550; Mar. = 21,100; Apr. = 13,550 **60.** x = number of miles on Thursday; $x + 106$ = number of miles on Friday; $x - 39$ = number of miles on Saturday; Thurs. = 263 miles; Fri. = 369 miles; Sat. = 224 miles **61.** x = width; $2x - 3$ = length; width = 13 inches; length = 23 inches **62.** x = width; $3x + 2$ = length; width = 22 meters; length = 68 meters **63.** x = length of the first side; $x + 7$ = length of the second side; $x - 4$ = length of the third side; first side = 32 meters; second side = 39 meters; third side = 28 meters **64.** x = number of degrees in angle B; $x + 74$ = number of degrees in angle A; $3x$ = number of degrees in angle C; angle A measures 95.2°; angle B measures 21.2°; angle C measures 63.6° **65.** x = cost of chair; $275 - 4x$ = cost of table; each chair costs \$35. The table costs \$135 **66.** x = length; $x - 44$ = width; the width is 50 feet. The length is 94 feet. **67.** $x + 88$ = number of miles on the first day; x = number of miles on the second day; the first day they drove 424 miles. The second day they drove 336 miles. **68.** x = number of applications in the first week; $x + 156$ = number of applications in the second week; $x - 142$ = number of applications in the third week; they received 262 the first week, 418 the second week, and 120 the third week. **69.** x = number of first-class passengers; $4x$ = number of coach passengers; there were 608 coach passengers and 152 first-class passengers. **70.** x = total sales; \$19,375

Chapter 10 Test

1. $-6a$ **2.** $\frac{2}{15}x + \frac{9}{8}y$ **3.** $-11x + 2y$ **4.** $a - 8b + 11$ **5.** $7x - 7z$ **6.** $-4x - 2y + 5$ **7.** $60x - 25y$ **8.** $-8x + 12y - 28$ **9.** $-4.5a + 3b - 1.5c + 12$ **10.** $-11a + 14b$ **11.** $x = -8$ **12.** $x = -6$ **13.** $x = 15$ **14.** $x = -\frac{1}{2}$ **15.** $x = 1$ **16.** $x = -\frac{5}{2}$ or $-2\frac{1}{2}$ **17.** $s = f + 15$ **18.** $n = s - 15,000$ **19.** $\frac{1}{2}s$ = measure of the first angle; s = measure of the second angle; $2s$ = measure of the third angle **20.** w = width; $2w - 5$ = length **21.** Prentice farm = 87 acres; Smithfield farm = 261 acres **22.** Marcia = \$24,000; Sam = \$22,500 **23.** Monday = 311 miles; Tuesday = 367 miles; Wednesday = 297 miles **24.** width = 25 feet; length = 34 feet

Cumulative Test for Chapters 1–10

1. 747 **2.** 10,815 **3.** 45,678,900 **4.** $\frac{5}{2}$ or $2\frac{1}{2}$ **5.** $\frac{65}{8}$ or $8\frac{1}{8}$ **6.** 0.07254 **7.** 30,550.301 **8.** $n = 16$ **9.** 1596 **10.** 5500 **11.** 0.345 meters **12.** 120 inches **13.** 37.7 yard **14.** 143 square meters **15.** −31 **16.** 30 **17.** $\frac{3}{4}a - \frac{1}{14}b$ **18.** $-6x + 2y + 3$ **19.** $21x - 7y + 56$ **20.** $-2x - 24y$ **21.** $x = 4$ **22.** $y = -\frac{5}{2}$ or $-2\frac{1}{2}$ **23.** $x = 26$ **24.** $x = -3$ **25.** p = weight of printer; $p + 322$ = weight of computer **26.** f = enrollment during fall; $f - 87$ = enrollment during summer **27.** Thursday = 376 miles; Friday = 424 miles; Saturday = 281 miles **28.** width = $13\frac{2}{3}$ feet; length = $35\frac{1}{3}$ feet

Practice Final Examination

1. eighty-two thousand, three hundred sixty-seven **2.** 30,333 **3.** 173 **4.** 34,103 **5.** 4212 **6.** 217,745 **7.** 158 **8.** 606 **9.** 116

10. 32 miles/gallon **11.** $\frac{7}{15}$ **12.** $\frac{42}{11}$ **13.** $\frac{33}{20}$ or $1\frac{13}{20}$ **14.** $\frac{89}{15}$ or $5\frac{14}{15}$ **15.** $\frac{31}{14}$ or $2\frac{3}{14}$ **16.** 4 **17.** $\frac{14}{5}$ or $2\frac{4}{5}$ **18.** $\frac{22}{13}$ or $1\frac{9}{13}$

19. $6\frac{17}{20}$ miles **20.** 5 packages **21.** 0.719 **22.** $\frac{43}{50}$ **23.** > **24.** 506.38 **25.** 21.77 **26.** 0.757 **27.** 0.492 **28.** 3.69 **29.** 0.8125

30. 0.7056 **31.** $\frac{1400 \text{ students}}{43 \text{ faculty}}$ **32.** no **33.** $n \approx 9.4$ **34.** $n \approx 7.7$ **35.** $n = 15$ **36.** $n = 9$ **37.** $3333.33 **38.** 9.75 inches

39. $294.12 **40.** 1.6 pounds **41.** 0.63% **42.** 21.25% **43.** 1.64 **44.** 17.3% **45.** 302.4 **46.** 250 **47.** 4284 **48.** $10,856

49. 4500 students **50.** 34.3% **51.** 4.25 gallons **52.** 6500 pounds **53.** 192 inches **54.** 5600 meters **55.** 0.0698 kilograms

56. 0.00248 liter **57.** 19.32 kilometers **58.** 6.3182×10^{-4} **59.** 1.264×10^{11} **60.** 1.36728 centimeters **61.** 14.4 meters

62. 206 centimeters **63.** 5.4 square feet **64.** 75 square meters **65.** 113.04 square meters **66.** 56.52 meters

67. 167.47 cubic centimeters **68.** 205.2 cubic feet **69.** 32.5 square meters **70.** $n = 32.5$ **71.** $8 million **72.** $1 million dollars

73. 50°F **74.** from 1990 to 2000 **75.** 600 students **76.** 1400 students **77.** mean ≈ 15.83; median = 16.5 **78.** 16 **79.** 11.091

80. 15 feet **81.** −13 **82.** $\frac{1}{8}$ **83.** −3 **84.** −17 **85.** 24 **86.** $-\frac{8}{3}$ or $-2\frac{2}{3}$ **87.** 4 **88.** 27 **89.** 8 **90.** $\frac{1}{2}$ **91.** $-3x - 7y$

92. $-7 - 4a - 17b$ **93.** $-2x + 6y + 10$ **94.** $-11x - 9y - 4$ **95.** $x = 2$ **96.** $x = -2$ **97.** $x = -\frac{1}{2}$ **98.** $x = -\frac{2}{5}$

99. 122 students are taking history; 110 students are taking math **100.** The length is 37 meters. The width is 16 meters.

Appendix B Scientific Calculator Exercises

1. 11,747 **3.** 352,554 **5.** 1.294 **7.** 254 **9.** 8934.7376; $123.45 + 45.9876 + 8765.3$ **11.** 16.81; $\frac{34}{8} + 12.56$ **13.** 10.03 **15.** 289.387

17. 378,224 **19.** 690,284 **21.** 1068.678 **23.** 1.58688 **25.** 1383.2 **27.** 0.7634 **29.** 16.03 **31.** $12,562.34 **33.** 85,163.7

35. $103,608.65 **37.** 15,782.038 **39.** 2.500344 **41.** 897 **43.** 22.5 **45.** 84.372 **47.** 4.435 **49.** 101.5673 **51.** 0.24685 **53.** 30.7572

55. −13.87849 **57.** −13.0384 **59.** 3.3174×10^{18} **61.** $4.630985915 \times 10^{11}$ **63.** −0.5517241379 **65.** 15.90974544 **67.** 1.378368601

69. 11.981752 **71.** 8.090444982 **73.** 8.325724507 **75.** 1.08120 **77.** 1.14773

Basic College Mathematics Applications Index

Basic College Mathematics Index

Photo Credits

CHAPTER 1   CO Stephen L. Saks/Photo Researchers, Inc.   **p. 30** Bob Winsett/Corbis
p. 55 Jose L. Pelaez/Corbis/Stock Market   **p. 61** Macduff Everton/Corbis
p. 63 Jeff Greenberg/PhotoEdit   **p. 69** Stocktrek/Corbis/Stock Market
p. 80 Jean Miele/Corbis/Stock Market   **p. 85** Rob & Sas/Corbis/Stock Market
p. 92 Ariel Skelley/Corbis/Stock Market

CHAPTER 2   CO Jeff Greenberg/PhotoEdit   **p. 113** Rob Lewine/Corbis/Stock Market
p. 133 John Paul Endress/Corbis/Stock Market   **p. 135** Bob Daemmrich/Stock Boston
p. 161 Stefan Lawrence/International Stock Photography Ltd.
p. 179 Index Stock Imagery, Inc.   **p. 181** Bill Stanton/International Stock Photography Ltd.

CHAPTER 3   CO Maria Stenzel/NGS Image Collection   **p. 215** Daemmrich/Stock Boston
p. 228 © J. Leonard/Weatherstock   **p. 237** Joyce Photographics/Photo Researchers, Inc.
p. 247 Richard Bickel/Corbis   **p. 248** Ramey/Woodfin Camp & Associates
p. 251 Catherine Ursillo/Photo Researchers, Inc.   **p. 252** David R. Frazier
Photolibrary/Photo Researchers, Inc.

CHAPTER 4   CO Ron Sanford/International Stock Photography Ltd.
p. 278 Lester Lefkowitz/Corbis/Stock Market   **p. 290** Charles Gupton/Stock Boston
p. 293 Greig Cranna/Stock Boston   **p. 297** SuperStock, Inc.

CHAPTER 5   CO SuperStock, Inc.   **p. 311** Robert Rathe/Stock Boston
p. 318 Hillary Wilkes/International Stock Photography Ltd.
p. 330 Mike Mazzaschi/Stock Boston   **p. 337** SuperStock, Inc.

CHAPTER 6   CO William Johnson/Stock Boston   **p. 371** William Taufic/Corbis/Stock
Market   **p. 384** Wernher Krutein/photovault.com   **p. 387** Gerard Lacz/Peter Arnold, Inc.
p. 391 Frank Grant/International Stock Photography Ltd.   **p. 392** Tom McHugh/Photo
Researchers, Inc.   **p. 399** Stan Ries/International Stock Photography, Ltd.

CHAPTER 7   CO Chuck Mason/International Stock Photography Ltd.   **p. 428** Clint
Clemens/International Stock Photography, Inc.   **p. 440** Rosenthal/SuperStock, Inc.
p. 458 Tom McCarthy/PhotoEdit   **p. 465** (top) Tom McCarthy/PhotoEdit
p. 465 (bottom) SuperStock, Inc.   **p. 475** Dick Durrance/Woodfin Camp & Associates
p. 489 John Elk III/Stock Boston   **p. 490** David Young-Wolff/PhotoEdit

CHAPTER 8   CO David Young-Wolff/PhotoEdit   **p. 539** Daemmrich/Stock Boston

CHAPTER 9   CO Alexander Tsiaras/Science Source/Photo Researchers, Inc.
p. 567 Gary Landsman/Corbis/Stock Market   **p. 579** Michael Giannechini/Photo
Researchers, Inc.   **p. 595** Alexander Tsiaras/Science Source/Photo Researchers, Inc.

CHAPTER 10   CO David Sailors/Corbis/Stock Market   **p. 651** Jeffry W. Myers/Stock
Boston